Renault Mégane

Service and Repair Manual

A K Legg LAE MIMI and Mark Coombs

(3395-328)

Models covered

Renault Mégane Hatchback, Saloon & Coupe models with petrol & diesel engines, including special/limited editions
1.4 litre (1390 cc) & 1.6 litre (1598 cc) petrol engines
1.9 litre (1870 cc) diesel engine (including Turbo)

Does not cover 2.0 litre (1998 cc) 8- or 16-valve petrol engines, Scenic or Cabriolet models

ABCDE
FGHIJ
KLMNO
PQRST

A book in the **Haynes Service and Repair Manual Series**

ISBN **1 85960 395 5**

British Library Cataloguing in Publication Data
A catalogue record for this book is available from the British Library.

Printed by **J H Haynes & Co. Ltd, Sparkford, Nr Yeovil, Somerset BA22 7JJ**

Haynes Publishing
Sparkford, Nr Yeovil, Somerset BA22 7JJ, England

Haynes North America, Inc
861 Lawrence Drive, Newbury Park, California 91320, USA

Editions Haynes S.A.
147/149, rue Saint Honoré, 75001 PARIS, France

Haynes Publishing Nordiska AB
Fyrisborgsgatan 5, 754 50 Uppsala, Sverige

Contents

Contents

REPAIRS AND OVERHAUL

Engine and associated systems

Transmission

Brakes and suspension

Body equipment

REFERENCE

The Renault Mégane model range was introduced into the UK in April 1996 to supersede the ageing Renault 19. 5-door Hatchback and Coupe models were available, with the Classic Saloon model following in November 1996. The Scenic miniature MPV was added to the range in early 1997, but this model is not covered in depth in this manual.

1390 cc and 1598 cc petrol engines together with 1870 cc non-Turbo and Turbo diesel engines are available, although the 1998 cc 8- and 16-valve engines are not covered in this manual. All engines are of overhead camshaft design, and are mounted transversely at the front of the vehicle.

Models may be fitted with five-speed manual, or four-speed automatic transmissions, mounted at the left-hand side of the engine.

All models have front-wheel-drive with fully-independent front and semi-independent rear suspension.

All models have a high trim level, which is very comprehensive in the upper model range. Central locking, electric windows, and electric sunroof, anti-lock brakes and air conditioning are all available.

For the home mechanic, the Renault Mégane is a straightforward vehicle to maintain and repair, since design features have been incorporated to reduce the actual cost of ownership to a minimum, and most of the items requiring frequent attention are easily accessible.

Your Renault Mégane manual

The aim of this manual is to help you get the best value from your vehicle. It can do so in several ways. It can help you decide what work must be done (even should you choose to get it done by a garage), provide information on routine maintenance and servicing, and give a logical course of action and diagnosis when random faults occur. However, it is hoped that you will use the manual by tackling the work yourself. On simpler jobs, it may even be quicker than booking the car into a garage and going there twice, to leave and collect it. Perhaps most important, a lot of money can be saved by avoiding the costs a garage must charge to cover its labour and overheads.

The manual has drawings and descriptions to show the function of the various components, so that their layout can be understood. Then the tasks are described and photographed in a clear step-by-step sequence.

Renault Mégane RT 1.6e

The Renault Mégane Team

Haynes manuals are produced by dedicated and enthusiastic people working in close co-operation. The team responsible for the creation of this book included:

Authors	**Andy Legg** **Mark Coombs**
Sub-editor	**Sophie Yar**
Editor & Page Make-up	**Bob Jex**
Workshop manager	**Paul Buckland**
Photo Scans	**John Martin** **Steve Tanswell**
Cover illustration & Line Art	**Roger Healing**
Wiring diagrams	**Matthew Marke**

We hope the book will help you to get the maximum enjoyment from your car. By carrying out routine maintenance as described you will ensure your car's reliability and preserve its resale value.

Acknowledgements

Thanks are due to the Champion Spark Plug Company who supplied the illustrations of various spark plug conditions. Certain other illustrations are the copyright of Renault (UK) Ltd, and are used with their permission. Thanks are also due to Kings of Taunton Ltd, who provided technical assistance, to Sykes-Pickavant Limited who provided some of the workshop tools, and to all those people at Sparkford who helped in the production of this manual.

This manual is not a direct reproduction of the vehicle manufacturers' data, and its publication should not be taken as implying any technical approval by the vehicle manufacturers or importers.

We take great pride in the accuracy of information given in this manual, but vehicle manufacturers make alterations and design changes during the production run of a particular vehicle of which they do not inform us. No liability can be accepted by the authors or publishers for loss, damage or injury caused by any errors in, or omissions from, the information given.

Project vehicles

The main vehicle used in the preparation of this manual, and which appears in many of the photographic sequences, was a petrol Renault Mégane RT 1.6e. Also used was a petrol Renault Mégane RT 1.4e and a diesel Renault Mégane 1.9D.

Working on your car can be dangerous. This page shows just some of the potential risks and hazards, with the aim of creating a safety-conscious attitude.

General hazards

Scalding

- Don't remove the radiator or expansion tank cap while the engine is hot.
- Engine oil, automatic transmission fluid or power steering fluid may also be dangerously hot if the engine has recently been running.

Burning

- Beware of burns from the exhaust system and from any part of the engine. Brake discs and drums can also be extremely hot immediately after use.

Crushing

- When working under or near a raised vehicle, always supplement the jack with axle stands, or use drive-on ramps. ***Never venture under a car which is only supported by a jack.***

- Take care if loosening or tightening high-torque nuts when the vehicle is on stands. Initial loosening and final tightening should be done with the wheels on the ground.

Fire

- Fuel is highly flammable; fuel vapour is explosive.
- Don't let fuel spill onto a hot engine.
- Do not smoke or allow naked lights (including pilot lights) anywhere near a vehicle being worked on. Also beware of creating sparks (electrically or by use of tools).
- Fuel vapour is heavier than air, so don't work on the fuel system with the vehicle over an inspection pit.
- Another cause of fire is an electrical overload or short-circuit. Take care when repairing or modifying the vehicle wiring.
- Keep a fire extinguisher handy, of a type suitable for use on fuel and electrical fires.

Electric shock

- Ignition HT voltage can be dangerous, especially to people with heart problems or a pacemaker. Don't work on or near the ignition system with the engine running or the ignition switched on.

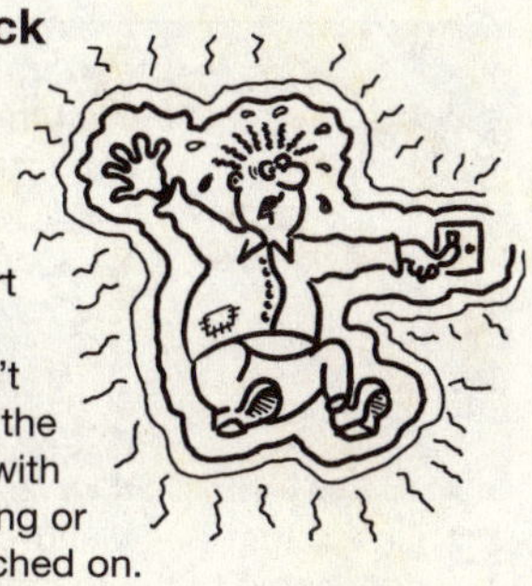

- Mains voltage is also dangerous. Make sure that any mains-operated equipment is correctly earthed. Mains power points should be protected by a residual current device (RCD) circuit breaker.

Fume or gas intoxication

- Exhaust fumes are poisonous; they often contain carbon monoxide, which is rapidly fatal if inhaled. Never run the engine in a confined space such as a garage with the doors shut.

- Fuel vapour is also poisonous, as are the vapours from some cleaning solvents and paint thinners.

Poisonous or irritant substances

- Avoid skin contact with battery acid and with any fuel, fluid or lubricant, especially antifreeze, brake hydraulic fluid and Diesel fuel. Don't syphon them by mouth. If such a substance is swallowed or gets into the eyes, seek medical advice.
- Prolonged contact with used engine oil can cause skin cancer. Wear gloves or use a barrier cream if necessary. Change out of oil-soaked clothes and do not keep oily rags in your pocket.
- Air conditioning refrigerant forms a poisonous gas if exposed to a naked flame (including a cigarette). It can also cause skin burns on contact.

Asbestos

- Asbestos dust can cause cancer if inhaled or swallowed. Asbestos may be found in gaskets and in brake and clutch linings. When dealing with such components it is safest to assume that they contain asbestos.

Special hazards

Hydrofluoric acid

- This extremely corrosive acid is formed when certain types of synthetic rubber, found in some O-rings, oil seals, fuel hoses etc, are exposed to temperatures above 400ºC. The rubber changes into a charred or sticky substance containing the acid. *Once formed, the acid remains dangerous for years. If it gets onto the skin, it may be necessary to amputate the limb concerned.*
- When dealing with a vehicle which has suffered a fire, or with components salvaged from such a vehicle, wear protective gloves and discard them after use.

The battery

- Batteries contain sulphuric acid, which attacks clothing, eyes and skin. Take care when topping-up or carrying the battery.
- The hydrogen gas given off by the battery is highly explosive. Never cause a spark or allow a naked light nearby. Be careful when connecting and disconnecting battery chargers or jump leads.

Air bags

- Air bags can cause injury if they go off accidentally. Take care when removing the steering wheel and/or facia. Special storage instructions may apply.

Diesel injection equipment

- Diesel injection pumps supply fuel at very high pressure. Take care when working on the fuel injectors and fuel pipes.

Warning: Never expose the hands, face or any other part of the body to injector spray; the fuel can penetrate the skin with potentially fatal results.

Remember...

DO

- Do use eye protection when using power tools, and when working under the vehicle.
- Do wear gloves or use barrier cream to protect your hands when necessary.
- Do get someone to check periodically that all is well when working alone on the vehicle.
- Do keep loose clothing and long hair well out of the way of moving mechanical parts.
- Do remove rings, wristwatch etc, before working on the vehicle – especially the electrical system.
- Do ensure that any lifting or jacking equipment has a safe working load rating adequate for the job.

DON'T

- Don't attempt to lift a heavy component which may be beyond your capability – get assistance.
- Don't rush to finish a job, or take unverified short cuts.
- Don't use ill-fitting tools which may slip and cause injury.
- Don't leave tools or parts lying around where someone can trip over them. Mop up oil and fuel spills at once.
- Don't allow children or pets to play in or near a vehicle being worked on.

The following pages are intended to help in dealing with common roadside emergencies and breakdowns. You will find more detailed fault finding information at the back of the manual, and repair information in the main chapters.

If your car won't start and the starter motor doesn't turn

- ☐ If it's a model with automatic transmission, make sure the selector is in 'P' or 'N'.
- ☐ Open the bonnet and make sure that the battery terminals are clean and tight.
- ☐ Switch on the headlights and try to start the engine. If the headlights go very dim when you're trying to start, the battery is probably flat. Get out of trouble by jump starting (see next page) using a friend's car.

If your car won't start even though the starter motor turns as normal

- ☐ Is there fuel in the tank?
- ☐ Is there moisture on electrical components under the bonnet? Switch off the ignition, then wipe off any obvious dampness with a dry cloth. Spray a water-repellent aerosol product (WD-40 or equivalent) on ignition and fuel system electrical connectors like those shown in the photos. Pay special attention to the ignition coil wiring connector and HT leads. (Note that diesel engines don't normally suffer from damp.)

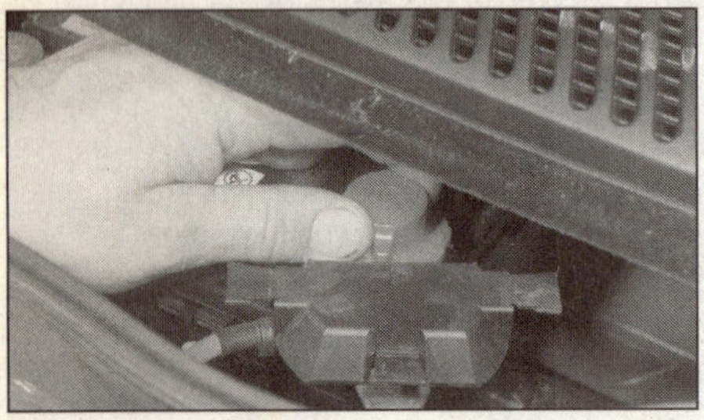

A Check the security and condition of the battery connections.

B Check that the spark plug HT leads are securely pushed onto the spark plugs.

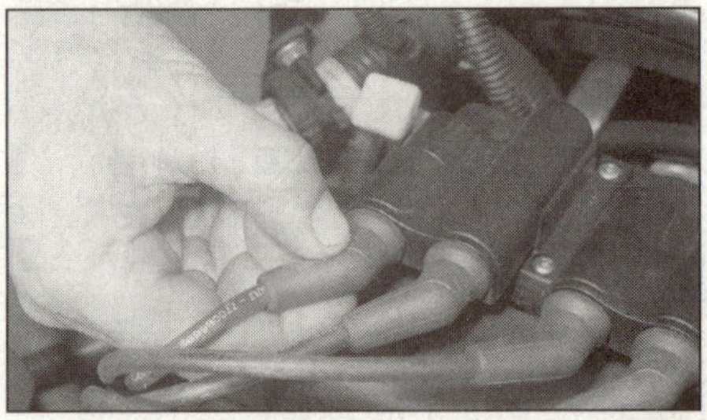

C Check that the spark plug HT leads are securely connected to the ignition coils.

Check that electrical connections are secure (with the ignition switched off) and spray them with a water dispersant spray like WD40 if you suspect a problem due to damp

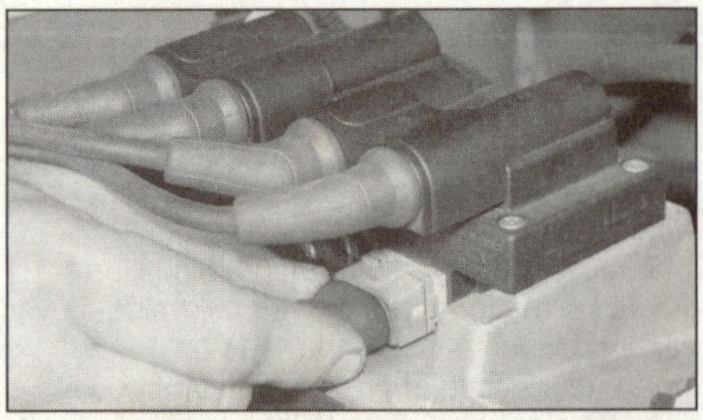

D Check that the wiring plugs are securely connected to the ignition coils.

E Check that the wiring plugs are securely connected to the throttle housing.

Jump starting

HAYNES HiNT ***Jump starting will get you out of trouble, but you must correct whatever made the battery go flat in the first place. There are three possibilities:***

1) The battery has been drained by repeated attempts to start, or by leaving the lights on.
2) The charging system is not working properly (alternator drivebelt slack or broken, alternator wiring fault or alternator itself faulty).
3) The battery itself is at fault (electrolyte low, or battery worn out).

When jump-starting a car using a booster battery, observe the following precautions:

- ✔ Before connecting the booster battery, make sure that the ignition is switched off.
- ✔ Ensure that all electrical equipment (lights, heater, wipers, etc) is switched off.
- ✔ Take note of any special precautions printed on the battery case.
- ✔ Make sure that the booster battery is the same voltage as the discharged one in the vehicle.
- ✔ If the battery is being jump-started from the battery in another vehicle, the two vehicles MUST NOT TOUCH each other.
- ✔ Make sure that the transmission is in neutral (or PARK, in the case of automatic transmission).

1 Connect one end of the red jump lead to the positive (+) terminal of the flat battery

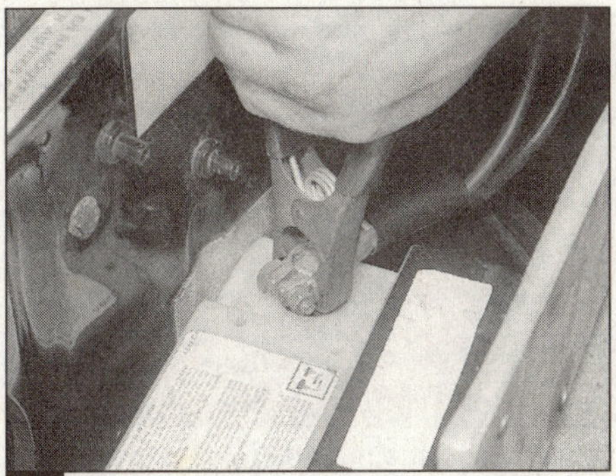

2 Connect the other end of the red lead to the positive (+) terminal of the booster battery

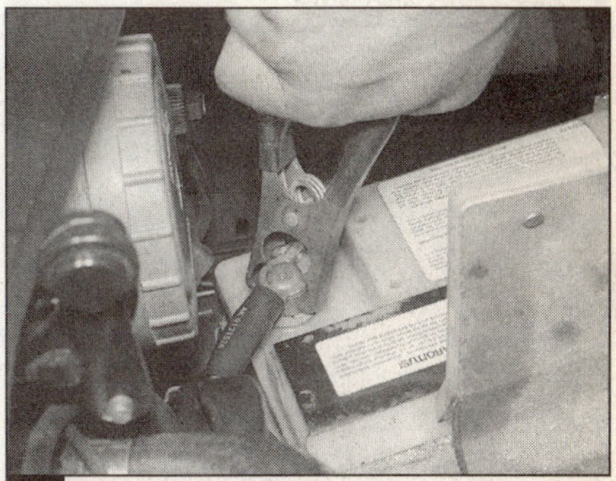

3 Connect one end of the black jump lead to the negative (-) terminal of the booster battery

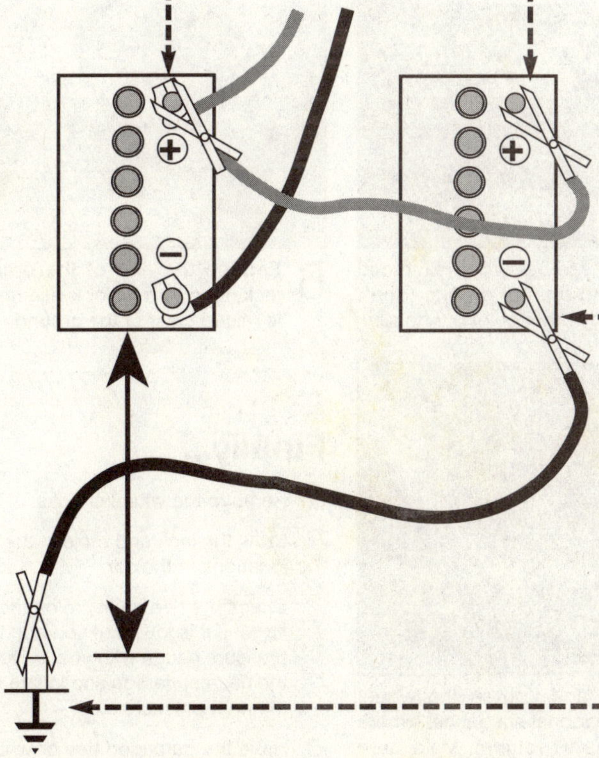

4 Connect the other end of the black jump lead to a bracket on the cylinder head on the vehicle to be started

5 Make sure that the jump leads will not come into contact with the fan, drivebelts or other moving parts of the engine

6 Start the engine using the booster battery, then with the engine running at idle speed, disconnect the jump leads in the reverse order of connection

Wheel changing

Some of the details shown here will vary according to model. For instance, the location of the spare wheel and jack is not the same on all cars. However, the basic principles apply to all vehicles.

Warning: Do not change a wheel in a situation where you risk being hit by other traffic. On busy roads, try to stop in a lay-by or a gateway. Be wary of passing traffic while changing the wheel – it is easy to become distracted by the job in hand.

Preparation

- ☐ When a puncture occurs, stop as soon as it is safe to do so.
- ☐ Park on firm level ground, if possible, and well out of the way of other traffic.
- ☐ Use hazard warning lights if necessary.
- ☐ If you have one, use a warning triangle to alert other drivers of your presence.
- ☐ Apply the handbrake and engage first or reverse gear (or Park on models with automatic transmission.
- ☐ Chock the wheel diagonally opposite the one being removed – a couple of large stones will do for this.
- ☐ If the ground is soft, use a flat piece of wood to spread the load under the jack.

Changing the wheel

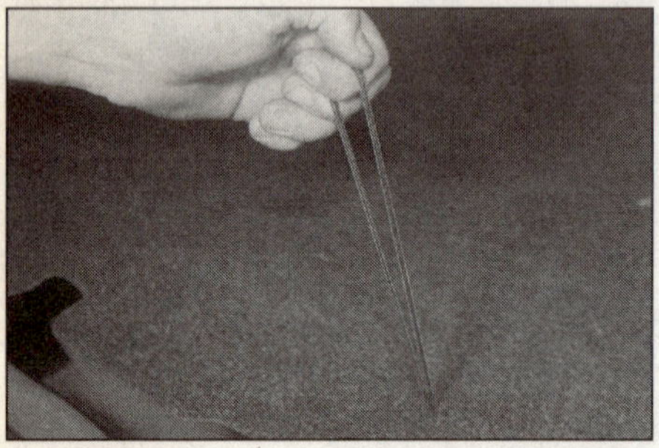

1 In the boot, lift the floor panel for access to the spare wheel. Attach the cord to the hook on the rear seat back.

2 Unscrew the nut and remove the spare wheel from the well. The tools are located beneath the spare wheel.

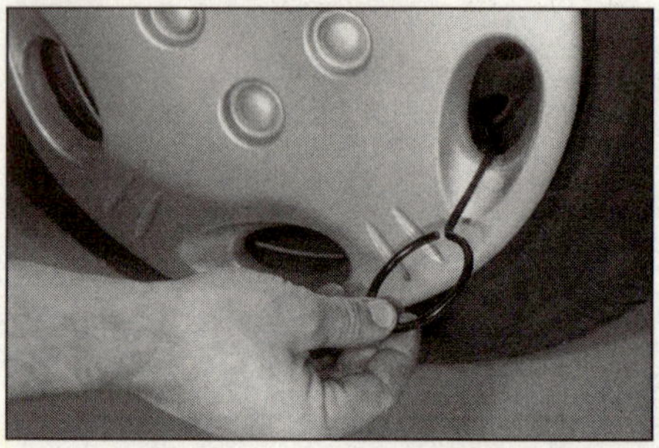

3 Use the hook provided to pull off the wheel trim. On some models, the trim is retained by the wheel bolts, and is removed together with the wheel.

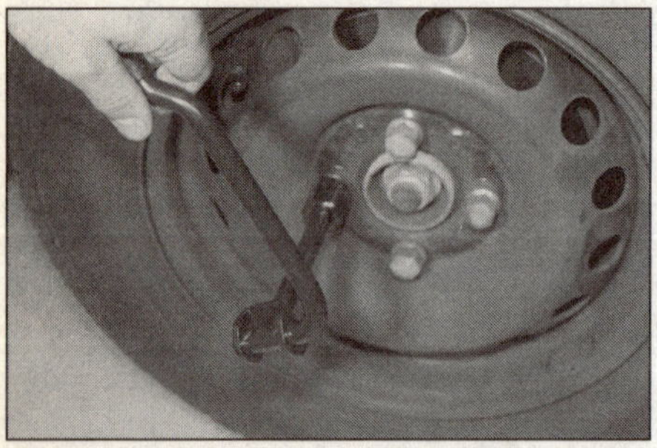

4 Use the brace to loosen each wheel bolt by half a turn.

5 Locate the jack head below the reinforced jacking point and on firm ground (don't jack the car at any other point on the sill).

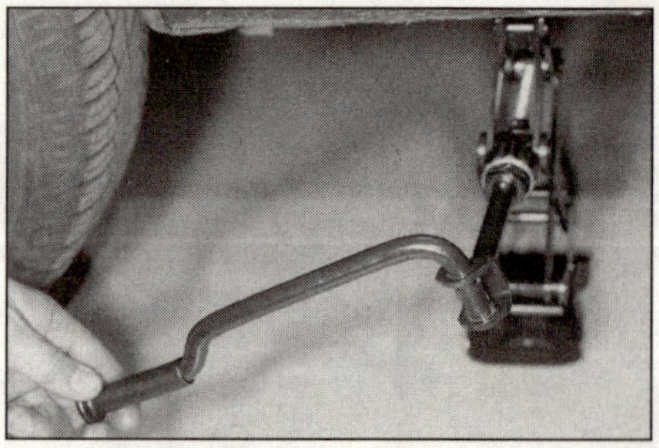

6 Engage the end of the brace with the jack, then turn it clockwise until the wheel is raised clear of the ground.

7 Remove the bolts (and wheel trim where necessary) and lift the wheel clear, then fit the spare wheel. Refit the wheel bolts and tighten moderately with the brace.

8 Lower the car, and tighten the wheel bolts fully in a diagonal sequence. Where applicable, fit the wheel trim. Make sure that the lug on the trim locates over the valve.

Finally...

- ☐ Remove the wheel chocks.
- ☐ Stow the jack and tools in the correct locations in the car.
- ☐ Check the tyre pressure on the wheel just fitted. If it is low, or if you don't have a pressure gauge with you, drive slowly to the nearest garage and inflate the tyre to the right pressure.
- ☐ Have the damaged tyre or wheel repaired as soon as possible.

Identifying leaks

Puddles on the garage floor or drive, or obvious wetness under the bonnet or underneath the car, suggest a leak that needs investigating. It can sometimes be difficult to decide where the leak is coming from, especially if the engine bay is very dirty already. Leaking oil or fluid can also be blown rearwards by the passage of air under the car, giving a false impression of where the problem lies.

Warning: Most automotive oils and fluids are poisonous. Wash them off skin, and change out of contaminated clothing, without delay.

HAYNES HiNT ***The smell of a fluid leaking from the car may provide a clue to what's leaking. Some fluids are distinctively coloured. It may help to clean the car and to park it over some clean paper as an aid to locating the source of the leak. Remember that some leaks may only occur while the engine is running.***

Sump oil

Engine oil may leak from the drain plug...

Oil from filter

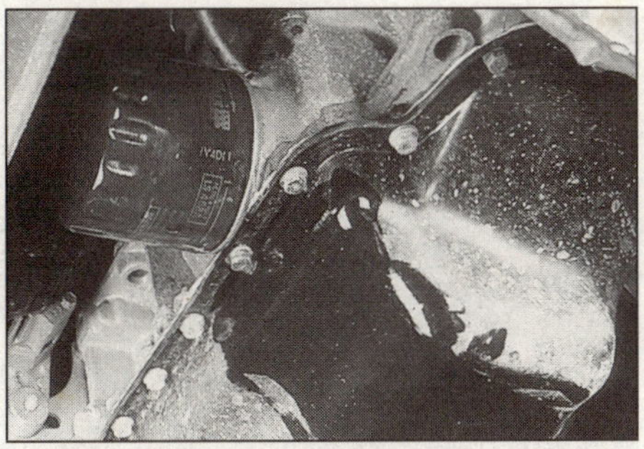

...or from the base of the oil filter.

Gearbox oil

Gearbox oil can leak from the seals at the inboard ends of the driveshafts.

Antifreeze

Leaking antifreeze often leaves a crystalline deposit like this.

Brake fluid

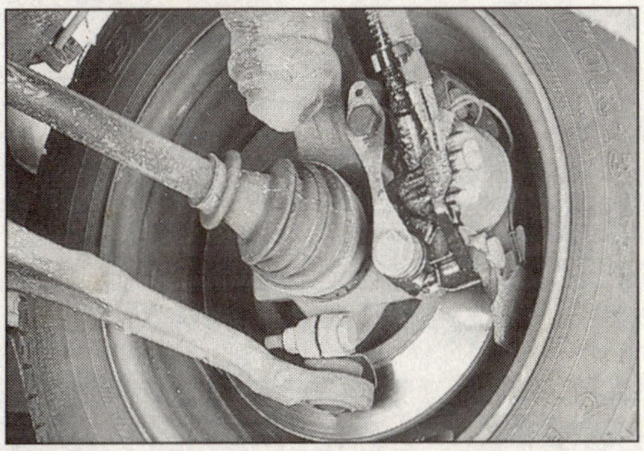

A leak occurring at a wheel is almost certainly brake fluid.

Power steering fluid

Power steering fluid may leak from the pipe connectors on the steering rack.

Towing

When all else fails, you may find yourself having to get a tow home – or of course you may be helping somebody else. Long-distance recovery should only be done by a garage or breakdown service. For shorter distances, DIY towing using another car is easy enough, but observe the following points:

☐ Use a proper tow-rope – they are not expensive. The vehicle being towed must display an 'ON TOW' sign in its rear window.

☐ Always turn the ignition key to the 'on' position when the vehicle is being towed, so that the steering lock is released, and that the direction indicator and brake lights will work.

☐ Towing eyes are provided at the front and rear of the vehicle, on the right-hand side. Do not attach the tow-rope to anything else.

☐ Before being towed, release the handbrake and select neutral on the transmission.

☐ Note that greater-than-usual pedal pressure will be required to operate the brakes, since the vacuum servo unit is only operational with the engine running.

☐ On models with power steering, greater-than-usual steering effort will also be required.

☐ The driver of the car being towed must keep the tow-rope taut at all times to avoid snatching.

☐ Make sure that both drivers know the route before setting off.

☐ Only drive at moderate speeds and keep the distance towed to a minimum. Drive smoothly and allow plenty of time for slowing down at junctions.

☐ On models with automatic transmission, special precautions apply. If in doubt, do not tow, or transmission damage may result.

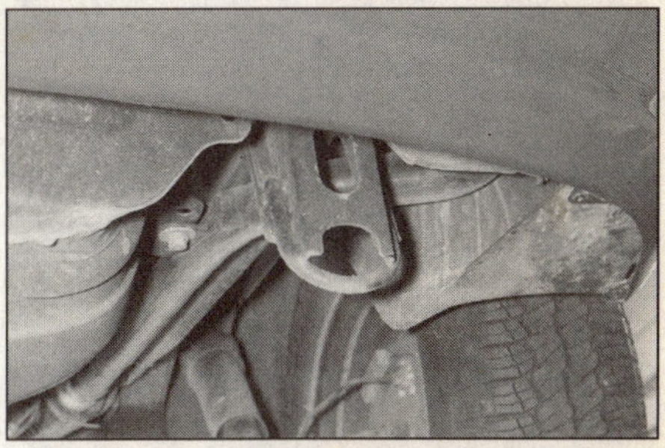

Towing eye at the rear of the vehicle

Introduction

There are some very simple checks which need only take a few minutes to carry out, but which could save you a lot of inconvenience and expense.

These "Weekly checks" require no great skill or special tools, and the small amount of time they take to perform could prove to be very well spent, for example;

☐ Keeping an eye on tyre condition and pressures, will not only help to stop them wearing out prematurely, but could also save your life.

☐ Many breakdowns are caused by electrical problems. Battery-related faults are particularly common, and a quick check on a regular basis will often prevent the majority of these.

☐ If your car develops a brake fluid leak, the first time you might know about it is when your brakes don't work properly. Checking the level regularly will give advance warning of this kind of problem.

☐ If the oil or coolant levels run low, the cost of repairing any engine damage will be far greater than fixing the leak, for example.

Underbonnet check points

◀ 1.4 litre petrol

A *Engine oil level dipstick*
B *Engine oil filler cap*
C *Coolant expansion tank*
D *Brake fluid reservoir*
E *Screen washer fluid reservoir*
F *Power steering fluid reservoir*
G *Battery*

◀ 1.6 litre petrol

A *Engine oil level dipstick*
B *Engine oil filler cap*
C *Coolant expansion tank*
D *Brake fluid reservoir*
E *Screen washer fluid reservoir*
F *Power steering fluid reservoir*
G *Battery*

◀ 1.9 litre diesel

A *Engine oil level dipstick*
B *Engine oil filler cap*
C *Coolant expansion tank*
D *Brake fluid reservoir*
E *Screen washer fluid reservoir*
F *Power steering fluid reservoir*
G *Battery*

Engine oil level

Before you start

✔ Make sure that your car is on level ground.
✔ Check the oil level before the car is driven, or at least 5 minutes after the engine has been switched off.

> **HAYNES HiNT** ***If the oil level is checked immediately after driving the vehicle, some of the oil will remain in the upper engine components, resulting in an inaccurate reading on the dipstick!***

The correct oil

Modern engines place great demands on their oil. It is very important that the correct oil for your car is used (See "Lubricants, fluids and tyre pressures").

Car care

- If you have to add oil frequently, you should check whether you have any oil leaks. Place some clean paper under the car overnight, and check for stains in the morning. If there are no leaks, the engine may be burning oil *(see "Fault finding").*

- Always maintain the level between the upper and lower dipstick marks (see photo 3). If the level is too low severe engine damage may occur. Oil seal failure may result if the engine is overfilled by adding too much oil.

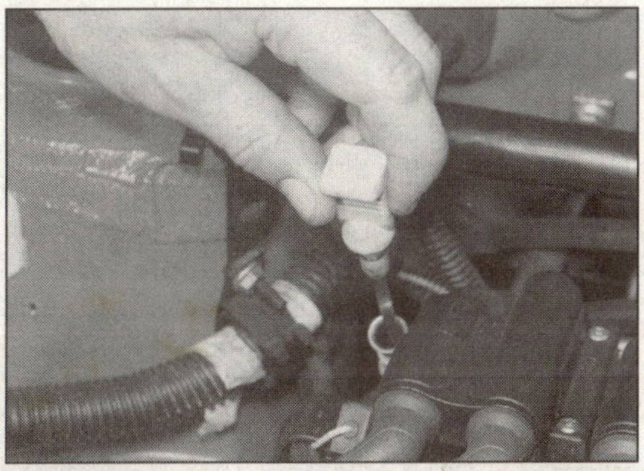

1 The dipstick is located on the right-hand rear of the engine on E7J and K7M petrol engines, and on the front of the engine on the F8Q diesel engine. Withdraw the dipstick.

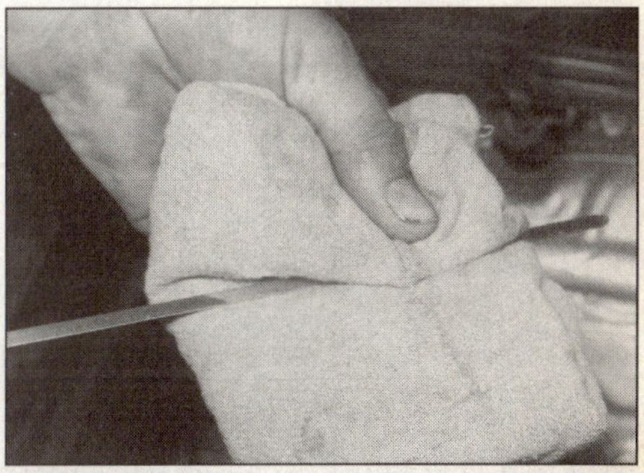

2 Using a clean rag or paper towel, wipe all the oil from the dipstick. Insert the clean dipstick into the tube as far as it will go, then withdraw it again.

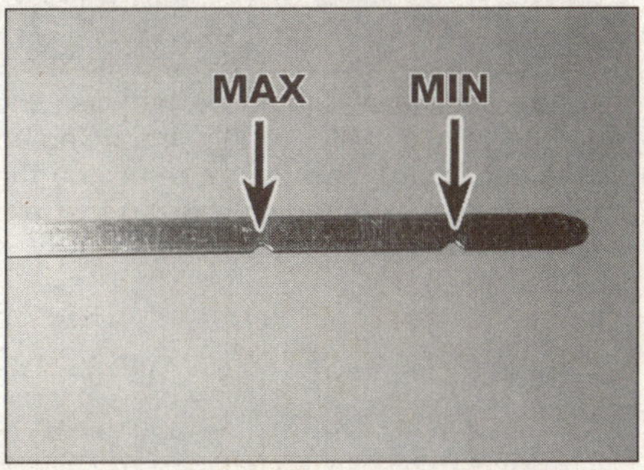

3 Note the oil level on the end of the dipstick, which should be between the upper (MAX) mark and lower (MIN) mark. Approximately 1.0 litre of oil will raise the level from the lower mark to the upper mark.

4 Oil is added through the filler cap. Unscrew the cap and top-up the level; a funnel may help to reduce spillage. Add the oil slowly, checking the level on the dipstick often. Don't overfill (see *"Car Care"* left).

Coolant level

Warning: DO NOT attempt to remove the expansion tank pressure cap when the engine is hot, as there is a very great risk of scalding. Do not leave open containers of coolant about, as it is poisonous.

Car care

● Adding coolant should not be necessary on a regular basis. If frequent topping-up is required, it is likely there is a leak. Check the radiator, all hoses and joint faces for signs of staining or wetness, and rectify as necessary.

● It is important that antifreeze is used in the cooling system all year round, not just during the winter months. Don't top-up with water alone, as the antifreeze will become too diluted.

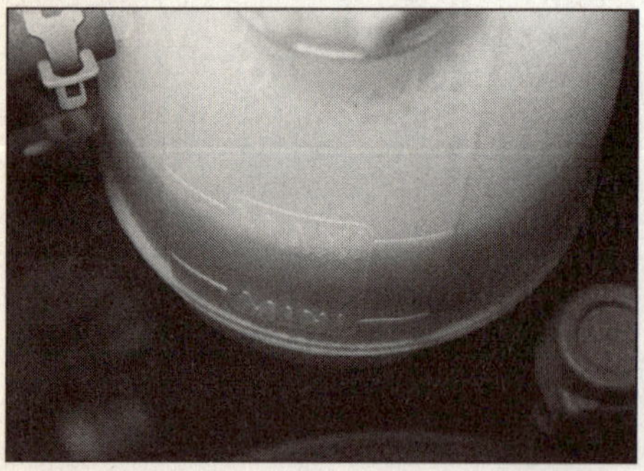

1 The coolant level varies with engine temperature. The level is checked in the expansion tank, which is in the left-hand side of the engine compartment. When the engine is cold, the coolant level should be between the "MAX" and "MIN" marks.

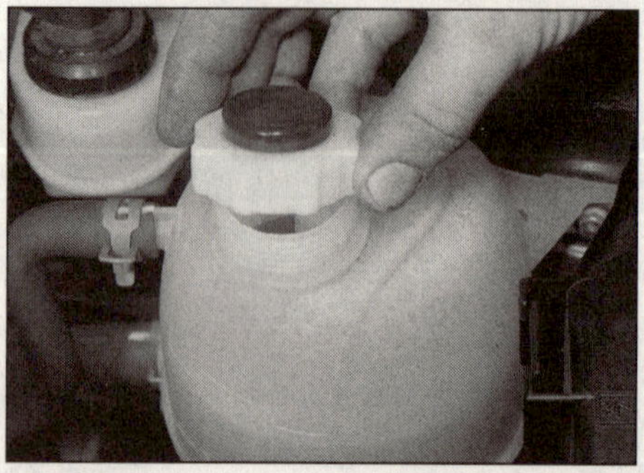

2 If topping-up is necessary, wait until the engine is cold then turn the pressure cap on the expansion tank slowly anti-clockwise, and pause until any pressure remaining in the system is released. Unscrew the cap and lift off.

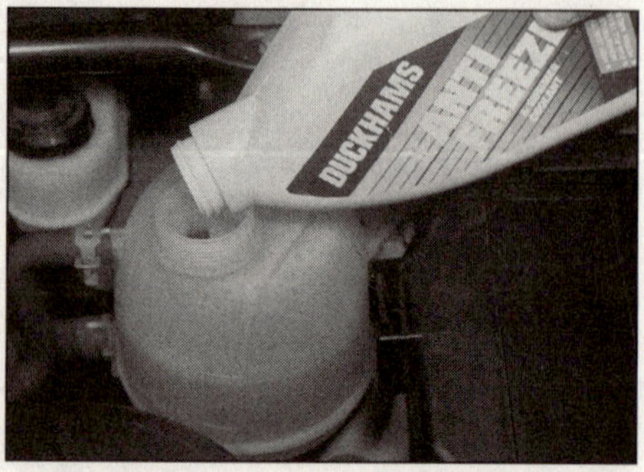

3 Add a mixture of water and antifreeze to the expansion tank, until the coolant level is up to the "MAX" level mark. Refit the cap, turning it clockwise as far as it will go until it is secure. Re-check that the cap is securely tightened once the engine is warm.

Power steering fluid level

Before you start:

- ✔ Park the vehicle on level ground.
- ✔ Set the steering wheel straight-ahead.
- ✔ The engine should be turned off.

HAYNES HiNT ***For the check to be accurate, the steering must not be turned once the engine has been stopped.***

Safety first!

● The need for frequent topping-up indicates a leak, which should be investigated immediately.

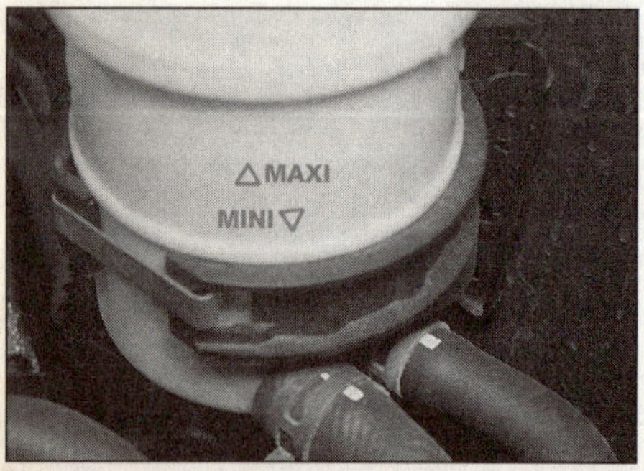

1 The power steering fluid reservoir is located on the right-hand side of the radiator. The fluid level should be checked with the engine stopped. A translucent reservoir is fitted, with "MAXI" and "MINI" markings on the side of the reservoir.

2 The fluid level should be between the "MAXI" and "MINI" marks. If topping-up is necessary, and before removing the cap, wipe the surrounding area so that dirt does not enter the reservoir.

3 Unscrew the cap, allowing the fluid to drain from the bottom of the cap as it is removed. Top up the fluid level to the "MAXI" mark, using the specified type of fluid (do not overfill the reservoir), then refit and tighten the filler cap.

Brake fluid level

Warning:

● ***Brake fluid can harm your eyes and will damage painted surfaces, so use extreme caution when handling and pouring it.***

● ***Do not use fluid that has been standing open for some time, as it absorbs moisture from the air, which can cause a dangerous loss of braking effectiveness.***

HAYNES HiNT

• Make sure that your car is on level ground.

• The fluid level in the reservoir will drop slightly as the brake pads wear down, but the fluid level must never be allowed to drop below the "MIN" mark.

Safety first!

● If the reservoir requires repeated topping-up this is an indication of a fluid leak somewhere in the system, which should be investigated immediately.

● If a leak is suspected, the car should not be driven until the braking system has been checked. Never take any risks where brakes are concerned.

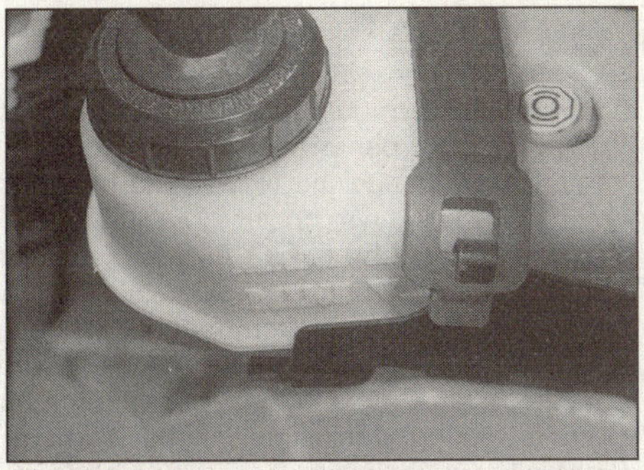

1 The "MAXI" and "MINI" marks are indicated on the side of the reservoir, which is located at the rear left-hand side of the engine compartment. The fluid level must be kept between these two marks.

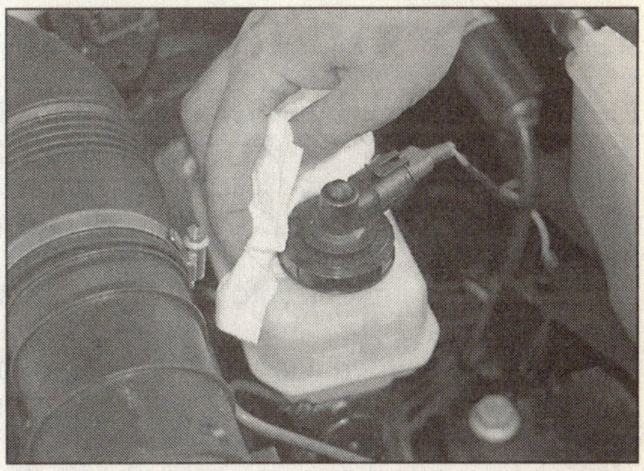

2 If topping-up is necessary, first wipe the area around the filler cap with a clean rag, then disconnect the fluid level sensor wires before removing the cap. When adding fluid, it's a good idea to inspect the reservoir. The system should be drained and refilled if dirt is seen in the fluid (see Chapter 9).

3 Carefully add fluid, avoiding spilling it on surrounding paintwork. Use only the specified hydraulic fluid; mixing different types of fluid can cause damage to the system and/or a loss of braking effectiveness. Bear in mind that the level in the reservoir will rise slightly when the cap/float assembly is refitted. After filling to the correct level, refit the cap securely and reconnect the sensor wires. Wipe off any spilt fluid.

4 When checking the fluid level it is a good idea to check the operation of the low fluid level warning light. Switch on the ignition and ask an assistant to press the test button on top of the brake fluid reservoir cap. When the button is pressed, the brake fluid level/ handbrake "on" warning light should come on - if not, the level switch, wiring or bulb may be faulty. If the warning light comes on and the fluid level is not low, check that the handbrake is not on (but note that certain models have a separate warning light for the handbrake). Switch off the ignition after testing.

Washer fluid level

Screenwash additives not only keep the winscreen clean during foul weather, they also prevent the washer system freezing in cold weather - which is when you are likely to need it most. Don't top up using plain water as the screenwash will become too diluted, and will freeze during cold weather. ***On no account use coolant antifreeze in the washer system - this could discolour or damage paintwork.***

1 The windscreen/tailgate/headlight washer fluid reservoir is located at the right-hand side of the engine compartment. If topping-up is necessary, open the cap.

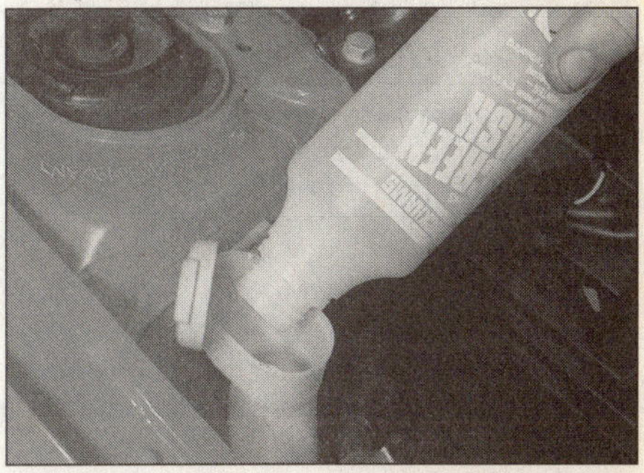

2 When topping-up the reservoir a screenwash additive should be added in the quantities recommended on the bottle.

Tyre condition and pressure

It is very important that tyres are in good condition, and at the correct pressure - having a tyre failure at any speed is highly dangerous. Tyre wear is influenced by driving style - harsh braking and acceleration, or fast cornering, will all produce more rapid tyre wear. As a general rule, the front tyres wear out faster than the rears. Interchanging the tyres from front to rear ("rotating" the tyres) may result in more even wear. However, if this is completely effective, you may have the expense of replacing all four tyres at once!

Remove any nails or stones embedded in the tread before they penetrate the tyre to cause deflation. If removal of a nail does reveal that the tyre has been punctured, refit the nail so that its point of penetration is marked. Then immediately change the wheel, and have the tyre repaired by a tyre dealer.

Regularly check the tyres for damage in the form of cuts or bulges, especially in the sidewalls. Periodically remove the wheels, and clean any dirt or mud from the inside and outside surfaces. Examine the wheel rims for signs of rusting, corrosion or other damage. Light alloy wheels are easily damaged by "kerbing" whilst parking; steel wheels may also become dented or buckled. A new wheel is very often the only way to overcome severe damage.

New tyres should be balanced when they are fitted, but it may become necessary to re-balance them as they wear, or if the balance weights fitted to the wheel rim should fall off. Unbalanced tyres will wear more quickly, as will the steering and suspension components. Wheel imbalance is normally signified by vibration, particularly at a certain speed (typically around 50 mph). If this vibration is felt only through the steering, then it is likely that just the front wheels need balancing. If, however, the vibration is felt through the whole car, the rear wheels could be out of balance. Wheel balancing should be carried out by a tyre dealer or garage.

1 *Tread Depth - visual check*

The original tyres have tread wear safety bands (B), which will appear when the tread depth reaches approximately 1.6 mm. The band positions are indicated by a triangular mark on the tyre sidewall (A).

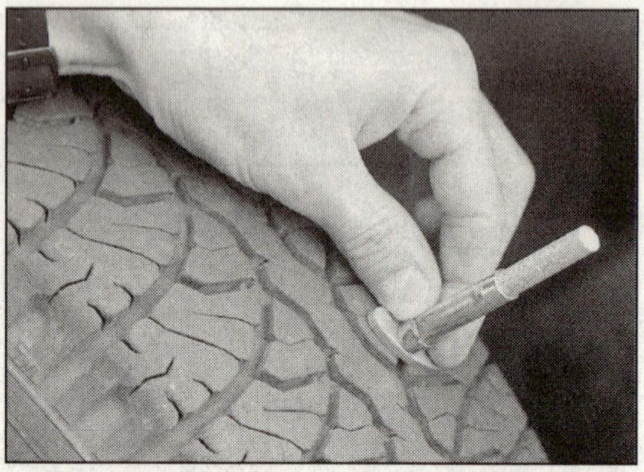

2 *Tread Depth - manual check*

Alternatively, tread wear can be monitored with a simple, inexpensive device known as a tread depth indicator gauge.

3 *Tyre Pressure Check*

Check the tyre pressures regularly with the tyres cold. Do not adjust the tyre pressures immediately after the vehicle has been used, or an inaccurate setting will result.

Tyre tread wear patterns

Shoulder Wear

Underinflation (wear on both sides)
Under-inflation will cause overheating of the tyre, because the tyre will flex too much, and the tread will not sit correctly on the road surface. This will cause a loss of grip and excessive wear, not to mention the danger of sudden tyre failure due to heat build-up.
Check and adjust pressures
Incorrect wheel camber (wear on one side)
Repair or renew suspension parts
Hard cornering
Reduce speed!

Centre Wear

Overinflation
Over-inflation will cause rapid wear of the centre part of the tyre tread, coupled with reduced grip, harsher ride, and the danger of shock damage occurring in the tyre casing.
Check and adjust pressures

If you sometimes have to inflate your car's tyres to the higher pressures specified for maximum load or sustained high speed, don't forget to reduce the pressures to normal afterwards.

Uneven Wear

Front tyres may wear unevenly as a result of wheel misalignment. Most tyre dealers and garages can check and adjust the wheel alignment (or "tracking") for a modest charge.
Incorrect camber or castor
Repair or renew suspension parts
Malfunctioning suspension
Repair or renew suspension parts
Unbalanced wheel
Balance tyres
Incorrect toe setting
Adjust front wheel alignment
Note: *The feathered edge of the tread which typifies toe wear is best checked by feel.*

Wiper blades

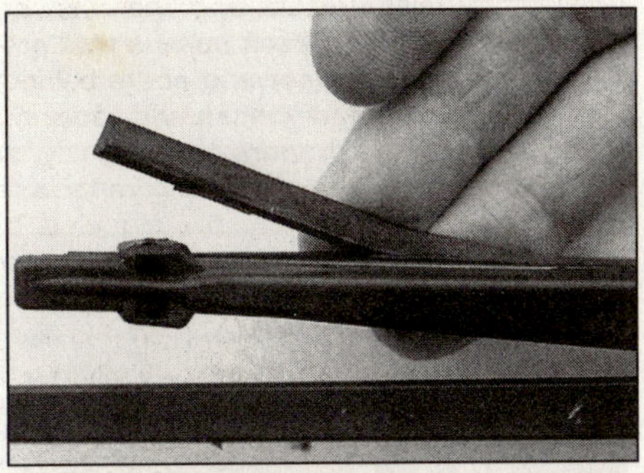

1 Check the condition of the wiper blades; if they are cracked or show any signs of deterioration, or if the glass swept area is smeared, renew them. Wiper blades should be renewed annually.

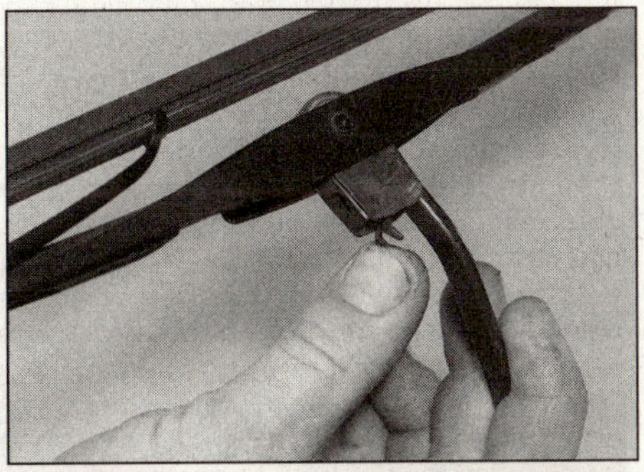

2 To remove a windscreen wiper blade, pull the arm fully away from the screen until it locks. Swivel the blade through 90°, then depress the locking clip at the base of the mounting block.

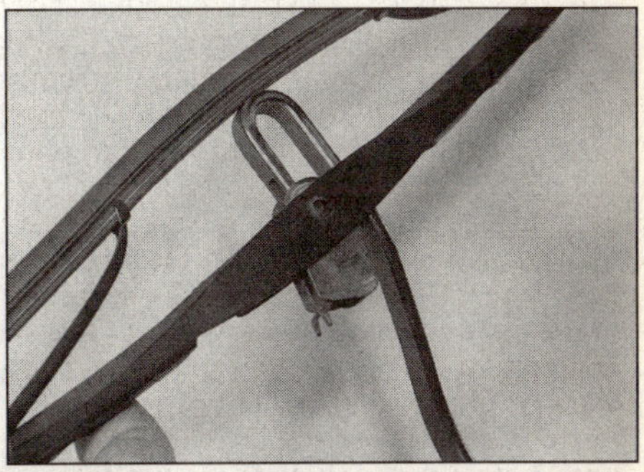

3 Slide the blade from the arm. Don't forget to check the tailgate wiper blade as well (where applicable).

Battery

Caution: *Before carrying out any work on the vehicle battery, read the precautions given in "Safety first" at the start of this manual.*

✔ Make sure that the battery tray is in good condition, and that the clamp is tight. Corrosion on the tray, retaining clamp and the battery itself can be removed with a solution of water and baking soda. Thoroughly rinse all cleaned areas with water. Any metal parts damaged by corrosion should be covered with a zinc-based primer, then painted.

✔ Periodically (approximately every three months), check the charge condition of the battery as described in Chapter 5A.

✔ If the battery is flat, and you need to jump start your vehicle, see ***Roadside Repairs***.

HAYNES HiNT

Battery corrosion can be kept to a minimum by applying a layer of petroleum jelly to the clamps and terminals after they are reconnected.

1 The battery is located in the right-hand rear bulkhead area of the engine compartment. The exterior of the battery should be inspected periodically for damage such as a cracked case or cover.

2 Check the tightness of the battery cable clamps to ensure good electrical connections. You should not be able to move them. Also check each cable for cracks and frayed conductors.

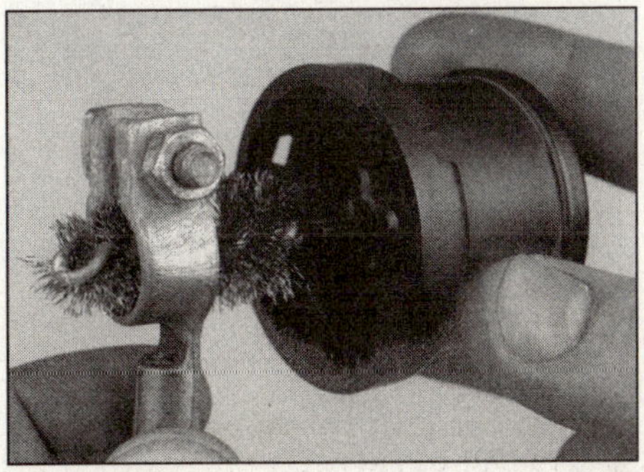

3 If corrosion (white, fluffy deposits) is evident, remove the cables from the battery terminals, clean them with a small wire brush, then refit them. Automotive stores sell a useful tool for cleaning the battery post and terminals.

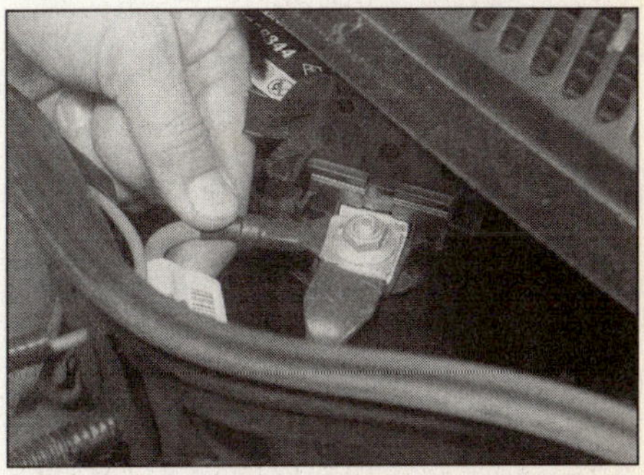

4 Note that the battery positive lead terminal can be disconnected by unscrewing the nut located under the plastic cover.

Bulbs and fuses

✔ Check all external lights and the horn. Refer to the appropriate Sections of Chapter 12 for details if any of the circuits are found to be inoperative.

✔ Visually check all accessible wiring connectors, harnesses and retaining clips for security, and for signs of chafing or damage.

HAYNES HiNT ***If you need to check your brake lights and indicators unaided, back up to a wall or garage door and operate the lights. The reflected light should show if they are working properly.***

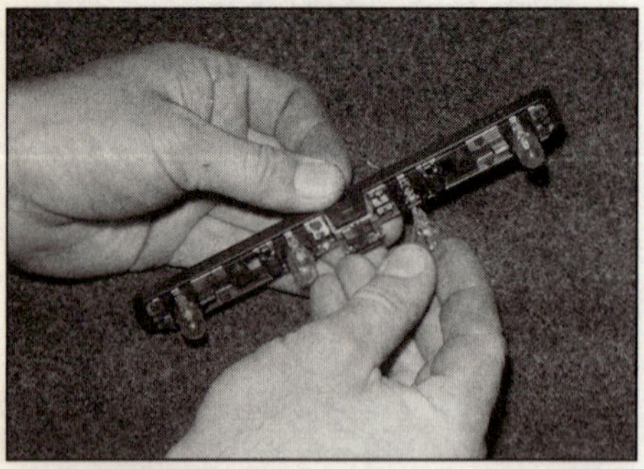

1 If a single indicator light, brake light or headlight has failed, it is likely that a bulb has blown and will need to be replaced. Refer to Chapter 12 for details. If both brake lights have failed, it is possible that the brake light switch operated by the brake pedal has failed. Refer to Chapter 9 for details.

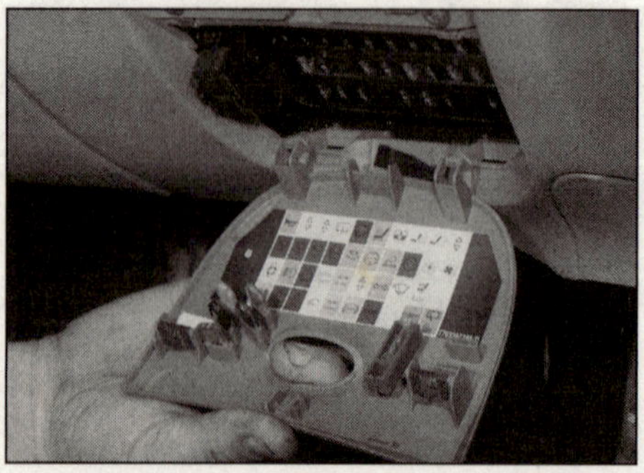

2 If more than one indicator light or tail light has failed it is likely that either a fuse has blown or that there is a fault in the circuit (see Chapter 12). The fuses are located under a cover on the driver's side of the facia panel (refer to Chapter 12).

3 To replace a blown fuse, remove it, where applicable, using the plastic tool provided. Fit a new fuse of the same rating, available from car accessory shops. It is important that you find the reason that the fuse blew (see *"Electrical fault finding"* in Chapter 12).

Lubricants and fluids

Engine:	
Petrol	Multigrade engine oil, viscosity SAE 10W/30 to 15W/50, to CCMC-G4 or better *(Duckhams QS, QXR, Hypergrade Plus, Hypergrade, or 10W-40 Motor Oil)*
Diesel	Multigrade engine oil, viscosity SAE 15W/40, to CCMC-PD2 or better *(Duckhams QS, QXR, Hypergrade Plus, Diesel, Hypergrade, or 10W-40 Motor Oil)*
Cooling system	Ethylene-glycol based antifreeze *(Duckhams Universal Antifreeze & Summer Coolant))*
Manual gearbox	SAE 75W-80W to API GL5 *(Duckhams Hypoid PT75W-80W)*
Automatic transmission:	
Main transmission	Elf Renaultmatic D2, Dexron type ATF *(Duckhams Unimatic)*
Final drive (sealed for life)	SAE 75W-80W to API GL5 *(Duckhams Hypoid PT75W-80W)*
Power steering reservoir	Elf Renaultmatic D2 or Mobil ATF 220 *(Duckhams Unimatic)*
Brake fluid reservoir	Hydraulic fluid to SAE J1703, DOT 3 or DOT 4 *(Duckhams Universal Brake and Clutch Fluid)*

Choosing your engine oil

Oils perform vital tasks in all engines. The higher the engine's performance, the greater the demand on lubricants to minimise wear as well as optimise power and economy. Duckhams tailors lubricants to the highest technical standards, meeting and exceeding the demands of all modern engines.

HOW ENGINE OIL WORKS

• Beating friction

Without oil, the surfaces inside your engine which rub together will heat, fuse and quickly cause engine seizure. Oil, and its special additives, forms a molecular barrier between moving parts, to stop wear and minimise heat build-up.

• Cooling hot spots

Oil cools parts that the engine's water-based coolant cannot reach, bathing the combustion chamber and pistons, where temperatures may exceed 1000°C. The oil assists in transferring the heat to the engine cooling system. Heat in the oil is also lost by air flow over the sump, and via any auxiliary oil cooler.

• Cleaning the inner engine

Oil washes away combustion by-products (mainly carbon) on pistons and cylinders, transporting them to the oil filter, and holding the smallest particles in suspension until they are flushed out by an oil change. Duckhams oils undergo extensive tests in the laboratory, and on the road.

OIL CARE
FOLLOW THE CODE
OIL BANK LINE
0800 66 33 66

***Note:** It is antisocial and illegal to dump oil down the drain. To find the location of your local oil recycling bank, call this number free.*

Engine oil types

Mineral oils are the "traditional" oils, generally suited to older engines and cars not used in harsh conditions. *Duckhams Hypergrade Plus* and *Hypergrade* are well suited for use in most popular family cars.
Diesel oils such as *Duckhams Diesel* are specially formulated for Diesel engines, including turbocharged models and 4x4s.
Synthetic oils are the state-of-the-art in lubricants, offering ultimate protection, but at a fairly high price. One such is *Duckhams QS*, for use in ultra-high performance engines.
Semi-synthetic oils offer high performance engine protection, but at less cost than full synthetic oils. *Duckhams QXR* is an ideal choice for hot hatches and hard-driven cars.

Tyre pressures (cold)

Note: *Pressures apply to original-equipment tyres, for normal use, and may vary if any other make or type of tyre is fitted, or for full-load/high-speed usage; check with a Renault dealer, or the tyre manufacturer or supplier for correct pressures if necessary.*

In the first instance, refer to the label attached to the driver's door for tyre pressures specific to your model, and always use these recommendations over the general information given for reference below.

	Front (psi)	Rear (psi)
Petrol engine models:		
Normal use	2.1 bars (30 psi)	2.0 bars (29 psi)
Fully laden or motorway driving	2.3 bars (33 psi)	2.2 bars (32 psi)
Non-Turbo diesel engine models:		
Normal use	2.1 bars (30 psi)	2.0 bars (29 psi)
Fully laden or motorway driving	2.3 bars (33 psi)	2.2 bars (32 psi)
Turbo diesel engine models:		
Normal use	2.2 bars (32 psi)	2.0 bars (32 psi)
Fully laden or motorway driving	2.4 bars (34 psi)	2.2 bars (32 psi)

Chapter 1 Part A:
Routine maintenance & servicing - petrol models

Contents

Degrees of difficulty

Easy, suitable for novice with little experience	**Fairly easy,** suitable for beginner with some experience	**Fairly difficult,** suitable for competent DIY mechanic	**Difficult,** suitable for experienced DIY mechanic	**Very difficult,** suitable for expert DIY or professional

Lubricants and fluids

Refer to "*Weekly checks*"

Capacities

Engine oil

Capacity (excluding filter):	
1390 cc (E7J) engines	3.5 litres
1598 cc (K7M) engines	3.5 litres
Oil filter capacity (all engines)	0.5 litre
Difference between MAX and MIN dipstick marks	1.5 litres

Cooling system

1390 cc (E7J) and 1598 cc (K7M) engines	6.0 litres

Fuel tank

All models	60 litres

Manual gearbox

JB1 and JB3 five-speed	3.4 litres

Automatic transmission (AD4)

Main section total capacity	5.7 litres
Main section drain and refill	3.5 litres
Final drive section	1.0 litre

Power-assisted steering reservoir	1.1 litres

Engine

Oil filter type:	
1390 cc (E7J)	Champion F103
1598 cc (K7M)	Champion F124

Cooling system

Antifreeze mixtures:	**Antifreeze**	**Water**
Protection to -23°C	35%	65%
Protection to -40°C	50%	50%
Auxiliary drivebelt tension:		
Models without air conditioning	84 SEEM units (see text) or 15 mm deflection	
Models with air conditioning	110 SEEM units (see text) or 15 mm deflection	

Fuel system

Specified idle speed (not adjustable - controlled by ECU):	
1390 cc/E7J 764 engine	750 ± 50 rpm
1598 cc/K7M 702 and 1598cc/K7M 720 engines	720 ± 50 rpm
1598 cc/K7M 703 engine	750 ± 50 rpm
Idle mixture CO content	Less than 0.3 % (controlled by ECU)
Air filter element:	
1390 cc (E7J) engine	Champion W190
1598 cc (K7M) engine	Champion W115
Fuel filter	Champion L224

Ignition system

Ignition timing	Controlled by the ECU - see Chapter 5B
Spark plugs:	
Type:	
1390 cc (E7J) engines	Champion RC9YCC
1598 cc (K7M) engines	Champion RC9YCC
Electrode gap	0.90 mm 0.0354"
Firing order	1-3-4-2
Location of No 1 cylinder	Flywheel end

Braking system

Front disc brakes:	
Pad thickness (including backing):	
New	18.0 mm
Minimum thickness	7.0 mm
Rear drum brakes:	
Shoe thickness (including backing):	
New	7.0 mm
Minimum thickness	2.5 mm

Electrical system

Windscreen wiper blade type:		
Driver's side	Champion X55	
Passenger side	Champion X51	
Torque wrench settings	**Nm**	**lbf ft**
Engine sump drain plug	15 to 25	11 to 18
Spark plugs	24 to 30	18 to 22
Roadwheel bolts	90	66

Maintenance schedule - petrol models

The maintenance intervals in this manual are provided with the assumption that you, not the dealer, will be carrying out the work. These are the minimum maintenance intervals recommended by us for vehicles driven daily. If you wish to keep your vehicle in peak condition at all times, you may wish to perform some of these procedures more often. We encourage frequent maintenance, because it enhances the efficiency, performance and resale value of your vehicle.

If the vehicle is driven in dusty areas, used to tow a trailer, or driven frequently at slow speeds (idling in traffic) or on short journeys, more frequent maintenance intervals are recommended.

When the vehicle is new, it should be serviced by a factory-authorised dealer service department, in order to preserve the factory warranty.

Every 250 miles (400 km) or weekly

- ☐ Refer to "*Weekly checks*"

Every 10 000 miles (16 000 km)

- ☐ Renew the engine oil and filter (Section 3)
- ☐ Check the front brake pad thickness and discs (Section 4)
- ☐ Check the condition of the exhaust system and mountings (Section 5)
- ☐ Check the suspension and steering components (Section 6)
- ☐ Check the driveshaft wear and gaiter (Section 7)
- ☐ Check all underbonnet components, hoses and brake lines for fluid leaks (Section 8)
- ☐ Check the tightness of the roadwheel bolts (Section 9)
- ☐ Renew the pollen filter (Section 10)
- ☐ Check the automatic transmission fluid level (Section 11)
- ☐ Check the operation of the clutch (Section 12)
- ☐ Check the exhaust emissions (Section 13)
- ☐ Check the operation of the air conditioning system (Section 14)
- ☐ Check the operation of all electrical systems (Section 15)
- ☐ Check the vehicle bodywork (Section 16)

Every 20 000 miles (32 000 km)

In addition to all the items listed previously, carry out the following:

- ☐ Check the condition of the seat belts (Section 17)
- ☐ Check the condition of the auxiliary drivebelts (Section 18)
- ☐ Check the headlight beam adjustment (Section 19)
- ☐ Renew the spark plugs and check the ignition system (Section 20)
- ☐ Renew the air filter element (Section 21)

Every 40 000 miles (64 000 km)

In addition to all the items listed previously, carry out the following:

- ☐ Check the operation of the handbrake (Section 22)
- ☐ Check the heating system (Section 23)
- ☐ Lubricate all hinges and locks (Section 24)
- ☐ Check the spare fuses are in place (Section 25)
- ☐ Check the front wheel alignment (Section 26)
- ☐ Check the rear brake shoe thickness (Section 27)
- ☐ Check the manual transmission oil level (Section 28)
- ☐ Renew the automatic transmission fluid (Section 29)
- ☐ Renew the fuel filter (Section 30)
- ☐ Renew the brake fluid (Section 31)
- ☐ Carry out a road test (Section 32)

Every 50 000 miles (80 000 km)

In addition to all the items listed previously, carry out the following:

- ☐ Clean the EGR valve (KTM petrol engine) (Section 33)

Every 70 000 miles (112 000 km)

In addition to all the items listed previously, carry out the following:

- ☐ Renew the timing belt (Section 34)

Every 2 years (regardless of mileage)

In addition to all the items listed previously, carry out the following:

- ☐ Renew the coolant (Section 35)

Underbonnet view of a 1.4 litre petrol model

1 Engine oil filler cap
2 Engine oil level dipstick
3 Battery
4 Brake fluid reservoir
5 Electrical connector box
6 Coolant expansion tank
7 Suspension strut upper mounting
8 Air filter housing
9 Engine oil filter
10 Power steering fluid reservoir
11 Alternator
12 Engine management electronic control unit
13 Washer fluid reservoir
14 Ignition coil and power module

Underbonnet view of a 1.6 litre petrol model

1 Engine oil filler cap
2 Engine oil level dipstick
3 Battery
4 Brake fluid reservoir
5 Electrical connector box
6 Coolant expansion tank
7 Suspension strut upper mounting
8 Air filter housing
9 Engine oil filter
10 Power steering fluid reservoir
11 Alternator
12 Engine management electronic control unit
13 Washer fluid reservoir
14 Ignition coils
15 MAP sensor

Front underbody view of a 1.6 litre petrol model

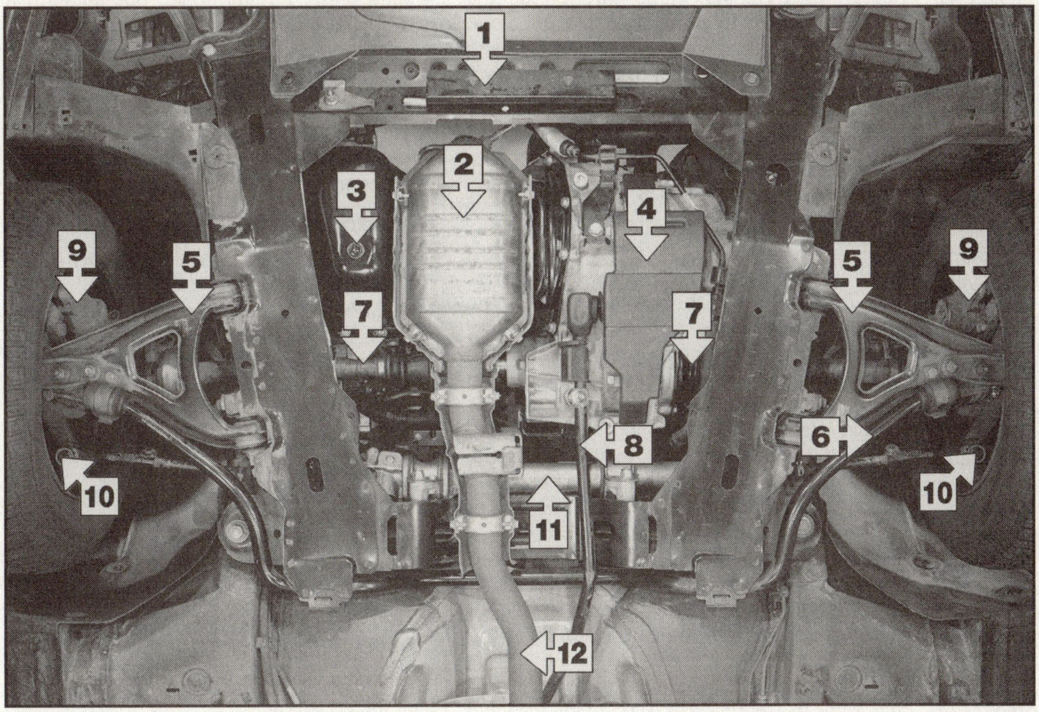

1 Radiator
2 Catalytic converter
3 Engine oil drain plug
4 Manual transmission
5 Front suspension lower arms
6 Front anti-roll arm
7 Front driveshafts
8 Gearchange rod
9 Front brake calipers
10 Steering track rod ends
11 Steering gear
12 Intermediate exhaust pipe

Rear underbody view of a 1.6 litre petrol model

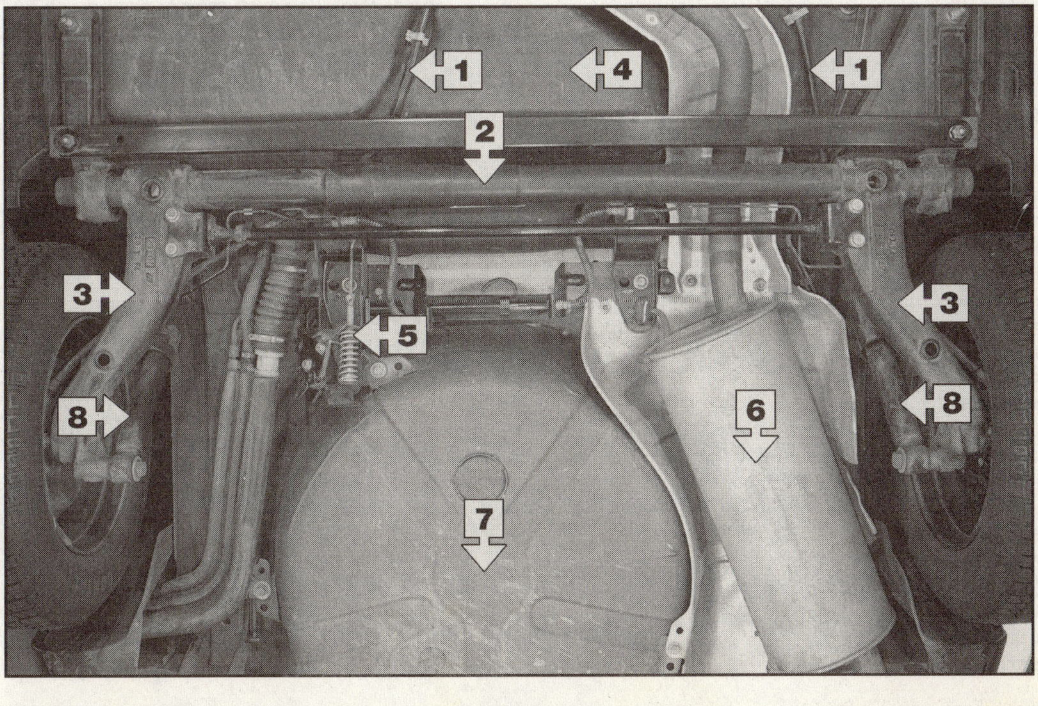

1 Handbrake cables
2 Rear axle assembly
3 Rear suspension trailing arms
4 Fuel tank
5 Rear brake pressure regulating valve
6 Exhaust tailpipe and silencer
7 Spare wheel well
8 Rear suspension shock absorbers

1 Introduction

This Chapter is designed to help the home mechanic maintain his/her vehicle for safety, economy, long life and peak performance.

The Chapter contains a master maintenance schedule, followed by Sections dealing specifically with each task in the schedule. Visual checks, adjustments, component renewal and other helpful items are included. Refer to the accompanying illustrations of the engine compartment and the underside of the vehicle for the locations of the various components.

Servicing your vehicle in accordance with the mileage/time maintenance schedule and the following Sections will provide a planned maintenance programme, which should result in a long and reliable service life. This is a comprehensive plan, so maintaining some items but not others at the specified service intervals, will not produce the same results.

As you service your vehicle, you will discover that many of the procedures can - and should - be grouped together, because of the particular procedure being performed, or because of the proximity of two otherwise-unrelated components to one another. For example, if the vehicle is raised for any reason, the exhaust can be inspected at the same time as the suspension and steering components.

The first step in this maintenance programme is to prepare yourself before the actual work begins. Read through all the Sections relevant to the work to be carried out, then make a list and gather all the parts and tools required. If a problem is encountered, seek advice from a parts specialist, or a dealer service department.

2 Intensive maintenance

1 If, from the time the vehicle is new, the routine maintenance schedule is followed closely, and frequent checks are made of fluid levels and high-wear items, as suggested throughout this manual, the engine will be kept in relatively good running condition, and the need for additional work will be minimised.

2 It is possible that there will be times when the engine is running poorly due to the lack of regular maintenance. This is even more likely if a used vehicle, which has not received regular and frequent maintenance checks, is purchased. In such cases, additional work may need to be carried out, outside of the regular maintenance intervals.

3 If engine wear is suspected, a compression test (refer to Chapter 2A) will provide valuable information regarding the overall performance of the main internal components. Such a test can be used as a basis to decide on the extent of the work to be carried out. If, for example, a compression test indicates serious internal engine wear, conventional maintenance as described in this Chapter will not greatly improve the performance of the engine, and may prove a waste of time and money, unless extensive overhaul work is carried out first.

4 The following series of operations are those usually required to improve the performance of a generally poor-running engine:

Primary operations

a) Clean, inspect and test the battery (see "Weekly checks").
b) Check all the engine-related fluids (see "Weekly checks").
c) Check the condition and tension of the auxiliary drivebelt (Section 18).
d) Check the condition of the air filter element, and renew if necessary (Section 21).
e) Check the fuel filter (Section 30).
f) Check the condition of all hoses, and check for fluid leaks (Section 8).

5 If the above operations do not prove fully effective, carry out the following secondary operations:

Secondary operations

All items listed under "Primary operations", plus the following:

a) Check the charging system (Chapter 5A).
b) Check the ignition system (Chapter 5B).
c) Check the fuel system (Chapter 4A).

Every 10 000 miles (16 000 km)

3 Engine oil and filter renewal

1 Frequent oil and filter changes are the most important preventative maintenance procedures which can be undertaken by the DIY owner. As engine oil ages, it becomes diluted and contaminated, which leads to premature engine wear.

2 Before starting this procedure, gather together all the necessary tools and materials. Also make sure that you have plenty of clean rags and newspapers handy, to mop up any spills. Ideally, the engine oil should be warm, as it will drain more easily, and more built-up sludge will be removed with it. Take care not to touch the exhaust or any other hot parts of the engine when working under the vehicle. To avoid any possibility of scalding, and to protect yourself from possible skin irritants and other harmful contaminants in used engine oils, it is advisable to wear gloves when carrying out this work.

3 Firmly apply the handbrake then jack up the front of the car and support it on axle stands (see *"Jacking and vehicle support"*). Where applicable, remove the engine undershield.

4 Remove the oil filler cap.

5 Using a spanner, or preferably a suitable socket and bar, slacken the drain plug about half a turn. Position the draining container under the drain plug, then remove the plug completely **(see Haynes Hint)**.

6 Allow some time for the oil to drain, noting that it may be necessary to reposition the container as the oil flow slows to a trickle.

7 After all the oil has drained, wipe the drain plug and the sealing washer with a clean rag. Examine the condition of the sealing washer, and renew it if it shows signs of scoring or other damage which may prevent an oil-tight seal. Clean the area around the drain plug opening, and refit the plug complete with the washer and tighten it securely.

8 Move the container into position under the oil filter which is located on the front of the cylinder block.

9 Use an oil filter removal tool to slacken the filter initially, then unscrew it by hand the rest of the way **(see illustrations)**. Empty the oil from the old filter into the container.

HAYNES HiNT

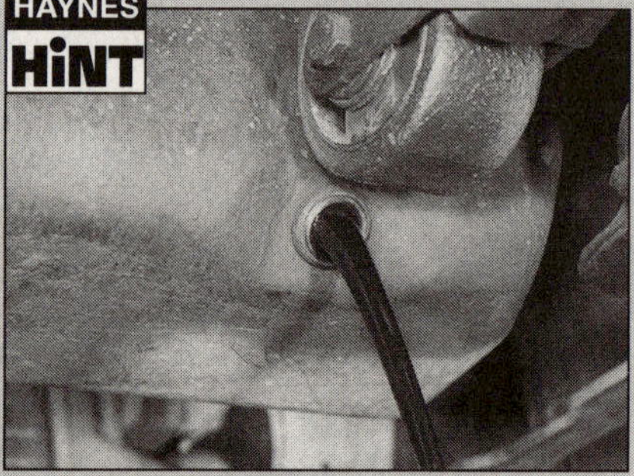

As the drain plug threads release, move it sharply away so the stream of oil issuing from the sump runs into the container, not up your sleeve!

3.9a Using an oil filter removal tool to slacken the oil filter

3.9b Removing the oil filter

3.11 Apply a light coating of clean engine oil to the sealing ring on the new filter

10 Use a clean rag to remove all oil, dirt and sludge from the filter sealing area on the engine. Check the old filter to make sure that the rubber sealing ring has not stuck to the engine. If it has, carefully remove it.

11 Apply a light coating of clean engine oil to the sealing ring on the new filter, then screw the filter into position on the engine **(see illustration)**. Tighten the filter firmly by hand only - **do not** use any tools.

12 Remove the old oil and all tools from under the vehicle then lower the vehicle to the ground.

13 Fill the engine through the filler hole, using the correct grade and type of oil (refer to "*Weekly checks*" for details of topping-up). Pour in half the specified quantity of oil first, then wait a few minutes for the oil to drain into the sump. Continue to add oil, a small quantity at a time, until the level is up to the lower mark on the dipstick. Adding approximately a further 1.0 litre will bring the level up to the upper mark on the dipstick.

14 Start the engine and run it for a few minutes, while checking for leaks around the oil filter seal and the sump drain plug. Note that there may be a delay of a few seconds before the low oil pressure warning light goes out when the engine is first started, as the oil circulates through the new oil filter and the engine oil galleries before the pressure builds up.

15 Stop the engine, and wait a few minutes for the oil to settle in the sump once more. With the new oil circulated and the filter now completely full, recheck the level on the dipstick, and add more oil as necessary.

16 Dispose of the used engine oil safely with reference to "*General repair procedures*".

OIL CARE
FOLLOW THE CODE
OIL BANK LINE
0800 66 33 66

Note: It is antisocial and illegal to dump oil down the drain. To find the location of your local oil recycling bank, call this number free.

4 Front brake pad and disc check

1 Apply the handbrake, then jack up the front of the car and support it securely on axle stands. Remove the front roadwheels.

2 For a comprehensive check, the brake pads should be removed and cleaned. The operation of the caliper can then also be checked, and the condition of the brake disc itself can be fully examined on both sides. Refer to Chapter 9 for further information **(see Haynes Hint)**.

3 On completion refit the roadwheels and lower the car to the ground.

5 Exhaust system check

1 With the engine cold (at least an hour after the vehicle has been driven), check the complete exhaust system from the engine to the end of the tailpipe. The exhaust system is most easily checked with the vehicle raised on a hoist, or suitably supported on axle stands, so that the exhaust components are readily visible and accessible.

2 Check the exhaust pipes and connections for evidence of leaks, severe corrosion and damage. Make sure that all brackets and mountings are in good condition, and that all relevant nuts and bolts are tight. Leakage at any of the joints or in other parts of the system will usually show up as a black sooty stain in the vicinity of the leak.

3 Rattles and other noises can often be traced to the exhaust system, especially the brackets and mountings. Try to move the pipes and silencers. If the components are able to come into contact with the body or suspension parts, secure the system with new mountings. Otherwise separate the joints (if possible) and twist the pipes as necessary to provide additional clearance.

HAYNES HiNT

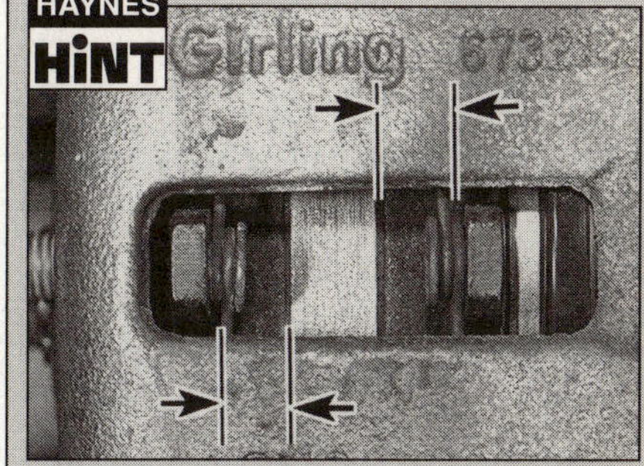

For a quick check, the thickness of each brake pad can be measured through the aperture in the caliper body

6 Suspension and steering check

Front suspension and steering check

1 Raise the front of the vehicle, and securely support it on axle stands.

2 Visually inspect the balljoint dust covers and the steering rack-and-pinion gaiters for splits, chafing or deterioration. Any wear of these components will cause loss of lubricant, together with dirt and water entry, resulting in rapid deterioration of the balljoints or steering gear.

3 On vehicles with power steering, check the fluid hoses for chafing or deterioration, and the pipe and hose unions for fluid leaks. Also check for signs of fluid leakage under pressure from the steering gear rubber gaiters, which would indicate failed fluid seals within the steering gear.

4 Grasp the roadwheel at the 12 o'clock and 6 o'clock positions, and try to rock it **(see illustration)**. Very slight free play may be felt, but if the movement is appreciable, further investigation is necessary to determine the source. Continue rocking the wheel while an assistant depresses the footbrake. If the movement is now eliminated or significantly reduced, it is likely that the hub bearings are at fault. If the free play is still evident with the

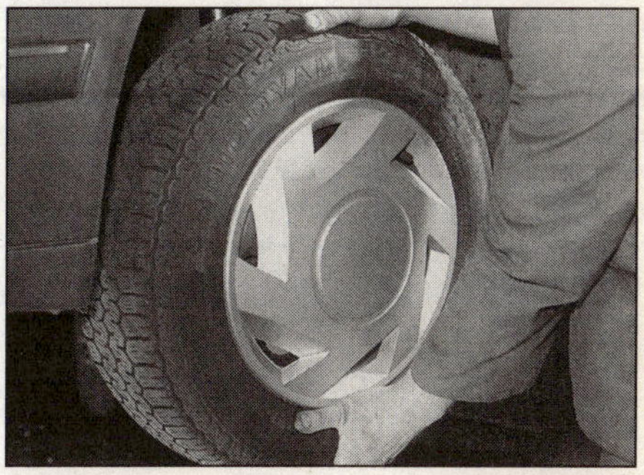
6.4 Check for wear in the hub bearings by grasping the wheel and trying to rock it

footbrake depressed, then there is wear in the suspension joints or mountings.

5 Now grasp the wheel at the 9 o'clock and 3 o'clock positions, and try to rock it as before. Any movement felt now may again be caused by wear in the hub bearings or the steering track-rod balljoints. If the outer balljoint is worn, the visual movement will be obvious. If the inner joint is suspect, it can be felt by placing a hand over the rack-and-pinion rubber gaiter and gripping the track-rod. If the wheel is now rocked, movement will be felt at the inner joint if wear has taken place.

6 Using a large screwdriver or flat bar, check for wear in the suspension mounting bushes by levering between the relevant suspension component and its attachment point. Some movement is to be expected, as the mountings are made of rubber, but excessive wear should be obvious. Also check the condition of any visible rubber bushes, looking for splits, cracks or contamination of the rubber.

7 With the car standing on its wheels, have an assistant turn the steering wheel back and forth, about an eighth of a turn each way. There should be very little, if any, lost movement between the steering wheel and roadwheels. If this is not the case, closely observe the joints and mountings previously described. In addition, check the steering column universal joints for wear, and also check the rack-and-pinion steering gear itself.

Rear suspension check

8 Chock the front wheels, then jack up the rear of the vehicle and support securely on axle stands.

9 Working as described previously for the front suspension, check the rear hub bearings, the suspension bushes and the shock absorber mountings for wear.

Suspension strut/shock absorber check

10 Check for any signs of fluid leakage around the strut/shock absorber body, or from the rubber gaiter around the piston rod. Should any fluid be noticed, the strut/shock absorber is defective internally, and should be renewed. **Note:** *Struts/shock absorbers should always be renewed in pairs on the same axle.*

11 The efficiency of the strut/shock absorber may be checked by bouncing the vehicle at each corner. Generally speaking, the body will return to its normal position and stop after being depressed. If it rises and returns on a rebound, the strut/shock absorber is probably suspect. Also examine the upper and lower mountings for any signs of wear.

7 Driveshaft rubber gaiter and CV joint check

1 With the car raised and securely supported on axle stands, turn the steering onto full-lock, then slowly rotate the roadwheel. Inspect the condition of the outer constant velocity (CV) joint rubber gaiters while squeezing the gaiters to open out the folds **(see illustration)**. Check for signs of cracking, splits or deterioration of the rubber, which may allow the grease to escape and lead to the entry of water and grit into the joint. Also check the security and condition of the retaining clips/fasteners. Repeat these checks on the inner CV joints. If any damage or deterioration is found, the gaiters should be renewed as described in Chapter 8.

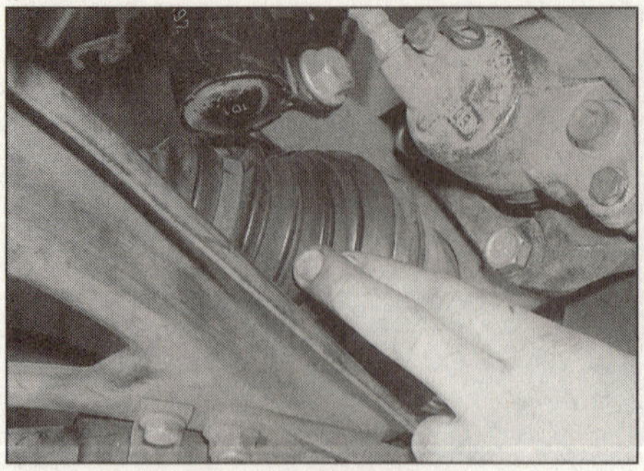

7.1 Checking a driveshaft outer gaiter for signs of damage

2 At the same time, check the general condition of the CV joints themselves by first holding the driveshaft and attempting to rotate the roadwheel. Repeat this check by holding the inner joint and attempting to rotate the driveshaft. Any appreciable movement indicates wear in the joints, in the driveshaft splines, or a loose driveshaft nut.

8 Hose, brake line and fluid leak check

1 Visually inspect the engine joint faces, gaskets and seals for any signs of water or oil leaks. Pay particular attention to the areas around the cylinder head cover, cylinder head, oil filter and sump joint faces. Bear in mind that, over a period of time, some very slight seepage from these areas is to be expected - what you are really looking for is any indication of a serious leak **(see Haynes Hint)**. Should a leak be found, renew the offending gasket or oil seal by referring to the appropriate Chapters in this manual.

HAYNES HiNT

A leak in the cooling system will usually show up as white- or rust-coloured deposits on the area adjoining the leak

2 Also check the security and condition of all the engine-related pipes and hoses, and all braking system pipes and hoses. Ensure that all cable-ties or securing clips are in place, and in good condition. Clips which are broken or missing can lead to chafing of the hoses, pipes or wiring, which could cause more serious problems in the future.

3 Carefully check the radiator hoses and heater hoses along their entire length. Renew any hose which is cracked, swollen or deteriorated. Cracks will show up better if the hose is squeezed. Pay close attention to the hose clips that secure the hoses to the cooling system components. Hose clips can pinch and puncture hoses, resulting in cooling system leaks. If the crimped-type hose clips are used, it may be a good idea to replace them with standard worm-drive clips.

4 Inspect all the cooling system components (hoses, joint faces, etc) for leaks.

5 Where any problems are found on system components, renew the component or gasket with reference to Chapter 3.

6 With the vehicle raised, inspect the fuel tank and filler neck for punctures, cracks and other damage. The connection between the filler neck and tank is especially critical. Sometimes a rubber filler neck or connecting hose will leak due to loose retaining clamps or deteriorated rubber.

7 Carefully check all rubber hoses and metal fuel lines leading away from the fuel tank. Check for loose connections, deteriorated hoses, crimped lines, and other damage. Pay particular attention to the vent pipes and hoses, which often loop up around the filler neck and can become blocked or crimped. Follow the lines to the front of the vehicle, carefully inspecting them all the way. Renew damaged sections as necessary. Similarly, whilst the vehicle is raised, take the opportunity to inspect all underbody brake fluid pipes and hoses.

8 From within the engine compartment, check the security of all fuel, vacuum and brake hose attachments and pipe unions, and inspect all hoses for kinks, chafing and deterioration.

9 Where applicable, check the condition of the power steering fluid pipes and hoses.

9 Roadwheel bolt check

1 Where applicable, remove the wheel trims, and slacken the roadwheel bolts slightly.

2 Tighten the bolts to the specified torque, using a torque wrench.

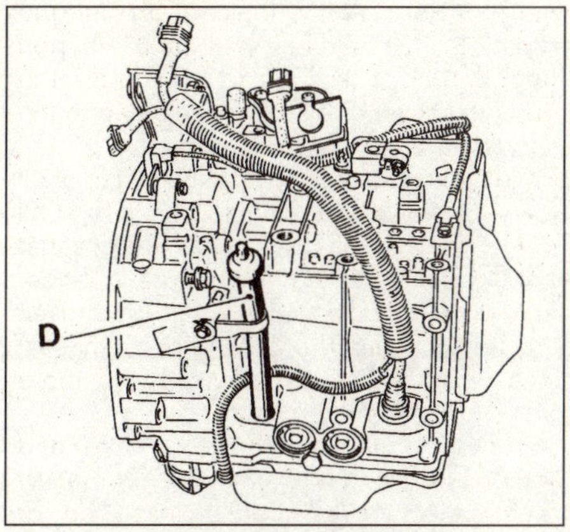

11.1 Automatic transmission fluid filler tube (D)

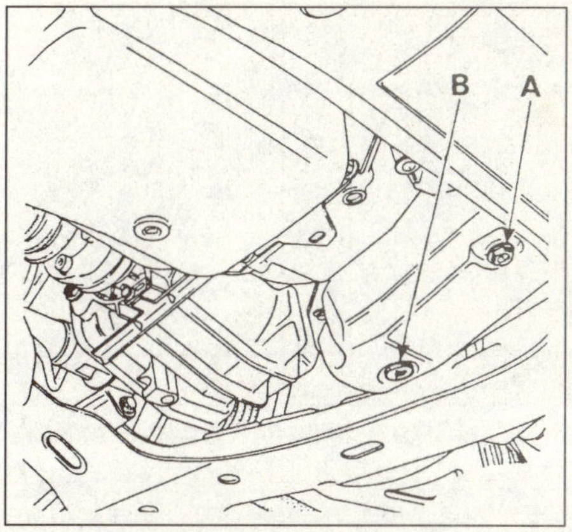

11.4 Automatic transmission level plug (A) and drain plug (B)

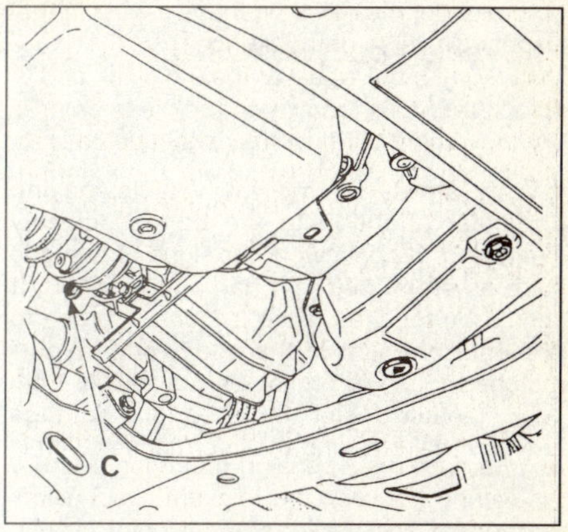

11.8 Automatic transmission final drive filler/level plug (C)

10 Pollen filter renewal

1 Open up the bonnet and unclip the small cover situated on the left-hand side of the intake vent panel.
2 Release the retaining clips and remove the pollen filter cover from the top of the blower motor housing.
3 Slide the pollen filter out of position and discard it.
4 Wipe clean the cover and housing and slide the new filter into position.
5 Securely clip the pollen filter cover onto the housing then clip the vent panel cover into position.

11 Automatic transmission fluid level check

Transmission fluid

Note: *The transmission fluid level checking procedure is particularly complicated, and the home mechanic would be well-advised to take the vehicle to a Renault dealer to have the work carried out, as special test equipment is necessary to carry out the check. However, the following procedure is given for those who may have access to this equipment.*
1 Remove the plug from the top of the filler tube 'D' on the front of the transmission **(see illustration)**. Add 0.5 litre of the specified fluid to the transmission via the filler tube, using a clean funnel with a fine-mesh filter, then refit the plug.
2 Position the vehicle over an inspection pit, on a ramp, or jack it up and support it on axle stands, ensuring that the vehicle remains level. Remove the engine undershield where fitted.
3 Connect the Renault XR25 test meter to the diagnostic socket, and enter 'DO4' then number '04'. With the selector lever in 'Park', run the engine at idle speed until the fluid temperature reaches 60°C.
4 With the engine still running, unscrew the level plug from the transmission **(see illustration)**. Allow the excess fluid to run out into a calibrated container for 20 seconds, then refit the plug. The amount of fluid should be more than 0.1 litre; if it is not, the fluid level in the transmission is incorrect.
5 If the level is incorrect, add an extra 1 litre of the specified fluid to the transmission, through the filler tube after removing the vent. Allow the transmission to cool down, then repeat the checking procedure again as described in the previous paragraphs. Repeat the procedure as required until more than the specified amount of fluid is drained as described in the previous paragraph, indicating that the transmission fluid level is correct, then securely tighten the level plug.

Final drive oil

6 This is not a routine operation, but it may be considered necessary if there is reason to suspect that the oil level is incorrect - for instance if there has been an oil leak.
7 Either position the vehicle over an inspection pit, or jack up the front and rear of the vehicle and support it on axle stands (see *"Jacking and vehicle support"*). The vehicle must be level for the check to be accurate. Remove the engine undershield where fitted.
8 Unscrew the final drive filler/level plug located behind the right-hand driveshaft on the right-hand side of the transmission **(see illustration)**.
9 Check that the level of the oil is up to the bottom of the plug hole. If not, inject oil of the correct grade into the hole until if overflows.
10 Clean the plug, then refit and tighten it.
11 Lower the vehicle to the ground.

12 Clutch check

1 Check and adjust the clutch cable as described in Chapter 6. Check that the cable moves freely and easily and lubricate its exposed section with multi-purpose grease.

13 Exhaust emission check

1 Both the idle speed and mixture (exhaust gas CO level) are automatically controlled by the ECU and cannot be adjusted. If either the idle speed or mixture settings are incorrect then a fault is present in the engine management system and the vehicle should be taken to a Renault dealer for testing.

14 Air conditioning system check

1 If the air conditioning system is functioning correctly, the only check needed is to ensure that the compressor drivebelt is correctly tensioned and in good condition (Section 18). If there is a problem with the air conditioning system the vehicle must be taken to a Renault dealer for testing.

15 Electrical systems check

1 Check the operation of all electrical equipment, ie, lights, direction indicators, horn, etc. Refer to the appropriate Sections of Chapter 12 for details if any of the circuits are found to be inoperative.
2 Note that the stop-light switch adjustment is described in Chapter 9.
3 Visually check all accessible wiring connectors, harnesses and retaining clips for security, and for signs of chafing or damage. Rectify any faults found.

16 Vehicle bodywork check

1 Work slowly around the vehicle, paying particular attention to the bonnet, closely examining the paintwork for signs of damage such as stone chips. If damage is found, it should be repaired using a touch-in brush to prevent corrosion of the bodywork; colour-matched touch-in brushes are available from your Renault dealer. More serious damage can be repaired using the information given in Chapter 11.

Every 20 000 miles (32 000 km)

17 Seat belt check

1 Carefully examine the seat belt webbing for cuts, or any signs of serious fraying or deterioration. If the belt is of the retractable type, pull the belt all the way out of the inertia reel, and examine the full extent of the webbing.
2 Fasten and unfasten the belt, ensuring that the locking mechanism holds securely, and releases properly when intended. If the belt is of the retractable type, check also that the retracting mechanism operates correctly when the belt is released.
3 Check the security of all seat belt mountings and attachments which are accessible without removing any trim or other components **(see illustration)**.

18 Auxiliary drivebelt(s) check and renewal

Checking

1 The auxiliary drivebelt is located at the right-hand side of the engine.
2 Two different drivebelt configurations may be encountered, depending on whether the vehicle is equipped with air conditioning.

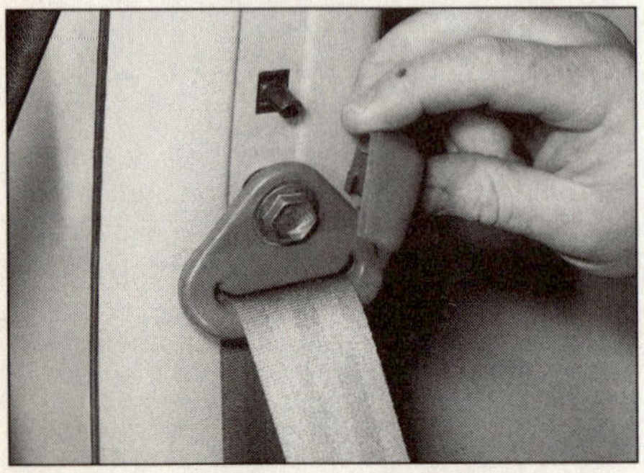

17.3 Check the security of the seat belt mountings

3 Due to their function and material makeup, drivebelts are prone to failure after a period of time and should therefore be inspected, and if necessary adjusted periodically.
4 Since the drivebelt is located very close to the right-hand side of the engine compartment, it is possible to gain better access by raising the front of the vehicle and removing the right-hand wheel, then removing the splash shield from inside the wheelarch.
5 With the engine stopped, inspect the full length of the drivebelt for cracks and separation of the belt plies. It will be necessary to turn the engine (using a spanner or socket and bar on the crankshaft pulley bolt) in order to move the belt from the pulleys so that the belt can be inspected thoroughly. Twist the belt between the pulleys so that both sides can be viewed. Also check for fraying, and glazing which gives the belt a shiny appearance. Check the pulleys for nicks, cracks, distortion and corrosion.

Tensioning

Note: *Renault recommend that a belt is renewed whenever it is slackened.*
6 The tension of the belt is checked by pushing midway between the pulleys at the point indicated **(see illustrations)**. Renault technicians use a special spring-tensioned tool which applies a force of 30 N (6.75 lbf) to the belt and then measures the deflection. An alternative arrangement can be made by using a straight-edge, steel rule and spring balance. Hold the straight-edge across the two pulleys, then position the steel rule on the belt, apply the force with the spring balance, and measure the deflection. Alternatively, depressing the belt under moderate finger-pressure should give an adequate approximation.
7 If adjustment is necessary on the power-assisted steering pump drivebelt, loosen the pump mounting and pivot bolts then loosen the adjustment locknut and turn the adjustment bolt as required until the tension is correct **(see illustrations)**.
8 If adjustment is necessary on the alternator drivebelt, jack up the front of the car and support on axle stands (see *"Jacking and vehicle support"*). Remove the right-hand front roadwheel and wheelarch liner, then loosen the alternator lower mounting bolt using a

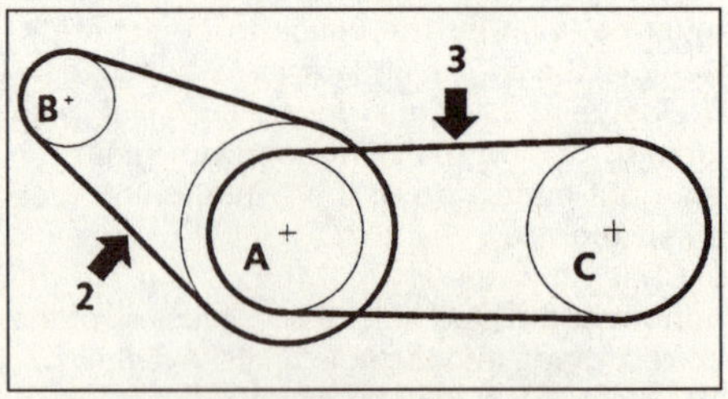

18.6a Auxiliary drivebelt tension checking - models without air conditioning

2 Tension checking point for the alternator drivebelt
3 Tension checking point for the power-assisted steering drivebelt
A Crankshaft pulley
B Alternator
C Power-assisted steering pump

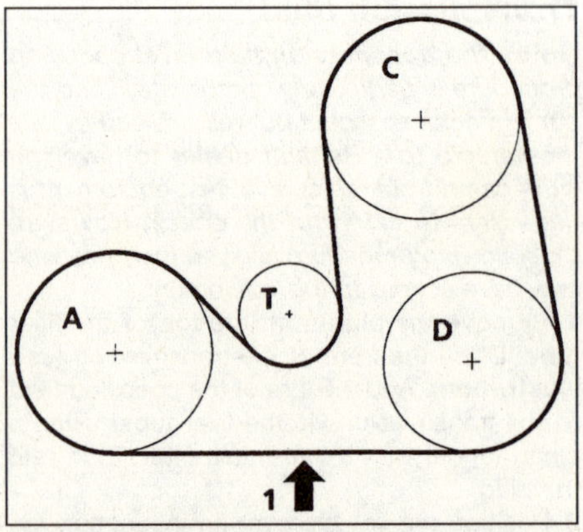

18.6b Auxiliary drivebelt tension checking - models with air conditioning

1 Tension checking point
A Crankshaft pulley
C Power-assisted steering pump
D Air conditioning compressor
T Tensioner

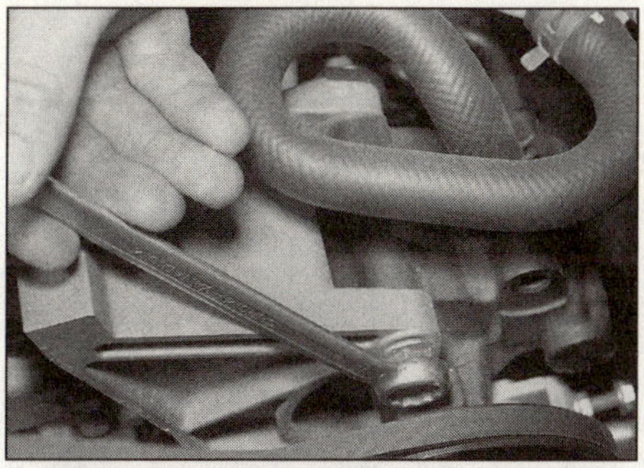

18.7a Loosen the power steering pump mounting bolts . . .

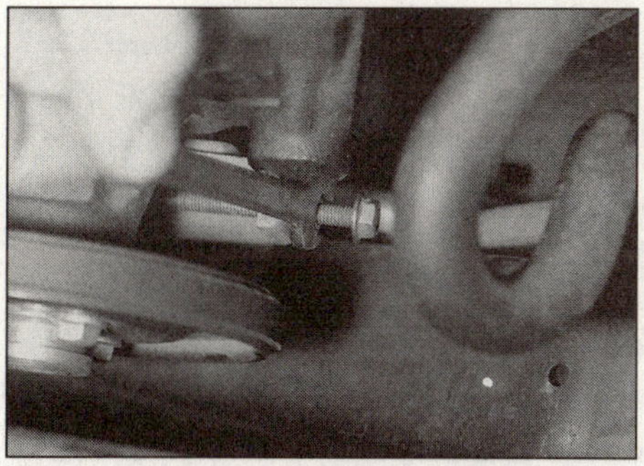

18.7b . . . then loosen the locknut and turn the adjustment bolt as required

18.8a Insert the socket through the hole in the inner panel . . .

18.8b . . . loosen the alternator lower mounting bolt . . .

18.8c . . . then loosen the adjustment bolt and re-position the alternator to tension the drivebelt

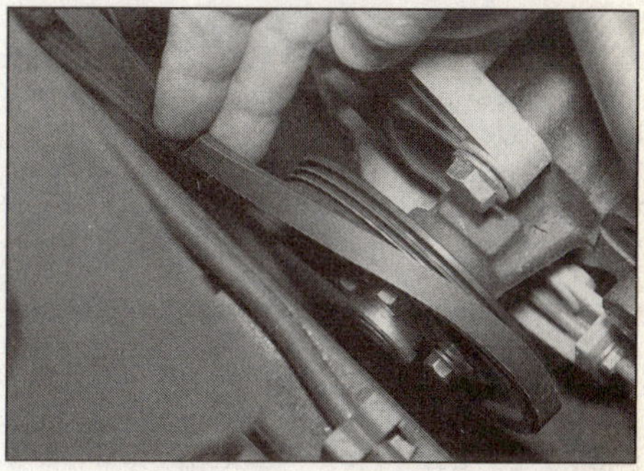

18.10 Removing the power steering pump drivebelt

socket through the hole in the inner panel. In the engine compartment, loosen the adjustment bolt, then move the alternator as required to tension the drivebelt and re-tighten the bolt. Refit the liner and roadwheel, and lower the car to the ground **(see illustrations)**.

9 Run the engine for about 5 minutes, then recheck the tension.

Renewal

10 To remove a belt, slacken the belt tension fully as described previously, then slip the belt off the pulleys. Note that the power steering pump drivebelt must be removed before the alternator drivebelt can be removed **(see illustration)**.

11 Fit the new belt ensuring that it is located correctly on the pulley grooves.

12 With the belt in position, adjust its tension as previously described.

19 Headlight beam adjustment

1 Refer to Chapter 12.

20 Spark plug renewal and ignition system check

Spark plug renewal

1 The correct functioning of the spark plugs is vital for the correct running and efficiency of the engine. It is essential that the plugs fitted are appropriate for the engine (a suitable type is specified at the beginning of this Chapter). If this type is used and the engine is in good condition, the spark plugs should not need attention between scheduled replacement intervals. Spark plug cleaning is rarely necessary, and should not be attempted unless specialised equipment is available, as damage can easily be caused to the firing ends.

2 If the marks on the original-equipment spark plug (HT) leads cannot be seen, mark the leads to correspond to the cylinder the lead serves. Pull the leads from the plugs by gripping the end fitting, not the lead, otherwise the lead connection may be fractured.

3 It is advisable to remove the dirt from the spark plug recesses using a clean brush, vacuum cleaner or compressed air before removing the plugs, to prevent dirt dropping into the cylinders.

4 Unscrew the plugs using a spark plug spanner, suitable box spanner or a deep socket and extension bar **(see illustrations)**. Keep the socket aligned with the spark plug - if it is forcibly moved to one side, the ceramic insulator may be broken off. As each plug is removed, examine it as follows.

5 Examination of the spark plugs will give a good indication of the condition of the engine. If the insulator nose of the spark plug is clean and white, with no deposits, this is indicative of a weak mixture or too hot a plug (a hot plug transfers heat away from the electrode slowly, a cold plug transfers heat away quickly).

6 If the tip and insulator nose are covered with hard black-looking deposits, then this is indicative that the mixture is too rich. Should the plug be black and oily, then it is likely that

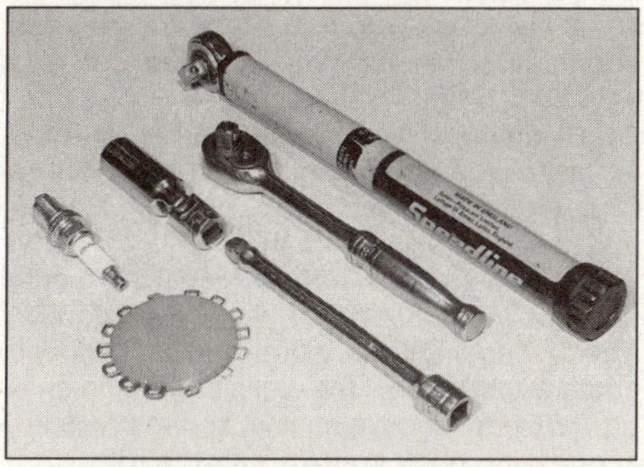

20.4a Tools required for spark plug removal, gap adjustment and refitting

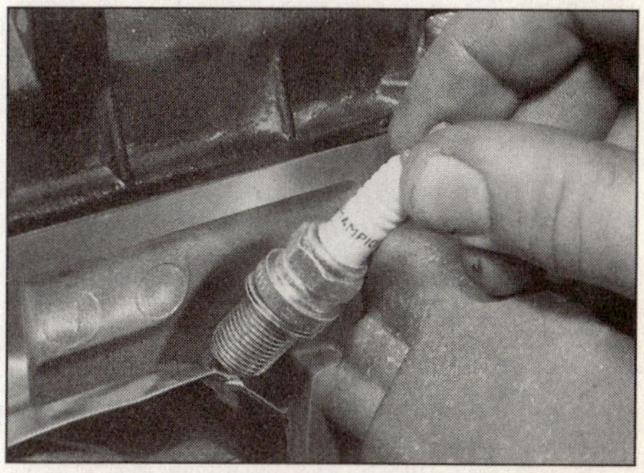

20.4b Removing a spark plug from the cylinder head

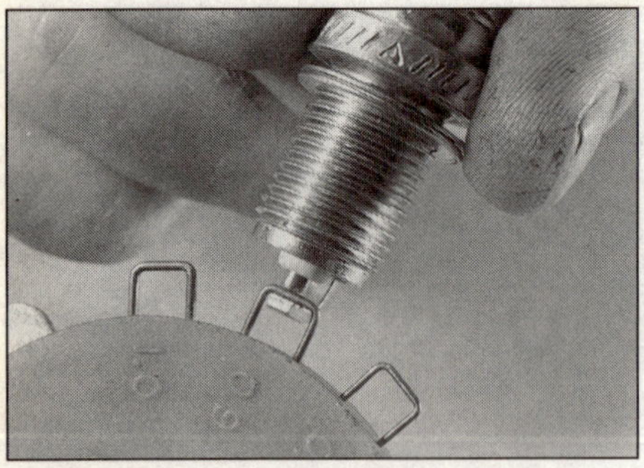

20.9a Measuring the spark plug gap with a wire gauge

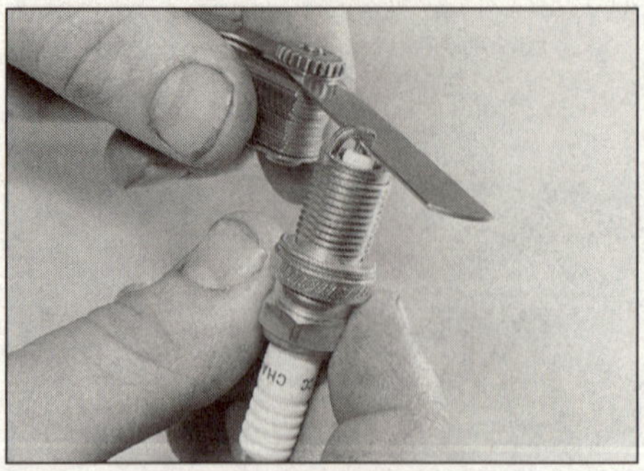

20.9b Measuring the spark plug gap with a feeler blade

the engine is fairly worn, as well as the mixture being too rich.

7 If the insulator nose is covered with light tan to greyish-brown deposits, then the mixture is correct and it is likely that the engine is in good condition.

8 The spark plug electrode gap is of considerable importance as, if it is too large or too small, the size of the spark and its efficiency will be seriously impaired. The gap should be set to the value given in the Specifications at the beginning of this Chapter.

9 To set the gap, measure it with a feeler blade and then bend open, or closed, the outer plug electrode until the correct gap is achieved. The centre electrode should never be bent, as this may crack the insulator and cause plug failure, if nothing worse. If using feeler blades, the gap is correct when the appropriate-size blade is a firm sliding fit **(see illustrations)**.

10 Special spark plug electrode gap adjusting tools are available from most motor accessory shops, or from some spark plug manufacturers.

11 Before fitting the spark plugs, check that the threaded connector sleeves are tight, and that the plug exterior surfaces and threads are clean **(see Haynes Hint)**.

12 Remove the rubber hose (if used), and tighten the plug to the specified torque using the spark plug socket and a torque wrench. Refit the remaining spark plugs in the same manner.

13 Connect the HT leads in their correct order.

Ignition system check

Warning: Voltages produced by an electronic ignition system are considerably higher than those produced by conventional ignition systems. Extreme care must be taken when working on the system with the ignition switched on. Persons with surgically-implanted cardiac pacemaker devices should keep well clear of the ignition circuits, components and test equipment.

Models with a distributor (E7J engine)

14 The spark plug (HT) leads should be checked whenever new spark plugs are fitted.

15 Ensure that the leads are numbered before removing them, to avoid confusion when refitting. Pull the leads from the plugs by gripping the end fitting, not the lead, otherwise the lead connection may be fractured.

16 Check inside the end fitting for signs of corrosion, which will look like a white crusty powder. Push the end fitting back onto the spark plug, ensuring that it is a tight fit on the plug. If not, remove the lead again and use pliers to carefully crimp the metal connector inside the end fitting until it fits securely on the end of the spark plug.

17 Using a clean rag, wipe the entire length of the lead to remove any built-up dirt and grease. Once the lead is clean, check for burns, cracks and other damage. Do not bend the lead excessively, nor pull the lead lengthwise - the conductor inside might break.

18 Disconnect the other end of the lead from the distributor cap. Again, pull only on the end fitting. Check for corrosion and a tight fit in the same manner as the spark plug end. If an ohmmeter is available, check the resistance of the lead by connecting the meter between the spark plug end of the lead and the segment inside the distributor cap. Refit the lead securely on completion.

19 Check the remaining leads one at a time, in the same way.

20 If new spark plug (HT) leads are required, purchase a set for your specific car and engine.

21 Remove the air cleaner ducting from the left-hand side of the engine with reference to Chapter 4A, then undo the retaining screws and remove the cap from the distributor body. Wipe it clean and carefully inspect it inside and out for signs of cracks, black carbon tracks (tracking) and worn, burned or loose contacts; check that the cap's carbon brush is unworn, free to move against spring pressure, and making good contact with the rotor arm. Remove the rotor arm and inspect it. It is common practice to renew the cap and rotor arm whenever new spark plug (HT) leads are fitted. When fitting a new cap, remove the leads from the old cap one at a time, and fit them to the new cap in the exact same location - do not simultaneously remove all the leads from the old cap, or firing order confusion may occur. When refitting, ensure that the arm is securely pressed onto the shaft, and tighten the cap retaining screws securely.

22 Even with the ignition system in first-class condition, some engines may still occasionally experience poor starting attributable to damp ignition components. To disperse moisture, a water-dispersant aerosol can be very effective.

Models with a static ignition system (K7M engine)

23 Check the condition of the HT leads as described above in paragraphs 14 to 19.

HAYNES HiNT

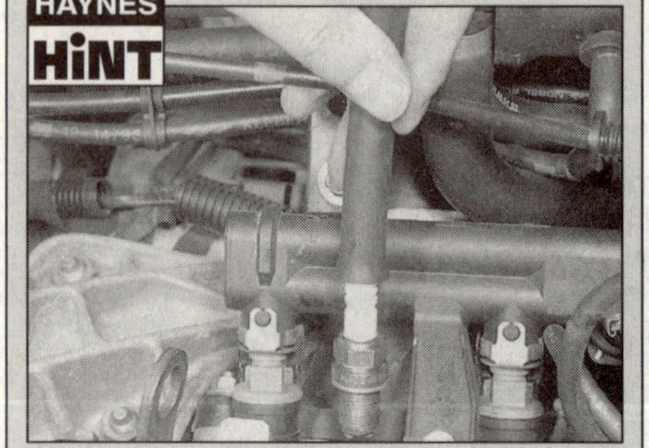

It's often difficult to insert spark plugs into their holes without cross-threading them. To avoid this possibility, fit a short piece of rubber hose over the end of the spark plug. The flexible hose acts as a universal joint, to help align the plug with the plug hole. Should the plug begin to cross-thread, the hose will slip on the spark plug, preventing thread damage to the aluminium cylinder head.

21 Air filter element renewal

E7J engine

1 Unscrew the air cleaner mounting bolts from the top of the air cleaner. Also unscrew the single cover retaining screw **(see illustration)**.

2 Release the retaining clips and remove the lid from the top of the air cleaner housing **(see illustrations)**.

3 Lift out the air filter element **(see illustration)**.

K7M engine

4 Undo the screws securing the cover to the air cleaner body and lift off the cover **(see illustration)**.

5 Lift out the air filter element **(see illustration)**.

21.1 Unscrewing the air cleaner mounting bolts (E7J engine)

21.2a Release the retaining clips . . .

21.2b . . . remove the lid . . .

21.3 . . . and lift out the air filter element (E7J engine)

21.4 Remove the air cleaner cover . . .

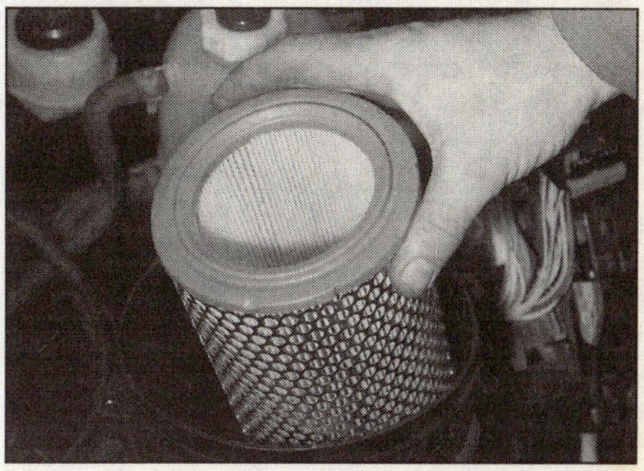

21.5 . . . and lift out the air filter element (K7M engine)

All engines

6 Wipe clean the filter housing, then fit the new filter. The remaining procedure is a reversal of removal.

Every 40 000 miles (64 000 km)

22 Handbrake check

1 Check and, if necessary, adjust the handbrake as described in Chapter 9.

23 Heating system check

1 Check that the heating controls move smoothly and easily with no sign of stiffness and check that the blower motor operates correctly. If there is a problem check the heating ventilation system components as described in Chapter 3.

2 If the heating system seems to be giving out insufficient heat, check that the coolant level is correct and that there are no air locks in the cooling system (see Section 35).

24 Hinge and lock lubrication

1 Lubricate the hinges of the bonnet, doors and boot lid/tailgate with a light general purpose oil. similarly, lubricate all latches, locks and lock strikers. At the same time, check the security and operation of all locks, adjusting them if necessary (see Chapter 11).

2 Lightly lubricate the bonnet release mechanism with a suitable grease.

25 Spare fuse check

1 Check that spare fuses are in place in the locations provided in the fusebox cover (see Chapter 12). It is advisable to carry at least one spare of each rating of fuse fitted. Spare fuses can be obtained from most car accessory shops, or from a Renault dealer.

26 Front wheel alignment check

1 Refer to the information in Chapter 10.

27 Rear brake shoe thickness check

1 Apply the handbrake, then jack up the rear of the car and support it securely on axle stands.

2 For a comprehensive check, remove the rear brake drums, and check the brake shoes for signs of wear or contamination. At the same time, also inspect the wheel cylinders for signs of leakage, and the brake drum for signs of wear. Refer to the relevant Sections of Chapter 9 for further information **(see Haynes Hint)**.

3 On completion lower the car to the ground.

HAYNES HiNT

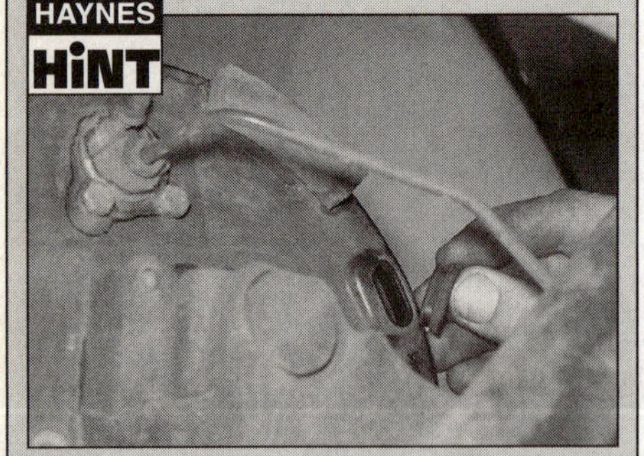

For a quick check, the thickness of the rear brake shoe can be checked through the aperture in the backplate once the rubber plug has been removed

28 Manual transmission oil level check

1 Either position the vehicle over an inspection pit, or jack up the front and rear of the vehicle and support it on axle stands (see *"Jacking and vehicle support"*). The vehicle must be level for the check to be accurate. Remove the engine undershield where fitted.

2 Clean the area around the filler/level plug on the left-hand side of the transmission, then slacken and remove the plug from the transmission **(see illustration)**.

3 The transmission oil level should be up to the lower edge of the filler/level plug aperture **(see illustration)**.

4 If necessary, top-up using the specified type of lubricant until the transmission oil level is correct. Fill the transmission until oil starts to flow out and allow excess oil to drain out **(see illustration)**.

5 Once the transmission oil level is correct, refit the filler/level plug and tighten it securely.

6 Lower the vehicle to the ground. Note that frequent need for topping-up indicates a leakage, possibly through an oil seal. The cause should be investigated and rectified.

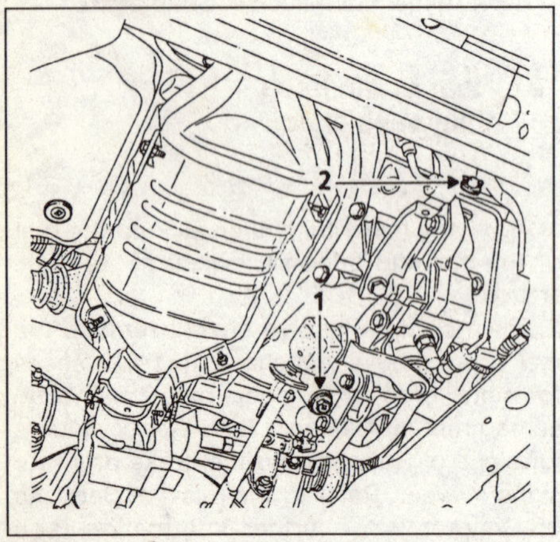

28.2 Manual transmission drain plug (1) and filler/level plug (2)

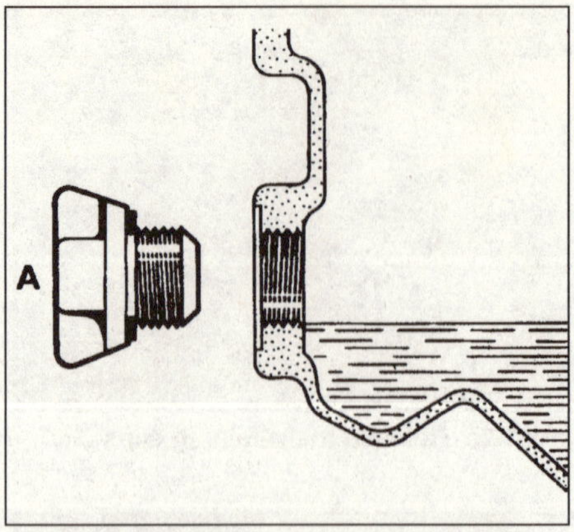

28.3 Manual transmission filler/level plug (A) - correct oil level shown

29 Automatic transmission fluid renewal

Note: *Refer to Section 11 to find out what is involved in checking the transmission fluid level before draining the transmission. Note that the final drive oil does not need to be renewed.*

1 Take the vehicle on a short run, to warm the transmission up to operating temperature.

2 Park the car on level ground, then switch off the ignition and apply the handbrake firmly. Jack up the front of the car and support it securely on axle stands (see *"Jacking and vehicle support"*). Note that, when refilling and checking the fluid level, the car must be level to ensure accuracy. Remove the engine undershield where fitted.

3 Position a suitable container under the transmission. Unscrew the transmission drain plug **(see illustration 11.4)** from the sump and allow the fluid to drain completely into the container.

Warning: If the fluid is hot, take precautions against scalding.

4 Clean the drain plug, being especially careful to wipe any metallic particles off the magnetic insert. Discard the original sealing washer; this should be renewed whenever it is disturbed.

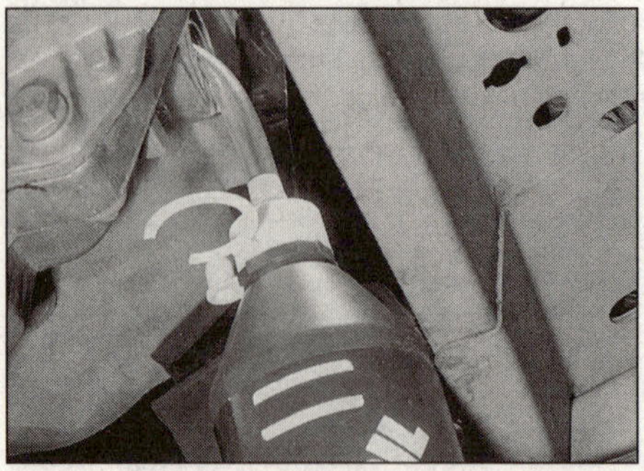

28.4 Topping-up the manual transmission oil level

5 When the fluid has finished draining, clean the drain plug threads and those of the transmission casing. Fit a new sealing washer to the drain plug, and refit the plug to the transmission, tightening it securely.

6 Make sure the vehicle is level then refill the transmission with the specified type and amount of fluid via the filler tube (Section 11). Refilling the transmission is an awkward operation, use a funnel with a fine mesh gauze, to avoid spillage, and to ensure that no foreign matter enters the transmission. Allow plenty of time for the fluid level to settle properly.

7 Check the transmission fluid level as described in Section 11.

30 Fuel filter renewal

Warning: Before carrying out the following operation refer to the precautions given in Safety first! at the beginning of this manual and follow them implicitly. Petrol is a highly dangerous and volatile liquid and the precautions necessary when handling it cannot be overstressed.

1 The fuel filter is located underneath the vehicle where it is mounted just in front of the fuel tank **(see illustration)**.

2 Bearing in mind the information given on depressurising the fuel system, disconnect the hoses from the fuel filter. The hoses are equipped with quick-release fittings to ease removal. To disconnect each hose, slide out the locking tab (where fitted) from the collar then depress the collar and detach each hose. Disconnect both hoses, noting the correct fitted position of the sealing rings and plug the hose ends to minimise fuel loss.

3 Slacken the clamp screw then slide the filter out of position, noting its correct fitted orientation. The arrow on the filter should point in the direction of fuel flow (towards the throttle body/fuel rail).

4 Slide the new filter into position making sure its arrow is pointing in the direction of fuel flow. Make sure the rubber mounting is correctly positioned then securely tighten the clamp bolt.

30.1 Fuel filter location just in front of the fuel tank

5 Ensure that the sealing rings are in position and reconnect the hoses to the fuel filter. Check the end fittings are clipped securely in position and (where necessary) refit the locking tabs to the collars.
6 Lower the vehicle to the ground then start the engine and check the filter for signs of fuel leakage.

31 Brake fluid renewal

Warning: Brake hydraulic fluid can harm your eyes and damage painted surfaces, so use extreme caution when handling and pouring it. Do not use fluid that has been standing open for some time, as it absorbs moisture from the air. Excess moisture can cause a dangerous loss of braking effectiveness.
Caution: On models equipped with ABS, disconnect the battery before carrying out the following operation and do not reconnect the battery until after the operation is complete. Failure to do this could lead to air entering the hydraulic unit. If air enters the hydraulic unit pump, it will prove very difficult (in some cases impossible) to bleed the unit.

1 The procedure is similar to that for the bleeding of the hydraulic system as described in Chapter 9, except that the brake fluid reservoir should be emptied by siphoning, using a clean poultry baster or similar before starting, and allowance should be made for the old fluid to be expelled when bleeding a section of the circuit.
2 Working as described in Chapter 9, open the first bleed screw in the sequence, and pump the brake pedal gently until nearly all the old fluid has been emptied from the master cylinder reservoir. Top-up to the "MAXI" level with new fluid, and continue pumping until only the new fluid remains in the reservoir, and new fluid can be seen emerging from the bleed screw. Tighten the screw, and top the reservoir level up to the "MAXI" level.

HAYNES HiNT ***Old brake fluid is invariably much darker in colour than the new, making it easy to distinguish the two.***

3 Work through all the remaining bleed screws in the sequence until new fluid can be seen at all of them. Be careful to keep the master cylinder reservoir topped-up to above the "MINI" level at all times, or air may enter the system and greatly increase the length of the task.
4 When the operation is complete, check that all bleed screws are securely tightened, and that their dust caps are refitted. Wash off all traces of spilt fluid, and recheck the master cylinder reservoir fluid level.
5 Check the operation of the brakes before taking the car on the road.

32 Road test

Instruments and electrical equipment

1 Check the operation of all instruments and electrical equipment.
2 Make sure that all instruments read correctly, and switch on all electrical equipment in turn, to check that it functions properly.

Steering and suspension

3 Check for any abnormalities in the steering, suspension, handling or road "feel".
4 Drive the vehicle, and check that there are no unusual vibrations or noises.
5 Check that the steering feels positive, with no excessive "sloppiness", or roughness, and check for any suspension noises when cornering and driving over bumps.

Drivetrain

6 Check the performance of the engine, clutch, transmission and driveshafts.
7 Listen for any unusual noises from the engine, clutch and transmission.
8 Make sure that the engine runs smoothly when idling, and that there is no hesitation when accelerating.
9 Check that, where applicable, the clutch action is smooth and progressive, that the drive is taken up smoothly, and that the pedal travel is not excessive. Also listen for any noises when the clutch pedal is depressed.
10 Check that all gears can be engaged smoothly without noise, and that the gear lever action is not vague or "notchy".
11 On automatic transmission models, make sure that all gearchanges occur smoothly, without snatching, and without an increase in engine speed between changes. Check that all of the gear positions can be selected with the vehicle at rest. If any problems are found, they should be referred to a Renault dealer.
12 Listen for a metallic clicking sound from the front of the vehicle, as the vehicle is driven slowly in a circle with the steering on full-lock. Carry out this check in both directions. If a clicking noise is heard, this indicates wear in a driveshaft joint (see Chapter 8).

Braking system

13 Make sure that the vehicle does not pull to one side when braking, and that the wheels do not lock prematurely when braking hard.
14 Check that there is no vibration through the steering when braking.
15 Check that the handbrake operates correctly, without excessive movement of the lever, and that it holds the vehicle stationary on a slope.
16 Test the operation of the brake servo unit as follows. Depress the footbrake four or five times to exhaust the vacuum, then start the engine. As the engine starts, there should be a noticeable "give" in the brake pedal as vacuum builds up. Allow the engine to run for at least two minutes, and then switch it off. If the brake pedal is now depressed again, it should be possible to detect a hiss from the servo as the pedal is depressed. After about four or five applications, no further hissing should be heard, and the pedal should feel considerably harder.

Every 50 000 miles (80 000 km)

33 EGR valve clean (K7M petrol engine)

1 Remove the EGR valve as described in Chapter 4C.
2 Remove all pipes from the valve, then blow air through the valve. Do not blow compressed air on the solenoid valve diaphragm as this may damage it.
3 Refit the EGR valve with reference to Chapter 4C.

34 Timing belt renewal

Refer to Chapter 2A.

Every 2 years (regardless of mileage)

35 Coolant renewal

Cooling system draining

Warning: Wait until the engine is cold before starting this procedure. Do not allow antifreeze to come in contact with your skin, or with the painted surfaces of the vehicle. Rinse off spills immediately with plenty of water. Never leave antifreeze lying around in an open container, or in a puddle in the driveway or on the garage floor. Children and pets are attracted by its sweet smell, but antifreeze can be fatal if ingested.

Note: *Renault do not specify renewal intervals for the coolant, but it is advisable to drain and refill the system every two years to ensure that the corrosion inhibiting properties of the coolant are maintained.*

1 With the engine completely cold, remove the expansion tank filler cap. Turn the cap anti-clockwise, wait until any pressure remaining in the system is released, then unscrew it and lift it off.

2 Where applicable, remove the undershield, then position a suitable container beneath the radiator bottom hose connection.

3 Slacken the hose clip, pull off the hose and allow the coolant to drain into the container.

4 To assist draining, open the cooling system bleed screw(s) in the heater hose(s) at the left-hand end of the cylinder head. Also open the bleed screw at the top right-hand side of the radiator. On the E7J engine there are two bleed screws, but on the K7M engine there is only one bleed screw **(see illustrations)**.

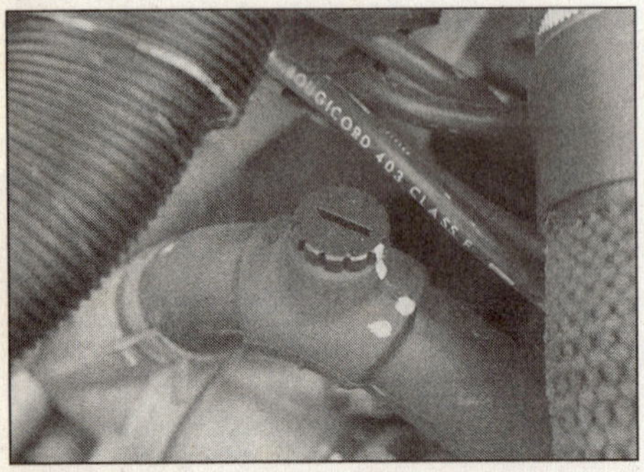

35.4a Cooling system bleed screw located in the heater inlet hose at the thermostat housing . . .

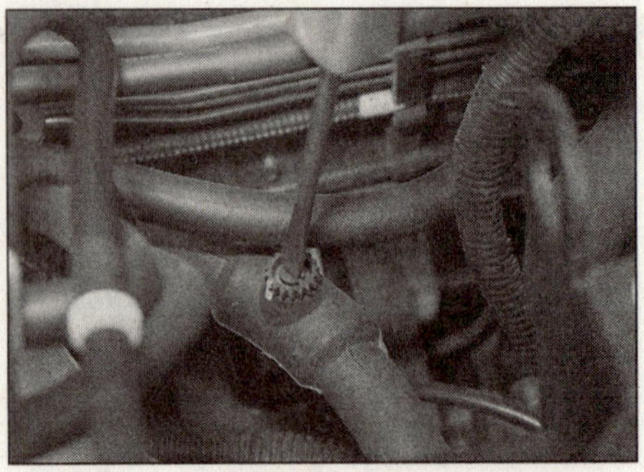

35.4b . . . and in the heater return hose (E7J engine)

5 If the coolant has been drained for a reason other than renewal, then provided it is clean and less than two years old, it can be re-used, though this is not recommended.

6 Refit the radiator bottom hose on completion of draining.

Cooling system flushing

7 If coolant renewal has been neglected, or if the antifreeze mixture has become diluted, then in time, the cooling system may gradually lose efficiency, as the coolant passages become restricted due to rust, scale deposits, and other sediment. The cooling system efficiency can be restored by flushing the system clean.

8 The radiator should be flushed independently of the engine, to avoid unnecessary contamination.

Radiator flushing

9 Disconnect the top and bottom hoses and any other relevant hoses from the radiator, with reference to Chapter 3.

10 Insert a garden hose into the radiator top inlet. Direct a flow of clean water through the radiator, and continue flushing until clean water emerges from the radiator bottom outlet.

11 If after a reasonable period, the water still does not run clear, the radiator can be flushed with a good proprietary cleaning agent. It is important that their manufacturer's instructions are followed carefully. If the contamination is particularly bad, insert the hose in the radiator bottom outlet, and reverse-flush the radiator.

Engine flushing

12 To flush the engine, remove the thermostat as described in Chapter 3, then temporarily refit the top hose at its engine connection.

13 With the top and bottom hoses disconnected from the radiator, insert a garden hose into the radiator top hose. Direct a clean flow of water through the engine, and continue flushing until clean water emerges from the radiator bottom hose.

14 On completion of flushing, refit the thermostat and reconnect the hoses with reference to Chapter 3.

Cooling system filling

15 Before attempting to fill the cooling system, make sure that all hoses and clips are in good condition, and that the clips are tight. Note that an antifreeze mixture must be used all year round, to prevent corrosion of the engine components.

16 Remove the expansion tank filler cap.

17 Open the cooling system bleed screw(s).

18 Place a wad of rags around the expansion tank.

19 Slowly fill the system until the coolant level reaches the top of the expansion tank filler neck.

20 Close the bleed screw(s) when coolant free from air bubbles emerges.

21 Start the engine, and run it at a fast idle speed (do not exceed 2500 rpm) for approximately 4 minutes. Keep the level topped up to the top of the expansion tank filler neck.

22 Refit and tighten the expansion tank filler cap.

23 Allow the engine to run for approximately 20 minutes at 2500 rpm until the cooling fan cuts in and out.

24 Stop the engine and check the coolant level, which should be up to the "MAX" mark on the side of the tank. Check that the expansion tank filler cap is tight.

25 Allow the engine to cool, then re-check the coolant level with reference to "*Weekly checks*". Top-up the level if necessary and refit the expansion tank filler cap. Where applicable, refit the undershield.

Antifreeze mixture

26 The antifreeze should always be renewed at the specified intervals. This is necessary not only to maintain the antifreeze properties, but also to prevent corrosion which would otherwise occur as the corrosion inhibitors become progressively less effective.

27 Always use an ethylene-glycol based antifreeze which is suitable for use in mixed-metal cooling systems. The quantity of antifreeze and levels of protection are given in the Specifications.

28 Before adding antifreeze, the cooling system should be completely drained, preferably flushed, and all hoses checked for condition and security.
29 After filling with antifreeze, a label should be attached to the expansion tank, stating the type and concentration of antifreeze used, and the date installed. Any subsequent topping-up should be made with the same type and concentration of antifreeze.
30 Do not use engine antifreeze in the windscreen/tailgate washer system, as it will cause damage to the vehicle paintwork. A screenwash additive should be added to the washer system in the quantities stated on the bottle.

Chapter 1 Part B:
Routine maintenance & servicing - diesel models

Contents

Degrees of difficulty

Easy, suitable for novice with little experience	**Fairly easy,** suitable for beginner with some experience	**Fairly difficult,** suitable for competent DIY mechanic	**Difficult,** suitable for experienced DIY mechanic	**Very difficult,** suitable for expert DIY or professional

Lubricants and fluids

Refer to "*Weekly checks*"

Capacities

Engine oil		
Capacity (excluding filter)	5.0 litres	
Oil filter capacity (all engines)	0.5 litre	
Difference between MAX and MIN dipstick marks	2.0 litres	
Cooling system		
1870 cc (F8Q) engine	7.5 litres	
Fuel tank		
All models	60 litres	
Manual gearbox		
JB1 five-speed	3.4 litres	
JC5 five-speed	3.1 litres	
Power-assisted steering reservoir		
All models	1.1 litres	

Engine

Oil filter type	Champion F105	

Cooling system

Antifreeze mixtures:	**Antifreeze**	**Water**
Protection to -23°C	35%	65%
Protection to -40°C	50%	50%
Auxiliary drivebelt tension:		
Models without air conditioning	97 SEEM units (see text) or 15 mm deflection	
Models with air conditioning	61 to 77 SEEM units (see text) or 15 mm deflection	

Fuel system

Idle speed	825 ± 50 rpm	
Fast idle speed:		
Bosch injection pump	950 ± 25 rpm	
Lucas injection pump	Not adjustable (factory set)	
Air filter element:		
Non-Turbo models	Champion W212	
Turbo models	No information at time of writing	
Fuel filter	No information at time of writing	

Braking system

Front disc brakes:		
Pad thickness (including backing):		
New	18.0 mm	
Minimum thickness	7.0 mm	
Rear drum brakes:		
Shoe thickness (including backing):		
New	7.0 mm	
Minimum thickness	2.5 mm	

Electrical system

Windscreen wiper blade type:		
Driver's side	Champion X55	
Passenger side	Champion X51	

Torque wrench settings

	Nm	**lbf ft**
Engine sump drain plug	15	11
Roadwheel bolts	90	66

Maintenance schedule - diesel models

The maintenance intervals in this manual are provided with the assumption that you, not the dealer, will be carrying out the work. These are the minimum maintenance intervals recommended by us for vehicles driven daily. If you wish to keep your vehicle in peak condition at all times, you may wish to perform some of these procedures more often. We encourage frequent maintenance, because it enhances the efficiency, performance and resale value of your vehicle.

If the vehicle is driven in dusty areas, used to tow a trailer, or driven frequently at slow speeds (idling in traffic) or on short journeys, more frequent maintenance intervals are recommended.

When the vehicle is new, it should be serviced by a factory-authorised dealer service department, in order to preserve the factory warranty.

Every 250 miles (400 km) or weekly

- ☐ Refer to "*Weekly checks*"

Every 6000 miles (10 000 km)

- ☐ Renew the engine oil and filter (Section 3)
- ☐ Drain any water from the fuel filter (Section 4)
- ☐ Check the front brake pad thickness and discs (Section 5)
- ☐ Check the condition of the exhaust system and mountings (Section 6)
- ☐ Check the suspension and steering components (Section 7)
- ☐ Check the driveshaft wear and gaiter (Section 8)
- ☐ Check all underbonnet components, hoses and brake lines for fluid leaks (Section 9)
- ☐ Check the tightness of the roadwheel bolts (Section 10)
- ☐ Check the operation of the clutch (Section 11)
- ☐ Check the exhaust emissions (Section 12)
- ☐ Check the operation of the air conditioning system (Section 13)
- ☐ Check the operation of all electrical systems (Section 14)
- ☐ Check the vehicle bodywork (Section 15)
- ☐ Check the boost pressure on Turbo engine only (Section 16)

Every 18 000 miles (30 000 km)

In addition to all the items listed previously, carry out the following:

- ☐ Check the condition of the seat belts (Section 17)
- ☐ Check the condition of the auxiliary drivebelt (Section 18)
- ☐ Check the headlight beam adjustment (Section 19)
- ☐ Renew the air filter element (Section 20)
- ☐ Renew the pollen filter (Section 21)
- ☐ Renew the fuel filter (Section 22)

Every 36 000 miles (60 000 km)

In addition to all the items listed previously, carry out the following:

- ☐ Check the operation of the handbrake (Section 23)
- ☐ Check the heating system (Section 24)
- ☐ Lubricate the hinges and locks (Section 25)
- ☐ Check the spare fuses are in place (Section 26)
- ☐ Check the front wheel alignment (Section 27)
- ☐ Check the rear brake shoe thickness (Section 28)
- ☐ Check the manual transmission oil level (Section 29)
- ☐ Renew the brake fluid (Section 30)
- ☐ Carry out a road test (Section 31)

Every 72 000 miles (120 000 km)

In addition to all the items listed previously, carry out the following:

- ☐ Renew the timing belt (Section 32)

Every 2 years (regardless of mileage)

In addition to all the items listed previously, carry out the following:

- ☐ Renew the coolant (Section 33)

Underbonnet view of a 1.9 litre non-Turbo diesel model

1 Engine oil filler cap
2 Engine oil level dipstick
3 Battery
4 Brake fluid reservoir
5 Electrical connector box
6 Coolant expansion tank
7 Suspension strut upper mountings
8 Air filter housing
9 Engine oil filter
10 Power steering fluid reservoir
11 Alternator
12 Washer fluid reservoir
13 Injection pump
14 Fuel injectors
15 Fuel filter
16 Pre/post-heating control unit

Front underbody view of a 1.9 litre non-Turbo diesel model

1 Radiator
2 Engine oil drain plug
3 Manual transmission
4 Front suspension lower arms
5 Front anti-roll arm
6 Front driveshafts
7 Gearchange rod
8 Front brake calipers
9 Steering track rod ends
10 Steering gear
11 Engine rear mounting
12 Exhaust front downpipe

Rear underbody view of a 1.9 litre diesel model

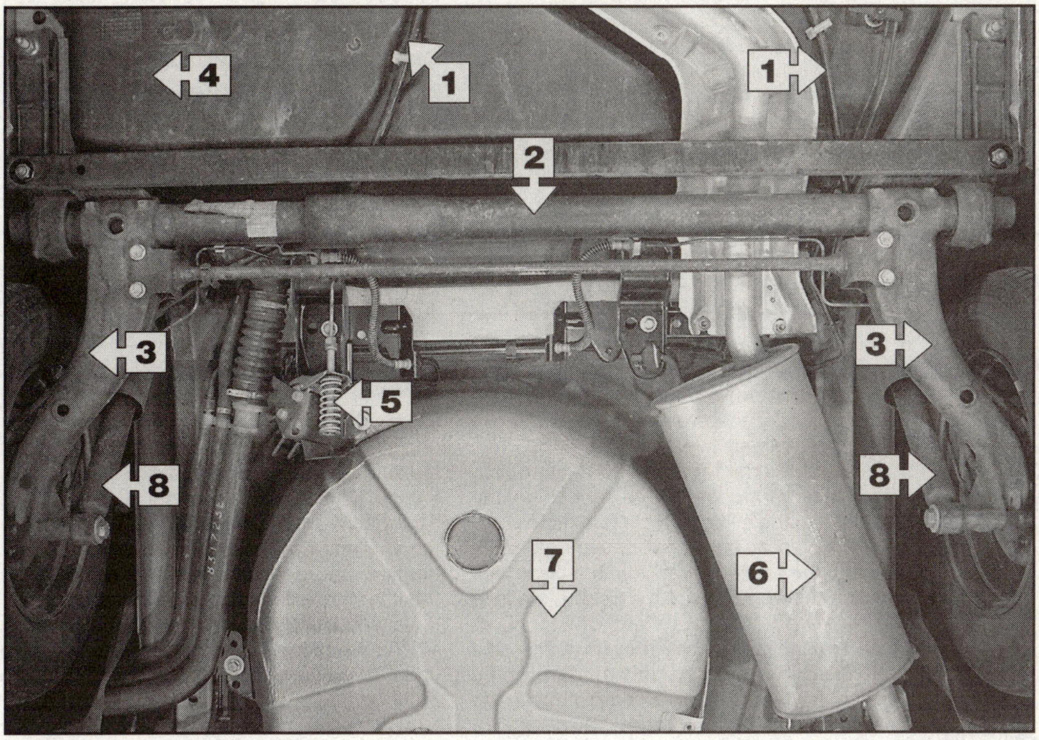

1 *Handbrake cables*
2 *Rear axle assembly*
3 *Rear suspension trailing arms*
4 *Fuel tank*
5 *Rear brake pressure regulating valve*
6 *Exhaust tailpipe and silencer*
7 *Spare wheel well*
8 *Rear suspension shock absorbers*

Maintenance procedures - diesel models

1 Introduction

1 This Chapter is designed to help the home mechanic maintain his/her vehicle for safety, economy, long life and peak performance.

2 The Chapter contains a master maintenance schedule, followed by Sections dealing specifically with each task in the schedule. Visual checks, adjustments, component renewal and other helpful items are included. Refer to the accompanying illustrations of the engine compartment and the underside of the vehicle for the locations of the various components.

3 Servicing your vehicle in accordance with the mileage/time maintenance schedule and the following Sections will provide a planned maintenance programme, which should result in a long and reliable service life. This is a comprehensive plan, so maintaining some items but not others at the specified service intervals, will not produce the same results.

4 As you service your vehicle, you will discover that many of the procedures can - and should - be grouped together, because of the particular procedure being performed, or because of the close proximity of two otherwise-unrelated components to one another. For example, if the vehicle is raised for any reason, the exhaust can be inspected at the same time as the suspension and steering components.

5 The first step in this maintenance programme is to prepare yourself before the actual work begins. Read through all the Sections relevant to the work to be carried out, then make a list and gather together all the parts and tools required. If a problem is encountered, seek advice from a parts specialist, or a dealer service department.

2 Intensive maintenance

1 If, from the time the vehicle is new, the routine maintenance schedule is followed closely, and frequent checks are made of fluid levels and high-wear items, as suggested throughout this manual, the engine will be kept in relatively good running condition, and the need for additional work will be minimised.

2 It is possible that there will be times when the engine is running poorly due to the lack of regular maintenance. This is even more likely if a used vehicle, which has not received regular and frequent maintenance checks, is purchased. In such cases, additional work may need to be carried out, outside of the regular maintenance intervals.

3 If engine wear is suspected, a compression test or leakdown test (refer to Chapter 2B) will provide valuable information regarding the overall performance of the main internal components. Such a test can be used as a basis to decide on the extent of the work to be carried out. If, for example, a compression test indicates serious internal engine wear, conventional maintenance as described in this Chapter will not greatly improve the performance of the engine, and may prove a waste of time and money, unless extensive overhaul work is carried out first.

4 The following series of operations are those usually required to improve the performance of a generally poor-running engine:

Primary operations

a) Clean, inspect and test the battery (See "Weekly checks").
b) Check all the engine-related fluids (See "Weekly checks").
c) Check the condition and tension of the auxiliary drivebelt (Section 18).
d) Check the condition of the air filter element, and renew if necessary (Section 20).
e) Check the fuel filter - drain off any water and renew the filter if necessary (Sections 4 and 22).
f) Check the condition of all hoses, and check for fluid leaks (Section 9).
g) Check the idle speed and anti-stall speed (Chapter 4B).

5 If the primary operations do not prove fully effective, carry out the following secondary operations:

Secondary operations

All items listed under "*Primary operations*", plus the following:

a) Check the charging system (Chapter 5A).
b) Check the preheating system (Chapter 5C).
c) Check the fuel system (Chapter 4B).

Every 6000 miles (10 000 km)

3 Engine oil and filter renewal

1 Frequent oil and filter changes are the most important preventative maintenance procedures which can be undertaken by the DIY owner. As engine oil ages, it becomes diluted and contaminated, which leads to premature engine wear.

2 Before starting this procedure, gather together all the necessary tools and materials **(see illustration)**. Also make sure that you have plenty of clean rags and newspapers handy, to mop up any spills. Ideally, the engine oil should be warm, as it will drain more easily, and more built-up sludge will be removed with it. Take care not to touch the exhaust or any other hot parts of the engine when working under the vehicle. To avoid any possibility of scalding, and to protect yourself from possible skin irritants and other harmful contaminants in used engine oils, it is advisable to wear gloves when carrying out this work.

3 Firmly apply the handbrake then jack up the front of the vehicle and support it on axle stands (see *"Jacking and vehicle support"*). Where applicable, remove the engine undershield.

4 Remove the oil filler cap.

5 Slacken the drain plug about half a turn using a special drain plug key (8 mm square) **(see illustration)**. Position the draining container under the drain plug, then remove the plug completely **(see Haynes Hint)**.

6 Allow some time for the oil to drain, noting that it may be necessary to reposition the container as the oil flow slows to a trickle.

7 After all the oil has drained, wipe the drain plug and the sealing washer with a clean rag. Examine the condition of the sealing washer, and renew it if it shows signs of scoring or other damage which may prevent an oil-tight seal. Clean the area around the drain plug opening, and refit the plug complete with the washer and tighten it securely.

8 Move the container into position under the oil filter which is located on the front of the cylinder block.

9 Use an oil filter removal tool to slacken the filter initially, then unscrew it by hand the rest of the way **(see illustration)**. Empty the oil from the old filter into the container.

10 Use a clean rag to remove all oil, dirt and sludge from the filter sealing area on the engine. Check the old filter to make sure that the rubber sealing ring has not stuck to the engine. If it has, carefully remove it.

11 Apply a light coating of clean engine oil to the sealing ring on the new filter, then screw the filter into position on the engine. Tighten the filter firmly by hand only - **do not** use any tools **(see illustration)**.

12 Remove the old oil and all tools from under the vehicle then lower the vehicle to the ground.

13 Fill the engine through the filler hole, using the correct grade and type of oil (refer to "*Weekly Checks*" for details of topping-up). Pour in half the specified quantity of oil first, then wait a few minutes for the oil to drain into the sump. Continue to add oil, a small

HAYNES HiNT

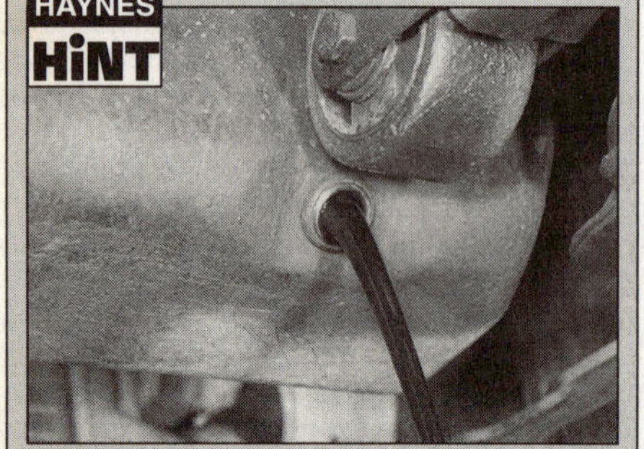

As the drain plug threads release, move it sharply away so the stream of oil issuing from the sump runs into the container, not up your sleeve!

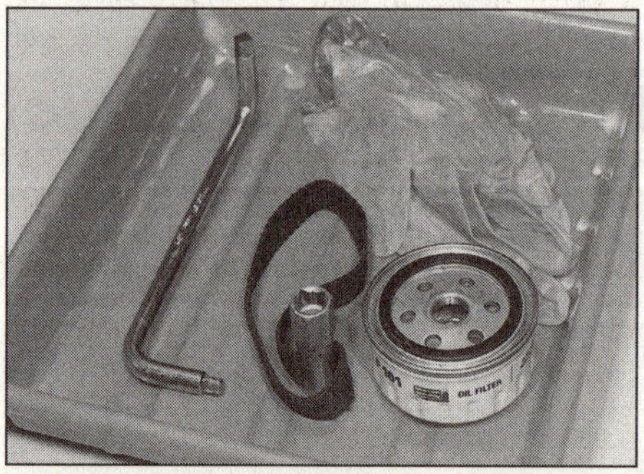

3.2 Tools and materials necessary for engine oil change and filter renewal

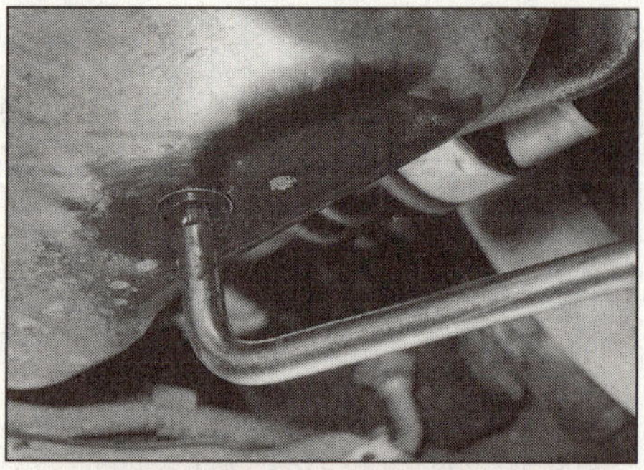

3.5 Using the special drain plug key to unscrew the sump drain plug

3.9 Using an oil filter removal tool to slacken the oil filter

3.11 Screw the new filter into position by hand

OIL CARE
FOLLOW THE CODE
OIL BANK LINE
0800 66 33 66

Note: It is antisocial and illegal to dump oil down the drain. To find the location of your local oil recycling bank, call this number free.

quantity at a time, until the level is up to the lower mark on the dipstick. Adding approximately a further 2.0 litres will bring the level up to the upper mark on the dipstick. Refit and tighten the oil filler cap.

14 Note that when the engine is first started, there will be a delay of a few seconds before the oil pressure warning light goes out while the new filter fills with oil (this does not apply to Turbo models, for which the procedure described in the following paragraph should be followed). Do not race the engine while the warning light is on.

15 On Turbo models, the following procedure must be observed before starting the engine.

a) Disconnect the wiring from the stop solenoid on the injection pump, and insulate the connector.

b) Crank the engine on the starter motor until the oil pressure warning light goes out (this may take several seconds).

c) Reconnect the wiring to the stop solenoid, then start the engine using the normal procedure.

d) Run the engine at idle speed, and check the turbocharger oil and coolant unions for leakage. Rectify any problems without delay.

16 Run the engine for a few minutes, and check that there are no leaks around the oil filter seal and the sump drain plug.

17 Stop the engine, and wait a few minutes for the oil to settle in the sump once more. With the new oil circulated and the filter now completely full, recheck the level on the dipstick, and add more oil as necessary.

18 Dispose of the used engine oil safely with reference to "*General repair procedures*".

4 Fuel filter water draining

1 A water drain screw is provided on the base of the fuel filter **(see illustration)**.

2 Place a suitable container beneath the drain screw. To make draining easier, a suitable length of tubing can be attached to the outlet on the screw to direct the fuel flow.

3 Loosen the fuel filter bleed screw, then open the drain screw by turning it anti-clockwise. On models where there is no bleed screw, loosen the fuel inlet union on the filter head.

4 Allow the entire contents of the filter to drain into the container, then securely tighten

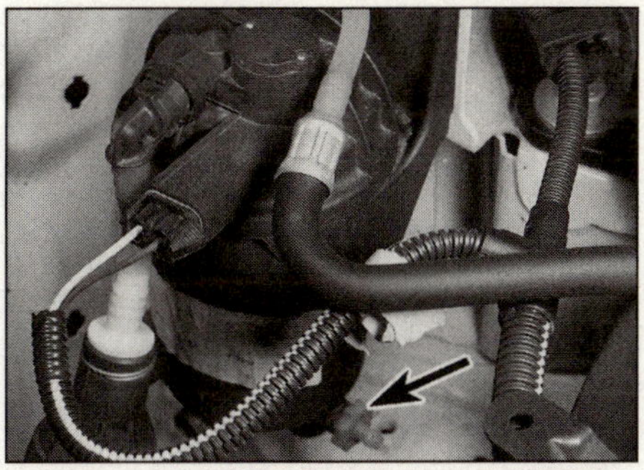

4.1 The water drain screw is on the base of the fuel filter

the drain screw and the bleed screw/fuel filter inlet union (as applicable).

5 Dispose of the drained fuel safely.

6 Prime and bleed the fuel system as described in Chapter 4B.

5 Front brake pad and disc check

1 Apply the handbrake, then jack up the front of the car and support it securely on axle stands. Remove the front roadwheels.

2 For a comprehensive check, the brake pads should be removed and cleaned. The operation of the caliper can then also be checked, and the condition of the brake disc itself can be fully examined on both sides. Refer to Chapter 9 for further information **(see Haynes Hint)**.

3 On completion refit the roadwheels and lower the car to the ground.

6 Exhaust system check

1 With the engine cold (at least an hour after the vehicle has been driven), check the complete exhaust system from the engine to the end of the tailpipe. The exhaust system is most easily checked with the vehicle raised on

HAYNES HiNT

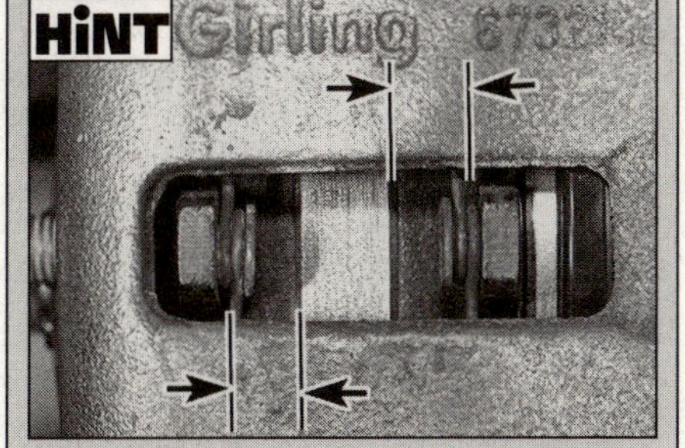

For a quick check, the thickness of each brake pad can be measured through the aperture in the caliper body

a hoist, or suitably supported on axle stands, so that the exhaust components are readily visible and accessible.

2 Check the exhaust pipes and connections for evidence of leaks, severe corrosion and damage. Make sure that all brackets and mountings are in good condition, and that all relevant nuts and bolts are tight. Leakage at any of the joints or in other parts of the system will usually show up as a black sooty stain in the vicinity of the leak.

3 Rattles and other noises can often be traced to the exhaust system, especially the brackets and mountings. Try to move the pipes and silencers. If the components are able to come into contact with the body or suspension parts, secure the system with new mountings. Otherwise separate the joints (if possible) and twist the pipes as necessary to provide additional clearance.

7 Suspension and steering check

Front suspension and steering check

1 Raise the front of the vehicle, and securely support it on axle stands.

2 Visually inspect the balljoint dust covers and the steering rack-and-pinion gaiters for splits, chafing or deterioration. Any wear of these components will cause loss of lubricant, together with dirt and water entry, resulting in rapid deterioration of the balljoints or steering gear.

3 On vehicles with power steering, check the fluid hoses for chafing or deterioration, and the pipe and hose unions for fluid leaks. Also check for signs of fluid leakage under pressure from the steering gear rubber gaiters, which would indicate failed fluid seals within the steering gear.

4 Grasp the roadwheel at the 12 o'clock and 6 o'clock positions, and try to rock it **(see illustration)**. Very slight free play may be felt, but if the movement is appreciable, further investigation is necessary to determine the source. Continue rocking the wheel while an assistant depresses the footbrake. If the

7.4 Check for wear in the hub bearings by grasping the wheel and trying to rock it

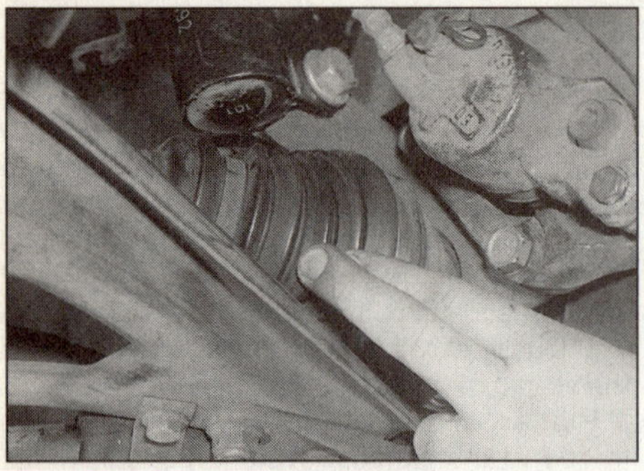

8.1 Checking a driveshaft outer gaiter for signs of damage

movement is now eliminated or significantly reduced, it is likely that the hub bearings are at fault. If the free play is still evident with the footbrake depressed, then there is wear in the suspension joints or mountings.

5 Now grasp the wheel at the 9 o'clock and 3 o'clock positions, and try to rock it as before. Any movement felt now may again be caused by wear in the hub bearings or the steering track-rod balljoints. If the outer balljoint is worn, the visual movement will be obvious. If the inner joint is suspect, it can be felt by placing a hand over the rack-and-pinion rubber gaiter and gripping the track-rod. If the wheel is now rocked, movement will be felt at the inner joint if wear has taken place.

6 Using a large screwdriver or flat bar, check for wear in the suspension mounting bushes by levering between the relevant suspension component and its attachment point. Some movement is to be expected, as the mountings are made of rubber, but excessive wear should be obvious. Also check the condition of any visible rubber bushes, looking for splits, cracks or contamination of the rubber.

7 With the car standing on its wheels, have an assistant turn the steering wheel back and forth, about an eighth of a turn each way. There should be very little, if any, lost movement between the steering wheel and roadwheels. If this is not the case, closely observe the joints and mountings previously described. In addition, check the steering column universal joints for wear, and also check the rack-and-pinion steering gear itself.

Rear suspension check

8 Chock the front wheels, then jack up the rear of the vehicle and support securely on axle stands.

9 Working as described previously for the front suspension, check the rear hub bearings, the suspension bushes and the shock absorber mountings for wear.

Suspension strut/shock absorber check

10 Check for any signs of fluid leakage around the strut/shock absorber body, or from the rubber gaiter around the piston rod. Should any fluid be noticed, the strut/shock absorber is defective internally, and should be renewed. **Note:** *Struts/shock absorbers should always be renewed in pairs on the same axle.*

11 The efficiency of the strut/shock absorber may be checked by bouncing the vehicle at each corner. Generally speaking, the body will return to its normal position and stop after being depressed. If it rises and returns on a rebound, the strut/shock absorber is probably suspect. Also examine the upper and lower mountings for any signs of wear.

8 Driveshaft rubber gaiter and CV joint check

1 With the car raised and securely supported on axle stands, turn the steering onto full-lock, then slowly rotate the roadwheel. Inspect the condition of the outer constant velocity (CV) joint rubber gaiters while squeezing the gaiters to open out the folds **(see illustration)**. Check for signs of cracking, splits or deterioration of the rubber, which may allow the grease to escape and lead to the entry of water and grit into the joint. Also check the security and condition of the retaining clips/fasteners. Repeat these checks on the inner CV joints. If any damage or deterioration is found, the gaiters should be renewed as described in Chapter 8.

2 At the same time, check the general condition of the CV joints themselves by first holding the driveshaft and attempting to rotate the roadwheel. Repeat this check by holding the inner joint and attempting to rotate the driveshaft. Any appreciable movement indicates wear in the joints, in the driveshaft splines, or a loose driveshaft nut.

9 Hose, brake line and fluid leak check

1 Visually inspect the engine joint faces, gaskets and seals for any signs of water or oil leaks. Pay particular attention to the areas around the cylinder head cover, cylinder head, oil filter and sump joint faces. Bear in mind that, over a period of time, some very slight seepage from these areas is to be expected - what you are really looking for is any indication of a serious leak **(see Haynes Hint)**. Should a leak be found, renew the offending gasket or oil seal by referring to the appropriate Chapters in this manual.

2 Also check the security and condition of all the engine-related pipes and hoses, and all braking system pipes and hoses. Ensure that all cable-ties or securing clips are in place, and in good condition. Clips which are broken or missing can lead to chafing of the hoses, pipes or wiring, which could cause more serious problems in the future.

3 Carefully check the radiator hoses and heater hoses along their entire length. Renew any hose which is cracked, swollen or deteriorated. Cracks will show up better if the hose is squeezed. Pay close attention to the hose clips that secure the hoses to the cooling system components. Hose clips can pinch and puncture hoses, resulting in cooling system leaks. If the crimped-type hose clips are used, it may be a good idea to replace them with standard worm-drive clips.

4 Inspect all the cooling system components (hoses, joint faces, etc) for leaks.

5 Where any problems are found on system components, renew the component or gasket with reference to Chapter 3.

6 With the vehicle raised, inspect the fuel tank and filler neck for punctures, cracks and other damage. The connection between the filler neck and tank is especially critical. Sometimes a rubber filler neck or connecting hose will leak due to loose retaining clamps or deteriorated rubber.

7 Carefully check all rubber hoses and metal fuel lines leading away from the fuel tank. Check for loose connections, deteriorated hoses, crimped lines, and other damage. Pay particular attention to the vent pipes and hoses, which often loop up around the filler neck and can become blocked or crimped. Follow the lines to the front of the vehicle, carefully inspecting them all the way. Renew damaged sections as necessary. Similarly, whilst the vehicle is raised, take the opportunity to inspect all underbody brake fluid pipes and hoses.

8 From within the engine compartment, check the security of all fuel, vacuum and brake hose attachments and pipe unions, and inspect all hoses for kinks, chafing and deterioration.

9 Where applicable, check the condition of the power steering fluid pipes and hoses.

HAYNES HiNT

A leak in the cooling system will usually show up as white- or rust-coloured deposits on the area adjoining the leak

10 Roadwheel bolt check

1 Where applicable, remove the wheel trims, and slacken the roadwheel bolts slightly.

2 Tighten the bolts to the specified torque, using a torque wrench.

11 Clutch check

1 Check and adjust the clutch cable as described in Chapter 6. Check that the cable moves freely and easily and lubricate its exposed section with multi-purpose grease.

12 Exhaust emission check

1 The exhaust emission control system can only be tested accurately using a suitable exhaust gas analyser (suitable for use with diesel engines). Have the test carried out by a Renault dealer.
2 Check the engine idle and anti-stall speeds as described in Chapter 4B.

13 Air conditioning system check

1 If the air conditioning system is functioning correctly, the only check needed is to ensure that the compressor drivebelt is correctly tensioned and in good condition (Section 18). If there is a problem with the air conditioning system the vehicle must be taken to a Renault dealer for testing.

14 Electrical systems check

1 Check the operation of all electrical equipment, ie, lights, direction indicators, horn, etc. Refer to the appropriate Sections of Chapter 12 for details if any of the circuits are found to be inoperative.
2 Note that the stop-light switch adjustment is described in Chapter 9.
3 Visually check all accessible wiring connectors, harnesses and retaining clips for security, and for signs of chafing or damage. Rectify any faults found.

15 Vehicle bodywork check

1 Work slowly around the vehicle, paying particular attention to the bonnet, closely examining the paintwork for signs of damage such as stone chips. If damage is found, it should be repaired using a touch-in brush to prevent corrosion of the bodywork; colour-matched touch-in brushes are available from your Renault dealer. More serious damage can be repaired with reference to Chapter 11.

16 Boost pressure check (Turbo engine only)

1 This work must be carried out by a Renault dealer.

Every 18 000 miles (30 000 km)

17 Seat belt check

1 Carefully examine the seat belt webbing for cuts, or any signs of serious fraying or deterioration. If the belt is of the retractable type, pull the belt all the way out of the inertia reel, and examine the full extent of the webbing.
2 Fasten and unfasten the belt, ensuring that the locking mechanism holds securely, and releases properly when intended. If the belt is of the retractable type, check also that the retracting mechanism operates correctly when the belt is released.
3 Check the security of all seat belt mountings and attachments which are accessible without removing any trim or other components **(see illustration)**.

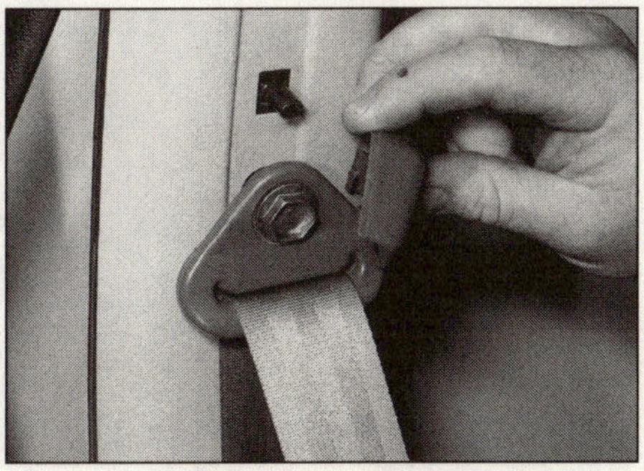

17.3 Check the security of the seat belt mountings

18 Auxiliary drivebelt check and renewal

Checking

1 The auxiliary drivebelt is located at the right-hand side of the engine.
2 Two different drivebelt configurations may be encountered, depending on whether the vehicle is equipped with air conditioning **(see illustrations)**.
3 Due to their function and material makeup, drivebelts are prone to failure after a period of time and should therefore be inspected, and if necessary adjusted periodically.

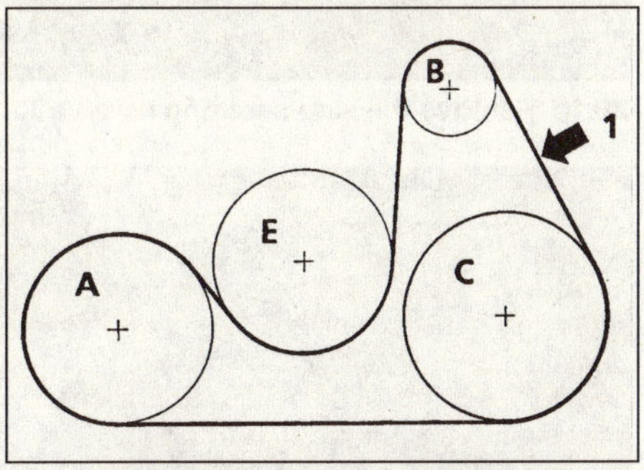

18.2a Auxiliary drivebelt tension checking - models without air conditioning

1 Tension checking point for the auxiliary drivebelt
A Crankshaft pulley
B Alternator
C Power-assisted steering pump
E Water pump

4 Since the drivebelt is located very close to the right-hand side of the engine compartment, it is possible to gain better access by raising the front of the vehicle and removing the right-hand wheel, then removing the splash shield from inside the wheelarch. On certain models it will also be necessary to remove the lower timing belt cover.
5 With the engine switched off, inspect the full length of the drivebelt for cracks and separation of the belt plies. It will be necessary to turn the crankshaft (using a socket or spanner on the crankshaft pulley

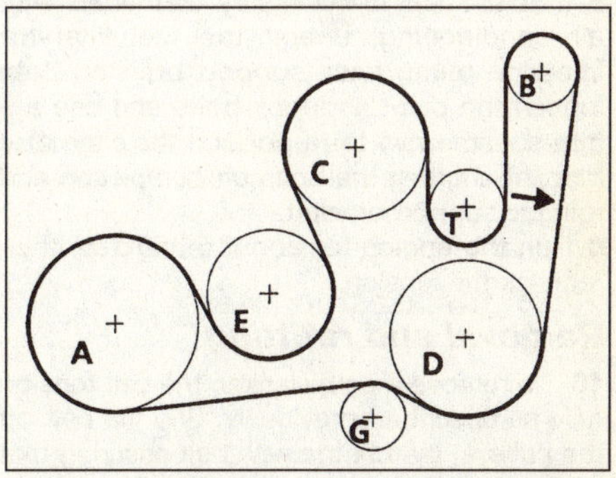

18.2b Auxiliary drivebelt tension checking - models with air conditioning

Arrow shows tension checking point for the auxiliary drivebelt
A Crankshaft pulley
B Alternator
C Power-assisted steering pump
D Air conditioning compressor
E Water pump
G Roller
T Tensioner

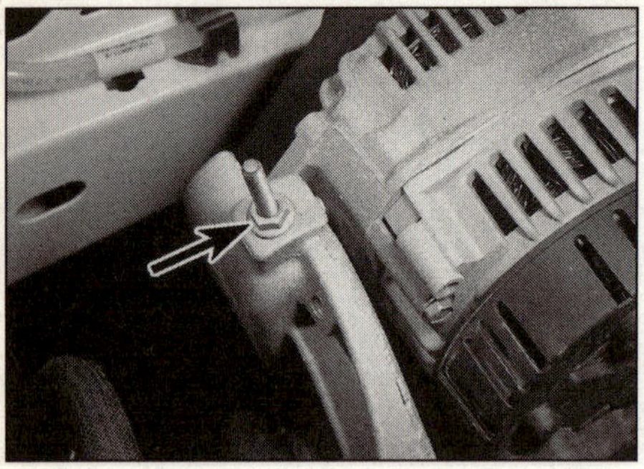

18.7 Auxiliary drivebelt adjustment bolt

bolt) in order to move the belt from the pulleys so that the full length of the belt can be inspected thoroughly. Twist the belt between the pulleys so that both sides can be viewed. Also check for fraying, and glazing which gives the belt a shiny appearance. Check the pulleys for nicks, cracks, distortion and corrosion.

Tensioning

6 The tension of the belt is checked by pushing midway between the pulleys at the point indicated **(see illustrations 18.2a and 18.2b)**. Renault technicians use a special spring-tensioned tool which applies a force of 30 N to the belt and then measures the deflection. An alternative arrangement can be made by using a straight-edge, steel rule and spring balance. Hold the straight-edge across the two pulleys, then position the steel rule on the belt, apply the force with the spring balance, and measure the deflection.
7 If adjustment is necessary on models without air conditioning, loosen the alternator mounting and pivot bolts, then loosen the adjustment locknut and turn the adjustment bolt as required until the tension is correct. Tighten the adjustment, mounting and pivot bolts on completion **(see illustration)**.
8 If adjustment is necessary on models with air conditioning, unbolt the fuel filter-to-injection pump hose support bracket, then loosen the pivot and lock bolts and use a 9 mm square drive to re-position the alternator bracket. Tighten the bolts on completion and refit the support bracket.
9 Run the engine for about 5 minutes, then recheck the tension.

Removal and refitting

10 To remove a belt, slacken the belt tension fully as described previously. Slip the belt off the pulleys, then fit the new belt ensuring that it is routed correctly.
11 With the belt in position, adjust its tension as previously described.

19 Headlight beam adjustment

1 Refer to Chapter 12.

20 Air filter element renewal

Non-Turbo models

1 Remove the securing screws (noting that one of the screws is located in the centre of the cover), and withdraw the air cleaner cover **(see illustrations)**.
2 Lift the filter element from the air cleaner casing **(see illustration)**.
3 Clean the inside of the air cleaner casing and the cover, and fit a new filter element.
4 Refit the cover using a reversal of the removal procedure.

Turbo models

5 Loosen the four captive securing bolts, and lift off the air cleaner cover.
6 Proceed as described in paragraphs 2 to 4 inclusive.

21 Pollen filter renewal

1 Open up the bonnet and unclip the small cover situated on the left-hand side of the intake vent panel.
2 Release the retaining clips and remove the pollen filter cover from the top of the blower motor housing.
3 Slide the pollen filter out of position and discard it.
4 Wipe clean the cover and housing and slide the new filter into position.
5 Securely clip the pollen filter cover onto the housing then clip the vent panel cover into position.

22 Fuel filter renewal

Caution: Do not allow dirt to enter the fuel system during this procedure.
1 Drain the contents of the fuel filter as described in Section 4.
2 Unscrew the filter and remove it from the base of the fuel filter housing. Recover the filter sealing ring. To improve access, unscrew the retaining nuts and free the filter housing from its mounting bracket. If the filter is tight, use an oil filter removal tool to unscrew it.
3 Smear the sealing ring of the new filter with fuel and screw the new filter onto the housing. Tighten the filter firmly by hand only - **do not** use any tools.
4 Refit the filter housing to its mounting bracket and securely tighten the retaining nuts.
5 Prime and bleed the fuel system as described in Chapter 4B.

20.1a Remove the side securing screws . . .

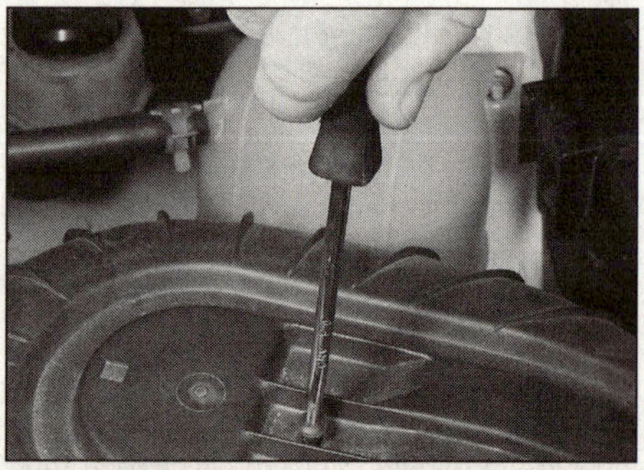

20.1b . . . and the centre securing screw . . .

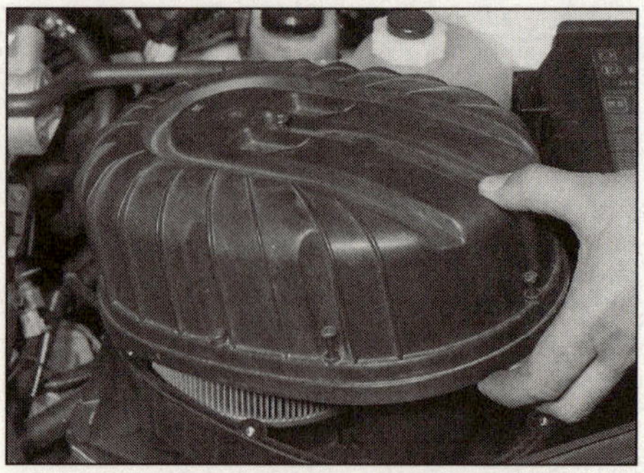

20.1c . . . then withdraw the air cleaner cover

20.2 Removing the filter element from the air cleaner casing

23 Handbrake check

1 Check and, if necessary, adjust the handbrake as described in Chapter 9.

24 Heating system check

1 Check that the heating controls move smoothly and easily with no sign of stiffness and check that the blower motor operates correctly. If there is a problem check the heating ventilation system components as described in Chapter 3.
2 If the heating system seems to be giving out insufficient heat, check that the coolant level is correct and that there are no air locks in the cooling system (see Section 33).

25 Hinge and lock lubrication

1 Lubricate the hinges of the bonnet, doors and boot lid/tailgate with a light general purpose oil. similarly, lubricate all latches, locks and lock strikers. At the same time, check the security and operation of all locks, adjusting them if necessary (see Chapter 11).
2 Lightly lubricate the bonnet release mechanism with a suitable grease.

26 Spare fuse check

1 Check that spare fuses are in place in the locations provided in the fusebox cover (see Chapter 12). It is advisable to carry at least one spare of each rating of fuse fitted. Spare fuses can be obtained from most car accessory shops, or from a Renault dealer.

27 Front wheel alignment check

1 Refer to the information in Chapter 10.

28 Rear brake shoe thickness check

1 Apply the handbrake, then jack up the rear of the car and support it securely on axle stands.
2 For a comprehensive check, remove the rear brake drums, and check the brake shoes for signs of wear or contamination. At the same time, also inspect the wheel cylinders for signs of leakage, and the brake drum for signs of wear. Refer to the relevant Sections of Chapter 9 for further information **(see Haynes Hint)**.
3 On completion lower the car to the ground.

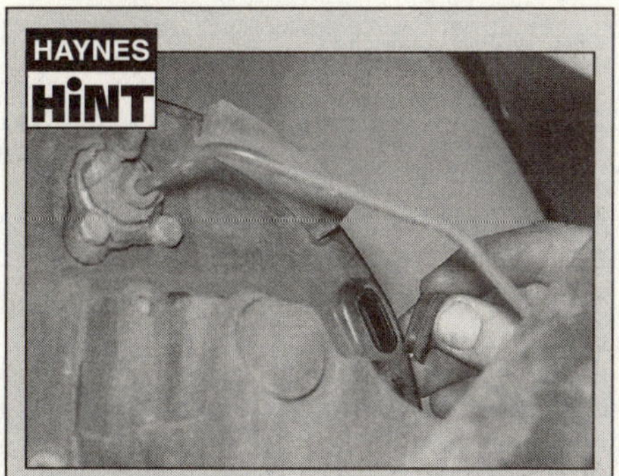

For a quick check, the thickness of the rear brake shoe can be checked through the aperture in the backplate once the rubber plug has been removed

29 Manual transmission oil level check

1 Either position the vehicle over an inspection pit, or jack up the front and rear of the vehicle and support it on axle stands (see *"Jacking and vehicle support"*). The vehicle must be level for the check to be accurate. Remove the engine undershield where fitted.
2 Clean the area around the filler/level plug on the left-hand side of the transmission, then slacken and remove the plug from the transmission **(see illustration)**.
3 The transmission oil level should be up to the lower edge of the filler/level plug aperture **(see illustration)**.
4 If necessary, top-up using the specified type of lubricant until the transmission oil level is correct. Fill the transmission until oil starts to flow out and allow excess oil to drain out **(see illustration)**.
5 Once the transmission oil level is correct, refit the filler/level plug and tighten it securely.

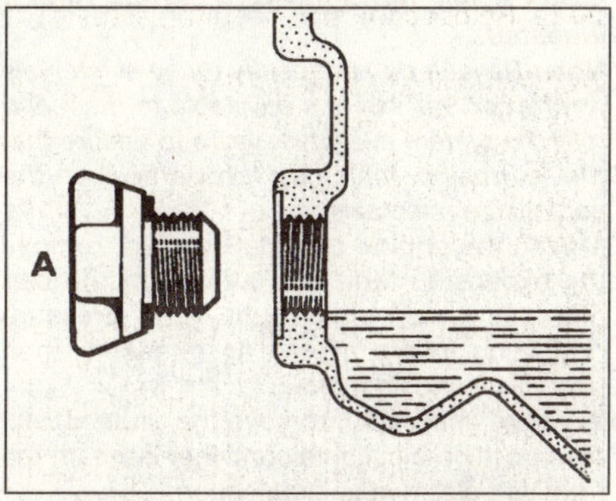

29.3 Manual transmission filler/level plug (A) - correct oil level shown

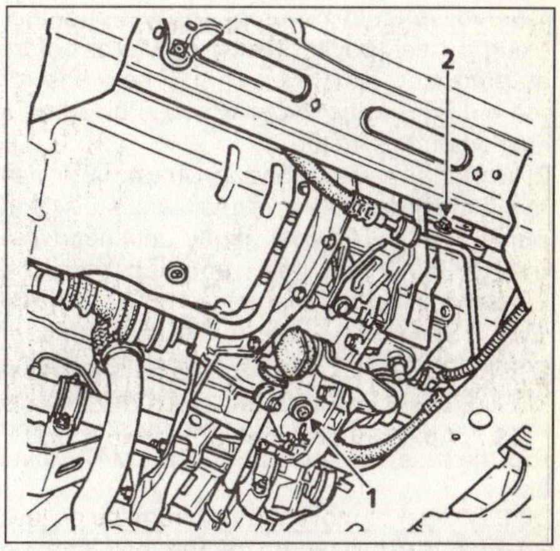

29.2 Manual transmission drain plug (1) and filler/level plug (2)

6 Lower the vehicle to the ground. Note that frequent need for topping-up indicates a leakage, possibly through an oil seal. The cause should be investigated and rectified.

30 Brake fluid renewal

Warning: Brake hydraulic fluid can harm your eyes and damage painted surfaces, so use extreme caution when handling and pouring it. Do not use fluid that has been standing open for some time, as it absorbs moisture from the air. Excess moisture can cause a dangerous loss of braking effectiveness.
Caution: On models equipped with ABS, disconnect the battery before carrying out the following operation and do not reconnect the battery until after the operation is complete. Failure to do this could lead to air entering the hydraulic unit. If air enters the hydraulic unit pump, it will prove very difficult (in some cases impossible) to bleed the unit.
1 The procedure is similar to that for the bleeding of the hydraulic system as described in Chapter 9, except that the brake fluid

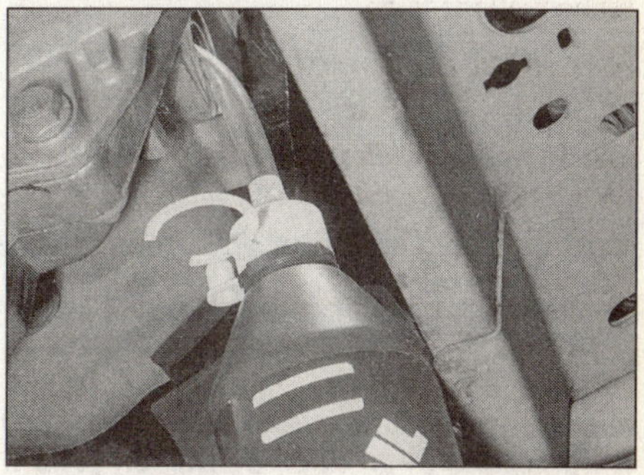

29.4 Topping-up the manual transmission oil level

1B

reservoir should be emptied by siphoning, using a clean poultry baster or similar before starting, and allowance should be made for the old fluid to be expelled when bleeding a section of the circuit.

2 Working as described in Chapter 9, open the first bleed screw in the sequence, and pump the brake pedal gently until nearly all the old fluid has been emptied from the master cylinder reservoir. Top-up to the "MAX" level with new fluid, and continue pumping until only the new fluid remains in the reservoir, and new fluid can be seen emerging from the bleed screw. Tighten the screw, and top the reservoir level up to the "MAX" level line.

HAYNES HiNT *Old brake fluid is invariably much darker in colour than the new, making it easy to distinguish the two.*

3 Work through all the remaining bleed screws in the sequence until new fluid can be seen at all of them. Be careful to keep the master cylinder reservoir topped-up to above the "MINI" level at all times, or air may enter the system and greatly increase the length of the task.

4 When the operation is complete, check that all bleed screws are securely tightened, and that their dust caps are refitted. Wash off all traces of spilt fluid, and recheck the master cylinder reservoir fluid level.

5 Check the operation of the brakes before taking the car on the road.

31 Road test

Instruments and electrical equipment

1 Check the operation of all instruments and electrical equipment.

2 Make sure that all instruments read correctly, and switch on all electrical equipment in turn, to check that it functions properly.

Steering and suspension

3 Check for any abnormalities in the steering, suspension, handling or road "feel".

4 Drive the vehicle, and check that there are no unusual vibrations or noises.

5 Check that the steering feels positive, with no excessive "sloppiness", or roughness, and check for any suspension noises when cornering and driving over bumps.

Drivetrain

6 Check the performance of the engine, clutch, transmission and driveshafts.

7 Listen for any unusual noises from the engine, clutch and transmission.

8 Make sure that the engine runs smoothly when idling, and that there is no hesitation when accelerating.

9 Check that, where applicable, the clutch action is smooth and progressive, that the drive is taken up smoothly, and that the pedal travel is not excessive. Also listen for any noises when the clutch pedal is depressed.

10 Check that all gears can be engaged smoothly without noise, and that the gear lever action is not vague or "notchy".

11 Listen for a metallic clicking sound from the front of the vehicle, as the vehicle is driven slowly in a circle with the steering on full-lock. Carry out this check in both directions. If a clicking noise is heard, this indicates wear in a driveshaft joint (see Chapter 8).

Braking system

12 Make sure that the vehicle does not pull to one side when braking, and that the wheels do not lock prematurely when braking hard.

13 Check that there is no vibration through the steering when braking.

14 Check that the handbrake operates correctly, without excessive movement of the lever, and that it holds the vehicle stationary on a slope.

15 Test the operation of the brake servo unit as follows. Depress the footbrake four or five times to exhaust the vacuum, then start the engine. As the engine starts, there should be a noticeable "give" in the brake pedal as vacuum builds up. Allow the engine to run for at least two minutes, and then switch it off. If the brake pedal is now depressed again, it should be possible to detect a hiss from the servo as the pedal is depressed. After about four or five applications, no further hissing should be heard, and the pedal should feel considerably harder.

Every 72 000 miles (120 000 km)

32 Timing belt renewal

Refer to Chapter 2B.

Every 2 years (regardless of mileage)

33 Coolant renewal

Cooling system draining

Warning: Wait until the engine is cold before starting this procedure. Do not allow antifreeze to come in contact with your skin, or with the painted surfaces of the vehicle. Rinse off spills immediately with plenty of water. Never leave antifreeze lying around in an open container, or in a puddle in the driveway or on the garage floor. Children and pets are attracted by its sweet smell, but antifreeze can be fatal if ingested.

Note: *Renault do not specify renewal intervals for the coolant, but it is advisable to drain and refill the system every two years to ensure that the corrosion inhibiting properties of the coolant are maintained.*

1 With the engine completely cold, remove the expansion tank filler cap. Turn the cap anti-clockwise, wait until any pressure remaining in the system is released, then unscrew it and lift it off.

2 Where applicable, remove the undershield, then position a suitable container beneath the radiator bottom hose connection.

3 Slacken the hose clip, pull off the hose and allow the coolant to drain into the container.

4 To assist draining, open the cooling system bleed screws in the heater hoses at the left-hand end of the cylinder head and on the bulkhead. Also open the bleed screw at the top right-hand side of the radiator.

5 If the coolant has been drained for a reason other than renewal, then provided it is clean and less than two years old, it can be re-used, though this is not recommended.

6 Refit the radiator bottom hose on completion of draining.

Cooling system flushing

7 If coolant renewal has been neglected, or if the antifreeze mixture has become diluted, then in time, the cooling system may gradually lose efficiency, as the coolant passages become restricted due to rust, scale deposits, and other sediment. The cooling system efficiency can be restored by flushing the system clean.

8 The radiator should be flushed independently of the engine, to avoid unnecessary contamination.

Radiator flushing

9 Disconnect the top and bottom hoses and any other relevant hoses from the radiator, with reference to Chapter 3.

10 Insert a garden hose into the radiator top inlet. Direct a flow of clean water through the radiator, and continue flushing until clean water emerges from the radiator bottom outlet.

11 If after a reasonable period, the water still does not run clear, the radiator can be flushed with a good proprietary cleaning agent. It is important that their manufacturer's instructions are followed carefully. If the contamination is particularly bad, insert the hose in the radiator bottom outlet, and reverse-flush the radiator.

Engine flushing

12 To flush the engine, remove the thermostat as described in Chapter 3, then temporarily refit the top hose at its engine connection.

13 With the top and bottom hoses disconnected from the radiator, insert a garden hose into the radiator top hose. Direct a clean flow of water through the engine, and continue flushing until clean water emerges from the radiator bottom hose.

14 On completion of flushing, refit the thermostat and reconnect the hoses with reference to Chapter 3.

Cooling system filling

15 Before attempting to fill the cooling system, make sure that all hoses and clips are in good condition, and that the clips are tight. Note that an antifreeze mixture must be used all year round, to prevent corrosion of the engine components.

16 Remove the expansion tank filler cap.

17 Open the cooling system bleed screws.

18 Place a wad of rags around the expansion tank.

19 Slowly fill the system until the coolant level reaches the top of the expansion tank filler neck.

20 Close the bleed screws when coolant free from air bubbles emerges.

21 Start the engine, and run it at a fast idle speed (do not exceed 2500 rpm) for approximately 4 minutes. Keep the level topped up to the top of the expansion tank filler neck.

22 Refit and tighten the expansion tank filler cap.

23 Allow the engine to run for approximately 20 minutes at 2500 rpm until the cooling fan cuts in and out.

24 Stop the engine and check the coolant level, which should be up to the "MAX" mark on the side of the tank. Check that the expansion tank filler cap is tight.

25 Allow the engine to cool, then re-check the coolant level with reference to "*Weekly checks*". Top-up the level if necessary and refit the expansion tank filler cap. Where applicable, refit the undershield.

Antifreeze mixture

26 The antifreeze should always be renewed at the specified intervals. This is necessary not only to maintain the antifreeze properties, but also to prevent corrosion which would otherwise occur as the corrosion inhibitors become progressively less effective.

27 Always use an ethylene-glycol based antifreeze which is suitable for use in mixed-metal cooling systems. The quantity of antifreeze and levels of protection are given in the Specifications.

28 Before adding antifreeze, the cooling system should be completely drained, preferably flushed, and all hoses checked for condition and security.

29 After filling with antifreeze, a label should be attached to the expansion tank, stating the type and concentration of antifreeze used, and the date installed. Any subsequent topping-up should be made with the same type and concentration of antifreeze.

30 Do not use engine antifreeze in the windscreen/tailgate washer system, as it will cause damage to the vehicle paintwork. A screenwash additive should be added to the washer system in the quantities stated on the bottle.

Chapter 2 Part A:
Petrol engine in-car repair procedures

Contents

Degrees of difficulty

Easy, suitable for novice with little experience	**Fairly easy,** suitable for beginner with some experience	**Fairly difficult,** suitable for competent DIY mechanic	**Difficult,** suitable for experienced DIY mechanic	**Very difficult,** suitable for expert DIY or professional

Specifications

General

Type	Four-cylinder, in-line, overhead camshaft, 8-valve
Designation:	
1390 cc	E7J
1598 cc	K7M
Bore:	
E7J	75.8 mm
K7M	79.5 mm
Stroke:	
E7J	77.0 mm
K7M	80.5 mm
Capacity:	
E7J	1390 cc
K7M	1598 cc
Firing order	1-3-4-2 (No 1 cylinder at flywheel/driveplate end)
Direction of crankshaft rotation	Clockwise viewed from pulley end
Compression ratio:	
E7J	9.5 : 1
K7M 803 and K7M 702	9.7 : 1
K7M 720	9.5 : 1

Cylinder block

Material	Cast-iron
Bore diameter:	
E7J (wet liners)	75.8 mm +0.03 -0.00 mm
K7M (dry liners)	79.5 mm +0.03 -0.00 mm
Liner height (E7J):	
Total	130.0 mm
From shoulder to top	91.5 mm +0.035 +0.005 mm
Cylinder block depth (E7J - from top to liner locating shoulder)	91.5 mm -0.015 -0.055 mm
Liner protrusion without O-ring (E7J)	0.02 to 0.09 mm
Maximum difference in protrusion between adjacent liners (E7J)	0.05 mm

Crankshaft

Number of main bearings	5
Main bearing journal diameter:	
E7J engine	
Standard	54.795 mm ± 0.01 mm
1st undersize	54.545 mm ± 0.01 mm
K7M engine	
Standard	48.010 +0 -0.02 mm
1st undersize	47.760 +0 -0.02 mm
Main bearing running clearance	0.020 to 0.058 mm
Crankpin (big-end) journal diameter:	
E7J engine	
Standard	43.980 mm +0 -0.02 mm
1st undersize	43.730 mm +0 -0.02 mm
K7M engine	
Standard	43.980 mm +0 -0.02 mm
1st undersize	43.790 mm +0 -0.02 mm
Crankshaft endfloat	0.045 to 0.852 mm

Pistons and piston rings

Piston ring end gaps	Pre-adjusted
Piston ring thickness:	
Top compression ring	1.5 mm
Second compression ring:	
E7J engine	1.75 mm
K7M engine	1.5 mm
Oil control ring:	
E7J engine	3.0 mm
K7M engine	2.5 mm
Piston clearance in liner	0.045 to 0.065 mm (suggested values)

Gudgeon pins

Length:	
E7J engine	60.0 mm
K7M engine	62.0 mm
Outside diameter	19.0 mm
Bore:	
E7J engine	11.0 mm
K7M engine	10.55 to 11.50 mm

Connecting rods

Big-end bearing running clearance	0.014 to 0.053 mm
Big-end cap side play:	
E7J engine	0.310 to 0.572 mm
K7M engine	0.310 to 0.604 mm

Cylinder head

Material	Aluminium
Height	113.0 mm ± 0.05 mm
Maximum acceptable gasket face distortion	0.05 mm
Refinishing limit	No refinishing permitted
Valve seat angle:	
Inlet	120°
Exhaust	90°
Valve seat width	1.7 mm

Camshaft

Drive	Toothed belt
Number of bearings	5
Camshaft endfloat:	
E7J engine	0.06 to 0.15 mm
K7M engine	0.01 to 0.15 mm

Valves

Head diameter:	
Inlet	37.5 ± 0.1 mm
Exhaust	33.5 ± 0.1 mm
Stem diameter	7.0 mm

Valve clearances (cold)

E7J engine

Inlet	0.10 mm
Exhaust	0.25 mm

K7M engine

Inlet	0.10 to 0.15 mm
Exhaust	0.25 to 0.30 mm (after fitting new valves 0.20 to 0.25 mm)

Lubrication system

System pressure:	
At idle	1.0 bar
At 4000 rpm (E7J) or 3000 rpm (K7M)	3.0 bars
Oil pump type	Two-gear
Oil pump clearances:	
Gear-to-body (minimum)	0.110 mm
Gear-to-body (maximum)	0.249 mm
Gear endfloat (minimum)	0.020 mm
Gear endfloat (maximum)	0.086 mm

Torque wrench settings	Nm	lbf ft
Camshaft sprocket bolt:		
EJ7 engine	55	41
K7M engine	45	33
Connecting rod (big-end) cap - oiled:		
EJ7 engine:		
Stage 1	10	7
Stage 2	45	33
K7M engine:		
Stage 1	10	7
Stage 2	43	32
Crankshaft pulley/sprocket bolt:		
EJ7 engine	85	63
K7M engine:		
Stage 1	20	15
Stage 2	Angle-tighten through 68° ± 6°	
Cylinder head bolts (E7J engine):		
Stage 1	30	22
Stage 2	60	44
Alternative Stage 1	20	15
Alternative Stage 2	Angle-tighten all bolts by 97° ± 2°	
Stage 3	Wait for 3 minutes	
Stage 4	Slacken bolts 1 and 2 fully	
Stage 5	Tighten bolts 1 and 2 to:	
	20	15
Stage 6	Angle-tighten bolts 1 and 2 by 97° ± 2°	
Stage 7	Slacken bolts 3, 4, 5 and 6 fully	
Stage 8	Tighten bolts 3, 4, 5 and 6 to:	
	20	15
Stage 9	Angle-tighten bolts 3, 4, 5 and 6 by 97° ± 2°	
Stage 10	Slacken bolts 7, 8, 9 and 10 fully	
Stage 11	Tighten bolts 7, 8, 9 and 10 to:	
	20	15
Stage 12	Angle-tighten bolts 7, 8, 9 and 10 by 97° ± 2°	

Torque wrench settings (continued)	Nm	lbf ft
Cylinder head bolts (K7M engine):		
Stage 1	Tighten bolts 1 and 2 to:	
	20	15
Stage 2	Angle-tighten bolts 1 and 2 by 100° ± 6°	
Stage 3	Tighten bolts 3, 4, 5 and 6 to:	
	20	15
Stage 4	Angle-tighten bolts 3,4, 5 and 6 by 100° ± 6°	
Stage 5	Tighten bolts 7, 8, 9 and 10 to:	
	20	15
Stage 6	Angle-tighten bolts 7,8, 9 and 10 by 100° ± 6°	
Stage 7	Wait for 3 minutes	
Stage 8	Slacken bolts 1 and 2 fully	
Stage 9	Tighten bolts 1 and 2 to:	
	20	15
Stage 10	Angle-tighten bolts 1 and 2 by 110° ± 6°	
Stage 11	Slacken bolts 3, 4, 5 and 6 fully	
Stage 12	Tighten bolts 3, 4, 5 and 6 to:	
	20	15
Stage 13	Angle-tighten bolts 3, 4, 5 and 6 by 110° ± 6°	
Stage 15	Slacken bolts 7, 8, 9 and 10 fully	
Stage 16	Tighten bolts 7, 8, 9 and 10 to:	
	20	15
Stage 17	Angle-tighten bolts 7, 8, 9 and 10 by 110° ± 6°	
Engine/gearbox mountings (see text):		
Rear mounting bracket to transmission	65	48
Rear mounting to subframe	75	55
RH and LH front engine mountings to subframe	45	33
Exhaust manifold:		
EJ7 engine	26	19
K7M engine	25	19
Flywheel/driveplate bolts	53	39
Front suspension strut to hub carrier bolts	170	126
Inlet manifold:		
EJ7 engine	26	19
K7M engine	25	19
Oil drain plug	20	15
Oil pump:		
Mounting bolts	25	19
Sprocket bolts	10	7
Main bearing cap:		
EJ7 engine	64	47
K7M engine:		
Stage 1	25	19
Stage 2	Angle-tighten through 47° ± 5°	
Rocker shaft bolts - oiled	23	17
Strengthening bracket:		
On engine	50	37
On transmission	25	19
Subframe-to-underbody bolts:		
Front	60	44
Rear	110	81
Sump:		
EJ7 engine	8	6
K7M engine	9	7
Timing belt tensioner pulley nut	50	37

1 General information

How to use this Chapter

This Part of Chapter 2 is devoted to in-car repair procedures on the petrol engine. All procedures concerning engine removal and refitting, and engine block/cylinder head overhaul can be found in Chapter 2C.

Refer to *"Buying spare parts and vehicle identification numbers"* in the Reference Section at the end of this manual for details of engine code locations.

Most of the operations in Chapter 2A are based on the assumption that the engine is still installed in the car. Therefore, if this information is being used during a complete engine overhaul, with the engine already removed, many of the steps included here will not apply.

Engine description

The engine is of four-cylinder, in-line, overhead camshaft type, mounted transversely in the front of the car.

The cylinder block is of cast iron. Replaceable wet liners are fitted to the E7J engine, while the K7M engine has conventional dry liners bored directly into the cylinder block. The crankshaft is supported within the cylinder block on five shell-type main bearings. Upper thrustwashers are fitted at the centre main bearing, to control crankshaft endfloat.

The connecting rods are attached to the crankshaft by horizontally-split shell-type big-end bearings, and to the pistons by gudgeon pins which are an interference-fit in the connecting rods. The aluminium alloy pistons are fitted with three piston rings, comprising two compression rings and a scraper-type oil control ring.

The overhead camshaft is mounted directly in the cylinder head, and is driven by the crankshaft via a toothed rubber timing belt, which also drives the water pump. The camshaft operates the valves via rocker arms located on a rocker shaft bolted to the top of the cylinder head.

A fully-enclosed crankcase ventilation system is employed.

Lubrication is by pressure feed from a gear-type oil pump, which is chain-driven direct from the front of the crankshaft.

On the E7J engine, the distributor rotor is driven directly from the rear end of the camshaft; on the K7M engine, the ignition is computer-controlled.

Repair operations possible with the engine in the vehicle

The following operations can be carried out without having to remove the engine from the car:

a) Removal and refitting of the cylinder head.
b) Removal and refitting of the timing belt and sprockets.
c) Renewal of the camshaft oil seal.
d) Removal and refitting of the camshaft.
e) Removal and refitting of the sump.
f) Removal and refitting of the connecting rods, pistons and liners.*
g) Removal and refitting of the oil pump.
h) Renewal of the front crankshaft oil seal.
i) Renewal of the engine mountings.

*Although the operation marked with an asterisk can be carried out with the engine in the car after removal of the sump, it is better for the engine to be removed, in the interests of cleanliness and improved access. For this reason, the procedure is described in Part C of this Chapter.

2 Compression test - description and interpretation

1 A compression check will tell you what mechanical condition the top end (pistons, rings, valves, head gaskets) of the engine is in. Specifically, it can tell you if the compression is down due to leakage caused by worn piston rings, defective valves and seats or a blown head gasket. **Note:** *The engine must be at normal operating temperature, and the battery must be fully charged, for this check.*

2 Begin by cleaning the area around the spark plugs before you remove them (compressed air should be used, if available, otherwise a small brush or even a bicycle tyre pump will work). The idea is to prevent dirt from getting into the cylinders as the compression check is being done.

3 Remove all of the spark plugs from the engine (see Chapter 1).

4 Disconnect the low tension wiring from the ignition power module (E7J engine) or wasted spark ignition coil (K7M engine). Also remove the fuse for the fuel injection system from the fusebox in the passenger compartment (refer to Chapter 12).

5 Fit the compression gauge into the No 1 spark plug hole - the type of tester which screws into the plug thread is to be preferred.

6 Have an assistant hold the accelerator pedal fully depressed, while at the same time cranking the engine over several times on the starter motor. Observe the compression gauge - the compression should build up quickly in a healthy engine. Low compression on the first stroke, followed by gradually-increasing pressure on successive strokes, indicates worn piston rings. A low compression reading on the first stroke, which does not build up during successive strokes, indicates leaking valves or a blown head gasket (a cracked head could also be the cause). Deposits on the undersides of the valve heads can also cause low compression. Record the highest gauge reading obtained, then repeat the procedure for the remaining cylinders.

7 Add some engine oil (about three squirts from a plunger-type oil can) to each cylinder, through the spark plug hole, and repeat the test.

8 If the compression increases after the oil is added, the piston rings are definitely worn. If the compression does not increase significantly, the leakage is occurring at the valves or head gasket. Leakage past the valves may be caused by burned valve seats and/or faces, or warped, cracked or bent valves.

9 If two adjacent cylinders have equally low compression, there is a strong possibility that the head gasket between them is blown. The appearance of coolant in the combustion chambers or the crankcase would verify this condition.

10 Actual compression pressures for the engines covered by this manual are not specified by the manufacturer. However, bearing in mind the information given in the preceding paragraphs, the results obtained should give a good indication of engine condition and what course of action, if any, to take.

3 Top Dead Centre (TDC) for No 1 piston - locating

1 Top dead centre (TDC) is the highest point in the cylinder that each piston reaches as the crankshaft turns. Each piston reaches TDC at the end of the compression stroke, and again at the end of the exhaust stroke; however, for the purpose of timing the engine, TDC refers to the position of No 1 piston at the end of its compression stroke. No 1 piston is at the flywheel end of the engine.

2 Disconnect the battery negative lead.

Caution: If the radio/cassette in your vehicle is equipped with an anti-theft system, make sure you have the correct activation code before disconnecting the battery.

3 Apply the handbrake, then jack up the front right-hand side of the car and support it on axle stands. Remove the right-hand roadwheel.

4 Remove the right-hand wheel arch liners after pulling out the plastic retainers and removing the screws.

5 Remove the spark plugs as described in Chapter 1.

6 Place a finger over the No 1 spark plug hole in the cylinder head (nearest the flywheel) - as the spark plugs are deeply recessed, alternatively use the handle of a screwdriver. Turn the engine in a clockwise direction, using a socket or spanner on the crankshaft pulley bolt, until pressure is felt in the No 1 cylinder. This indicates that No 1 piston is rising on its compression stroke.

7 Look through the aperture at the flywheel end of the engine, and continue turning the crankshaft until the TDC timing mark on the flywheel/driveplate is aligned with the TDC mark on the gearbox bellhousing.

8 On the E7J engine, remove the distributor cap and check that the rotor arm is aligned with the No 1 HT lead segment.

9 If it is wished to view the timing mark on the camshaft sprocket, refer to Section 5 and remove the timing cover. With No 1 piston at TDC, the timing mark on the camshaft sprocket must be in alignment with the mark on the end of the valve cover.

4 Valve clearances - checking and adjustment

1 On the E7J engine, remove the air cleaner assembly with reference to Chapter 4, then disconnect the accelerator cable from the throttle housing. If necessary, unbolt the accelerator cable support from the valve cover.

2 On the K7M engine, remove the ignition coil assembly from the valve cover with reference to Chapter 5B, then disconnect the HT leads from the spark plugs and unbolt the HT lead support from the valve cover.

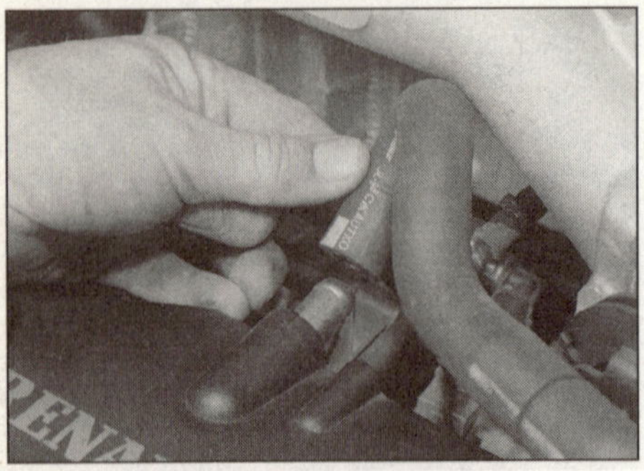

4.3a Disconnecting the large (to air cleaner) . . .

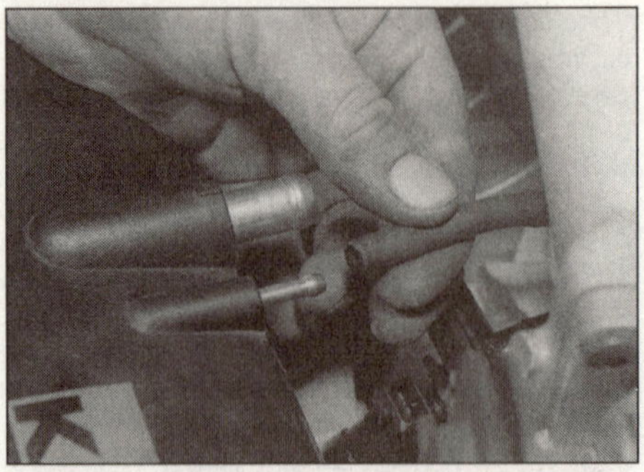

4.3b . . . and small (to inlet manifold) crankcase ventilation hoses from the camshaft cover

4.4a Removing the valve cover bolts

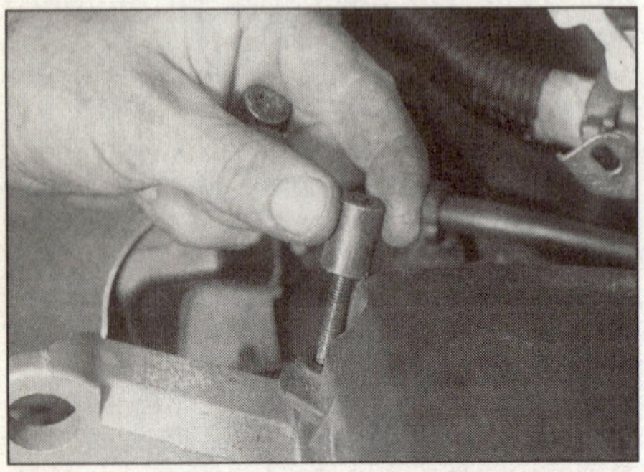

4.4b The right-hand valve cover bolts have threads in their tops for the ignition coil

4.4c Removing the valve cover . . .

4.4d . . . and gasket

3 Disconnect the crankcase ventilation hoses from the valve cover **(see illustrations)**.

4 Unscrew the bolts, and remove the valve cover and gasket. On the K7M engine, the right-hand bolts have threads in their tops, as the ignition coil is mounted on them **(see illustrations)**.

5 Remove the spark plugs (Chapter 1) in order to make turning the engine easier.

6 Draw the valve positions on a piece of paper, numbering them 1 to 4 inlet and exhaust according to their cylinders, from the flywheel/driveplate (left-hand) end of the engine (ie 1E, 1I, 2E, 2I and so on). The inlet valves are on the inlet manifold side of the cylinder head, and the exhaust valves are on the exhaust manifold side. As the valve clearances are adjusted, cross them off.

7 Using a socket or spanner on the crankshaft pulley bolt, turn the engine in a clockwise direction until No 1 exhaust valve is completely open (ie the valve spring is completely compressed).

8 Insert a feeler blade of the correct thickness (see Specifications) between the No 3 cylinder inlet valve stem and the end of the rocker arm. It should be a firm sliding fit. If adjustment is necessary, loosen the locknut on the rocker arm using a ring spanner, and turn the adjustment screw with a small open-ended spanner or pliers until the fit is correct. Hold the adjustment screw, tighten the locknut (preferably using a torque wrench) and recheck the adjustment **(see illustrations)**. Repeat the adjustment procedure on No 4 cylinder exhaust valve. Note that the valve clearances for the inlet and exhaust valves are different.

9 Turn the engine in a clockwise direction until No 3 exhaust valve is completely open. Adjust the valve clearances on No 4 inlet and No 2 exhaust valves. Continue to adjust the valve clearances in the following sequence.

Exhaust valve fully open	Inlet valve to adjust	Exhaust valve to adjust
1E	*3I*	*4E*
3E	*4I*	*2E*
4E	*2I*	*1E*
2E	*1I*	*3E*

10 Remove the socket or spanner from the crankshaft pulley bolt.

11 Refit the spark plugs with reference to Chapter 1. Refit the valve cover, using a new gasket where necessary. Reconnect the crankcase ventilation hose.

12 On the K7M engine, refit the ignition coil

4.8a Loosening the adjustment locknut on the rocker arm

4.8b Turning the adjustment screw with pliers

4.8c Tightening the adjustment locknut with a torque wrench

5.6a Loosening the crankshaft pulley bolt

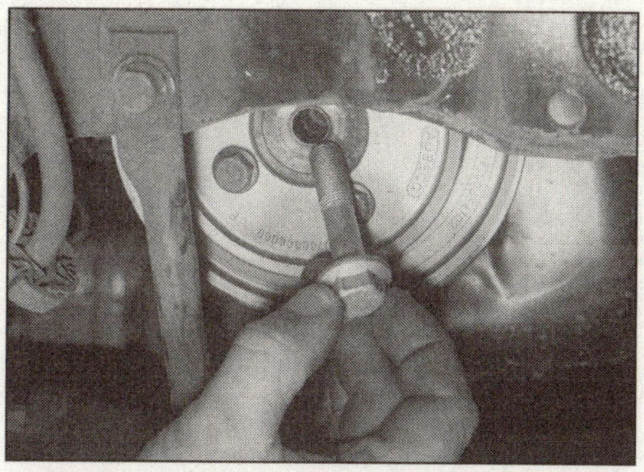
5.6b Removing the crankshaft pulley bolt

5.6c Removing the crankshaft pulley

5.7a Timing cover lower mounting nut

5.7b Removing the timing cover

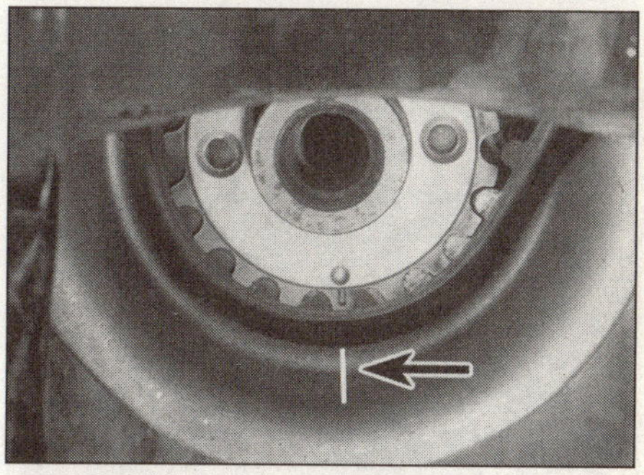
5.8 Mark the front oil seal housing in line with the TDC mark on the crankshaft pulley

assembly and HT lead support, then reconnect the HT leads to the spark plugs.

13 On the E7J engine, refit the accelerator cable and support, followed by the air cleaner assembly.

5 Timing belt - removal, inspection and refitting

Caution: If the timing belt breaks in service, extensive engine damage may result. Renew the belt at the intervals specified in Chapter 1, or earlier if its condition is at all doubtful.

Removal

1 Disconnect the battery negative lead.

Caution: If the radio/cassette in your vehicle is equipped with an anti-theft system, make sure you have the correct activation code before disconnecting the battery.

2 Apply the handbrake, then jack up the front right-hand side of the vehicle and support on axle stands (see *"Jacking and vehicle support"*). Remove the right-hand roadwheel. Where fitted, remove the engine compartment undertray.

3 Remove the right-hand wheel arch liners, after pulling out the plastic retainers and removing the screws.

4 Remove the power-assisted steering drivebelt, followed by the alternator drivebelt, as described in Chapter 1.

5 Set the engine at TDC for No 1 piston as described in Section 3.

6 Have an assistant engage 4th gear and press the brake pedal firmly. Unscrew and remove the crankshaft pulley bolt, then slide the pulley and hub from the key on the end of the crankshaft **(see illustrations)**. **Note:** *The bolt is very tight.* Alternatively, the crankshaft may be held stationary by unbolting the ignition timing sensor from the top of the gearbox/automatic transmission, and inserting a screwdriver in the starter ring gear teeth through the access hole.

7 Unbolt and remove the timing cover **(see illustrations)**.

8 Check that the TDC mark on the camshaft sprocket is pointing upwards and in line with the mark on the valve cover. On some models there is also an alignment mark on the front oil seal housing - on later models, mark the housing as an aid to refitting **(see illustration)**.

5.10a The running direction arrows on the timing belt

9 Loosen the nut on the timing belt tensioner, turn the tensioner pulley anti-clockwise to release the tension, then temporarily tighten the nut to hold the pulley away from the timing belt.

10 Check if the belt is marked with arrows to indicate its running direction; mark it if necessary. Release the belt from the camshaft sprocket, water pump sprocket and crankshaft sprocket, and remove it from the engine. Be careful not to kink or otherwise damage the belt if it is to be re-used **(see illustrations)**.

11 Clean the sprockets and tensioner, and wipe them dry. Also clean the cylinder head and block behind the timing belt running area.

Inspection

12 Examine the timing belt carefully for any signs of cracking, fraying or general wear, particularly at the roots of the teeth. Renew

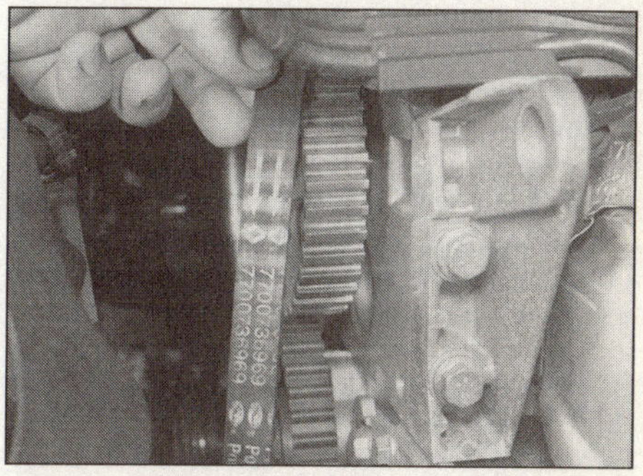
5.10b Removing the timing belt from the camshaft sprocket

5.13a Align the timing mark on the inside of the belt with the TDC mark on the crankshaft sprocket

5.13b The timing belt is also marked on its outside surface

5.13c Align the upper mark on the timing belt with the timing mark on the valve cover

5.14 Tightening the tensioner nut with a torque wrench

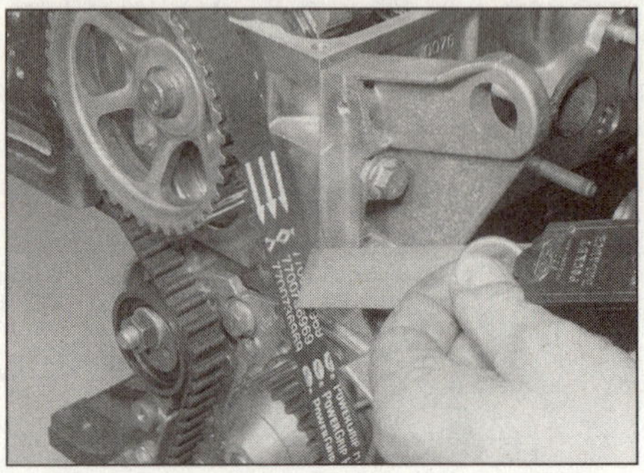

5.15 Checking the tension of the timing belt with a spring balance and steel rule

5.17 Angle-tightening the crankshaft pulley on the K7M engine

the belt if there is any sign of deterioration of this nature, or if there is any oil or grease contamination. Renew any leaking oil seals. The belt **must** be renewed if it has completed the maximum mileage given in Chapter 1. Also examine the sprockets and tensioner as described in the next Section.

Refitting

13 Check the directional mark and timing bands on the timing belt. Fit the belt on the crankshaft sprocket so that one of the timing bands is aligned with the marks on the sprocket and cover. The other timing band should be positioned so that it will locate on the camshaft sprocket in alignment with the other timing mark. After engaging the belt with the crankshaft sprocket, pull it taut over the water pump sprocket and onto the camshaft sprocket, then position it over the tensioner pulley **(see illustrations)**.

14 With the belt fully engaged with the sprockets, slacken the tensioner nut and tension the belt, then retighten the nut **(see illustration)**. The tensioner is not spring-loaded, so it will have to be rotated manually to tension the belt; a screwdriver can be used as a lever between bolts in the holes in the tensioner hub, or use a tool like the one shown in illustration 6.2. Check that the timing marks are still aligned correctly.

15 The belt deflection must now be checked. To do this, first make a mark on the cylinder head in line with the timing belt, mid-way between the camshaft and water pump sprockets. A force of 30 N (7 lbf) must now be applied to the timing belt. Belt deflection under this force must be 6.0 ± 0.5 mm (0.236 ± 0.02 in). Renault technicians use a special tool to do this, but an alternative arrangement can be made by using a spring balance and steel rule. Apply the force with the spring balance, and read off the deflection on the steel rule **(see illustration)**.

16 If the adjustment is incorrect, the tensioner hub will have to be repositioned. The tensioner nut must be tightened to the specified torque, since if it were to come loose, considerable engine damage would result.

17 Refit the crankshaft pulley and its bolt. Using a socket on the pulley bolt, turn the engine through two complete revolutions in its normal direction, then recheck the timing belt tension and make sure that the timing marks are still in alignment. The timing marks on the sprockets must align every two turns, however the timing bands on the timing belt will only align after several more turns of the engine. Tighten the crankshaft pulley bolt to the specified torque (and angle on the K7M engine) while the assistant holds the crankshaft stationary as described earlier in this Section **(see illustration)**.

18 Refit the timing cover and tighten the retaining bolts.

19 Refit the alternator drivebelt followed by the power-assisted steering drivebelt and tension them as described in Chapter 1.

20 Refit the right-hand wheel arch liners, then refit the roadwheel and lower the vehicle to the ground.

21 Reconnect the battery negative lead.

6 Timing belt sprockets and tensioner - removal, inspection and refitting

Removal

1 Remove the timing belt as described in Section 5

2 To remove the camshaft sprocket, hold the sprocket stationary using a metal bar with two bolts tightened onto it inserted into the holes in the sprocket, then unscrew the bolt. The Renault tool is also shown **(see illustrations)**.

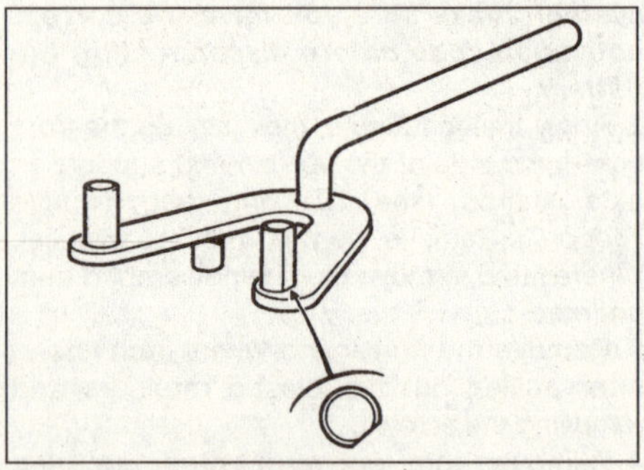

6.2a Renault tool for holding the camshaft sprocket stationary

6.2b Home-made tool for holding the camshaft sprocket stationary

6.2c Removing the camshaft sprocket bolt

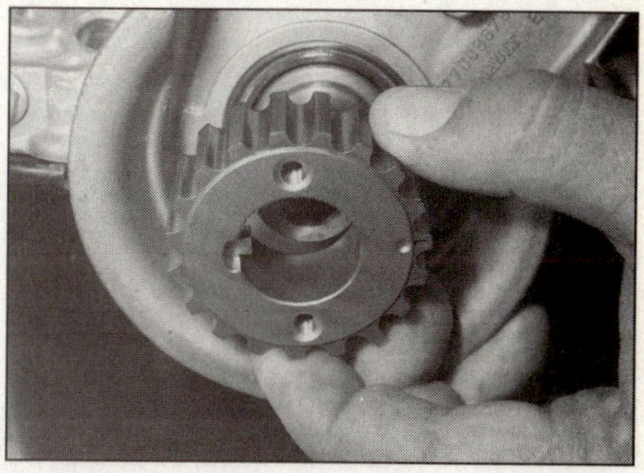

6.4 Removing the crankshaft sprocket

3 Remove the sprocket from the end of the camshaft. Note that it has a tab on its inner face which locates in a slot in the end of the camshaft.

4 A puller may be necessary to remove the crankshaft sprocket if it is tight **(see Tool Tip)**. Note that the key is incorporated in the sprocket, and is not separate **(see illustration)**.

5 Unscrew the nut, and withdraw the washer and tensioner pulley from the stud on the front oil seal housing **(see illustrations)**.

Inspection

6 Inspect the teeth of the sprockets for signs of nicks and damage. Also examine the water pump teeth. The teeth are not prone to wear, and should normally last the life of the engine.

7 Spin the tensioner pulley by hand, and check it for any roughness or tightness. Do not attempt to clean it with solvent, as this may enter the bearing. If wear is evident, renew the tensioner.

Refitting

8 Locate the tensioner on the stud on the front housing, then refit the nut and washer and tighten it finger-tight at this stage.

9 Slide the sprocket fully onto the crankshaft engaging the key with the groove in the crankshaft; use a metal tube if necessary to tap it into position.

TOOL TIP

It is easy to make up a crankshaft sprocket puller using two bolts, a metal bar and the existing crankshaft pulley bolt. By unscrewing the crankshaft pulley bolt, the sprocket is pulled from the end of the crankshaft.

6.5a Unscrew the nut . . .

6.5b . . . and withdraw the washer and tensioner pulley from the stud on the front oil seal housing

10 Locate the sprocket on the end of the camshaft, making sure that the tab locates in the special slot, then screw in the bolt. Tighten the bolt to the specified torque, holding the sprocket stationary using the method described in paragraph 2.

11 Refit the timing belt as described in Section 5.

7 Camshaft oil seal - renewal

1 Remove the camshaft sprocket as described in Section 6.

2 Note the fitted position of the old oil seal. Using a small screwdriver, prise out the oil seal from the cylinder head.

3 Wipe clean the seating in the cylinder head.

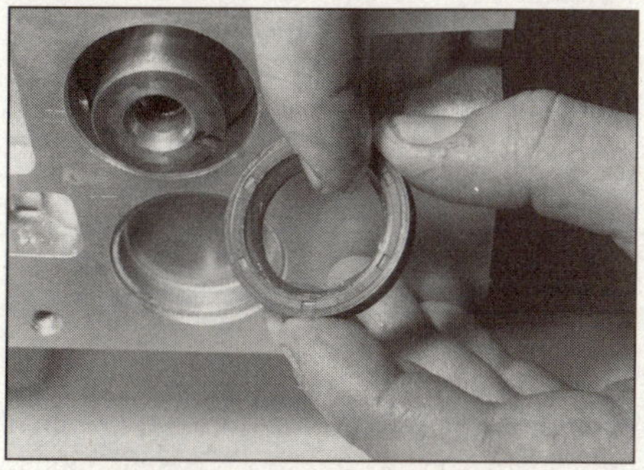

7.4a Smear a little grease on the oil seal . . .

4 Smear a little oil on the outer perimeter and sealing lip of the new oil seal. Locate the seal squarely in the cylinder head. Drive the seal into position using a metal tube or socket which has an external diameter slightly less than that of the bore in the cylinder head **(see illustrations)**. Make sure that the oil seal is the correct way round, with the lip facing inwards.

5 Refit the camshaft sprocket as described in Section 6.

8 Camshaft - removal, inspection and refitting

Removal

1 Remove the cylinder head as described in Section 9, and place it on the workbench.

7.4b . . . before driving it into the cylinder head with a suitable socket

8.2a Unscrew and remove the bolts . . .

8.2b . . . then withdraw the rocker shaft from the cylinder head

8.5a Unscrew the bolts . . .

8.5b . . . and remove the cover from the left-hand end of the cylinder head on the K7M engine

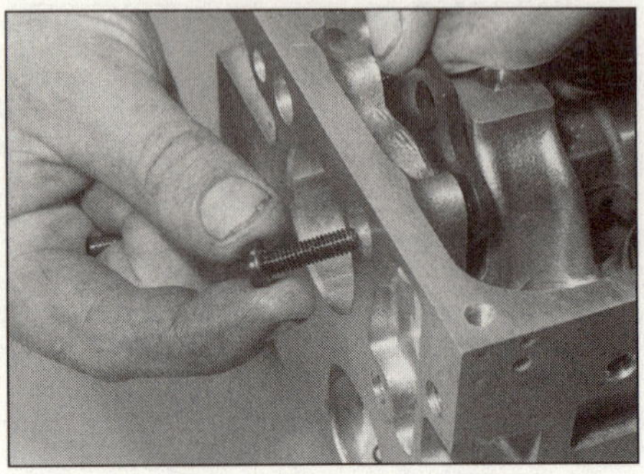

8.7a Unscrew the two bolts . . .

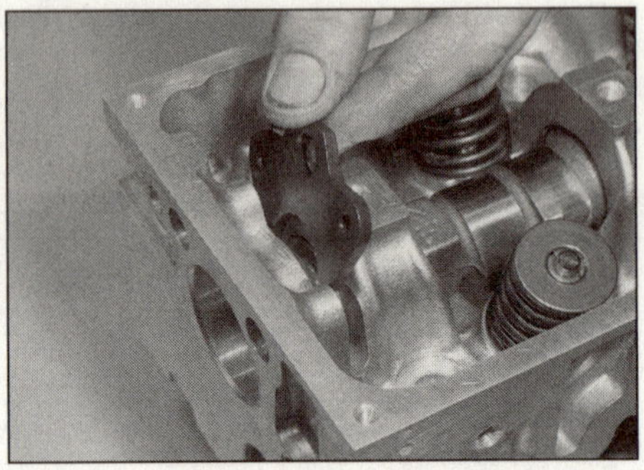

8.7b . . . and lift the thrustplate from the slot in the camshaft

2 Progressively unscrew the bolts holding the rocker shaft and retaining plate to the cylinder head, and withdraw the shaft **(see illustrations)**.

3 Hold the camshaft stationary using a spanner on the special flats provided on the camshaft, or with a suitable tool inserted through the sprocket holes, then unscrew the bolt and withdraw the sprocket.

4 On the E7J engine, using a Torx key, unscrew the two bolts and remove the distributor. There is no need to mark the distributor, as it is not possible to adjust its position. Although there is an elongated slot for one of the bolts, the other bolt locates in a single hole.

5 On the K7M engine, unbolt the cover and remove the gasket from the left-hand end of the cylinder head **(see illustrations)**.

6 Using a dial gauge, measure the endfloat of the camshaft, and compare with that given in the Specifications. This will give an indication of the amount of wear in the thrustplate.

7 Unscrew the two bolts and lift the thrustplate out from the slot in the camshaft **(see illustrations)**.

8 Withdraw the camshaft from the sprocket end of the cylinder head, taking care not to damage the bearing surfaces **(see illustration)**.

Inspection

9 Examine the camshaft bearing surfaces and cam lobes for wear ridges and scoring. Also examine the rocker shaft for wear. If necessary, the rocker shaft may be dismantled for inspection **(see illustration)**. Renew worn components as necessary.

8.8 Removing the camshaft from the cylinder head

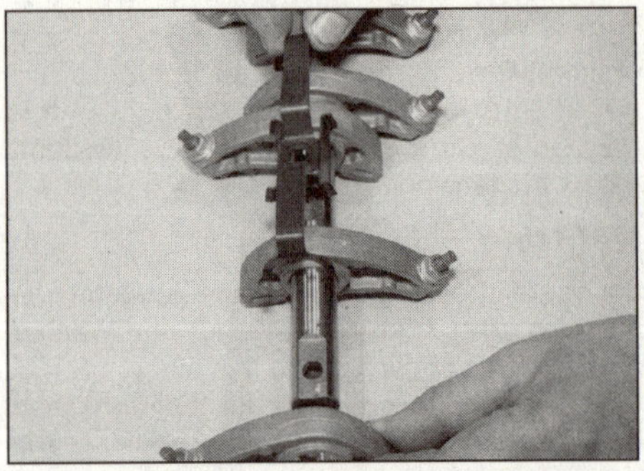

8.9 If necessary, the rockers may be removed from the rocker shaft, but keep them in order

10 Examine the condition of the bearing surfaces both on the camshaft and in the cylinder head. If the head bearing surfaces are worn excessively, a new cylinder head will be required.

Refitting

11 Lubricate the bearing surfaces in the cylinder head and the camshaft journals, then insert the camshaft into the head.

12 Refit the thrustplate, then insert and tighten the bolts.

13 Measure the endfloat as described in paragraph 6, and make sure that it is within the limits given in the Specifications. Excessive endfloat can only be due to wear of the thrustplate or the camshaft.

14 On the E7J engine, refit the distributor and tighten the two bolts using a Torx key.

15 On the K7M engine, refit the end cover together with a new gasket, and tighten the bolts.

16 Renew the oil seal (Section 7) then refit the camshaft sprocket, making sure that the tab engages with the cut-out in the end of the camshaft. Hold the camshaft stationary with a spanner on the special flats, then insert the bolt and tighten it to the specified torque.

17 Refit the rocker shaft and retaining plate, then insert the bolts in their original positions and tighten them to the specified torque **(see illustration)**. Note that the bolt threads and head contact surfaces must be oiled before inserting them. The hollow bolts are located at each end and in the middle.

18 Refit the cylinder head (see Section 9).

8.17 Tightening the rocker shaft mounting bolts

9 Cylinder head - removal, inspection and refitting

Note: *The cylinder head is removed from the cylinder block together with the inlet and exhaust manifolds, then dismantled on the workbench.*

Removal

1 Disconnect the battery negative lead.

Caution: If the radio/cassette in your vehicle is equipped with an anti-theft system, make sure you have the correct activation code before disconnecting the battery.

2 Drain the cooling system, including the cylinder block on the E7J engine, with reference to Chapter 1. The block drain plug on the E7J engine is located on the right-hand rear face of the cylinder block, beneath the inlet manifold. It is important to drain the block on the E7J engine, because if the wet cylinder liners are disturbed, the coolant will drain into the sump.

3 Drain the engine oil with reference to Chapter 1.

4 Remove the timing belt with reference to Section 5 of this Chapter.

5 Remove the air cleaner assembly as described in Chapter 4. On the K7M engine, remove the strengthening bar from between the front suspension turrets, and also disconnect the air duct from the throttle housing **(see illustrations)**.

6 Disconnect the accelerator cable from the throttle body housing.

7 On the E7J engine, unbolt the accelerator cable support from the valve cover.

8 On the K7M engine, remove the ignition coil assembly from the valve cover with reference to Chapter 5B; disconnect the HT leads from the spark plugs, and unbolt the HT lead support from the valve cover.

9 Disconnect the crankcase ventilation hoses from the valve cover.

10 Unscrew the bolts, and remove the valve cover and gasket.

11 Unbolt the earth strap from the right-hand side of the bulkhead **(see illustration)**.

12 Loosen the clips, and disconnect the fuel supply and return hoses from the throttle housing (E7J engine) or fuel rail (K7M engine).

13 Disconnect the evaporation emission control solenoid valve hose.

14 Disconnect the wiring and hoses from the absolute pressure sensor on the bulkhead. Also unbolt the wiring loom from the inlet manifold, and disconnect all wiring from the cylinder head **(see illustration)**.

15 Loosen the clip and disconnect the brake vacuum servo hose from the inlet manifold.

16 Unscrew and remove the oil level dipstick tube upper mounting bolt from the cylinder head.

17 Remove the spark plugs as described in Chapter 1.

18 Release the clip and disconnect the radiator top hose from the thermostat housing. Also disconnect the heater hoses and expansion tank hose, and unbolt the heater hose support from the left-hand side of the cylinder head **(see illustrations)**.

19 Unscrew and remove the alternator upper mounting/adjustment bolt from the bracket on the rear of the cylinder head.

20 Unscrew the nuts and remove the heat shield from the exhaust manifold. Remove the exhaust front downpipe with reference to Chapter 4.

21 Unbolt and remove the rocker shaft and retaining plate, then progressively slacken the cylinder head bolts in the reverse order to that shown in illustration 9.36a **(see illustrations)**. Remove all the bolts except the one positioned on the front right-hand corner, which should be unscrewed by only three or four threads.

9.5a On the K7M engine, unclip the accelerator cable support, then . . .

9.5b . . . unbolt the strengthening bar from between the suspension turrets

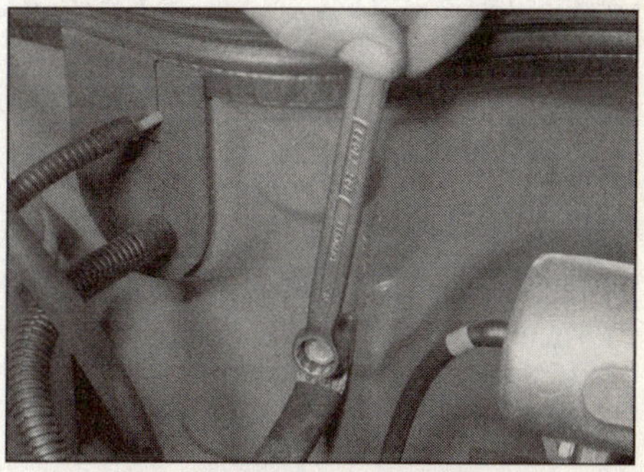

9.11 Unbolting the earth strap from the bulkhead

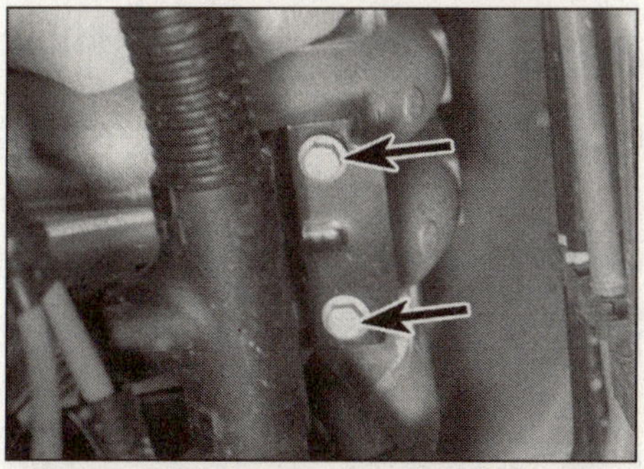

9.14 Engine wiring loom support bracket on the bottom of the inlet manifold on the K7M engine

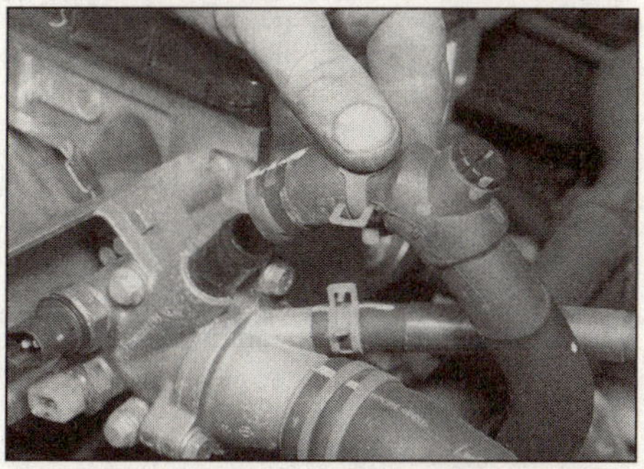

9.18a Disconnecting the heater hose from the thermostat housing

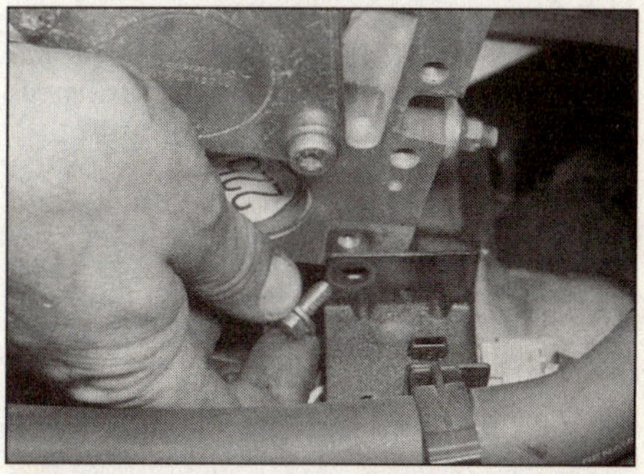

9.18b Unbolting the heater hose support from the cylinder head

9.21a Removing a rocker shaft end bolt - the bolt is hollow for oil supply to the rockers

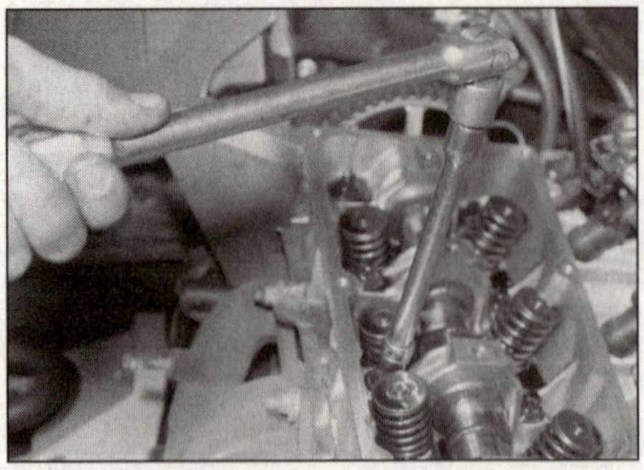

9.21b Slackening the cylinder head bolts

9.21c Removing the cylinder head bolts

9.23a Lifting the cylinder head from the block

9.23b Removing the cylinder head gasket

9.25 Clamps (arrowed) holding the liners in place on the E7J engine

22 The joint between the cylinder head, gasket and cylinder block must now be broken. On the E7J engine, it is important not to lift or disturb the 'wet' cylinder liners as the head is removed. To avoid this, pull the left-hand end of the cylinder head forward so as to swivel it around the single bolt still fitted, then move the head back to its original position. If this procedure is not followed on the E7J engine, there is a possibility of the cylinder liners moving and their bottom seals being disturbed, causing leakage after refitting the head.

23 Remove the remaining bolt, and lift the head from the cylinder block, followed by the gasket. Note the location dowel on the front right-hand corner of the block **(see illustrations)**.

24 Remove the inlet and exhaust manifolds with reference to Chapter 4A.

25 Note that on the E7J engine, the crankshaft must **not** be rotated with the cylinder head removed, otherwise the cylinder liners may be displaced. If it is necessary to turn the crankshaft (eg to clean the piston crowns), clamp the liners using bolts and washers, or make up some retaining clamps out of flat metal bar, held in place with bolts screwed into the block **(see illustration)**.

Inspection

26 The mating faces of the cylinder head and block must be perfectly clean before refitting the head. Use a scraper to remove all traces of gasket and carbon, and also clean the tops of the pistons. Take particular care with the aluminium cylinder head, as the soft metal is easily damaged. Also, make sure that debris is not allowed to enter the oil and water channels - this is particularly important for the oil circuit, as carbon could block the oil supply to the camshaft and rocker arms or crankshaft bearings. Using adhesive tape and paper, seal the water, oil and bolt holes in the cylinder block. Clean the piston crowns in the same way.

HAYNES HiNT ***To prevent carbon entering the gap between the pistons and bores, smear a little grease in the gap. After cleaning the piston, rotate the crankshaft so that the piston moves down the bore, then wipe out the grease and carbon with a cloth rag.***

27 Check the block and head for nicks, deep scratches and other damage. If slight, they may be removed carefully with a file. It may be possible to repair more serious damage by machining, but this is a specialist job.

28 If warpage of the cylinder head is suspected, use a straight-edge to check it for distortion. On the E7J engine, also check the protrusion of the cylinder liners. Either of these items can be associated with the head gasket blowing. Refer to Part C of this Chapter for further information.

29 Clean out all the bolt holes in the block using a pipe cleaner, or a rag and screwdriver. Make sure that all oil is removed, otherwise there is a possibility of the block being cracked by hydraulic pressure when the bolts are tightened.

30 Examine the bolt threads and the threads in the cylinder block for damage. If necessary, use the correct-size tap to chase out the threads in the block, and use a die to clean the threads on the bolts.

Refitting

31 Refit the inlet and exhaust manifolds to the cylinder head, referring to Chapter 4A.

32 Check that No 1 piston is positioned at TDC, then wipe clean the faces of the head and block. On the E7J engine, remove the cylinder liner clamps.

33 Check that the location dowel is in place on the front right-hand corner of the block.

34 Position the new gasket on the block and over the dowel - it can only be fitted one way round **(see illustration)**.

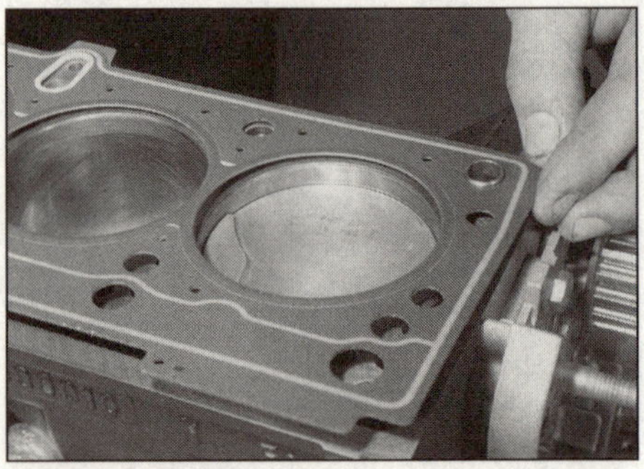

9.34 Locate the new cylinder head gasket on the block and over the dowel

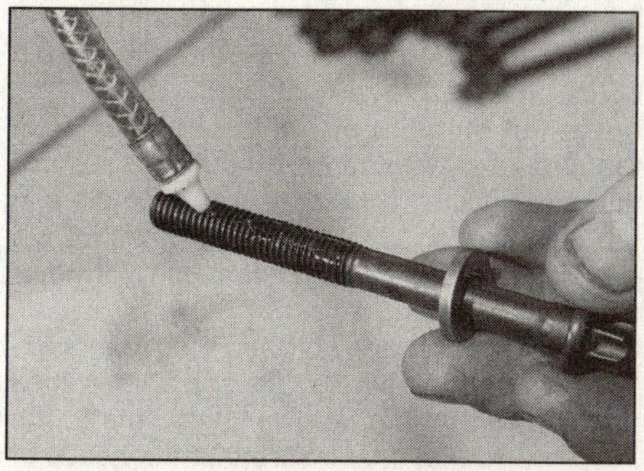

9.35a Oil the cylinder head bolts before . . .

9.35b . . . inserting them

35 Lower the cylinder head onto the block. Oil the threads and the head undersides of the cylinder head bolts, then insert them and initially screw them in finger-tight **(see illustrations)**. Make sure that the washers are positioned on the bolts. Note that the shorter bolts are located on the inlet side of the head.
36 Tighten the cylinder head bolts to the specified torques in the sequence shown **(see illustrations)**, and in the stages given in the Specifications at the beginning of this Chapter. The initial stages pre-compress the gasket, and the remaining stages are the main tightening procedure. When angle-tightening the bolts, put paint marks on the bolt heads and cylinder head as a guide for the correct angle, or obtain a special angle-tightening tool.
37 Refit the rocker shaft and retaining plate with reference to Section 8. If the cylinder head has been overhauled, it is worthwhile checking the valve clearances at this stage, to prevent any possibility of the valves touching the pistons when the timing belt is being fitted. Turn the crankshaft so that there are no pistons at TDC. Use a socket on the camshaft sprocket to turn the camshaft and check the valve clearances.
38 Refit the exhaust front downpipe with reference to Chapter 4, then refit the heat shield to the exhaust manifold and tighten the nuts.
39 Insert the alternator upper mounting/ adjustment bolt to the bracket on the rear of the cylinder head, then adjust the tension of the drivebelt with reference to Chapter 1 and tighten the bolt.
40 Reconnect the expansion tank hose, the heater hoses and the radiator top hose to the thermostat housing and make sure the clips are correctly positioned.
41 Refit and tighten the oil level dipstick tube upper mounting bolt.
42 Reconnect the brake vacuum servo hose to the inlet manifold.
43 Refit the spark plugs with reference to Chapter 1.
44 Reconnect the wiring and hoses to the absolute pressure sensor.
45 Reconnect the evaporation emission control solenoid valve hose.
46 Reconnect the fuel supply and return hoses to the throttle housing (E7J engine) or fuel rail (K7M engine).
47 Refit the earth strap to the bulkhead and tighten the bolt.
48 If not already done, adjust the valve clearances (see Section 4 of this Chapter).
49 Refit the valve cover together with a new gasket, and tighten the bolts to the specified torque.
50 Reconnect the crankcase ventilation hose to the valve cover.
51 On the K7M engine, refit the ignition coil assembly to the valve cover, and reconnect the HT leads to the spark plugs. Refit the HT lead support to the valve cover.
52 On the E7J engine, refit the accelerator cable support to the valve cover.
53 Reconnect the accelerator cable and adjust it as described in Chapter 4.
54 Refit the air cleaner assembly as described in Chapter 4. On the K7M engine, reconnect the air duct to the throttle housing.
55 Refit the timing belt with reference to Section 5 of this Chapter.
56 Refill the engine with fresh oil, with reference to Chapter 1.
57 Refit and tighten the cylinder block drain plug on the E7J engine.
58 Reconnect the battery negative lead.
59 Refill and bleed the cooling system with reference to Chapter 1.

10 Sump - removal and refitting

Note: *An engine lifting hoist is required during this procedure. On models fitted with the K7M engine with an* ***aluminium*** *sump, the engine and transmission assembly must be completely removed from the engine compartment as described in Part C of this Chapter.*

E7J engine

Removal

1 Disconnect the battery negative lead.
Caution: If the radio/cassette in your vehicle is equipped with an anti-theft system, make sure you have the correct activation code before disconnecting the battery.
2 Drain the engine oil referring to Chapter 1, then refit and tighten the drain plug using a new washer.
3 Jack up the front of the vehicle and support on axle stands. Remove the right-hand front roadwheel.
4 Remove the exhaust front downpipe and catalytic converter, referring to Chapter 4A.
5 Disconnect the track rod end from the steering arm on the hub carrier, with reference to Chapter 10.
6 Unscrew and remove the upper bolt securing the right-hand suspension strut to the hub carrier. Note that the head end of the bolt has location splines machined on it, and it is necessary to tap out the bolt after removing the nut.
7 Loosen only the lower bolt securing the strut to the hub carrier.
8 Using a suitable drift, drive out the roll pin securing the right-hand driveshaft to the

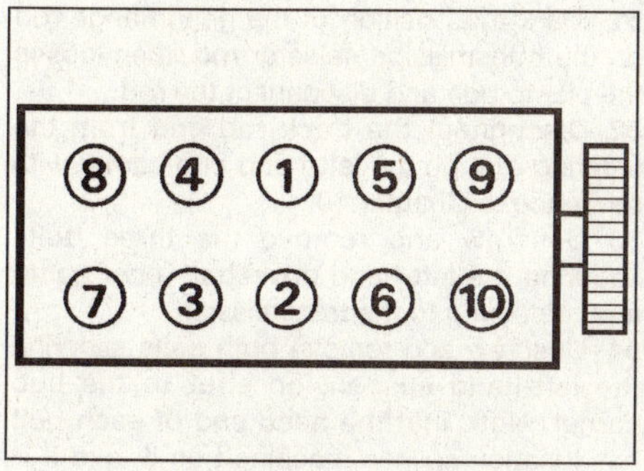

9.36a Cylinder head bolt tightening sequence

9.36b Torque-tightening the cylinder head bolts

9.36c Angle-tightening the cylinder head bolts

transmission output shaft. Tilt the hub carrier while releasing the driveshaft from the splines on the output shaft. Support the driveshaft by tying it to the steering gear.
9 Unbolt the engine-to-transmission strengthening bracket/flywheel cover.
10 Unscrew and remove the two nuts securing the front engine mountings to the subframe, and recover the spacer from the right-hand mounting.
11 At the rear of the engine, unscrew and remove the nuts from the bolts securing the rear engine mounting bracket to the transmission, and the nuts from the rear mounting through-bolt. Do not remove the bolts at this stage.
12 Remove the air filter as described in Chapter 4A.
13 Attach a hoist to the engine using the lifting eyes on the inlet manifold at the rear and at the front right-hand end of the cylinder head. Alternatively, use a support bar across the engine compartment.
14 Disconnect the carbon canister hose from the solenoid valve on the inlet manifold.
15 Release the fuel supply and return hoses from the inlet manifold and side member.
16 Unbolt the power steering fluid reservoir, and position it to one side. Do not disconnect the hoses.
17 Disconnect the speedometer sender unit from the transmission by pulling out the retaining clip.
18 Carefully lift the engine and transmission assembly until the rear mounting bolts and bracket can be removed.
19 Continue to lift the assembly until the power-assisted steering fluid pipe can be released from the supports on the subframe.
20 Lift the assembly until the gap between the lower edge of the cylinder block and subframe is 25 cm.
Caution: Make sure the engine/transmission assembly is supported adequately before removing the sump.
21 Where applicable remove the tie-bar and cover assembly from under the sump, then unscrew the bolts securing the sump to the cylinder block, but leave two diagonally-opposite bolts entered on a few threads until the sump joint has been released **(see illustration)**.

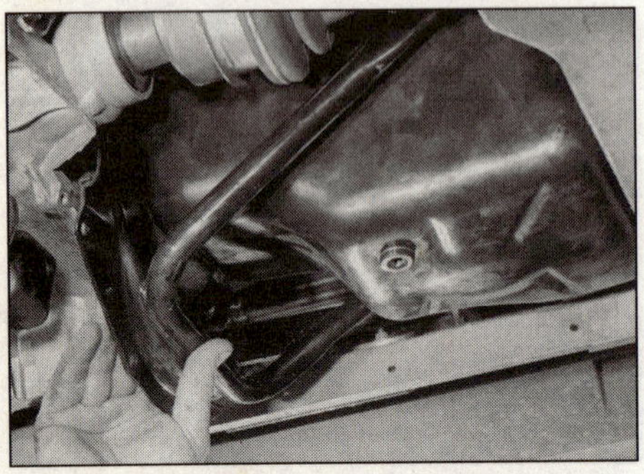

10.21 Removing the tie-bar and cover assembly (E7J engine)

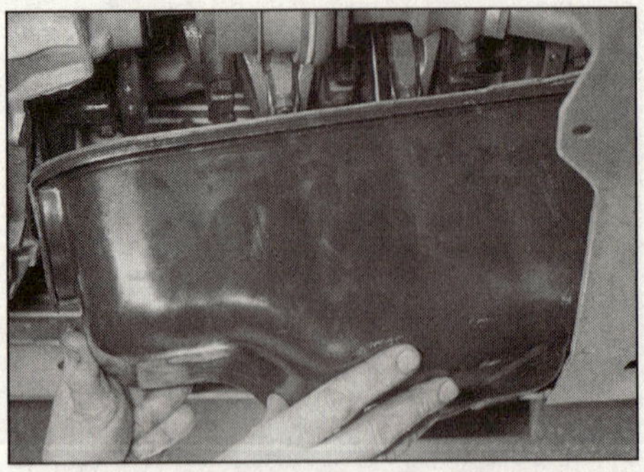

10.23 Lowering the sump from the engine (E7J engine)

22 Using a palette knife or similar tool, release the sump from the bottom of the cylinder block. Do not use a screwdriver as this may damage the mating surfaces.
23 Unscrew the two remaining bolts and lower the sump from the engine. It will be necessary to lower it first, then move it towards the transmission to clear the subframe **(see illustration)**.
24 Remove the rubber sealing strips from the grooves in the front and rear main bearing caps, or from the curved sections at each end of the sump. Clean all traces of sealant from the cylinder block and sump, and wipe them dry.

Refitting

25 Apply a 3 mm wide bead of suitable sealant to the mating face of the sump, making sure that the bead is around the inner edge of each bolt hole. Renault recommend Rhodorseal 5661.
26 Locate new rubber sealing strips in the grooves on the front and rear main bearing caps. Note that the strips are different.
27 Offer the sump to the cylinder block, and insert the bolts. Progressively tighten the bolts in diagonal sequence to the specified torque.
28 Lower the engine/transmission assembly until the power-assisted steering fluid pipe can be refitted to the supports on the subframe.
29 Lower the assembly, and refit the rear mounting bolts and bracket. Locate the front mounting studs in the subframe, and completely lower the assembly. Remove the hoist from the engine.
30 Refit the nuts to the front and rear mounting bolts, and tighten to the specified torque. Make sure the spacer is fitted to the right-hand mounting.
31 Refit the speedometer sender unit to the transmission, and retain with the clip.
32 Reposition the power steering fluid reservoir and tighten the bolts.
33 Refit the fuel supply and return hoses to the inlet manifold and side member.
34 Reconnect the evaporative charcoal canister hose to the inlet manifold.
35 Refit the air filter, referring to Chapter 4A.
36 Refit the engine-to-transmission strengthening bracket/flywheel cover, and tighten the bolts.
37 Locate the right-hand driveshaft on the transmission output shaft so that the roll pin holes are correctly aligned. Slide the driveshaft on and push in the hub carrier at the same time. Drive in the outer section of the roll pin using a suitable drift, then drive in the inner section. Seal both ends of the roll pin with suitable sealant to prevent entry of water and dirt.
38 Insert the upper bolt securing the right-hand suspension strut to the hub carrier, then tighten both the upper and lower nuts to the specified torque. Make sure that the heads of the bolts face the front of the vehicle.
39 Refit the track rod end to the steering arm on the hub carrier, referring to Chapter 10.
40 Refit the exhaust front downpipe and catalytic converter, referring to Chapter 4A.
41 Refit the right-hand front roadwheel, and lower the vehicle to the ground.
42 Refill the engine with oil with reference to Chapter 1.
43 Reconnect the battery negative lead.

K7M engine (with steel sump)

Note: *See note at beginning of this Section for engines with an aluminium sump.*

Removal

44 Disconnect the battery negative lead.
Caution: If the radio/cassette in your vehicle is equipped with an anti-theft system, make sure you have the correct activation code before disconnecting the battery.
45 Drain the engine oil as described in Chapter 1, then refit and tighten the drain plug using a new washer.
46 Jack up the front of the vehicle and support on axle stands. Remove both front roadwheels.
47 Drain the cooling system as described in Chapter 1.
48 Loosen the clip and disconnect the radiator top hose from the thermostat housing.
49 Remove the clips and retaining screws, and remove the wheel arch liners from beneath the right-hand front wheel arch.
50 Disconnect the oxygen sensor wiring then remove the catalytic converter and front downpipe with reference to Chapter 4A.
51 Mark the position of the gearchange rod on the transmission selector rod, then loosen the clamp bolt and disconnect the rod.
52 Disconnect the track rod end from the steering arm on the left-hand hub carrier with reference to Chapter 10.
53 Unscrew and remove the three bolts securing the left-hand driveshaft inner gaiter and metal ring to the transmission.
54 Unscrew and remove both bolts securing the left-hand suspension strut to the hub carrier. Note that the head end of each bolt has location splines machined on it, and it is necessary to tap out the bolt after removing the nut.

55 Carefully withdraw the left-hand driveshaft inner joint from the sun wheel in the transmission. Tie the driveshaft to one side.
56 Unscrew and remove the upper bolt securing the right-hand suspension strut to the hub carrier. Note that the head end of the bolt has location splines machined on it and it is necessary to tap out the bolt after removing the nut.
57 Loosen (do not remove) the lower bolt securing the strut to the hub carrier.
58 Using a suitable drift, drive out the roll pin securing the right-hand driveshaft to the transmission output shaft. Tilt the hub carrier while releasing the driveshaft from the splines on the sun wheel. Support the driveshaft by tying it to the steering gear.
59 Unbolt and remove the engine-to-transmission strengthening bracket/flywheel cover.
60 At the front of the subframe, unscrew the nuts from the front engine mountings, and recover the plate from the right-hand mounting.
61 At the rear of the engine, unscrew and remove the nuts from the bolts securing the rear engine mounting bracket to the transmission, and the nuts from the rear mounting through-bolt. Do not remove the bolts at this stage.
62 Unbolt and remove the strengthening bar from between the front suspension strut turrets.
63 Unbolt the earth strap from the right-hand side of the bulkhead.
64 Extract the spring clip and remove the speedometer sender unit from the rear of the transmission.
65 Unbolt or release the power-assisted steering reservoir from the right-hand front of the engine compartment, and position it on the engine. On the K7M engine, the reservoir is removed by sliding it up from the support channel on the right-hand side of the radiator.
66 Disconnect the brake vacuum servo hose from the inlet manifold.
67 Disconnect the carbon canister hose from the inlet manifold.
68 Where necessary, disconnect the crankcase ventilation hose.
69 Disconnect the wiring from the throttle potentiometer.
70 Disconnect the fuel supply and return hoses from the fuel pressure regulator and fuel rail. Refer to Chapter 4A if necessary.
71 Attach a hoist to the engine and transmission assembly, and take the weight of the engine.
72 Where applicable, release the clutch cable from the side of the engine compartment.
73 Carefully lift the engine and transmission assembly until the rear mounting bolts and bracket can be removed.
74 Continue to lift the assembly until the power-assisted steering fluid pipe can be released from the supports on the subframe.
75 Lift the assembly until the gap between the lower edge of the cylinder block and subframe is 26 cm. Make sure the engine wiring is not trapped between the transmission and the side member.
Caution: Make sure the engine/transmission assembly is supported adequately before removing the sump.
76 Unscrew the bolts securing the sump to the cylinder block, but leave two diagonally-opposite bolts entered on a few threads until the sump joint has been released **(see illustration)**.
77 Using a palette knife or similar tool, release the sump from the bottom of the cylinder block. Do not use a screwdriver, as this may damage the mating surfaces.
78 Unscrew the two remaining bolts, and lower the sump from the engine. It will be necessary to lower it first, then move it towards the transmission to clear the subframe **(see illustration)**.
79 Remove the rubber sealing strips from the grooves in the front and rear main bearing caps or from the curved sections at each end of the sump. Clean all traces of sealant from the cylinder block and sump, and wipe them dry **(see illustrations)**.

Refitting

80 Apply a 3 mm wide bead of suitable sealant to the mating face of the sump, making sure that the bead is around the inner edge of each bolt hole. Renault recommend Rhodorseal 5661 **(see illustration)**.
81 Locate new rubber sealing strips in the grooves on the front and rear main bearing caps. Note that the strips are different.
82 Offer the sump to the cylinder block, and insert the bolts. Progressively tighten the bolts in diagonal sequence to the specified torque.
83 Lower the engine/transmission assembly until the power-assisted steering fluid pipe can be refitted to the supports on the subframe.
84 Lower the assembly and refit the rear mounting bolts and bracket. Locate the front mounting studs in the subframe, and completely lower the assembly. Remove the hoist from the engine.
85 Refit the nuts to the front and rear mounting bolts and tighten to the specified torque. Make sure the spacer is fitted to the right-hand mounting.
86 Where applicable, refit the clutch cable to the side of the engine compartment.
87 Reconnect the fuel supply and return hoses to the fuel pressure regulator and fuel rail and tighten the clips.

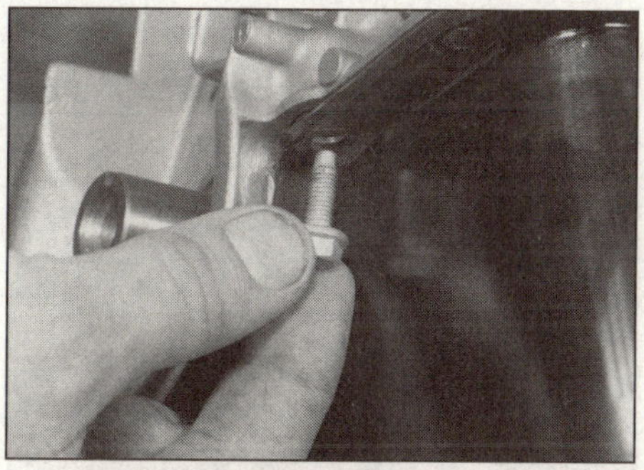
10.76 Removing the sump bolts (K7M engine)

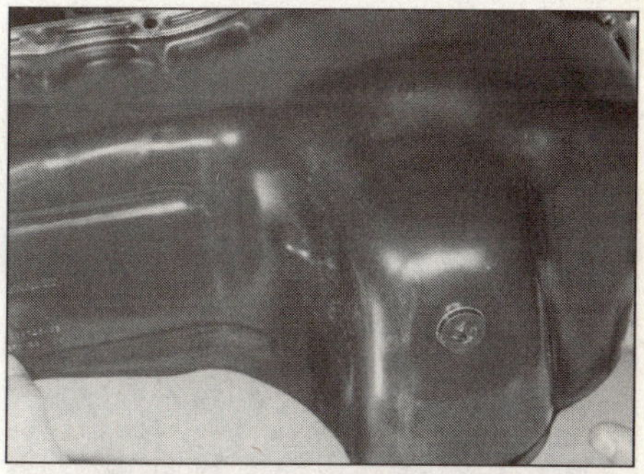
10.78 Lowering the sump from the engine (K7M engine)

10.79a Removing the rubber sealing strip from the rear main bearing cap . . .

10.79b . . . and front main bearing cap

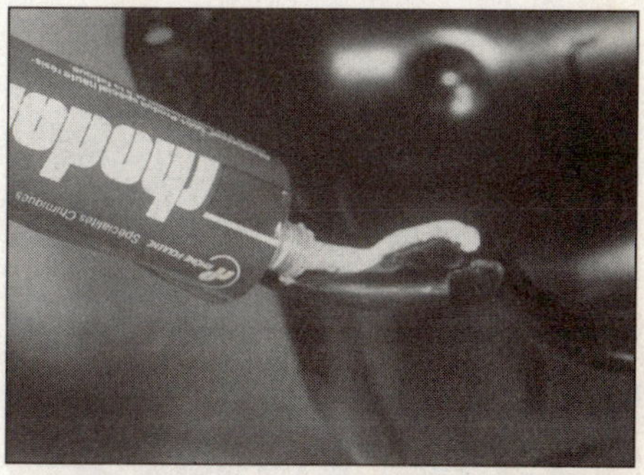
10.80 Apply a 3 mm wide bead of sealant to the mating flange of the sump

88 Reconnect the wiring to the throttle potentiometer.
89 Reconnect the crankcase ventilation hose.
90 Reconnect the carbon canister hose and brake vacuum servo hose to the inlet manifold, and tighten the clips.
91 Reposition the power steering fluid reservoir and tighten the bolts.
92 Refit the speedometer sender unit to the transmission, and retain with the clip.
93 Refit the earth strap to the right-hand side of the bulkhead and tighten the bolt.
94 Refit the strengthening bar to the strut turrets and tighten the bolts.
95 Refit the nuts to the front and rear mounting bolts and tighten to the specified torque. Make sure the spacer is fitted to the right-hand mounting.
96 Refit the engine-to-transmission strengthening bracket/flywheel cover.
97 Locate the right-hand driveshaft on the transmission output shaft so that the roll pin holes are correctly aligned. Slide the driveshaft on and push in the hub carrier at the same time. Drive in the outer section of the roll pin using a suitable drift, then drive in the inner section. Seal both ends of the roll pin with sealant to prevent entry of water and dirt.
98 Insert the upper bolt securing the right-hand suspension strut to the hub carrier, then tighten both the upper and lower nuts to the specified torque. Make sure that the heads of the bolts face the front of the vehicle.
99 Insert the left-hand driveshaft inner joint in the transmission sun wheel, then refit the bolts securing the left-hand suspension strut to the hub carrier. Make sure the heads of the bolts are facing the front of the vehicle, and tighten their nuts to the specified torque.
100 Refit the three bolts securing the left-hand driveshaft inner gaiter to the transmission and tighten to the specified torque.
101 Refit the track rod end to the steering arm on the left-hand hub carrier (see Chapter 10).
102 Refit the gearchange rod to the transmission selector rod, position it as noted during removal, then tighten the clamp bolt.
103 Refit the catalytic converter and front downpipe with reference to Chapter 4A and reconnect the oxygen sensor wiring.
104 Refit the wheel arch liners beneath the right-hand front wheel arch.
105 Refit the top hose to the thermostat housing.
106 Refit the front roadwheels and lower the vehicle to the ground.
107 Reconnect the battery negative lead.
108 Refill the engine with oil (see Chapter 1).
109 Refill the cooling system (see Chapter 1).

11 Oil pump and sprockets - removal, inspection and refitting

Removal

1 Remove the timing belt and the crankshaft sprocket, with reference to Sections 5 and 6.
2 Remove the sump referring to Section 10.
3 On the E7J engine, remove the Woodruff key (when fitted) from its slot in the crankshaft.
4 Unbolt the front cover from the cylinder block **(see illustration)**.
5 Unscrew the bolts securing the sprocket to the oil pump hub. Use a screwdriver through one of the holes in the sprocket to hold it stationary **(see illustrations)**.
6 Remove the sprocket from the oil pump.
7 Slide off the oil seal spacer **(see illustration)**.
8 Slide the drive sprocket from the crankshaft, then release both sprockets from the chain **(see illustration)**. Note that the sprocket is not keyed to the crankshaft, but relies on the pulley bolt being tightened correctly in order to clamp the sprocket. It is most important that the pulley bolt is correctly tightened otherwise there is the possibility of the oil pump not functioning.
9 Unscrew the two mounting bolts, and withdraw the oil pump from the crankcase. If the two locating dowels are displaced, refit them in the crankcase **(see illustrations)**.

11.4 Removing the front cover from the cyinder block

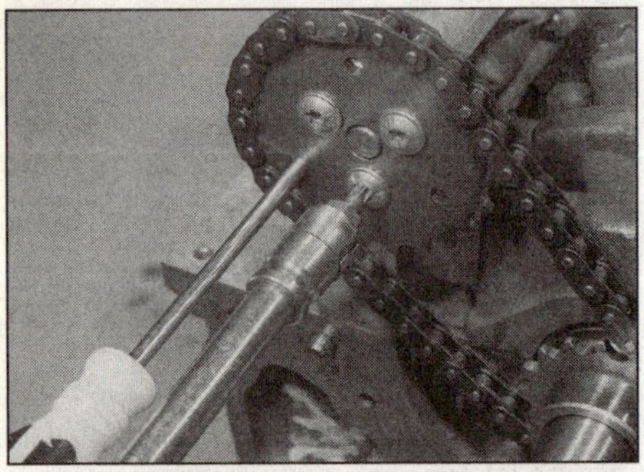

11.5a Using a screwdriver to hold the oil pump sprocket when loosening the bolts

11.5b Removing the sprocket bolts from the oil pump

11.7 Slide off the oil seal spacer

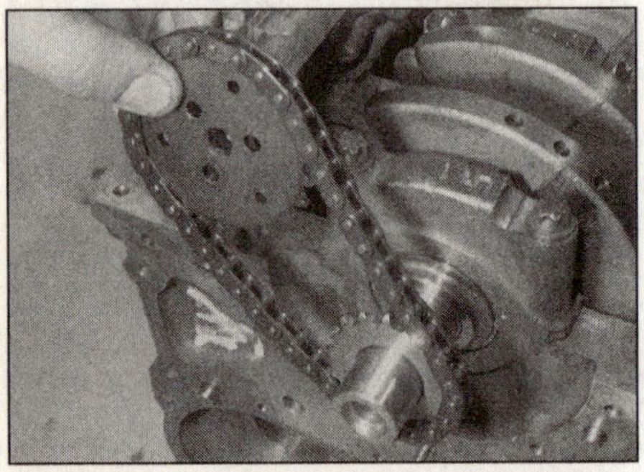

11.8 Removing the oil pump sprockets and chain from the front of the crankshaft

11.9a Unscrew the bolts . . .

11.9b . . . and remove the oil pump from the crankcase

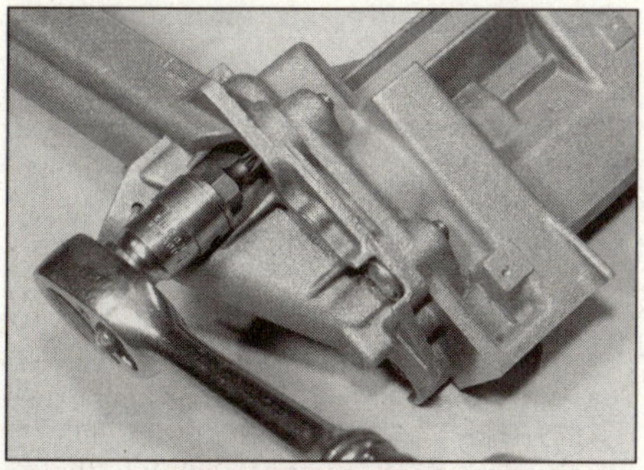

11.10a Unscrewing the bolts securing the cover and pick-up tube to the oil pump

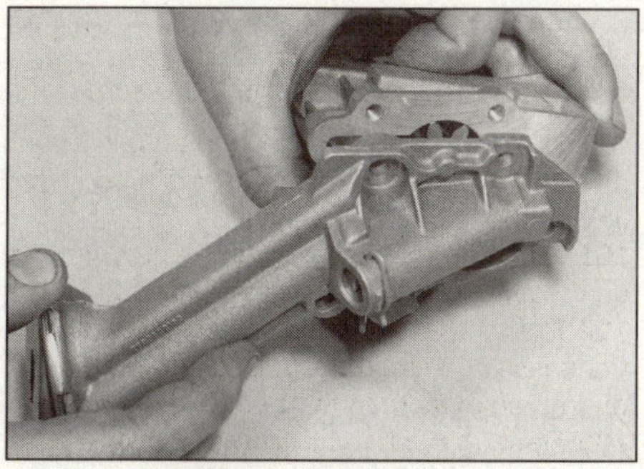

11.10b Removing the cover and pick-up tube

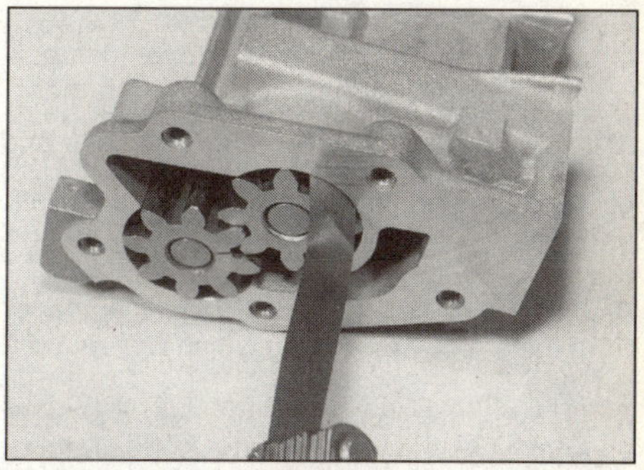

11.11a Using a feeler blade to check the clearance between the gears and the oil pump body

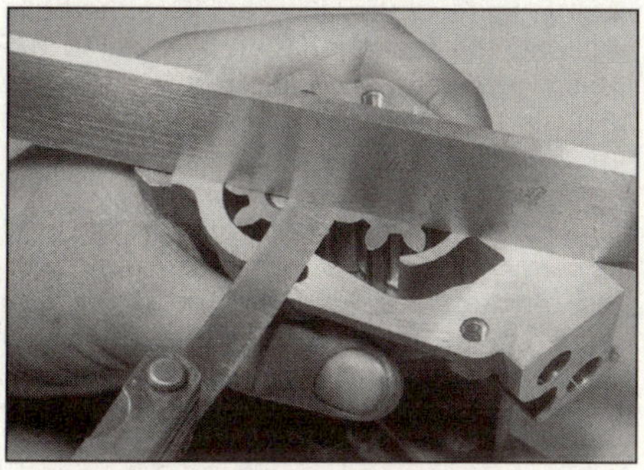

11.11b Checking the endfloat of the oil pump gears

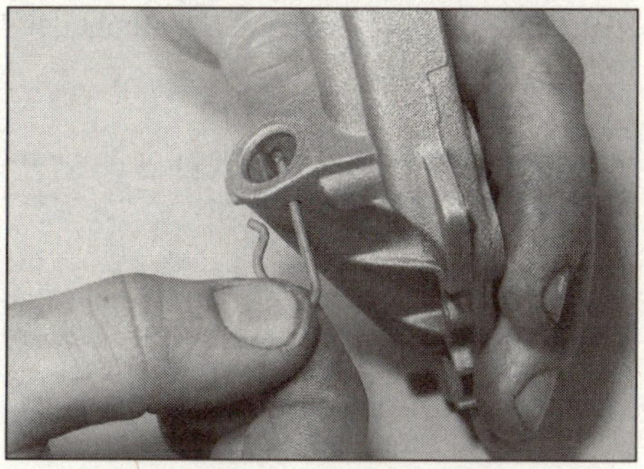

11.12a Extract the spring clip . . .

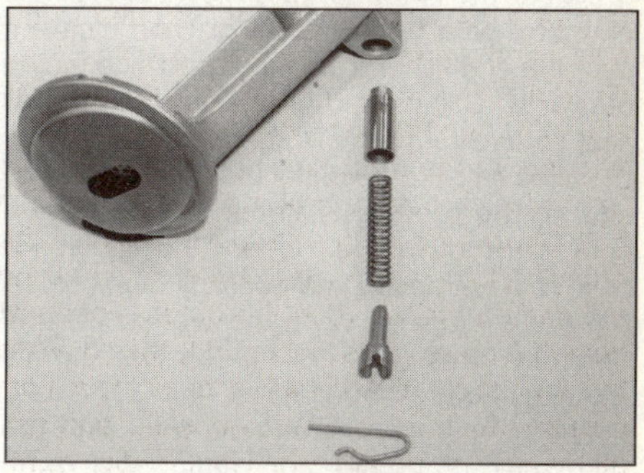

11.12b . . . and remove the end stop, spring and piston

Inspection

10 Unscrew the retaining bolts, and lift off the pump cover and pick-up tube **(see illustrations)**.

11 Using a feeler blade, check the clearance between each of the gears and the oil pump body. Also check the endfloat by measuring the clearance between the gears and the cover joint face. If any clearance is outside the tolerances given in the Specifications, the oil pump must be renewed **(see illustrations)**.

12 Depress the relief valve end stop and extract the spring clip (with the piston at rest there is very little spring tension on the end stop, since the oil pressure release hole is located some way up the piston bore). Release the end stop, and remove the spring and piston **(see illustrations)**.

13 Examine the relief valve piston and bore for signs of wear and damage. If evident, renew the oil pump complete.

14 If the components are serviceable, clean them and reassemble in the reverse order to dismantling. Before refitting the cover, fill the pump with fresh engine oil to assist circulation when the engine is first started. A new pump should also be primed with oil.

15 Examine the chain for excessive wear, and renew it if necessary. Similarly check the sprockets.

Refitting

16 Wipe clean the oil pump and crankcase mating surfaces.

17 Check that the two locating dowels are fitted in the crankcase, then position the oil pump on them and insert the two mounting bolts. Tighten the bolts to the specified torque.

18 Slide the drive sprocket onto the crankshaft.

19 Engage the oil pump sprocket with the chain, then engage the chain with the crankshaft drive sprocket, and locate the sprocket on the oil pump hub.

20 Align the holes, then insert the sprocket bolts and tighten them to the specified torque while holding the sprocket stationary with a screwdriver.

21 The oil seal in the front cover should be renewed whenever the cover is removed. Note the fitted position of the old seal, then prise it out with a screwdriver and wipe clean the seating. Smear the outer perimeter of the new seal with fresh engine oil, and locate it squarely on the cover with its closed side facing outwards. Place the cover on a block of wood, then use a socket or metal tube to drive in the oil seal.

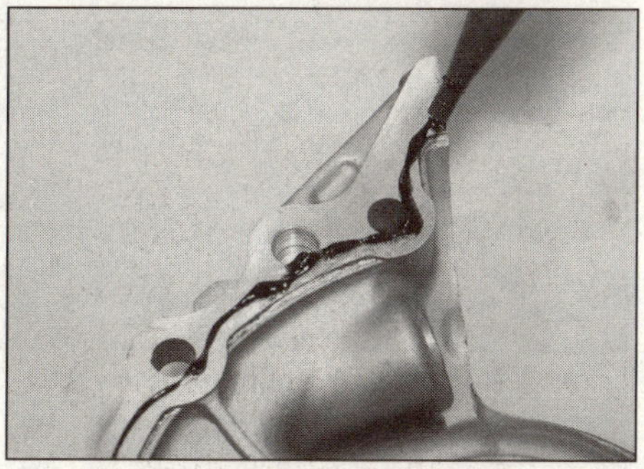

11.22 Apply a 0.6 to 1.0 mm diameter bead of sealant to the front cover

22 Clean all traces of sealant from the front cover and block mating faces. Apply a 0.6 to 1.0 mm diameter bead of sealant to the front cover, then refit it to the cylinder block and tighten the bolts securely. The sealant must be applied around the inner edges of the bolt holes **(see illustration)**.

23 Smear the oil seal with a little engine oil, then slide the spacer onto the front of the crankshaft. Turn the spacer slightly as it enters the oil seal, to prevent damage to the seal lip. If the spacer is worn where the old oil seal contacted it, it can be turned around so that the new oil seal contacts the unworn area.

24 On the E7J engine, refit the Woodruff key (when fitted) to its slot in the crankshaft.

25 Refit the sump (refer to Section 10).

26 Refit the crankshaft sprocket and timing belt with reference to Sections 6 and 5.

12 Crankshaft oil seals - renewal

Right-hand (timing end) oil seal

1 Remove the timing belt and the crankshaft sprocket with reference to Sections 5 and 6.

2 Note the fitted position of the old seal, then prise it out from the front cover using a screwdriver or suitable hooked instrument. Take care not to damage the surface of the spacer. An alternative method of removing the oil seal is to carefully drill two small holes in it and insert self-tapping screws, then pull on the screws with grips.

3 With the oil seal removed, slide the spacer from the crankshaft, noting which way round it is fitted.

4 Examine the spacer for excessive oil seal wear. If necessary, the spacer can be refitted so that the new oil seal contacts an unworn area. Clean the oil seal seating in the front cover.

5 Smear the lips and outer perimeter of the new seal with fresh engine oil, and locate it over the crankshaft with its closed side facing outwards. Using hand pressure, press the oil seal squarely into the front cover a little way, then use a socket or metal tube to drive the oil seal to the previously noted position. Take care not to drive it in too far otherwise it will have to be removed and possibly renewed.

6 Slide the spacer onto the crankshaft and carefully press it into the oil seal, while twisting it to prevent damage.

7 Wipe away any excess oil, then refit the crankshaft sprocket and timing belt with reference to Sections 6 and 5.

Left-hand (flywheel/driveplate end) oil seal

8 Renewal of the rear crankshaft oil seal requires the engine and transmission assembly to be removed from the engine compartment and the transmission and flywheel/driveplate separated on the bench. The procedure is therefore described in Chapter 2C.

13 Engine/transmission mountings - inspection and renewal

Inspection

1 Apply the handbrake, then jack up the front of the car and support it on axle stands (see *"Jacking and vehicle support"*). Where fitted, remove the engine compartment undertray.

2 Visually inspect the rubber pads on the two front and one rear engine/transmission mountings for signs of cracking and deterioration. Careful use of a lever will help to determine the condition of the rubber pads. If there is excessive movement in the mounting, or if the rubber has deteriorated, the mounting should be renewed.

Renewal

Right-hand front mounting

3 Apply the handbrake, then jack up the front of the car and support it on axle stands (see *"Jacking and vehicle support"*)

4 Unscrew the nut securing the mounting lower stud to the subframe and remove the spacer plate **(see illustration)**.

5 Unscrew the two upper nuts securing the mounting to the bracket.

6 Using a hoist or trolley jack and block of wood, raise the right-hand end of the engine sufficient to remove the mounting.

7 If necessary the bracket may be unbolted from the cylinder block. On the E7J engine, recover the spacer from the upper bolt. Where power steering is fitted, it will be necessary to remove the power steering pump before removing the bracket.

8 Fit the new mounting using a reversal of the removal procedure, but tighten the nuts/bolts to the specified torque.

Left-hand front mounting

9 Apply the handbrake, then jack up the front of the vehicle and support it on axle stands (see *"Jacking and vehicle support"*).

10 Unscrew and remove the nut securing the mounting lower stud to the subframe. Take care not to allow the nut to drop back inside the subframe **(see illustration)**.

11 Unscrew and remove the upper nut securing the mounting to the transmission bracket.

12 Using a hoist or trolley jack and block of wood, raise the left-hand end of the engine/transmission sufficient to remove the mounting.

13 If necessary, the bracket may be unbolted from the transmission.

14 Fit the new mounting using a reversal of the removal procedure, but tighten the nuts/bolts to the specified torque.

Rear mounting

15 For improved access, apply the handbrake then jack up the front of the vehicle and support it on axle stands (see *"Jacking and vehicle support"*).

16 On the K7M engine, remove the air cleaner assembly as described in Chapter 4A.

17 From within the engine compartment, reach down behind the left-hand side of the engine, and unscrew the nuts from the bolts securing the rear mounting bracket to the transmission **(see illustration)**.

18 Using a hoist or trolley jack and block of wood, take the weight of the engine/transmission off the rear mounting.

19 Extract the bolts from the bracket and manoeuvre the bracket from the transmission. The bracket is a tight fit, and it may be necessary to raise the engine/transmission further to provide room to remove it.

20 Unbolt and remove the mounting from the subframe **(see illustration)**.

21 Fit the new mounting using a reversal of the removal procedure, but tighten the bolts to the specified torque.

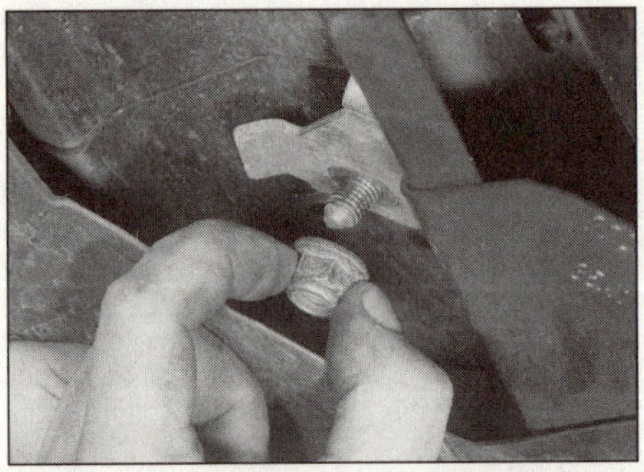

13.4 Removing the right-hand engine mounting lower nut and spacer

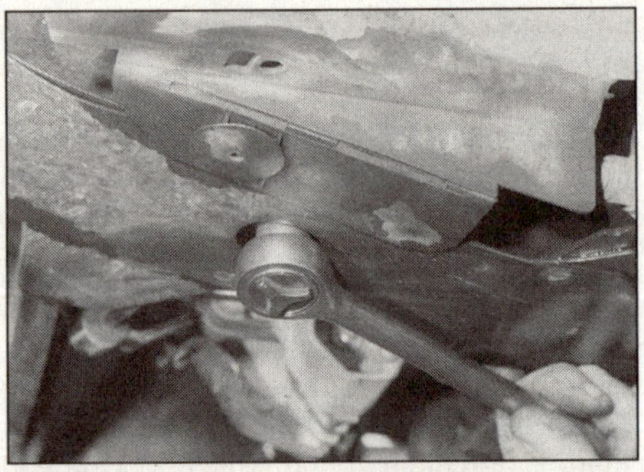

13.10 Removing the left-hand engine mounting lower nut

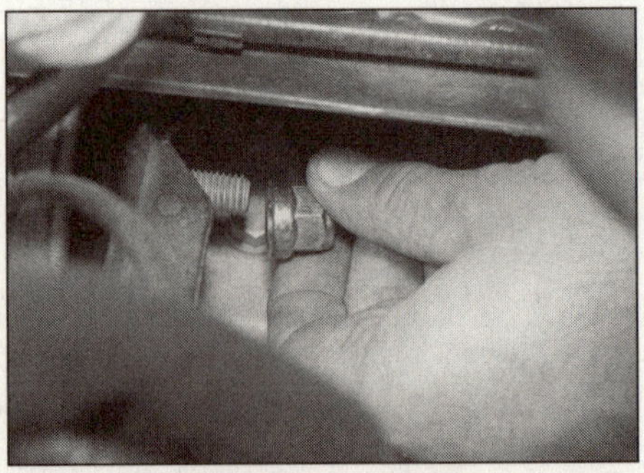

13.17 Removing the nuts from the rear engine mounting

13.20 The rear engine mounting is bolted to the subframe

Chapter 2 Part B:
Diesel engine in-car repair procedures

Contents

Degrees of difficulty

Easy, suitable for novice with little experience	**Fairly easy,** suitable for beginner with some experience	**Fairly difficult,** suitable for competent DIY mechanic	**Difficult,** suitable for experienced DIY mechanic	**Very difficult,** suitable for expert DIY or professional

Specifications

General

Type	Four-cylinder, in-line, overhead camshaft
Designation:	
Non-turbo engine	F8Q 620
Turbo engine	F8Q 784, F8Q 704
Capacity	1870 cc
Bore	80.0 mm
Stroke	93.0 mm
Firing order	1-3-4-2 (No 1 cylinder at flywheel end)
Direction of crankshaft rotation	Clockwise viewed from timing belt end
Compression ratio	21.5:1
Maximum power output (typical):	
Non-turbo engine	47.0 kW (61 bhp) at 4500 rpm
Turbo engine	68.5 kW (89 bhp) at 4250 rpm
Maximum torque (typical):	
Non-turbo engine	118 Nm (87 lbf ft) at 2250 rpm
Turbo engine	175 Nm (129 lbf ft) at 2250 rpm

Compression pressures (engine warm - approximately 80°C)

Minimum pressure	20 bars
Maximum difference between cylinders	4 bars

Camshaft

Drive	Toothed belt
Number of bearings	5
Camshaft endfloat	0.05 to 0.13 mm

Valve clearances

Inlet	0.20 mm
Exhaust	0.40 mm

Timing belt

Tension setting using Renault tool Elé. 346.04 (see text)	7.0 to 8.0 mm deflection under a load of 30 N

Cylinder block

Material	Cast iron	
Cylinder bore diameter:		
Class A or 1	80.000 to 81.015 mm	
Class B or 2	80.015 to 80.030 mm	

Auxiliary shaft

Endfloat	0.07 to 0.15 mm

Crankshaft

Number of main bearings	5
Main bearing journal diameter:	
Standard	54.795 ± 0.01 mm
Undersize	54.545 ± 0.01 mm
Main bearing running clearance	0.020 to 0.058 mm
Crankpin (big-end) journal diameter:	
Standard	48.00 mm
Undersize	47.75 mm +0.02 mm - 0.00 mm
Big-end bearing running clearance	0.014 to 0.053 mm
Big-end cap side play	0.220 to 0.400 mm
Crankshaft endfloat	0.070 to 0.230 mm
Thrustwasher thicknesses	2.30, 2.35, 2.40 and 2.45 mm

Pistons and piston rings

Piston ring end gaps	Rings supplied pre-set
Piston ring thickness:	
Top compression ring:	
Non-turbo engines	2.0 mm
Turbo engines	2.5 mm
Second compression ring	2.0 mm
Oil control ring	3.0 mm
Piston clearance in bore	0.015 to 0.030 mm (suggested values)

Connecting rods

Big-end side play	0.220 to 0.400 mm

Cylinder head

Material	Aluminium	
Height	159.50 mm	
Maximum acceptable gasket face distortion	0.05 mm	
Swirl chamber protrusion	0.01 to 0.04 mm	
Valve depth below cylinder head gasket face:	**Inlet**	**Exhaust**
Non-turbo engines	0.85 ± 0.09 mm	0.97 ± 0.09 mm
Turbo engines	0.65 ± 0.09 mm	0.57 ± 0.09 mm
Valve seat angle	120°	90°
Valve seat width	1.8 mm	
Valve guide bore	8.0 mm	
Valve guide outer diameter:		
Standard	13.00 mm	
Oversize	13.30 mm	

Valves

Head diameter:	**Inlet**	**Exhaust**
Non-turbo engines:		
Conventional inlet valve seat	36.1 mm	31.5 mm
Stellite inlet valve seat*	36.35 mm	31.5 mm
Turbo engines	36.225 ± 0.125 mm	31.620 ± 0.120 mm
Stem diameter:		
Non-turbo engines	8.0 mm	
Turbo engines	7.985 ± 0.04 mm	
Valve spring free length:		
Non-turbo engines	43.41 mm	
Turbo engines	47.57 mm	

**Stellite valve seats are used with steel inlet valves.*

Lubrication system	Non-turbo engines	Turbo engines
Minimum oil pressure at 80°C	2 bars at 1000 rpm 3.5 bars at 3000 rpm	1 bar at idle speed 3 bars at 3500 rpm
Oil pump type	Two-gear	
Oil pump clearances:		
Gear-to-body (minimum)	0.10 mm	
Gear-to-body (maximum)	0.24 mm	
Gear endfloat (minimum)	0.02 mm	
Gear endfloat (maximum)	0.085 mm	

Torque wrench settings	Nm	lbf ft
Auxiliary shaft sprocket bolt	50	37
Camshaft sprocket	50	37
Camshaft bearing caps:		
8 mm diameter bolts	20	15
6 mm diameter bolts	10	7
Connecting rod (big-end) cap bolts	45 to 50	33 to 37
Crankshaft pulley bolt:		
Pre-February 1997	90 to 100	66 to 74
February 1997-on (locking fluid on threads):		
Stage 1	20	15
Stage 2	Angle-tighten a further 115°	
Cylinder head bolts (new bolts may be required - see Section 9):		
Non-turbo engines with hexagon socket-head bolts:		
Stage 1	30	22
Stage 2	70	52
Stage 3	Wait for 3 minutes minimum	
Stage 4	Loosen all the bolts completely	
Stage 5	20	15
Stage 6	Angle-tighten a further 123° ± 2°	
Turbo engines, and non-turbo engines with T55 Torx bolts:		
Stage 1	30	22
Stage 2	Angle-tighten a further 50° ± 4°	
Stage 3	Wait for 3 minutes minimum	
Stage 4	Loosen all the bolts completely	
Stage 5	25	18
Stage 6 (final stage for non-turbo)	Angle-tighten a further 213° ± 7°	
Turbo engines only:		
Stage 7	Run engine until it reaches normal operating temperature (cooling fan cuts in), then allow the engine to cool for a minimum of 2 1/2 hours	
Stage 8	Angle-tighten a further 120° ± 7°	
Engine mounting bolt on transmission	55	41
Engine mounting bolt on front left-hand side member	25	19
Flywheel bolts*	50 to 55	37 to 41
Front right-hand engine mounting cover bolt on engine	40	30
Front right-hand engine mounting movement limiter bolt	60	44
Injection pump sprocket securing nut	50	37
Main bearing caps	60 to 65	44 to 48
Sump bolts	12 to 15	9 to 11
Timing belt tensioner roller nut	50	37
Valve cover nuts/bolts	12	9

***Note:** *Use new bolts.*

2B

1 General information

How to use this Chapter

This Part of Chapter 2 is devoted to in-car repair procedures. Details on engine removal and refitting, and engine block/ cylinder head overhaul, can be found in Chapter 2C.

Refer to *"Buying spare parts and vehicle identification numbers"* in the Reference Section of this manual for details of engine code locations.

Most of the operations in Chapter 2B assume that the engine is still installed in the car. Therefore, if this information is being used during a complete engine overhaul, with the engine already removed, many of the steps included here will not apply.

Engine description

The engine is of four-cylinder, in-line, single overhead camshaft type, mounted transversely at the front of the vehicle.

The crankshaft is supported in five shell-type main bearings. Thrustwashers are fitted to No 2 main bearing to control crankshaft endfloat.

The connecting rods are attached to the crankshaft by horizontally-split shell-type big-end bearings and to the pistons by gudgeon pins. The gudgeon pins are fully floating and are retained by circlips. The aluminium alloy pistons are of the slipper type, and are fitted with three piston rings; two compression rings and a scraper-type oil control ring.

The single overhead camshaft is mounted directly in the cylinder head, and is driven by the crankshaft via a toothed timing belt.

The camshaft operates the valves via inverted bucket-type tappets, which operate in bores machined directly in the cylinder head. Valve clearance adjustment is by shims located externally between the tappet bucket and the cam lobe. The inlet and exhaust valves are mounted vertically in the cylinder head and are each closed by a single valve spring.

An auxiliary shaft located alongside the crankshaft is also driven by the timing belt, and actuates the oil pump via a skew gear.

The fuel injection pump is driven by the timing belt, and is described in further detail in Chapter 4B.

A semi-closed crankcase ventilation system is employed, and crankcase fumes are drawn from an oil separator on the cylinder block and passed via a hose (and in certain cases, a second oil separator) to the inlet tract (see Chapter 4B for further details).

Engine lubrication is by pressure feed from a gear-type oil pump located beneath the crankshaft. Engine oil is fed through an externally-mounted oil filter to the main oil gallery feeding the crankshaft, auxiliary shaft and camshaft. On turbo models, oil spray jets are fitted to the cylinder block to supply oil to the underside of the pistons. Certain models are fitted with an oil cooler mounted on the cylinder block.

Repair operations possible with the engine in the vehicle

The following operations can be carried out without having to remove the engine from the vehicle:

a) Removal and refitting of the cylinder head.
b) Removal and refitting of the timing belt and sprockets.
c) Renewal of the camshaft oil seals.
d) Removal and refitting of the camshaft.
e) Removal and refitting of the sump.
f) Removal and refitting of the connecting rods and pistons.*
g) Removal and refitting of the oil pump.
h) Renewal of the crankshaft oil seals.
i) Renewal of the engine mountings.

**Although the operation marked with an asterisk can be carried out with the engine in the car after removal of the sump, it is better for the engine to be removed in the interests of cleanliness and improved access. For this reason, the procedure is described in Chapter 2C.*

2 Compression and leakdown tests - description and interpretation

Compression test

Note: *A compression tester specifically designed for diesel engines must be used for this test.*

2.2 Carrying out a compression test

1 When engine performance is down, or if misfiring occurs which cannot be attributed to a fault in the fuel system, a compression test can provide diagnostic clues as to the engine's condition. If the test is performed regularly it can give warning of trouble before any other symptoms become apparent.

2 A compression tester specifically intended for diesel engines must be used, because of the higher pressures involved. The tester is connected to an adapter which screws into the glow plug or injector hole **(see illustration)**. It is unlikely to be worthwhile buying such a tester for occasional use, but it may be possible to borrow or hire one - if not, have the test performed by a garage.

3 Unless specific instructions to the contrary are supplied with the tester, observe the following points:

a) The battery must be in a good state of charge, the air filter must be clean and the engine should be at normal operating temperature.
b) All the injectors or glow plugs should be removed before starting the test. If removing the injectors, also remove the fire seal washers (which must be renewed when the injectors are refitted - see Chapter 5C), otherwise they may be blown out.
c) It is advisable to disconnect the stop solenoid on the pump to reduce the amount of fuel discharged as the engine is cranked.

4 There is no need to hold the accelerator pedal down during the test because the diesel engine air inlet is not throttled.

5 The actual compression pressures measured are not so important as the balance between cylinders. Values are given in the Specifications.

6 The cause of poor compression is less easy to establish on a diesel engine than on a petrol one. The effect of introducing oil into the cylinders ("wet" testing) is not conclusive, because there is a risk that the oil will sit in the swirl chamber or in the recess on the piston crown instead of passing to the rings. However, the following can be used as a rough guide to diagnosis.

7 All cylinders should produce very similar pressures; any difference greater than that specified indicates the existence of a fault. Note that the compression should build up quickly in a healthy engine; low compression on the first stroke, followed by gradually increasing pressure on successive strokes, indicates worn piston rings. A low compression reading on the first stroke, which does not build up during successive strokes, indicates leaking valves or a blown head gasket (a cracked head could also be the cause).

8 A low reading from two adjacent cylinders is almost certainly due to the head gasket having blown between them.

9 If the compression reading is unusually high, the cylinder head surfaces, valves and pistons are probably coated with carbon deposits. If this is the case, the cylinder head should be removed and decarbonised (see Chapter 2C).

Leakdown test

10 A leakdown test measures the rate at which compressed air fed into the cylinder is lost. It is an alternative to a compression test and in many ways it is better, since the escaping air provides easy identification of where pressure loss is occurring (piston rings, valves or head gasket).

11 The equipment needed for leakdown testing is unlikely to be available to the home mechanic. If poor compression is suspected, have the test performed by a suitably equipped garage.

3 Top dead centre (TDC) for No 1 piston - locating

Note: *An 8 mm diameter rod or drill will be required if it is desired to lock the crankshaft in the TDC position.*

1 Top dead centre (TDC) is the highest point in the cylinder that each piston reaches as the crankshaft turns. Each piston reaches TDC at the end of the compression stroke and again at the end of the exhaust stroke; however, for the purpose of timing the engine, TDC refers to the position of No 1 piston at the end of its compression stroke. No 1 piston is at the flywheel end of the engine.

2 When No 1 piston is at TDC, the timing mark on the camshaft sprocket should be aligned with the pointer on the outer timing belt cover (the sprocket mark can be viewed through the cut-out in the timing belt cover, below the pointer). Additionally, the timing mark on the flywheel should be aligned with the TDC mark on the gearbox bellhousing.

3 To align the timing marks, the crankshaft must be turned. This should be done by using a spanner on the crankshaft pulley bolt. Improved access to the pulley bolt can be obtained by jacking up the front right-hand corner of the vehicle and removing the roadwheel and the lower wheel arch cover (secured by plastic clips). If desired, to enable the engine to be turned more easily, remove

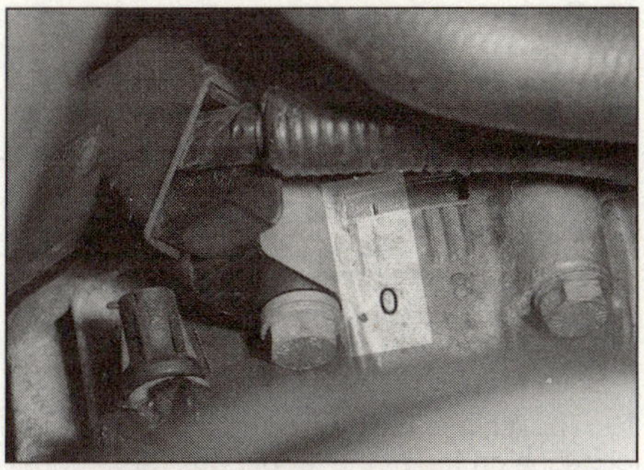

3.4 Flywheel timing mark aligned with TDC (0°) mark on bellhousing

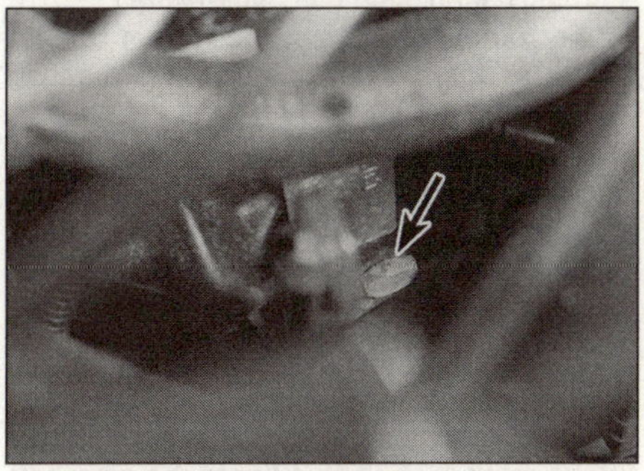

3.8a Remove the plug (arrowed) from the cylinder block . . .

the glow plugs (Chapter 5C) or the fuel injectors (Chapter 4B).

4 Look through the timing aperture in the gearbox bellhousing, and turn the crankshaft until the timing mark on the flywheel is aligned with the TDC (0°) mark on the bellhousing **(see illustration)**.

5 Unscrew the three securing bolts, and remove the plastic cover from the upper right-hand engine mounting bracket. Note the locations of any brackets which may be secured by the bolts.

6 Check that the timing mark on the camshaft sprocket is aligned with the pointer on the outer timing belt cover **(see illustration)**.

7 It is possible to lock the crankshaft in the TDC position as follows. If necessary, remove the air cleaner housing assembly (Chapter 4B).

8 Remove the plug on the lower front-facing side of the cylinder block, at the flywheel end, and obtain a metal rod which is a snug fit in the plug hole (8 mm diameter). Turn the crankshaft slightly if necessary to the TDC position, then push the rod through the hole to locate in the slot in the crankshaft web. Make sure that the crankshaft is exactly at TDC for No 1 piston (flywheel end) by aligning the timing mark on the flywheel with the TDC (0°) mark on the gearbox bellhousing as described previously. If the crankshaft is not positioned accurately, it is possible to engage the rod with a balance hole in the crankshaft web by mistake, instead of the TDC slot. It is a good idea to put a notice on the engine warning that the timing rod is in position **(see illustrations)**.

9 On completion, remove the metal rod and all removed components.

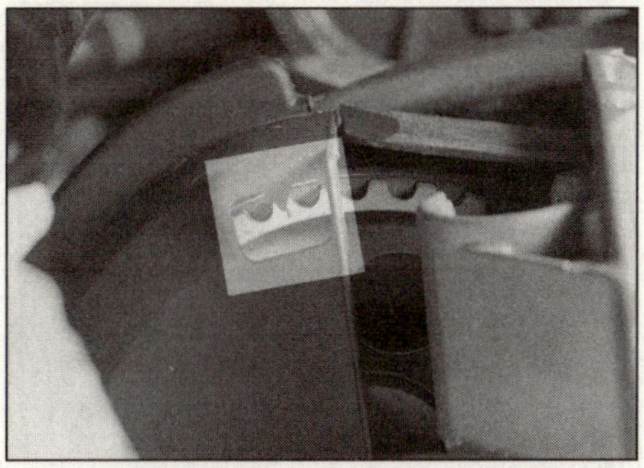

3.6 Camshaft sprocket timing mark aligned with pointer on timing belt cover

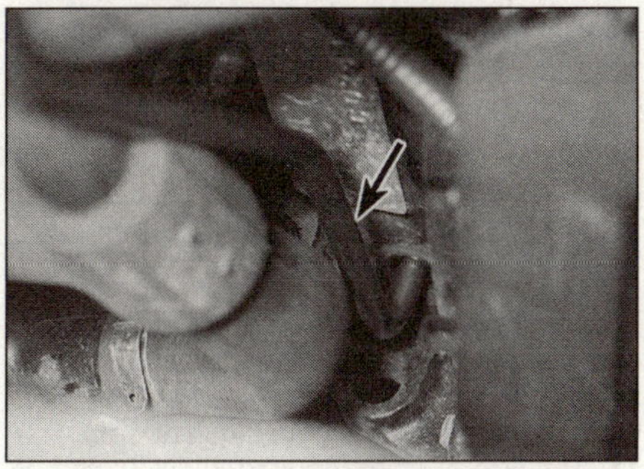

3.8b . . . and insert a suitable 8 mm rod (arrowed)

4 Valve clearances - checking and adjustment

Note: *This operation is not part of the maintenance schedule. It should be undertaken if noise from the valvegear becomes evident, or if loss of performance gives cause to suspect that the clearances may be incorrect. A new valve cover gasket may be required on refitting.*

Checking

1 Where necessary for improved access, unclip any hoses which are routed across the top of the valve cover, and move them to one side out of the way. If fuel lines are disconnected, cover open unions to prevent dirt ingress.

2 Where applicable, unscrew the securing bolts, and remove the upper timing belt/ engine mounting cover. Note the locations of any brackets secured by the bolts.

3 Unscrew the nuts from the valve cover, and withdraw the cover from the engine. Recover the gasket.

4 During the following procedure, the crankshaft must be turned, using a spanner on the crankshaft pulley bolt. Improved access to the pulley bolt can be obtained by jacking up the front right-hand corner of the vehicle and removing the roadwheel and the lower wheel arch cover (secured by plastic clips).

5 If desired, to enable the crankshaft to be

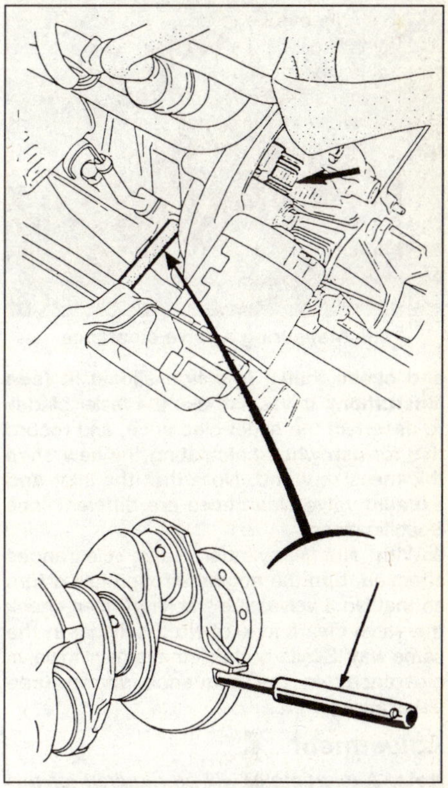

3.8c TDC mark on flywheel viewed through bellhousing aperture, and TDC locking tool engaged with crankshaft (arrowed)

2B

turned more easily, remove the glow plugs (Chapter 5C) or the fuel injectors (Chapter 4B).

6 Draw the valve positions on a piece of paper, numbering them 1 to 8 from the flywheel end of the engine. Identify them as inlet or exhaust (ie 1E, 2I, 3E, 4I, 5I, 6E, 7I, 8E) **(see illustration)**.

7 Turn the crankshaft until the valves of No 1 cylinder (flywheel end) are "rocking". The exhaust valve will be closing and the inlet valve will be opening. The piston of No 4 cylinder will be at the top of its compression stroke, with both valves fully closed. The clearances for both valves of No 4 cylinder may be checked at the same time.

8 Insert a feeler blade of the correct thickness (see Specifications) between the cam lobe and the shim on the top of the tappet bucket,

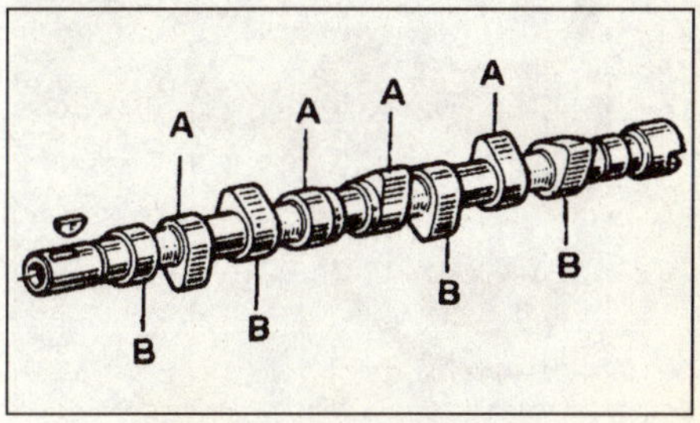

4.6 Cam lobe identification
A Inlet B Exhaust

4.8 Measuring a valve clearance

and check that it is a firm sliding fit **(see illustration).** If it is not, use the feeler blades to ascertain the exact clearance, and record this for use when calculating the new shim thickness required. Note that the inlet and exhaust valve clearances are different (see Specifications).

9 With No 4 cylinder valve clearances checked, turn the engine through half a turn so that No 3 valves are "rocking", then check the valve clearances of No 2 cylinder in the same way. Similarly check the remaining valve clearances in the sequence shown **(see illustration).**

Adjustment

Note: *A micrometer will be required for this operation.*

10 Where a valve clearance differs from the specified value, then the shim for that valve must be replaced with a thinner or thicker shim accordingly. The shim size is stamped on the bottom face of the shim, but it is prudent to use a micrometer to measure the true thickness of any shim removed, as it may have been reduced by wear **(see illustrations).**

4.10a Shim thickness engraved on the underside

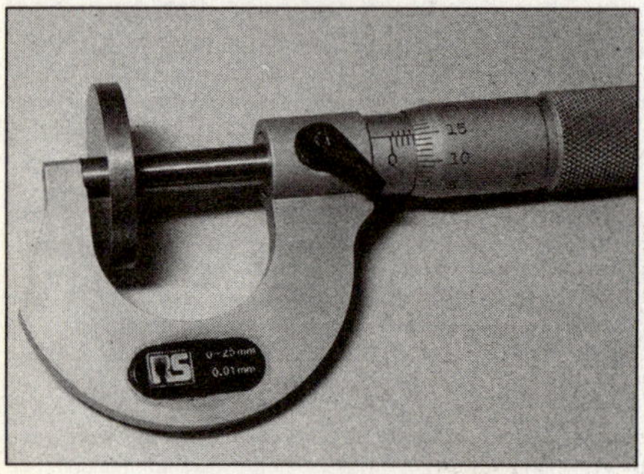

4.10b Measuring a shim using a micrometer

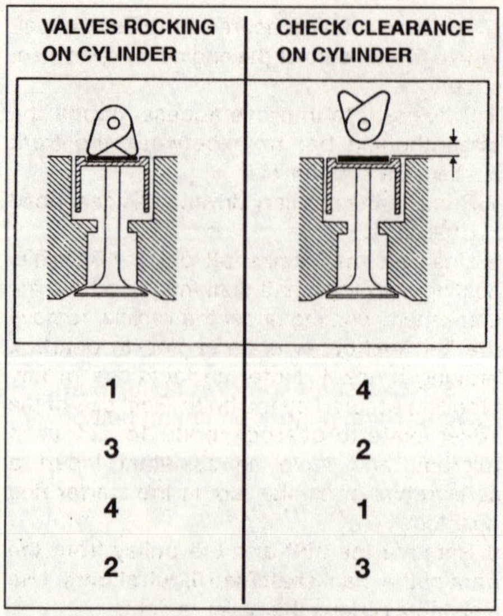

4.9 Valve clearance adjustment

11 The size of shim required is calculated as follows. If the measured clearance is less than specified, subtract the measured clearance from the specified clearance, and deduct the result from the thickness of the existing shim. For example:

Sample calculation - clearance too small
Clearance measured (A) = 0.15 mm
Desired clearance (B) = 0.20 mm
Difference (B - A) = 0.05 mm
Shim thickness fitted = 3.70 mm
Shim required = 3.70 - 0.05 = 3.65 mm

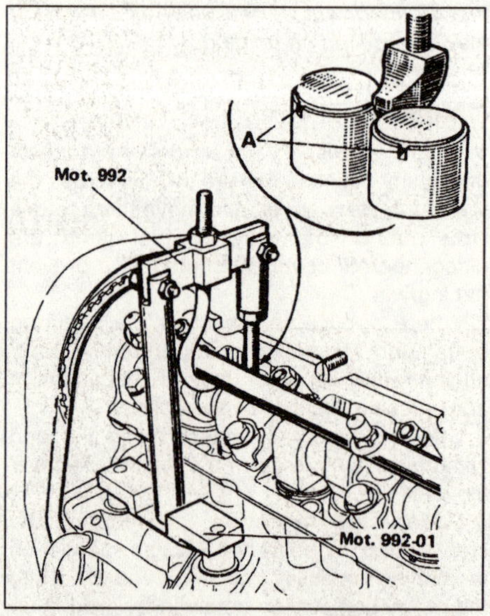

4.13 Renault tool for compressing tappet buckets to change shims
A Notches positioned at right-angles to camshaft centre-line

12 If the measured clearance is greater than specified, subtract the specified clearance from the measured clearance, and add the result to the thickness of the existing shim. For example:

Sample calculation - clearance too big
Clearance measured (A) = 0.50 mm
Desired clearance (B) = 0.40 mm
Difference (A - B) = 0.10 mm
Shim thickness fitted = 3.45 mm
Shim required = 3.45 + 0.10 = 3.55 mm

13 The shims can be removed from their locations on top of the tappet buckets without removing the camshaft if the Renault tool shown can be borrowed, or a suitable alternative fabricated **(see illustration).**

14 To remove the shim, the tappet bucket has to be pressed down against valve spring pressure just far enough to allow the shim to be slid out. Theoretically, this could be done by levering against the camshaft between the cam lobes with a suitable screwdriver or similar tool to push the bucket down, but this is not recommended by the manufacturers. If this method is to be used, take great care not to damage the camshaft, cylinder head, or tappet bucket **(see illustration).**

15 An arrangement similar to the Renault tool can be made by bolting a bar to the camshaft bearing studs and levering down against this with a stout screwdriver. The contact pad should be a triangular-shaped metal block with a lip filed along each side to contact the edge of the buckets. Levering down against this will open the valve and allow the shim to be withdrawn.

16 Make sure that the cam lobe peaks are uppermost when depressing a tappet, and rotate the buckets so that the notches are at right-angles to the camshaft centre-line. When refitting the shims, ensure that the size markings face the tappet buckets (ie face downwards).

17 If the Renault tool cannot be borrowed or a suitable alternative improvised, then it will be necessary to remove the camshaft to gain access to the shims, as described in Section 8.

18 Remove the spanner from the crankshaft pulley bolt.

19 Refit the valve cover, using a new gasket where necessary.

4.14 Removing a shim from a tappet bucket using screwdrivers

20 Where applicable, refit the fuel injectors (as described in Chapter 4B), or the glow plugs (Chapter 5C).

21 Refit/reconnect any hoses which were moved for access. If fuel lines were disconnected, reconnect them, then prime and bleed the fuel system as described in Chapter 4B.

22 Where applicable, refit the upper timing belt/engine mounting cover.

5 Timing belt - removal, inspection and refitting

Caution: If the timing belt breaks or slips in service, extensive engine damage may result. Renew the belt at the intervals specified in Chapter 1, or earlier if its condition is at all doubtful.

Note: *A suitable tool will be required to check the timing belt tension on completion of refitting - see text. A suitable puller may be required to remove the crankshaft pulley.*

Removal

1 Disconnect the battery negative lead.

Caution: If the radio/cassette in your vehicle is equipped with an anti-theft system, make sure you have the correct activation code before disconnecting the battery.

2 Jack up the front right-hand corner of the vehicle (see *"Jacking and vehicle support"*) and remove the roadwheel and the lower wheel arch cover (secured by plastic clips). Where fitted, remove the engine compartment undertray.

3 If desired, to improve access, unbolt the strengthening bar from between the front suspension strut turrets.

4 Remove the auxiliary drivebelt as described in Chapter 1.

5 Unscrew the crankshaft pulley bolt while holding the crankshaft stationary. To hold the crankshaft, working under the vehicle, remove the flywheel cover plate/engine-to-gearbox bracing bracket. Note the locations of any brackets secured by the bolts. Refit one of the cover plate-to-gearbox bolts to act as a fulcrum, and have an assistant insert a screwdriver or similar tool in the starter ring gear teeth.

6 Remove the bolt and the pulley from the front of the crankshaft **(see illustrations)**. Use a suitable puller if the pulley is tight.

7 To improve access, unscrew the two bolts securing the fuel filter assembly to the body panel, then unclip the filter hoses from the brackets on the engine. The assembly can be moved to one side (on top of the engine) until completion of refitting.

8 Temporarily refit the crankshaft pulley bolt. Turn the crankshaft to position No 1 piston at TDC on the compression stoke, and lock the crankshaft in position, as described in Section 3.

9 The right-hand upper engine mounting bracket must be removed to enable the timing belt to be removed, therefore the engine must be supported. The assembly can be supported using a jack and a suitable block of wood to spread the load under the sump. Alternatively, connect a hoist and suitable lifting tackle to the engine lifting brackets (in this case, the fuel return hose will have to be disconnected from the injection pump, and withdrawn from the hole in the rear engine lifting bracket).

10 Ensure that the engine/gearbox assembly is adequately supported, then unscrew the nut securing the upper mounting bracket to the mounting on the body. On certain models it is necessary to counterhold the mounting threaded rod using a suitable Allen key or hexagon bit, whilst loosening the nut with an open-ended spanner.

11 Unscrew the three bolts securing the upper mounting bracket to the engine, then withdraw the bracket.

12 Unbolt the engine mounting rubber/movement limiter assembly from the body.

13 Unscrew the securing bolts, and withdraw the timing belt cover which covers the fuel injection pump sprocket **(see illustration)**.

14 Unscrew the securing bolts, noting the locations of any brackets secured by the bolts, and remove the timing belt cover which covers the camshaft sprocket **(see illustrations)**.

15 With the crankshaft locked in position with No 1 piston at TDC (see paragraph 8), note the position of the timing mark on the fuel injection pump sprocket **(see illustrations)**. **Note:** *On models fitted with a two-piece adjustable injection pump sprocket, the*

5.6a Remove the crankshaft pulley bolt . . .

5.6b . . . and the pulley

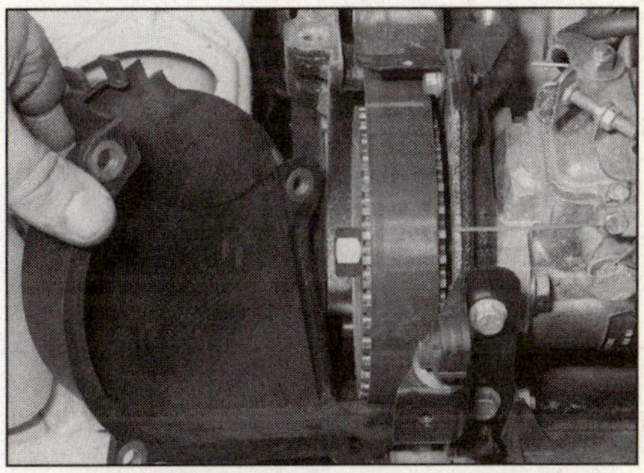

5.13 Removing the timing belt cover from the fuel injection pump sprocket

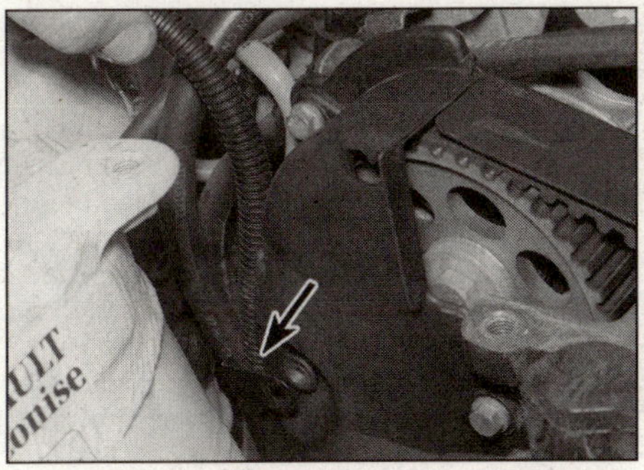

5.14a Note the location of any brackets (arrowed) secured by the bolts . . .

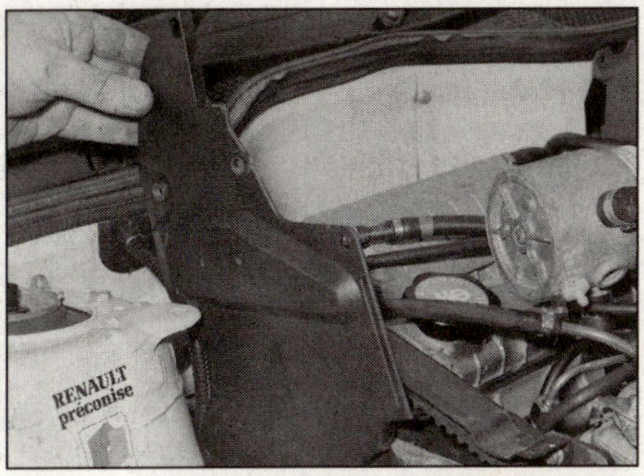

5.14b . . . and remove the timing belt cover from the camshaft sprocket

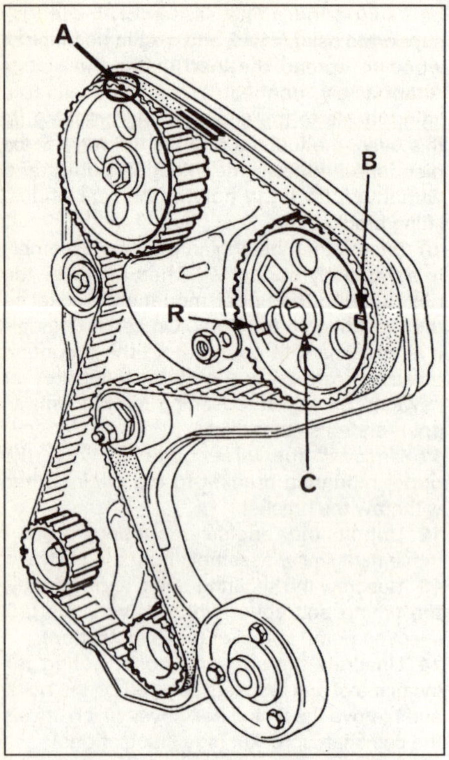

5.15a Camshaft sprocket and fuel injection pump sprocket timing mark positions with No 1 piston at TDC - models with one-piece fuel injection pump sprocket and Bosch injection pump

A Camshaft sprocket mark (aligned with pointer on outer timing belt cover - cover removed in this view)
B Injection pump sprocket mark for use with Bosch pump
C Woodruff key
R Injection pump sprocket mark for use with Lucas pump

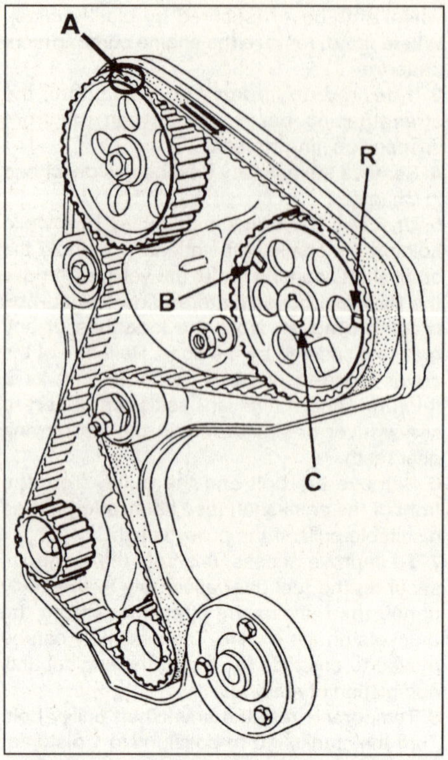

5.15b Camshaft sprocket and fuel injection pump sprocket timing mark positions with No 1 piston at TDC - models with one-piece fuel injection pump sprocket and Roto-Diesel/Lucas injection pump

A Camshaft sprocket mark (aligned with pointer on outer timing belt cover - cover removed in this view)
B Injection pump sprocket mark for use with Bosch pump
C Woodruff key
R Injection pump sprocket mark for use with Lucas pump

sprocket timing mark may differ from those shown for the one-piece sprocket. If this is the case, take note of the type of mark found, and its location, for use when refitting. Note that on models with a single-piece sprocket, there are two timing marks on the sprocket. The mark used depends on whether a Bosch or Lucas injection pump is fitted.

16 If the original belt is to be re-used, check if the belt is marked with arrows to indicate its running direction, and if necessary mark it. Similarly, make accurate alignment marks on the belt, corresponding to the timing marks on the camshaft, fuel injection pump and crankshaft sprockets.

17 Loosen the adjuster nut, and slide the timing belt tensioner back to relieve the tension from the belt **(see illustration)**. Re-tighten the nut.

18 Release the belt from the camshaft sprocket, fuel injection pump sprocket, idler wheel, crankshaft sprocket and auxiliary shaft sprocket, and remove it from the engine.

19 Do not turn the camshaft or the crankshaft whilst the timing belt is removed, as there is a risk of piston-to-valve contact. If it is necessary to turn the camshaft for any reason, remove the TDC locking tool from the cylinder block, before doing so, and turn the crankshaft anti-clockwise (viewed from the timing belt end of the engine) by a quarter turn to position all four pistons half way down their bores.

5.17 Loosen the timing belt tensioner adjuster nut

5.15c Fuel injection pump sprocket timing mark location with No 1 piston at TDC - Lucas injection pump

Inspection

20 Clean the sprockets, idler wheel and tensioner and wipe them dry, although do not apply excessive amounts of solvent to the idler and tensioner wheels otherwise the bearing lubricant may be contaminated. Also clean the rear timing belt cover, and the front of the cylinder head and block.

21 Examine the timing belt carefully for any signs of cracking, fraying or general wear, particularly at the roots of the teeth. Renew the belt if there is any sign of deterioration of this nature, or if there is any oil or grease contamination. The belt must, of course, be renewed if it has completed the mileage given in the maintenance schedule in Chapter 1.

Refitting

22 Check that the crankshaft is positioned with No 1 piston at TDC, and locked in position using the tool through the hole in the cylinder block as described previously. If the pistons have been positioned halfway down their bores (see paragraph 19), temporarily refit the outer timing belt cover which covers the camshaft sprocket, and check that the TDC mark on the camshaft sprocket is aligned with the pointer on the timing belt cover, then turn the crankshaft clockwise (viewed from the timing belt end of the engine) until the TDC locking tool can be refitted.

23 Align the timing marks on the belt with those on the crankshaft, camshaft and fuel injection pump sprockets, ensuring that the running direction arrows on the belt are pointing clockwise (viewed from the timing belt end of the engine). Note that the belt should be marked with lines across its width to act as timing marks **(see illustrations)**. If the original belt is to be refitted, and the timing marks have deteriorated (in which case, it is likely that the belt is in need of renewal in any case), use the alignment marks made on the belt before removal. Fit the timing belt over the crankshaft sprocket first, followed by the idler wheel, fuel injection pump sprocket, camshaft sprocket, tensioner, and auxiliary shaft sprocket.

24 Check that all the timing marks are still aligned.

25 The belt must now be tensioned as follows.

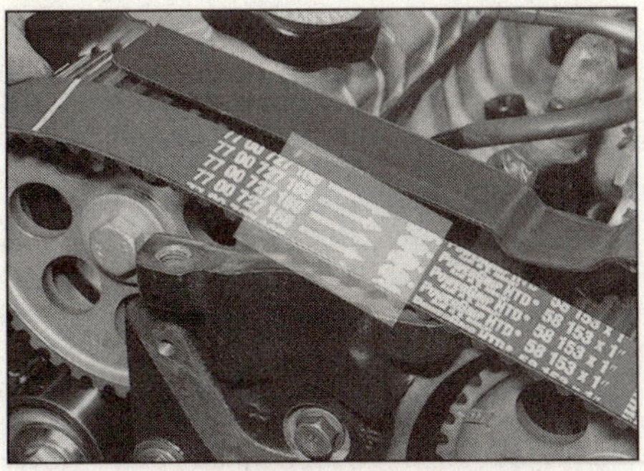

5.23a The running direction arrows on the belt must point clockwise

5.23b Align the timing marks on the belt with the crankshaft . . .

5.23c . . . camshaft . . .

5.23d . . . and injection pump sprocket marks (model with Lucas pump shown)

Adjustment using Renault special tool Ele. 346.04

Note: *The tension of the belt must be adjusted with the engine cold. The manufacturers specify the use of a special gauge, Renault tool Ele. 346.04 for checking the timing belt tension. If access to a suitable tool cannot be obtained, it is strongly recommended that the vehicle is taken to a Renault dealer to have the belt tension checked at the earliest opportunity.*

26 Fit the gauge to the engine, as shown **(see illustration)**.

27 Screw an M6 bolt through the threaded hole in the rear timing belt cover next to the tensioner. Loosen the tensioner nut, and tighten the bolt so that it pushes against the tensioner roller bracket to tension the belt **(see illustration)**.

5.27 M6 bolt (arrowed) fitted to rear timing belt cover

28 Adjust the bolt pushing against the tensioner roller bracket to give the specified reading on the gauge (**see illustration 5.26**).

29 With the correct tension applied, re-tighten the tensioner nut to the specified torque. This torque is critical, since if the nut were to come loose, considerable engine damage would result. Loosen the bolt fitted to the rear timing belt cover so that it no longer rests on the tensioner roller bracket.

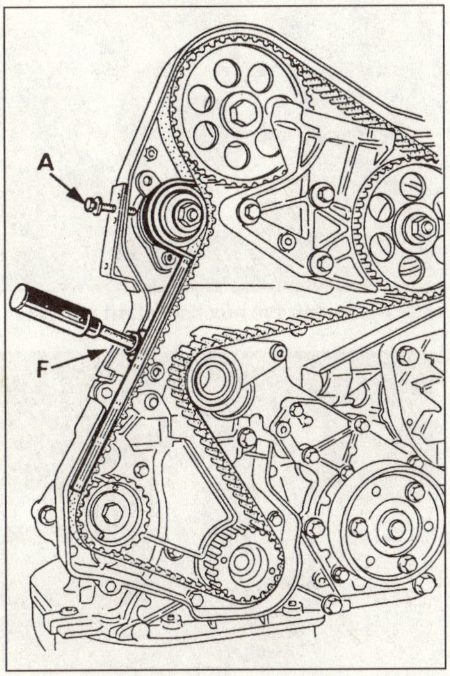

5.26 Renault tool (F) and M6 bolt (A) used to tension timing belt

30 Remove the TDC locking tool from the cylinder block, then refit the crankshaft pulley and securing bolt. Prevent the crankshaft turning using the method described previously, and tighten the bolt to the specified torque. Refit the flywheel cover plate/engine-to-gearbox bracing bracket.

31 Check that the crankshaft is still positioned with No 1 piston at TDC (by temporarily refitting the locking tool to the cylinder block), then remove the TDC locking tool, and turn the crankshaft two complete turns in the normal direction of rotation, returning it to the TDC position again. Re-insert the TDC locking tool in the cylinder block.

32 Temporarily refit the timing belt cover which covers the camshaft sprocket, and check that the camshaft sprocket timing mark still aligns with the pointer on the cover, as noted before removal (see Section 3).

33 Re-check the belt tension as described previously. If the tension is incorrect, the setting and checking procedure must be repeated until the correct tension is achieved.

34 With the belt tensioned correctly, remove the M6 bolt from the rear timing belt cover, and remove the TDC locking tool from the cylinder block, if not already done. Refit the plug to the cylinder block TDC hole.

35 Check the fuel injection pump timing as described in Chapter 4B.

36 Refit the upper timing belt covers, ensuring that any brackets secured by the bolts are in position as noted before removal.

37 Refit the engine mounting rubber/movement limiter assembly to the body, and refit the upper mounting bracket to the engine, as described in Section 14.

38 Refit the fuel filter assembly to the body, and clip the hoses into position.

39 Refit the auxiliary drivebelt as described in Chapter 1.

40 Refit the strengthening bar to the front suspension turrets.

41 Withdraw the jack, or the lifting tackle, as applicable, used to support the engine.

42 Refit the wheel arch cover and the roadwheel, and lower the vehicle to the ground.

43 Reconnect the battery negative lead.

Approximate adjustment

44 Refer to the note at the beginning of paragraph 26 before proceeding.

45 If the special gauge is not available, the timing belt tension can be checked approximately by twisting the belt between the thumb and forefinger, at the centre of the run between the tensioner pulley and the auxiliary shaft pulley. It should be just possible to twist the belt through 90° using moderate pressure **(see illustration)**.

5.45 Twisting the timing belt as an approximate tension check

46 The checking, adjustment and final refitting procedures are as described previously in paragraphs 27 to 43, but have the belt tension checked by a Renault dealer at the earliest opportunity.

6 Timing belt sprockets, idler roller and tensioner - removal and refitting

Crankshaft sprocket

Note: *A suitable puller may be required for this operation.*

Removal

1 Remove the timing belt as described in Section 5.

2 It should be possible to simply pull the sprocket from the front of the crankshaft, however in some cases a puller may be required **(see illustration)**. It is a simple matter to make up a puller using two bolts, a metal bar and the existing crankshaft pulley bolt. By unscrewing the crankshaft pulley bolt, the sprocket is pulled from the end of the crankshaft.

3 Recover the Woodruff key if it is loose.

Refitting

4 Refitting is a reversal of removal. Note that the sprocket fits with the flange against the end of the cylinder block/rear timing belt cover.

5 Refit the timing belt as described in Section 5.

Auxiliary shaft sprocket

Removal

6 Remove the timing belt as described in Section 5.

7 Hold the sprocket stationary using a suitable gear-holding tool. Alternatively, an old timing belt can be wrapped around the sprocket and held firmly with a pair of grips.

8 Unscrew the securing bolt, then pull the sprocket from the end of the shaft **(see illustration)**. If necessary, use two levers or screwdrivers to free the sprocket (if necessary, a puller can be used as described previously for the crankshaft sprocket).

Refitting

9 Refitting is a reversal of removal. Tighten the sprocket securing bolt to the specified torque.

10 Refit the timing belt as described in Section 5.

Fuel injection pump sprocket

Note: *A suitable puller will be required for this operation.*

Removal

11 Remove the timing belt (see Section 5).

12 Hold the sprocket stationary using a suitable gear-holding tool. Alternatively, an old timing belt can be wrapped around the sprocket and held firmly with a pair of grips.

13 Unscrew the sprocket securing nut, as far as the end of the pump shaft. Do not remove the nut at this stage **(see illustration)**. **Note:** *On models with a two-piece adjustable sprocket, do not loosen the three sprocket adjuster bolts, as this will alter the injection pump timing.*

14 On models with a single-piece sprocket, note the location of the Woodruff key in the end of the injection pump shaft. There are two key slots in the sprocket (for use with Bosch and Lucas injection pumps - **(see illustrations 5.15a and 5.15b)**. Mark the relevant key slot in the sprocket to ensure correct refitting.

15 Using a suitable puller, acting on the sprocket nut (not on the end of the pump shaft), release the sprocket from the taper on the pump shaft **(see illustration)**.

16 Remove the puller, then remove the sprocket securing nut, and recover the washer.

17 Make a mark on the timing belt corresponding to the location of the fuel injection pump sprocket timing mark **(see illustrations 5.15a and 5.15b)** to aid alignment when refitting. Disengage the sprocket from the timing belt, remove the sprocket, and recover the Woodruff key from the end of the pump shaft if it is loose **(see illustration)**.

Refitting

18 Refitting is a reversal of removal, bearing in mind the following points:

a) *On models with a single-piece sprocket, ensure that the Woodruff key is engaged with the correct slot in the sprocket.*
b) *Tighten the sprocket securing nut to the specified torque.*
c) *Refit and tension the timing belt as described in Section 5.*
d) *Before refitting the timing belt cover over the injection pump sprocket, check the injection timing (see Chapter 4B).*

6.2 Removing the crankshaft sprocket

6.8 Removing the auxiliary shaft sprocket

6.13 Unscrewing the fuel injection pump sprocket nut

6.15 Using a puller to release the fuel injection pump sprocket

6.17 Make an alignment mark on the timing belt

Camshaft sprocket - models with a rear timing belt cover which surrounds both the camshaft and fuel injection pump sprockets

Note: *This is an involved procedure due to the location of the upper engine mounting/fuel injection pump mounting bracket, which leaves insufficient clearance to remove the sprocket from the end of the camshaft. This means that the fuel injection pump and mounting bracket must be removed, as described in the following procedure. It is suggested that this procedure is read through thoroughly before beginning work.*

Removal

19 Disconnect the battery negative lead.
Caution: If the radio/cassette in your vehicle is equipped with an anti-theft system, make sure you have the correct activation code before disconnecting the battery.
20 Remove the timing belt as described in Section 5.
21 Remove the plastic alternator shield.
22 Remove the alternator, with reference to Chapter 5A.
23 Unbolt the alternator mounting bracket from the cylinder block **(see illustration)**.
24 Remove the fuel injection pump sprocket as described previously in this Section.
25 Remove the fuel injection pump with reference to Chapter 4B.
26 Remove the bolt(s) securing the rear timing belt cover to the fuel injection pump bracket.
27 Remove the three securing bolts, and withdraw the upper engine mounting/fuel injection pump mounting bracket from the engine. Note that it may not be possible to fully withdraw the upper two bolts from the bracket (due to limited clearance), in which case slide out the bracket with the bolts in position in their holes.
28 If not already done, remove the crankshaft TDC locking tool, and turn the engine a quarter of a turn anti-clockwise (viewed from the timing belt end of the engine) to position the pistons halfway down their bores. This is to avoid any possibility of piston-to-valve contact if the camshaft is inadvertently turned when loosening the sprocket bolt.

6.23 Removing the alternator mounting bracket

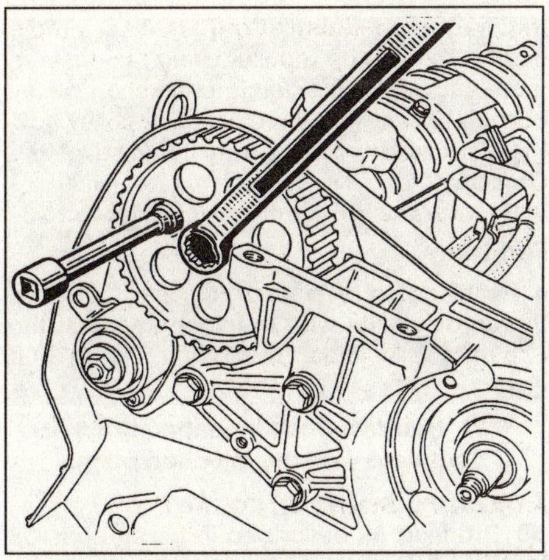

6.29 Using a socket and extension bar to counterhold the camshaft sprocket whilst unscrewing the sprocket bolt

29 Unscrew the camshaft sprocket bolt. The sprocket can be held using a suitable socket and extension bar engaged with one of the rear timing belt cover securing bolts as shown **(see illustration)**. Alternatively, use an old timing belt wrapped around the sprocket. Recover the washer.
30 Remove the bolt and washer, and the sprocket from the front of the camshaft. A suitable puller may be required, in which case ensure that the legs of the puller act on the holes in the sprocket, **not** on the sprocket teeth.
31 Recover the Woodruff key from the end of the camshaft if it is loose **(see illustration)**.

Refitting

32 Ensure that the Woodruff key is in place in the end of the camshaft, then refit the camshaft sprocket, noting that the projecting hub fits towards the cylinder head.
33 Ensure that the washer is in place, then refit the sprocket bolt, and tighten it to the specified torque, holding the sprocket as during removal.
34 Refit the upper engine mounting/fuel injection pump mounting bracket, and tighten the securing bolts. Where applicable, refit the upper two bolts to the holes in the bracket before the bracket is refitted.
35 Refit the bolt(s) securing the rear timing belt cover to the upper engine mounting/fuel injection pump mounting bracket.
36 Refit the fuel injection pump with reference to Chapter 4B.
37 Refit the fuel injection pump sprocket as described previously in this Section.
38 Refit the plastic alternator shield.
39 Refit the alternator mounting bracket to the cylinder block, and tighten the securing bolts.
40 Refit the alternator, with reference to Chapter 5A.
41 Temporarily refit the outer timing belt cover which covers the camshaft sprocket, and check that the TDC mark on the camshaft sprocket is aligned with the pointer on the timing belt cover.

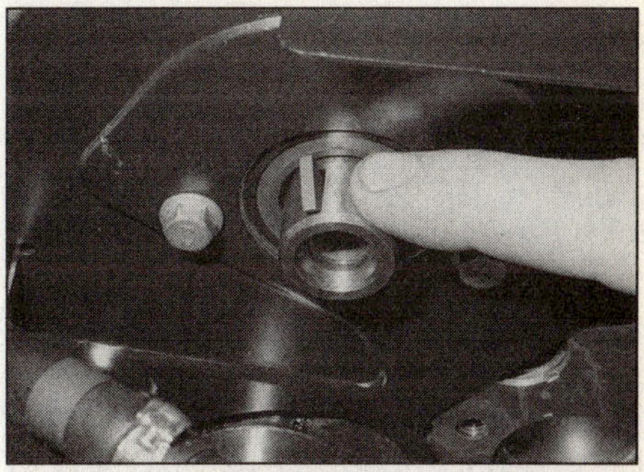

6.31 Recover the Woodruff key from the end of the camshaft if it is loose

42 Turn the crankshaft clockwise (viewed from the timing belt end of the engine) until the crankshaft locking tool can be refitted (to position No 1 piston at TDC).
43 Refit and tension the timing belt, as described in Section 5.
44 Reconnect the battery negative lead.
45 Bleed the fuel system as described in Chapter 4B.

Camshaft sprocket - models with a rear timing belt cover which surrounds the camshaft sprocket, but not the fuel injection pump sprocket

Note: *Refer to the note at the beginning of the previous sub-Section before proceeding.*

2B

Removal

46 Proceed as described in paragraphs 19 to 23.
47 Disconnect all cables, hoses, pipes and wiring from the fuel injection pump, to facilitate removal, referring to Chapter 4B.
48 Remove the bolts securing the injection pump rear mounting bracket to the cylinder head **(see illustration)**.
49 Remove the nuts securing the rear mounting bracket to the injection pump, then withdraw the bracket.

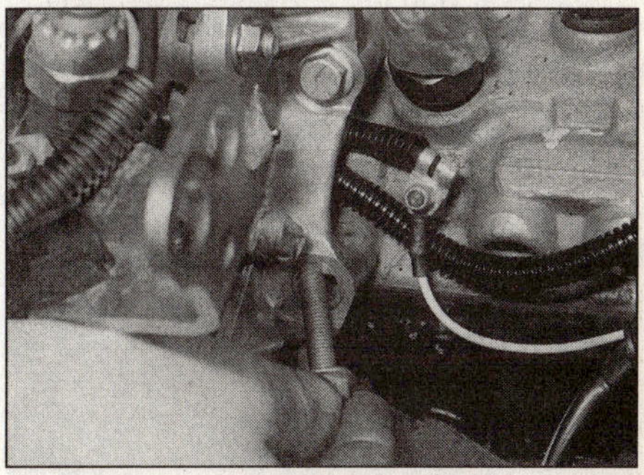

6.48 Removing one of the bolts securing the injection pump rear mounting bracket to the cylinder head

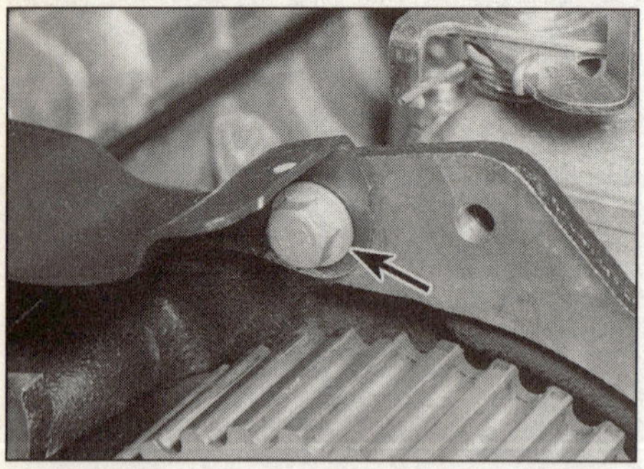

6.50 Remove the bolt (arrowed) securing the rear timing belt cover to the bracket

6.51 Removing the mounting bracket complete with the injection pump

50 Remove the bolt securing the rear timing belt cover to the upper engine mounting/fuel injection pump mounting bracket **(see illustration)**.

51 Unscrew the three securing bolts, and remove the upper engine mounting/fuel injection pump mounting bracket, complete with the injection pump from the engine **(see illustrations)**. The assembly is removed by sliding it out towards the front of the vehicle. Note that it may not be possible to fully withdraw the upper two bolts from the bracket (due to limited clearance), in which case slide out the assembly with the bolts in position in their holes.

52 Proceed as described in paragraphs 28 to 31.

Refitting

53 Proceed as described in paragraphs 32 and 33.

54 Refit the upper engine mounting/fuel injection pump mounting bracket, complete with the injection pump, and tighten the securing bolts. Where applicable, refit the upper two bolts to the holes in the bracket before the assembly is refitted.

55 Refit the bolt securing the rear timing belt cover to the upper engine mounting/fuel injection pump mounting bracket.

56 Refit the injection pump rear mounting bracket and tighten the securing bolts and nuts.

57 Reconnect all relevant cables, hoses, pipes and wiring to the fuel injection pump with reference to Chapter 4B, Section 13.

58 Proceed as described in paragraphs 38 to 45.

Idler roller

Removal

59 Remove the timing belt as described in Section 5.

60 Unscrew the two securing bolts, and withdraw the idler roller assembly, manipulating it out from the rear timing belt covers.

Refitting

61 Refitting is a reversal of removal, but check that the roller turns freely without binding or excessive play.

62 Refit and tension the timing belt as described in Section 5.

Tensioner

Removal

63 Remove the timing belt as described in Section 5.

64 Remove the securing nut and the washer, and withdraw the tensioner assembly from the engine.

Refitting

65 Refitting is a reversal of removal, but check that the roller turns freely without binding or excessive play. Ensure that the peg on the cylinder block engages with the hole in the tensioner bracket.

66 Refit and tension the timing belt as described in Section 5.

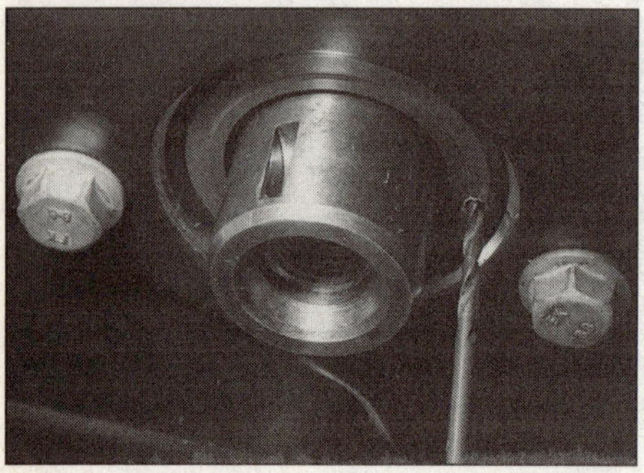

7.3a Drill a small hole . . .

7.3b . . . and use a screw and pliers to pull out the seal

7 Camshaft oil seals - renewal

Front (timing belt) end oil seal

1 Remove the camshaft sprocket as described in Section 6.

2 Remove the Woodruff key from the end of the camshaft, if not already done.

3 Using a small screwdriver, prise out the oil seal from the cylinder head, taking care not to damage the surface of the camshaft. Alternatively, the oil seal can be removed by drilling a small hole and inserting a self-tapping screw. A pair of grips can then be used to pull out the oil seal, by pulling on the screw **(see illustrations)**. If difficulty is experienced, insert two screws diagonally opposite each other. To ensure correct fitting, note the fitted position of the old oil seal.

4 Wipe clean the oil seal seating in the cylinder head, then dip the new seal in fresh engine oil, and locate it over the camshaft with its closed side facing outwards. Make sure that the oil seal lip is not damaged as it is located on the camshaft.

5 Using a tube of suitable diameter, drive the oil seal squarely into the housing to the previously noted depth. A block of wood cut to pass over the end of the camshaft may be used instead.

6 Refit the camshaft sprocket as described in Section 6.

Rear (flywheel end) oil seal

7 No oil seal is fitted to the flywheel end of the camshaft. The sealing is provided by a gasket between the cylinder head and the brake vacuum pump housing, and on certain models by an O-ring fitted between the pump and the housing. The gasket and the O-ring, where applicable, can be renewed after unbolting the pump from the cylinder head (see Chapter 9).

8 Camshaft and tappets - removal, inspection and refitting

Note: *A new camshaft front oil seal should be fitted, and a new valve cover gasket may be required on refitting. Suitable sealant will be required for the camshaft bearing caps and the bearing cap bolts.*

Removal

1 Remove the camshaft sprocket as described in Section 6.

2 Remove the securing nut and bolt, and withdraw the timing belt tensioner roller from the engine **(see illustration)**.

3 Unscrew and remove the two bolts securing the upper rear timing belt cover to the cylinder head **(see illustration)**.

8.2 Withdraw the timing belt tensioner roller

4 Unscrew the lower bolt(s) securing the upper rear timing belt cover to the cylinder block.

5 Remove the timing belt idler pulley securing bolt which also passes through the rear timing belt cover.

6 Manipulate the rear timing belt cover from the front of the camshaft and, where possible, withdraw the cover from the engine.

7 Remove the brake vacuum pump as described in Chapter 9.

8 Where necessary for improved access, unclip any hoses which are routed across the top of the valve cover, and move them to one side out of the way. If any fuel lines are disconnected, cover the open unions to prevent dirt ingress.

9 Unscrew the nuts from the valve cover, and withdraw the cover from the engine. Recover the gasket.

10 Using a dial gauge, measure the camshaft endfloat, and compare with the value given in the Specifications **(see illustration)**. This will give an indication of the amount of wear present on the thrust surfaces.

11 If the original camshaft is to be refitted, it is advisable to measure the valve clearances at this stage, as described in Section 4, so that any shims required can be obtained before the camshaft is refitted.

12 Check the camshaft bearing caps for identification marks, and if none are present, make identifying marks so that they can be refitted in their original positions and the same way round. Number the caps from the flywheel end of the engine.

13 Progressively slacken the bearing cap bolts and studs until the valve spring pressure is relieved. Remove the bolts and studs (noting their locations to ensure correct refitting), and the bearing caps themselves. Note that No 1 bearing cap is secured by two studs and two additional bolts.

14 Lift out the camshaft together with the oil seal **(see illustration)**.

15 Remove the tappets, keeping each with its shim **(see illustration)**. Place them in a compartmented box, or on a sheet of card marked into eight sections, so that they may be refitted to their original locations. Write down the shim thicknesses - they will be needed later if any of the valve clearances are

8.3 Remove the bolts securing the rear timing belt cover to the cylinder head

8.14 Lifting out the camshaft

incorrect. The shim size is stamped on the bottom face of the shim, but it is prudent to use a micrometer to measure the true thickness of any shim removed, as it may have been reduced by wear.

Inspection

16 Examine the camshaft bearing surfaces and cam lobes for wear ridges, pitting or scoring. Renew the camshaft if evident.

17 Renew the oil seal at the front end of the camshaft as a matter of course. Lubricate the lips of the new seal before fitting, and store the camshaft so that its weight is not resting on the seal.

18 Examine the camshaft bearing surfaces in the cylinder head and bearing caps. Deep scoring or other damage means that the cylinder head must be renewed.

19 Inspect the tappet buckets and shims for scoring, pitting and wear ridges. Renew as necessary.

Refitting

20 Ensure that the pistons are positioned half way down their bores, as described for sprocket removal in Section 6.

21 Oil the tappets and fit them to the bores from which they were removed. Fit the correct shim, numbered side downwards, to each tappet.

22 Oil the camshaft bearings. Place the camshaft with its oil seal onto the cylinder head. The oil seal must be positioned so that it is flush with the cylinder head face.

8.10 Measuring the camshaft endfloat using a dial gauge

8.15 Lift out the tappets

23 Apply sealant (CAF 4/60 THIXO, or a suitable equivalent) to the cylinder head mating faces of the rear and front camshaft bearing caps (Nos 1 and 5).

24 Refit the camshaft bearing caps to their original locations, ensuring that the front oil seal is correctly located in the bearing cap.

25 Apply sealant to the threads of the bearing cap bolts and studs. Fit the bolts and studs, and tighten them progressively to the specified torque.

26 If a new camshaft has been fitted, measure the endfloat using a dial gauge, and check that it is within the specified limits.

27 Refit the brake vacuum pump with reference to Chapter 9.

28 Refit the rear timing belt cover, then refit and tighten the securing bolts.

29 Refit and tighten the bolt securing the timing belt idler pulley assembly.

30 Refit the timing belt tensioner roller, ensuring that the peg on the cylinder block engages with the hole in the tensioner roller bracket.

31 Refit the camshaft sprocket as described in Section 6, ignoring the reference to bleeding the fuel system at this stage.

32 Check the valve clearances as described in Section 4, and take any corrective action necessary.

33 Refit the valve cover, using a new gasket if necessary, and tighten the securing nuts.

34 Refit/reconnect any hoses which were moved for access.

35 Reconnect the battery negative lead.

36 Bleed the fuel system as described in Chapter 4B.

9 Cylinder head - removal, inspection and refitting

Note: *A new cylinder head gasket must be fitted, and a new valve cover gasket and new cylinder head bolts may be required on refitting - see text.*

Removal

1 The following procedure describes removal and refitting of the cylinder head complete with manifolds and the fuel injection pump.

2 Disconnect the battery negative lead.

Caution: If the radio/cassette in your vehicle is equipped with an anti-theft system, make sure you have the correct activation code before disconnecting the battery.

3 Drain the cooling system with reference to Chapter 1. Also drain the cylinder block by unscrewing the drain plug located on the right-hand rear face of the engine. Refit the plug after draining.

4 To improve access, unbolt the strengthening bar from between the front suspension strut turrets.

5 Remove the alternator drivebelt (Chapter 1), then unbolt and remove the alternator drivebelt tensioner and bracket.

6 Unbolt and remove the earth strap from the rear of the engine.

7 Disconnect the accelerator cable from the fuel injection pump, and move the cable clear of the engine, noting its routing, with reference to Chapter 4B if necessary.

8 Undo the banjo union and disconnect the fuel supply pipe from the injection pump. Recover the sealing washers from the banjo union. Cover the open end of the pipe, and plug the opening in the injection pump to keep dirt out (the banjo bolt can be refitted to the pump and covered) **(see illustrations)**.

9 Remove the securing clip, and disconnect the main fuel return hose from the pipe on the fuel injection pump. Feed the hose through the rear engine lifting bracket, located at the rear of the cylinder head, and move the hose clear of the engine. Cover the open ends of the pipe and the hose to prevent dirt ingress.

10 Disconnect the air trunking running from the air cleaner to the inlet manifold, or turbocharger, as applicable, and remove the trunking (note that, where applicable, the breather hoses which connect to the trunking will also have to be disconnected).

11 On turbo models, disconnect the air trunking running from the intercooler to the turbocharger, and remove the trunking.

12 Remove the timing belt as described in Section 5. Since the cylinder head will be removed, it is not possible to support the engine/transmission with a hoist. Renault technicians use special support tools located beneath the right-hand front of the cylinder block and the subframe, and an additional support bracket bolted beneath the water pump, however blocks of wood may be used instead. Note that the engine **must** be supported adequately as the cylinder head bolts are tightened to a high torque.

13 Unbolt the hose bracket from the front of the fuel injection pump mounting bracket, and move the hoses and bracket clear of the injection pump.

14 Disconnect the breather hose(s) from the crankcase oil separator(s), and where applicable (on turbo models), disconnect the hose from the boost pressure fuel delivery corrector on the injection pump. Similarly, disconnect the corresponding ends of the hoses from the manifold.

15 Unclip the hose bracket from the front engine lifting bracket, unbolt the remaining hose bracket from the brake vacuum pump, and remove the hose assembly from the engine.

16 On non-turbo models, disconnect the coolant hose from the thermostat housing.

17 On turbo models, unbolt the thermostat housing from the cylinder head, and move it to one side, leaving the hoses and sensor wiring connected **(see illustration)**.

18 Disconnect the coolant hose from the left-hand rear corner of the cylinder head, and move the hose clear.

19 Disconnect the vacuum hose from the brake vacuum pump **(see illustration)**.

20 Disconnect all relevant wiring from the fuel injection pump. Note that on certain pumps, this can be achieved by simply disconnecting the wiring connectors at the brackets on the pump. On some pumps it will be necessary to disconnect the wiring from the individual components (some connections may be protected by rubber covers). Label all connections to aid correct refitting.

21 Disconnect the electrical feed wires from the relevant glow plugs.

22 On non-turbo models, disconnect the wiring plug from the temperature gauge/warning light sender unit, located at the left-hand front or rear of the cylinder head.

23 Disconnect all relevant pipes and hoses from the manifolds, and the turbocharger where applicable, with reference to the relevant Section(s) of Chapter 4B. Label all pipes and hoses to aid correct refitting.

24 Where applicable, unbolt any hose brackets from the manifolds and surrounding area, and move the hoses to one side.

25 On turbo models, remove the two bolts securing the turbocharger inlet elbow to the bracket on the gearbox. Remove the nut and bolt securing the bracing bracket to the turbocharger and the inlet elbow, and remove the elbow.

26 Remove the bolt(s) securing the upper rear timing belt cover to the cylinder block.

27 Remove the timing belt idler pulley securing bolt which also passes through the rear timing belt cover.

28 Remove the exhaust front section as described in Chapter 4B.

29 On turbo models, unscrew the union nut securing the turbocharger oil feed pipe to the union on the cylinder block, and remove the pipe/hose assembly.

30 On turbo models, remove the bolts securing the turbocharger support bracket to the turbocharger and the engine, and remove the support bracket.

31 If not already done, remove the TDC locking tool from the cylinder block, and turn the crankshaft anti-clockwise (viewed from the timing belt end of the engine) by a quarter turn to position all four pistons half way down their bores.

32 Progressively slacken the cylinder head bolts in the reverse sequence to that shown in illustration 9.53 **(see illustration)**. When the tension has been relieved, remove all the bolts.

9.8 Unscrewing the fuel supply pipe banjo union

9.17 Unbolt the thermostat housing and move it to one side - turbo models

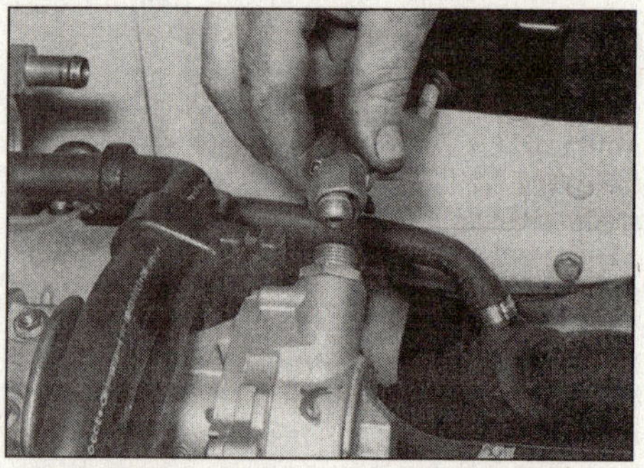

9.19 Disconnecting the brake vacuum pump hose

9.32 Slackening a cylinder head bolt

9.34 Lifting the cylinder head assembly from the engine - turbo model shown

9.42 Measuring piston protrusion using a dial test indicator

33 The cylinder head assembly complete with ancillaries is heavy, and it is advisable to attach a hoist and suitable lifting tackle to the lifting brackets on the cylinder head in order to lift it from the engine.

34 Lift the cylinder head (complete with manifolds, injection pump, and upper rear timing belt cover) upwards and off the cylinder block **(see illustration)**. If it is stuck, tap it upwards using a hammer and block of wood (taking care not to damage the fuel injection pump). **Do not** try to turn the cylinder head (it is located by two dowels), nor attempt to prise it free using a screwdriver inserted between the block and head faces.

35 If desired, the manifolds, turbocharger (where applicable) and fuel injection pump can be removed from the cylinder head with reference to the relevant Sections of Chapter 4.

Inspection

36 The mating faces of the cylinder head and block must be perfectly clean before refitting the head. Use a scraper to remove all traces of gasket and carbon, and also clean the tops of the pistons. Take particular care with the aluminium cylinder head, as the soft metal is damaged easily. Also, make sure that debris is not allowed to enter the oil and water channels - this is particularly important for the oil circuit, as carbon could block the oil supply to the camshaft or crankshaft bearings. Using adhesive tape and paper, seal the water, oil and bolt holes in the cylinder block. Clean the piston crowns in the same way.

> **HAYNES HiNT** ***To prevent carbon entering the gap between the pistons and bores, smear a little grease in the gap. After cleaning the piston, rotate the crankshaft so that the piston moves down the bore, then wipe out the grease and carbon with a cloth rag.***

37 Check the block and head for nicks, deep scratches and other damage. If slight, they may be removed carefully with a file. More serious damage may be repaired by machining, but this is a specialist job.

38 If warpage of the cylinder head is suspected, use a straight-edge to check it for distortion. Refer to Chapter 2C if necessary.

39 Clean out the bolt holes in the block using a pipe cleaner, or a rag and screwdriver. Make sure that all oil is removed, otherwise there is a possibility of the block being cracked by hydraulic pressure when the bolts are tightened.

40 Examine the bolt threads and the threads in the cylinder block for damage. If necessary, use the correct-size tap to chase out the threads in the block, and use a die to clean the threads on the bolts.

Gasket selection

41 Turn the crankshaft to bring piston Nos 1 and 4 to just below the TDC position (just below the top face of the cylinder block). Position a dial test indicator (DTI) on the cylinder block, and zero it on the block face. Transfer the probe to the centre of No 1 piston, then slowly turn the crankshaft back and forth past TDC, noting the highest reading produced on the indicator. Record this reading.

42 Repeat this measurement procedure on No 4 piston, then turn the crankshaft half a turn (180°) and repeat the procedure on Nos 2 and 3 pistons **(see illustration)**. Ensure that all measurements are taken along the longitudinal centreline of the crankshaft (this will eliminate errors due to piston slant).

43 If a dial test indicator is not available, piston protrusion may be measured using a straight-edge and feeler blades or vernier calipers. However, these methods are inevitably less accurate and cannot therefore be recommended.

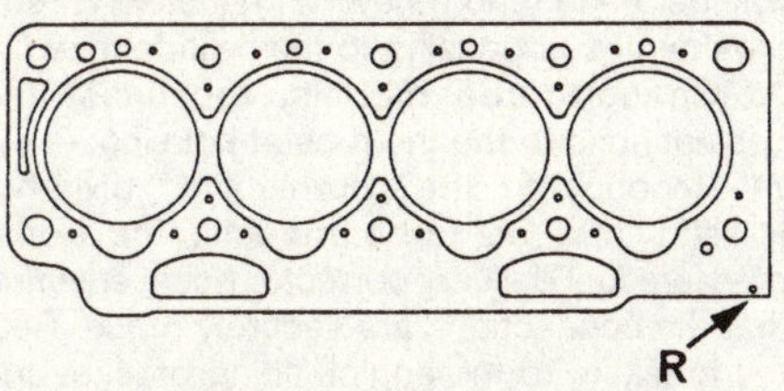

9.44a Cylinder head gasket thickness identification marking location (R)

44 Ascertain the greatest piston protrusion measurement, and use this to determine the correct cylinder head gasket from the following table. The identification holes are located at the front corner of the gasket, at the flywheel end **(see illustrations)**. **Note:** *The gasket thickness identification holes are located in an area 25 mm from the flywheel end of the gasket. Do not take into account any other holes outside this area.*

Piston protrusion	**Gasket identification**
Non-turbo engines:	
Less than 0.868 mm	*2 holes*
0.868 to 1.000 mm	*1 hole*
More than 1.000 mm	*3 holes*
Turbo engines:	
Less than 0.073 mm	*2 holes*
0.073 to 0.206 mm	*1 hole*
More than 0.206 mm	*3 holes*

Cylinder head bolt examination

45 On models fitted with T55 Torx cylinder head bolts, the manufacturers recommend that the bolts are measured to determine whether renewal is necessary, however some owners may wish to renew the bolts as a matter of course. The manufacturers do not specify checking or renewal for the hexagon socket-head-type bolts.

46 Where applicable, measure the length of each bolt from the bottom of the bolt head (without a washer fitted) to the end of the threaded shank.

47 If the length of any bolt is greater than 120.5 mm, all ten bolts **must** be renewed.

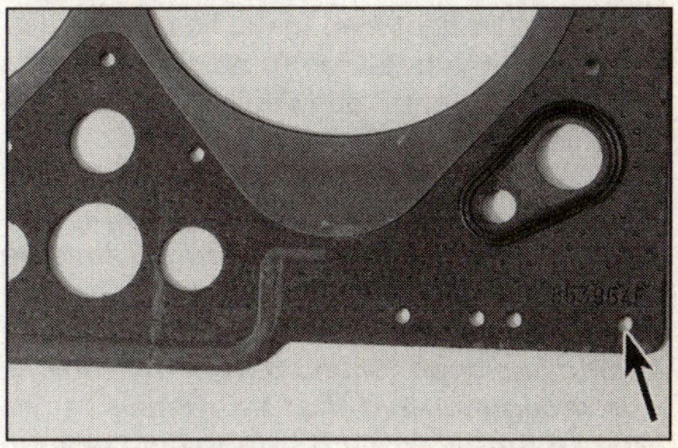

9.44b Cylinder head gasket thickness marking (arrowed) - "1 hole" type shown (ignore remaining holes - see text)

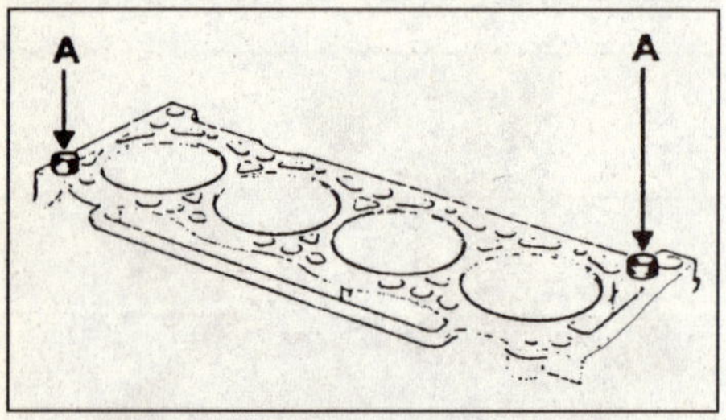

9.50 Cylinder head locating dowel locations (A)

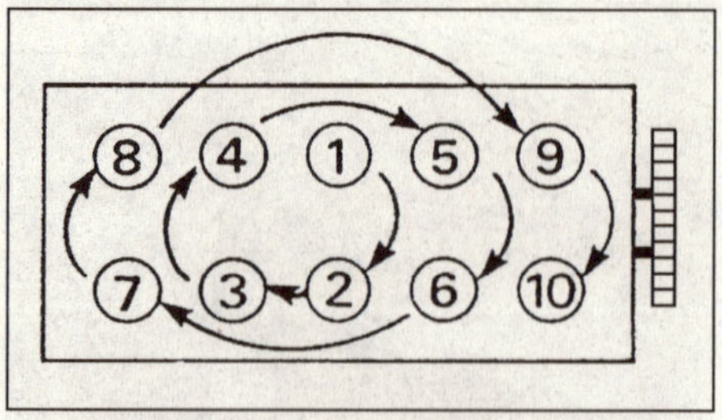

9.53 Cylinder head bolt tightening sequence

Refitting

48 Where applicable, refit the manifolds, turbocharger (where applicable) and fuel injection pump to the cylinder head, with reference to the relevant Sections of Chapter 4.
49 Turn the crankshaft clockwise (viewed from the timing belt end) until Nos 1 and 4 pistons pass bottom dead centre (BDC) and begin to rise, then position them halfway up their bores. Nos 2 and 3 pistons will also be at their mid-way positions, but descending their bores. Do not turn the crankshaft again until the timing belt is to be refitted (this is to prevent the possibility of piston-to-valve contact).
50 Ensure that the cylinder head locating dowels are fitted to the cylinder block, then fit the correct gasket the right way round on the cylinder block with the identification mark(s) at the front corner of the engine at the flywheel end **(see illustration)**.
51 Lower the cylinder head onto the block. Ensure that the upper rear timing belt cover engages correctly with the lower rear timing belt cover on the cylinder block. Where applicable, disconnect the lifting tackle and hoist. Ensure that the swirl chambers do not drop out of their locations in the cylinder head as it is lowered into position.
52 Lubricate the cylinder head bolt threads, and the undersides of the bolt heads with a little engine oil, then insert them, together with their washers. Screw the bolts into their threads as far as possible by hand.
53 Tighten the bolts in the order shown, and in the stages given in the Specifications - ie, tighten all bolts in sequence to the Stage 1 torque, then tighten all bolts in sequence to the Stage 2 torque, and so on **(see illustration)**. Note that on turbo models, the bolts must be tightened to the final stage after the engine has been run to normal operating temperature and allowed to cool (see Specifications).

Warning: On models with Torx type cylinder head bolts, the final tightening stages involve very high forces. Ensure that the tools used are in good condition. If the engine has been removed from the vehicle, it is recommended that tightening to the Stage 6 torque wrench setting (and beyond for turbo models) is carried out with the engine refitted to the vehicle (it may be necessary to remove the right-hand upper engine mounting bracket for access to one of the bolts with the engine in the vehicle).

54 On turbo models, refit the turbocharger-to-engine bracket, and tighten the securing bolts.
55 On turbo models, refit the turbocharger oil feed pipe, and tighten the union to the cylinder block.
56 Refit the exhaust front section with reference to Chapter 4B.
57 Refit the timing belt idler pulley securing bolt.
58 Refit the bolts securing the upper rear timing belt cover to the cylinder block.
59 Refit the timing belt as described in Section 5.
60 On turbo models, examine the sealing ring in the turbocharger inlet elbow, and renew it if necessary. Refit the elbow and the bracing bracket.
61 Refit any hose brackets to the manifolds, as noted before removal.
62 Reconnect all relevant pipes and hoses to the manifolds, and the turbocharger where applicable, as noted before removal.
63 On non-turbo models, reconnect the wiring plug to the temperature gauge/warning light sender unit.
64 Reconnect the feed wires to the relevant glow plugs.
65 Reconnect all wiring to the fuel injection pump.
66 Reconnect the engine earth lead to the rear engine lifting bracket.
67 Reconnect the coolant hose to the cylinder head and the brake vacuum hose to the vacuum pump.
68 Reconnect the air trunking between the air cleaner, inlet manifold, turbocharger and intercooler, as applicable. Ensure that any breather hoses are correctly reconnected.
69 On turbo models, examine the sealing ring between the thermostat housing and the cylinder head, and renew it if necessary. Refit the thermostat housing to the cylinder head.
70 On non-turbo models, reconnect the coolant hose to the thermostat housing.
71 Reconnect the crankcase breather hose(s), and on turbo models, the boost pressure fuel delivery corrector hose, ensuring that the connections are securely made. Refit the brackets to the engine lifting bracket and the brake vacuum pump.
72 Refit the hose bracket to the fuel injection pump mounting bracket.
73 Feed the main fuel return hose through the engine lifting bracket, and reconnect it to the fuel injection pump pipe.
74 Reconnect the fuel supply hose to the injection pump.
75 Reconnect the accelerator cable with reference to Chapter 4B.
76 Refit the strengthening bar to the front suspension strut turrets.
77 Refill and bleed the cooling system as described in Chapter 1.
78 Refit the alternator drivebelt tensioner and bracket, followed by the drivebelt (refer to Chapter 1 if necessary).
79 Reconnect the battery negative lead.
80 Prime and bleed the fuel system as described in Chapter 4B.
81 On turbo models, follow the procedure described in Chapter 4B (priming the turbocharger oil circuit) before starting the engine.
82 On turbo models, tighten the cylinder head bolts to the final stage (refer to the Specifications) after running the engine up to normal operating temperature and allowing it to cool. It may be necessary to remove the right-hand upper engine mounting bracket, with reference to Section 14, for access to one of the bolts.

10 Sump - removal and refitting

Note: *A new sump gasket or suitable sealant (as applicable) must be used on refitting.*

Removal

1 Disconnect the battery negative lead, and place it to one side.
Caution: If the radio/cassette in your vehicle is equipped with an anti-theft system, make sure you have the correct activation code before disconnecting the battery.
2 Drain the engine oil (see Chapter 1), then refit and tighten the drain plug using a new washer.
3 Jack up the front of the vehicle and support on axle stands (see *"Jacking and vehicle support"*). Where fitted, remove the engine compartment undertray.
4 Unbolt and remove the strengthening bar from between the front suspension turrets.
5 Remove the cover from the right-hand engine mounting.
6 Using a hoist, support the weight of the engine/transmission.
7 Unscrew and remove the bolts securing the exhaust downpipe to the exhaust manifold. Release the downpipe and support it.
8 Unbolt and remove the engine-to-transmission bellhousing bracing bracket/ flywheel cover. Note that one of the bolts secures the rear engine mounting bracket to the transmission, and that the bolts securing the bracket to the engine require a Torx key to unscrew them.

9 Unbolt the supports securing the power-assisted steering fluid pipe to the right-hand side of the subframe.
10 Unscrew the nut from the right-hand engine mounting.
11 Lift the engine only sufficient to provide room to remove the sump. Remember that the driveshafts and remaining engine mountings are still connected.
Caution: Make sure the engine/transmission assembly is supported adequately before removing the sump.
12 Unscrew the bolts securing the sump to the cylinder block, but leave two diagonally opposite bolts entered on a few threads until the sump joint has been released.
13 Using a palette knife or similar tool, release the sump from the bottom of the cylinder block. Do not use a screwdriver as this may damage the mating surfaces.
14 Unscrew the two remaining bolts and lower the sump from the engine. Recover the gasket where fitted.
15 Clean all traces of gasket or sealant from the cylinder block and sump, and wipe them dry.

Refitting

16 Clean all traces of gasket or sealing compound from the mating faces of the sump and cylinder block.
17 Where fitted, locate a new gasket on the sump, otherwise apply a bead of CAF 4/60 THIXO sealant (or equivalent) to the sump face.
18 If a gasket is not fitted, it is important that the sump is positioned correctly the first time and not moved around after the sealant has touched the cylinder block. Temporary long bolts or dowel rods may be used to help achieve this.
19 To prevent oil dripping from the oil pump and crankcase, wipe these areas clean before refitting the sump.
20 Lift the sump into position, then insert the bolts and tighten them progressively to the specified torque.
21 Lower the engine, then refit and tighten the right-hand engine mounting nut.
22 Refit the supports for the power-assisted steering fluid pipe.
23 Refit the flywheel cover plate/engine-to-transmission bracing bracket and tighten the securing bolts.
24 Refit the exhaust downpipe and tighten the bolts.
25 Remove the hoist and refit the cover to the right-hand engine mounting.
26 Refit the strengthening bar between the front suspension turrets, and tighten the bolts.
27 Reconnect the battery, then lower the vehicle to the ground.
28 Fill the engine with fresh oil with reference to Chapter 1.

11.2 Removing the oil pump

11.7a Measuring the oil pump-to-body clearance

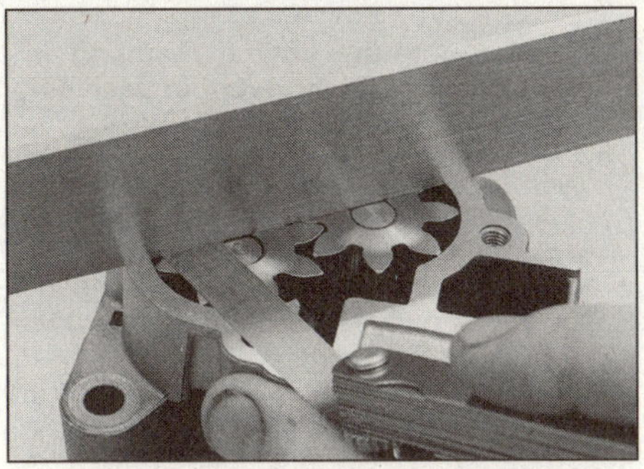

11.7b Measuring the oil pump gear endfloat

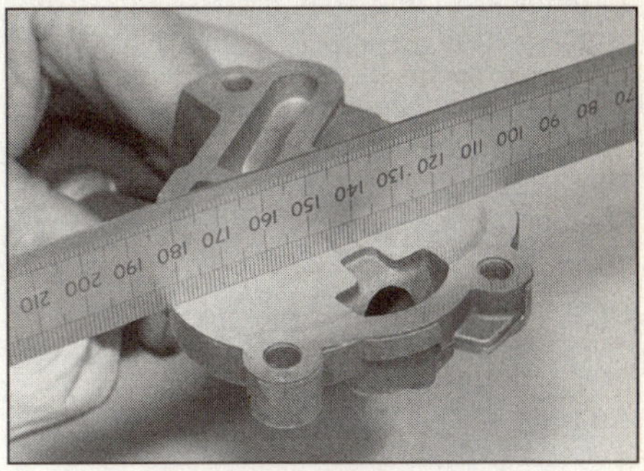

11.7c Checking the flatness of the oil pump cover

11 Oil pump - removal, inspection and refitting

Removal

1 Remove the sump (see Section 10).
2 Unscrew the four retaining bolts, and withdraw the pump from the cylinder block and drivegear **(see illustration)**.

Inspection

3 Unscrew the retaining bolts and lift off the pump cover.
4 Withdraw the idler gear and the drivegear/shaft. Mark the idler gear before removal, so that it can be refitted in its original position.
5 Extract the retaining clip, and remove the oil pressure relief valve spring retainer, spring, and plunger.
6 Clean the components, and carefully examine the gears, pump body and relief valve plunger for any signs of scoring or wear. Renew the complete pump assembly if excessive wear is evident (no spare parts are available).
7 If the components appear serviceable, measure the clearance between the pump body and the gears using feeler blades. Also measure the gear endfloat, and check the flatness of the end cover **(see illustrations)**. If the clearances exceed the specified tolerances, the pump must be renewed. There should be no perceivable distortion of the end cover.
8 If the pump is satisfactory, reassemble the components in the reverse order of removal. Fill the pump with oil, then refit the cover and tighten the bolts securely. (A new pump should also be primed with oil.) Check that the locating dowel is in position where the driveshaft enters the oil pump body.

Refitting

9 Wipe clean the mating faces of the oil pump and cylinder block.
10 Lift the oil pump into position, engaging the driveshaft with the drivegear. Insert and fully tighten the retaining bolts.
11 Refit the sump as described in Section 10.

12 Crankshaft oil seals - renewal

Front (timing belt) oil seal

1 Remove the crankshaft sprocket, as described in Section 6.
2 Prise out the old oil seal using a small screwdriver, taking care not to damage the surface of the crankshaft. Alternatively, the oil seal can be removed by drilling two small holes diagonally opposite each other and inserting self-tapping screws in them. A pair of grips can then be used to pull out the oil seal, by pulling on each side in turn.
3 Inspect the seal rubbing surface on the crankshaft. If it is grooved or rough in the area where the old seal was fitted, the new seal should be fitted slightly less deeply, so that it

rubs on an unworn part of the crankshaft surface.

4 Wipe clean the oil seal seating, then dip the new seal in fresh engine oil, and locate it over the crankshaft with its closed side facing outwards. Make sure that the oil seal lip is not damaged as it is located on the crankshaft.

5 Using a tube of suitable diameter, drive the oil seal squarely into the housing until flush. A block of wood cut to pass over the end of the crankshaft may be used instead.

6 Refit the crankshaft sprocket as described in Section 6.

Rear (flywheel end) oil seal

7 Renewal of the rear crankshaft oil seal requires the engine and transmission assembly to be removed from the engine compartment and the transmission and flywheel/driveplate separated on the bench. The procedure is therefore described in Chapter 2C.

13 Engine mountings - renewal

Inspection

1 Apply the handbrake, then jack up the front of the car and support it on axle stands (see *"Jacking and vehicle support"*). Where fitted, remove the engine compartment undertray.

2 Visually inspect the rubber pads on the two front and one rear engine/transmission mountings for signs of cracking and deterioration. Careful use of a lever will help to determine the condition of the rubber pads. If there is excessive movement in the mounting, or if the rubber has deteriorated, the mounting should be renewed.

Renewal

Right-hand front mounting

3 Connect a hoist and suitable lifting tackle to the engine lifting brackets to support the engine/transmission assembly while the mounting is removed. Alternatively, the assembly can be supported using a jack and a suitable block of wood to spread the load under the sump.

4 Remove the plastic cover from the upper engine mounting bracket. Note the location of any brackets secured by the bolts.

5 Ensure that the engine/gearbox assembly is adequately supported, then unscrew the nut securing the upper mounting bracket to the mounting on the body.

6 Unscrew the three bolts securing the upper mounting bracket to the main bracket on the engine, then withdraw the bracket **(see illustration)**.

7 Unscrew the bolts securing the mounting plate and the engine movement limiting rubber assembly to the body. Lift the movement limiting rubber assembly and the mounting plate from the body.

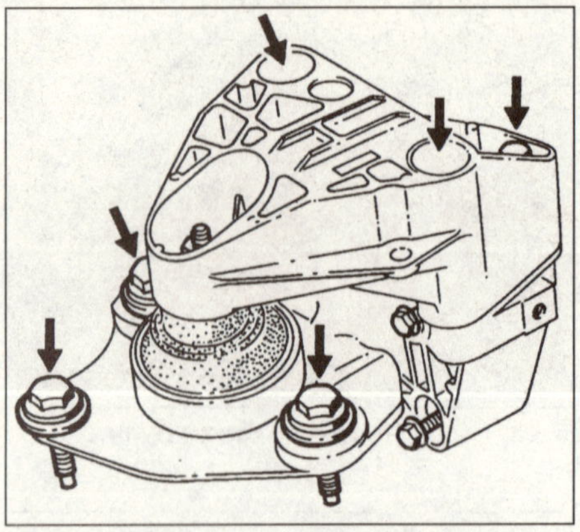

13.6 Right-hand engine mounting plate/movement limiting rubber assembly mounting bolt locations

8 The main engine mounting bracket is integral with the fuel injection pump mounting bracket. Removal and refitting of this bracket is an involved procedure, and is described in Section 6 as part of the camshaft sprocket removal and refitting procedure.

9 Commence refitting by refitting the movement limiting rubber assembly and the mounting plate. Do not fully tighten the securing bolts at this stage.

10 Refit the upper mounting bracket, and tighten the bolts. At this stage, do not fully tighten the nut securing the bracket to the mounting on the body.

11 Remove the hoist or jack from the engine/transmission assembly.

12 The upper mounting bracket must now be centred before tightening the body mounting bolts. Renault technicians use a special forked tool which is inserted through the slots in the bracket. If this tool is not available it may be possible to make up a similar tool out of strips of metal.

13 With the mounting centred, tighten the body mounting bolts to the specified torque. Remove the tool after making the adjustment, and also refit the plastic cover.

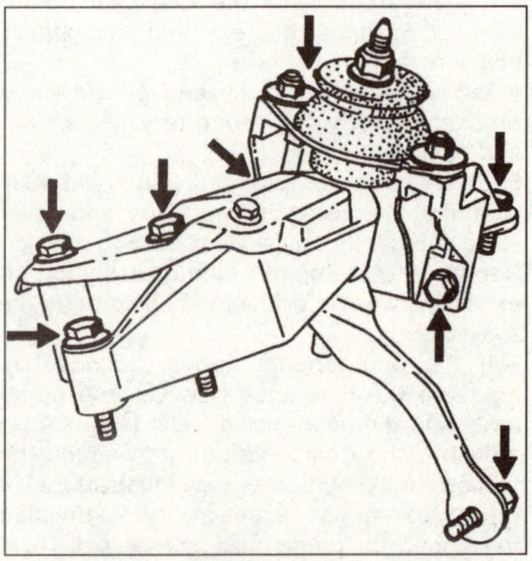

13.17 Left-hand front mounting bolt locations

Left-hand front mounting

14 Remove the left-hand roadwheel.

15 Using a jack and block of wood, support the weight of the transmission.

16 Unscrew and remove the nut securing the lower mounting stud to the upper bracket. Using a soft-faced mallet, tap the stud to release it from the upper bracket.

17 Unbolt the upper mounting rubber and bracket **(see illustration)**.

18 Unbolt the lower mounting bracket.

19 Fit the new mounting components using a reversal of the removal procedure, but tighten the nuts/bolts to the specified torque.

Rear mounting

20 Remove the air cleaner assembly as described in Chapter 4B.

21 From within the engine compartment, reach down behind the left-hand side of the engine and unscrew the nuts from the bolts securing the rear mounting bracket to the transmission. At the same time unbolt the shock absorber from the top of the mounting.

22 Using a hoist or trolley jack and block of wood, take the weight of the engine/transmission off the rear mounting.

23 Unscrew and remove the bolts from the bracket and manoeuvre the bracket from the transmission and subframe **(see illustration)**. The bracket is a tight fit, and it may be necessary to raise the engine/transmission further to provide room to remove it.

24 Fit the new mounting using a reversal of the removal procedure, but tighten the bolts to the specified torque.

14 Engine oil cooler - removal and refitting

Removal

1 Drain the cooling system as described in Chapter 1.

2 Remove the oil filter (refer to Chapter 1).

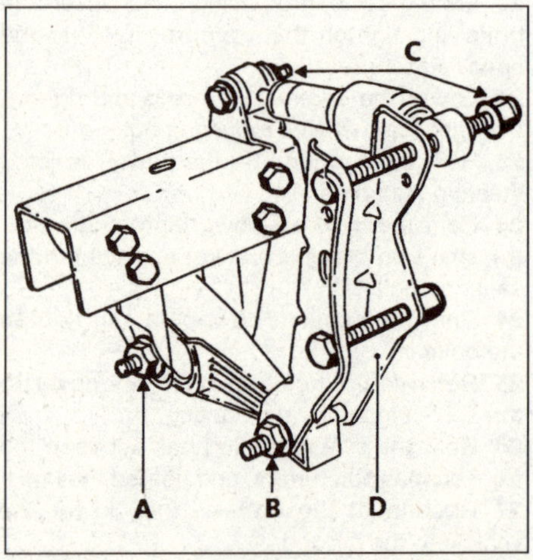

13.23 Rear mounting removal sequence (A to D)

3 Loosen the clips, and disconnect the coolant hoses from the oil cooler.
4 Unscrew the oil filter mounting stud, which also secures the oil cooler, and withdraw the oil cooler from the engine. Recover the sealing ring **(see illustrations)**.

Refitting

5 Refitting is a reversal of removal, but use a new sealing ring.

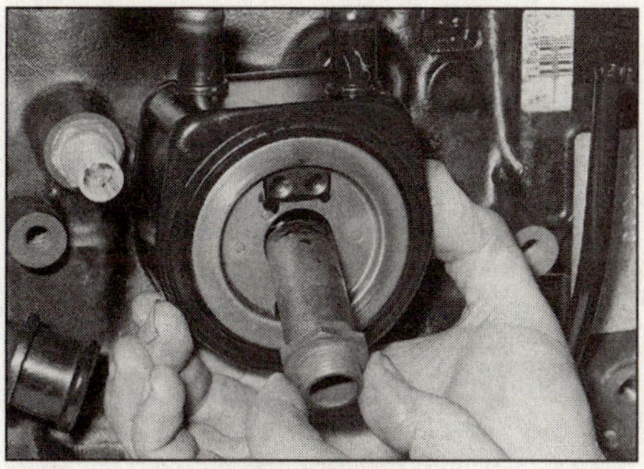

14.4a Remove the oil filter mounting stud . . .

14.4b . . . then withdraw the oil cooler and recover the sealing ring

Notes

Chapter 2 Part C: Engine removal and overhaul procedures

Contents

Degrees of difficulty

Easy, suitable for novice with little experience	**Fairly easy,** suitable for beginner with some experience	**Fairly difficult,** suitable for competent DIY mechanic	**Difficult,** suitable for experienced DIY mechanic	**Very difficult,** suitable for expert DIY or professional

2C

Specifications

Torque wrench settings

Refer to Specifications in Chapter 2A or 2B.

1 General information

Included in this part of Chapter 2 are the general overhaul procedures for the cylinder head, cylinder block/crankcase and internal engine components.

The information ranges from advice concerning preparation for an overhaul and the purchase of replacement parts, to detailed step-by-step procedures covering removal, inspection, renovation and refitting of internal engine parts.

The following Sections have been compiled based on the assumption that the engine has been removed from the car. For information concerning in-car engine repair, as well as the removal and refitting of the external components necessary for the overhaul, refer to Part A (petrol engines) or Part B (diesel engine) of this Chapter, and to Section 7 of this Part.

2 Engine overhaul - general information

It is not always easy to determine when, or if, an engine should be completely overhauled, as a number of factors must be considered.

High mileage is not necessarily an indication that an overhaul is needed, while low mileage does not preclude the need for an overhaul. Frequency of servicing is probably the most important consideration. An engine which has had regular and frequent oil and filter changes, as well as other required maintenance, will most likely give many thousands of miles of reliable service. Conversely, a neglected engine may require an overhaul very early in its life.

Excessive oil consumption is an indication that piston rings, valve stem oil seals and/or valves and valve guides are in need of attention. Make sure that oil leaks are not responsible before deciding that the rings and/or guides are bad. Perform a cylinder compression check to determine the extent of the work required.

Check the oil pressure with a gauge fitted in place of the oil pressure sender, and compare it with the value given in the Specifications. If it is extremely low, the main and big-end bearings and/or the oil pump are probably worn out.

Loss of power, rough running, knocking or metallic engine noises, excessive valve gear noise and high fuel consumption may also point to the need for an overhaul, especially if they are all present at the same time. If a complete tune-up does not remedy the situation, major mechanical work is the only solution.

An engine overhaul involves restoring the internal parts to the specifications of a new engine. During an overhaul, the pistons and rings are renewed, and the cylinder bores are reconditioned. New main bearings, connecting rod bearings and camshaft bearings are generally fitted, and if necessary, the crankshaft may be reground to restore the journals. The valves are also serviced as well, since they are usually in less-than-perfect condition at this point. While the engine is being overhauled, other components, such as the distributor, starter and alternator, can be overhauled as well. The end result should be a like-new engine that will give many trouble-free miles. **Note:** *Critical cooling system components such as the hoses, drivebelts, thermostat and water pump MUST be renewed when an engine is overhauled. The radiator should be checked carefully, to ensure that it is not clogged or leaking. Also, it is a good idea to renew the oil pump whenever the engine is overhauled.*

Before beginning the engine overhaul, read through the entire procedure to familiarise yourself with the scope and requirements of the job. Overhauling an engine is not difficult if you follow all of the instructions carefully, have the necessary tools and equipment, and pay close attention to all specifications; however, it can be time-consuming. Plan on the vehicle being tied up for a minimum of two weeks, especially if parts must be taken to an engineering works for repair or reconditioning. Check on the availability of parts, and make sure that any necessary special tools and equipment are obtained in advance. Most work can be done with typical hand tools, although a number of precision measuring tools are required for inspecting parts to determine if they must be renewed. Often the engineering works will handle the inspection of parts, and offer advice concerning reconditioning and renewal. **Note:** *Always wait until the engine has been completely disassembled, and all components, especially the engine block, have been inspected before deciding what service and repair operations must be performed by an engineering works. Since the condition of the block will be the major factor to consider when determining whether to overhaul the original engine or buy a reconditioned unit, do not purchase parts or have overhaul work done on other components until the block has been thoroughly inspected.* As a general rule, time is the primary cost of an overhaul, so it does not pay to fit worn or substandard parts.

As a final note, to ensure maximum life and minimum trouble from a reconditioned engine, everything must be assembled with care, and in a spotlessly-clean environment.

3 Engine removal - methods and precautions

If you have decided that an engine must be removed for overhaul or major repair work, several preliminary steps should be taken.

Locating a suitable place to work is extremely important. Adequate work space, along with storage space for the vehicle, will be needed. If a garage is not available, at the very least a flat, level, clean work surface is required.

Cleaning the engine compartment and engine before beginning the removal procedure will help keep tools clean and organised.

An engine hoist or A-frame will also be necessary. Make sure the equipment is rated in excess of the combined weight of the engine and transmission. Safety is of primary importance, considering the potential hazards involved in lifting the engine out of the vehicle.

If the engine is being removed by a novice, an assistant should be available. Advice and aid from someone more experienced would also be helpful. There are many instances when one person cannot simultaneously perform all of the operations required when lifting the engine out of the vehicle.

Plan the operation ahead of time. Arrange for, or obtain, all of the tools and equipment you will need, prior to beginning the job. Some of the equipment necessary to perform engine removal and installation safely and with relative ease are (in addition to an engine hoist) a heavy-duty floor jack, complete sets of spanners and sockets as described in the front of this manual, wooden blocks, and plenty of rags and cleaning solvent for mopping up spilled oil, coolant and fuel. If the hoist must be hired, make sure that you arrange for it in advance, and perform all of the operations possible without it beforehand. This will save you money and time.

Plan for the vehicle to be out of use for quite a while. An engineering works will be required to perform some of the work which the do-it-yourselfer cannot accomplish without special equipment. These places often have a busy schedule, so it would be a good idea to consult them before removing the engine, in order to accurately estimate the amount of time required to rebuild or repair components that may need work.

Always be extremely careful when removing and refitting the engine. Serious injury can result from careless actions. Plan ahead, take your time, and you will find that a job of this nature, although major, can be accomplished successfully.

4 Petrol engine and manual transmission - removal, separation and refitting

Note: *The engine is removed upwards from the engine compartment together with the manual gearbox, then separated on the bench. Downwards removal together with the subframe and front suspension struts is possible, but this method is not described, as it requires the use of a special wheeled cradle.*

Removal

1 Disconnect the battery negative lead.
Caution: If the radio/cassette in your vehicle is equipped with an anti-theft system, make sure you have the correct activation code before disconnecting the battery.
2 Remove the radiator with reference to Chapter 3. Also disconnect the top hose from the thermostat housing, and the bottom hose from the coolant pipe **(see illustration)**.
3 Drain the manual gearbox oil with reference to Chapter 7A **(see illustration)**.
4 If necessary, drain the engine oil with reference to Chapter 1.
5 Remove the bonnet with reference to Chapter 11.
6 On the K7M engine, unbolt and remove the strengthening bar from between the front suspension strut turrets.
7 Remove the air cleaner assembly with reference to Chapter 4A.

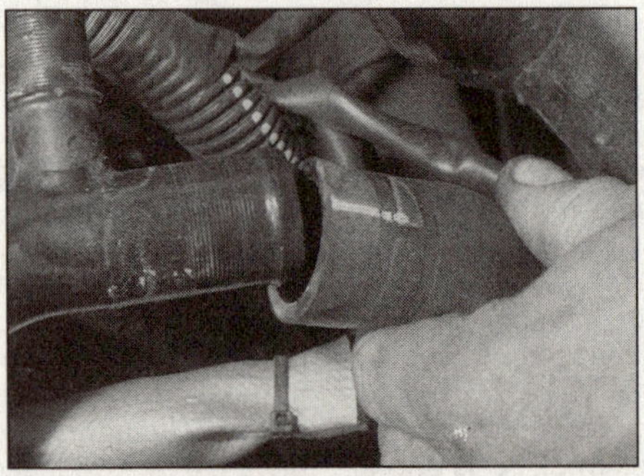

4.2 Disconnecting the bottom hose from the coolant pipe

4.3 Unscrewing the drain plug from the manual transmission

4.9 Disconnecting the track rod end from the steering arm on the stub axle carrier

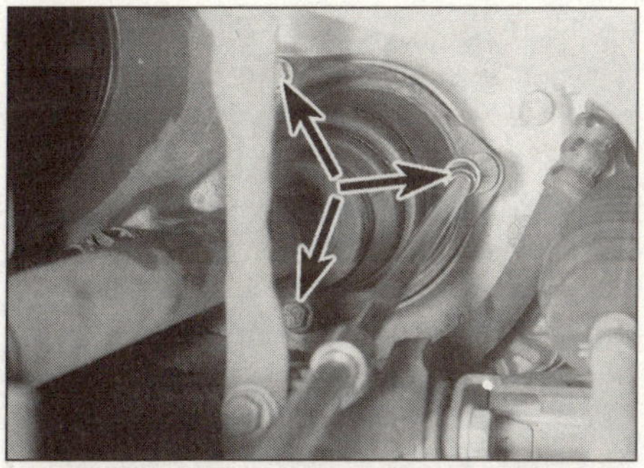

4.10 Remove the three bolts (arrowed) to remove the metal ring securing the left-hand driveshaft inner rubber boot to the transmission

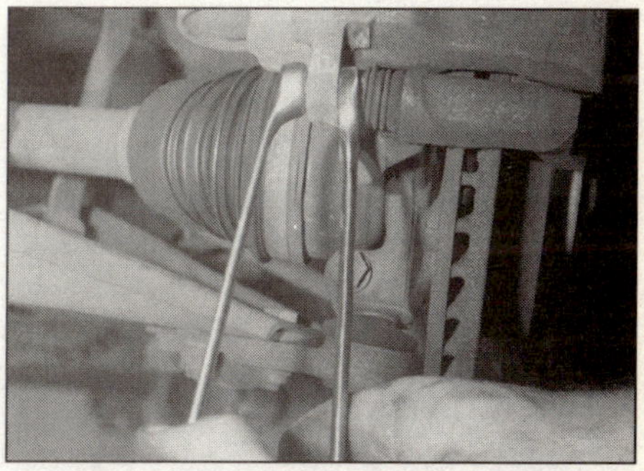

4.11a Unscrew the mounting bolts . . .

4.11b . . . remove the left-hand brake caliper . . .

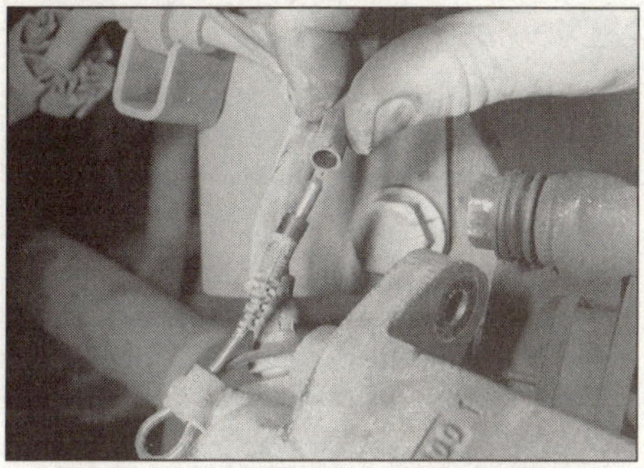

4.11c . . . and disconnect the pad wear warning wiring

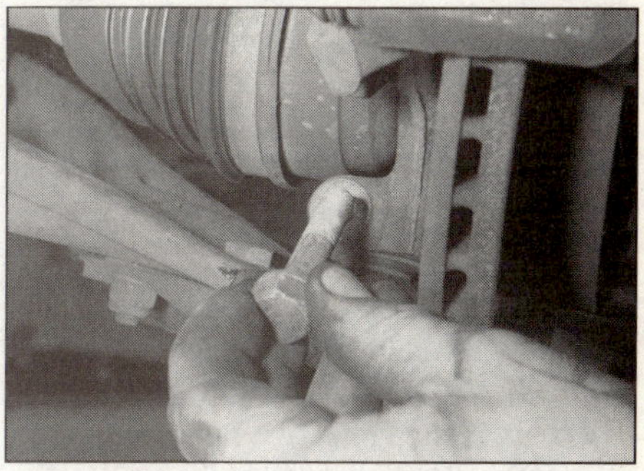

4.12 Removing the pinch-bolt securing the front left lower balljoint to the bottom of the stub axle carrier

4.13a Unscrew the nuts . . .

4.13b . . . and tap out the bolts securing the stub axle carrier to the bottom of the suspension strut

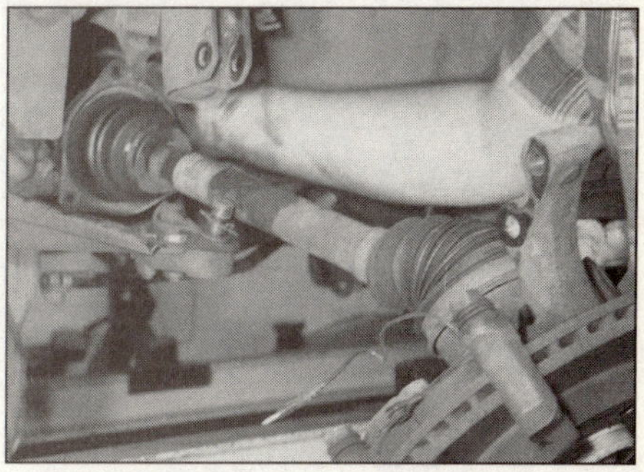

4.14 Withdrawing the left-hand driveshaft from the transmission

8 Apply the handbrake, then jack up the front of the car and support it on axle stands (see *"Jacking and vehicle support"*). Remove both front roadwheels. Where fitted, remove the engine compartment undertray.

9 Unscrew the nut from the left-hand track rod end, and disconnect it from the stub axle carrier steering arm using a balljoint separator tool **(see illustration)**.

10 Unscrew the bolts securing the left-hand driveshaft inner rubber boot and metal ring to the transmission **(see illustration)**.

11 Unscrew the two mounting bolts securing the left-hand brake caliper to the stub axle carrier, and disconnect the pad wear warning wiring. Tie the caliper to the coil spring, taking care not to strain the hose **(see illustrations)**.

12 Unscrew and remove the pinch-bolt securing the front left lower balljoint to the bottom of the stub axle carrier **(see illustration)**.

13 Unscrew and remove both bolts securing the left-hand front suspension strut to the stub axle carrier (note that the nuts are on the rear of the carrier) **(see illustrations)**.

14 Lever the suspension lower arm downwards, then pull the driveshaft from the sun gear and withdraw it together with the stub axle carrier. Make sure the tripod components remain in position on the inner end of the driveshaft, otherwise they may fall into the transmission. There may be some oil loss from the gearbox, so place a small container on the floor to catch it **(see illustration)**.

15 Working beneath the car, drive out the roll pin securing the right-hand driveshaft to the differential sun gear shaft. Note that if the original double roll pin is renewed, the new one will be of single coiled type **(see illustrations)**.

16 Unscrew the nut from the right-hand track rod end, and disconnect it from the stub axle carrier steering arm using a balljoint separator tool.

17 Unscrew and remove the upper bolt securing the right-hand front suspension strut to the stub axle carrier (note that the nut is on the rear of the carrier). Loosen only the lower bolt.

18 Pull out the driveshaft and disconnect the inner end from the splines on the transmission

4.15a Roll pin (arrowed) securing the driveshaft to the transmission sun gear

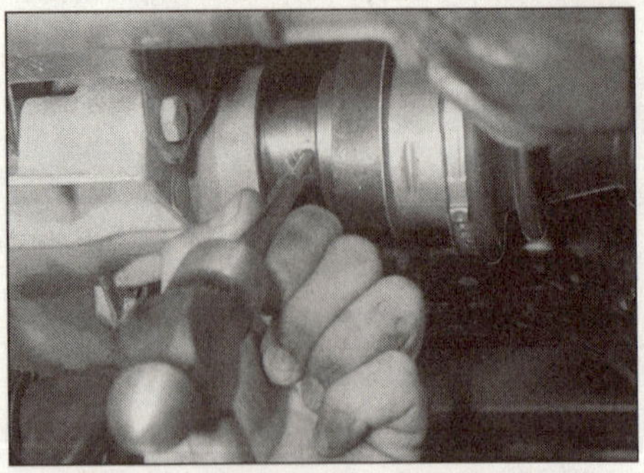
4.15b Using a punch to drive out the driveshaft roll pin

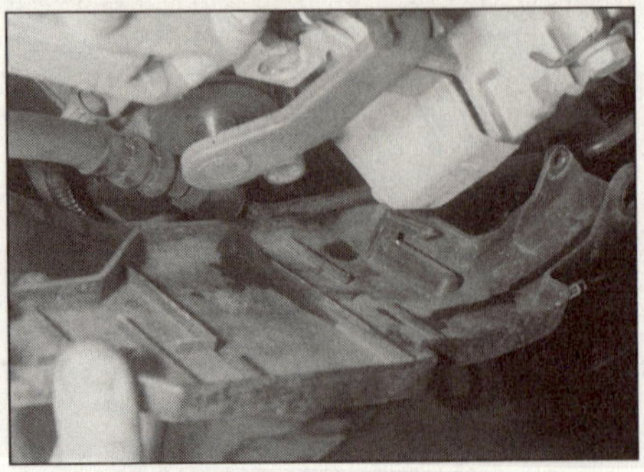
4.20a Remove the plastic cover from the bottom of the transmission . . .

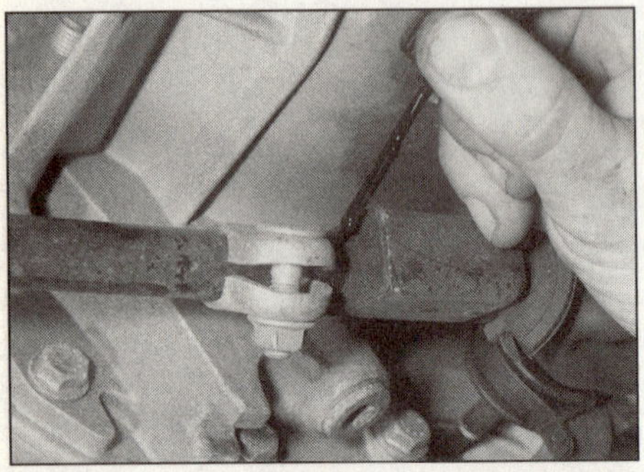
4.20b . . . then mark the position of the gearchange rod clamp with paint . . .

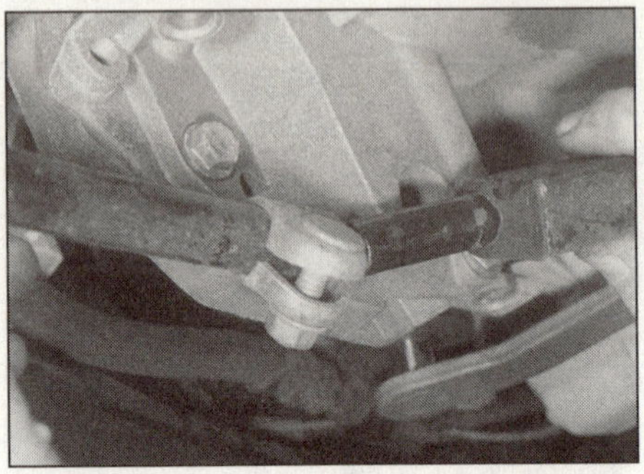
4.20c . . . loosen the clamp, and disconnect the gearchange rod

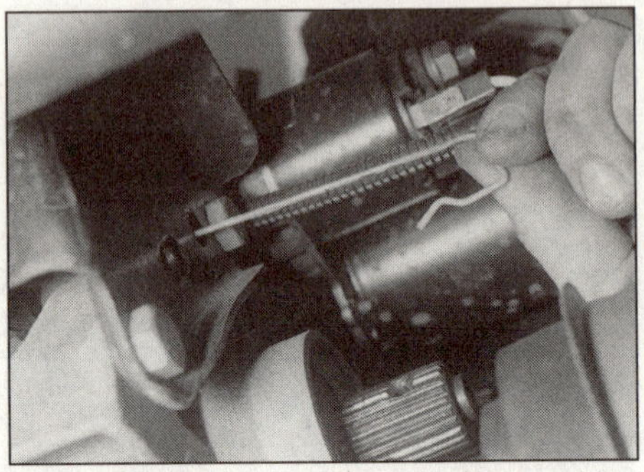
4.22a Pull out the spring clip . . .

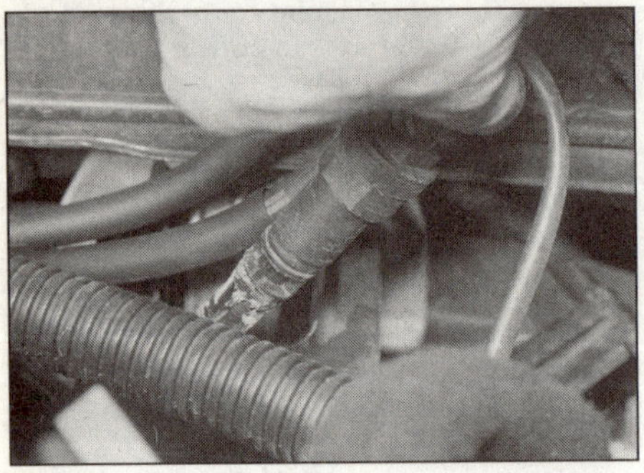
4.22b . . . and remove the speedometer sender unit from the transmission

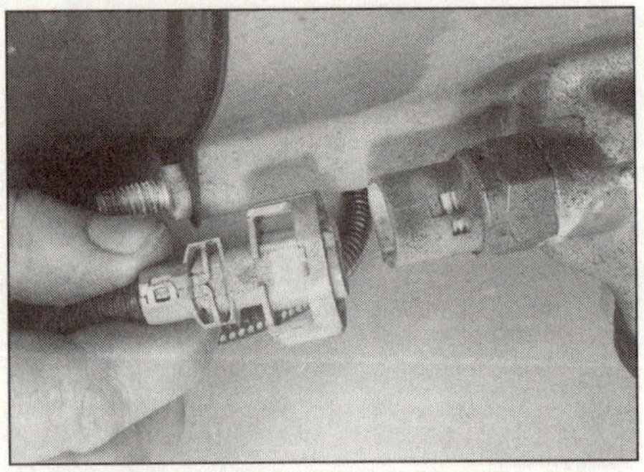
4.23 Disconnecting the wiring from the reversing light switch

4.24 Removing the power-assisted steering pump and fluid reservoir from the bracket on the engine

sun gear shaft. Tie the driveshaft to the steering gear.

19 Remove the exhaust front downpipe and catalytic converter as described in Chapter 4.

20 Remove the plastic cover from the bottom of the transmission, then mark the position of the gearchange rod clamp, loosen the clamp bolt and disconnect the rod at the transmission **(see illustrations)**.

21 Where applicable, detach the power-assisted steering hydraulic fluid pipes from the transmission.

22 Pull out the spring clip and remove the speedometer sender unit from the rear of the transmission **(see illustrations)**.

23 Disconnect the wiring from the reversing light switch on the transmission **(see illustration)**.

24 Remove the power-assisted steering pump and fluid reservoir from the engine with reference to Chapter 10, but do not disconnect the fluid pipes. Position the pump to one side **(see illustration)**.

25 On models fitted with air conditioning, remove the compressor with reference to Chapter 3, but leave the hoses attached, and position the pump to one side. **Do not** open any refrigerant lines.

26 Disconnect the accelerator cable from the throttle body with reference to Chapter 4.

27 Disconnect the clutch cable from the transmission with reference to Chapter 6.

28 Disconnect the remaining hoses from the thermostat housing and/or the coolant pipe (as applicable) on the left-hand side of the engine. These include the coolant expansion tank hose and the heater hoses **(see illustration)**.

29 On the K7M engine, remove the coolant expansion tank and tie it to the left-hand side of the engine compartment.

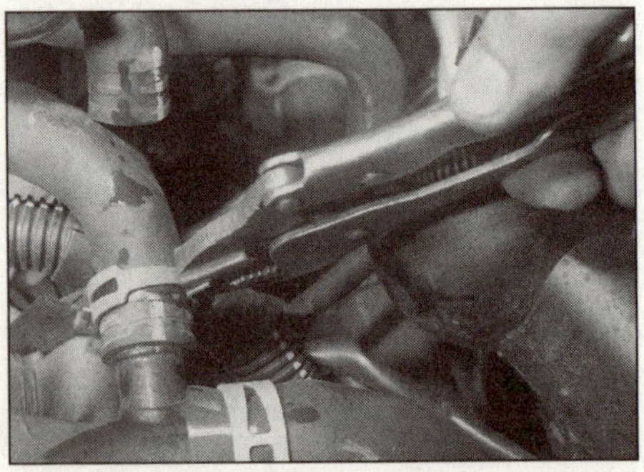
4.28 Disconnecting the heater hose from the coolant pipe - note the use of pliers to compress the hose retaining clip

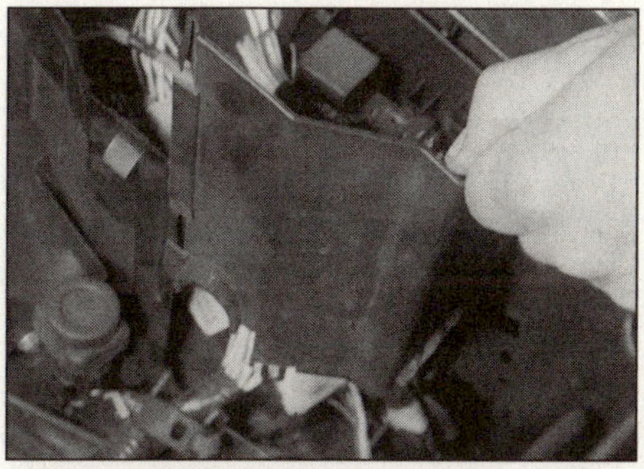

4.30a Remove the covers . . .

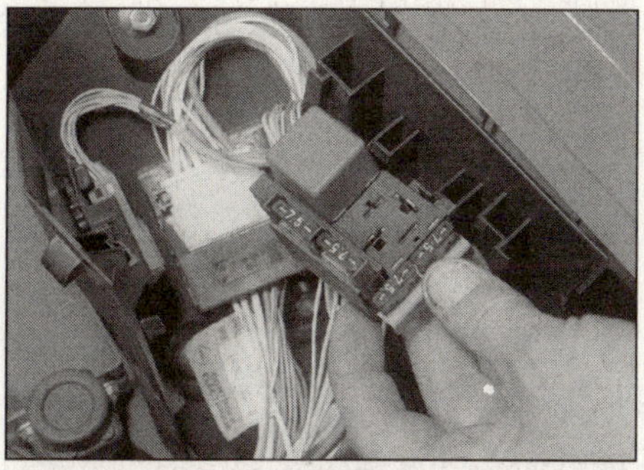

4.30b . . . and disconnect the wiring from the relay box

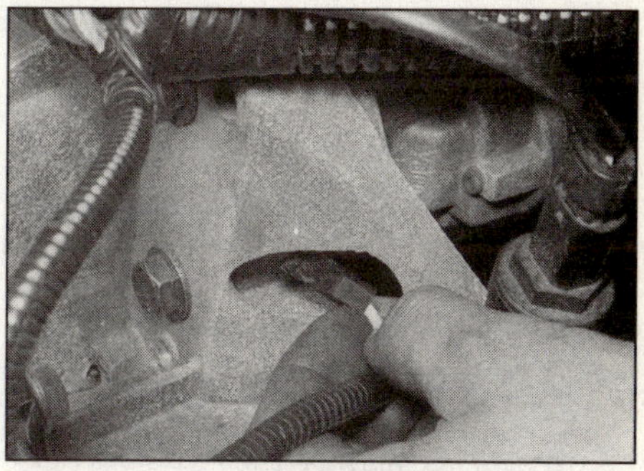

4.44a Disconnecting the wiring from the starter motor . . .

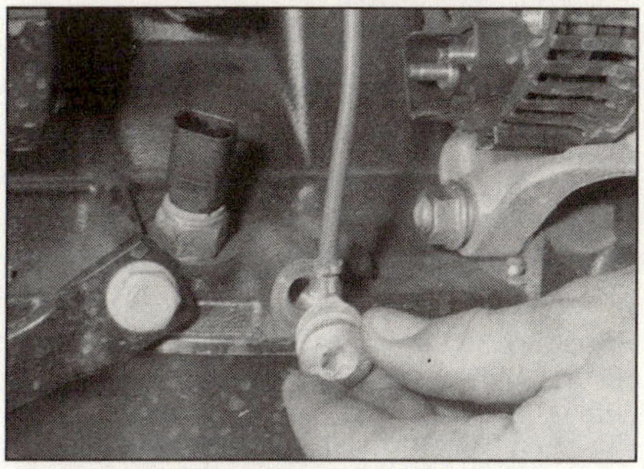

4.44b . . . the earth wire from the cylinder block . . .

4.44c . . . and the loom support from the transmission

30 Remove the top and side covers from the relay box on the left-hand side of the engine compartment, then release the relay plate and wiring cable-ties and position the wiring on the engine **(see illustrations)**.

31 Disconnect the wiring from the radiator fan and thermostatic switch.

32 Disconnect the brake vacuum servo unit hose from the inlet manifold.

33 Disconnect the map sensor hose from the inlet manifold, and disconnect the wiring from the map sensor.

34 Unscrew the mounting nuts and remove the ECU from the right-hand side of the engine compartment. Tie the ECU on top of the engine to prevent it being damaged when the assembly is lifted out.

35 Loosen the clips and disconnect the fuel supply and return hoses from the throttle body housing (E7J engine) or fuel rail (K7M engine).

36 Unbolt the earth cable from the bulkhead.

37 Disconnect the evaporative carbon canister hose from the solenoid valve on the inlet manifold.

38 Disconnect the wiring from the ignition module on the bulkhead.

39 Disconnect the starter main supply cable from the battery. To do this, remove the battery (Chapter 5) and disconnect the wiring at the connector box, then release the cable from the cover/grommet in the rear engine compartment panel. Also disconnect the fuel injection feed wire.

40 Unscrew and remove the engine/transmission front lower mounting nuts, and recover the plate from the right-hand mounting.

41 Connect a hoist to the engine lifting eyes, and take the weight of the engine assembly.

42 Unscrew and remove the bracket bolts and through-bolt from the engine/transmission rear mounting. Lift the assembly slightly and remove the rear mounting bracket.

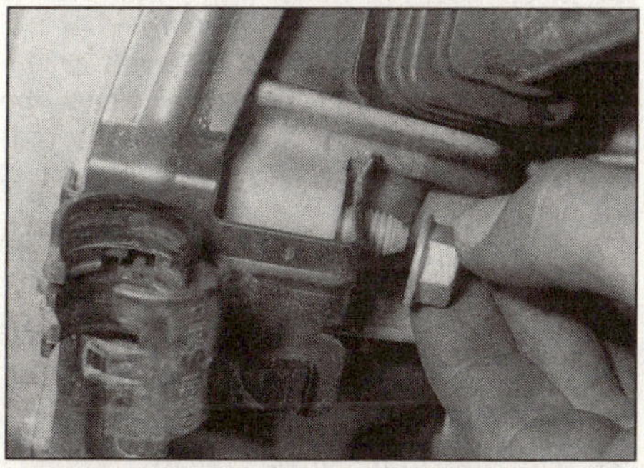

4.47a Unscrew the front lower nut . . .

4.47c Upper bolts securing the transmission to the engine

43 With the help of an assistant, slowly lift the engine/transmission assembly from the engine compartment, taking care not to damage any components on the surrounding panels. When high enough, lift the assembly over the front body panel, and lower to the ground.

44 If the engine is to be overhauled, remove the wiring loom from the engine, noting the location and routing of each part **(see illustrations)**.

45 To separate the transmission from the engine, first remove the starter motor as described in Chapter 5A.

46 Unbolt and remove the strengthening bracket between the engine cylinder block and transmission.

47 Support the transmission then unscrew and remove the bolts and nuts securing it to the engine, noting the location of brackets and supports **(see illustrations)**.

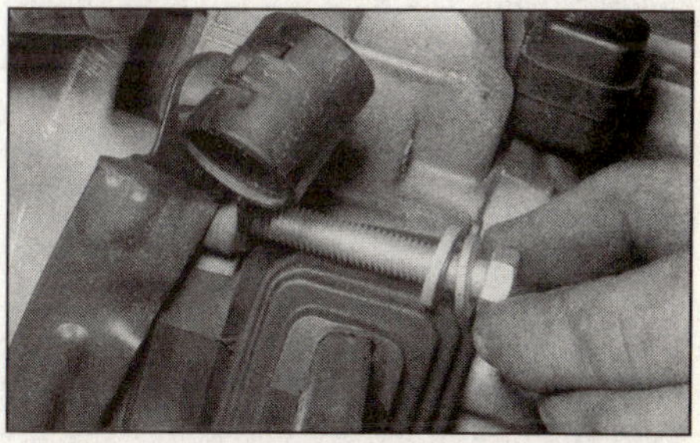

4.47b . . . and upper bolt securing the transmission to the engine, noting the location of the wiring support

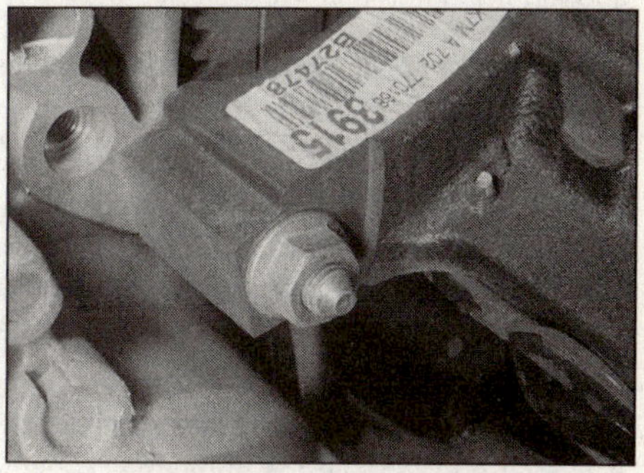

4.47d Rear lower nut securing the transmission to the engine

48 With the help of an assistant, withdraw the transmission from the engine making sure that it is not allowed to hang on the input shaft while engaged with the clutch components.

Refitting

49 Refitting is a reversal of removal, noting the following additional points:

a) *Apply a little high melting-point grease to the splines of the transmission input shaft.*
b) *Make sure the clutch release bearing and fork are correctly positioned before refitting the transmission.*
c) *Tighten all nuts and bolts to the specified torque.*
d) *Refer to the applicable Chapters as for removal.*
e) *Fill the engine and gearbox with oil, with reference to Chapter 1.*
f) *Fill the cooling system with reference to Chapter 1.*
g) *Depress the brake pedal several times to set the front brake pads in their normal position.*

5 Petrol engine and automatic transmission - removal, separation and refitting

Note: *Renault recommend that the engine is removed downwards from the engine compartment together with the automatic transmission, subframe and front suspension struts. Ideally, a purpose-made cradle should be positioned beneath the assembly to support it after lowering from the engine compartment. The help of several assistants will be required to lift the body upwards over the assembly. This procedure is not recommended for the home mechanic unless adequate assistance is available.*

Removal

1 Disconnect the battery negative lead.

Caution: If the radio/cassette in your vehicle is equipped with an anti-theft system, make sure you have the correct activation code before disconnecting the battery.

2 Remove the radiator with reference to Chapter 3.

3 Drain the automatic transmission fluid with reference to Chapter 7B.

4 If necessary, drain the engine oil with reference to Chapter 1.

5 Remove the bonnet with reference to Chapter 11.

6 Apply the handbrake, then jack up the front of the car and support it on axle stands (see *"Jacking and vehicle support"*). Remove both front roadwheels. Where fitted, remove the engine compartment undertray.

7 Remove the left and right-hand side wheel arch liners.

8 Unbolt and remove the subframe-to-body support rods from each side of the vehicle.

9 Unbolt the front brake calipers from the hub carriers with reference to Chapter 9, and support them on axle stands or tie them to the body.

10 Remove the front bumper as described in Chapter 11.

11 Remove the exhaust front downpipe and catalytic converter, referring to Chapter 4A.

12 Remove the horn with reference to Chapter 12.

13 Unbolt and remove the strengthening bar from between the front suspension strut turrets.

14 Remove the air cleaner assembly with reference to Chapter 4A.

15 Unbolt the earth cable from the bulkhead.

16 Disconnect the map sensor hose from the inlet manifold, and disconnect the wiring from the map sensor.

17 Remove the coolant expansion tank and attach it to the engine/transmission assembly.

18 Disconnect the accelerator cable from the throttle housing.

19 Disconnect the wiring from the kickdown switch.

20 Disconnect the brake vacuum servo unit hose from the inlet manifold.

21 Remove the top and side covers from the relay box on the left-hand side of the engine compartment, then release the relay plate and wiring cable-ties and position the wiring on the engine.

22 Disconnect the wiring from the automatic transmission ECU on the left-hand side of the engine compartment.

23 Disconnect the evaporative carbon canister hose from the solenoid valve on the inlet manifold.

24 Disconnect the starter main supply cable from the battery. To do this, remove the battery (Chapter 5A) and disconnect the wiring at the connector box, then release the cable from the cover/grommet in the rear engine compartment panel. Also disconnect the fuel injection feed wire.

25 Loosen the clips and disconnect the fuel supply and return hoses from the fuel rail.

26 Disconnect the heater hoses from the thermostat housing.

27 Disconnect the selector control from the transmission.

28 With the front wheels in the straight-ahead position, working in the engine compartment pull back the rubber grommet then unscrew and remove the eccentric bolt securing the bottom of the steering inner column to the steering gear, and separate the column (refer to Chapter 10 for more information) **(see illustration)**. **Note:** *It is important that the airbag rotary switch beneath the steering wheel is not damaged - make sure that the column is held in its central position until it is reconnected to the steering gear.*

29 The left-hand transmission mounting must now be removed. First wedge a block of wood between the automatic transmission and the left-hand side of the subframe to support the weight of the transmission. Unscrew the nut from the top of the mounting then using a soft-faced mallet tap the mounting stud to release it.

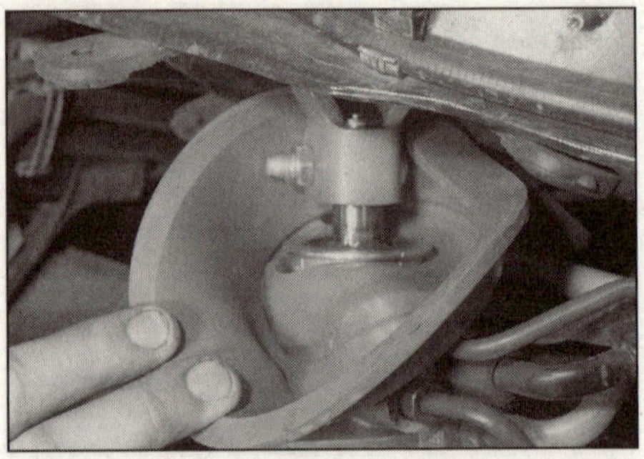

5.28 Pull back the rubber grommet, then unscrew and remove the bolt securing the bottom of the steering inner column to the steering gear

30 Support the subframe and engine/transmission adequately on axle stands. Alternatively attach a hoist to the assembly and support it.

31 Unscrew and remove the front suspension strut upper mounting bolts from the turrets.

32 Unscrew and remove the subframe-to-underbody bolts from the corners of the subframe, noting the location of spacers and washers.

33 If not already supported on axle stands, lower the assembly to the ground and support.

34 With the help of several assistants, lift the vehicle body to clear the assembly, then release the handbrake and move the vehicle away. Two of the assistants must support the struts, and care must be taken not to damage any components on the surrounding panels. With the vehicle clear of the assembly, carefully lower it to the ground and apply the handbrake.

35 To remove the engine/transmission from the subframe, first disconnect the radiator top hose.

36 Remove the power-assisted steering fluid reservoir and pump, referring to Chapter 10.

37 Release the wiring loom from the subframe and attach it to the engine.

38 Connect the hoist to the engine lifting eyes, and take the weight of the assembly.

39 Detach the right-hand engine mounting and the rear mounting with reference to Chapter 2A.

40 Have an assistant move the tops of the suspension struts outwards while the driveshafts are released from the transmission. Support the driveshafts making sure that the CV joints are not angled excessively.

41 Carefully lift the engine/transmission assembly from the subframe.

42 Remove the starter motor as described in Chapter 5A.

43 Release the clips and disconnect the hoses from the fluid cooler on top of the transmission.

44 Unbolt the cover from the automatic transmission sump/bellhousing.
45 Using a wide-bladed screwdriver in the starter ring gear teeth, hold the driveplate stationary while the driveplate-to-torque converter bolts are removed. It will be necessary to turn the crankshaft for access to all of the bolts.
45 Support the transmission, then unscrew and remove the bolts securing it to the engine.
47 Carefully withdraw the transmission from the engine, making sure that the torque converter remains fully engaged with the input shaft/oil pump shaft. With the transmission removed, a bent metal bar can be fitted to the bellhousing to keep the torque converter in place.

Refitting

48 Refitting is a reversal of removal, with reference to the Chapters needed for removal, but tighten all nuts and bolts to the specified torque. When attaching the subframe to the underbody, the use of 100 mm long rods temporarily screwed into the nuts will help alignment of the bolt holes. With the front of the subframe aligned, the rear bolts can be inserted, then the temporary rods can be removed and the front bolts inserted. Refer to Chapter 10 when connecting the steering column to the steering gear. Top-up the power steering fluid level as described in *"Weekly checks"*. Fill the engine with fresh oil and fill the cooling system with coolant with reference to Chapter 1. Fill the automatic transmission with fresh fluid with reference to Chapter 7B. Before taking the vehicle on the road, depress the brake pedal several times to bring the brake pads to their normal position.

6 Diesel engine and manual transmission - removal, separation and refitting

Note: *The engine is removed upwards from the engine compartment together with the manual gearbox, then separated on the bench. Downwards removal together with the subframe and front suspension struts is possible, but this method is not described, as it requires the use of a special cradle.*

Removal

1 Disconnect the battery negative lead.
Caution: If the radio/cassette in your vehicle is equipped with an anti-theft system, make sure you have the correct activation code before disconnecting the battery.
2 Remove the radiator with reference to Chapter 3. Also disconnect the top hose from the thermostat housing, and the bottom hose from the coolant pipe.
3 Drain the manual gearbox oil with reference to Chapter 7A.
4 If necessary, drain the engine oil with reference to Chapter 1.
5 Remove the bonnet with reference to Chapter 11.
6 Remove the air cleaner assembly and air duct with reference to Chapter 4B. On turbo models, remove the air ducting connecting the intercooler to the turbocharger and inlet manifold.
7 Apply the handbrake, then jack up the front of the car and support it on axle stands (see *"Jacking and vehicle support"*). Remove both front roadwheels. Where fitted, remove the engine compartment undertray.
8 Remove the front wheel arch liners from both sides.
9 Remove the alternator drivebelt as described in Chapter 1.
10 On models fitted with air conditioning, remove the compressor with reference to Chapter 3, but leave the hoses attached, and position the pump to one side. **Do not** open any refrigerant lines.
11 Remove the oil filter as described in Chapter 1.
12 Remove the power-assisted steering pump and fluid reservoir from the engine with reference to Chapter 10, but do not disconnect the fluid pipes. Position the pump to one side..
13 Identify then disconnect the heater hoses from the thermostat housing on the left-hand side of the cylinder head.
14 Disconnect the brake vacuum servo unit hose from the vacuum pump on the left-hand side of the cylinder head.
15 Disconnect the pipes and wiring from the fast idle and EGR solenoid valves.
16 Remove the pre-heating unit, and tie it to the engine (refer to Chapter 5C).
17 Disconnect the accelerator cable and clutch cable from the engine with reference to Chapters 4 and 6.
18 Disconnect the wiring from the altitude sensor corrector.
19 Remove the top and side covers from the relay box on the left-hand side of the engine compartment; release the relay plate and wiring cable-ties, and place the wiring on the engine.
20 Squeeze the connector and disconnect the fuel supply pipe from the fuel filter on the bulkhead. Disconnect the wiring from the fuel filter.
21 Detach the fuel return union from the injection pump.
22 Unbolt the earth strap from the bulkhead.
23 Unbolt and remove the strengthening bar from between the front suspension strut turrets.
24 Remove the battery as described in Chapter 5A.
25 Disconnect the starter main cable and the pre-heating cable from the wiring connector on the right-hand side of the battery compartment.
26 Unscrew the nut from the left-hand track rod end, and disconnect it from the stub axle carrier steering arm using a balljoint separator tool.
27 Unscrew the bolts securing the left-hand driveshaft inner rubber boot and metal ring to the transmission
28 Unscrew the two mounting bolts securing the left-hand brake caliper to the stub axle carrier. Tie the caliper to the coil spring, taking care not to strain the hose.
29 Unscrew and remove the pinch-bolt securing the front left lower balljoint to the bottom of the stub axle carrier.
30 Unscrew both bolts securing the left-hand front suspension strut to the stub axle carrier (note that the nuts are on the rear of the carrier).
31 Lever the suspension lower arm downwards, then pull the driveshaft from the sun gear and withdraw it together with the stub axle carrier.
32 Make sure that the tripod components remain in position on the inner end of the driveshaft, otherwise they may fall into the transmission. There may be some loss of oil from the gearbox, so position a small container on the floor to catch it.
33 Working beneath the car, drive out the roll pin securing the right-hand driveshaft to the differential sun gear shaft. Note that if the original double roll pin is renewed, the new one will be of single coiled type.
34 Unscrew the nut from the right-hand track rod end, and disconnect it from the stub axle carrier steering arm using a balljoint separator tool.
35 Unscrew and remove the upper bolt securing the right-hand front suspension strut to the stub axle carrier (note that the nut is on the rear of the carrier). Loosen (do not remove) the lower bolt.
36 Pull out the driveshaft and disconnect the inner end from the splines on the transmission sun gear shaft. Tie the driveshaft to the steering gear.
37 Remove the exhaust front downpipe as described in Chapter 4.
38 Mark the position of the gearchange rod clamp, then loosen the clamp bolt and disconnect the rod at the transmission.
39 Where applicable, detach the power-assisted steering hydraulic fluid pipes from the transmission.
40 Pull out the spring clip and remove the speedometer sender unit from the rear of the transmission.
41 Disconnect the wiring from the reversing light switch on the transmission.
42 Disconnect the oil cooler pipes.
43 Connect a hoist to the engine lifting eyes, and take the weight of the engine/transmission assembly.
44 Loosen (do not remove) the lower rear bolt on the rear engine mounting, then unscrew and remove the lower front bolt. Note which way round the shock absorber is, then unbolt it from the top of the rear mounting. Also unbolt the rear mounting bracket from the transmission.
45 Unscrew and remove the bolts securing the right-hand engine mounting brackets to

the engine and body bracket. It is not necessary to separate the two sections.

46 Unbolt and transmission mounting brackets from the transmission and body. Again there is no need to separate the two sections of the mounting.

47 With the help of an assistant, slowly lift the engine/transmission assembly from the engine compartment, taking care not to damage any components on the surrounding panels. When high enough, lift the assembly over the front body panel, and lower to the ground.

48 To separate the transmission from the engine, first remove the starter motor as described in Chapter 5A.

49 Unbolt and remove the strengthening bracket between the engine cylinder block and transmission.

50 Support the transmission then unscrew and remove the bolts and nuts securing it to the engine.

51 With the help of an assistant, withdraw the transmission from the engine, making sure that it is not allowed to hang on the input shaft while engaged with the clutch components.

Refitting

52 Refitting is a reversal of removal, noting the following additional points.

a) Apply a little high melting-point grease to the splines of the transmission input shaft.

b) Make sure the clutch release bearing and fork are correctly positioned before refitting the transmission.

c) Tighten all nuts and bolts to the specified torque, however, before fully tightening the right-hand engine mounting-to-body bolts, make sure that the upper bracket is centralised.

d) Refer to the applicable Chapters as for removal.

e) Fill the engine and gearbox with oil and fit a new oil filter, with reference to Chapter 1.

f) Fill the cooling system with reference to Chapter 1.

g) Check and top-up the power steering fluid reservoir, and if necessary bleed the system as described in Chapter 10.

h) Prime and bleed the fuel system as described in Chapter 4B.

i) On turbo models, before starting the engine, ensure that the oil circuit has been primed as described in Chapter 1.

j) Depress the brake pedal several times to set the front brake pads in their normal position.

7 Engine overhaul - dismantling sequence

1 It is much easier to disassemble and work on the engine if it is mounted on a portable engine stand. These stands can often be hired from a tool hire shop. Before the engine is mounted on a stand, the flywheel/driveplate should be removed from the engine, so that the engine stand bolts can be tightened into the end of the cylinder block.

2 If a stand is not available, it is possible to disassemble the engine with it blocked up on a sturdy workbench or on the floor. Be extra-careful not to tip or drop the engine when working without a stand.

3 If you are going to obtain a reconditioned engine, all the external components must come off first, in order to be transferred to the replacement engine (just as they will if you are doing a complete engine overhaul yourself). Check with the engine supplier for details. Normally these components include:

Petrol engine models

a) Alternator and brackets.

b) Distributor (where applicable), HT leads and spark plugs.

c) Thermostat and cover.

d) Fuel injection equipment.

e) Inlet and exhaust manifolds.

f) Oil filter.

*g) Engine mountings and lifting brackets **(see illustration).***

*h) Ancillary (power steering pump, air conditioning compressor) brackets **(see illustration).***

*i) Oil filler tube and dipstick **(see illustrations).***

*j) Coolant pipes and hoses **(see illustration).***

k) Flywheel/driveplate.

Diesel engine models

a) Alternator mounting bracket.

b) Fuel injection pump and mounting bracket, and fuel injectors and glow plugs (see Chapter 4B).

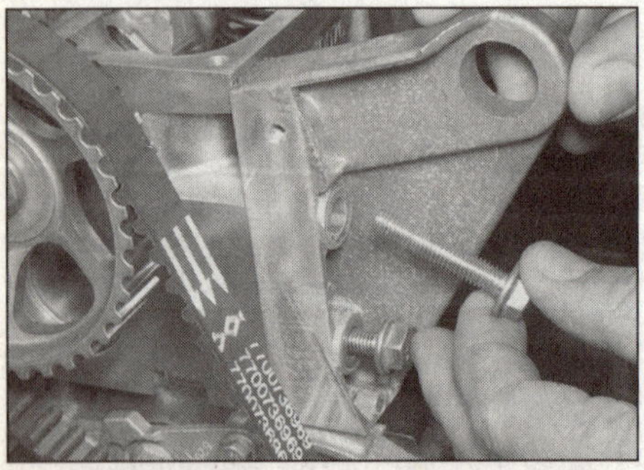

7.3a Removing the front engine lifting bracket (K7M engine)

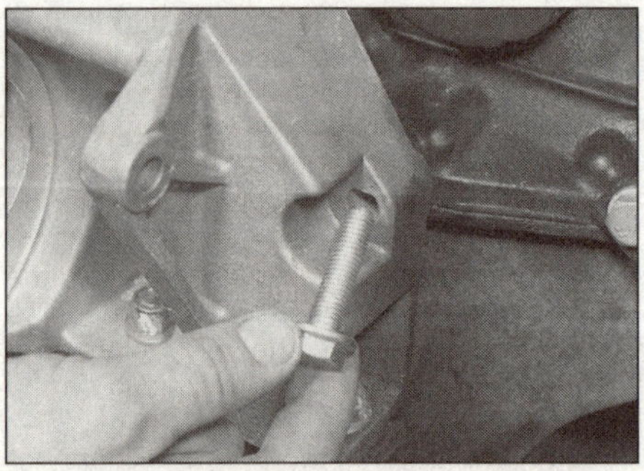

7.3b Removing the power-assisted steering pump bracket from the engine (K7M engine)

7.3c On the K7M engine, unscrew the bolt . . .

7.3d . . . lift the oil filler tube and dipstick from the front cover . . .

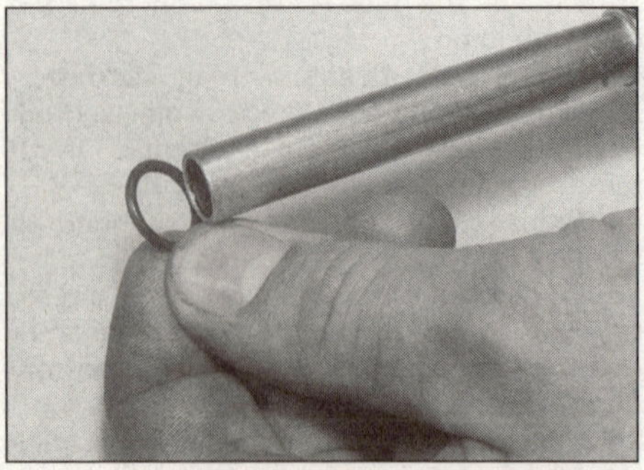

7.3e . . . and remove the sealing O-ring

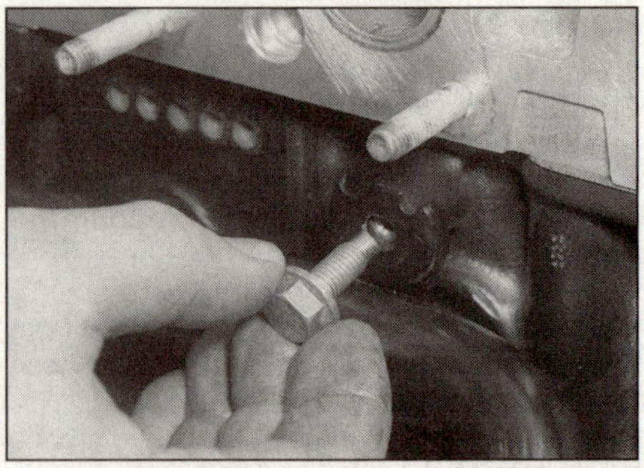

7.3f Unbolting the coolant pipe from the cylinder block

7.5a Unbolting the water pump from the cylinder block (K7M engine)

7.5b Removing the cylinder head (K7M engine)

c) Thermostat and cover.
d) Turbocharger (where applicable).
e) Inlet and exhaust manifolds.
f) Oil cooler.
g) Engine lifting brackets, hose brackets and wiring brackets.
h) Ancillary (power steering pump, air conditioning compressor) brackets.
i) Oil pressure warning light switch and oil level sensor (where applicable) (see Chapter 5A).
j) Coolant temperature sensors (see Chapter 3).
k) Wiring harnesses and brackets.
l) Coolant pipes and hoses.
m) Oil filler tube and dipstick.
n) Clutch (Chapter 6).

Note: *When removing the external components from the engine, pay close attention to details that may be helpful or important during refitting. Note the fitted position of gaskets, seals, spacers, pins, washers, bolts and other small items.*

4 If you are obtaining a 'short' motor (which, when available, consists of the engine cylinder block, crankshaft, pistons and connecting rods all assembled), then the cylinder head, sump, oil pump, and timing belt (where applicable) will have to be removed also.

5 If you are planning a complete overhaul, the engine can be disassembled and the internal components removed in the following order:

a) Inlet and exhaust manifolds.
*b) Timing belt and sprockets, and water pump **(see illustration)**.*
*c) Cylinder head **(see illustration)**.*
d) Flywheel/driveplate.
e) Sump.
f) Oil pump.
g) Pistons.
h) Crankshaft.

6 Before beginning the disassembly and overhaul procedures, make sure that you have all of the correct tools necessary. Refer to the introductory pages at the beginning of this manual for further information.

8 Cylinder head - dismantling

Note: *New and reconditioned cylinder heads are available from the manufacturers and from engine overhaul specialists. Due to the fact that some specialist tools are required for the dismantling and inspection procedures, and new components may not be readily available, it may be more practical and economical for the home mechanic to purchase a reconditioned head rather than dismantle, inspect and recondition the original head.*

1390 cc (E7J) and 1598 cc (K7M) engines

1 Remove the rocker shaft and camshaft (Chapter 2A).

1870 cc (F8Q) engine

2 Remove the brake vacuum pump (see Chapter 9), the fuel injectors (Chapter 4), the glow plugs (Chapter 5C), and the camshaft and tappets (Chapter 2B).

All engines

3 Using a valve spring compressor, compress each valve spring in turn until the split collets can be removed. Release the compressor and lift off the cap, spring and spring seat. If, when the valve spring compressor is screwed down, the valve spring cap refuses to free and expose the split collets, gently tap the top of the tool, directly over the cap, with a light hammer. This will free the cap **(see illustrations)**.

4 Withdraw the oil seal from the top of the valve guide, then remove the valve through the combustion chamber **(see illustrations)**.

5 It is essential that the valves and associated components are kept in their correct

2C

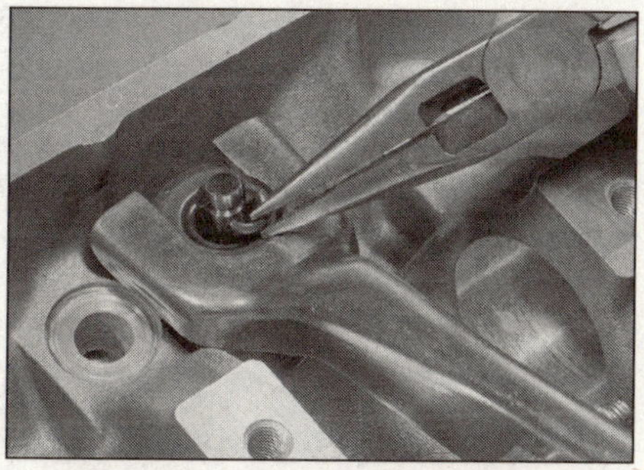

8.3a Removing the split collets . . .

8.3b . . . and lift off the cap and valve spring . . .

8.3c . . . followed by the spring seat

8.4a Removing the oil seal from the top of the valve guide

8.4b Withdrawing a valve from the combustion chamber

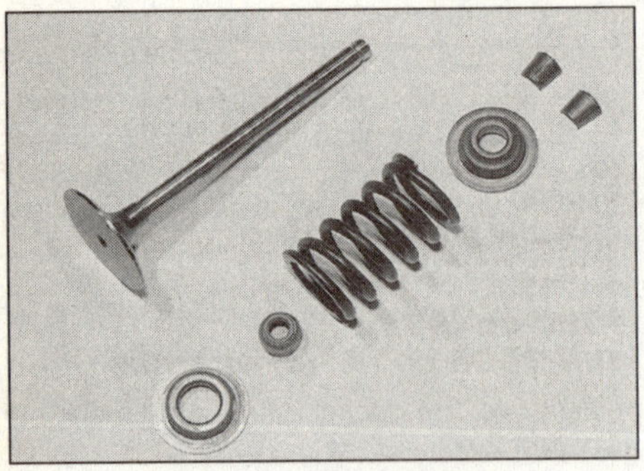
8.5a Valve components

8.5b Store the valve components in a labelled polythene bag

sequence, unless they are so badly worn that they are to be renewed. If they are going to be kept and used again, place them in labelled polythene bags, or in a compartmented box **(see illustrations)**.

9 Cylinder head and valves - cleaning, inspection and renovation

1 Thorough cleaning of the cylinder head and valve components, followed by a detailed inspection, will enable you to decide how much valve service work must be carried out during the engine overhaul.

Cleaning

2 Scrape away all traces of old gasket material and sealing compound from the cylinder head. Take care not to damage the cylinder head surfaces.

3 Scrape away the carbon from the combustion chambers and ports, then wash the cylinder head thoroughly with paraffin or a suitable solvent.

4 Scrape off any heavy carbon deposits that may have formed on the valves, then use a power-operated wire brush to remove deposits from the valve heads and stems.

5 On the F8Q diesel engine, the swirl chambers should be removed from their locations if they are loose (if this is done, mark the swirl chambers so that they can be refitted in their original locations) **(see illustration)**.

6 If the head is extremely dirty, it should be steam-cleaned. On completion, make sure that all oil holes and oil galleries are cleaned.

Inspection and renovation

Note: *Be sure to perform all the following inspection procedures before concluding that the services of an engine overhaul specialist are required. Make a list of all items that require attention.*

Cylinder head

7 Inspect the head very carefully for cracks, evidence of coolant leakage and other damage. If cracks are found, a new cylinder head should be obtained.

8 Use a straight-edge and feeler blade to check that the cylinder head surface is not distorted. On the F8Q diesel engine, do not position the straight-edge over the swirl chambers, as these may be proud of the cylinder head face. If the specified distortion limit is exceeded, machining of the gasket face is not recommended by the manufacturers, so the only course of action is to renew the cylinder head. On the F8Q engine, check that the overall height of the cylinder head is as specified, which will indicate if the head has been machined in a mistaken attempt to compensate for surface distortion **(see illustrations)**.

9 Examine the valve seats in each of the combustion chambers. If they are severely pitted, cracked or burned, then they will need to be renewed or recut by an engine overhaul specialist. If they are only slightly pitted, this can be removed by grinding the valve heads and seats together with coarse, then fine, grinding paste as described below. Note on the F8Q engine the valve seats can only be recut to a limited depth, to avoid decreasing the compression ratio. Using a dial test indicator, check that valve depth below the cylinder head gasket surface is within the limits given in the Specifications **(see illustration)**.

10 If the valve guides are worn, indicated by a side-to-side motion of the valve in the guide, new guides must be fitted. A dial gauge may be used to determine the amount of side play of the valve. Recheck the fit using a new valve if in doubt, to decide whether it is the valve or the guide which is worn. If new guides are to be fitted, the valves must be renewed in any case. Valve guides may be renewed using a press and a suitable mandrel, making sure that they are at the correct height. The work is best carried out by an engine overhaul specialist, since if it is not done skilfully, there is a risk of damaging the cylinder head.

11 On the F8Q diesel engine, inspect the swirl chambers for burning or cracks. If required, the chambers can be renewed by an engine overhaul specialist. Using a dial test indicator check that the swirl chamber protrusion is within the limits given in the Specifications. Zero the dial test indicator on the gasket surface of the cylinder head, then measure the protrusion of the swirl chamber **(see illustrations)**.

12 On the F8Q diesel engine, check the tappet bores in the cylinder head for wear. If excessive wear is evident, the cylinder head must be renewed.

13 Examine the camshaft bearing surfaces in the cylinder head (and bearing caps on the

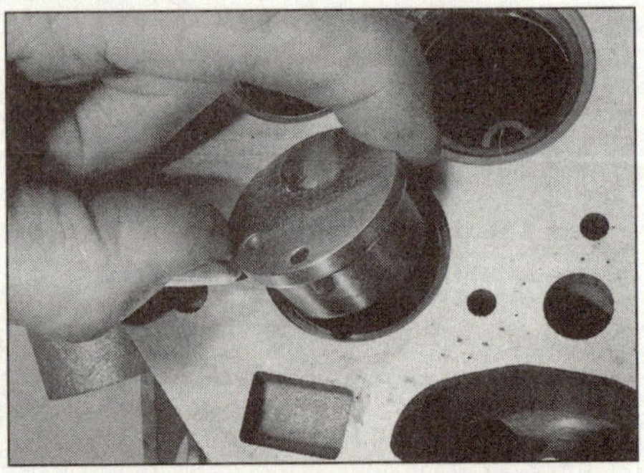
9.5 Removing a swirl chamber (F8Q engine)

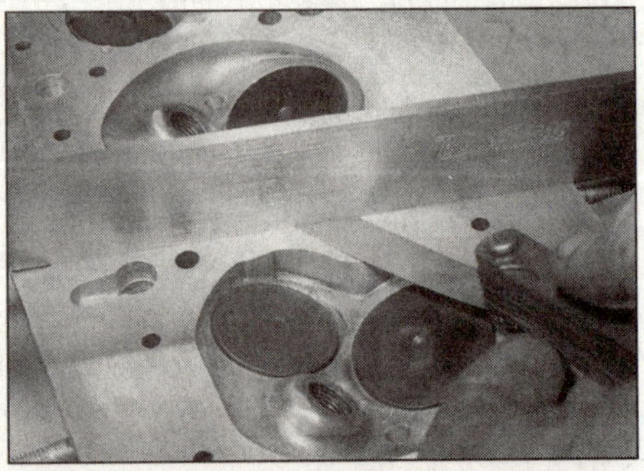
9.8a Checking the cylinder head surface for distortion with feeler blades

9.8b Checking the overall height of the cylinder head (F8Q engine)

9.9 Measuring the valve depth (F8Q engine)

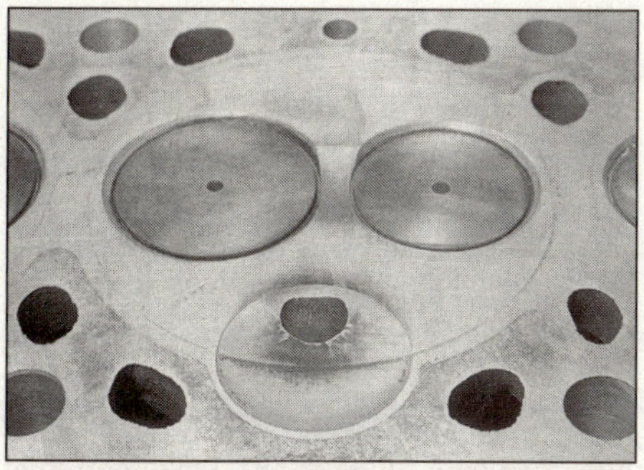

9.11a This swirl chamber shows the initial stages of cracking and burning (F8Q engine)

9.16 Grinding a valve to its seat - lift the valve to redistribute the paste

F8Q diesel engine). Wear here can only be corrected by renewing the head. Also examine the camshaft as described in Chapter 2A or 2B.

Valves

14 Examine the head of each valve for pitting, burning, cracks and general wear, and check the valve stem for scoring and wear ridges. Rotate the valve, and check for any obvious indication that it is bent. Look for pits and excessive wear on the end of each valve stem. If the valve appears satisfactory at this stage, measure the valve stem diameter at several points using a micrometer **(see illustration)**. Any significant difference in the readings obtained indicates wear of the valve stem. Should any of these conditions be apparent, the valve(s) must be renewed. If the valves are in satisfactory condition, or if new valves are being fitted, they should be ground (lapped) into their respective seats to ensure a smooth gas-tight seal.

15 Valve grinding is carried out as follows. Place the cylinder head upside-down on a bench, with a block of wood at each end to give clearance for the valve stems.

16 Smear a trace of coarse carborundum paste on the seat face, and press a suction grinding tool onto the valve head. With a semi-rotary action, grind the valve head to its seat, lifting the valve occasionally to redistribute the grinding paste **(see illustration)**. When a dull-matt even surface is produced on both the valve seat and the valve, wipe off the paste and repeat the process with fine carborundum paste. A light spring placed under the valve head will greatly ease this operation. When a smooth unbroken ring of light grey matt finish is produced on both the valve and seat, the grinding operation is complete. Be sure to remove all traces of grinding paste, using paraffin or a suitable solvent, before reassembly of the cylinder head.

9.11b Measuring the swirl chamber protrusion using a dial gauge (F8Q engine)

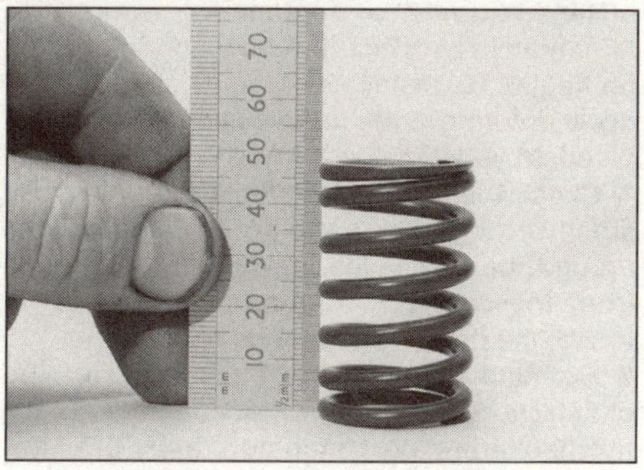

9.17 Checking a valve spring free length

Valve components

17 Examine the valve springs for signs of damage and discoloration, and also measure their free length using vernier calipers or a steel rule **(see illustration)** or by comparing the existing spring with a new component.

18 Stand each spring on a flat surface, and check it for squareness. If any of the springs are damaged, distorted or have lost their tension, obtain a complete new set of springs. It is normal to renew the springs as a matter of course during a major overhaul.

10.2a Fitting the valve stem oil seals to the valve guides

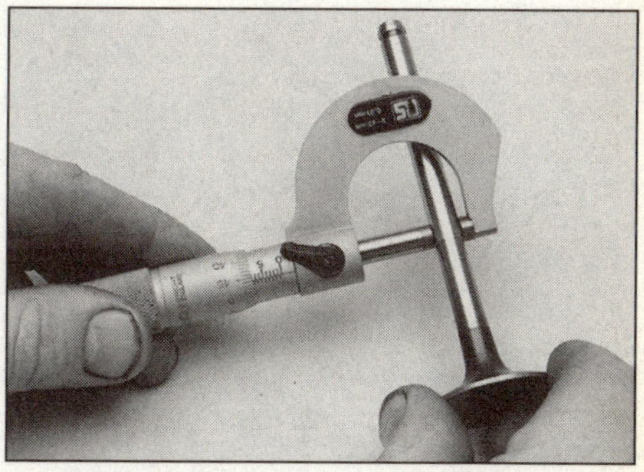

9.14 Measuring a valve stem using a micrometer

19 On the F8Q diesel engine, check the tappet buckets and the shims for scoring, pitting (especially on the shims), and wear ridges. Renew any components as necessary. Some scuffing is to be expected, and is acceptable provided that the tappets are not scored.

Rocker arm components - E7J and K7M engines

20 Check the rocker arm contact surfaces for pits, wear, score marks or any indication that the surface-hardening has worn through. Dismantle the rocker shaft, and check the rocker arm and rocker shaft pivot and contact areas in the same way. Measure the internal diameter of each rocker, and check their fit on the shaft. Clean out the oil spill holes in each rocker using a length of wire. Renew the rocker arm or the rocker shaft itself if any are suspect.

Valve stem oil seals

21 The valve stem oil seals should be renewed as a matter of course.

10 Cylinder head - reassembly

1 On the F8Q diesel engine, if the swirl chambers have been removed, refit them to their original locations.

2 Lubricate the valve stem oil seals with clean engine oil, then fit them by pushing into position in the cylinder head using a suitable socket or special tool **(see illustrations)**. Ensure that the seals are fully engaged with the valve guide.

10.2b Using a special tool to fit the valve stem oil seals

10.3 Lubricate the valve stems before inserting the valves

3 Insert the valves into their original locations. If new valves are being fitted, insert them into the locations to which they have been ground. Take care not to damage the valve stem oil seal as each valve is fitted **(see illustration)**.

4 Locate the spring seat on the guide, followed by the spring and cap.

5 Compress the valve spring, and locate the split collets in the recess in the valve stem. Release the compressor, then repeat the procedure on the remaining valves. Use a little grease to hold the collets in place **(see illustration)**.

6 With all the valves installed, place the cylinder head flat on the bench and, using a hammer and interposed block of wood, tap the end of each valve stem to settle the components.

7 The previously removed components can now be refitted with reference to paragraphs 1 or 2 (as applicable), Section 8.

11 Flywheel/driveplate - removal, inspection and refitting

Removal

1 Remove the manual or automatic transmission as described in Section 4, 5 or 6 of this Chapter.

2 On manual transmission models, remove the clutch as described in Chapter 6.

3 Mark the flywheel/driveplate in relation to the crankshaft to aid refitting. Note that the flywheel/driveplate can only be refitted in one position, as the bolts are unequally spaced.

4 The flywheel/driveplate must now be held stationary while the bolts are loosened. To do this, locate a long bolt in one of the transmission-to-engine mounting bolt holes, and either insert a wide-bladed screwdriver in the starter ring gear, or use a piece of bent metal bar engaged with the ring gear.

5 Unscrew the mounting bolts, and withdraw the flywheel/driveplate from the crankshaft.

Inspection

6 Examine the flywheel/driveplate for wear or chipping of the ring gear teeth. If the ring gear is worn or damaged, it may be possible to renew it separately, but this job is best left to a Renault dealer or engineering works. The temperature to which the new ring gear must be heated for installation is critical and, if not done accurately, the hardness of the teeth will be destroyed.

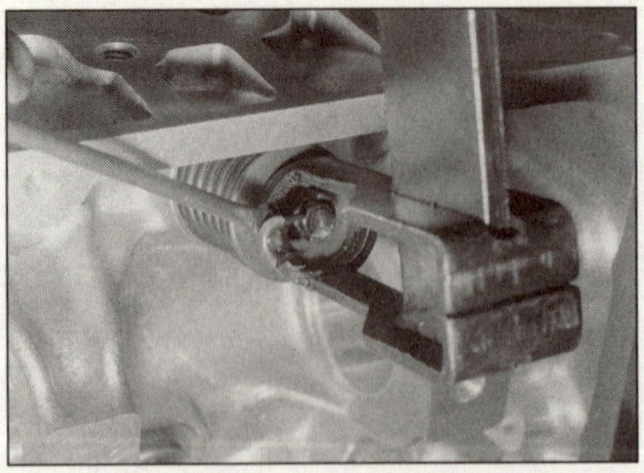

10.5 Use a little grease to hold the collets in place

7 Check the flywheel/driveplate carefully for signs of distortion, and for hairline cracks around the bolt holes, or radiating outwards from the centre. If damage of this sort is found, the driveplate must be renewed.

8 Examine the flywheel for scoring of the clutch face. If the clutch face is scored, the flywheel may be machined until flat, but renewal is preferable.

Refitting

9 Clean the flywheel/driveplate and crankshaft faces, then locate the unit on the crankshaft, making sure that any previously-made marks are aligned.

10 Apply a few drops of locking fluid to the mounting bolt threads, fit the bolts and tighten them in a diagonal sequence to the specified torque.

11 Refit the transmission with reference to Section 4, 5 or 6 of this Chapter.

12 Left-hand (flywheel/driveplate end) crankshaft oil seal - renewal

1 Remove the flywheel/driveplate as described in Section 11.

2 Prise out the old oil seal using a small screwdriver, taking care not to damage the surface on the crankshaft. Alternatively, the oil seal can be removed by drilling two small holes diagonally opposite each other and inserting self-tapping screws in them. A pair of grips can then be used to pull out the oil seal, by pulling on each side in turn.

12.4 Fitting a new crankshaft rear oil seal

3 Inspect the seal rubbing surface on the crankshaft. If it is grooved or rough in the area where the old seal was fitted, the new seal should be fitted slightly less deeply, so that it rubs on an unworn part of the surface.

4 Wipe clean the oil seal seating, then dip the new seal in fresh engine oil, and locate it over the crankshaft with its closed side facing outwards **(see illustration)**. Make sure that the oil seal lip is not damaged as it is located on the crankshaft.

5 Using a metal tube, drive the oil seal squarely into the bore until flush. A block of wood cut to pass over the end of the crankshaft may be used instead.

6 Refit the flywheel/driveplate with reference to Section 11.

13 Sump (K7M petrol engine with aluminium sump only) - removal and refitting

Removal

1 Remove the engine as described in Section 4 of this Chapter, then drain the engine oil as described in Chapter 1.

2 Unbolt and remove the engine-to-transmission strengthening bracket/flywheel cover.

3 Unscrew the bolts securing the sump to the cylinder block, but leave two diagonally-opposite bolts entered on a few threads until the sump joint has been released.

4 Using a palette knife or similar tool, release the sump from the bottom of the cylinder block. Do not use a screwdriver, as this may damage the mating surfaces.

5 Unscrew the two remaining bolts, and lower the sump from the engine.

6 Remove the rubber sealing strips from the grooves in the front and rear main bearing caps, or from the curved sections at each end of the sump. Clean all traces of sealant from the cylinder block and sump, and wipe them dry.

Refitting

7 Apply a 3 mm wide bead of suitable sealant to the mating face of the sump, making sure that the bead is around the inner edge of each bolt hole. Renault recommend Rhodorseal 5661.

8 Locate new rubber sealing strips in the grooves on the front and rear main bearing caps. Note that the strips are different. If necessary, use a little sealant to hold the strips in position until the sump is refitted.

9 Offer the sump to the cylinder block, and insert the bolts. Progressively tighten the bolts in diagonal sequence to the specified torque.

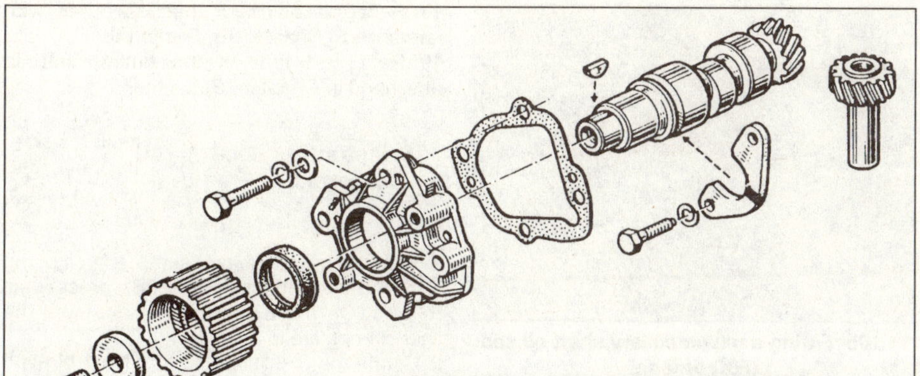

14.5a Auxiliary shaft components (F8Q engine)

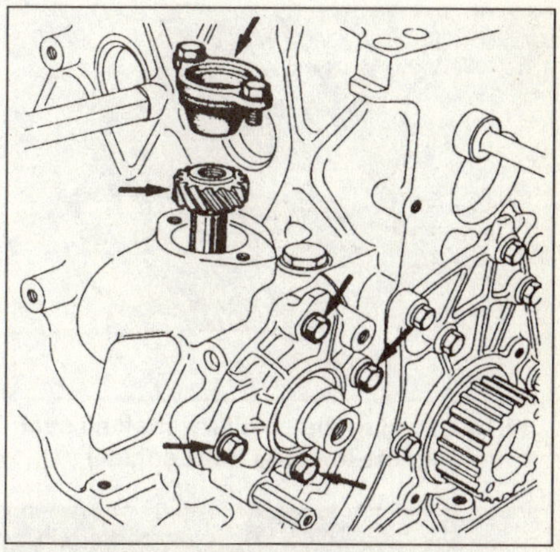

14.5b Auxiliary shaft housing retaining bolt locations, oil pump driveshaft and cover plate (F8Q engine)

14 Auxiliary shaft (F8Q diesel engine) - removal and refitting

Note: *A new auxiliary shaft oil seal, housing gasket (or suitable sealant, as applicable), and oil pump drivegear cover plate O-ring will be required on refitting.*

Removal

1 With the engine removed from the vehicle, proceed as follows.
2 Remove the timing belt (see Chapter 2B).
3 Remove the auxiliary shaft sprocket with reference to Chapter 2B, Section 6.
4 Unbolt the timing belt rear lower cover from the cylinder block, and remove it.
5 Unscrew the four bolts and withdraw the auxiliary shaft housing **(see illustrations)**, then remove the gasket (if fitted). Note that the housing locates on two dowels.
6 Unscrew the two bolts and withdraw the oil pump drivegear cover plate/breather noting the location of the O-ring. Withdraw the drivegear from its location. Use a screwdriver or similar tool to hook the drivegear out if necessary **(see illustrations)**.
7 Unscrew the two bolts and washers, and lift out the auxiliary shaft thrustplate and the auxiliary shaft **(see illustrations)**.

Inspection

8 Examine the auxiliary shaft and oil pump driveshaft for pitting, scoring or wear ridges on the bearing journals, and for chipping or wear of the gear teeth. Renew as necessary. Check the auxiliary shaft bearings in the cylinder block for wear and, if worn, have these renewed by your Renault dealer or suitably-equipped engineering works. Wipe them clean if they are still serviceable.
9 Temporarily fit the thrustplate to its position on the auxiliary shaft, and use a feeler blade to check that the endfloat is as given in the Specifications **(see illustration)**. If it is greater than the upper tolerance, a new thrustplate should be obtained, but first check the thrust surfaces on the shaft to ascertain if wear has occurred here.

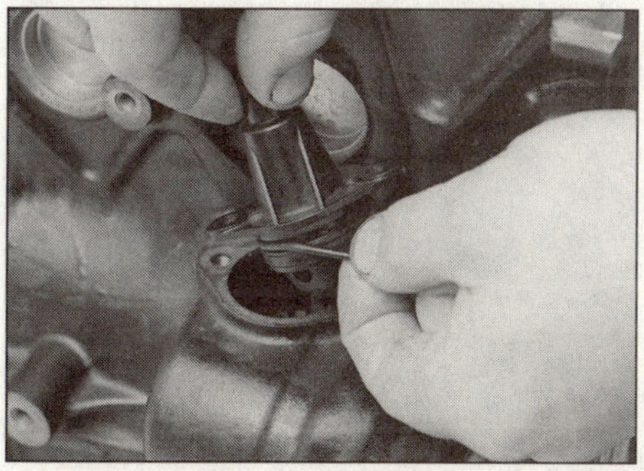

14.6a Removing oil pump drivegear cover plate/breather and O-ring (F8Q engine)

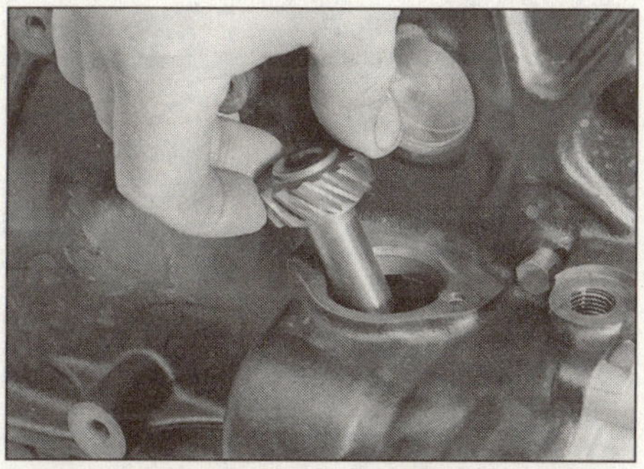

14.6b Removing the oil pump drivegear (F8Q engine)

14.7a Removing the auxiliary shaft thrustplate . . .

14.7b . . . and the auxiliary shaft (F8Q engine)

14.9 Measuring the auxiliary shaft endfloat with a feeler blade (F8Q engine)

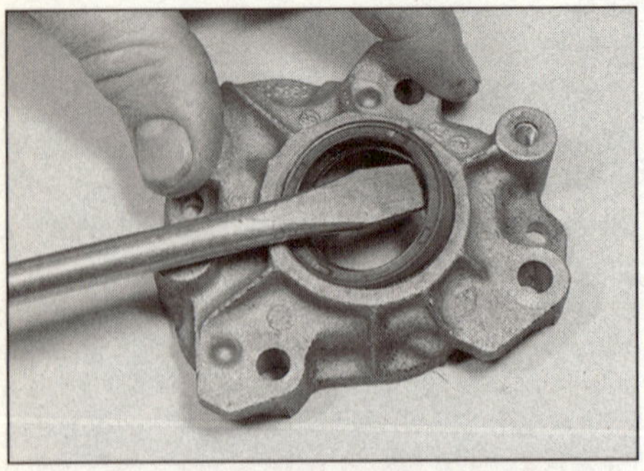

14.10a Prising the auxiliary shaft oil seal from the housing (F8Q engine)

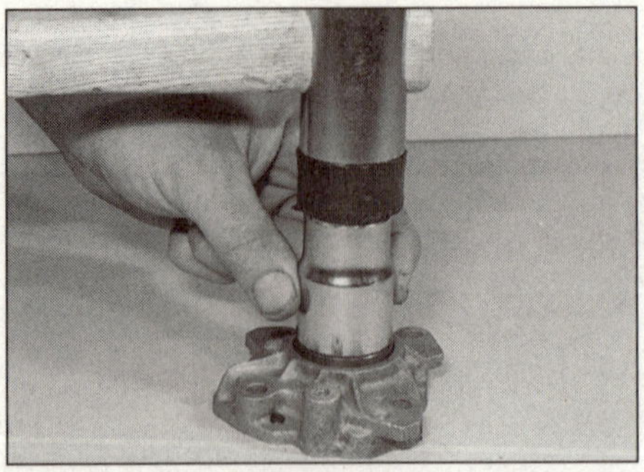

14.10b Fitting a new auxiliary shaft oil seal (F8Q engine)

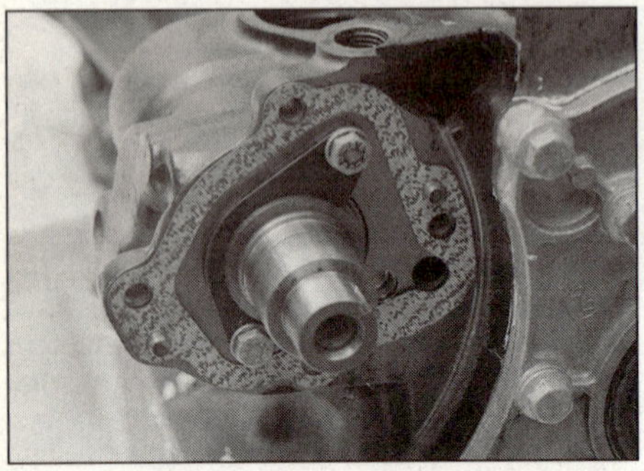

14.13 Auxiliary shaft gasket positioned over the dowels (F8Q engine)

14.14 Refitting the auxiliary shaft housing. Note tape around front of shaft (F8Q engine)

Refitting

10 Clean off all traces of the old gasket or sealant from the auxiliary shaft housing, and prise out the oil seal with a screwdriver. Install the new oil seal using a block of wood, or a suitable socket, and tap it in until it is flush with the outer face of the housing. The open side of the seal must be towards the engine **(see illustrations)**.

11 Liberally lubricate the auxiliary shaft, and slide it into its bearings.

12 Place the thrustplate in position with its curved edge away from the crankshaft, and refit the two retaining bolts, tightening them securely.

13 Place a new housing gasket in position over the dowels of the cylinder block **(see illustration)**. If a gasket was not used previously, apply a bead of CAF 4/60 THIXO sealant (or an alternative) to the housing mating face.

14 Wind a length of tape around the end of the auxiliary shaft, to prevent damage to the oil seal as the housing is refitted. Liberally lubricate the oil seal lips, and then locate the housing in place, engaging it with the dowels **(see illustration)**. Refit and tighten the housing retaining bolts progressively in a diagonal sequence. Remove the tape from the end of the shaft.

15 Lubricate the oil pump drivegear, and lower the gear into its location. Ensure that the splines on the drivegear engage with the oil pump.

16 Inspect the O-ring seal on the oil pump drivegear cover plate/breather, and renew it if necessary. Fit the cover plate/breather and secure with the two retaining bolts.

17 Refit the rear lower timing belt cover to the cylinder block, and tighten the bolts.

18 Refit the auxiliary shaft sprocket with reference to Chapter 2B, Section 6.

19 Refit and tension the timing belt as described in Chapter 2B.

15 Piston/connecting rod assemblies - removal

1 With the cylinder head, sump and oil pump removed (see Chapter 2A or 2B), proceed as follows. On the E7J engine, make sure the liner clamps are in position.

2 Rotate the crankshaft so that No 1 big-end cap (nearest the flywheel/driveplate position) is at the lowest point of its travel. If the big-end cap and rod are not already numbered, mark them with a centre-punch **(see illustration)**. Mark both cap and rod to identify the cylinder they operate in.

3 Before removing the big-end caps, check the amount of side play between the caps and the crankshaft webs **(see illustration)**.

4 Unscrew and remove the big-end bearing cap nuts (E7J and K7M petrol engines) or bolts (F8Q diesel engine). Withdraw the cap, complete with shell bearing, from the connecting rod. Strike the cap with a wooden or copper mallet if it is stuck **(see illustration)**.

5 If only the bearing shells are being attended to, push the connecting rod up and off the crankpin, and remove the upper bearing shell **(see illustration)**. Keep the bearing shells and cap together in their correct sequence if they are to be refitted.

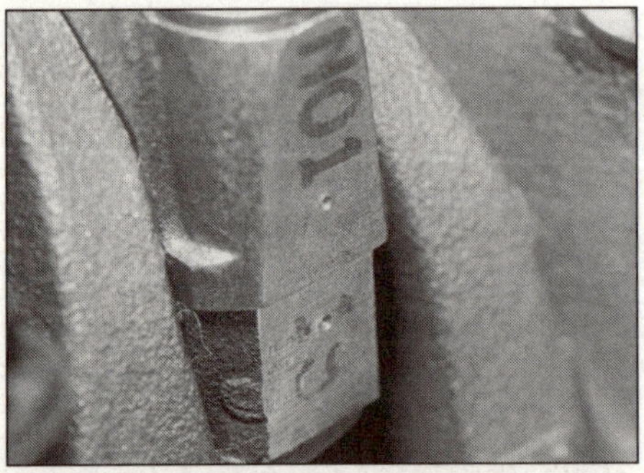

15.2 Big-end caps marked with a centre-punch

15.3 Checking the side play between a big-end cap and the crankshaft web with a feeler blade

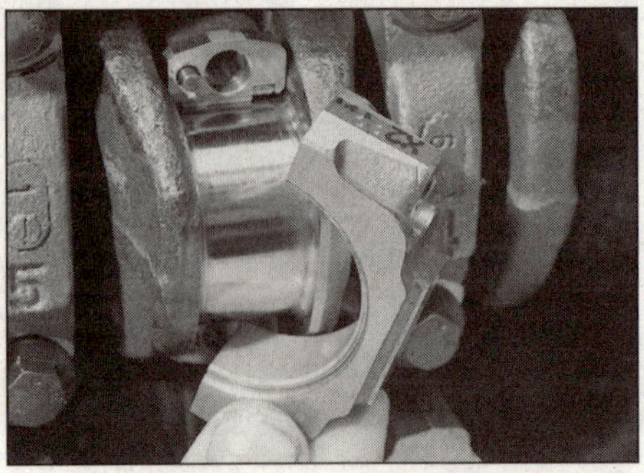

15.4 Removing a big-end bearing cap

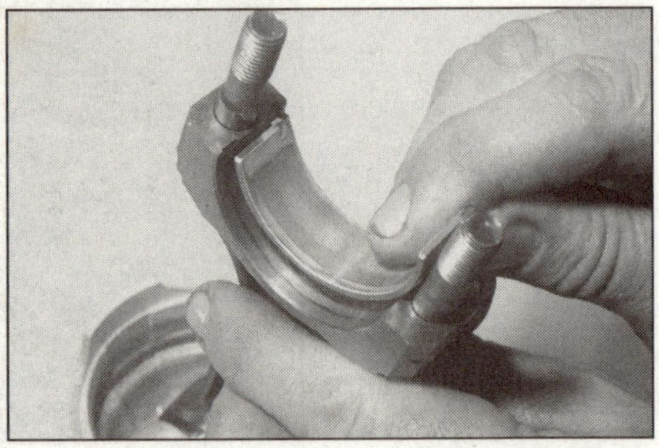

15.5 Removing a big-end bearing upper shell

E7J petrol engine

6 Remove the liner clamps and withdraw each liner, together with piston and connecting rod, from the top of the cylinder block. Mark the liners using masking tape, so that they may be refitted in their original locations.
7 Withdraw the piston from the bottom of the liner. Keep each piston with its respective liner if they are to be re-used.

K7M petrol and F8Q diesel engines

8 Push the connecting rod up, and remove the piston and rod from the bore. Note that if there is a pronounced wear ridge at the top of the bore, there is a risk of damaging the piston as the rings foul the ridge. However, it is reasonable to assume that a rebore and new pistons will be required in any case if the ridge is so pronounced.

All engines

9 Repeat the procedure for the remaining piston/connecting rod assemblies. Ensure that the caps and rods are marked before removal, as described previously, and keep all components in order.

16 Crankshaft - removal

1 Remove the timing belt, crankshaft sprocket, oil pump, and flywheel/driveplate. The pistons/connecting rods must be free of the crankshaft journals, however it is not essential to remove them completely from the cylinder block.
2 Unscrew the nuts/bolts and remove the lower rear timing cover, then unscrew the securing bolts, and remove the crankshaft front oil seal housing, along with the oil seal. Note that the housing is fitted using sealant **(see illustration)**.
3 Before the crankshaft is removed, check the endfloat using a dial gauge in contact with the end of the crankshaft **(see illustration)**. Push the crankshaft fully one way, and then zero the gauge. Push the crankshaft fully the other way, and check the endfloat. The result can be compared with the specified amount, and will give an indication as to whether new thrustwashers are required.
4 If a dial gauge is not available, feeler blades can be used. First push the crankshaft fully towards the flywheel end of the engine, then slip the feeler blade between the web of No 2 crankpin and the thrustwasher of the centre main bearing (E7J and K7M engines), or between the web of No 1 crankpin and the thrustwasher of No 2 main bearing (F8Q engine).
5 Identification numbers should already be cast onto the base of each main bearing cap. If not, number the cap and crankcase using a centre-punch, as was done for the connecting rods and caps **(see illustration)**.
6 Unscrew and remove the main bearing cap retaining bolts, and withdraw the caps, complete with bearing shells **(see illustration)**. Tap the caps with a wooden or copper mallet if they are stuck. Note on the F8Q diesel engine No 1 main bearing cap (flywheel end) is secured by Allen bolts and sealed to the cylinder block using silicon sealant or butyl seals.
7 Carefully lift the crankshaft from the crankcase **(see illustration)**.
8 Remove the thrustwashers at each side of the centre main bearing (E7J and K7M engines) or No 2 main bearing (F8Q engine), then remove the bearing shell upper halves from the crankcase **(see illustration)**. Place each shell with its respective bearing cap.
9 Remove the oil seal from the rear of the crankshaft.

17 Cylinder block/crankcase and bores - cleaning and inspection

Cleaning

1 For complete cleaning, the core plugs should be removed. Drill a small hole in them, then insert a self-tapping screw and pull out the plugs using a pair of grips or a slide-hammer. Also remove all external components and senders (if not already done), noting their locations. As applicable unbolt the drivebelt tensioner bracket, coolant pipe and oil cooler from the cylinder block. On the F8Q turbo engine, remove the oil sprayers from the bottom of each bore by unscrewing the retaining bolts **(see illustration)**.

2C

16.2 Unscrewing the nuts/bolts securing the front oil seal housing to the cylinder block (K7M engine)

16.3 Checking the crankshaft endfloat with a dial gauge

16.5 The main bearing caps are numbered for position

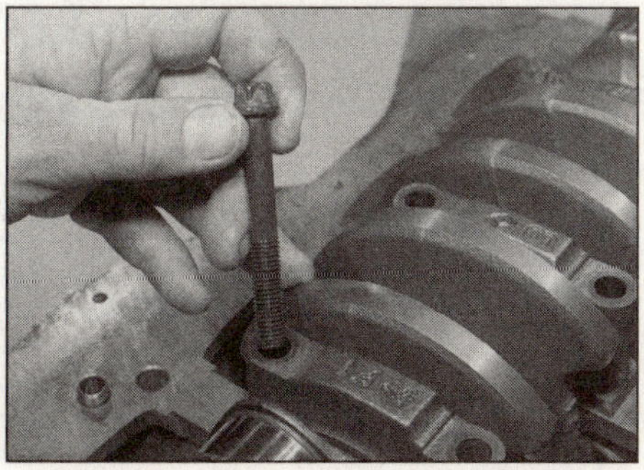

16.6 Removing a main bearing cap bolt

16.7 Lifting the crankshaft from the crankcase

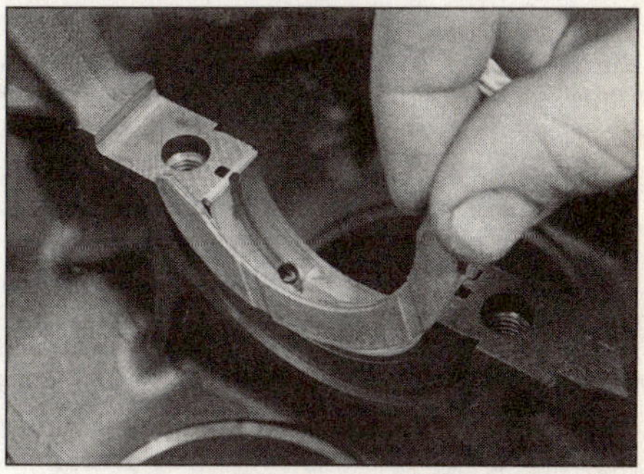

16.8 Removing the crankshaft thrustwashers

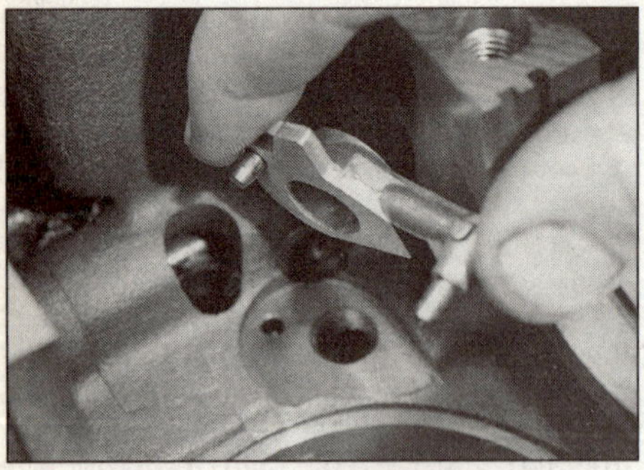

17.1 Removing a piston oil spray jet - F8Q turbo engine

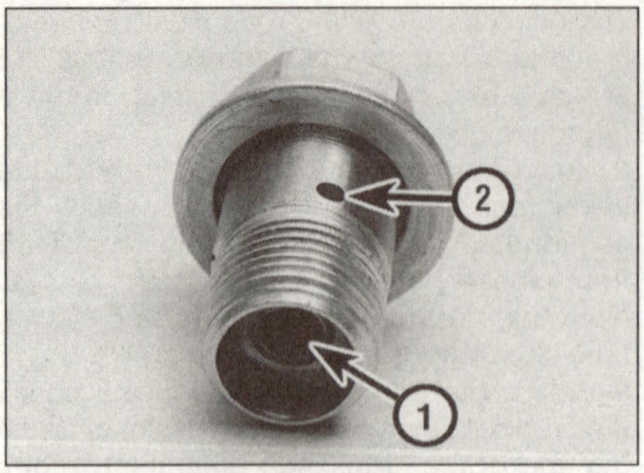

17.9a Check the gauze filters (1) and the oil holes (2) in the oil spray jet bolts . . .

17.9b . . . and the oil holes (arrowed) in the jets for blockage (F8Q engine)

2 Scrape all traces of gasket or sealant from the cylinder block, taking care not to damage the head and sump mating faces.

3 If the block is extremely dirty, it should be steam-cleaned.

4 After the block has been steam-cleaned, clean all oil holes and oil galleries one more time. Flush all internal passages with warm water until the water runs clear, dry the block thoroughly and wipe all machined surfaces with a light rust-preventative oil. If you have access to compressed air, use it to speed up the drying process and to blow out all the oil holes and galleries.

Warning: Wear eye protection when using compressed air!

5 If the block is not very dirty, you can do an adequate cleaning job with hot soapy water and a stiff brush. Take plenty of time, and do a thorough job. Regardless of the cleaning method used, be sure to clean all oil holes and galleries very thoroughly, dry the block completely and coat all machined surfaces with light oil.

6 The threaded holes in the block must be clean to ensure accurate torque wrench readings during reassembly. Run the proper-size tap into each of the holes to remove rust, corrosion, thread sealant or sludge, and to restore damaged threads. If possible, use compressed air to clear the holes of debris produced by this operation. Now is a good time to clean the threads on the head bolts and the main bearing cap bolts as well.

7 Refit the main bearing caps, and tighten the bolts finger-tight.

8 After coating the mating surfaces of the new core plugs with suitable sealant, refit them in the cylinder block. Make sure that they are driven in straight and seated properly, or leakage could result. Special tools are available for this purpose, but a large socket, with an outside diameter that will just slip into the core plug, will work just as well.

9 On F8Q turbo diesel models, check the gauze filters and the oil holes in the piston oil spray jet securing bolts, and the oil holes in the jets themselves for blockage **(see illustrations)**. Clean if necessary, then refit the jets and tighten the securing bolts. Ensure the locating pegs on the jets engage with the corresponding holes in the cylinder block.

10 If the engine is not going to be reassembled right away, cover it with a large plastic bag to keep it clean and prevent it rusting.

Inspection

11 Visually check the block for cracks, rust and corrosion. Look for stripped threads in the threaded holes. If there has been any history of internal water leakage, it may be worthwhile having an engine overhaul specialist check the block with special equipment. If defects are found, have the block repaired, if possible, or renewed.

12 Check the cylinder bores/liners for scuffing and scoring. Normally, bore wear will show up in the form of a wear ridge at the top of the bore. This ridge marks the limit of piston travel.

13 Measure the diameter of each cylinder at the top (just under the ridge area), centre and bottom of the cylinder bore, parallel to the crankshaft axis.

14 Next measure each cylinder's diameter at the same three locations across the crankshaft axis. If the difference between any of the measurements is greater than 0.20 mm, indicating that the cylinder is excessively out-of-round or tapered, then remedial action must be considered.

15 Repeat this procedure for the remaining cylinders.

16 If the cylinder walls are badly scuffed or scored, or if they are excessively out-of-round or tapered, have the cylinder block rebored (K7M petrol and F8Q diesel engines) or obtain new cylinder liners (E7J engine). New pistons (oversize in the case of a rebore) will also be required.

17.19a Checking the liner protrusion with a dial gauge . . .

17 If the cylinders are in reasonably good condition, then it may only be necessary to renew the piston rings.

18 If this is the case, the bores should be honed in order to allow the new rings to bed in correctly and provide the best possible seal. The conventional type of hone has spring-loaded stones, and is used with a power drill. You will also need some paraffin or honing oil and rags. The hone should be moved up and down the cylinder to produce a crosshatch pattern, and plenty of honing oil should be used. Ideally, the crosshatch lines should intersect at approximately a 60° angle. Do not take off more material than is necessary to produce the required finish. If new pistons are being fitted, the piston manufacturers may specify a finish with a different angle, so their instructions should be followed. Do not withdraw the hone from the cylinder while it is still being turned, but stop it first. After honing a cylinder, wipe out all traces of the honing oil. If equipment of this type is not available, or if you are not sure whether you are competent to undertake the task yourself, an engine overhaul specialist will carry out the work at a moderate cost.

19 Before refitting the cylinder liners to the E7J engine, their protrusions must be checked as follows, and new base O-rings fitted. Place the liner without a base O-ring in the cylinder block, and press down to make sure that it is seated correctly. Using a dial gauge or straight-edge and feeler blade, check that the protrusion of the liner above the upper surface of the cylinder block is within the specified limits **(see illustrations)**.

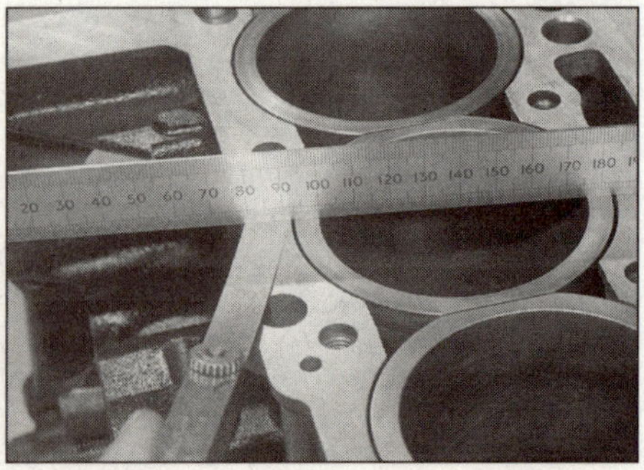

17.19b . . . and with feeler blades (E7J engine)

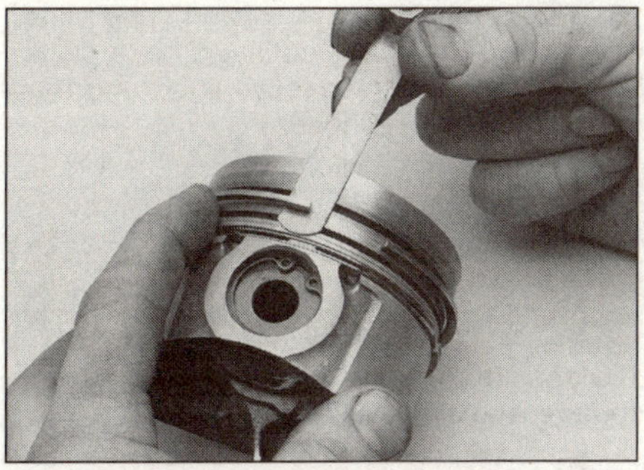

18.2 Removing a piston ring with the aid of a feeler blade

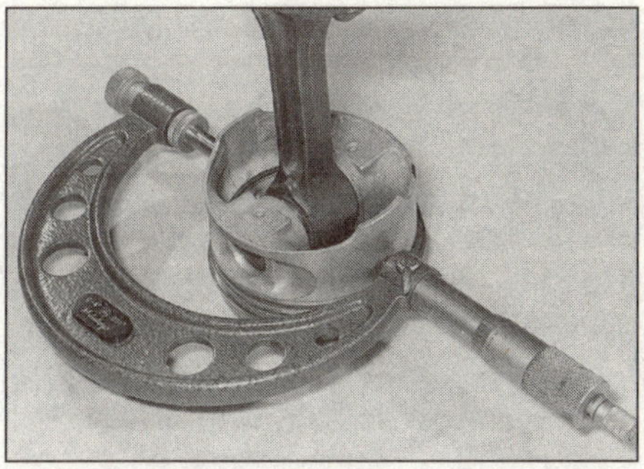

18.11a Measuring a piston diameter using a micrometer

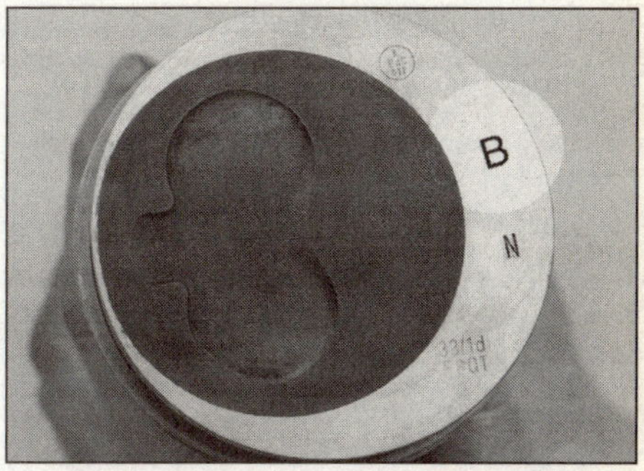

18.11b Piston diameter grade marked on piston crown (F8Q engine)

Check all of the liners in the same manner, and record the protrusions. Note that there is also a limit specified for the difference of protrusion between two adjacent liners. If new liners are being fitted, it is permitted to interchange them to bring this difference within limits. The protrusions may be stepped upwards or downwards from the flywheel end of the engine.

20 Refit all external components and senders in their correct locations, as noted before removal.

18 Piston/connecting rod assemblies - inspection and reassembly

Inspection

1 Before the inspection process can begin, the piston/connecting rod assemblies must be cleaned, and the original piston rings removed from the pistons.

2 Carefully expand the old rings over the top of the pistons. The use of two or three old feeler blades will be helpful in preventing the rings dropping into empty grooves **(see illustration)**. Note that the oil control ring is in two sections.

3 Scrape away all traces of carbon from the top of the piston. A hand-held wire brush or a piece of fine emery cloth can be used once the majority of the deposits have been scraped away.

4 Remove the carbon from the ring grooves in the piston by cleaning them using an old ring. Break the ring in half to do this. Be very careful to remove only the carbon deposits; do not remove any metal, nor nick or scratch the sides of the ring grooves. Protect your fingers - piston rings are sharp.

5 Once the deposits have been removed, clean the piston/connecting rod assembly with paraffin or a suitable solvent, and dry thoroughly. Make sure the oil return holes in the ring grooves are clear.

6 If the pistons and cylinder bores are not damaged or worn excessively, and if the cylinder block does not need to be rebored or the liners renewed, the original pistons can be re-used. Normal piston wear appears as even vertical wear on the piston thrust surfaces, and slight looseness of the top ring in its groove. New piston rings, however, should always be used when the engine is reassembled.

7 Carefully inspect each piston for cracks around the skirt, at the gudgeon pin bosses, and at the piston ring lands (between the piston ring grooves).

8 Look for scoring and scuffing on the sides of the skirt, holes in the piston crown, and burned areas at the edge of the crown. If the skirt is scored or scuffed, the engine may have been suffering from overheating and/or abnormal combustion, which caused excessively-high operating temperatures. The cooling and lubricating systems should be checked thoroughly. Scorch marks on the sides of the pistons show that blow-by has occurred and the rings are not sealing correctly. A hole in the piston crown is an indication that abnormal combustion (pre-ignition, knocking or detonation) has been occurring. If any of the above problems exist, the causes must be corrected, or the damage will occur again. On petrol engines, the causes may include inlet air leaks, incorrect fuel/air mixture or incorrect ignition timing on petrol engines. On F8Q diesel engines incorrect injection pump timing or a faulty injector may be the cause.

9 Corrosion of the piston, in the form of small pits, indicates that coolant is leaking into the combustion chamber and/or the crankcase. Again, the cause must be corrected, or the problem may persist in the rebuilt engine.

10 If new rings are being fitted to old pistons, measure the piston ring-to-groove clearance by placing a new piston ring in each ring groove and measuring the clearance with a feeler blade. Check the clearance at three or four places around each groove. No values are specified, but if the measured clearance is excessive - say greater than 0.10 mm - new pistons will be required. If the new ring is excessively tight, the most likely cause is dirt remaining in the groove.

11 Check the piston-to-bore/liner clearance by measuring the cylinder bore/liner (see Section 16) and the piston diameter. Measure the piston across the skirt, at a 90° angle to the gudgeon pin, approximately half way down the skirt **(see illustrations)**. Subtract the piston diameter from the bore/liner diameter to obtain the clearance. On K7M and F8Q engines, if this is greater than the figures given in the Specifications, the block will have to be rebored and new pistons and rings fitted. On E7J engines, new pistons and liners are supplied in matched pairs.

12 Check the fit of the gudgeon pin by twisting the piston and connecting rod in opposite directions. Any noticeable play indicates excessive wear, which must be corrected. If the pistons or connecting rods are to be renewed, the work should be carried out by a Renault garage or engine overhaul specialist. Note in the case of the F8Q diesel engine, the gudgeon pins are secured by circlips, so the pistons and connecting rods can be separated without difficulty. Note the position of the piston relative to the rod before dismantling, and use new circlips on reassembly **(see illustrations)**.

18.12a Gudgeon pin retaining circlip - arrowed (F8Q engine)

18.12b Oil hole (1) in connecting rod small-end should face away from combustion chamber (2) in piston crown (F8Q engine)

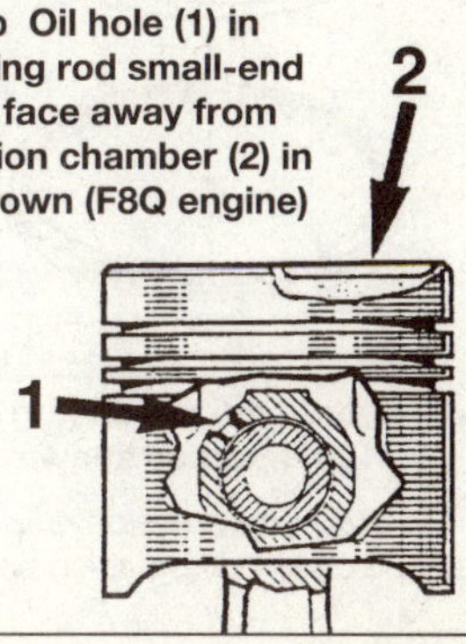

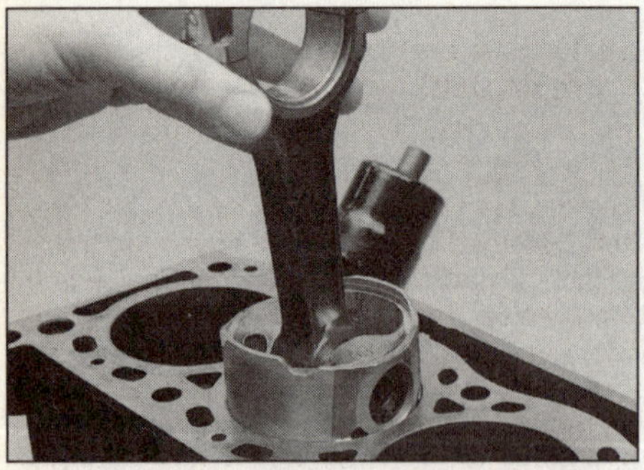

18.13a Use the piston to push the rings into the cylinder bores . . .

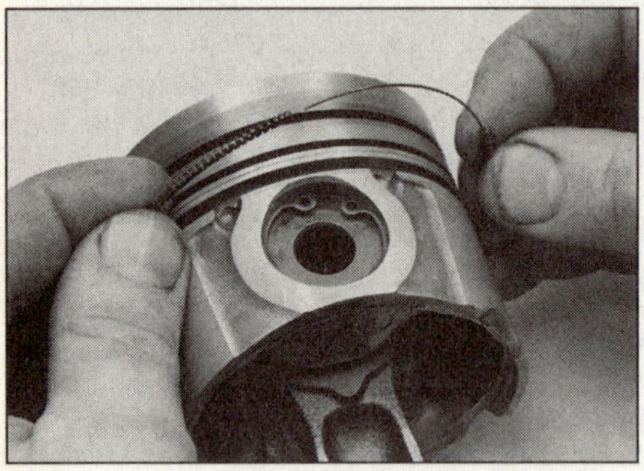

18.14a Fit the oil control ring expander . . .

13 Before refitting the rings to the pistons, check their end gaps by inserting each of them in their cylinder bores. Use the piston to make sure that they are square **(see illustrations)**. No values are specified, but typical gaps would be of the order of 0.50 mm for compression rings, perhaps somewhat greater for the oil control rings. Renault rings are supplied pre-gapped; no attempt should be made to adjust the gaps by filing.

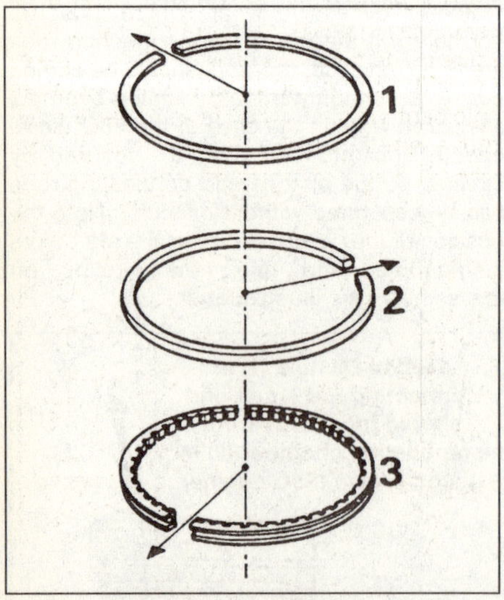

18.14c Position the piston ring end gaps 120° apart as shown

1 Top compression ring
2 Lower compression ring
3 Oil control ring

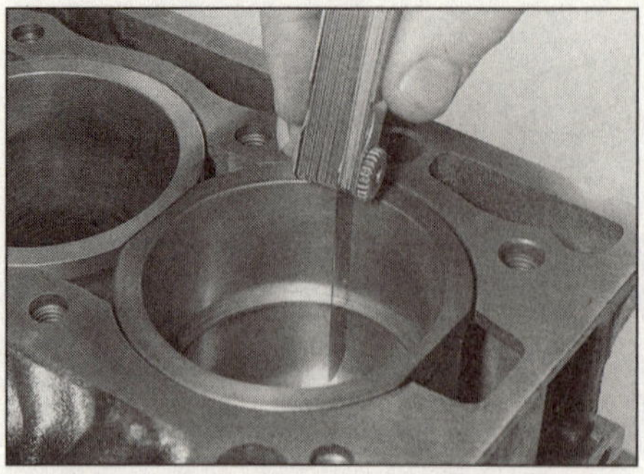

18.13b . . . then measure the ring end gaps

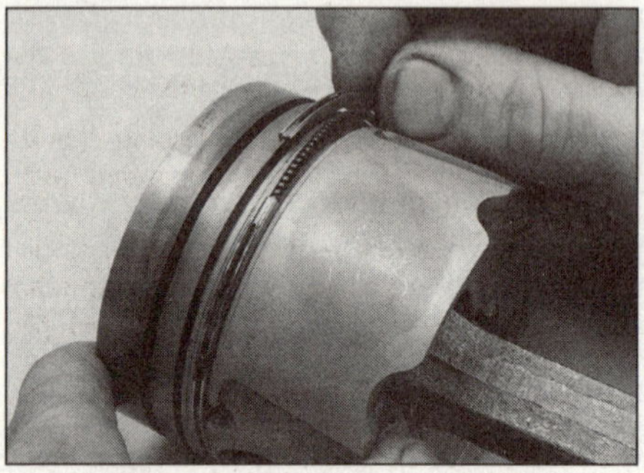

18.14b . . . followed by the ring

Reassembly

14 Install the new rings by fitting them over the top of the piston, starting with the oil control scraper ring **(see illustrations)**. Use feeler blades in the same way as when removing the old rings. Note that the second compression ring is tapered, and additionally stepped in the case of the K7M engine. Both compression rings must be fitted with the word 'TOP' uppermost. Be careful when handling the compression rings; they will break if they are handled roughly or expanded too far. With all the rings in position, space the ring gaps at 120° to each other.

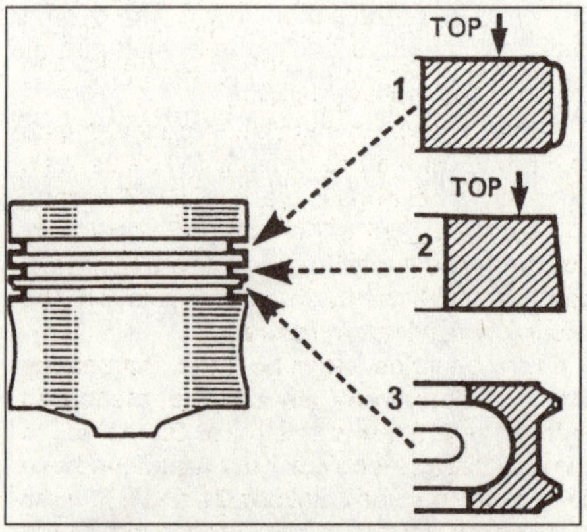

18.14d Piston ring profiles (F8Q engine)

1 Top compression ring
2 Lower compression ring
3 Oil control ring
Position the "TOP" markings as shown

15 Note that if new piston and liner assemblies have been obtained, each piston is matched to its respective liner, and they must not be interchanged.

19 Crankshaft - inspection

1 Clean the crankshaft and dry it with compressed air if available.

Warning: Wear eye protection when using compressed air! Be sure to clean the oil holes with a pipe cleaner or similar probe.

2 Check the main and big-end bearing journals for uneven wear, scoring, pitting and cracking.

3 If the crankshaft has been reground, check for burrs around the crankshaft oil holes (the holes are usually chamfered, so burrs should not be a problem unless regrinding has been carried out carelessly). Remove any burrs with a fine file or scraper, and thoroughly clean the oil holes as described previously.

4 Using a micrometer, measure the diameter of the main bearing and connecting rod journals, and compare the results with the Specifications **(see illustration)**. By measuring the diameter at a number of points around each journal's circumference, you will be able to determine whether or not the journal is out-of-round. Take the measurement at each end of the journal, near the webs, to determine if the journal is tapered. If any of the measurements vary by more than 0.025 mm, the crankshaft will have to be reground, and undersize bearings fitted.

5 Check the oil seal contact surfaces at each end of the crankshaft for wear and damage. If the seal has worn an excessive groove in the surface of the crankshaft, consult an engine overhaul specialist who will be able to advise whether a repair is possible or if a new crankshaft is necessary.

19.4 Measuring a main bearing journal diameter using a micrometer

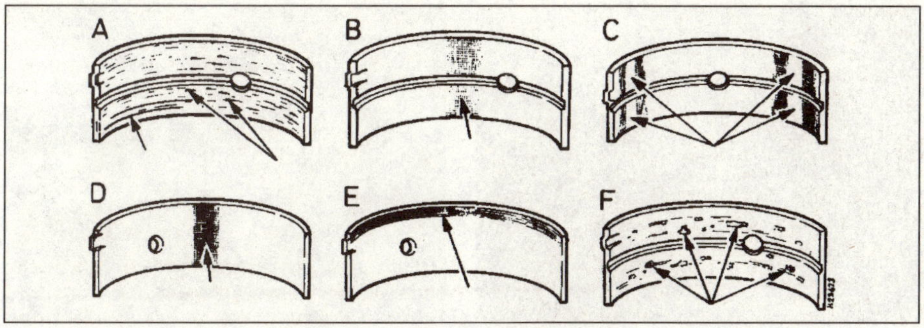

20.2 Typical bearing failures

A Scratched by dirt; dirt embedded into bearing material
B Lack of oil; overlay wiped out
C Improper seating; bright (polished) sections
D Tapered journal; overlay gone from entire surface
E Radius ride
F Fatigue failure; craters or pockets

20 Main and big-end bearings - inspection

1 Even though the main and big-end bearings should be renewed during the engine overhaul, the old bearings should be retained for close examination, as they may reveal valuable information about the condition of the engine. The size of the bearing shells is stamped on the back metal, and this information should be given to the supplier of the new shells.

2 Bearing failure occurs because of lack of lubrication, the presence of dirt or other foreign particles, overloading the engine, and corrosion. Regardless of the cause of bearing failure, it must be corrected before the engine is reassembled, to prevent it from happening again **(see illustration)**.

3 When examining the bearings, remove them from the engine block, the main bearing caps, the connecting rods and the rod caps, and lay them out on a clean surface in the same general position as their location in the engine. This will enable you to match any bearing problems with the corresponding crankshaft journal.

4 Dirt and other foreign particles get into the engine in a variety of ways. Dirt may be left in the engine during assembly, or it may pass through filters or the crankcase ventilation system. It may get into the oil, and from there into the bearings. Metal chips from machining operations and normal engine wear are often present. Abrasives are sometimes left in engine components after reconditioning, especially when parts are not thoroughly cleaned using the proper cleaning methods. Whatever the source, these foreign objects often end up embedded in the soft bearing material, and are easily recognised. Large particles will not embed in the bearing, and will score or gouge the bearing and journal. The best prevention for this cause of bearing failure is to clean all parts thoroughly, and keep everything spotlessly-clean during engine assembly. Frequent and regular engine oil and filter changes are also recommended.

5 Lack of lubrication (or oil breakdown) has a number of interrelated causes. Excessive heat (which thins the oil), overloading (which squeezes the oil from the bearing face) and oil leakage (from excessive bearing clearances, worn oil pump or high engine speeds) all contribute to lubrication breakdown. Blocked oil passages, which usually are the result of misaligned oil holes in a bearing shell, will also oil-starve a bearing and destroy it. When lack of lubrication is the cause of bearing failure, the bearing material is wiped or extruded from the steel backing of the bearing. Temperatures may increase to the point where the steel backing turns blue from overheating.

6 Driving habits can have a definite effect on bearing life. Full-throttle, low-speed operation (labouring the engine) puts very high loads on bearings, which tends to squeeze out the oil film. These loads cause the bearings to flex, which produces fine cracks in the bearing face (fatigue failure). Eventually, the bearing material will loosen in pieces and tear away from the steel backing. Short-trip driving leads to corrosion of bearings, because insufficient engine heat is produced to drive off the condensed water and corrosive gases. These products collect in the engine oil, forming acid and sludge. As the oil is carried to the engine bearings, the acid attacks and corrodes the bearing material.

7 Incorrect bearing installation during engine assembly will lead to bearing failure as well. Tight-fitting bearings leave insufficient bearing oil clearance, and will result in oil starvation. Dirt or foreign particles trapped behind a bearing shell result in high spots on the bearing which lead to failure.

8 If new bearings are to be fitted, the bearing running clearances should be measured before the engine is finally reassembled, to ensure that the correct bearing shells have been obtained (see Sections 21 and 22). If the crankshaft has been reground, the engineering works which carried out the work will advise on the correct size bearing shells to suit the work carried out. If there is any doubt as to which bearing shells should be used, seek advice from a Renault dealer.

21 Engine overhaul - reassembly sequence

1 Before starting, ensure all new parts have been obtained, and all necessary tools are available. Read through the entire procedure to familiarise yourself with the work involved, and to ensure all items necessary for engine reassembly are at hand. In addition to all normal tools and materials, a thread-locking compound will be needed. A tube of RTV sealing compound will also be required for the joint faces that are fitted without gaskets; it is recommended that Rhodorseal 5661 paste is used for the crankshaft main bearing cap and sump, and Loctite 518 for the water pump and cylinder head end cover (K7M engine). Both sealants are available from Renault dealers.

2 To save time and avoid problems, assembly can be carried out in the following order.

a) Crankshaft.
b) Pistons/connecting rod assemblies.
c) Oil pump and sump.
d) Flywheel/driveplate.
e) Cylinder head.
f) Timing belt and sprockets.
g) Engine external components.

22 Crankshaft - main bearing running clearance check and refitting

1 Before fitting the crankshaft and main bearings on the E7J and F8Q engines, decide whether the No 1 main bearing cap is to be sealed using butyl seals or silicone sealant. If butyl seals are to be used, it is necessary to determine the correct thickness of the seals to obtain from Renault. To do this, place the bearing cap in position without any seals, and secure it with the two retaining bolts. Locate a twist drill, dowel rod or any other suitable implement which will just fit in the side seal groove **(see illustration)**. Now measure the

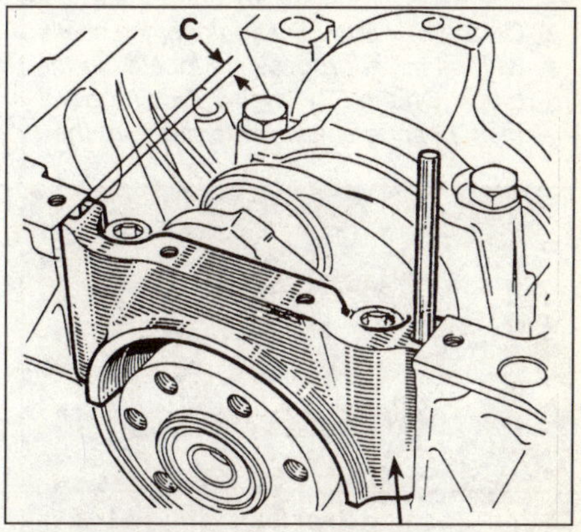

22.1 Measuring No 1 main bearing cap side seal grooves using a dowel rod

Bearing cap (arrowed)
C Seal groove measurement

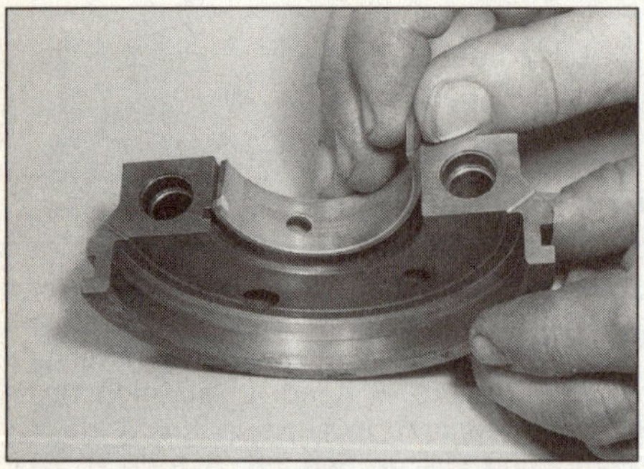
22.3 Fitting a main bearing shell to No 1 cap on the K7M engine

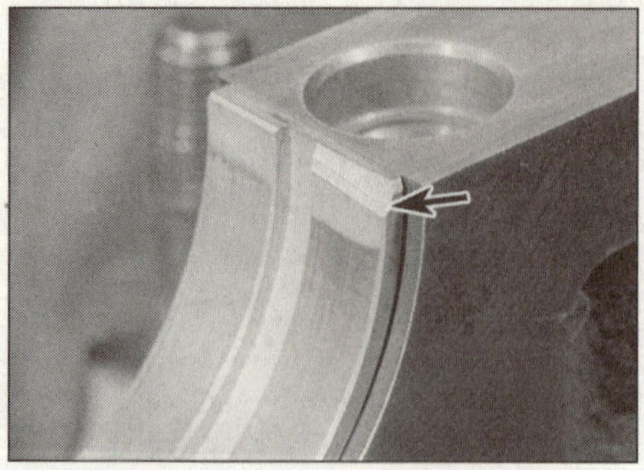
22.4a On the E7J engine, No 5 upper main bearing shell has an oil pump lubricating chamfer

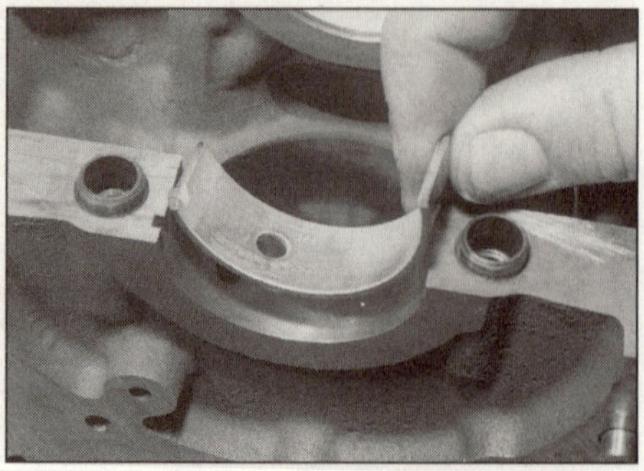
22.4b Fitting a main bearing shell to No 5 position on the K7M engine

implement - this dimension is the side seal groove size. If the dimension is less than or equal to 5 mm, a 5.10 mm thick side seal is needed. If the dimension is more than 5 mm, a 5.4 mm thick side seal is required. Having determined the side seal size and obtained the necessary seals, proceed as follows for the other types of engine as well. Note that silicone sealant is used instead of side seals when the engine is originally assembled at the factory.

Main bearing running clearance check

2 Clean the backs of the bearing shells and the bearing recesses in both the cylinder block and main bearing caps.

3 Press the bearing shells into the caps, ensuring that the tag on the shell engages in the notch in the cap. Note the following points.

- a) *On the E7J and F8Q engines, the shells without oil holes are located in the caps.*
- b) *On the K7M engine, the shells with grooves are located in main bearing caps 2 and 4, and the shells without grooves are located in main bearing caps 1, 3 and 5* ***(see illustration).***

4 Press the bearing shells into the recesses in the cylinder block. Note the following points.

- a) *On the E7J and F8Q engines, the shells with oil holes/grooves are located in the block. On the E7J engine, No 5 upper main bearing shell is different from the rest, in that it has an oil pump pinion lubricating chamfer which must face the timing belt end of the crankshaft* ***(see illustration).***
- b) *On the K7M engine, the shells with grooves are located in main bearing caps 2 and 4, and the shells without grooves are located in main bearing caps 1, 3 and 5* ***(see illustration).***
- c) *On all engines, if the original main bearing shells are being re-used, these must be refitted to their original locations in the block and caps.*

5 Before the crankshaft can be permanently installed, the main bearing running clearance should be checked; this can be done in either of two ways. One method is to fit the main bearing caps to the cylinder block, with the bearing shells in place. With the cap retaining bolts tightened to the specified torque, measure the internal diameter of each assembled pair of bearing shells using a vernier dial indicator or internal micrometer. If the diameter of each corresponding crankshaft journal is measured and then subtracted from the bearing internal diameter, the result will be the main bearing running clearance. The second (and more accurate) method is to use a product known as 'Plastigauge'. This consists of a fine thread of perfectly-round plastic which is compressed between the bearing cap and the journal. When the cap is removed, the deformation of the plastic thread is measured with a special card gauge supplied with the kit. The running clearance is determined from this gauge. The procedure for using Plastigauge is as follows.

6 With the upper main bearing shells in place, carefully lay the crankshaft in position **(see illustration)**. Do not use any lubricant; the crankshaft journals and bearing shells must be perfectly clean and dry.

7 Cut several pieces of the appropriate-size Plastigauge (they should be slightly shorter than the width of the main bearings), and place one piece on each crankshaft journal axis **(see illustration)**.

8 With the bearing shells in position in the caps, fit the caps to their numbered or previously-noted locations. Take care not to disturb the Plastigauge.

9 Starting with the centre main bearing and working outward, tighten the main bearing cap bolts progressively to their specified torque setting. Don't rotate the crankshaft at any time during this operation.

10 Remove the bolts and carefully lift off the main bearing caps, keeping them in order. Don't disturb the Plastigauge or rotate the crankshaft. If any of the bearing caps are difficult to remove, tap them from side-to-side with a soft-faced mallet.

11 Compare the width of the crushed Plastigauge on each journal to the scale printed on the gauge to obtain the main bearing running clearance **(see illustration)**.

12 If the clearance is not as specified, the bearing shells may be the wrong size (or badly worn if the original shells are being re-used). Before deciding that different size shells are

22.6 Lay the crankshaft in position in the crankcase

22.7 Strip of Plastigauge placed on No 5 main bearing journal

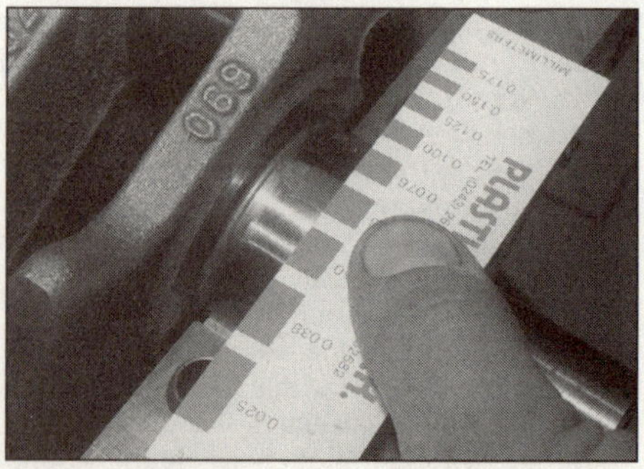
22.11 Using the gauge to check the main bearing running clearance

22.15a Smear a little grease on the crankshaft thrustwashers . . .

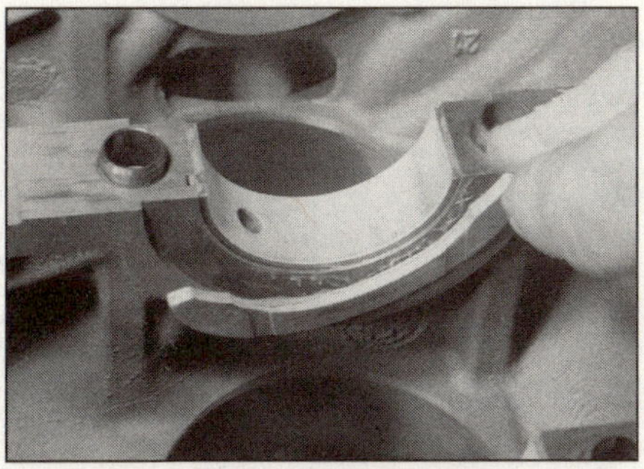

22.15b . . . and stick them to the centre main bearing (K7M engine)

22.16 Lubricate the main bearing shells before fitting the crankshaft

22.17a Fitting No 5 main bearing cap (K7M engine)

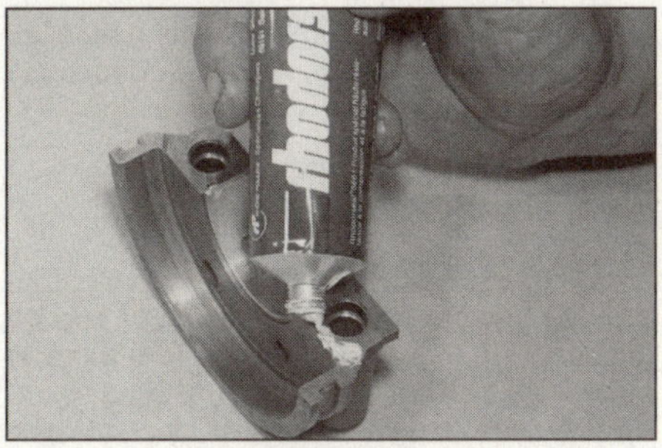

22.17b On the K7M engine, apply a thin coating of sealant to the contact faces of No 1 main bearing cap . . .

22.17c . . . before refitting the cap

needed, make sure no dirt or oil was trapped between the bearing shells and the caps or block when the clearance was measured. If the Plastigauge was wider at one end than at the other, the journal may be tapered.

13 Carefully scrape away all traces of the Plastigauge material from the crankshaft and bearing shells, using a fingernail or something similar which is unlikely to score the shells.

Final refitting

14 Carefully lift the crankshaft out of the cylinder block once more.

15 Using a little grease, stick the thrustwashers to each side of the centre main bearing (E7J and K7M engines) or No 2 main bearing (F8Q engine). Ensure that the oilway grooves on each thrustwasher face outwards from the bearing location, towards the crankshaft webs **(see illustrations)**.

16 Liberally lubricate each bearing shell in the cylinder block, and lower the crankshaft into position **(see illustration)**.

17 Lubricate the bearing shells, then fit the bearing caps in their numbered or previously-noted locations **(see illustration)**. Note the following points.

a) *On the E7J engine, apply a thin coating of sealant to the outer corners of No 1 main bearing cap.*

b) *On the K7M engine, apply a thin coating of sealant to the complete contact faces of the No 1 main bearing cap **(see illustrations)**.*

c) *If fitting butyl seals to No 1 main bearing cap on the E7J and F8Q engines, fit the seals with their grooves facing outwards. Position the seals so that approximately 0.2 mm of seal protrudes at the bottom-facing side (the side towards the crankcase). Lubricate the seals with a little oil, and apply a little sealant to the bottom corners of the cap prior to fitting it. When the cap is being fitted, use the bolts as a guide by just starting them in their threads, then pressing the cap firmly into position. When the cap is almost fully home, check that the seals still protrude slightly at the cylinder block mating face **(see illustration)**.*

d) *If silicone sealant is to be used on the No 1 main bearing on the E7J and F8Q engines, do not press the cap right down onto the crankshaft, but leave it raised so that the first few threads of the main bearing bolts can just be entered. Now inject the sealant into each of the cap side grooves until it enters the space below the cap and completely fills the grooves **(see illustrations)**.*

2C

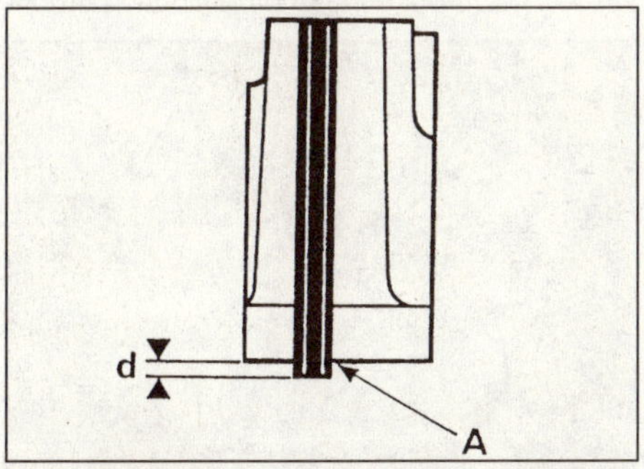

22.17d On E7J and F8Q engines, the butyl seals should protrude (d = 0.2 mm) beyond the lower face (A) of No 1 main bearing cap

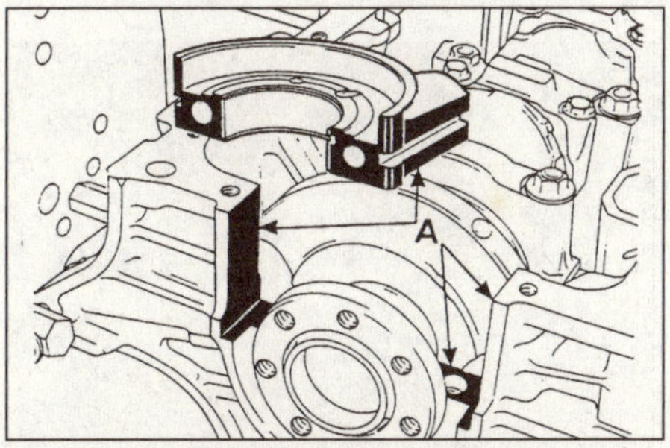

22.17e On E7J and F8Q engines, clean the cylinder block and bearing cap mating faces (A) when using silicone sealant to seal No 1 main bearing cap . . .

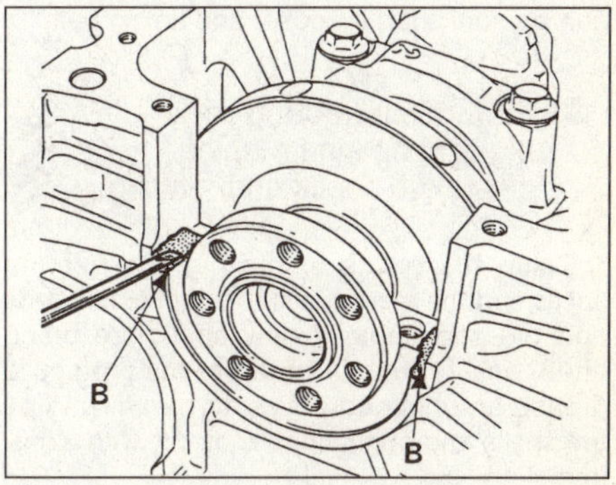

22.17f . . . and coat the lower faces of the cylinder block (B) with sealant

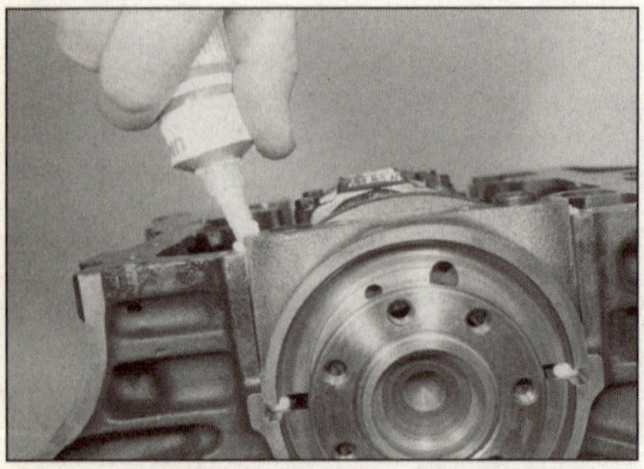

22.17g Injecting sealant into the side grooves of No 1 main bearing cap

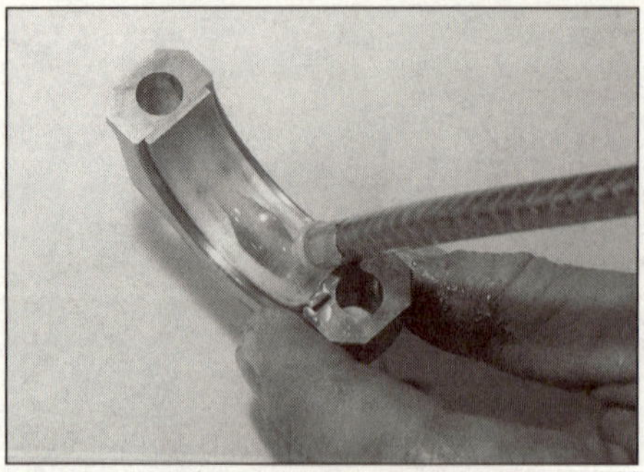

22.18a Tighten the main bearing cap bolts to the specified torque . . .

22.18b . . . and angle (K7M engine)

18 Fit the main bearing cap bolts, and tighten them progressively to the specified torque (and angle on the K7M engine) **(see illustrations)**.

19 Where butyl seals have been fitted to seal No 1 bearing cap, trim the protruding ends flush with the surface of the cylinder block sump mating face.

20 Where silicone sealant is to be used to seal No 1 main bearing cap, mix the sealant and the hardener as described in the instructions supplied with the kit. Inject the mixture into the bearing cap grooves, allowing the mixture to flow out slightly either side of the grooves, and at the cylinder block sump mating face, to ensure that the grooves are completely filled. Allow the sealant to dry for a few minutes, then cut away surplus sealant from the joint face, both inside and outside the cylinder block, and at the sump mating face.

21 Check that the crankshaft is free to turn. Some stiffness is normal if new components have been fitted, but there must be no jamming or tight spots.

22 Check the crankshaft endfloat with reference to Section 16.

23 Fit a new oil seal to the crankshaft front oil seal housing, and refit the housing with reference to Chapter 2A or 2B.

24 Fit a new crankshaft rear oil seal, with reference to Chapter 2A or 2B.

25 On the F8Q engine, refit the lower rear timing belt cover.

26 On completion, refit the piston/connecting rod assemblies, the flywheel/driveplate, oil pump, crankshaft sprocket and timing belt.

23 Piston/connecting rods - big-end bearing running clearance check and refitting

1 Clean the backs of the big-end bearing shells and the recesses in the connecting rods and big-end caps. If new shells are being fitted, ensure that all traces of the protective grease are cleaned off using paraffin. Wipe the shells and connecting rods dry with a lint-free cloth.

2 Press the big-end bearing shells into the connecting rods and caps in their correct positions. Make sure that the location tabs are engaged with the cut-outs in the connecting rods **(see illustration)**

Big-end bearing running clearance check

E7J petrol engine

3 Place the four liners face down in a row on the bench, in their correct order. Turn them as necessary so that the flats on the edges of liners 1 and 2 are towards each other, and the flats on liners 3 and 4 are towards each other also. Fit the O-rings to the base of each liner **(see illustration)**.

4 Lubricate the pistons and piston rings, then place each piston and connecting rod assembly with its respective liner.

5 Starting with assembly No 1, make sure that the piston ring gaps are still spaced at 120° to each other. Clamp the piston rings using a piston ring compressor.

6 Insert the piston and connecting rod assembly into the bottom of the liner, ensuring that the arrow on the piston crown will be facing the flywheel/driveplate end of the engine. Using a block of wood or a hammer handle against the end of the connecting rod, tap the piston into the liner until the top of the piston is approximately 25 mm away from the top of the liner.

7 Repeat the procedure for the remaining three piston-and-liner assemblies.

8 Turn the crankshaft so that No 1 crankpin is at the bottom of its travel.

9 With the liner seal/O-ring in position, place No 1 liner, piston and connecting rod assembly into its location in the cylinder block. Ensure that the arrow on the piston crown faces the flywheel/driveplate end of the engine, and that the flat on the liner is positioned as described previously.

10 To measure the big-end bearing running clearance, refer to the information contained in Section 21; the same general procedures apply. If the Plastigauge method is being used, ensure that the crankpin journal and the big-end bearing shells are clean and dry, then pull the connecting rod down and engage it with the crankpin. Place the Plastigauge strip on the crankpin, check that the marks made on the cap and rod during removal are next to each other, then refit the cap and retaining nuts. Tighten the nuts to the specified torque in two stages. Do not rotate the crankshaft during this operation. Remove the cap and check the running clearance by measuring the Plastigauge as previously described.

11 With the liner/piston assembly installed, retain the liner using a bolt and washer screwed into the cylinder head bolt holes, or using liner clamps **(see illustration)**.

12 Repeat the above procedures for the remaining piston-and-liner assemblies.

K7M petrol and F8Q diesel engines

13 Lubricate No 1 piston and piston rings, and check that the ring gaps are spaced at 120° intervals to each other **(see illustration)**.

14 Fit a ring compressor to No 1 piston, then insert the piston and connecting rod into No 1 cylinder. On the K7M engine, the 'V' arrow must point to the flywheel end of the engine. On the F8Q engine, the combustion chamber

23.3 Fitting a liner base O-ring (E7J engine)

23.11 Liner clamp to hold the liners in place (E7J engine)

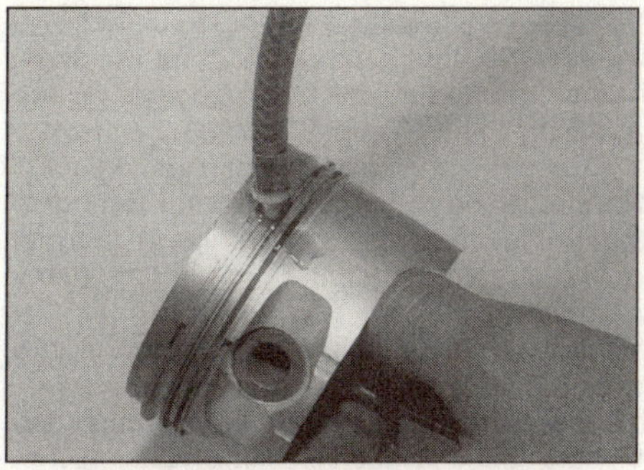

23.13 Lubricating the piston rings

23.14a Markings on the top of the piston (K7M engine)

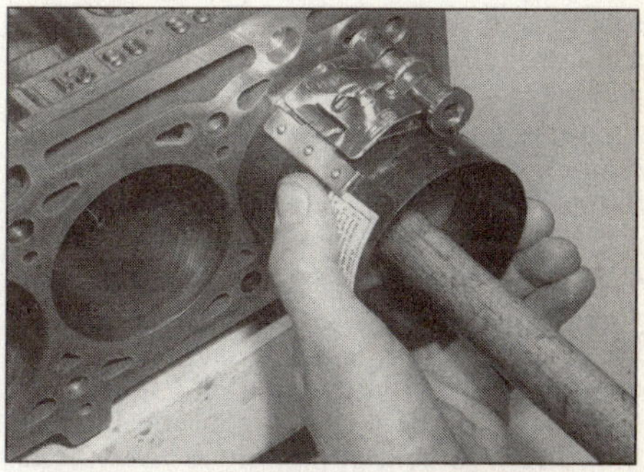

23.14b Using the wooden handle of a hammer to drive the piston into the bore

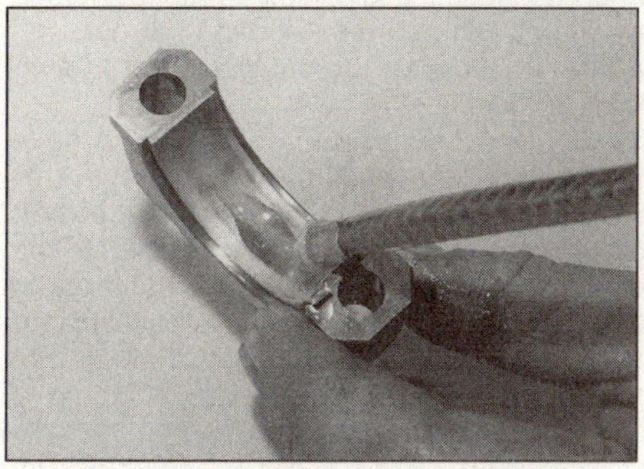

23.18a Lubricate both the cap shells . . .

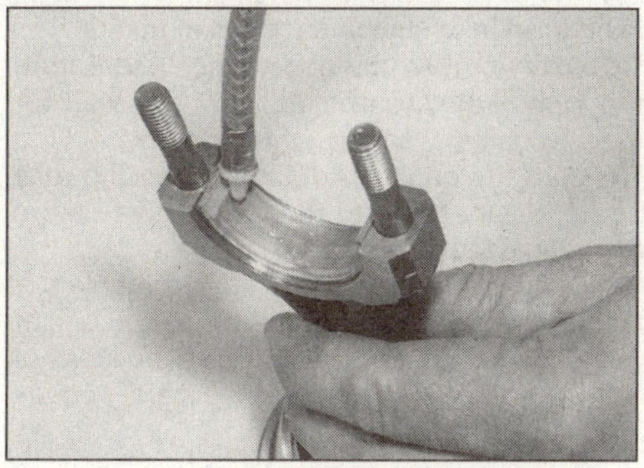

23.18b . . . and connecting rod shells before . . .

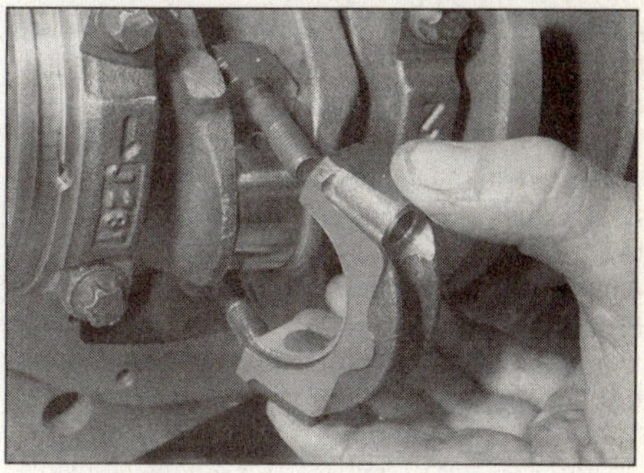

23.18c . . . refitting the cap . . .

recess in the piston crown should be on the oil filter side of the engine. With No 1 crankpin at its lowest point, drive the piston carefully into the cylinder with the wooden handle of a hammer, at the same time guiding the connecting rod onto the crankpin **(see illustrations)**.

15 To measure the big-end bearing running clearance, refer to the information contained in Section 21; the same general procedures apply. If the Plastigauge method is being used, ensure that the crankpin journal and the big-end bearing shells are clean and dry, then engage the connecting rod with the crankpin. Place the Plastigauge strip on the crankpin, fit the bearing cap in its previously-noted position, then tighten the nuts/bolts to the specified torque. Do not rotate the crankshaft during this operation. Remove the cap and check the running clearance by measuring the Plastigauge as previously described.

16 Repeat the above procedures on the remaining piston/connecting rod assemblies.

Final refitting

17 Having checked the running clearance of all the crankpin journals and taken any corrective action necessary, clean off all traces of Plastigauge from the bearing shells and crankpin.

18 Liberally lubricate the crankpin journals and big-end bearing shells. Refit the bearing caps once more, ensuring correct positioning as previously described. Tighten the bearing cap nuts/bolts to the specified torque, and turn the crankshaft each time to make sure that it is free before moving on to the next assembly. On the E7J engine, check that all the liners are positioned relative to each other, so that a 0.1 mm feeler blade can pass freely through the gaps between the liners. If this is not the case, it may be necessary to interchange one or more of the complete piston/liner assemblies to achieve this clearance **(see illustrations)**.

19 On completion, refit the oil pump, sump and cylinder head as described in Chapter 2A or 2B.

24 Engine - initial start-up after overhaul

1 With the engine refitted in the vehicle, double-check the engine oil and coolant levels. Make a final check that everything has been reconnected, and that there are no tools or rags left in the engine compartment.

Petrol engine models

2 With the spark plugs removed and the ignition and fuel injection systems disabled by

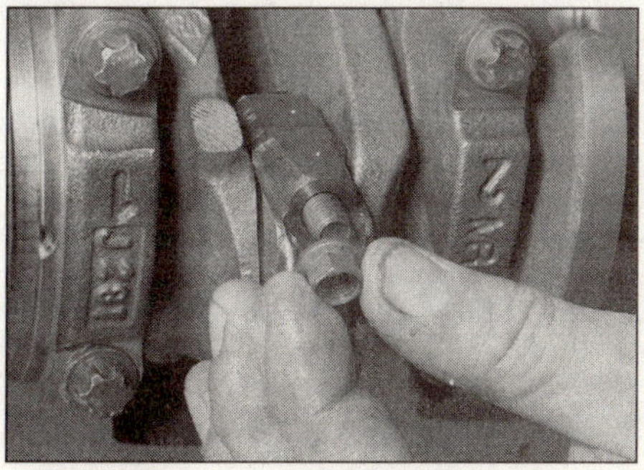

23.18d . . . screwing on the nuts . . .

23.18e . . . and tightening the nuts

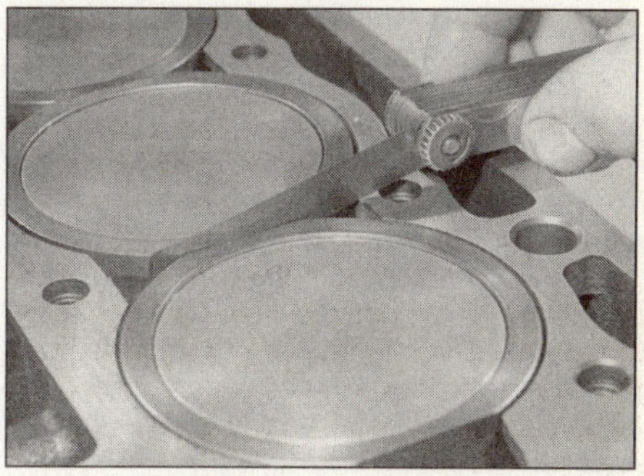

23.18f Checking the gap between the liners (E7J engine)

disconnecting the wiring from the coil(s) (Chapter 5B) and depressurising the fuel system by removing the fuel pump relay (Chapter 4A) crank the engine on the starter motor until the oil pressure light goes out.

3 Refit the spark plugs and connect the coil wiring. Refit the fuel pump relay.

4 Start the engine, noting that this may take a little longer than usual, due to the fuel pump being empty.

5 While the engine is idling, check for fuel, water and oil leaks. Where applicable, check the power steering pipe/hose unions for leakage. Do not be alarmed if there are some odd smells and smoke from parts getting hot and burning off oil deposits.

6 Keep the engine idling until hot water is felt circulating through the top hose, then switch it off.

7 After a few minutes, recheck the oil and water levels, and top-up as necessary.

8 There is no requirement to retighten the cylinder head bolts.

9 If new pistons, rings or crankshaft bearings have been fitted, the engine must be run-in for the first 500 miles (800 km). Do not operate the engine at full-throttle, nor allow it to labour in any gear during this period. It is recommended that the oil and filter be changed at the end of this period.

Diesel engine models

10 Prime the fuel system as described in Chapter 4B.

11 On turbo models, prime the turbocharger lubrication circuit by disconnecting the stop solenoid and cranking the engine on the starter motor in several ten-second bursts, pausing for half a minute or so between each burst. Reconnect the solenoid when satisfied that oil pressure has been established (ensure that the oil pressure warning light on the facia extinguishes when the engine is cranked).

12 Fully depress the accelerator pedal, turn the ignition key to position "M" and wait for the preheating warning light to go out.

13 Start the engine. Additional cranking may be necessary to bleed the fuel system before the engine starts.

14 Once started, keep the engine running at fast tickover. Check that the oil pressure light goes out, then check that there are no leaks of oil, fuel and coolant. Where applicable, check the power steering pipe/hose unions for leakage. Do not be alarmed if there are some odd smells and smoke from parts getting hot and burning off oil deposits.

15 Keep the engine idling until hot coolant is felt circulating through the radiator top hose, indicating that the engine is at normal operating temperature, then stop the engine and allow it to cool.

16 Recheck the oil and coolant levels and top-up if necessary.

17 Check the fuel injection pump timing and the idle speed as described in Chapter 4B.

18 On turbo models, tighten the cylinder head bolts to the final stage (see Chapter 2B) after running the engine until it reaches normal operating temperature (cooling fan cuts in), and then allowing the engine to cool for at least 2 hours.

19 If new pistons, rings or bearings have been fitted, the engine must be run-in at reduced speeds and loads for the first 500 miles (800 km) or so. Do not operate the engine at full throttle, or allow it to labour in any gear during this period. It is beneficial to change the engine oil and filter at the end of this period.

Chapter 3
Cooling, heating and ventilation systems

Contents

Degrees of difficulty

Easy, suitable for novice with little experience	**Fairly easy,** suitable for beginner with some experience	**Fairly difficult,** suitable for competent DIY mechanic	**Difficult,** suitable for experienced DIY mechanic	**Very difficult,** suitable for expert DIY or professional

Specifications

General

Maximum system pressure:	
Petrol models:	
1.4 litre engine	1.2 bar
1.6 litre engine	1.6 bar
Diesel models	1.2 bar

Thermostat

Opening temperatures:	
Starts to open	89°C
Fully open	101°C
Valve lift	7.5 mm

1 General information and precautions

General information

The cooling system is of pressurised type, comprising a coolant pump (driven by the timing belt on petrol engines and the auxiliary drivebelt on diesel engines), an aluminium crossflow radiator, electric cooling fan(s), a thermostat, heater matrix, and all associated hoses and switches.

The system functions as follows. Cold coolant in the bottom of the radiator passes through the bottom hose to the coolant pump, where it is pumped around the cylinder block and head passages, and through the oil cooler(s) (where fitted). After cooling the cylinder bores, combustion surfaces and valve seats, the coolant reaches the underside of the thermostat, which is initially closed. The coolant passes through the heater, and is returned via the cylinder block to the coolant pump.

When the engine is cold, the coolant circulates only through the cylinder block, cylinder head, and heater. When the coolant reaches a predetermined temperature, the thermostat opens, and the coolant passes through the top hose to the radiator. As the coolant circulates through the radiator, it is cooled by the inrush of air when the car is in forward motion. The airflow is supplemented by the action of the electric cooling fan(s) when necessary. Upon reaching the bottom of the radiator, the coolant has now cooled, and the cycle is repeated.

On diesel engine models, the coolant is also circulated through the engine oil cooler and around the turbocharger (where fitted). On models with automatic transmission, the coolant is also passed through the fluid cooler on the top of the transmission.

When the engine is at normal operating temperature, the coolant expands, and some of it is displaced into the expansion tank. Coolant collects in the tank, and is returned to the radiator when the system cools.

The electric cooling fan mounted behind the radiator are controlled by a thermostatic switch. At a predetermined coolant temperature, the switch actuates the fan.

Precautions

Warning: Do not attempt to remove the expansion tank filler cap, or to disturb any part of the cooling system, while the engine is hot, as there is a high risk of scalding. If the expansion tank filler cap must be removed before the engine and radiator

have fully cooled (even though this is not recommended), the pressure in the cooling system must first be relieved. Cover the cap with a thick layer of cloth to avoid scalding, and slowly unscrew the filler cap until a hissing sound is heard. When the hissing has stopped, indicating that the pressure has reduced, slowly unscrew the filler cap until it can be removed; if more hissing sounds are heard, wait until they have stopped before unscrewing the cap completely. At all times, keep well away from the filler cap opening, and protect your hands.

Warning: Do not allow antifreeze to come into contact with your skin, or with the painted surfaces of the vehicle. Rinse off spills immediately, with plenty of water. Never leave antifreeze lying around in an open container, or in a puddle in the driveway or on the garage floor. Children and pets are attracted by its sweet smell, but antifreeze can be fatal if ingested.

Warning: If the engine is hot, the electric cooling fan may start rotating even if the engine is not running. Be careful to keep your hands, hair, and any loose clothing well clear when working in the engine compartment.

Warning: Refer to Section 10 for precautions to be observed when working on models equipped with air conditioning.

2 Cooling system hoses - disconnection and renewal

Note: *Refer to the warnings given in Section 1 of this Chapter before proceeding. Hoses should only be disconnected once the engine has cooled sufficiently to avoid scalding.*

1 If the checks described in Chapter 1 reveal a faulty hose, it must be renewed as follows.

2 First drain the cooling system (Chapter 1). If the coolant is not due for renewal, it may be re-used, providing it is collected in a clean container.

3 Release the hose clips from the hose concerned. Three types of clip are used: worm-drive, spring and crimped-type. The worm-drive clip is released by turning its screw anti-clockwise. The spring clip is released by squeezing its tags together with pliers, at the same time working the clip away from the hose stub **(see illustration)**. The crimped-type clip is not re-usable, and is best cut off with snips or side cutters and replaced with a standard worm-drive clip. Move the clips along the hose, and position them clear of the union.

4 Carefully work the hose free. The hoses can be removed with relative ease when new - on an older car, they may have stuck.

5 If a hose proves to be difficult to remove, try to release it by rotating its ends before attempting to free it. Gently prise the end of the hose with a blunt instrument (such as a flat-bladed screwdriver), but do not apply too much force, and take care not to damage the pipe stubs or hoses. Note in particular that the radiator inlet stub is fragile; do not use excessive force when attempting to remove the hose. If all else fails, cut the hose with a sharp knife, then slit it so that it can be peeled off in two pieces. Although this may prove expensive if the hose is otherwise undamaged, it is preferable to buying a new radiator. Check first, however, that a new hose is readily available.

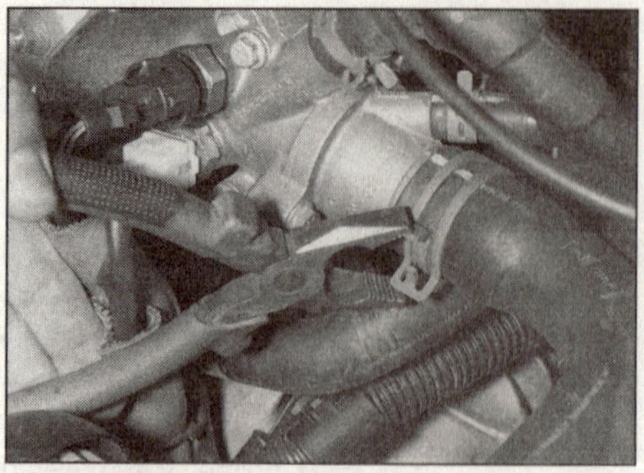

2.3 Removing a coolant hose spring clip using pliers

6 When fitting a hose, first slide the clips onto the hose, then work the hose into position. If crimped-type clips were originally fitted, use standard worm-drive clips when refitting the hose.

HAYNES HiNT ***If the hose is stiff, use a little soapy water as a lubricant, or soften the hose by soaking it in hot water. Do not use oil or grease, which may attack the rubber.***

7 Work the hose into position, checking that it is correctly routed, then slide each clip back along the hose until it passes over the flared end of the relevant inlet/outlet, before securing it in position with the retaining clips.

8 Refill the cooling system with reference to Chapter 1.

9 Check thoroughly for leaks as soon as possible after disturbing any part of the cooling system.

3 Radiator - removal, inspection and refitting

Note: *If leakage is the reason for removing the radiator, bear in mind that minor leaks can often be cured using a radiator sealant with the radiator in situ.*

Removal

1 Disconnect the battery negative lead.

2 Drain the cooling system as described in Chapter 1.

3 Disconnect the wiring connector from the cooling fan and cooling fan switch, which is screwed into the left-hand side of the radiator.

4 Release the retaining clips (see Section 2) and disconnect the coolant hoses from the radiator.

5 Remove the left-hand front direction indicator light unit as described in Chapter 12.

6 Slacken and remove the retaining screw from the centre of the radiator grille and remove the joining section.

7 Undo the retaining screws, then release the lower retaining clips and slide the left-and right-hand sections of the radiator grille forwards and away from the vehicle.

8 Slacken and remove the nuts/bolts (as applicable) securing the radiator upper mounting brackets to the bonnet crossmember. If access cannot easily be gained to the mounting nut/bolt, it may prove sufficient to simply unclip the radiator from the bracket.

9 Slacken and remove the six bonnet crossmember mounting bolts, then remove the crossmember from the vehicle. Remove the upper mounting brackets and rubbers from the top of the radiator.

10 Where necessary, unclip the power steering fluid reservoir and position it clear of the radiator **(see illustration)**.

11 On models with air conditioning, slacken and remove the retaining bolts and nuts, and free the drier reservoir from the rear of the radiator. Slacken and remove the bolts securing the condenser to the front of the radiator, and separate the two.

Warning: Do not attempt to open the refrigerant circuit. Refer to the precautions in Section 10.

12 Check that all pipes/wiring is released, then lift the radiator and cooling fan assembly out from the engine compartment, taking care not to lose the radiator lower mounting rubbers.

Inspection

13 If the radiator has been removed due to suspected blockage, reverse-flush it as described in Chapter 1. Clean dirt and debris

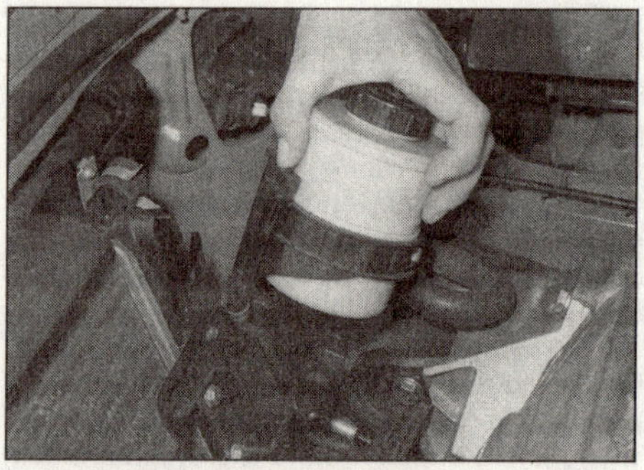

3.10 On models with power steering, unclip the reservoir and position it clear of the radiator

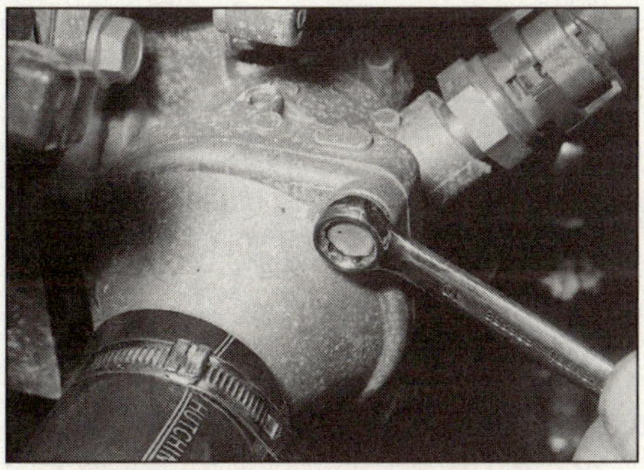

4.5a On diesel models, slacken and remove the retaining bolts . . .

from the radiator fins, using an air line (in which case, wear eye protection) or a soft brush. Be careful, as the fins are sharp, and easily damaged.

14 If necessary, a radiator specialist can perform a "flow test" on the radiator, to establish whether an internal blockage exists.

15 A leaking radiator must be referred to a specialist for permanent repair. Do not attempt to weld or solder a leaking radiator, as damage to the plastic components may result.

16 In an emergency, minor leaks from the radiator can be cured by using a suitable radiator sealant, in accordance with its manufacturer's instructions, with the radiator *in situ*.

17 If the radiator is to be sent for repair or renewed, remove all hoses and the cooling fan switch (where fitted).

18 Inspect the condition of the radiator mounting rubbers, and renew them if necessary.

Refitting

19 Refitting is a reversal of removal, ensuring that all the mounting rubbers are correctly positioned, and that the upper mounting brackets are correctly engaged with the radiator. On completion, refill the cooling system as described in Chapter 1.

4 Thermostat - removal, testing and refitting

Note: *A new thermostat sealing ring will be required on refitting.*

Removal

1 Disconnect the battery negative lead.

2 Drain the cooling system as described in Chapter 1. The thermostat housing is mounted on the left-hand end of the cylinder head.

3 Where necessary, release any relevant wiring and hoses from the retaining clips, and position clear of the thermostat housing to improve access.

4 To improve access, disconnect the coolant hose(s) from the thermostat housing cover (see Section 2).

4.5b . . . then remove the thermostat housing cover (a gasket maybe fitted on some engines) . . .

5 Unscrew the nuts/bolts (as applicable), and carefully withdraw the thermostat housing cover to expose the thermostat **(see illustrations)**. Where necessary, remove the housing gasket and discard it; a new one should be used on refitting.

6 Lift the thermostat from the housing, and recover the sealing ring **(see illustration)**.

Testing

7 A rough test of the thermostat may be made by suspending it with a piece of string in a container full of water. Heat the water to bring it to the boil - the thermostat must open by the time the water boils. If not, renew it.

8 If a thermometer is available, the precise opening temperature of the thermostat may be determined; compare with the figures given in the Specifications. The opening temperature is also marked on the thermostat.

9 A thermostat which fails to close as the water cools must also be renewed.

Refitting

10 Refitting is a reversal of removal, bearing in mind the following points:

a) *Always fit a new sealing ring and ensure that the thermostat is fitted the correct way round, with the spring(s) facing into the housing. The housing gasket (where fitted) should also be renewed.*

b) *On completion, refill the cooling system as described in Chapter 1.*

5 Electric cooling fan - testing, removal and refitting

Testing

1 Current supply to the cooling fan is via the ignition switch and a fuse (see Chapter 12) the circuit is completed by the cooling fan thermostatic switch, which is mounted in the left-hand side of the radiator.

2 If a fan does not appear to work, first check the fuse. Run the engine until normal operating temperature is reached, then allow it to idle. The fan should cut in within a few minutes (before the temperature gauge needle enters the red section, or before the coolant

4.6 . . . and lift out the thermostat and sealing ring

temperature warning light comes on). If not, switch off the ignition and disconnect the wiring plug from the cooling fan switch. Bridge the two contacts in the wiring plug using a length of spare wire, and switch on the ignition. If the fan now operates, the switch is probably faulty, and should be renewed.

3 If the fan still fails to operate, check that battery voltage is available at the feed wire to the switch; if not, then there is a fault in the feed wire. If there is no problem with the feed, check that there is continuity between the switch earth terminal and a good earth point on the body; if not, then the earth connection is faulty, and must be re-made.

4 If the switch and the wiring are in good condition, the fault must lie in the motor itself. The motor can be checked by disconnecting it from the wiring loom and connecting a 12-volt supply directly to it.

Removal

5 Disconnect the battery negative lead.

6 Where necessary, unclip/unbolt the power steering fluid reservoir and position it clear of the cooling fan.

7 On models with air conditioning, slacken and remove the retaining bolts and nuts, and free the drier reservoir from the rear of the radiator.

Warning: Do not attempt to open the refrigerant circuit. Refer to the precautions in Section 10.

8 Disconnect the wiring connector from the cooling fan motor, and free the wiring from the fan shroud **(see illustration)**.

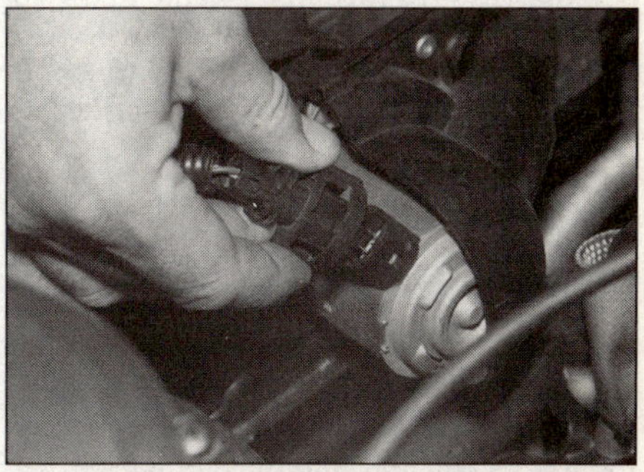

5.8 Disconnecting the cooling fan motor wiring connector

9 Slacken and remove the bolts securing the fan shroud to the rear of the radiator then manoeuvre the assembly out from the engine compartment.
10 If necessary, slide off retaining clip/undo the retaining screw (as applicable) and remove the fan from the motor spindle. The motor can then be unbolted and removed from the shroud. If the motor is faulty, the complete unit must be renewed, as no spares are available.

Refitting

11 Refitting is a reversal of removal, noting the following points.

a) Prior to refitting, inspect the fan shroud mountings, renewing them if they show signs of wear or damage.
b) Ensure that the wiring is correctly routed and secured in position with all the relevant clips and ties, so that it is no danger of contacting the fan.
c) Refit the radiator as described in Section 3.

6 Cooling system electrical switches and sensors - testing, removal and refitting

Electric cooling fan thermostatic switch

Testing

1 Testing of the switch is described in Section 5, as part of the electric cooling fan test procedure.

Removal

2 The switch is located in the left-hand side of the radiator. The engine and radiator should be cold before removing the switch.
3 Disconnect the battery negative lead.
4 Partially drain the cooling system to just below the level of the switch (as described in Chapter 1). Alternatively, have ready a suitable bung to plug the switch aperture in the radiator when the switch is removed. If this method is used, take great care not to damage the radiator, and do not use anything which will allow foreign matter to enter the radiator.
5 Disconnect the wiring plug from the switch.
6 Carefully unscrew the switch from the radiator, and recover the sealing ring (where fitted). If the system has not been drained, plug the switch aperture to prevent further coolant loss.

Refitting

7 If the switch was originally fitted using a sealing ring, use a new sealing ring on refitting. Where no sealing ring was fitted, clean the switch threads thoroughly, and coat them with fresh sealing compound.
8 Refitting is a reversal of removal. Tighten the switch securely and refill (or top-up) the cooling system as described in Chapter 1.
9 On completion, start the engine and run it until it reaches normal operating temperature. Continue to run the engine, and check that the cooling fan cuts in and out correctly.

Coolant temperature gauge/ temperature warning light sender

Testing

10 The coolant temperature gauge/warning light sender is fitted to the thermostat housing on the left-hand end of the cylinder head. On petrol engine models, the sender is the upper of the two senders on the front of the thermostat housing; on diesel engine models, the sender is fitted to the end of the housing.
11 The temperature gauge is fed with a stabilised voltage from the instrument panel feed (via the ignition switch and a fuse). The gauge earth is controlled by the sender. The sender contains a thermistor - an electronic component whose electrical resistance decreases at a predetermined rate as its temperature rises. When the coolant is cold, the sender resistance is high, current flow through the gauge is reduced, and the gauge needle points towards the cold end of the scale. As the coolant temperature rises and the sender resistance falls, current flow increases, and the gauge needle moves towards the upper end of the scale. If the sender is faulty, it must be renewed.
12 The temperature warning light, is fed with a voltage from the instrument panel. The light earth is controlled by the sender. The sender is effectively a switch, which operates at a predetermined temperature to earth the light and complete the circuit.
13 If the gauge develops a fault, first check the other instruments; if they do not work at all, check the instrument panel electrical feed. If the readings are erratic, there may be a fault in the voltage stabiliser, which will necessitate renewal of the stabiliser (the stabiliser is integral with the instrument panel printed circuit board - see Chapter 12). If the fault lies in the temperature gauge alone, check it as follows.
14 If the gauge needle remains at the "cold" end of the scale when the engine is hot, disconnect the sender wiring plug, and earth the relevant wire to the cylinder head. If the needle then deflects when the ignition is switched on, the sender unit is proved faulty, and should be renewed. If the needle still does not move, remove the instrument panel (Chapter 12) and check the continuity of the wire between the sender unit and the gauge, and the feed to the gauge unit. If continuity is shown, and the fault still exists, then the gauge is faulty, and the gauge unit should be renewed.
15 If the gauge needle remains at the "hot" end of the scale when the engine is cold, disconnect the sender wire. If the needle then returns to the "cold" end of the scale when the ignition is switched on, the sender unit is proved faulty, and should be renewed. If the needle still does not move, check the remainder of the circuit as described previously.
16 The same basic principles apply to testing the warning light. The light should illuminate when the relevant sender wire is earthed.

Removal and refitting

17 On petrol engine models, remove and refit the sender using the information given for the cooling fan switch (paragraphs 2 to 9) **(see illustration)**.
18 On diesel engine models, refer to the information given in paragraphs 2 to 9 for the cooling fan switch, noting that the sender is clipped into the housing. Press the sender into the housing then, using a small flat-bladed screwdriver, lever out the retaining clip and remove the sender and sealing ring. On refitting, fit a new sealing ring to the sender, then press the sender into the thermostat housing and secure it in position with the retaining clip.

Fuel injection system coolant temperature sensor - petrol models

Testing

19 The fuel injection system coolant temperature sensor is located in the thermostat housing on the left-hand end of the cylinder head. The sensor is the lower of the two sensors which are screwed into the front of the housing **(see illustration)**.

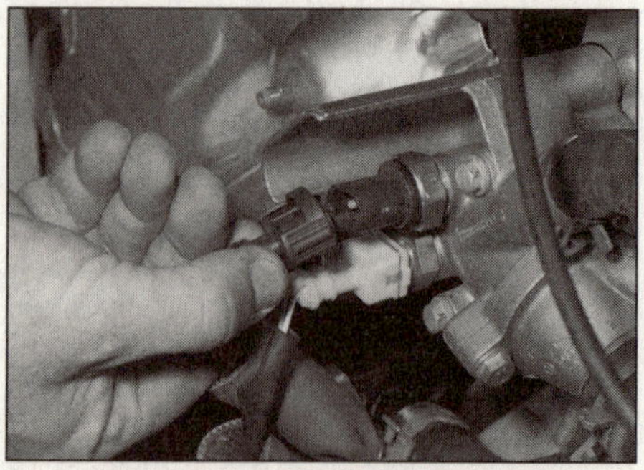

6.17 Disconnecting the coolant temperature gauge/temperature warning light sender wiring connector (1.6 petrol engine shown)

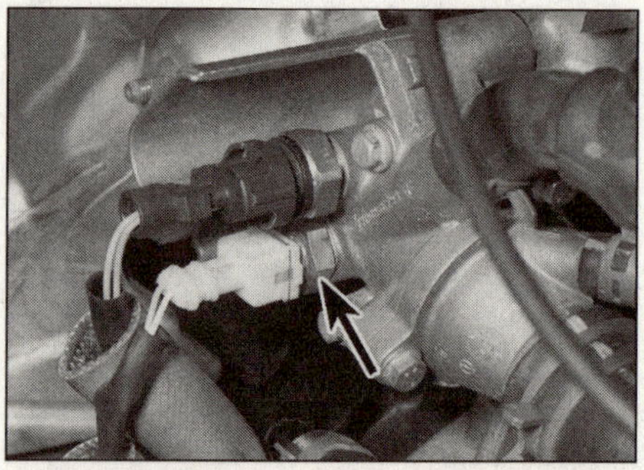

6.19 On petrol models, the fuel injection system coolant temperature sensor (arrowed) is the lower of the sensors on the thermostat housing (1.6 engine shown)

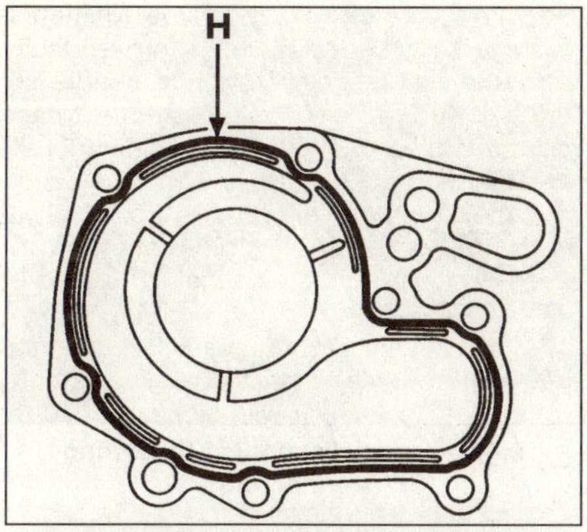

7.4a On petrol models, apply a bead of sealant (H) to the coolant mating surface as shown (1.4 litre engine shown - 1.6 litre similar)

20 The sensor is a thermistor (refer to paragraph 11). The fuel injection/engine management electronic control unit (ECU) supplies the sensor with a set voltage and then, by measuring the current flowing in the sensor circuit, it determines the engine's temperature. This information is then used, in conjunction with other inputs, to control the injector opening time (pulse width). On some models, the idle speed and/or ignition timing settings are also temperature-dependent.

21 If the sensor circuit should fail to provide adequate information, the ECU's back-up facility will override the sensor signal. In this event, the ECU assumes a predetermined setting which will allow the fuel injection/engine management system to run, albeit at reduced efficiency. When this occurs, the warning light on the instrument panel will come on, and the advice of a Renault dealer should be sought. The sensor itself can only be tested using special Renault diagnostic equipment. *Do not* attempt to test the circuit using any other equipment, as there is a high risk of damaging the ECU.

Removal and refitting

22 Remove and refit the sensor using the information given for the cooling fan switch (paragraphs 2 to 9).

7 Coolant pump - removal and refitting

Petrol models

Removal

1 Drain the cooling system (see Chapter 1).

2 Remove the timing belt as described in Chapter 2.

3 Slacken and remove the retaining bolts, and withdraw the pump assembly from the engine. Note the correct fitted location of the pump locating dowels, and remove them for safe-keeping if they are loose.

7.4b Applying sealant to the coolant pump - 1.6 litre engine

Refitting

4 Ensure that the mating surfaces of the pump and cylinder block are clean and dry, then apply a bead of sealant 0.6 to 1.0 mm thick (Renault recommend the use of Loctite 518 - available from your Renault dealer) to the pump sealing face as shown **(see illustrations)**.

5 Ensure that the locating dowels are correctly positioned then manoeuvre the pump into position. Refit the pump bolts and tighten them securely.

6 Refit the timing belt (Chapter 2).

7 Refill the cooling system as described in Chapter 1.

Diesel models

Removal

8 Remove the auxiliary drivebelt as described in Chapter 1, noting that the coolant pump pulley retaining bolts should be slackened before the belt is removed.

9 Drain the cooling system (see Chapter 1).

10 Unscrew the retaining bolts, and remove the drivebelt pulley from the coolant pump **(see illustration)**.

11 Unscrew the retaining bolts, then manoeuvre the coolant pump out of position **(see illustrations)**. Note the correct fitted location of the pump locating dowels, and remove them for safe-keeping if they are loose. Recover the pump gasket and discard it; a new one must be used on refitting.

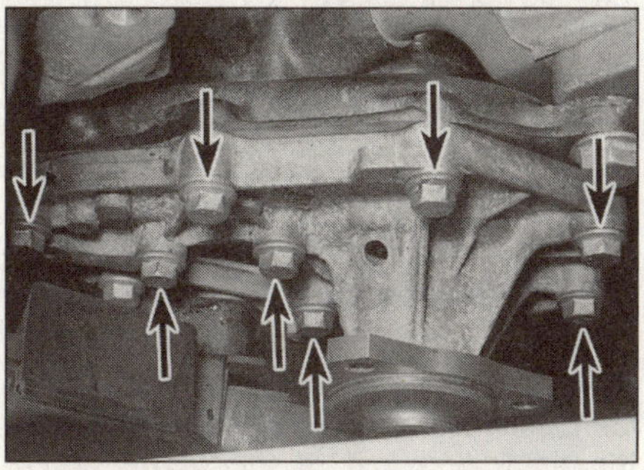

7.11a Slacken and remove the retaining bolts (arrows - not all bolts visible) . . .

7.10 On diesel models, remove the auxiliary drivebelt, then undo the retaining bolts and remove the coolant pump drive pulley

Refitting

12 Ensure that pump and cylinder block/housing mating faces are clean and dry, and that the locating dowels are correctly positioned.

13 Offer up the new gasket and fit the pump assembly, tightening its retaining bolts securely.

14 Refit the coolant pump pulley, refit the drivebelt and tension as described in Chapter 1, then securely tighten the pulley retaining bolts.

15 Refill the cooling system as described in Chapter 1.

8 Heating and ventilation system - general information

The heating/ventilation system consists of a fully-adjustable blower motor, face level vents in the centre and at each end of the facia, and air ducts to the front footwells.

The control unit is located in the facia, and the controls operate flap valves to deflect and mix the air flowing through the various parts of the heating/ventilation system. The flap valves are contained in the air distribution housing, which acts as a central distribution unit, passing air to the various ducts and vents.

Cold air enters the system through the grille at the rear of the engine compartment. If required, the airflow is boosted by the blower,

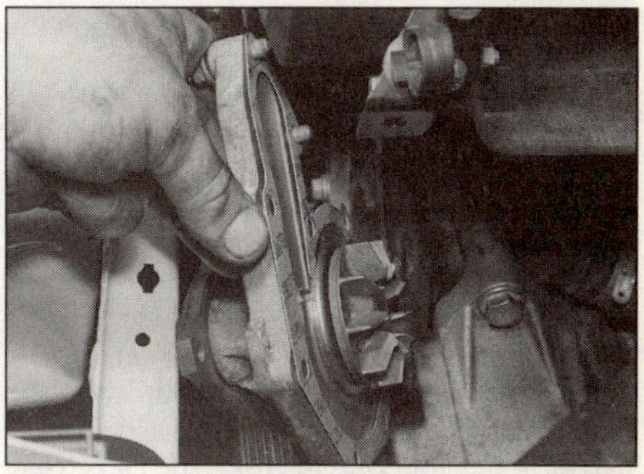

7.11b . . . then manoeuvre the pump out of position

and then flows through the various ducts, according to the settings of the controls. Stale air is expelled through ducts at the rear of the vehicle. If warm air is required, the cold air is passed over the heater matrix, which is heated by the engine coolant.

On some models, a recirculation switch enables the outside air supply to be closed off, while the air inside the vehicle is recirculated. This can be useful to prevent unpleasant odours entering from outside the vehicle, but should only be used briefly, as the recirculated air inside the vehicle will soon become stale.

9 Heater/ventilation components - removal and refitting

Models without air conditioning

Heater/ventilation control unit

1 Disconnect the battery negative lead.
2 Remove the cigarette lighter as described in Chapter 12.
3 Unclip the control unit and withdraw it from the facia **(see illustration)**.
4 Disconnect the wiring connectors, then release the outer cable clips from the unit and detach the inner cables from their levers; note that the control cables are colour-coded for identification **(see illustrations)**. If necessary, unclip the control unit from the facia panel.
5 Refitting is a reversal of removal, but ensure that the control cables (where fitted) are securely reconnected to their original locations; the air temperature (hot/cold) cable is red, and the air distribution cable is blue. Check the operation of the controls prior to refitting the cigarette lighter.

Heater/ventilation control cables

6 Remove the control unit as described in paragraphs 1 to 4. Access to the cable fixings on the right-hand side of the air distribution housing can be gained both from beneath the facia and through the control unit aperture.
7 Make alignment marks between the control cables and their retaining clips on the side of the air distribution housing. The marks can then be used on refitting to ensure that the cables are correctly positioned.
8 Unhook the retaining clip(s) then detach the cable(s) from the housing and lever and withdraw from the facia; note that the control cables are colour-coded for identification.
9 On refitting, manoeuvre the cable(s) into position, noting that they are colour-coded; the air temperature (hot/cold) cable is red, and the air distribution cable is blue.
10 Engage the inner cable(s) with the housing levers, and seat the outer cable(s) on the housing. Align the retaining clips with the marks made prior to removal, and secure both cables in position with the retaining clips. Refit the control unit and check the operation of the cables.

9.3 Unclip the heater control unit from the facia . . .

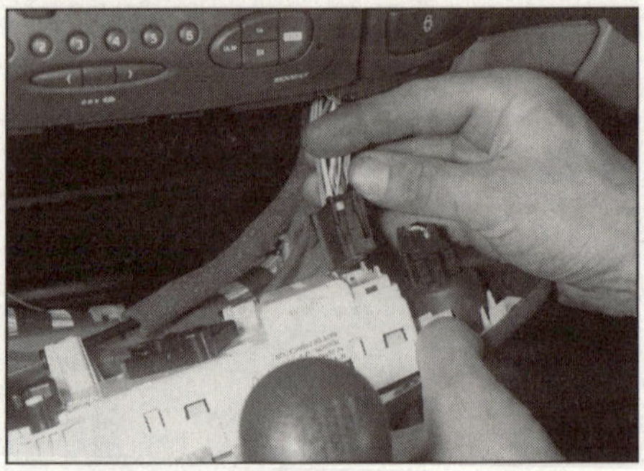

9.4a . . . and disconnect the wiring connectors

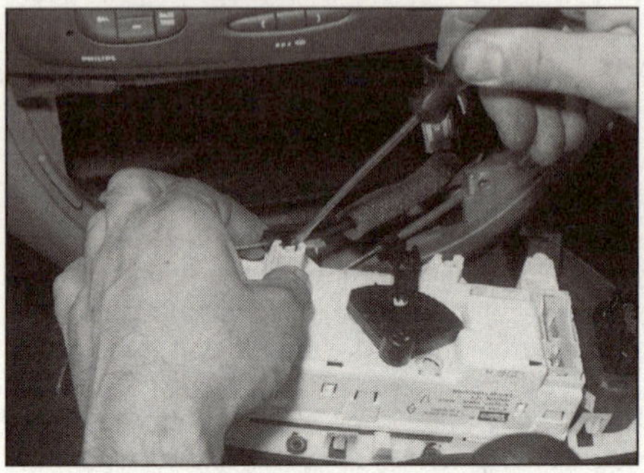

9.4b Release the outer cable retaining clips . . .

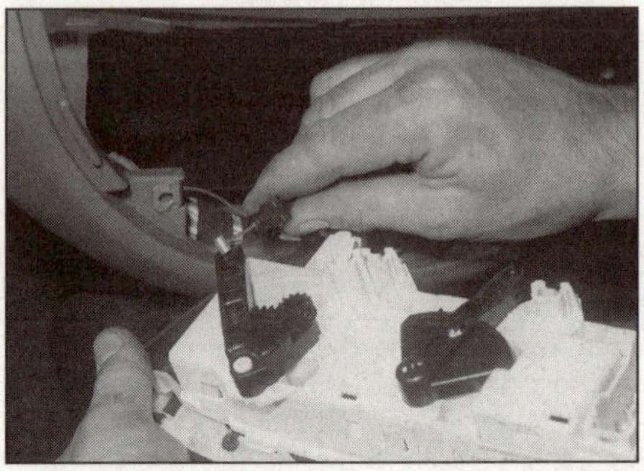

9.4c . . . then detach the inner cables from the levers, and remove the control unit from the vehicle

Heater blower motor - models with standard heating/ventilation system

11 Disconnect the battery negative terminal. Referring to Chapter 12, on right-hand-drive models remove the passenger side wiper arm, and on left-hand-drive models remove both windscreen wiper arms.
12 Remove the sealing strip from the top of the engine compartment bulkhead.
13 Undo the retaining screws and remove the inlet vent panel from beneath the left-hand side of the windscreen.
14 Undo the retaining screws/nuts and remove the plastic cover from the left-hand side of the blower motor.
15 Disconnect the wiring connectors from the right-hand side of the blower motor assembly.
16 Slacken and remove the blower motor mounting bolts, then manoeuvre the motor out from the left-hand side.
17 Refitting is the reverse of removal, ensuring that the blower motor seal is in good condition.

Heater blower motor - models with standard heating/ventilation system incorporating an air recirculation switch

18 Carry out the operations described in paragraphs 11 to 13.
19 Disconnect the wiring connectors from the right-hand end of the blower motor housing, then unscrew the end mounting bolt.
20 Unscrew the mounting bolts and remove the strengthening bar which is fitted between the front suspension strut turrets.
21 On petrol engine models, in order to allow the left-hand section of the engine compartment bulkhead to be removed, carry out the following operations.

a) *On 1.4 litre models, remove the air cleaner housing and free the MAP sensor and the ignition control unit from the bulkhead (see Chapters 4 and 5).*
b) *On 1.6 litre models, remove the air cleaner inlet duct and free the MAP sensor from the bulkhead (see Chapter 4)* **(see illustration)**.

22 On diesel engine models, in order to allow the left-hand section of the engine compartment bulkhead to be removed, carry out the following operations.

a) *Free the preheating unit from the bulkhead (see Chapter 5).*

9.21 On 1.6 litre petrol models, remove the MAP sensor from the bulkhead panel

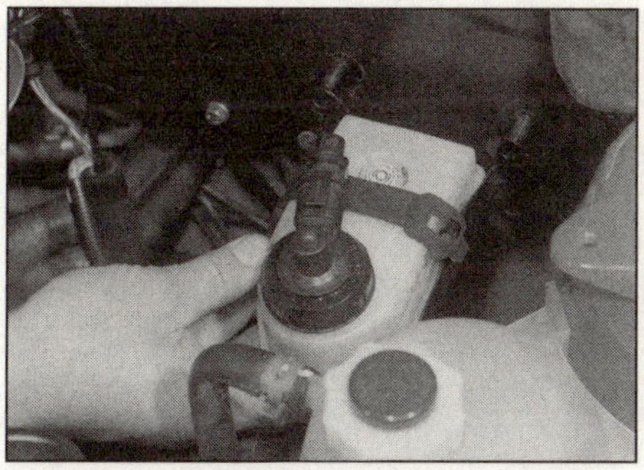

9.23 Unscrew the retaining nuts, and free the braking system master cylinder reservoir from the bulkhead

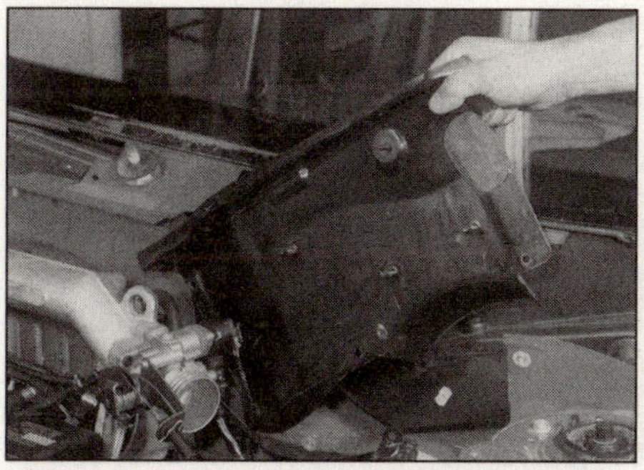

9.24 Undo the retaining bolts and remove the bulkhead panel to gain access to the blower motor housing

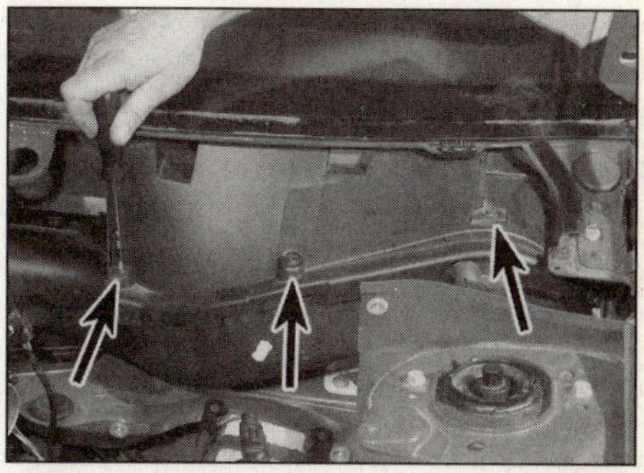

9.26a On models with a one-piece blower motor housing, undo the retaining screws (arrowed) . . .

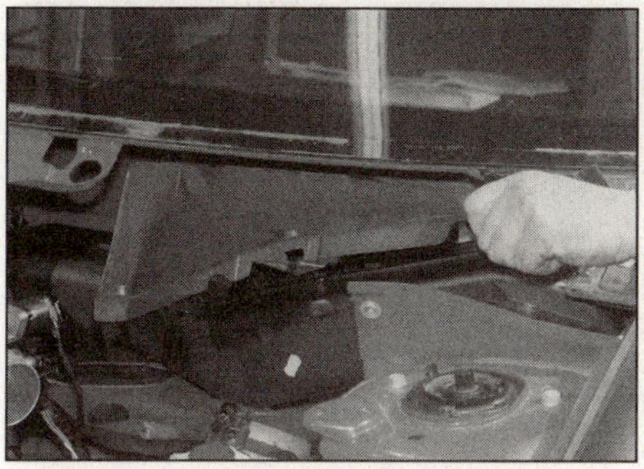

9.26b . . . and remove the top cover from the housing

9.26c Unscrew the retaining bolts and remove the blower motor housing assembly from the vehicle

9.32 On some models, it will be necessary to disconnect the control module wiring from the motor (arrow) and release it from its clips to allow the module to be removed

b) Free the solenoid valves (where fitted) from the bulkhead.

c) Release all pipes/hoses and their mounting brackets from the bulkhead panel.

23 Slacken the retaining nuts, and free the braking system master cylinder fluid reservoir bracket from the bulkhead panel **(see illustration)**. Position the reservoir clear, ensuring it stays upright to avoid fluid spillage.

24 On all models, slacken and remove the retaining bolts and remove the heatshield (where fitted). Slacken and remove the five retaining bolts and manoeuvre the bulkhead panel out from the engine compartment **(see illustration)**.

25 On models with a two-piece blower motor housing, undo the retaining screws and remove the top cover. Unscrew the bolts securing the two halves of the housing together, then separate the two halves and remove the right-hand end section from the vehicle. Disconnect the wiring connector from the blower motor, then undo the two mounting bolts and remove the left-hand section of the blower motor housing. Undo the retaining screws and remove the blower motor assembly from its housing.

26 On models with a one-piece blower motor housing, undo the retaining screws and remove the top cover from the housing. Slacken and remove the retaining bolts, and manoeuvre the assembly out of position **(see illustrations)**. At the time of writing, it was unclear whether the blower motor was available separately from the housing; refer to your Renault dealer for the latest parts information.

27 Refitting is the reverse of removal, ensuring all the housing seals are in good condition. Check the operation of the blower motor before refitting the bulkhead.

Heater blower motor control module

28 Disconnect the battery negative terminal.

29 Referring to Chapter 12, on right-hand drive-models remove the passenger side wiper arm, and on left-hand-drive models remove both windscreen wiper arms.

30 Remove the sealing strip from the top of the engine compartment bulkhead.

31 Undo the retaining screws and remove the inlet vent panel from beneath the left-hand side of the windscreen.

32 On models where an air recirculation switch is incorporated into the system, disconnect the wiring connectors from the control module which is mounted on the right-hand end of the blower motor housing; on some models it will also be necessary to disconnect the wiring connector from the blower motor and free it from its retaining clips so that its free to be removed with the module housing **(see illustration)**. Undo the retaining screws then release the retaining clips and remove the control module from the.

33 On models without an air recirculation switch, the control module is mounted onto the right-hand side of the blower motor. Disconnect the wiring, then undo the retaining screws (where fitted) and unclip the unit from the blower motor housing **(see illustration)**.

34 Refitting is the reverse of removal.

Air distribution housing

35 Drain the cooling system as described in Chapter 1. Alternatively, clamp the heater matrix hoses as close as possible to the engine compartment bulkhead.

36 Remove the facia assembly as described in Chapter 11.

37 Release the retaining clips, and disconnect the coolant hoses from the heater matrix unions **(see illustration)**.

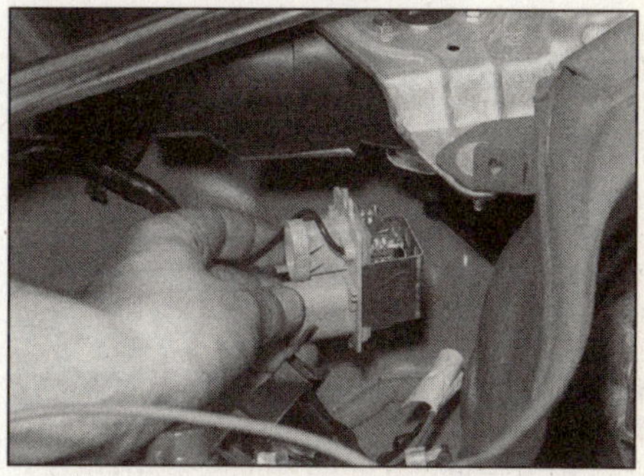

9.33 Removing the blower motor control module from the housing

9.37 Release the retaining clips and disconnect the coolant hoses (arrowed) from the heater matrix unions

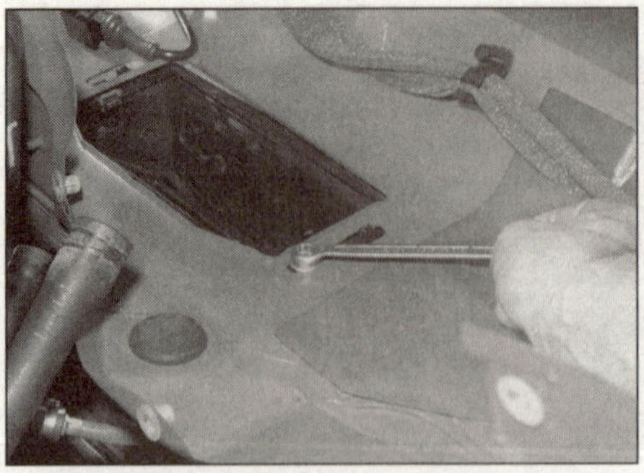

9.39 With the blower motor housing removed, slacken and remove the air distribution housing bolt

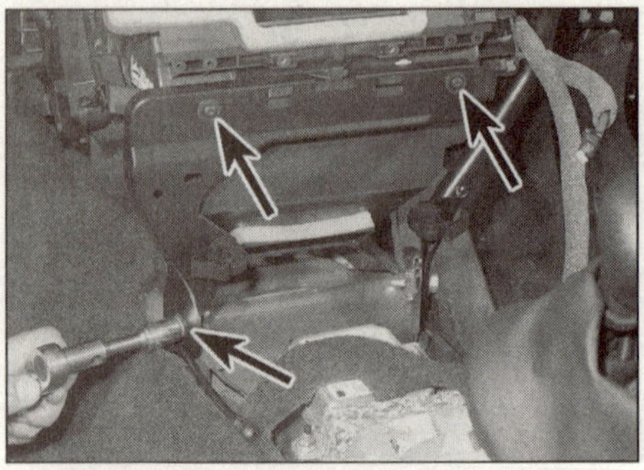

9.40 Slacken and remove the retaining screws and lower bolt (arrowed) and remove the mounting bracket . . .

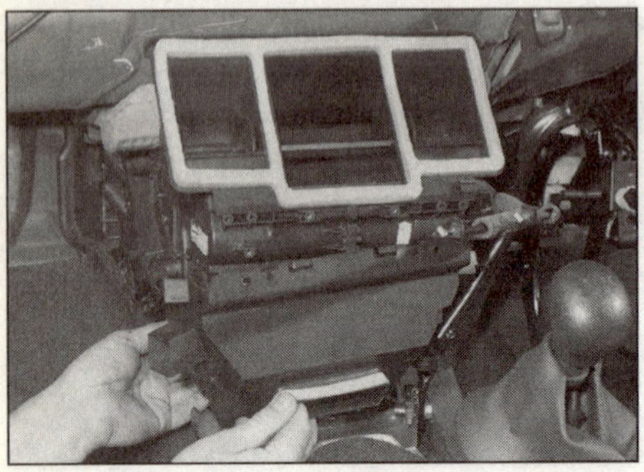

9.41 . . . then manoeuvre the lower duct and housing assembly out from the vehicle

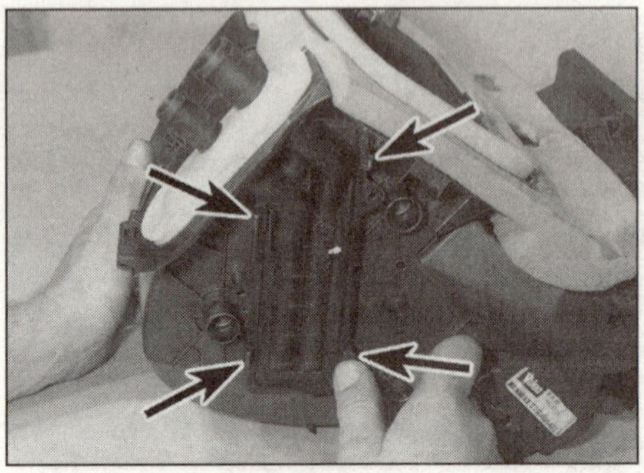

9.44a Release the retaining clips (arrowed) . . .

9.44b . . . and carefully slide the heater matrix out from the housing

38 Remove the blower motor/housing as described earlier in this Section to gain access to the air distribution housing retaining bolt. **Note:** *On models with a two-piece housing, it will only be necessary to remove the right-hand housing section.*

39 Slacken and remove the air distribution housing bolt **(see illustration)**.

40 From inside the vehicle, slacken and remove the housing mounting screws, then unscrew the bolt securing the lower mounting bracket to the floor and remove the bracket **(see illustration)**.

41 Carefully ease housing and lower duct assembly clear of the bulkhead, and remove it from the vehicle **(see illustration)**. As the housing is removed, try and keep the heater matrix unions uppermost, to prevent coolant spillage on the interior. Place absorbent rags around the housing as a precaution, and mop up any spilt coolant immediately with a damp cloth to prevent staining.

42 Refitting is the reverse of removal, noting that the mounting bracket lower bolt should be tightened by hand only, and should only be fully tightened once the blower motor housing has been refitted. On completion, top-up/refill the cooling system as described in Chapter 1.

Heater matrix

Note: *If leakage is the reason for removing the matrix, bear in mind that minor leaks can often be cured using a radiator sealant with the matrix in situ.*

43 Remove the air distribution housing as described earlier in this Section.

44 Slacken and remove the retaining screws (where fitted), then release the retaining clips and slide the matrix out from the housing, taking great care not to damage the matrix fins **(see illustrations)**.

45 Refitting is the reverse of removal. If the retaining clips were damaged on removal, secure the matrix in position with screws using the holes provided.

Recirculation valve flap motor - models incorporating an air recirculation switch

46 Remove the blower motor housing as described earlier in this Section.

47 Trace the wiring back from the flap motor, freeing it from the retaining clips, and disconnect its wiring connector.

48 Undo the retaining nuts and remove the flap housing assembly from the vehicle **(see illustration)**.

49 With the assembly on the bench, slacken and remove the retaining screws, and remove the motor and the flap drivegear from the housing **(see illustrations)**.

50 Refitting is the reverse of removal, ensuring

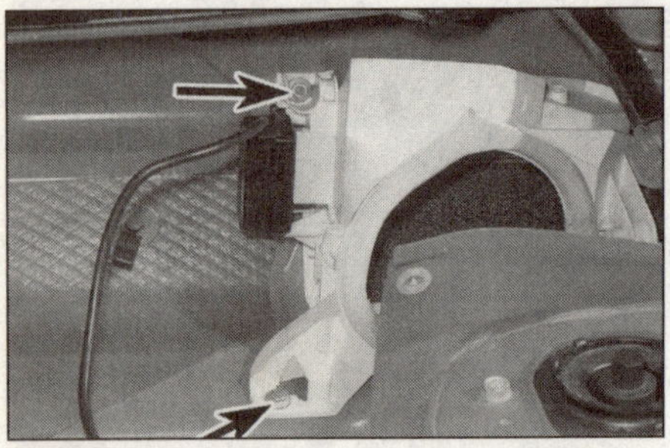

9.48 Slacken and remove the nuts (arrowed), and remove the recirculation flap valve assembly from the vehicle

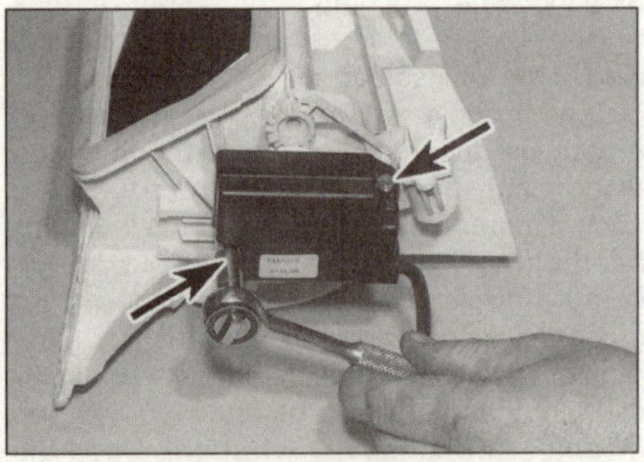

9.49a Undo the retaining screws (arrowed) . . .

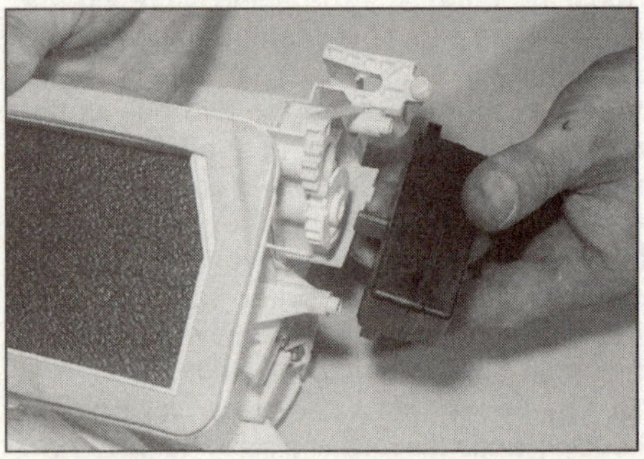

9.49b . . . and remove the motor and drivegears from the housing

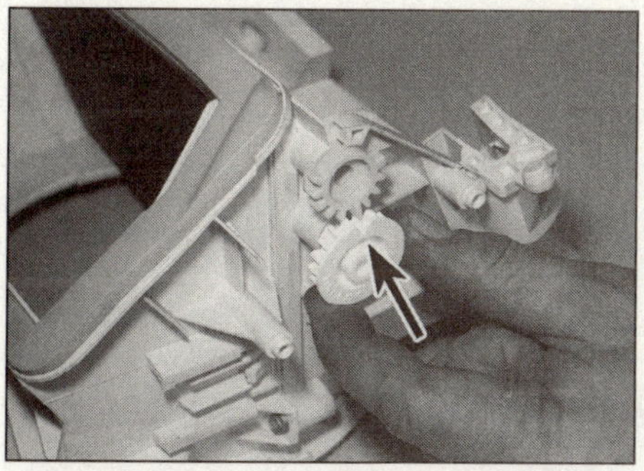

9.50 On refitting, ensure that the slot on the motor gear (arrowed) is correctly aligned with the lug on the flap gear

that the drivegears are correctly engaged so that the slot on the motor gear aligns with the lug on the flap gear as they rotate **(see illustration)**. Engage the motor with the gear, and securely tighten its retaining screws.

Models with air conditioning

51 On models equipped with air conditioning, the only operations which can be easily performed by the home mechanic are removal/refitting of the heater/ventilation control panel and cables, and of the blower motor control module. These operations can be carried out using the information given earlier for standard models.

52 Removal of the blower motor, air distribution housing, heater matrix and recirculation valve flap motor will have to be entrusted to a Renault dealer. This is since none of these operations are possible without discharging and disconnecting the air conditioning system. Therefore, the task should be entrusted to a Renault dealer with access to the necessary equipment required to discharge/recharge the system safely (see Section 10).

Warning: Do not attempt to open the refrigerant circuit. Refer to the precautions in Section 10.

10 Air conditioning system - general information and precautions

General information

An air conditioning system is available on certain models. It enables the temperature of incoming air to be lowered, and also dehumidifies the air, which makes for rapid demisting and increased comfort.

The cooling side of the system works in the same way as a domestic refrigerator. Refrigerant gas is drawn into a belt-driven compressor, and passes into a condenser mounted on the front of the radiator, where it loses heat and becomes liquid. The liquid passes through an expansion valve to an evaporator, where it changes from liquid under high pressure to gas under low pressure. This change is accompanied by a drop in temperature, which cools the evaporator. The refrigerant returns to the compressor, and the cycle begins again.

Air blown through the evaporator passes to the air distribution unit, where it is mixed with hot air blown through the heater matrix to achieve the desired temperature in the passenger compartment.

The heating side of the system works in the same way as on models without air conditioning (see Section 8).

The operation of the system is controlled electronically by the control unit, which controls the electric cooling fan(s), the compressor and the facia-mounted warning light. Any problems with the system should be referred to a Renault dealer.

Precautions

When an air conditioning system is fitted, it is necessary to observe special precautions whenever dealing with any part of the system, or its associated components. If for any reason the system must be disconnected, entrust this task to your Renault dealer or a refrigeration engineer.

Warning: The refrigeration circuit may contain a liquid refrigerant (Freon), and it is therefore dangerous to disconnect any part of the system without specialised knowledge and equipment.

The refrigerant is potentially dangerous, and should only be handled by qualified persons. If it is splashed onto the skin, it can cause frostbite. It is not itself poisonous, but in the presence of a naked flame (including a cigarette) it forms a poisonous gas. Uncontrolled discharging of the refrigerant is dangerous, and potentially damaging to the environment.

Do not operate the air conditioning system if it is known to be short of refrigerant, as this may damage the compressor.

11 Air conditioning system components - removal and refitting

Warning: Do not attempt to open the refrigerant circuit. Refer to the precautions in Section 10.

1 The only operation which can be carried out easily without discharging the refrigerant is the renewal of the compressor drivebelt. This is described in Chapter 1. All other operations must be referred to a Renault dealer or an air conditioning specialist.

2 If necessary, the compressor can be unbolted and moved aside, without disconnecting its flexible hoses, after removing the drivebelt.

Chapter 4 Part A:
Fuel/exhaust systems - petrol engine models

Contents

Degrees of difficulty

Easy, suitable for novice with little experience	**Fairly easy,** suitable for beginner with some experience	**Fairly difficult,** suitable for competent DIY mechanic	**Difficult,** suitable for experienced DIY mechanic	**Very difficult,** suitable for expert DIY or professional

Specifications

System type

Single-point injection models (1390 cc E7J engine)	Seimens - Fenix 3 Monopoint with Bosch throttle body
Multi-point injection models (1598 cc K7M engine)	Siemens - Fenix 5 Semi-sequential Multipoint with Peirburg throttle body

Fuel system data

Fuel pump type	Electric, immersed in tank
Fuel pump regulated constant pressure:	
Single-point injection models	1.06 bars
Multi-point injection models:	
With fuel pressure regulator vacuum hose attached	2.5 ± 0.2 bars
With fuel pressure vacuum hose disconnected	3.0 ± 0.2 bars
Air temperature sensor resistance:	
At 0°C	7470 to 11970 ohms
At 20°C	3060 to 4045 ohms
At 40°C	1315 to 1600 ohms
Coolant temperature sensor resistance:	
At 20°C	3060 to 4045 ohms
At 40°C	1315 to 1600 ohms
At 80°C	300 to 370 ohms
At 90°C	210 to 270 ohms
Injector resistance:	
Single-point injection	2.0 ohms
Multi-point injection	14.5 ± 1 ohms
Specified idle speed (not adjustable - controlled by ECU):	
1390 cc/E7J 764 engine	750 ± 50 rpm
1598 cc/K7M 702 and 1598cc/K7M 720 engines	720 ± 50 rpm
1598 cc/K7M 703 engine	750 ± 50 rpm
Idle mixture CO content	Less than 0.3 % (controlled by ECU)

Recommended fuel

Minimum octane rating	95 RON unleaded (UK unleaded premium). Leaded fuel must **not** be used

4A

Torque wrench settings

Torque wrench settings	Nm	lbf ft
Exhaust manifold retaining nuts	20	15
Fuel pump/gauge sender locking ring	35	26
Fuel rail mounting nuts - multi-point injection models	8	6
Inlet manifold retaining nuts/bolts	25	19
Throttle body/housing bolts (K7M engine - multi-point injection)	20	15

1 General information and precautions

The fuel system consists of a fuel tank which is mounted under the rear of the vehicle with an electric fuel pump immersed in it, a fuel filter and the fuel feed and return lines. On single-point injection models, the fuel is supplied by a throttle body assembly which incorporates the single fuel injector and the fuel pressure regulator. On multi-point injection models, the fuel pump supplies fuel to the fuel rail, which acts as a reservoir for the four fuel injectors which inject fuel into the inlet tracts - a semi-sequential system is used whereby the injectors are controlled in pairs (1 and 4, and 2 and 3). In addition, there is an Electronic Control Unit (ECU) and various sensors, electrical components and related wiring.

Refer to Section 7 for further information on the operation of each fuel injection system, and to Section 17 for information on the exhaust system.

Warning: Many of the procedures in this Chapter require the removal of fuel lines and connections, which may result in some fuel spillage. Before carrying out any operation on the fuel system, refer to the precautions given in "Safety first!" at the beginning of this manual, and follow them implicitly. Petrol is a highly-dangerous and volatile liquid, and the precautions necessary when handling it cannot be overstressed.

Note: *Residual pressure will remain in the fuel lines long after the vehicle was last used. When disconnecting any fuel line, first depressurise the fuel system as described in Section 8.*

2 Air cleaner assembly and inlet ducts - removal and refitting

Removal

Single-point injection engine

1 Unscrew and remove the bolts and cover screw from the top of the air cleaner, then release the clips and remove the cover from the body. Remove the element from the body **(see illustrations)**.

2 Disconnect the warm air hose from the exhaust manifold shroud, then disconnect the ambient air hose from the inlet elbow on the air cleaner body **(see illustrations)**.

3 Release the rubber strap, then lift the body from the throttle housing and disconnect the crankcase ventilation hose **(see illustrations)**.

4 If necessary, unbolt and remove the support bracket. Remove the hose and unbolt the air inlet duct from the front of the engine compartment.

Multi-point injection engine

5 For improved access, unbolt and remove

2.1a Removing the bolts . . .

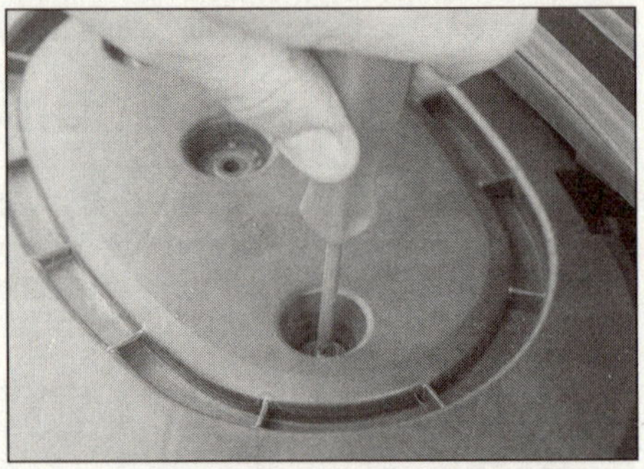

2.1b . . . and cover screw (E7J engine)

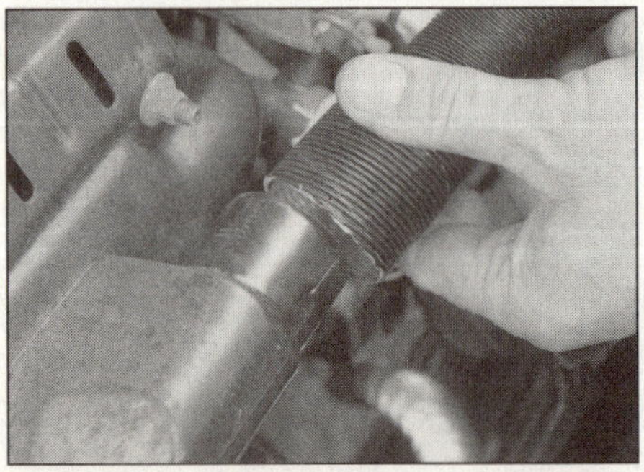

2.2a Disconnecting warm air hose from the exhaust manifold shroud (E7J engine)

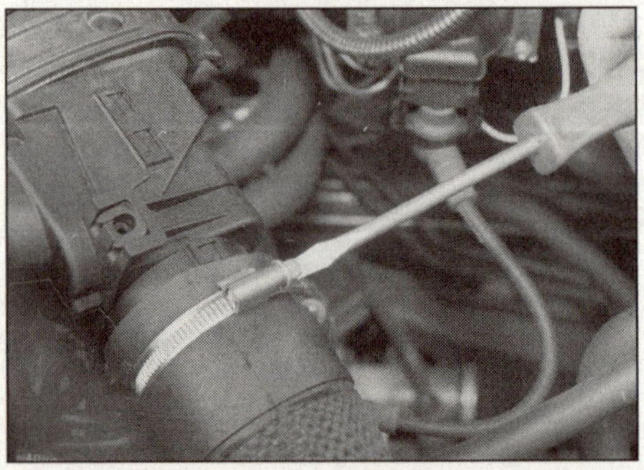

2.2b Loosen the clip . . .

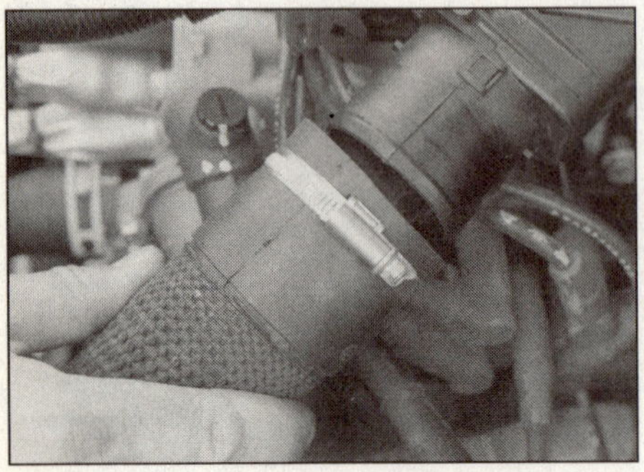

2.2c . . . and disconnect the ambient air hose (E7J engine)

2.3a Release the rubber strap . . .

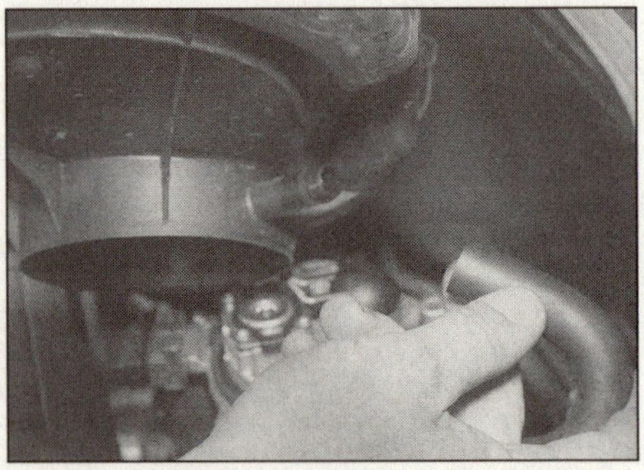

2.3b . . . and disconnect the crankcase ventilation hose (E7J engine)

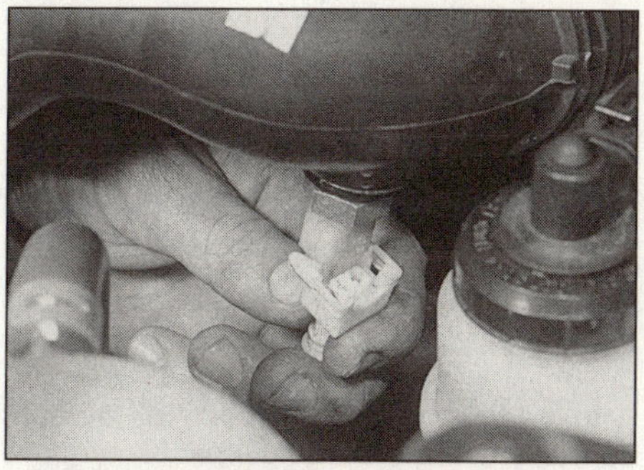

2.6 Disconnecting the wiring from the inlet air temperature sensor (K7M engine)

2.7a Disconnecting the intermediate air duct from the air cleaner (K7M engine)

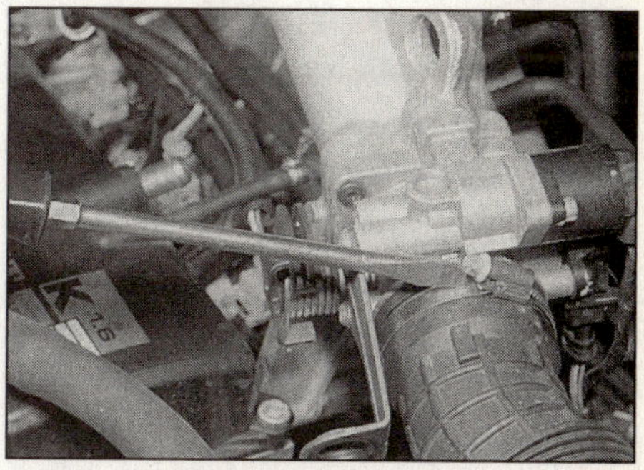

2.7b Loosen the clip . . .

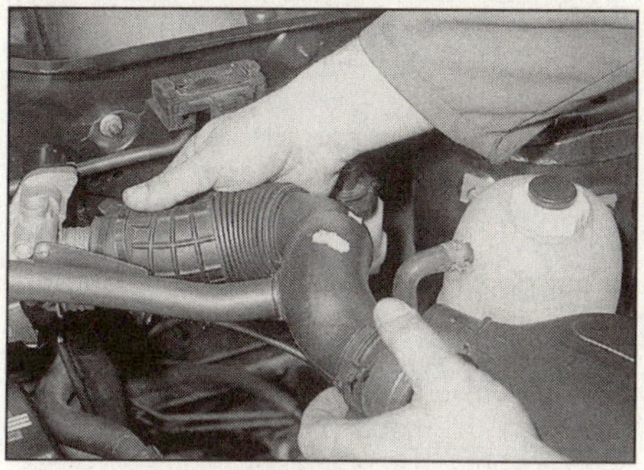

2.7c . . . and disconnect the intermediate air duct from the throttle housing (K7M engine)

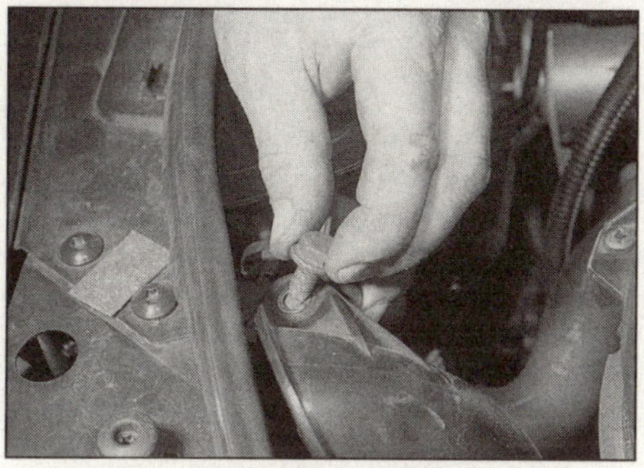

2.8 Unscrew the inlet air elbow mounting bolt . . .

2.9 . . . and lift the air cleaner body and primary hose from the engine compartment (K7M engine)

the tie bar from the front suspension strut turrets.

6 Disconnect the wiring from the inlet air temperature sensor on the intermediate air duct **(see illustration)**.

7 Loosen the clip and disconnect the intermediate air duct from the air cleaner, then loosen the clip and disconnect the duct from the throttle housing. Disconnect the crankcase ventilation hose and remove the air duct **(see illustrations)**.

8 Unscrew and remove the bolt securing the inlet air elbow to the front of the engine compartment **(see illustration)**.

9 Lift the air cleaner body and primary hose from the holder in the engine compartment **(see illustration)**.

10 If necessary, disconnect the primary air inlet hose from the bottom of the air cleaner body and remove the element with reference to Chapter 1A.

Refitting

11 Refitting is a reversal of removal.

3 Air cleaner air temperature control system (single-point injection) - general information

The system is controlled by a heat-sensitive wax capsule located in the air cleaner body inlet. The capsule moves a flap which controls the mix of hot and cold air admitted to the engine to ensure maximum efficiency. When the ambient air is below a certain temperature, the flap is positioned to admit warm air from the exhaust manifold shroud, however, with warm ambient air the flap moves to admit air from the front of the vehicle and the warm air inlet is closed.

If the system does not function correctly, the air cleaner body must be renewed; it is not possible to renew the flap valve separately.

4 Accelerator cable - removal, refitting and adjustment

Removal

1 On single-point injection engines, remove the air cleaner assembly as described in Section 2. On multi-point injection engines, unbolt and remove the tie bar from between the front suspension strut turrets.

2 On the throttle housing, turn the throttle quadrant by hand to release the cable tension. Disconnect the inner cable from the quadrant **(see illustration)**.

3 Remove the outer cable and grommet from the support bracket **(see illustration)**. On automatic transmission models, disconnect the kickdown sensor wiring from the end of the outer cable. If necessary, remove the spring clip from the end of the cable noting its position in the groove.

4 Working inside the car, remove the lower trim panel from under the steering column. Disconnect the cable from the accelerator pedal by squeezing the lugs of the cable end fitting **(see illustration)**.

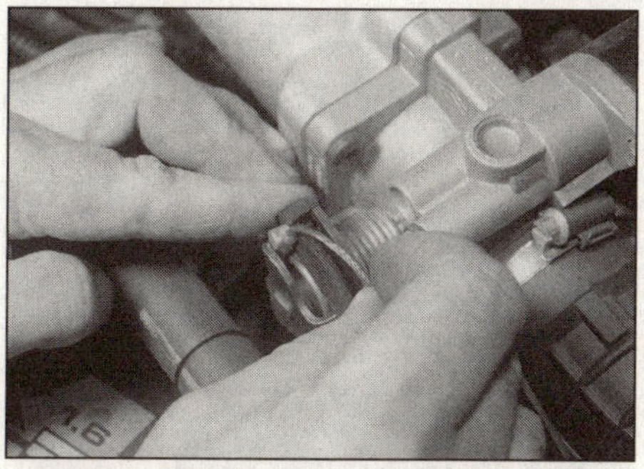

4.2 Disconnecting the accelerator cable from the quadrant on the throttle housing (K7M engine)

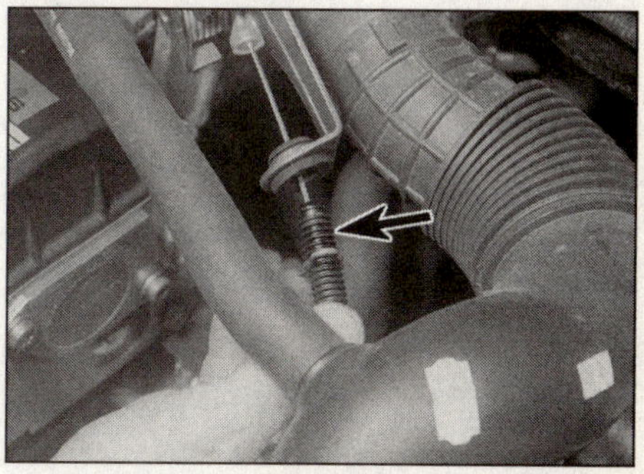

4.3 Removing the outer cable and grommet from the support bracket (K7M engine)

4.4 Accelerator cable end fitting on the top of the pedal

4.8b Accelerator cable adjustment ferrule . . .

5 Return to the engine compartment, release the outer cable from the bulkhead and withdraw the cable.

Refitting

6 Refitting is a reversal of removal, but adjust the cable if necessary as follows. Make the adjustment before refitting the tie bar on multi-point injection engines or the air cleaner assembly on single-point engines.

Adjustment

7 With the spring clip removed from the accelerator outer cable, ensure that the throttle cam is fully against its stop. Gently pull the cable out of its grommet until all free play is removed from the inner cable.

8 With the cable held in this position, refit the spring clip to the last exposed outer cable groove in front of the rubber grommet and washer **(see illustrations)**. When the clip is refitted and the outer cable is released, there should be only a small amount of free play in the inner cable. **Note:** *The idle speed control motor opens the throttle slightly when the ignition is switched off. There must be enough slack in the cable to allow the throttle to close past this position, otherwise a stable idle speed will not be obtained.*

9 Have an assistant depress the accelerator pedal, and check that the throttle cam opens fully and returns smoothly to its stop.

4.8a Accelerator cable adjustment ferrule and support (E7J engine)

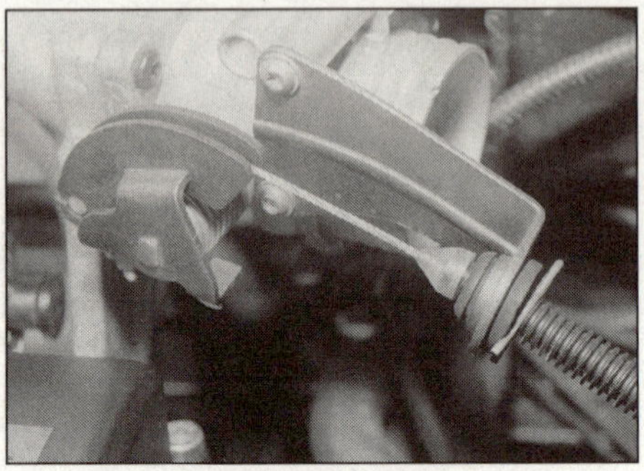

4.8c . . . and support bracket (K7M engine)

5 Accelerator pedal - removal and refitting

Removal

1 Remove the lower trim panel from under the steering column.

2 Disconnect the cable from the top of the accelerator pedal by squeezing the lugs of the cable end fitting.

3 Remove the nut which secures the accelerator pedal pivot and bush to the bulkhead **(see illustration)**.

4 Withdraw the accelerator pedal.

5 Examine the pedal and pivot for signs of wear and renew as necessary.

Refitting

6 Refitting is a reversal of removal. Check the adjustment of the accelerator cable as described in Section 4.

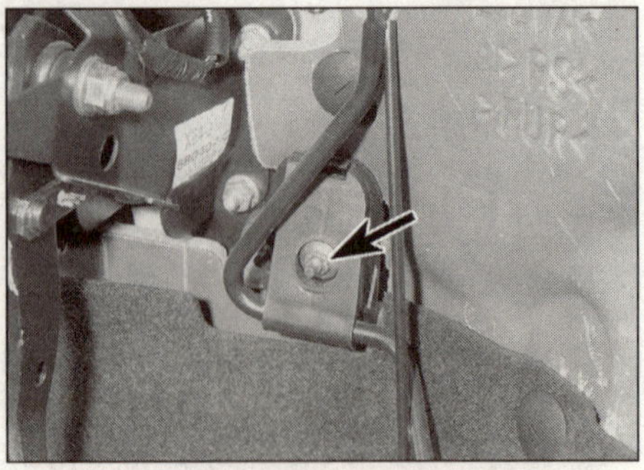

5.3 Accelerator pedal pivot and bush mounting nut

6 Unleaded petrol - general information and usage

Note: *The information given in this Chapter is correct at the time of writing. If updated information is thought to be required, check with a Renault dealer. If travelling abroad, consult one of the motoring organisations (or a similar authority) for advice on the fuel available.*

The fuel recommended by Renault is given in the *Specifications* section of this Chapter, followed by the equivalent petrol currently on sale in the UK.

All petrol models are designed to run on fuel with a minimum octane rating of 95 (RON). All models have a catalytic converter, and so must be run on unleaded fuel only. Under no circumstances should leaded fuel (UK "4-star") be used, as this may damage the converter.

Super unleaded petrol (98 octane) can also be used in all models if wished, though there is no advantage in doing so.

7 Fuel injection systems - general information

Single-point injection system (Seimens-Fenix 3 Monopoint with Bosch throttle body)

1 Single-point injection models are equipped with a Seimens-Fenix 3 Monopoint fuel injection/ignition system with a Bosch throttle body. The system incorporates a closed-loop catalytic converter and an evaporative emission control system, and complies with the latest emission control standards. The fuel injection side of the system operates as follows; refer to Chapter 5 for information on the ignition system.

2 The fuel pump, immersed in the fuel tank, pumps fuel from the fuel tank to the fuel injector, via a filter mounted underneath the rear of the vehicle. Fuel supply pressure is controlled by the pressure regulator in the throttle body assembly. The regulator operates by allowing excess fuel to return to the tank.

3 The electrical control system consists of the ECU, along with the following sensors.

a) Throttle potentiometer - informs the ECU of the throttle position, and the rate of throttle opening or closing.

b) Coolant temperature sensor - informs the ECU of engine temperature.

c) Inlet air temperature sensor - informs the ECU of the temperature of the air passing through the throttle body.

d) *Lambda sensor - informs the ECU of the oxygen content of the exhaust gases (explained in greater detail in Part C of this Chapter).*

e) *Microswitch (built into idle speed stepper motor) - informs the ECU when the throttle valve is closed (ie when the accelerator pedal is released).*

f) *Crankshaft sensor - informs the ECU of engine speed and crankshaft position.*

g) *Power steering pressure switch - informs the ECU when the power steering pump is working so the engine idle speed can be increased to prevent stalling.*

h) *Knock sensor - informs the ECU when pre-ignition ("pinking") is occurring.*

i) *Manifold absolute pressure (MAP) sensor - informs the ECU of the engine load by monitoring the pressure in the inlet manifold.*

j) *Fuel vapour recirculation valve - operates the fuel evaporative control system (explained in greater detail in Part C of this Chapter).*

4 All the above information is analysed by the ECU and, based on this, the ECU determines the appropriate ignition and fuelling requirements for the engine. The ECU controls the fuel injector by varying its pulse width - the length of time the injector is held open - to provide a richer or weaker mixture, as appropriate. The mixture is constantly varied by the ECU, to provide the best setting for cranking, starting (with either a hot or cold engine), warm-up, idle, cruising, and acceleration.

5 The ECU also has full control over the engine idle speed, via a stepper motor which is fitted to the throttle body. The motor pushrod rests against a cam on the throttle valve spindle. When the throttle valve is closed (accelerator pedal released), the ECU uses the motor to vary the opening of the throttle valve and so control the idle speed.

6 The ECU also controls the exhaust and evaporative emission control systems, which are described in detail in Part C of this Chapter.

7 If there is an abnormality in any of the readings obtained from either the coolant temperature sensor, the inlet air temperature sensor or the Lambda sensor, the ECU enters its back-up mode. In this event, the ECU ignores the abnormal sensor signal, and assumes a pre-programmed value which will allow the engine to continue running (albeit at reduced efficiency). If the ECU enters this back-up mode, the warning light on the instrument panel will come on, and the relevant fault code will be stored in the ECU memory.

8 If the warning light comes on, the vehicle should be taken to a Renault dealer at the earliest opportunity. A complete test of the engine management system can then be carried out, using a special electronic diagnostic test unit which is simply plugged into the system's diagnostic connector (located on the fusebox on the facia panel).

9 A fuel cut-off inertia switch is incorporated into the fuel injection system. In the event of an impact the switch cuts off the electrical supply to the fuel pump and so prevents fuel being expelled should the fuel pipes/hoses be damaged in an accident.

Multi-point injection system (Seimens - Fenix 5 Semi-sequential Multipoint with Peirburg throttle body)

10 Multi-point injection models are equipped with a Siemens-Fenix 5 semi-sequential multi-point fuel injection/ignition system with a Peirburg throttle body. In many respects the multi-point system is similar to the single-point system described in the previous paragraphs, however, there are four injectors (one per cylinder) located in the inlet manifold downstream of the throttle butterfly. All the injectors are fed from a common rail, which also carries the fuel pressure regulator.

11 The system is of semi-sequential design with the injectors operating in pairs (1 and 4, then 2 and 3). There is one injection per revolution of the engine for each cylinder, and for the complete four-stroke cycle these take place during the compression and exhaust phases.

12 An EGR valve is also incorporated into the system. Information on this is given in Part C of this Chapter.

13 On automatic transmission models, information from sensors on the transmission is sent to the ECU for processing to determine the most efficient settings for the engine.

8 Fuel injection system - depressurisation

Warning: Refer to the warning note in Section 1 before proceeding. The following procedure will merely relieve the pressure in the fuel system - remember that fuel will still be present in the system components, and take precautions accordingly before disconnecting any of them.

1 The fuel system referred to in this Section includes the tank-mounted fuel pump, the fuel filter, the fuel injector(s) and the pressure regulator in the injector housing/fuel rail, and the metal pipes and flexible hoses of the fuel lines between these components. All these contain fuel which will be under pressure while the engine is running, and/or while the ignition is switched on. The pressure will remain for some time after the ignition has been switched off, and it must be relieved when any of these components are disturbed for servicing work.

Method 1

2 Disconnect the battery negative lead.

Caution: If the radio/cassette in your vehicle is equipped with an anti-theft system, make sure you have the correct activation code before disconnecting the battery.

3 Place a suitable container beneath the connection or union to be disconnected, and have a large rag ready to soak up any escaping fuel not being caught by the container.

4 Slowly loosen the connection or union nut to avoid a sudden release of pressure, and position the rag around the connection, to catch any fuel spray which may be expelled. Once the pressure is released, disconnect the fuel line. Plug the pipe ends, to minimise fuel loss and prevent the entry of dirt into the fuel system.

Method 2

5 Remove the fuel pump relay located in the left-hand front corner of the engine compartment (see Chapter 12).

6 Start the engine and allow it to idle until it stops due to lack of fuel. Operate the starter motor a couple more times, to ensure that all fuel pressure has been relieved.

7 Switch off the ignition and refit the fuel pump relay.

9 Fuel pump - removal and refitting

Warning: Refer to the warning note in Section 1 before proceeding.

Removal

1 Disconnect the battery negative lead.

Caution: If the radio/cassette in your vehicle is equipped with an anti-theft system, make sure you have the correct activation code before disconnecting the battery.

2 Remove the rear seat cushion as described in Chapter 11, for access to the fuel pump cover.

3 Using a screwdriver, carefully prise the plastic access cover from the floor to expose the fuel pump **(see illustration)**.

4 Disconnect the wiring connector from the fuel pump, and tape the connector to the

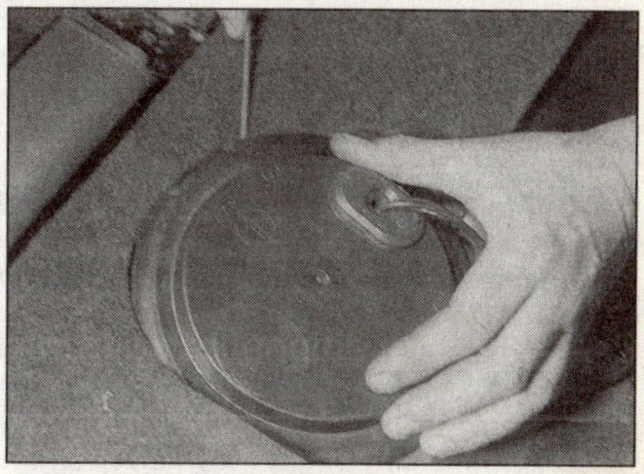

9.3 Use a screwdriver to prise up the fuel pump access cover

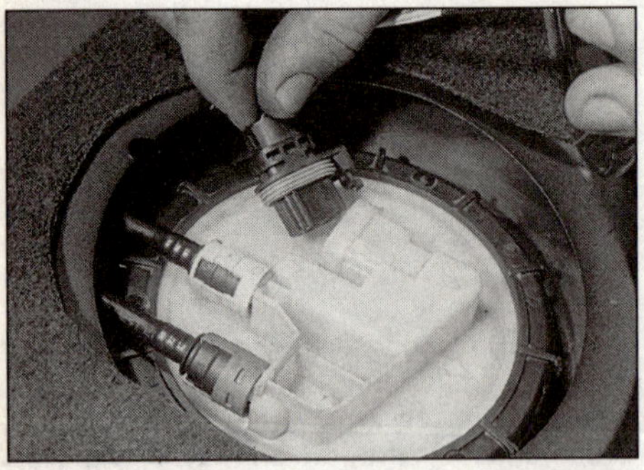

9.4 Disconnecting the fuel pump wiring

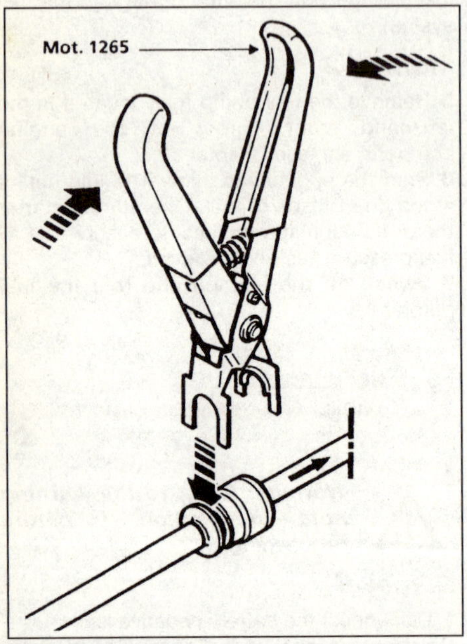

9.5a Renault special pliers (Mot.1265) for releasing the fuel hose quick-release fittings

vehicle body, to prevent it disappearing behind the tank **(see illustration)**.

5 Identify the fuel hoses for position; the green hose on the left-hand side is the supply to the engine compartment, and the red hose next to the wiring connector is the return (arrows on the top of the pump indicate the direction of fuel flow). The hoses are equipped with quick-release fittings to ease removal. To disconnect each hose, slide out the locking tab from the collar then compress the collar and detach the hose from the pump. Renault technicians use a special tool to compress the collar however it is possible to use a small screwdriver to **carefully** press back the locking collar inside the end of the special end fitting. Disconnect both hoses from the top of the pump, then plug the hose ends to minimise fuel loss **(see illustrations)**.

6 Noting the alignment arrows on the pump cover and fuel tank, unscrew the locking ring and remove it from the tank. This can be accomplished by using a screwdriver on the raised ribs of the locking ring - carefully tap the screwdriver to turn the ring anti-clockwise

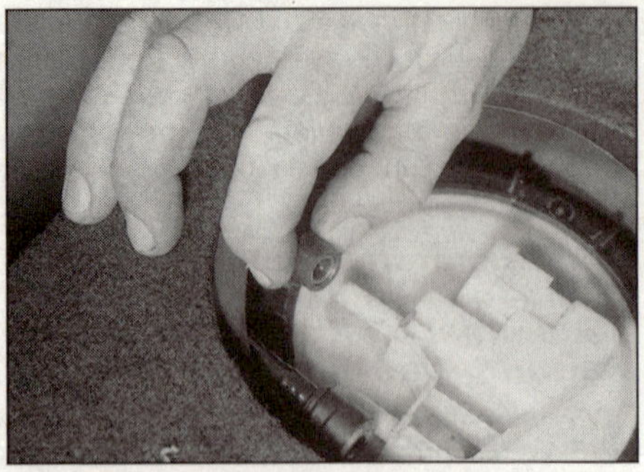

9.5b Disconnecting the fuel return hose from the fuel pump

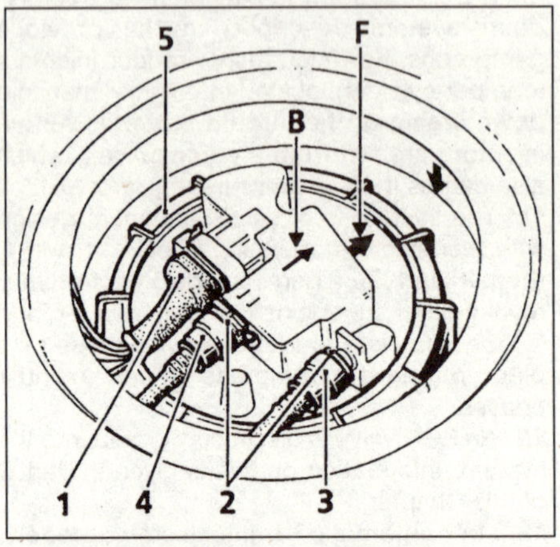

9.5c Fuel pump connections

1 *Wiring connector*
2 *Quick-release clips*
3 *Fuel supply hose*
4 *Fuel return hose*
5 *Locking ring*
B *Arrow indicating flow of return fuel*
F *The arrow must be pointing to the rear and aligned with the longitudinal axis of the car*

until it can be unscrewed by hand. Alternatively a removal tool can be fabricated out of metal bar as shown **(see illustrations)**.

7 Carefully lift the fuel pump assembly out of the fuel tank, taking great care not to damage the fuel level gauge sender arm, or to spill fuel onto the interior of the vehicle. Remove the rubber sealing ring and check it for deterioration; if it is in good condition, it may

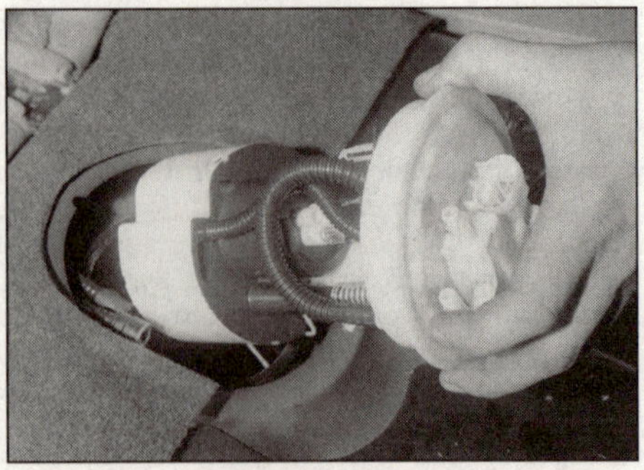

9.7a Lifting the fuel pump assembly from the fuel tank

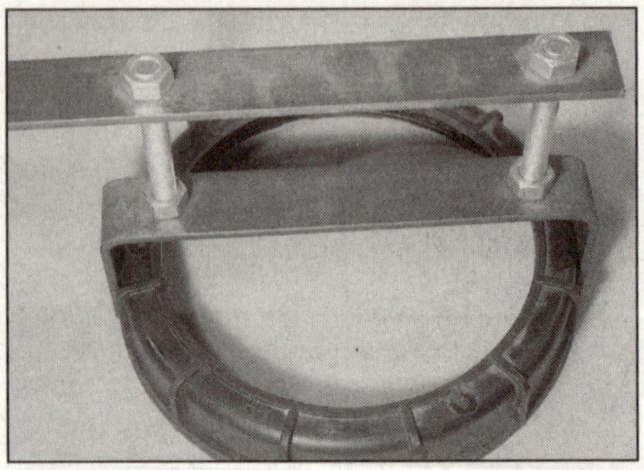

9.6a Fuel pump locking ring removal tool fabricated out of metal bar and two bolts

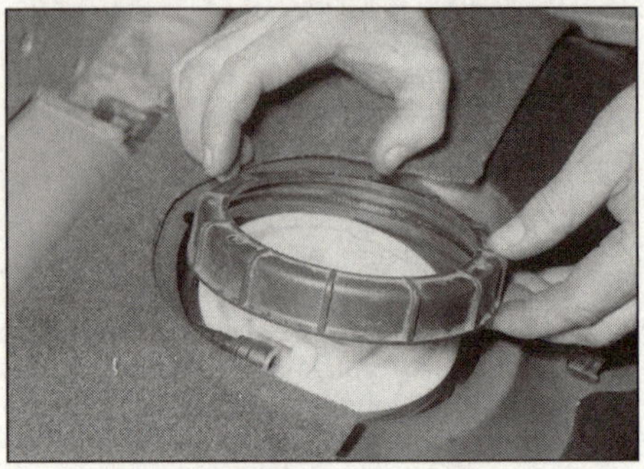

9.6b Removing the locking ring

be reused, however if the pump is to remain out of the fuel tank for several hours, the locking ring should be refitted temporarily to prevent the sealing ring from distorting. If the sealing ring is unserviceable, obtain a new one **(see illustrations)**.

8 Note that the fuel pump/fuel gauge sender unit is only available as a complete assembly - no components are available separately.

Refitting

9 Ensure that the fuel pump pick-up filter is clean and free of debris. Fit the sealing ring to the top of the fuel tank.

10 Carefully manoeuvre the pump assembly into the fuel tank.

11 Align the arrow on the fuel pump cover with the arrow on the fuel tank (the arrow must point to the rear of the vehicle), then refit the locking ring. Securely tighten the locking ring,

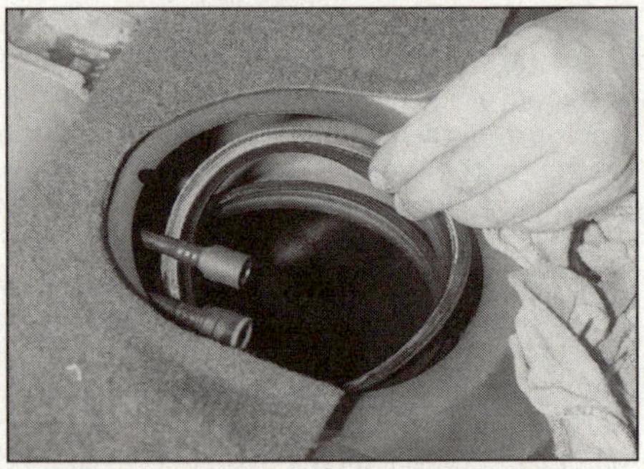

9.7b Removing the sealing ring

then recheck that the pump cover and tank marks are all correctly aligned. **Note:** *If a suitable adapter is available tighten the locking ring to the specified torque.*

12 Reconnect the feed and return hoses to the top of the fuel pump; it is not necessary to depress the collars when refitting the hoses. Check the hoses are securely in position.

13 Reconnect the wiring connector.

14 Reconnect the battery negative lead, and start the engine. Check the fuel pump feed and return hoses for signs of leakage.

15 Refit the plastic access cover and the rear seat cushion.

10 Fuel gauge sender unit - removal and refitting

The fuel gauge sender unit is an integral part of the fuel pump assembly (see Section 9).

11 Fuel tank - removal and refitting

Warning: Refer to the warning note in Section 1 before proceeding.

Removal

1 Before removing the fuel tank, all fuel must be drained from it. Since a drain plug is not provided, it is preferable to carry out the removal operation when the tank is nearly empty.

2 Remove the rear seat cushion (Chapter 11), for access to the fuel pump cover.

3 Using a screwdriver, carefully prise the plastic access cover from the floor to expose the fuel pump.

4 If there is any fuel remaining in the fuel tank, it can be removed by disconnecting the fuel delivery hose and connecting a suitable hose leading to a container outside the vehicle (refer to Section 9 for disconnecting the hose). Remove the fuel pump relay located at the left-hand front of the engine compartment, and connect a bridging wire between terminals 3 and 5 (the terminals with the thick wires). Allow the fuel pump to operate until the fuel flow is intermittent, then disconnect the bridging wire and refit the relay.

5 Disconnect the wiring connector from the fuel pump, and tape the connector to the vehicle body, to prevent it disappearing behind the tank.

6 Disconnect the battery negative lead.

Caution: If the radio/cassette in your car is equipped with an anti-theft system, make sure you have the correct activation code before disconnecting the battery.

7 Disconnect the return hose from the fuel pump with reference to Section 9.

8 Chock the front wheels then jack up the rear of the vehicle and support on axle stands (see *"Jacking and vehicle support"*). Remove the right-hand rear wheel.

9 Remove the exhaust system and relevant heat shield(s) with reference to Section 17.

10 Note the position of the adjustment nut on the rear of the handbrake lever equaliser rod, then unscrew and remove it and detach the handbrake cables from the supports on the underbody. Position the cables to one side away from the fuel tank.

11 Disconnect the hoses from the fuel filler neck (see Section 9 when disconnecting the special collar).

12 Place a trolley jack with an interposed block of wood beneath the tank, then raise the jack until it is supporting the weight of the tank.

13 Unscrew and remove the mounting bolts, then slowly lower the fuel tank out of position, disconnecting any other relevant vent pipes as they become accessible (where necessary), and remove the tank from underneath the vehicle. Note that it may be necessary to carefully bend the brake pipes to provide sufficient clearance when lowering the tank.

14 If the tank is contaminated with sediment or water, remove the fuel pump/sender unit (Section 9), and swill the tank out with clean fuel. The tank is injection-moulded from a synthetic material - if seriously damaged, it should be renewed. However, in certain cases, it may be possible to have small leaks or minor damage repaired. Seek the advice of a specialist before attempting to repair the fuel tank.

Refitting

15 Refitting is the reverse of the removal procedure, noting the following points:

a) *When lifting the tank back into position, take care to ensure that the hoses are not trapped between the tank and vehicle body.*

b) *Ensure that all pipes and hoses are correctly routed. Make sure the sealing rings are in position in the quick-release fittings prior to fitting and make sure they are securely clipped in position.*

c) *On completion, refill the tank with a small amount of fuel, and check for signs of leakage prior to taking the vehicle out on the road.*

12 Throttle body/housing - removal and refitting

Warning: Refer to the warning note in Section 1 before proceeding.

Removal

1 Depressurize the fuel system with reference to Section 8.

2 Disconnect the battery negative lead and proceed as described under the relevant heading.

Caution: If the radio/cassette in your vehicle is equipped with an anti-theft system, make sure you have the correct activation code before disconnecting the battery.

Single-point injection models

3 Remove the air cleaner assembly as described in Section 2.

4 Depress the retaining clips and disconnect the wiring connectors from the air temperature sensor/injector, throttle potentiometer and idle control stepper motor **(see illustrations)**.

5 Loosen the clips and disconnect the fuel feed and return hoses. Plug the hoses to prevent leakage and entry of dust and dirt.

6 Turn the throttle lever cam and disconnect the accelerator inner cable.

7 Unscrew and remove the bolts securing the throttle body assembly to the inlet manifold, then remove the assembly and gasket. Discard the gasket; a new one should be used on refitting.

Multi-point injection models

8 Unbolt and remove the tie bar from between the front suspension strut turrets.

9 Remove the air duct from between the throttle housing and air cleaner with reference to Section 2.

10 Turn the throttle lever cam and disconnect the accelerator inner cable. On automatic transmission models also disconnect the kick-down sensor wiring. Remove the accelerator cable from the support bracket making sure that the adjustment hairpin clip remains in its groove.

12.4a Wiring connector for the air temperature sensor/injector (E7J engine)

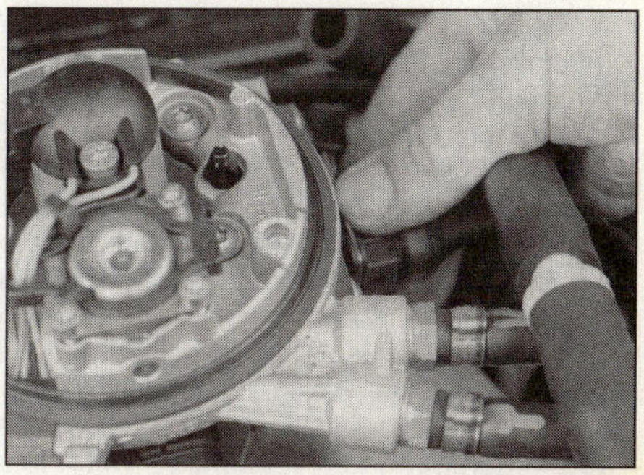

12.4b Disconnecting the wiring for the throttle potentiometer (E7J engine)

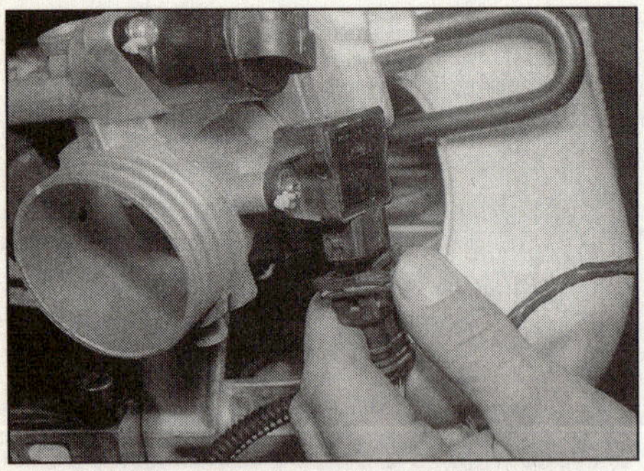

12.11a Disconnecting the throttle potentiometer wiring (K7M engine)

11 Disconnect the wiring from the throttle potentiometer and idle control stepper motor **(see illustrations)**. Where applicable, disconnect the EGR valve control vacuum hose.

12 Unscrew and remove the mounting bolts and remove the throttle housing from the inlet manifold. Remove and discard the gasket/spacer; a new one should be used on refitting.

Refitting

Single- and multi-point injection models

13 Clean the mating faces of the throttle housing and inlet manifold.

14 Refitting is a reversal of removal, but fit a new gasket and check the adjustment of the accelerator cable with reference to Section 4.

13 Fuel injection system - testing and adjustment

Testing

1 If a fault appears in the fuel injection system, first ensure that all the system wiring connectors are securely connected and free of corrosion. Ensure that the fault is not due to poor maintenance; ie, check that the air cleaner filter element is clean, the spark plugs are in good condition and correctly gapped, the cylinder compression pressures are correct, the ignition timing is correct, and that the engine breather hoses are clear and undamaged, referring to Chapters 1, 2 and 5 for further information.

2 If these checks fail to reveal the cause of the problem, the vehicle should be taken to a suitably-equipped Renault dealer for testing. A diagnostic connector is incorporated in the engine management circuit, into which a special electronic diagnostic tester can be plugged; the connector is located on the facia fusebox. The tester will locate the fault quickly and simply, alleviating the need to test all the system components individually, which is a time-consuming operation that carries a risk of damaging the ECU. The special Renault

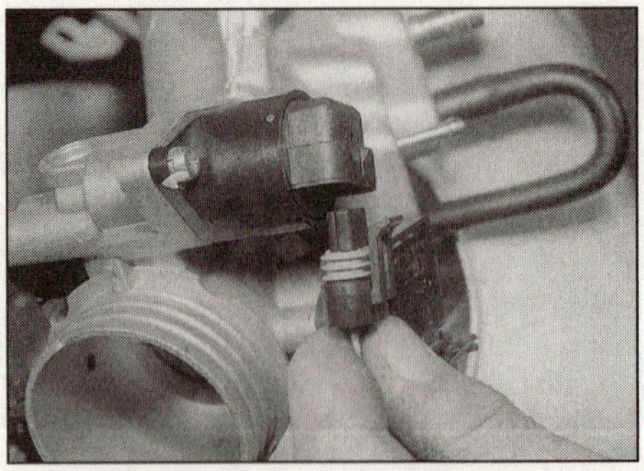

12.11b Disconnecting the idle control stepper motor wiring (K7M engine)

XR25 diagnostic tester is specific for Renault dealerships, and at the time of writing there was no other similar tester available. The tester uses a barchart configuration on a LCD screen; a fiche for the particular model is placed on the screen and each circuit can be checked instantly. There is no code output as such, so it is not possible for the home mechanic to determine a faulty area of the fuel injection system.

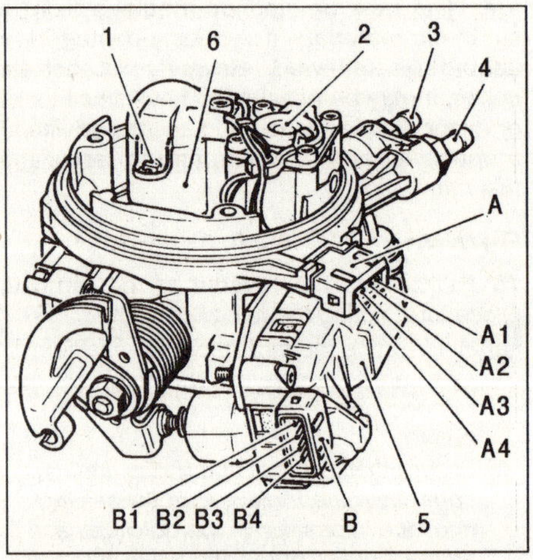

13.4a Throttle body/housing components and terminals - right-hand side (E7J engine)

1 *Air temperature sensor*
2 *Fuel pressure regulator*
3 *Fuel return to fuel tank*
4 *Fuel supply*
5 *Stepper motor*
6 *Injector*
A *Injector and air temperature sensor wiring connector*
A1 *Air temperature sensor*
A2 *+ Monopoint injector*
A3 *- Monopoint injector*
A4 *Air temperature injector*
B *Idle speed and low load switch wiring connector*
B1 *+ or - motor feed*
B2 *- or + motor feed*
B3 *Low load switch*
B4 *Low load switch*

3 If the "electronic incident" warning light illuminates on the instrument panel whilst driving, or remains illuminated longer than 3 seconds after switching on the ignition, a fault is indicated in one or more of the following components.

a) Absolute pressure sensor
b) Throttle position potentiometer
c) Injectors
d) Idle speed stepper motor (K7M engine only)
e) Vehicle speed sensor when the vehicle is moving
f) EGR solenoid valve (K7M 703 engine only)
g) Automatic transmission computer (K7M engine only)

4 Some individual components may be tested for resistance after removal using the information given in the Specifications, however other items (such as the idle speed stepper motor) cannot be checked and are not adjustable **(see illustrations)**.

Adjustment

5 Experienced home mechanics with a considerable amount of skill and equipment (including a tachometer and an accurately calibrated exhaust gas analyser) may be able to check the exhaust CO level and the idle speed. However, if these are found to be in need of adjustment, the car *must* be taken to a suitably-equipped Renault dealer for further testing. Neither the mixture adjustment (exhaust gas CO level) nor the idle speed are adjustable, and should either be incorrect, a fault must be present in the fuel injection system.

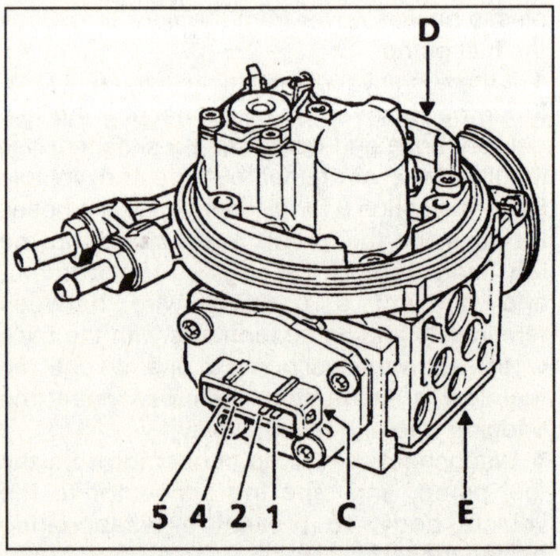

13.4b Throttle body/housing components and terminals - left-hand side (E7J engine)

C *Throttle potentiometer wiring connector*
C1 *Earth*
C2 *+ 5 volt feed*
C4 *Throttle position information (0 to 5 volts)*
C5 *Unused*
D *Throttle body/housing upper section*
E *Throttle body/housing lower section*

14 Single-point injection system components - removal and refitting

Warning: *Refer to the warning note in Section 1 before proceeding.*

Fuel injector

Note: *If a faulty injector is suspected, before condemning the injector, it is worth trying the effect of one of the proprietary injector-cleaning treatments.*

Removal

1 Remove the inlet air temperature sensor as described later in this Section.

2 Note its fitted position, then lift out the injector and recover its sealing rings **(see illustration)**.

Refitting

3 Refitting is a reversal of the removal procedure ensuring that the injector sealing rings and injector cap O-ring are in good condition. When refitting the injector cap ensure that the injector pins are correctly aligned with the cap terminals, the terminals are marked '+' and '-' for identification.

Fuel pressure regulator

4 The fuel pressure regulator is an integral part of the upper half of the throttle housing. If it is defective, the upper half of the housing must be renewed.

Inlet air temperature sensor

Removal

5 Remove the air cleaner assembly as described in Section 2.

6 Disconnect the sensor multi-plug.

7 Remove the single securing screw, and free the sensor cover **(see illustration)**.

8 Free the connector lugs, and withdraw the sensor together with the cover, wiring and connector.

Refitting

9 Refit by reversing the removal operations.

Idle speed control stepper motor

Removal

10 Remove the air cleaner assembly as described in Section 2.

11 Remove the throttle housing securing screws. Carefully lift the throttle housing to improve access to the motor.

12 Disconnect the motor multi-plug.

13 Remove the three mounting screws and withdraw the motor.

Refitting

14 Refit by reversing the removal operations, then perform an initial setting operation as follows.

15 Place a shim or feeler blades of thickness approximately 5 mm between the throttle lever and the motor plunger. Switch the ignition on for a few seconds, then switch it off again. Remove the shim or blades, then switch on and off again.

16 Run the engine and check that the idle speed and quality are satisfactory.

Throttle potentiometer

17 The throttle potentiometer is an integral part of the lower part of the throttle housing. If it is defective, the lower half of the housing must be renewed. The idle speed control motor can be transferred from the old half-housing to the new one.

18 After renewal, the initial setting of the potentiometer must be verified using Renault test box XR25.

Coolant temperature sensor

19 The coolant temperature sensor is located on the front of the thermostat housing. Refer to Chapter 3 for removal and refitting details.

Electronic control unit (ECU)

Removal

20 The ECU is located in the front right-hand corner of the engine compartment.

21 To remove the ECU first disconnect the battery negative lead.

Caution: If the radio/cassette in your vehicle is equipped with an anti-theft system, make sure you have the correct activation code before disconnecting the battery.

22 Unbolt the bracket from the top of the ECU and release the strap. Alternatively the bracket can remain on the ECU until the assembly is removed.

23 Undo the mounting screws and remove the ECU and mounting bracket **(see illustration)**.

24 Disconnect the wiring connector and remove the ECU from the engine compartment.

Refitting

25 Refitting is a reverse of the removal procedure ensuring that the wiring connector is securely reconnected.

Manifold absolute pressure (MAP) sensor

Removal

26 The MAP sensor is mounted onto the left-hand side of the engine compartment bulkhead.

27 Disconnect the wiring from the sensor.

28 Unscrew the mounting nuts and remove the sensor.

Refitting

29 Refitting is a reversal of removal.

Fuel injection system relay and fuel pump relay

Removal

30 The injection relay and fuel pump relay are located in a plastic box in the left-hand front corner of the engine compartment.

31 Remove the cover from the box.

32 Remove the relevant relay from the relay box.

Refitting

33 Refitting is the reverse of removal.

Crankshaft position sensor

Removal

34 The crankshaft position sensor is mounted onto the top of the transmission

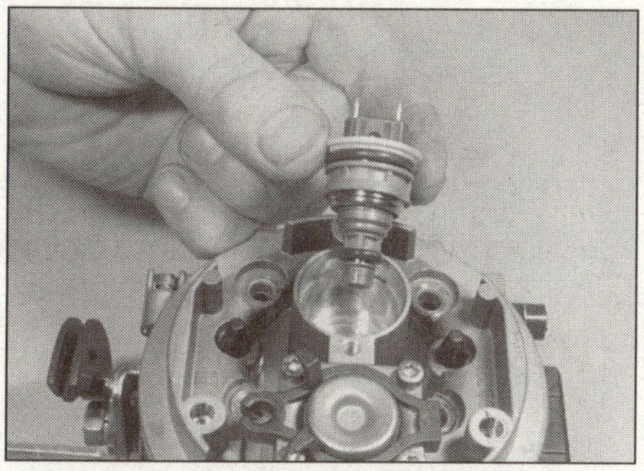

14.2 Removing the fuel injector (E7J engine)

14.7 Inlet air temperature sensor cover retaining screw (E7J engine)

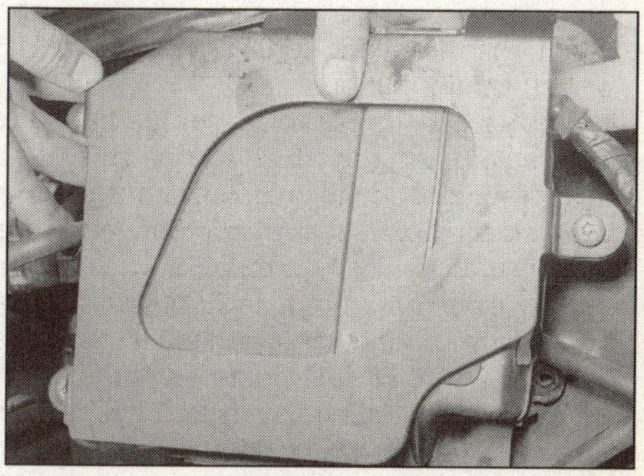

14.23 Removing the ECU complete with mounting bracket

housing at the left-hand end of the cylinder block.

35 To remove the sensor, disconnect the battery negative terminal then remove the air cleaner housing as described in Section 2.

36 Trace the wiring back from the sensor to the wiring connector, and disconnect it from the main harness.

37 Unscrew and remove the retaining bolts and remove the sensor from the top of the transmission.

Refitting

38 Refitting is a reverse of the removal procedure. Ensure that the sensor retaining bolts are securely tightened.

Fuel cut-off inertia switch

Removal

39 The fuel cut-off inertia switch is located in the left-hand side of the engine compartment.

40 Unscrew and remove the switch retaining screws then disconnect its wiring connector and remove the switch from the engine compartment.

Refitting

41 Refitting is the reverse of removal. On completion, reset the switch by depressing its button.

Power steering pressure switch

Removal

42 The pressure switch is screwed into the power steering feed pipe from the pump to the steering gear.

43 To remove the switch, disconnect the wiring connector.

44 Wipe clean the area around the switch then unscrew the switch and remove it from the pipe. Plug the pipe aperture to prevent excess fluid leakage and prevent dirt entry into the hydraulic system.

Refitting

45 Refitting is the reverse of removal. On completion check the power steering fluid level as described in "*Weekly checks*".

Knock sensor

Removal

46 The knock sensor is located on the rear of the cylinder block. Apply the handbrake then jack up the front of the vehicle and support on axle stands (see "*Jacking and vehicle support*").

47 From below the vehicle, reach up and disconnect the wiring.

48 Unscrew the knock sensor from the cylinder block.

Refitting

49 Refitting is a reversal of removal.

15 Multi-point injection system components - removal and refitting

Fuel rail and injectors

Warning: Refer to the warning note in Section 1 before proceeding.

Note: *If a faulty injector is suspected, before condemning the injector, it is worth trying the effect of one of the proprietary injector-cleaning treatments.*

Removal

1 Disconnect the battery negative lead.

Caution: If the radio/cassette in your vehicle is equipped with an anti-theft system, make sure you have the correct activation code before disconnecting the battery.

2 Unbolt the tie bar from between the front suspension strut turrets.

3 Disconnect the crankcase ventilation hoses from the valve cover.

4 Disconnect the wiring from the injectors and the fuel evaporative solenoid valve and move the wiring loom to one side **(see illustration)**.

5 Disconnect the vacuum hose connecting the pressure regulator (on the fuel rail) to the inlet manifold.

6 Disconnect the fuel feed and return hoses from the fuel rail **(see illustration)**. The hoses have the same connectors as fitted to the fuel pump (see Section 9).

7 Unscrew and remove the mounting bolts and carefully ease the fuel rail together with the injectors from the inlet manifold **(see illustration)**.

8 Note the fitted positions of the injectors, then remove the clips and ease the injectors from the fuel rail **(see illustrations)**.

9 Remove the sealing rings from the grooves at each end of the injectors and obtain new ones.

Refitting

10 Refitting is a reversal of the removal procedure, noting the following points.

a) *Renew all sealing rings, using a smear of engine oil to aid installation.*
b) *Refit the fuel rail assembly to the manifold, making sure the sealing rings remain correctly positioned, and tighten the retaining bolts to the specified torque.*
c) *On completion start the engine and check for fuel leaks.*

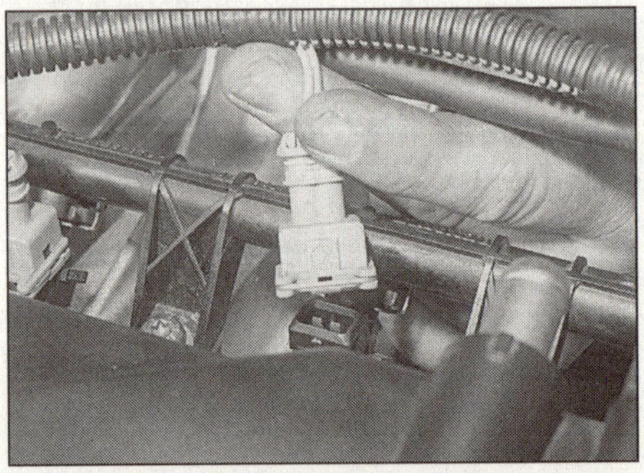

15.4 Disconnecting the injector wiring (K7M engine)

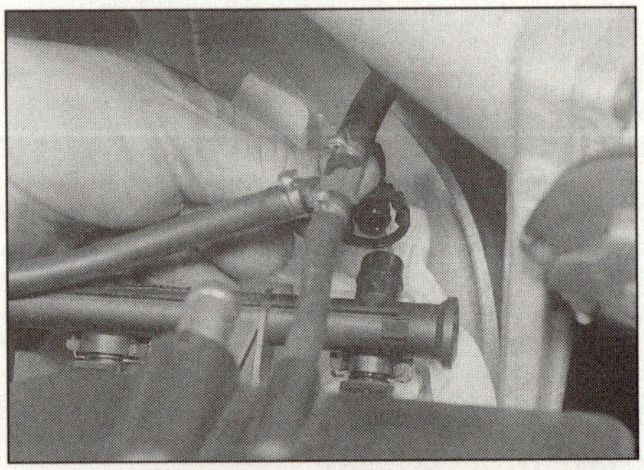

15.6 Disconnecting the fuel return hose from the fuel rail (K7M engine)

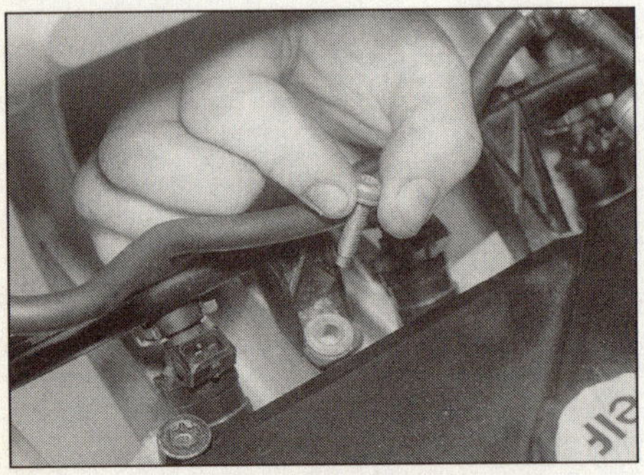

15.7 Removing the fuel rail mounting bolts (K7M engine)

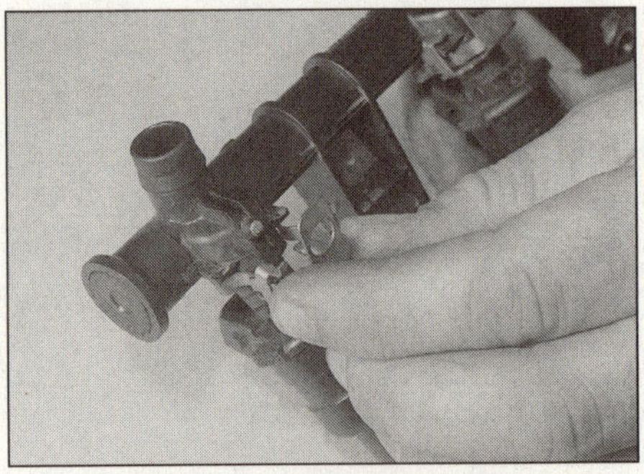

15.8a Remove the retaining clips . . .

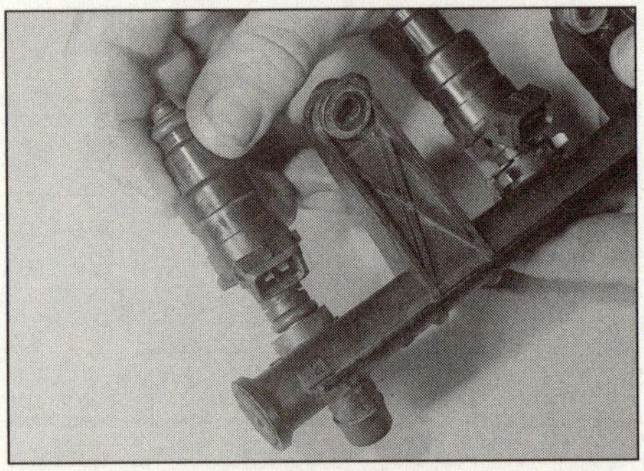

15.8b . . . and ease the injectors from the fuel rail (K7M engine)

15.11 Fuel pressure regulator vacuum pipe (K7M engine)

Fuel pressure regulator

Warning: Refer to the warning note in Section 1 before proceeding.

Removal

11 Disconnect the vacuum pipe from the regulator **(see illustration)**.

12 Place a wad of rag over the regulator to catch any spilled fuel, then extract the retaining spring and ease the regulator from the fuel rail.

13 Remove the sealing rings from the grooves in the pressure regulator and obtain new ones.

Refitting

14 On refitting, fit new sealing rings to the regulator grooves and apply a smear of engine oil to them to ease installation. Ease the regulator back into the end of the fuel rail and refit the retaining spring and vacuum pipe.

Throttle potentiometer

Removal

15 Remove the throttle housing as described in Section 12.

16 Undo the retaining screws and remove the potentiometer from the throttle housing.

Refitting

17 Refitting is a reverse of the removal procedure ensuring that the potentiometer is correctly engaged with the throttle valve spindle. Note that Renault recommend that the potentiometer operation should be checked using the special electronic (XR25) tester whenever it is disturbed.

Inlet air temperature sensor

Removal

18 Remove the air inlet duct as described in Section 2.

19 Unscrew and remove the inlet air temperature sensor.

Refitting

20 Refitting is a reversal of removal.

Coolant temperature sensor

21 The sensor is located on the thermostat housing at the left-hand end of the cylinder head. Refer to Chapter 3 for removal and refitting details.

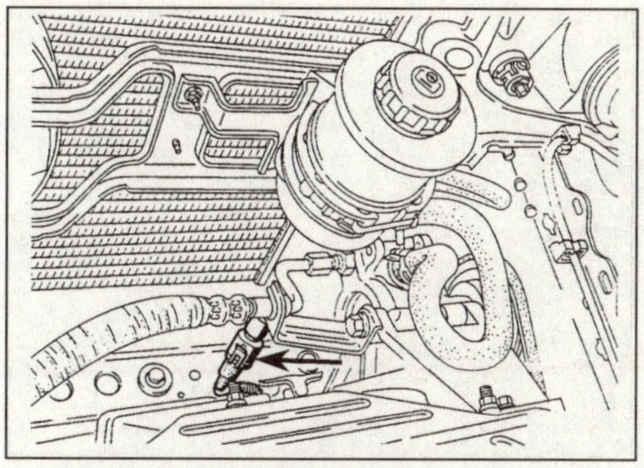

15.22 The power steering pressure switch is screwed into the power steering feed pipe (K7M engine)

Power steering pressure switch

Removal

22 The switch is screwed into the power steering feed pipe from the pump to the steering gear **(see illustration)**.

23 To remove the switch, disconnect the wiring connector.

24 Wipe clean the area around the switch then unscrew the switch and remove it from the pipe. Plug the pipe aperture to prevent excess fluid leakage and prevent dirt entry into the hydraulic system.

Refitting

25 Refitting is the reverse of removal. On completion check the power steering fluid level as described in "*Weekly checks*".

Knock sensor

Removal

26 The knock sensor is located on the rear of the cylinder block. Apply the handbrake then jack up the front of the vehicle and support on axle stands (see *"Jacking and vehicle support"*).

27 From below the vehicle, reach up and disconnect the wiring **(see illustration)**.

28 Unscrew the knock sensor from the cylinder block.

Refitting

29 Refitting is a reversal of removal.

Idle speed control stepper motor

Removal

30 The idle speed control stepper motor is mounted on the top of the throttle housing. To remove it first remove the throttle housing as described in Section 12.

31 Undo the retaining screws and remove the stepper motor from the throttle housing. Recover the gasket and discard it; a new one should be used on refitting.

Refitting

32 Refitting is a reversal of the removal procedure using a new gasket.

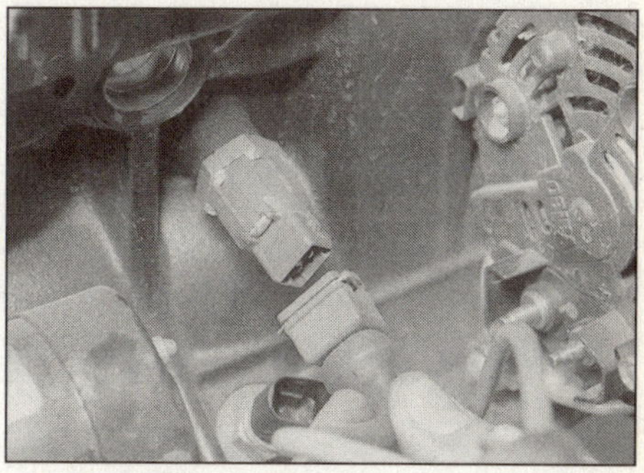

15.27 Disconnecting the wiring from the knock sensor (K7M engine)

Manifold absolute pressure (MAP) sensor

Removal

33 The MAP sensor is mounted onto the left-hand side of the engine compartment bulkhead.

34 Disconnect the wiring from the sensor.

35 Unscrew the mounting nuts and remove the sensor.

Refitting

36 Refitting is a reversal of removal.

Fuel injection system relay and fuel pump relay

Removal

37 The injection relay and fuel pump relay are located in a plastic box in the left-hand front corner of the engine compartment.

38 Remove the cover from the box.

39 Remove the relevant relay from the relay box.

Refitting

40 Refitting is the reverse of removal.

Crankshaft position sensor

Removal

41 The sensor is mounted onto the top of the transmission housing at the left-hand end of the cylinder block **(see illustration)**.

42 To remove the sensor, first remove the air cleaner housing as described in Section 2.

15.41 The crankshaft position sensor is mounted onto the top of the transmission

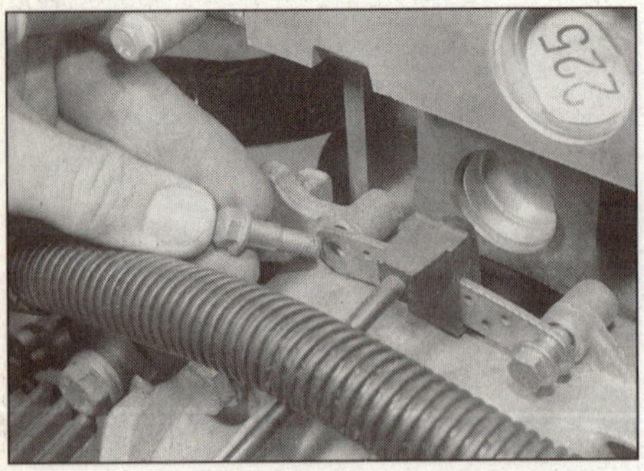

15.44 Removing the crankshaft position sensor retaining bolts

43 Trace the wiring back from the sensor to the wiring connector, and disconnect it from the main harness.
44 Unscrew and remove the retaining bolts and remove the sensor from the top of the transmission **(see illustration)**.

Refitting

45 Refitting is a reverse of the removal procedure. Ensure that the sensor retaining bolts are securely tightened.

Fuel cut-off inertia switch

Removal

46 The switch is located in the left-hand side of the engine compartment **(see illustration)**.
47 Unscrew and remove the switch retaining screws then disconnect its wiring connector and remove the switch from the engine compartment.

Refitting

48 Refitting is the reverse of removal. On completion, reset the switch by depressing its button.

Electronic control unit (ECU)

Removal

49 The ECU is located in the front right-hand corner of the engine compartment.
50 To remove the ECU first disconnect the battery negative lead.
Caution: If the radio/cassette in your vehicle is equipped with an anti-theft system, make sure you have the correct activation code before disconnecting the battery.
51 Unbolt the bracket from the top of the ECU and release the strap. Alternatively the bracket can remain on the ECU until the assembly is removed.
52 Undo the mounting screws and remove the ECU and mounting bracket.
53 Disconnect the wiring connector and remove the ECU from the engine compartment.

Refitting

54 Refitting is a reverse of the removal procedure ensuring that the wiring connector is securely reconnected.

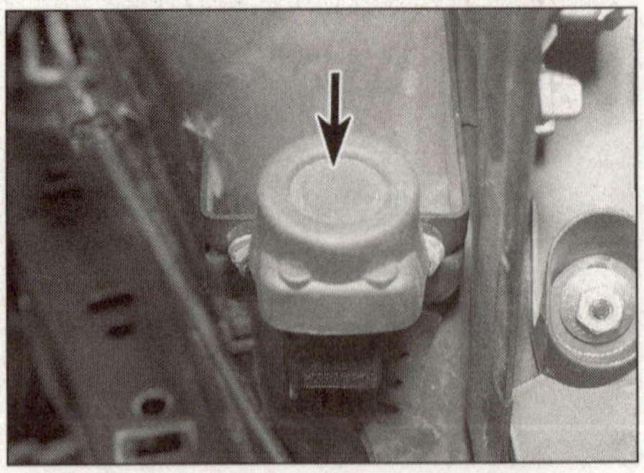
15.46 Fuel cut-off inertia switch

16 Manifolds - removal and refitting

Inlet manifold

E7J (single-point) engine

Removal

1 Remove the throttle body/housing as described in Section 12.
2 Drain the cooling system as described in Chapter 1A, then disconnect the coolant hoses from the inlet manifold.
3 Disconnect the brake servo vacuum hose and the wiring connector from the heater on the bottom of the inlet manifold.
4 Progressively unscrew the nuts, then withdraw the inlet manifold from the studs on the cylinder head. Remove the gasket.
5 If necessary, the heater may be removed from the bottom of the inlet manifold, by first extracting the circlip. Note the location of the seal **(see illustration)**.

Refitting

6 Refitting is a reversal of removal but use a new gasket and tighten the mounting nuts to the specified torque. Ensure that the cylinder head and manifold mating surfaces are clean, and use a new gasket. Refill the cooling system with reference to Chapter 1A.

K7M (multi-point engine)

Removal

7 Remove the fuel rail and injector assembly as described in Section 15. Also disconnect the fuel pressure regulator hose from the right-hand end of the inlet manifold **(see illustration)**.
8 Remove the air inlet duct as described in Section 2. For improved access, apply the handbrake then jack up the front of the vehicle and support on axle stands (see *"Jacking and vehicle support"*). Where applicable, remove the engine undershield.
9 Remove the throttle body/housing as described in Section 12. Position the fuel supply and return hoses to one side and plug them to prevent loss of fuel.
10 Where applicable, disconnect the wiring from the EGR valve, then unbolt and remove the valve. Release the clip then unscrew the steel pipe leading to the manifold from the EGR valve.

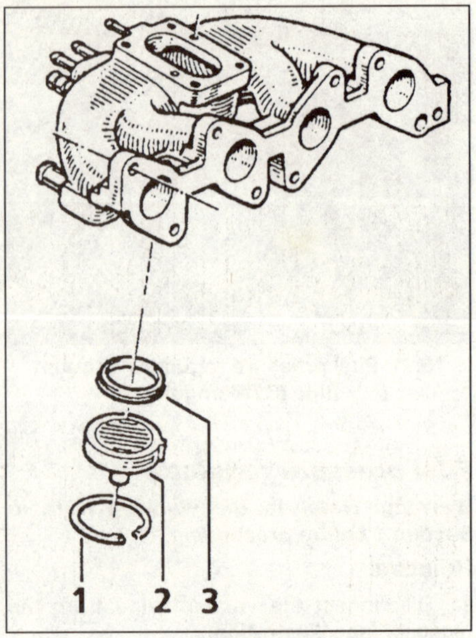

16.5 Inlet manifold heater components (E7J engine)
1 Circlip 2 Heater element 3 Seal

11 Unscrew the bolts securing the wiring loom support bracket to the underside of the inlet manifold.
12 Disconnect the manifold absolute pressure vacuum hose and the brake servo vacuum hose.
13 Disconnect the fuel evaporative canister solenoid valve hoses.
14 Unscrew the lower mounting nuts followed by the upper mounting bolts, then carefully withdraw the inlet manifold from the lower studs and recover the gasket **(see illustrations)**.

Refitting

15 Refitting is a reversal of removal but use a new gasket and tighten the mounting nuts and bolts to the specified torque. Ensure that the cylinder head and manifold mating surfaces are clean, and use a new gasket.

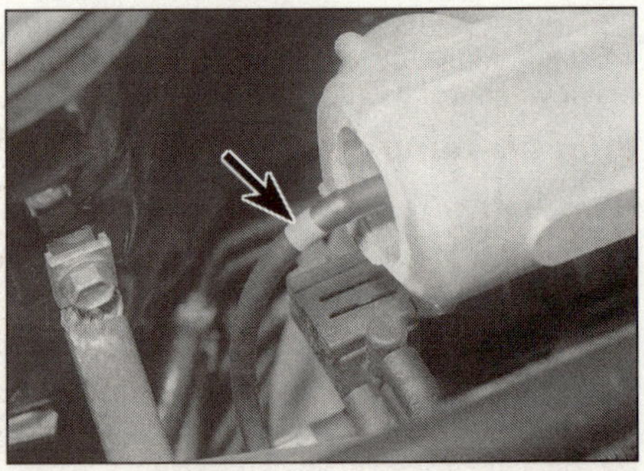
16.7 Fuel pressure regulator vacuum hose on the right-hand end of the inlet manifold (K7M engine)

16.14a Inlet manifold lower mounting nuts (K7M engine)

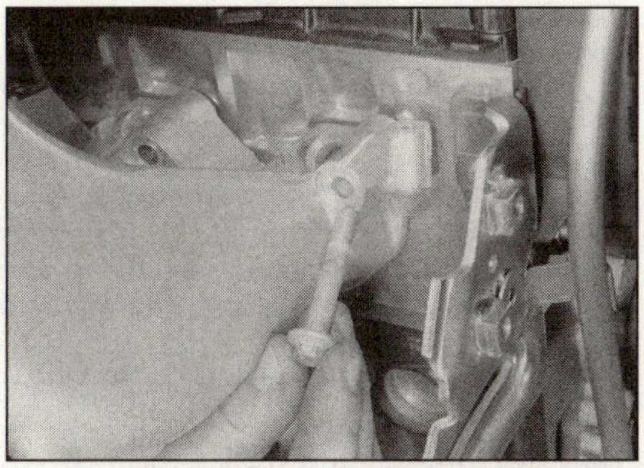

16.14b Removing the inlet manifold upper mounting bolts (K7M engine)

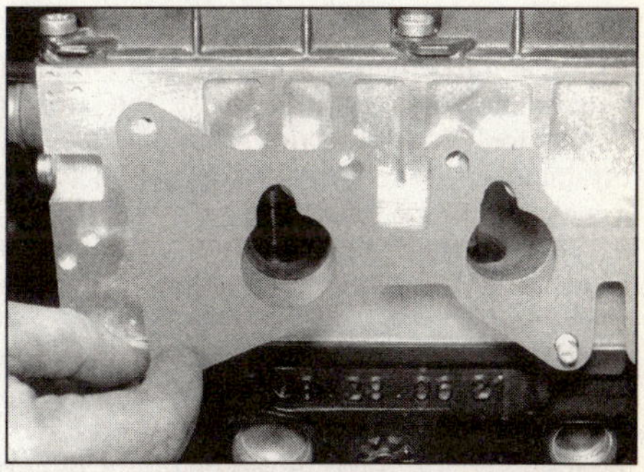

16.14c Removing the inlet manifold gasket (K7M engine)

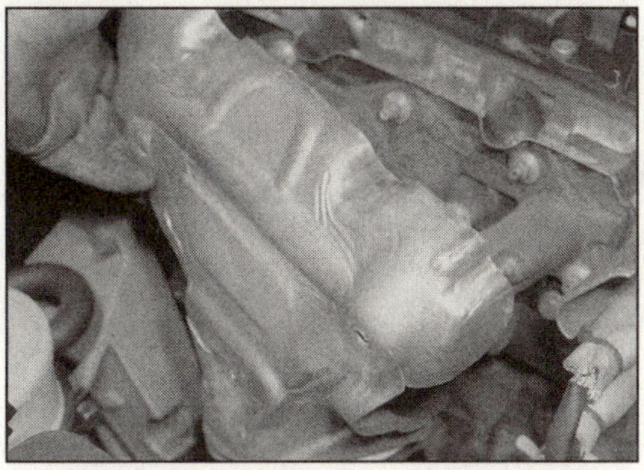

16.18 Removing the hot air shroud from the exhaust manifold (K7M engine)

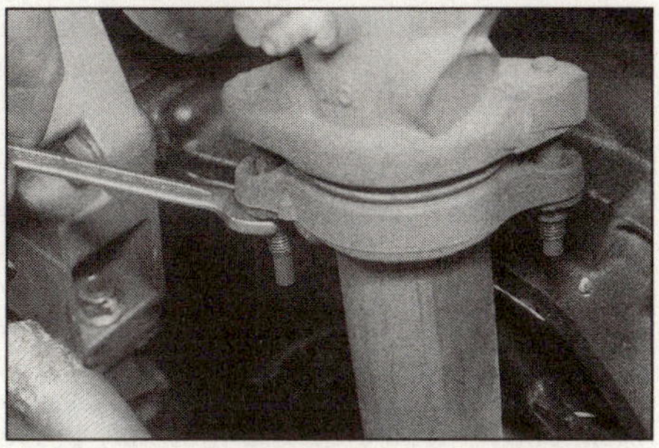

16.19 Unscrewing the nuts securing the downpipe to the exhaust manifold (K7M engine)

16.21a Unscrew the mounting nuts . . .

Exhaust manifold

Removal

16 Apply the handbrake then jack up the front of the vehicle and support on axle stands (see *"Jacking and vehicle support"*). Where applicable, remove the engine undershield.

17 On the E7J engine, disconnect the hot air hose from the exhaust manifold shroud.

18 Unscrew the nuts and remove the hot air shroud from the manifold **(see illustration)**.

19 Where necessary, support the exhaust front downpipe (take care not to damage the oxygen sensor), then unscrew the nuts/bolts securing the downpipe to the exhaust manifold. Recover the gasket **(see illustration)**.

20 Where applicable, unscrew the nut and release the clips securing the EGR valve steel pipe to the manifold.

21 Unscrew the mounting nuts and withdraw the exhaust manifold from the studs on the cylinder head **(see illustrations)**.

22 Recover the gasket **(see illustration)**.

Refitting

23 Refitting is a reversal of removal but tighten the nuts to the specified torque. Ensure that the cylinder head and manifold mating surfaces are clean, and use a new gasket.

17 Exhaust system - general information, removal and refitting

General information

1 On new vehicles the exhaust system consists of just two sections; the front downpipe (and catalytic converter where applicable) and the remaining system consisting of a resonator, tailpipe and silencer. The downpipe is attached to the rear section by a flanged joint with gasket or a cone-and-socket joint.

2 The rear section of the exhaust is located above the rear suspension, however the intermediate and tailpipe/silencer may be renewed separately by cutting the intermediate pipe with a hacksaw. The system is suspended throughout its entire length by rubber mountings **(see illustration)**.

Removal

3 To remove a part of the system, first jack up the front or rear of the car, and support it on axle stands (see *"Jacking and vehicle support"*). Alternatively, position the car over an inspection pit, or on car ramps. Where fitted, remove the engine compartment undershield.

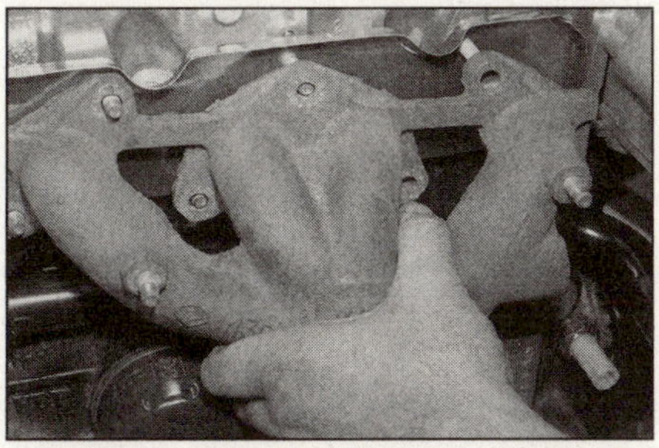

16.21b . . . and withdraw the exhaust manifold from the cylinder head (K7M engine)

16.22 Removing the exhaust manifold gasket (K7M engine)

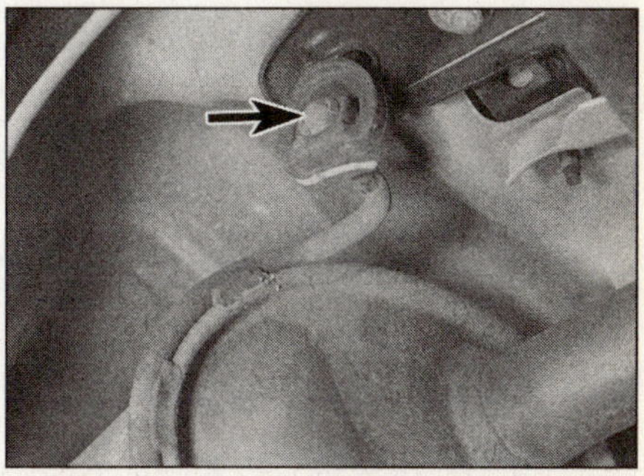

17.2 Rear silencer mounting rubber (K7M engine)

Front pipe and catalytic converter

4 Trace the wiring back from the Lambda sensor and disconnect it at the wiring connector. Free the wiring from any relevant retaining clips so the sensor is free to be removed with the front pipe.

5 Where applicable, unbolt the front pipe from the mounting bracket on the transmission.

6 Unscrew and remove the nuts/bolts securing the front pipe flange joint to the manifold, and recover the gasket.

7 Unscrew and remove the clamp or flange bolts (as applicable), and disconnect the front pipe and catalytic converter from the rear section. Withdraw the pipe from under the vehicle.

Intermediate pipe and resonator

8 Unscrew and remove the clamp or flange bolts (as applicable) attaching the front pipe and catalytic converter to the rear section. Recover the gasket where applicable.

9 If the original rear section is fitted, it must be cut in half using either a hacksaw or pipe cutter. Locate the cutting area which is situated approximately midway between the resonator and rear silencer. The cutting point is marked with two circular punch marks on the side of the pipe. The punch marks are 90 mm apart and the exhaust section should be cut at the mid-point between the two punch marks. **Note:** *Ensure that the exhaust pipe is cut squarely else it will be difficult to obtain a gas-tight seal when the exhaust is refitted.*

10 With the intermediate pipe cut, withdraw the intermediate exhaust section from under the vehicle.

11 If the rear section is in two halves, unscrew the bolt and slide the clamp sleeve onto the rear section then release the rubber mountings and withdraw the intermediate section from under the vehicle.

Rear tailpipe and silencer

12 If the original rear section is fitted, follow the instructions given in paragraph 9.

13 If the rear section is in two halves, unscrew the bolt and slide the clamp sleeve onto the intermediate section then release the rubber mountings and withdraw the tailpipe and silencer from under the vehicle.

Heat shield(s)

14 The heat shields are secured to the underside of the body by various nuts and bolts. Each shield can be removed separately but note that they overlap making it necessary to loosen another section first. If a shield is being removed to gain access to a component located behind it, it may prove sufficient in some cases to remove the retaining nuts and/or bolts, and simply lower the shield, without disturbing the exhaust system. Otherwise remove the exhaust section as described earlier.

Refitting

15 Each section is refitted by reversing the removal sequence, noting the following points:

a) *Ensure that all traces of corrosion have been removed from the flanges, and renew all necessary gaskets.*

b) *Inspect the rubber mountings for signs of damage or deterioration, and renew as necessary.*

c) *When reconnecting the intermediate pipe to the tailpipe, apply a smear of exhaust system jointing paste (Renault recommend the use of Sodicam) to the sleeve inner surface, to ensure a gas-tight seal. Make sure both inner ends of the cut pipe are positioned squarely against the stop of the clamp sleeve. Position the sleeve bolt vertically on the left-hand side of the pipe and securely tighten the nut until it is heard to click; the clamp bolt has a groove in it to ensure that the nut is correctly tightened (equivalent to a tightening torque of approximately 25 Nm).*

d) *Prior to tightening the exhaust system fasteners, ensure that all rubber mountings are correctly located, and that there is adequate clearance between the exhaust system and vehicle underbody.*

Chapter 4 Part B:
Fuel/exhaust systems - diesel engine models

Contents

Degrees of difficulty

Easy, suitable for novice with little experience	**Fairly easy,** suitable for beginner with some experience	**Fairly difficult,** suitable for competent DIY mechanic	**Difficult,** suitable for experienced DIY mechanic	**Very difficult,** suitable for expert DIY or professional

Specifications

General

System type	Rear-mounted fuel tank, distributor fuel injection pump with integral transfer pump, indirect injection. Turbocharger on F8Q 784 engine
Firing order	1-3-4-2 (No 1 at flywheel end)

Maximum speed

Bosch injection pump:	
F8Q 620 non-Turbo engine	5100 to 5300 rpm
Lucas injection pump:	
F8Q 784 Turbo engine	5100 to 5300 rpm

Idle speed

All engines	825 ± 50 rpm

Fast idle speed

Bosch injection pump	950 ± 25 rpm
Lucas injection pump	Not adjustable (factory set)

Anti-stall speed

Non-Turbo engines (with 4 mm shim - see text)	1150 ± 25 rpm
Turbo engines (with 4 mm shim - see text)	1250 ± 50 rpm

Injection pump (Bosch)

Direction of rotation	Clockwise viewed from sprocket end
Static timing:	
Engine position	No 1 cylinder at TDC (see Section 16)
Pump position	0.82 ± 0.04 mm
Dynamic timing	No information available at time of writing

Injection pump (Lucas)

Direction of rotation	Clockwise viewed from sprocket end	
Static timing:		
Engine position	No 1 cylinder at TDC (see Section 15)	
Pump position	Value shown on pump (see Section 15)	
Dynamic timing	No information available at time of writing	

Injectors

Type	Pintle	
Opening pressure	130 + 8 bars - 5 bars	

Turbocharger

Type	Garrett T2	
Boost pressure (full load at 2500 to 2600 rpm)	900 mbar + 50 - 100 mbar	
Speed of rotation	Approximately 150 000 rpm	
Wastegate pressure settings:	**Control pressure**	**Adjustment**
Actuator rod movement 0.38 mm	1020 to 1080 mbars	1050 to 1080 mbars
Actuator rod movement 4.0 mm	1180 to 1260 mbars	1220 to 1260 mbars

Recommended fuel

All engines	Commercial diesel fuel for road vehicles (DERV)	

Torque wrench settings

	Nm	lbf ft
Coolant pipe-to-turbocharger union nuts	20 to 30	15 to 22
Exhaust front section-to-turbocharger nuts	45	33
Fast idle thermostatic actuator-to-cylinder head	35	26
Fuel gauge sender unit locking ring	35	26
Fuel injectors to cylinder head	70	52
Fuel pipe union nuts and bolts	25	19
Injection pump mounting nuts and bolts	25	18
Injection pump sprocket adjuster bolts (two-piece adjustable sprocket)	20	15
Injection pump sprocket securing nut	50	37
Injection pump timing hole blanking plug:		
Lucas pump	5	4
Bosch pump	10	7
Inlet/exhaust manifold mounting nuts	25	19
Lucas injection pump sprocket nut (LH thread) with micrometric adjustment	90	66
Oil feed pipe-to-turbocharger union nut	15 to 25	11 to 18
Oil return pipe-to-turbocharger union nut	40	30
Turbocharger-to-manifold nuts	45	33
Wastegate actuator rod locknut	6 to 8	4 to 6
Wastegate bracket-to-turbocharger bolts*	15	11

**Coat the bolt threads with thread-locking compound*

1 General information and precautions

General information

1 The fuel system consists of a rear-mounted fuel tank, a fuel filter with integral water separator, a fuel injection pump, injectors and associated components. As it passes through the filter, the fuel is heated by an electric heater. A turbocharger is fitted to the F8Q 784 engine. The exhaust system is conventional, but on certain models, an exhaust gas recirculation (EGR) system and/or an unregulated catalytic converter may be fitted to reduce exhaust gas emissions (see Chapter 4C for further details).

2 Fuel is drawn from the fuel tank to the fuel injection pump by a vane-type transfer pump incorporated in the fuel injection pump. Before reaching the pump the fuel passes through a fuel filter where foreign matter and water are removed. Excess fuel lubricates the moving components of the injection pump and is then returned to the tank **(see illustration)**.

3 The fuel injection pump is driven at half crankshaft speed by the timing belt. The high pressure required to inject the fuel into the compressed air in the swirl chambers is achieved by a cam plate acting on a single piston on the Bosch pump, or by two opposed pistons forced together by rollers running in a cam ring on the Lucas/CAV pump. The fuel passes through a central rotor with a single outlet drilling which aligns with ports leading to the injector pipes.

4 Fuel metering is controlled by a centrifugal governor which reacts to accelerator pedal position and engine speed. The governor is linked to a metering valve which increases or decreases the amount of fuel delivered at each pumping stroke. On turbocharged models a separate device also increases fuel delivery with increasing boost pressure.

5 Basic injection timing is determined when the pump is fitted. When the engine is running it is varied automatically to suit the prevailing engine speed by a mechanism which turns the cam plate or ring.

6 A solenoid valve in the pump is used to advance the injection timing briefly after a cold start. The system operates by increasing the pump transfer pressure via a restrictor valve.

7 The four fuel injectors produce a homogeneous spray of fuel into the swirl chambers located in the cylinder head. The injectors are calibrated to open and close at critical pressures to provide efficient and even combustion. Each injector needle is lubricated by fuel which accumulates in the spring chamber and is channelled to the injection pump return hose by leak-off pipes.

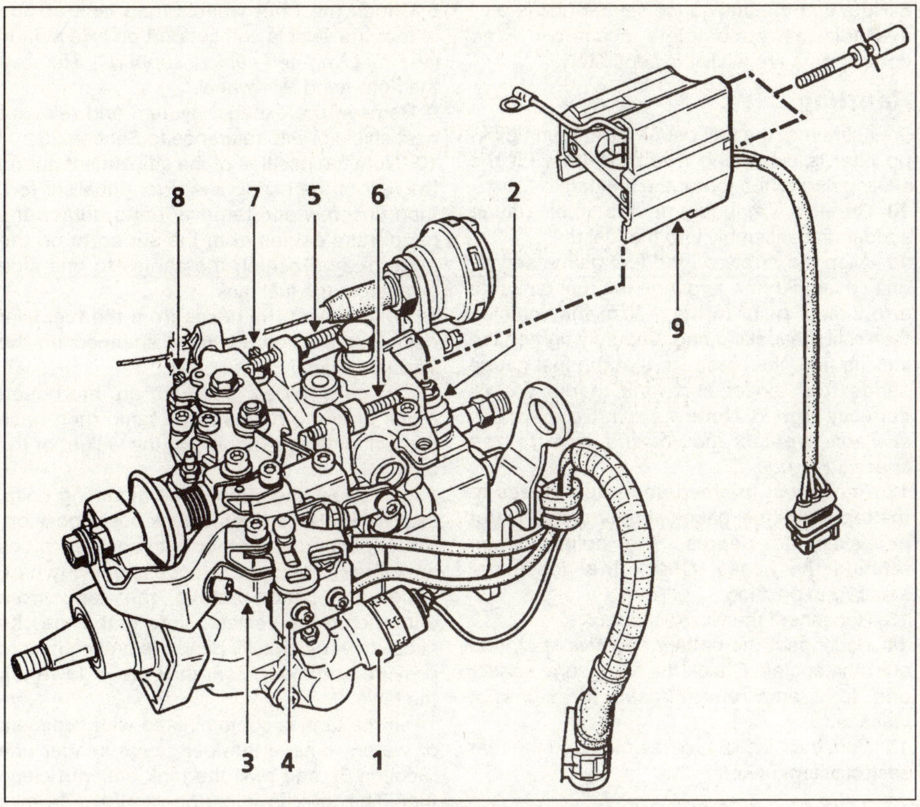

1.2 Fuel injection pump components

1 *Load dependence control solenoid valve (ALFB)*
2 *Fuel cut-off solenoid*
3 *Post-heating/EGR valve cutting microswitch*
4 *Air conditioning microswitch*
5 *Residual flow stop screw (anti-stall)*
6 *Maximum engine speed stop screw*
7 *Idle speed adjustment stop screw*
8 *Fast idle speed adjustment stop screw*
9 *Coded solenoid valve electronic unit*

8 A Bosch injection pump is fitted to the F8Q 620 engine, and a Lucas injection pump is fitted to the F8Q 784 turbocharged engine. Replacement components must be of the same make as those originally fitted.
9 Cold starting is assisted by preheater or "glow" plugs fitted to each swirl chamber (see Chapter 5C for further details).
10 The fast idle system is operated either by a thermostatic actuator, or a vacuum actuator, depending on model.
11 On models with a thermostatic fast idle actuator, a thermostatic sensor in the cooling system operates a fast idle lever on the injection pump, via a cable, to increase the idling speed when the engine is cold.
12 On models with a vacuum actuator, the fast idle cable is operated by a vacuum-controlled actuator, via a solenoid valve. The solenoid valve is controlled by the preheating system control unit, and on models with air conditioning, an additional input is provided by the air conditioning compressor clutch relay.
13 A stop solenoid cuts the fuel supply to the injection pump rotor when the ignition is switched off, and there is also a hand-operated stop lever for use in an emergency **(see illustration)**.
14 Provided that the specified maintenance is carried out, the fuel injection equipment will give long and trouble-free service. The injection pump itself may well outlast the engine. The main potential cause of damage to the injection pump and injectors is dirt or water in the fuel.
15 Servicing of the injection pump and injectors is very limited for the home

1.13 Hand-operated stop lever (arrowed) - Lucas injection pump

mechanic, and any dismantling or adjustment other than that described in this Chapter must be entrusted to a Renault dealer or fuel injection specialist.

Precautions

Warning: It is necessary to take certain precautions when working on the fuel system components, particularly the fuel injectors. Before carrying out any operations on the fuel system, refer to the precautions given in "Safety first!" at the beginning of this manual, and to any additional warning notes at the start of the relevant Sections.

2 Air cleaner housing assembly - removal and refitting

Non-Turbo models

Removal

1 Remove the securing screws (noting that one of the screws is located in the centre of the cover), and withdraw the air cleaner cover.
2 Lift the filter element from the air cleaner body.
3 Disconnect the inlet manifold air hose from the air cleaner body.
4 Disconnect the fresh air hose from the bottom of the air cleaner body and from the front left-hand corner of the engine compartment.
5 Unscrew the mounting nuts and withdraw the air cleaner body from the bracket in the engine compartment. If necessary, unbolt the air inlet from the front crossmember, and also unbolt the mounting bracket from the engine compartment.
6 Check the condition of the rubber mountings and renew them as necessary. Also check the hoses and hose clips for condition.

Refitting

7 Refitting is a reversal of removal.

Turbo models

Removal

8 Undo the screws and lift off the air cleaner cover.
9 Remove the filter element from the air cleaner body.
10 Release the strap securing the air cleaner body to the support bracket.
11 Disconnect the air trunking connecting the air cleaner body to the turbocharger.
12 Disconnect the fresh air hose from the bottom of the air cleaner, and lift the air cleaner body from its mounting bracket.
13 If necessary, unbolt the air inlet from the front crossmember, and also unbolt the mounting bracket from the engine compartment.
14 Check the condition of the securing strap and the hoses and hose clips.

Refitting

15 Refitting is a reversal of removal.

3 Fuel gauge sender unit - removal and refitting

Warning: Refer to the warning note in Section 1 before proceeding.

Removal

1 Disconnect the battery negative lead.
Caution: If the radio/cassette in your vehicle is equipped with an anti-theft system, make sure you have the correct activation code before disconnecting the battery.

2 Remove the rear seat cushion as described in Chapter 11, for access to the fuel gauge sender unit cover.

3 Using a screwdriver, carefully prise the plastic access cover from the floor to expose the fuel gauge sender unit.

4 Disconnect the wiring connector from the fuel gauge sender unit, and tape the connector to the vehicle body, to prevent it disappearing behind the tank.

5 Identify the fuel hoses for position; the green hose on the left-hand side is the supply to the engine compartment, and the red hose next to the wiring connector is the return (arrows on the top of the fuel gauge sender unit indicate the direction of fuel flow). The hoses are equipped with quick-release fittings to ease removal. To disconnect each hose, slide out the locking tab from the collar then compress the collar and detach the hose from the fuel gauge sender unit. Renault technicians use a special tool to compress the collar, however it is possible to use a small screwdriver to carefully press back the locking collar inside the end of the special end fitting. Disconnect both hoses from the top of the fuel gauge sender unit, then plug the hose ends to minimise fuel loss.

6 Noting the alignment arrows on the fuel gauge sender unit cover and fuel tank, unscrew the locking ring and remove it from the tank. This can be accomplished by using a screwdriver on the raised ribs of the locking ring - carefully tap the screwdriver to turn the ring anti-clockwise until it can be unscrewed by hand. Alternatively a removal tool can be fabricated out of metal bar as shown in Chapter 4A.

7 Carefully lift the fuel gauge sender unit assembly out of the fuel tank, taking great care not to damage the fuel gauge sender arm, or to spill fuel onto the interior of the vehicle. Remove the rubber sealing ring and check it for deterioration; if it is in good condition, it may be reused, however if the fuel gauge sender unit is to remain out of the fuel tank for several hours, the locking ring should be refitted temporarily to prevent the sealing ring from distorting. If the sealing ring is unserviceable, obtain a new one.

8 Note that the fuel gauge sender unit is only available as a complete assembly - no components are available separately.

Refitting

9 Ensure that the fuel gauge sender unit pick-up filter is clean and free of debris. Fit the sealing ring to the top of the fuel tank.

10 Carefully manoeuvre the fuel gauge sender unit assembly into the fuel tank.

11 Align the arrow on the fuel gauge sender unit cover with the arrow on the fuel tank (the arrow must point to the rear of the vehicle), then refit the locking ring. Securely tighten the locking ring, then recheck that the fuel gauge sender unit cover and tank marks are all correctly aligned. **Note:** *If a suitable adapter is available tighten the locking ring to the specified torque.*

12 Reconnect the feed and return hoses to the top of the fuel gauge sender unit; it is not necessary to depress the collars when refitting the hoses. Check the hoses are securely in position.

13 Reconnect the wiring connector.

14 Reconnect the battery negative lead, and start the engine. Check the fuel gauge sender unit feed and return hoses for signs of leakage.

15 Refit the plastic access cover and the rear seat cushion.

4 Fuel tank - removal and refitting

Warning: Refer to the warning note in Section 1 before proceeding.

Removal

1 Before removing the fuel tank, all fuel must be drained from it. Since a drain plug is not provided, it is preferable to carry out the removal operation when the tank is nearly empty.

2 Disconnect the battery negative lead.
Caution: If the radio/cassette in your vehicle is equipped with an anti-theft system, make sure you have the correct activation code before disconnecting the battery.

3 Remove the rear seat cushion as described in Chapter 11, for access to the fuel gauge sender unit cover.

4 Using a screwdriver, carefully prise the plastic access cover from the floor to expose the sender unit.

5 If there is any fuel remaining in the fuel tank at this stage, the sender unit may be removed as described in Section 3 and the fuel siphoned out.

6 Disconnect the wiring connector from the fuel gauge sender unit, and tape the connector to the vehicle body, to prevent it disappearing behind the tank.

7 Disconnect the feed and return hoses with reference to Section 3.

8 Chock the front wheels then jack up the rear of the vehicle and support on axle stands (see *"Jacking and vehicle support"*). Remove the right-hand rear wheel.

9 Remove the exhaust system and relevant heat shield(s) with reference to Section 25.

10 Note the position of the adjustment nut on the rear of the handbrake lever equaliser rod, then unscrew and remove it and detach the handbrake cables from the supports on the underbody. Position the cables to one side away from the fuel tank.

11 Disconnect the hoses from the fuel filler neck (see Section 3 when disconnecting the special collar).

12 Place a trolley jack with an interposed block of wood beneath the tank, then raise the jack until it is supporting the weight of the tank.

13 Unscrew and remove the mounting bolts, then slowly lower the fuel tank out of position, disconnecting any other relevant vent pipes as they become accessible (where necessary), and remove the tank from underneath the vehicle. Note that it may be necessary to carefully bend the brake pipes to provide sufficient clearance when lowering the tank.

14 If the tank is contaminated with sediment or water, remove the fuel gauge sender unit (Section 3), and swill the tank out with clean fuel. The tank is injection-moulded from a synthetic material - if seriously damaged, it should be renewed. However, in certain cases, it may be possible to have small leaks or minor damage repaired. Seek the advice of a specialist before attempting to repair the fuel tank.

Refitting

15 Refitting is the reverse of the removal procedure, noting the following points:

a) *When lifting the tank back into position, take care to ensure that the hoses are not trapped between the tank and vehicle body.*
b) *Ensure that all pipes and hoses are correctly routed. Make sure the sealing rings are in position in the quick-release fittings prior to fitting and make sure they are securely clipped in position.*
c) *On completion, refill the tank with a small amount of fuel, and check for signs of leakage prior to taking the vehicle out on the road.*

5 Accelerator cable - removal, refitting and adjustment

Removal

1 Working in the engine compartment, operate the accelerator lever on the fuel injection pump, and release the inner cable from the lever. Alternatively, on models with a press-fit balljoint cable end fitting, pull the

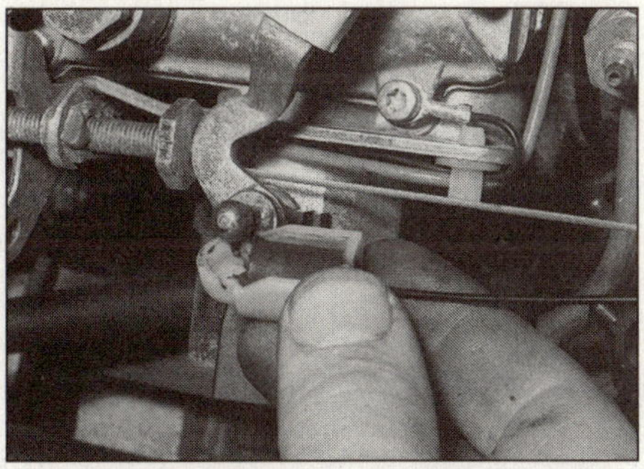

5.1 Disconnecting the accelerator cable from the injection pump

5.2b . . . to enable the cable to pass through the bracket (Lucas injection pump)

cable end from the lever **(see illustration)**.

2 Pull the outer cable ferrule from the grommet in the fuel injection pump bracket. Note that on models with a balljoint end fitting, it will be necessary to prise the balljoint from the rubber cable end fitting to enable the cable to pass through the pump bracket **(see illustrations)**.

3 Working inside the vehicle, remove the lower trim panel from under the steering column. Disconnect the cable from the accelerator pedal by squeezing the lugs of the cable end fitting.

4 Return to the engine compartment, release the outer cable from the bulkhead and withdraw the cable.

Refitting and adjustment

5 Refitting is a reversal of removal, ensuring that the cable is routed as noted before removal, and on completion, check the cable adjustment as follows.

6 Have an assistant fully depress the accelerator pedal, then check that the accelerator lever on the injection pump is touching the maximum speed adjustment screw. If adjustment is required, remove the spring clip from the adjustment ferrule, reposition the ferrule as necessary, then insert the clip in the next free groove on the ferrule.

7 With the accelerator pedal fully released, check that the accelerator lever is touching the anti-stall adjustment screw.

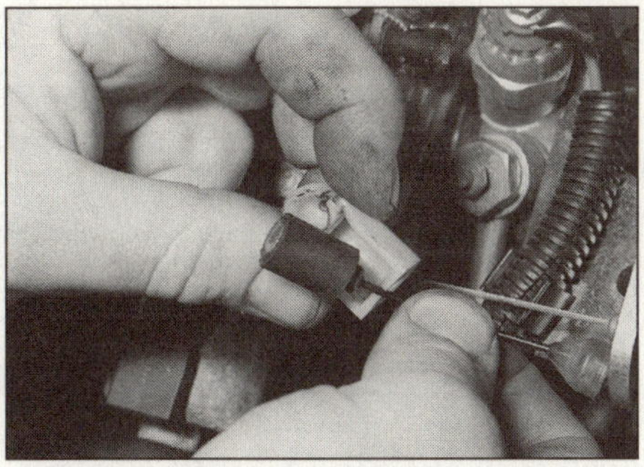

5.2a Remove the balljoint . . .

5.2c Accelerator cable end fitting (Bosch injection pump)

6 Accelerator pedal - removal and refitting

Removal

1 Remove the lower trim panel from under the steering column.

2 Disconnect the cable from the top of the accelerator pedal by squeezing the lugs of the cable end fitting.

3 Remove the nut which secures the accelerator pedal pivot and bush to the bulkhead.

4 Withdraw the accelerator pedal.

5 Examine the pedal and pivot for signs of wear and renew as necessary.

Refitting

6 Refitting is a reversal of removal. Check the adjustment of the accelerator cable as described in Section 5.

7 Fuel system - priming and bleeding

Warning: Refer to the warning note in Section 1 before proceeding.

1 After disconnecting part of the fuel supply system or running out of fuel, it is necessary to prime the system and bleed off any air which may have entered the system components.

2 All models are fitted with a hand-operated priming bulb located in the right-hand corner of the engine compartment, in the fuel feed line to the fuel filter.

3 To prime the system, loosen the bleed screw located on the top of the injection pump near the fuel inlet union, then squeeze the priming bulb several times until fuel free from air bubbles emerges from the bleed screw **(see illustrations)**.

4 Retighten the bleed screw and continue pumping until firm resistance is felt.

5 Attempt to start the engine at this stage, by fully depressing the accelerator pedal and operating the starter motor. Do not operate the heater plugs. If it refuses to start after 15 seconds, operate the heater plugs as normal then attempt to start the engine again.

6 If a large amount of air has entered the pump, place a wad of rag around the fuel return union on top of the pump (to absorb spilt fuel), then slacken the union. Operate the priming bulb until fuel free from air bubbles emerges from the fuel union. Tighten the union and mop up spilt fuel.

7 If air has entered the injector pipes, place wads of rag around the injector pipe unions at the injectors (to absorb spilt fuel), then slacken the unions. Crank the engine on the starter motor until fuel emerges from the unions, then stop cranking the engine and retighten the unions. Mop up spilt fuel.

Warning: Be prepared to stop the engine if it should fire, to avoid fuel spray and spillage.

7.3a Loosen the fuel system bleed screw located on the fuel pump inlet union . . .

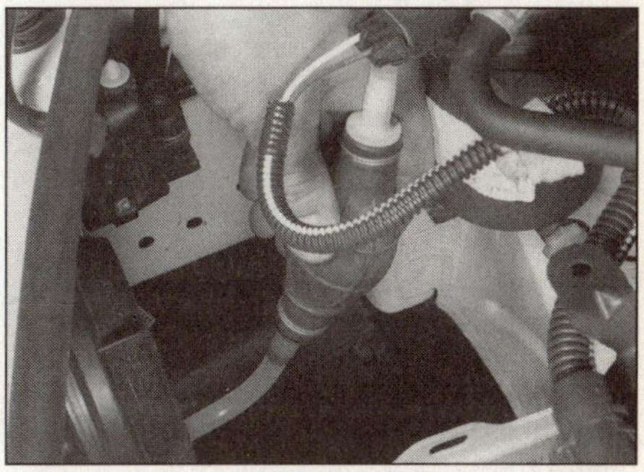

7.3b . . . and squeeze the priming bulb

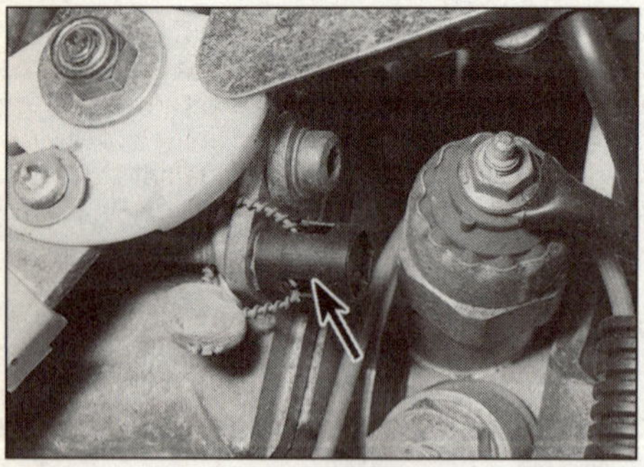

8.3 Maximum speed adjustment screw (arrowed) - Lucas injection pump

8 Attempt to start the engine by fully depressing the accelerator pedal and operating the starter motor. Initially do not operate the heater plugs, however, if it refuses to start after 15 seconds, operate the heater plugs as normal then start the engine.

8 Maximum speed - checking and adjustment

Caution: The maximum speed adjustment screw is sealed by the manufacturers at the factory using paint or a locking wire and a lead seal. There is no reason why it should require adjustment. Do not disturb the screw if the vehicle is still within the warranty period otherwise the warranty will be invalidated. This adjustment requires the use of a tachometer - refer to Chapter 1B for alternative methods.

1 Run the engine to normal operating temperature.

2 Have an assistant fully depress the accelerator pedal and check that the maximum engine speed is as given in the Specifications. Do not keep the engine at maximum speed for more than two or three seconds.

3 If adjustment is necessary, stop the engine then loosen the locknut, turn the maximum speed adjustment screw as necessary, and retighten the locknut **(see illustration)**.

4 Repeat the procedure in paragraph 2 to check the adjustment.

5 Stop the engine and disconnect the tachometer.

9 Fast idle thermostatic actuator and cable - removal, refitting, testing and adjustment

Note: *A new sealing ring will be required on refitting.*

Removal

1 The thermostatic actuator is located at the left-hand end of the cylinder head **(see illustration)**.

2 Partially drain the cooling system as described in Chapter 1B.

9.1 Thermostatic actuator on the left-hand end of the cylinder head

3 Loosen the clamp screw or nut (as applicable) and slide the fast idle cable end fitting from the inner cable at the fast idle lever on the injection pump **(see illustration)**.

4 Slide the outer cable from the bracket on the fuel injection pump.

5 Unscrew the thermostatic actuator from the cylinder head, and withdraw the actuator complete with the cable. Recover the sealing ring **(see illustration)**.

Refitting

6 Fit the actuator, using a new sealing ring, and tighten it.

7 Refill the cooling system as described in Chapter 1.

8 Insert the outer cable through the bracket on the injection pump.

9 Insert the inner cable through the fast idle lever, and position the end fitting on the cable, but do not tighten the clamp screw or nut (as applicable).

10 Adjust the cable as described in the following paragraphs.

Testing and adjustment

Bosch injection pump

Note: *During this procedure, it is necessary to measure the engine speed. The usual type of tachometer (rev counter), which works from ignition system pulses, cannot be used on diesel engines. A diagnostic socket is provided for the use of Renault test equipment, but this will not normally be available to the home mechanic. If it is not felt that adjusting the idle speed "by ear" is satisfactory, or that (where fitted) the instrument panel tachometer is not sufficiently reliable, one of the following alternatives may be used.*

a) Purchase or hire of an appropriate tachometer.

b) Delegation of the job to a Renault dealer or other specialist.

c) Timing light (strobe) operated by a petrol engine running at the desired speed. If the timing light is pointed at a mark on the

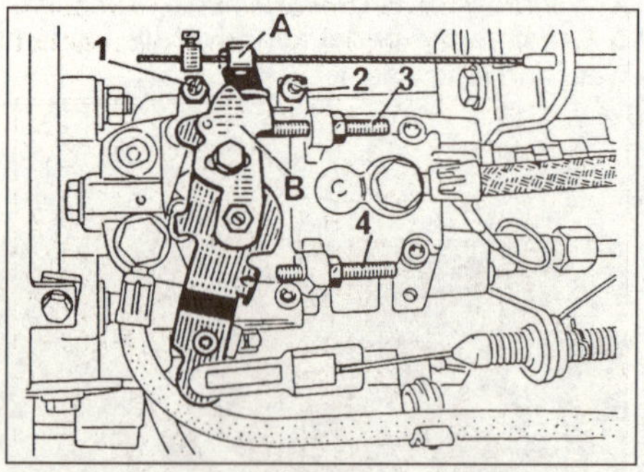

9.12 Bosch injection pump adjustment screws

A Fast idle lever
B Accelerator lever
1 Fast idle adjustment screw
2 Idle adjustment screw
3 Anti-stall adjustment screw
4 Maximum speed adjustment screw

9.3 Thermostatic actuator fast idle cable adjustment clamp

9.5 Unscrew the thermostatic actuator from the cylinder head. Note the sealing ring (arrowed)

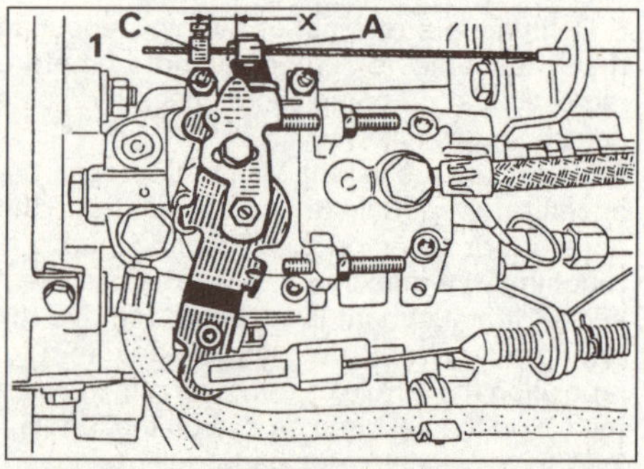

9.17 Bosch injection pump fast idle cable adjustment (thermostatic actuator)

A Fast idle lever
C Cable end fitting
1 Fast idle adjustment screw
x = 6.0 mm

camshaft or injection pump sprocket, the mark will appear stationary when the two engines are running at the same speed (or multiples of that speed). The sprocket will be rotating at half the crankshaft speed but this will not affect the adjustment. (In practice it was found impossible to use this method on the crankshaft pulley due to the acute viewing angle.)

11 With the engine warm (the cooling fan should have cut in and out once) and running at the correct idle speed (see Chapter 1B), proceed as follows.

12 Move the fast idle lever on the injection pump towards the flywheel end of the engine so that it contacts the fast idle adjustment screw **(see illustration)**.

13 Check the fast idle speed. If necessary, loosen the locknut and turn the adjustment screw to give the specified fast idle speed. Retighten the locknut on completion.

14 Stop the engine, and disconnect the tachometer where applicable.

15 The fast idle cable should now be adjusted as follows.

16 With the engine still at normal operating temperature, and the fast idle lever in its rest position (resting against the idle speed adjustment screw, **not** the fast idle adjustment screw), gently pull the fast idle cable taut.

17 Position the end fitting on the cable so that the dimension between the end of the fast idle lever and the end fitting is 6.0 ± 1 mm on non-air conditioning models or 2.0 ± 1 mm on air conditioning models **(see illustration)**.

18 With the end fitting correctly positioned, tighten the clamp screw.

19 When the engine has cooled, check that the cable has pulled the fast idle lever so that it rests against the fast idle adjustment screw. Re-check that with the engine at normal operating temperature the dimension "x" between the end of the fast idle lever and the cable end fitting is as specified. If not, it is likely that the thermostatic actuator is faulty.

Lucas injection pump

20 The fast idle speed is set at the factory on a test bench and cannot be adjusted.

21 With the engine cold, note the length of the exposed inner fast idle cable, measured from the end of the cable to the point where it enters the cable outer.

22 Run the engine until it is at normal operating temperature (the cooling fan should have cut in and out once), then stop the engine.

23 Again measure the length of the exposed inner fast idle cable, which should have increased by 7.0 to 8.5 mm. If the cable travel between cold and normal operating temperature is not as specified, it is likely that the thermostatic actuator is faulty.

24 With the engine still at normal operating temperature, and the fast idle lever in the rest position, gently pull the fast idle cable taught.

25 Position the fast idle cable end fitting on the cable so that the dimension between the end of the fast idle lever and the end fitting is 3.0 ± 1.0 mm.

26 With the end fitting correctly positioned, tighten the clamp nut.

10 Fast idle system components - removal, refitting, testing and adjustment

Vacuum actuator and cable

Removal

1 The vacuum actuator is attached to a bracket located on the end of the injection pump.

2 Loosen the clamp screw or nut (as applicable) and slide the fast idle cable end fitting from the inner cable at the fast idle lever on the injection pump.

3 Disconnect the vacuum hose from the actuator.

4 Unscrew the nut securing the actuator to the bracket, and withdraw the actuator, passing the cable through the bracket as it is withdrawn **(see illustrations)**.

Refitting

5 Refit the actuator to the bracket, and tighten the securing nut.

6 Reconnect the vacuum hose.

7 Insert the inner cable through the fast idle lever, and position the end fitting on the cable, but do not tighten the clamp screw or nut (as applicable).

8 Adjust the cable as described in the following paragraphs.

Testing and adjustment - Bosch injection pump

9 Proceed as described in Section 9, paragraphs 11 to 18 inclusive, with reference to the note at the beginning of paragraph 11.

10 If a faulty vacuum actuator is suspected, first test the solenoid valve as described later in this Section.

11 The actuator can be tested using a vacuum pump. With the engine stopped, disconnect the vacuum hose from the actuator, and connect the vacuum pump in its place. With a vacuum of 500 mbars applied, the actuator should operate the cable sufficiently to pull the fast idle lever so that it rests against the fast idle adjustment screw. If not, it is likely that the vacuum actuator is faulty.

Testing and adjustment - Lucas injection pump

12 The fast idle speed is set at the factory on a test bench and cannot be adjusted.

13 With the engine stopped, and the fast idle lever in the rest position, gently pull the fast idle cable taught.

14 Position the fast idle cable end fitting on the cable so that the dimension between the end of the fast idle lever and the end fitting is 3.0 ± 1.0 mm.

15 With the end fitting correctly positioned, tighten the clamp nut.

16 The actuator can be tested with reference to paragraphs 10 and 11, but note that the fast idle lever should rest against the lever stop rather than the fast idle adjustment screw.

Solenoid valve

Removal

Note: *A special kit is available from Renault dealers to drill out the shear-head screws which retain the valve on the injection pump. New shear-head screws will be required for refitting the valve.*

17 The solenoid valve is located on the left-hand side of the injection pump and is retained with shear-screws (screws which are tightened until their heads break off). The ignition must be switched off before starting work.

18 Remove the crankcase ventilation oil separator from the left-hand front of the engine for access to the solenoid valve mounting screws.

19 The heads of the mounting screws must now be drilled out. If the Renault kit is used, locate the drilling tube over one of the screws and drill to a depth of 3.0 mm using a 4.0 mm

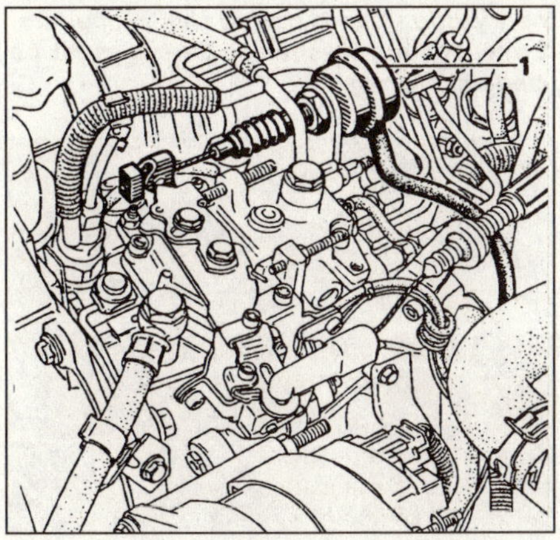

10.4a Fast idle vacuum actuator (1) - Bosch injection pump

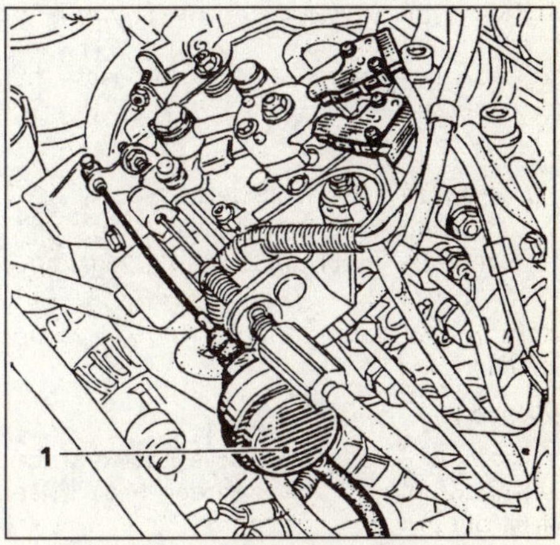

10.4b Fast idle vacuum actuator (1) - Lucas injection pump

4B

drill (lightly oil the drill to facilitate drilling). Using a No 6 "Easy-out" (or a stud extractor tool), unscrew the mounting screw. Repeat the procedure on the remaining screw.

20 Disconnect the wiring and vacuum hoses from the solenoid valve, and withdraw it from the injection pump.

Refitting

21 Refitting is a reversal of removal ensuring that the vacuum hoses are securely reconnected, but tighten the new shear-head screws until the heads break off.

Testing

22 Remove the solenoid valve as described previously.

23 Attempt to blow gently through one of the vacuum hose connections. No air should pass through the valve.

24 With a 12-volt supply connected across the solenoid terminals, again attempt to blow through the valve. Air should now pass through.

25 If the valve proves to be faulty, it should be renewed.

11 Cold start timing advance solenoid - testing, removal and refitting

Testing

1 To check the operation of the system, start the engine from cold, and listen for a knocking or harshness, disappearing after between 30 seconds and 2 minutes 45 seconds. This shows that the system is working correctly - if dynamic timing equipment is available, this can be used to check the advance.

2 If the system does not seem to be working, check for voltage at the solenoid feed wire with the starter motor cranking, and for 5 to 6 seconds after start-up (the feed to the solenoid is controlled by the preheating system control unit).

3 If voltage is present, but the system is still not working, the solenoid valve is probably faulty, and should be renewed.

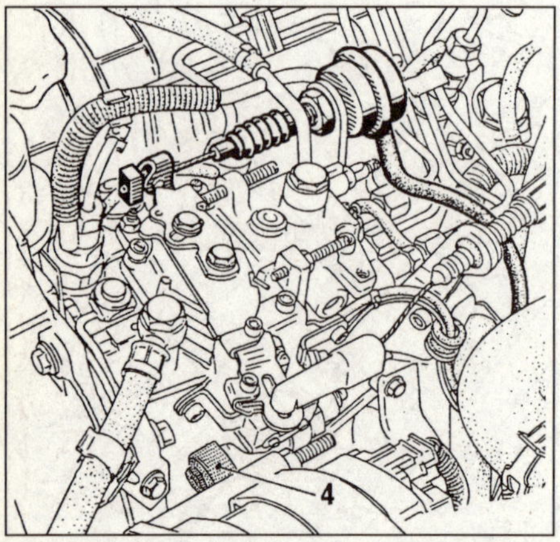

11.4a Cold start timing advance solenoid location (4) - Bosch injection pump

Removal

Note: *Be careful not to allow dirt into the injection pump during this procedure. A new sealing washer must be used on refitting.*

4 Disconnect the battery negative lead, then disconnect the solenoid wiring connector **(see illustrations)**.

Caution: If the radio/cassette in your vehicle is equipped with an anti-theft system, make sure you have the correct activation code before disconnecting the battery.

5 Unscrew the solenoid valve from the pump and recover the sealing washer. Take great care not to allow dirt to enter the pump.

Refitting

6 Refitting is a reversal of removal, using a new sealing washer.

12 Stop solenoid - description, removal and refitting

Description

1 The stop solenoid is located on the end of the fuel injection pump. Its purpose is to cut the fuel supply when the ignition is switched off. If an open circuit occurs in the solenoid or supply wiring it will be impossible to start the engine, as the fuel will not reach the injectors. The same applies if the solenoid plunger jams in the "stop" position. If the solenoid jams in the "run" position, the engine will not stop when the ignition is switched off.

2 If the solenoid has failed and the engine will not run, a temporary repair may be made by removing the solenoid as described in the following paragraphs. Refit the solenoid body without the plunger and spring. Tape up the wire so that it cannot touch earth. The engine can now be started as usual, but it will be necessary to use the manual stop lever on the fuel injection pump (or to stall the engine in gear) to stop it.

Removal

Caution: Be careful not to allow dirt into the injection pump during this procedure. A new sealing washer or O-ring must be used on refitting.

11.4b Cold start timing advance solenoid location (arrowed) - Lucas injection pump

3 Disconnect the battery negative lead.

Caution: If the radio/cassette in your vehicle is equipped with an anti-theft system, make sure you have the correct activation code before disconnecting the battery.

4 Withdraw the rubber boot (where applicable), then unscrew the terminal nut and disconnect the wire from the top of the solenoid.

5 Carefully clean around the solenoid, then unscrew and withdraw the solenoid, and recover the sealing washer or O-ring (as applicable). Recover the solenoid plunger and spring if they remain in the pump. Operate the hand priming pump as the solenoid is removed to flush away any dirt.

Refitting

6 Refitting is a reversal of removal, using a new sealing washer or O-ring.

13 Fuel injection pump - removal and refitting

Note: *Refer to the precautions given in Section 1 of this Chapter before proceeding. Be careful not to allow dirt into the pump or injector pipes during this procedure. New sealing rings should be used on the fuel pipe banjo unions when refitting.*

Removal

Non-turbo models with air conditioning

1 In order to gain access to the injection pump securing bolts, the alternator and its mounting bracket must be removed.

2 Disconnect the battery negative lead.

Caution: If the radio/cassette in your vehicle is equipped with an anti-theft system, make sure you have the correct activation code before disconnecting the battery.

3 Remove the alternator as described in Chapter 5A, then proceed as follows.

4 Unscrew the securing bolts, and remove the power steering pump pulley from the pump drive flange. Note that it will be necessary to counterhold the pulley (eg, using an old drivebelt) in order to loosen the bolts.

5 Remove the securing bolts, and withdraw the auxiliary drivebelt guide roller/bracket assembly.

6 Remove the two upper bolts securing the air conditioning compressor to the alternator mounting bracket.

7 Apply the handbrake, then jack up the front of the vehicle and support securely on axle stands (see *"Jacking and vehicle support"*). Where fitted, remove the engine compartment undershield.

8 Remove the right-hand front roadwheel and the wheel arch liner.

9 Loosen the lower air conditioning compressor mounting nuts and bolts, then pivot the compressor downwards to enable the alternator mounting bracket to be removed from above.

10 Proceed as described in the following paragraphs.

All models

11 If not already done, disconnect the battery negative lead.

Caution: If the radio/cassette in your vehicle is equipped with an anti-theft system, make sure you have the correct activation code before disconnecting the battery.

12 Unbolt and remove the right-hand engine mounting cover and the injection pump sprocket cover.

13 If not already done, remove the two bolts securing the fuel filter assembly to the body panel, unbolt the hose bracket from the fuel injection pump mounting bracket, and move the fuel filter assembly to one side, leaving the hoses connected. Take care not to strain the hoses.

14 Turn the crankshaft to bring No 1 piston to TDC on the compression stroke, and fit the tool to lock the crankshaft in position, as described in Chapter 2B.

15 Unscrew the securing bolts, and remove the timing belt cover which covers the fuel injection pump sprocket.

16 If an injection pump sprocket locking tool (Renault tool Mot.1200) is available, remove the crankshaft locking tool, turn the engine back from TDC by one camshaft sprocket tooth, and fit the injection pump sprocket locking tool (turning the engine back by one sprocket tooth will ensure that sufficient adjustment is available to set the pump timing on refitting).

17 If a locking tool is not available, remove the timing belt as described in Chapter 2B.

18 Disconnect the accelerator cable from the injection pump, with reference to Section 5 if necessary.

19 On models with a fast idle thermostatic actuator, disconnect the fast idle cable from the injection pump, with reference to Section 9 if necessary.

20 On models with a fast idle vacuum actuator, disconnect the vacuum hose from the actuator (see Section 10).

21 On Turbo models, disconnect the hose connecting the boost pressure fuel delivery corrector to the inlet manifold **(see illustration)**.

22 Loosen the clip or undo the banjo union, as applicable and disconnect the fuel supply hose from the injection pump **(see illustration)**. Recover the sealing washers from the banjo union, where applicable. Cover the open end of the hose or pipe, and plug the opening in the injection pump to keep dirt out (on models with a banjo union, the banjo bolt can be refitted to the pump and covered).

23 Disconnect the main fuel leak-off return hose (connected to the main fuel return pipe) from the relevant fuel injector **(see illustration)**.

24 Where applicable, disconnect the auxiliary fuel return hose from the pump. On Turbo models, the auxiliary hose is connected to the boost pressure fuel delivery corrector on the pump (also unclip it from the accelerator cable bracket) **(see illustration)**.

25 Disconnect the main fuel return pipe banjo union from the injection pump. Recover the sealing washers from the banjo union. Again, cover the open end of the hose and the banjo bolt to keep dirt out **(see illustration)**. Take care not to get the inlet and outlet banjo unions mixed up.

26 Disconnect all relevant wiring from the pump. Note that on certain pumps, this can be achieved by simply disconnecting the wiring connectors at the brackets on the pump. On some pumps it will be necessary to disconnect the wiring from the individual components (some connections may be protected by rubber covers) **(see illustration)**.

27 Unscrew the union nuts securing the injector pipes to the injection pump and injectors. Counterhold the unions on the pump, when unscrewing the pipe-to-pump union nuts. Remove the pipes as a set. Cover open unions to keep dirt out, using small plastic bags or fingers cut from discarded (but clean!) rubber gloves. Note that the leak-off hoses will have to be removed from the fuel injectors to enable the injectors to be covered **(see illustration)**.

28 Where applicable, remove the alternator plastic shield from below the injection pump. Note the locations of any brackets secured by the shield securing nut and bolt.

13.21 Disconnect the hose connecting the boost pressure fuel delivery corrector to the inlet manifold - Turbo models

13.22 Fuel supply hose (1) and main return hose (2) - Bosch injection pump

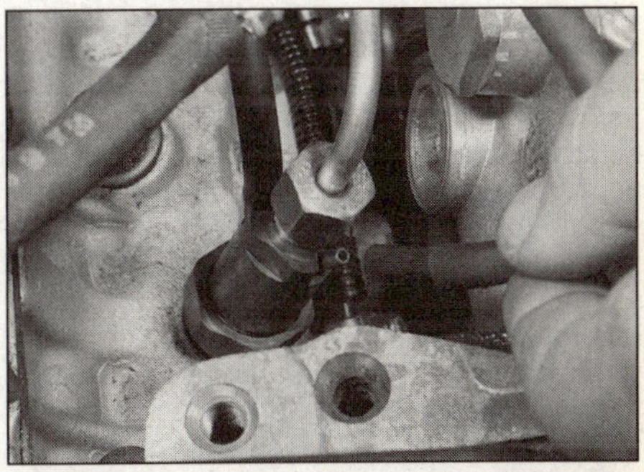

13.23 Disconnect the main fuel leak-off return hose from the relevant fuel injector

13.24 Disconnecting the auxiliary fuel return hose - Turbo model

13.25 Disconnecting the main fuel return pipe union - Turbo model

13.26 Wiring terminal on the ALFB solenoid valve - Bosch injection pump

13.27 Unscrewing an injector pipe-to-injection pump union

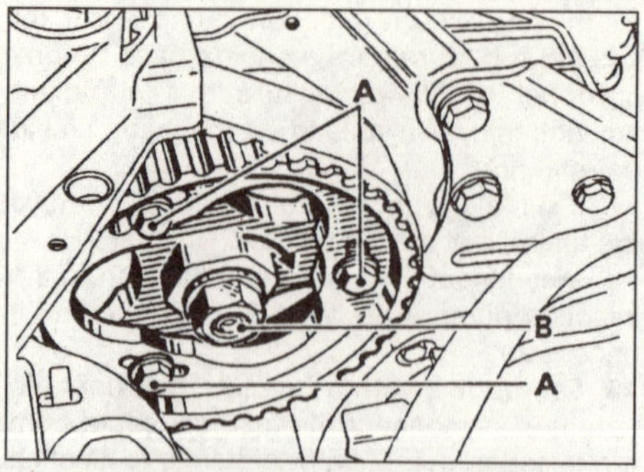

13.30 Two-piece sprocket adjuster bolts (A) and sprocket securing nut (B). Turn sprocket boss in direction arrowed before removing pump

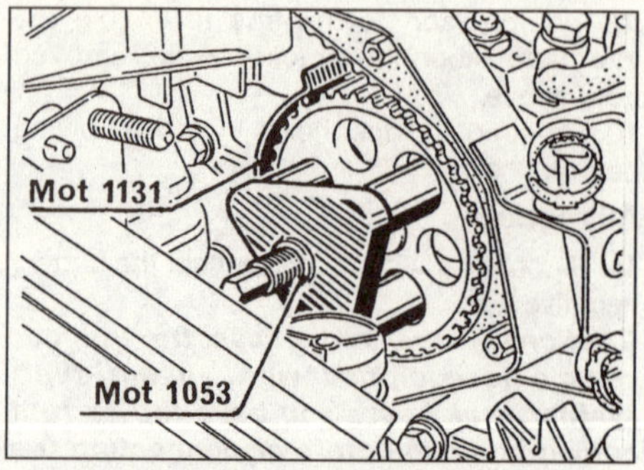

13.31a Renault sprocket locking tool and puller in position - one-piece sprocket

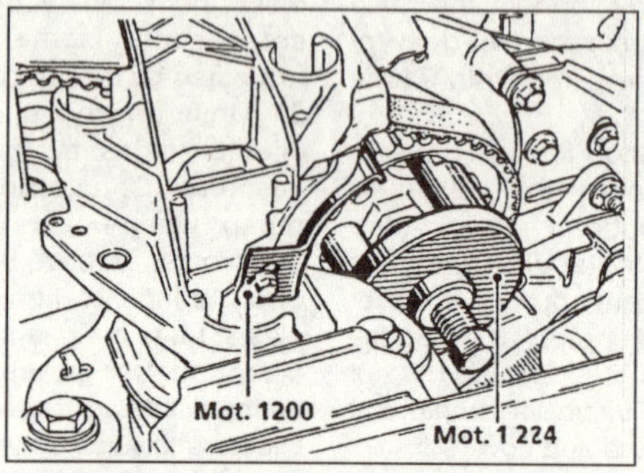

13.31b Renault sprocket locking tool and puller in position - two-piece adjustable sprocket

13.33 Unscrew the bolts securing the rear injection pump mounting bracket to the cylinder head - Turbo model shown

29 Unscrew the injection pump sprocket securing nut, and remove the washer. If a Renault sprocket locking tool is not available (see paragraph 17), counterhold the sprocket using an improvised tool engaged with the holes in the sprocket (**not** the sprocket teeth).

30 Mark the sprocket in relation to the end of the pump shaft to ensure correct refitting. On models with a Bosch injection pump fitted with a one-piece sprocket, this is necessary regardless of whether the timing belt has been removed, because there are two keyways in the sprocket (for use with different injection pump types), and it is possible to refit the sprocket to the pump shaft incorrectly (see Chapter 2B) **(see illustration)**.

31 Fit a puller to the pump sprocket, and free the sprocket from the taper on the pump shaft **(see illustrations)**. **Note:** *The puller must locate in the sprocket holes,* **not** *on the teeth. On the Lucas two-piece (adjustable) sprocket a special puller will be required which locates on the hub of the sprocket.* **Do not** hammer on the end of the pump shaft to free it, as this will damage the internal components of the pump.

32 Remove the puller. Unless an injection pump sprocket locking tool has been fitted and the timing belt is still in position (see paragraph 17), remove the sprocket.

33 Unscrew the bolts securing the rear injection pump mounting bracket to the cylinder head **(see illustration)**.

34 Make a final check to ensure that all relevant pipes, hoses and wires have been disconnected to facilitate pump removal.

35 Make alignment marks between the pump and the mounting bracket. This will aid pump timing on refitting.

36 Unscrew the three pump securing bolts, and withdraw the pump from its mounting bracket, leaving the sprocket engaged with the timing belt, where applicable **(see illustrations)**. Access to the lower pump mounting bolt is most easily obtained from the rear of the pump, using a deep socket and extension.

37 Recover the Woodruff key from the end of the pump shaft if it is loose.

38 If desired, the rear pump mounting bracket and the accelerator cable bracket can be unbolted from the rear of the pump.

Refitting

39 Where applicable, refit the rear mounting bracket and the accelerator cable bracket to the rear of the pump.

40 Commence refitting by fitting the Woodruff key to the end of the pump shaft, where applicable.

41 Offer the pump to the mounting bracket. If the sprocket is still engaged with the timing belt, engage the pump shaft with the sprocket. Align the marks made on the sprocket and the pump shaft before removal (on the one-piece sprocket, this will ensure that the key engages with the correct keyway

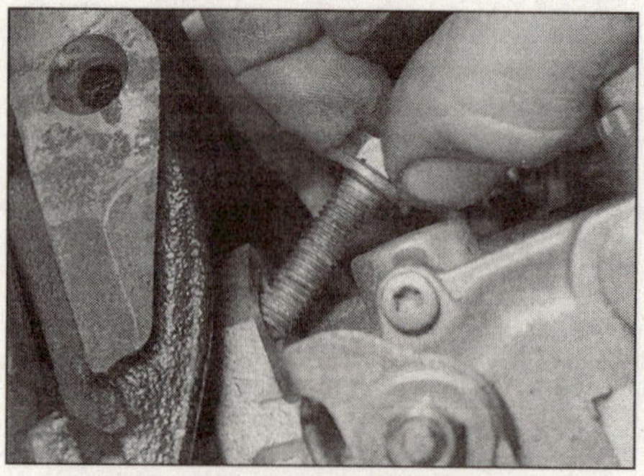

13.36a Remove the injection pump securing bolts . . .

13.36b . . . and withdraw the fuel injection pump - Turbo model shown

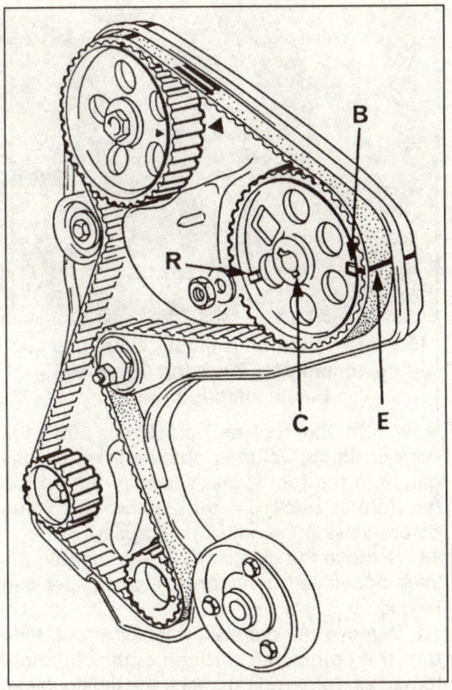

13.41a One-piece injection pump sprocket orientation - Bosch injection pump (engine shown with No 1 piston at TDC)

B Sprocket alignment mark for Bosch injection pump
C Position of injection pump shaft Woodruff key
E Sprocket alignment mark on timing belt
R Sprocket alignment mark for Lucas injection pump

in the sprocket) **(see illustrations)**. Ensure that the Woodruff key does not fall out of the shaft as the sprocket is engaged.

42 Where applicable, align the marks made on the pump and the mounting bracket before removal. If a new pump is being fitted, transfer the mark from the old pump to give an approximate setting.

43 Refit and lightly tighten the front pump mounting bolts.

44 Refit and tighten the bolts securing the rear pump mounting bracket to the cylinder head.

45 Where applicable, refit the sprocket to the pump shaft, ensuring the Woodruff key engages correctly. Align the marks made on the sprocket and the pump shaft before removal.

46 Refit the washer, then tighten the pump sprocket securing nut to the specified torque, counterholding the sprocket as during removal **(see illustration)**.

47 If the timing belt has been removed, refit and tension it as described in Chapter 2B.

48 Where applicable, remove the sprocket locking tool, then turn the crankshaft to bring No 1 piston to TDC on the compression stroke, and fit the tool to lock the crankshaft in position, as described in Chapter 2B.

49 Carry out injection timing as described in Sections 14, 15 or 16, as applicable.

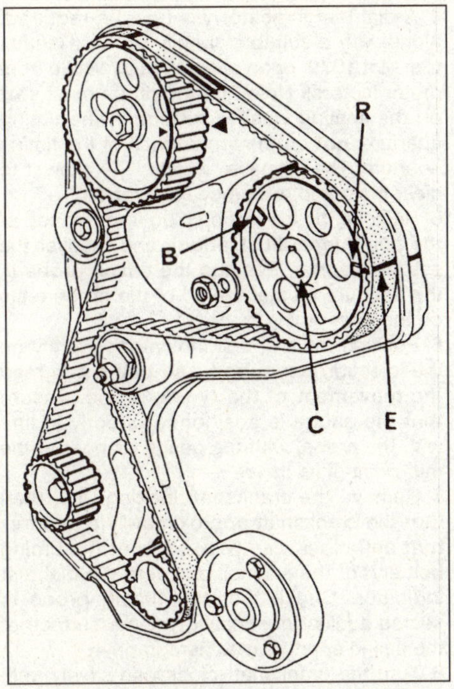

13.41b One-piece injection pump sprocket orientation - Lucas injection pump (engine shown with No 1 cylinder at TDC)

B Sprocket alignment mark for Bosch injection pump
C Position of injection pump shaft Woodruff key
E Sprocket alignment mark on timing belt
R Sprocket alignment mark for Lucas injection pump

50 Refit the plastic shield under the pump, ensuring that any brackets noted during removal are in place.

51 Refit and reconnect the injector fuel pipes, and tighten the unions. Counterhold the unions on the pump when tightening the pipe-to-pump union nuts.

52 Reconnect all relevant wiring to the pump.

53 Reconnect the fuel supply and return pipes and hoses, and tighten the unions, as applicable. Use new sealing washers on the banjo unions. Also reconnect the leak-off hoses.

54 On Turbo models, reconnect the hose connecting the boost pressure fuel delivery corrector to the inlet manifold.

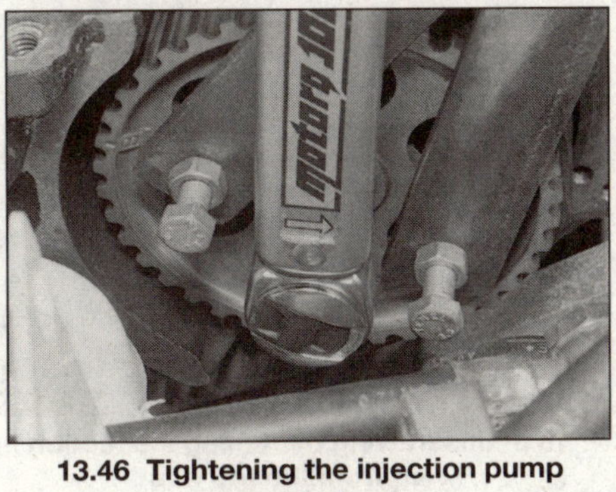

13.46 Tightening the injection pump sprocket nut

55 Reconnect the accelerator cable and adjust as described in Section 5.

56 Refit the fuel filter assembly and tighten the securing bolts, and refit the hose bracket.

57 On non-Turbo models with air conditioning, reverse the operations carried out in paragraphs 1 to 10. Refit and tighten the auxiliary drivebelt, referring to Chapter 1B.

58 On models with a fast idle thermostatic actuator, reconnect and adjust the fast idle cable as described in Section 9, noting that final adjustment must be carried out after the engine has been started and reached normal operating temperature.

59 On models with a fast idle vacuum actuator, reconnect the hose to the actuator, and check the fast idle cable adjustment as described in Section 10.

60 Refit the timing belt cover over the injection pump sprocket and tighten the bolts.

61 Refit the right-hand engine mounting cover and the injection pump sprocket cover and tighten the bolts.

62 Check the no-load switch adjustment as described in Chapter 5C.

63 Reconnect the battery negative lead.

64 Prime and bleed the fuel system as described in Section 7.

65 Start the engine, and check the idle speed and anti-stall speed, referring to Chapter 1B.

14 Injection timing - checking methods and adjustment

1 Checking the injection timing is not a routine operation. It is only necessary after the injection pump has been disturbed.

2 Dynamic timing equipment does exist, but it is unlikely to be available to the home mechanic. The equipment works by converting pressure pulses in an injector pipe into electrical signals. If such equipment is available, use it in accordance with its maker's instructions.

3 Static timing as described in this Chapter gives good results if carried out carefully. A dial test indicator will be needed, with probes and adaptors appropriate to the type of injection pump **(see illustration)**. Read through the

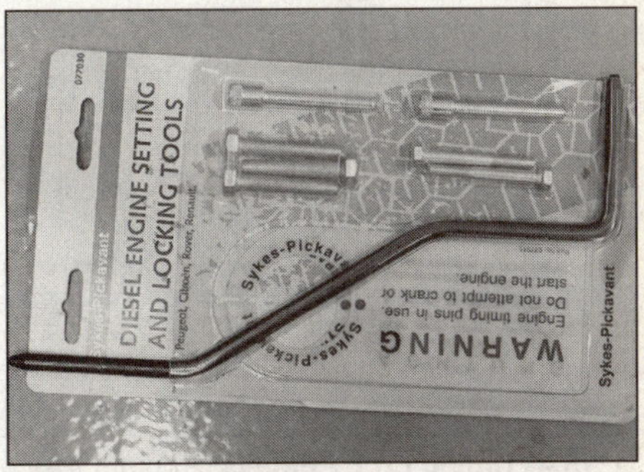

14.3 TDC locking tools for setting injection timing on Renault diesel engines

procedures before starting work to find out what is involved.

15 Injection timing (Lucas injection pump) - checking and adjustment

Caution: The maximum engine speed and transfer pressure settings, together with timing access plugs, are sealed by the manufacturers at the factory using locking wire and lead seals. Do not disturb the wire if the vehicle is still within the warranty period otherwise the warranty will be invalidated. Also do not attempt the timing procedure unless accurate instrumentation is available. Suitable special tools for carrying out pump timing are available from motor factors, and a dial test indicator will be required regardless of the method used. Refer to the precautions given in Section 1 of this Chapter before proceeding.

1 Disconnect the battery negative lead.

Caution: If the radio/cassette in your vehicle is equipped with an anti-theft system, make sure you have the correct activation code before disconnecting the battery.

2 Apply the handbrake, then jack up the front right-hand corner of the vehicle until the wheel is just clear of the ground. Support the vehicle on an axle stand and engage 4th or 5th gear. This will enable the crankshaft to be turned easily by turning the right-hand wheel. Alternatively, the engine can be turned using an open-ended spanner on the crankshaft pulley bolt.

3 Turn the crankshaft to bring No 1 piston to TDC on the compression stroke, and fit the tool to lock the crankshaft in position, as described in Chapter 2B.

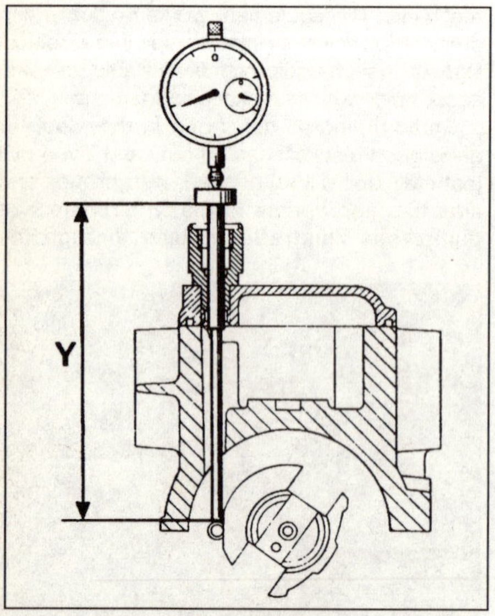

15.4 Timing probe details (Renault tool Mot.1079) - Lucas injection pump
Y = 95.5 ± 0.1 mm

4 A dial test indicator will now be required, along with a suitable special probe (Renault tool Mot.1079, or an alternative available from motor factors). Note that the probe must seat on the sealing washer surface in the timing aperture, **not** on the top surface of the timing aperture. The probe is "waisted" to allow it to clear the pump rotor **(see illustration)**.

5 Remove the inspection plug from the top of the pump **(see illustration)**, and recover the sealing washer. Position the timing probe in the aperture so that the tip of the probe rests on the rotor timing piece.

6 Position the dial test indicator securely on the injection pump body, so that it can read the movement of the timing probe. Ensure that the gauge is positioned directly in line with the probe, with the gauge plunger at the mid-point of its travel.

7 Remove the crankshaft locking tool, then turn the crankshaft approximately a quarter-turn anti-clockwise (viewed from the timing belt end of the engine), and zero the dial test indicator. Check that the timing probe is seated against the sealing washer surface of the timing aperture (see paragraph 4).

8 Turn the crankshaft clockwise slowly until the crankshaft locking tool can be re-inserted (bringing the engine back to TDC).

9 Read the dial test indicator; the reading should correspond to the value marked on the pump load lever. The timing value may also be marked on a plastic disc on the front of the pump, on a label attached to the top of the pump, or on a tag attached to the accelerator pump lever **(see illustration)**.

10 If the reading is not as specified, proceed as follows.

11 Unbolt and remove the timing belt cover from over the injection pump sprocket.

12 The injection pump sprocket consists of a hub/flange and outer gear which are locked together by an internal bolt (left-hand thread). The flange incorporates three holes into which a special tool (Renault tool Mot. 1358) is inserted in order to turn the pump shaft. First, the internal bolt must be loosened using Renault tool Mot. 1359 or a suitable hexagon tool.

13 Fit the special tool in the three holes and turn the tool clockwise as far as possible.

15.5 Unscrewing the timing inspection plug from the top of a Lucas injection pump

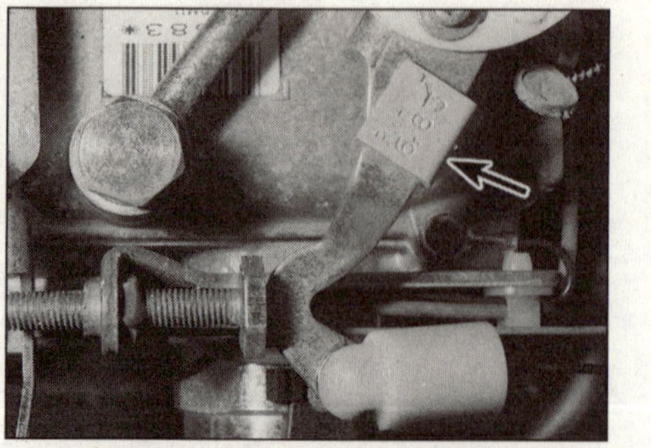

15.9 Injection pump timing value marked on accelerator lever tag (arrowed) - Lucas injection pump

Now turn the tool anti-clockwise until the correct timing value is obtained on the dial gauge. If the tool is inadvertently turned too far, turn it back 0.7 mm below the value before making the adjustment again.

14 Remove the special tool, then tighten the internal bolt (left-hand thread) to the specified torque.

15 Remove the crankshaft locking tool, then turn the crankshaft through eight complete turns clockwise, and recheck the timing value.

16 If adjustment is necessary, repeat the procedure described in the previous paragraphs.

17 When the timing is correct, remove the dial test indicator. Remove the probe from the inspection hole, and refit the inspection plug, ensuring that the sealing washer is in place.

18 Remove the crankshaft locking tool.

19 Lower the vehicle to the ground and reconnect the battery negative lead.

20 Bleed the fuel system (see Section 7).

21 Check and if necessary adjust the idle speed and anti-stall speed as described in Section 19.

16 Injection timing (Bosch injection pump) - checking and adjustment

Caution: Some of the injection pump settings and access plugs may be sealed by the manufacturers at the factory using paint or locking wire and lead seals. Do not disturb the seals if the vehicle is still within the warranty period, otherwise the warranty will be invalidated. Also do not attempt the timing procedure unless accurate instrumentation is available. Suitable special tools for carrying out pump timing are available from motor factors, and a dial test indicator will be required regardless of the method used. Refer to the precautions given in Section 1 of this Chapter before proceeding.

1 Disconnect the battery negative lead.

Caution: If the radio/cassette in your vehicle is equipped with an anti-theft system, make sure you have the correct activation code before disconnecting the battery.

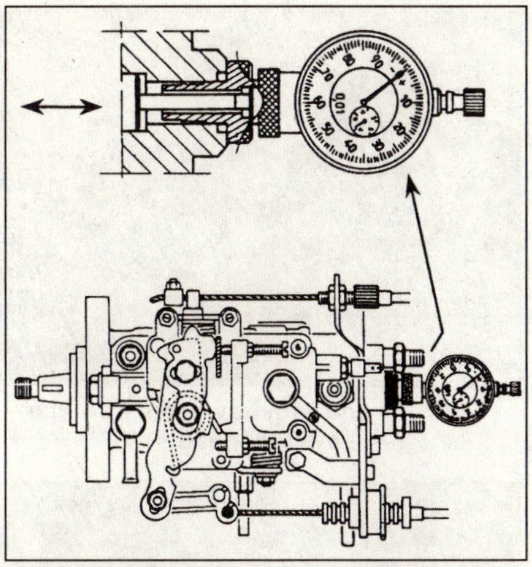

16.5 Dial test indicator and timing probe for use with Bosch injection pump
Note: *Pump cable linkages may differ from that shown*

2 Apply the handbrake, then jack up the front right-hand corner of the vehicle until the wheel is just clear of the ground. Support the vehicle on an axle stand and engage 4th or 5th gear. This will enable the engine to be turned easily by turning the right-hand wheel. Alternatively, the engine can be turned using an open ended spanner on the crankshaft pulley bolt.
3 Turn the crankshaft to bring No 1 piston to TDC on the compression stroke, and fit the tool to lock the crankshaft in position, as described in Chapter 2B.
4 Cover the alternator with a plastic bag or rags as a precaution against spillage of fuel.
5 A dial test indicator will now be required, along with a special probe and adaptor to screw into the hole in the rear of the pump (designed specifically for the Bosch pump and available from motor factors) **(see illustration)**.
6 Unscrew the union nuts securing the injector pipes to the fuel injection pump. Counterhold the unions on the pump, when unscrewing the nuts. Cover open unions to keep dirt out, using small plastic bags or fingers cut from discarded (but clean!) rubber gloves.
7 Unscrew the blanking plug from the end of the injection pump between the injector pipe connections. Be prepared for the loss of some fuel.
8 Insert the probe and connect it to the dial test indicator positioned directly over the inspection hole.
9 Remove the locking tool from the flywheel, and turn the engine approximately a quarter-turn anti-clockwise (viewed from the timing belt end of the engine), and zero the dial test indicator.
10 Turn the crankshaft clockwise slowly until the crankshaft locking tool can be re-inserted (bringing the engine back to TDC).
11 Read the dial test indicator; the reading should correspond to the value given in the Specifications. Note that on later pumps, the timing value is marked on the pump accelerator lever.
12 If the reading is not as specified, proceed as follows.
13 Slacken the three front pump mounting bolts, and the rear pump mounting nuts, and slowly rotate the pump body until the point is found where the specified reading is obtained on the dial gauge. When the pump is correctly positioned, tighten the mounting nuts and bolts, ensuring that the reading on the dial gauge does not change as the fixings are tightened.
14 Remove the crankshaft locking tool, and rotate the crankshaft through one and three quarter turns clockwise. Check that the dial test indicator is reading zero.
15 Rotate the crankshaft slowly clockwise until the crankshaft locking tool can be reinserted (bringing No 1 piston back to TDC). Recheck the timing measurement.
16 If adjustment is necessary, slacken the pump mounting nuts and bolts, and repeat the operations described in paragraphs 13 to 15.
17 When the timing is correct, remove the dial test indicator, remove the probe from the inspection hole, and refit the blanking plug.
18 Reconnect the fuel injector pipes to the pump.
19 Where applicable, remove the plastic bag or rags from the alternator.
20 Remove the crankshaft locking tool.
21 Lower the vehicle to the ground and reconnect the battery negative lead.
22 Bleed the fuel system as described in Section 7.
23 Check and if necessary adjust the idle speed and anti-stall speed as described in Section 19.

17 Fuel injectors - testing, removal and refitting

Warning: Exercise extreme caution when working on the fuel injectors. Never expose the hands or any part of the body to injector spray, as the high working pressure can cause the fuel to penetrate the skin, with possibly fatal results. You are strongly advised to have any work which involves testing the injectors under pressure carried out by a dealer or fuel injection specialist. Refer to the precautions given in Section 1 of this Chapter before proceeding.

Testing

1 Injectors do deteriorate with prolonged use and it is reasonable to expect them to need reconditioning or renewal after 60 000 miles (100 000 km) or so. Accurate testing, overhaul and calibration of the injectors must be left to a specialist. A defective injector which is causing knocking or smoking can be located without dismantling as follows.
2 Run the engine at a fast idle. Slacken each injector union in turn, placing rag around the union to catch spilt fuel and being careful not to expose the skin to any spray. When the union on the defective injector is slackened, the knocking or smoking will stop.

Removal

Note: *Take great care not to allow dirt into the injectors or fuel pipes during this procedure.*
3 Carefully clean around the injectors and injector pipe union nuts.
4 Pull the leak-off pipes from the injectors **(see illustration)**.
5 Unscrew the union nuts securing the injector pipes to the fuel injection pump. Counterhold the unions on the pump, when unscrewing the nuts. Cover open unions to keep dirt out, using small plastic bags or fingers cut from discarded (but clean!) rubber gloves.
6 Unscrew the union nuts and disconnect the pipes from the injectors **(see illustration)**. If necessary the injector pipes may be completely removed. Note the locations of any clips attached to the pipes. Cover the ends of the injectors to prevent dirt ingress.
7 Unscrew the injectors using a deep socket or box spanner (27 mm across flats) and remove them from the cylinder head **(see illustrations)**.
8 Recover the copper washers and fire seal washers from the cylinder head. Also recover the sleeves if they are loose **(see illustrations)**.

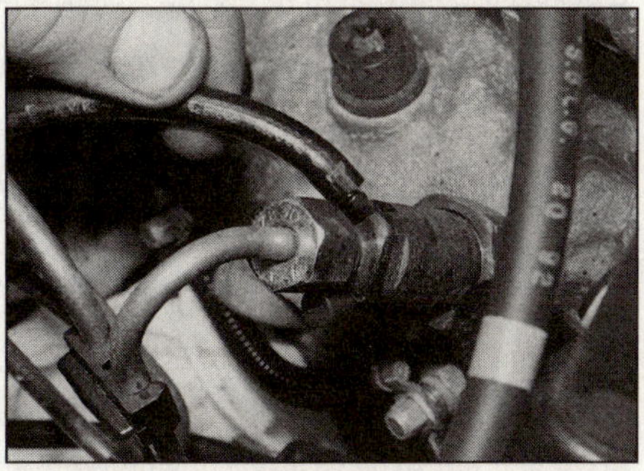

17.4 Pull the leak-off pipes from the fuel injectors

17.6 Disconnect the fuel pipes from the injectors

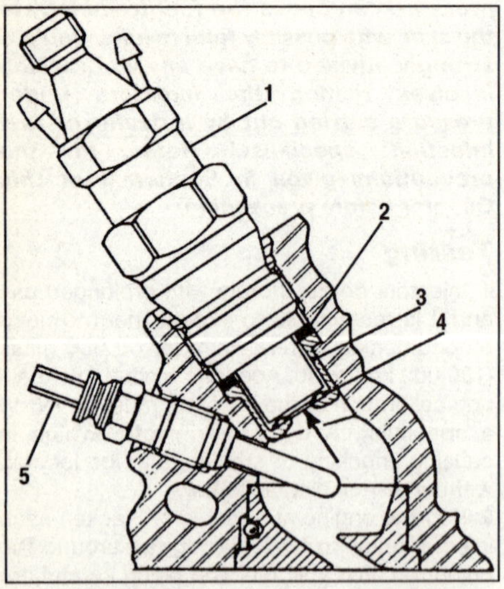

17.7a Cross-section of fuel injector

1 Fuel injector
2 Copper washer
3 Sleeve
4 Fire seal washer
5 Glow plug

Refitting

9 Obtain new copper washers and fire seal washers. Also renew the sleeves if they are damaged.

10 Take care not to drop the injectors or allow the needles at their tips to become damaged. The injectors are precision-made to fine limits and must not be handled roughly. In particular, do not mount them in a bench vice.

11 Commence refitting by inserting the sleeves (if removed) into the cylinder head.

12 Fit the new fire seal washers to the cylinder head. Note that the fire seal washers must be fitted with the convex side downwards (towards the cylinder head) **(see illustration)**.

13 Fit the copper washers to the cylinder head.

14 Insert the injectors and tighten them to the specified torque.

15 Refit the injector pipes and tighten the union nuts. Position any clips attached to the pipes as noted before removal.

16 Reconnect the leak-off pipes.

17 Start the engine. If difficulty is experienced, bleed the fuel system, referring to Section 7.

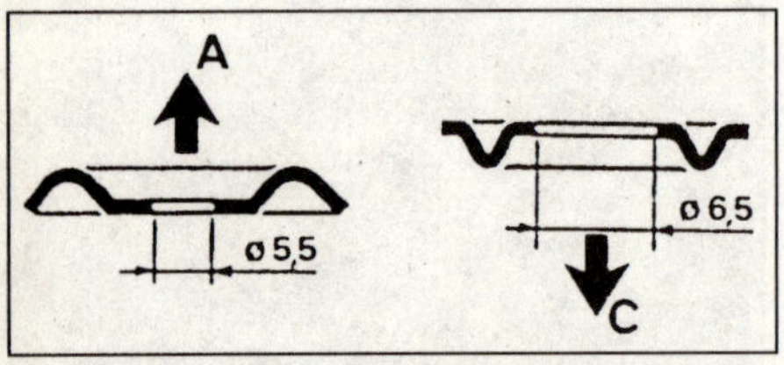

17.12 Fire seal washer fitting details

A Early type washer - convex side upwards
C Later type washer - convex side downwards

17.7b Unscrew the injectors . . .

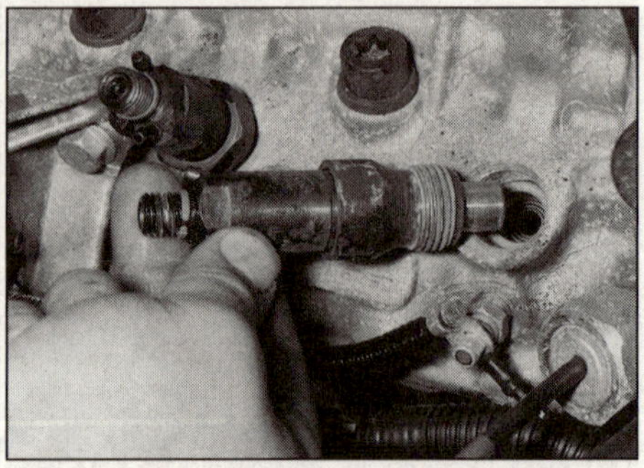

17.7c . . . and withdraw them from the cylinder head

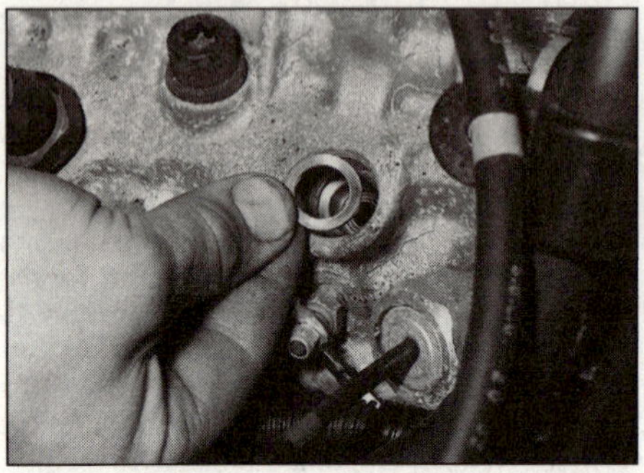

17.8a Recover the copper washers . . .

17.8b . . . and the fire seal washers

18 Manifolds - removal and refitting

Note: *A new gasket must be used on refitting.*

Non-Turbo models

Removal

1 Although the manifolds are separate, they are retained by the same nuts, since the stud holes are split between the manifold flanges.

2 To improve access, unscrew the securing bolts, and remove the acoustic bar from between the front suspension strut turrets. Also jack up the front of the vehicle and support on axle stands (see *"Jacking and vehicle support"*). Where fitted, remove the engine compartment undershield.

3 Note the location of any wiring or hose brackets/clips attached to the manifolds, and remove them.

4 Loosen the securing clip, and disconnect the air inlet hose from the inlet manifold.

5 Where applicable, disconnect the hoses from the oil separator mounted on the inlet manifold. If desired, the oil separator can be unbolted from its bracket.

6 Where applicable, disconnect the breather hose(s) from the inlet manifold.

7 On models fitted with an EGR system, remove the recirculation valve and pipe as described in Chapter 4C.

8 Detach the exhaust front downpipe from the exhaust manifold with reference to Section 25.

9 Progressively unscrew the nuts securing the inlet and exhaust manifolds and withdraw them from the cylinder head. Recover the manifold gasket.

Refitting

10 Refitting is a reversal of removal, bearing in mind the following points.

a) Ensure that the cylinder head and manifold mating surfaces are clean and use a new gasket.
b) Reconnect the exhaust front section to the manifold with reference to Section 25.
c) Where applicable, refit the EGR recirculation valve and pipe with reference to Chapter 4C.
d) Ensure that any wiring or hose brackets/clips are positioned as noted before removal.

Turbo models

Removal

11 Although the manifolds are separate, they are retained by the same nuts, since the stud holes are split between the manifold flanges.

12 To improve access, unscrew the securing bolts, and remove the tie-bar from between the front suspension strut turrets. Also jack up the front of the vehicle and support on axle stands (see *"Jacking and vehicle support"*). Where fitted, remove the engine compartment undershield.

13 Note the location of any wiring or hose brackets/clips attached to the manifolds, and remove them.

14 Disconnect the breather hoses from the

18.18 Manifold securing nuts (arrowed) - Turbo model (viewed with engine removed for clarity)

inlet manifold, noting their locations to aid refitting.

15 Remove the turbocharger as described in Section 21.

16 On models fitted with an EGR system, remove the recirculation valve and pipe as described in Chapter 4C.

17 Unbolt the engine earth strap from the engine lifting bracket at the right-hand end of the inlet manifold, then unbolt the bracket from the manifold.

18 Progressively unscrew the nuts securing the inlet and exhaust manifolds and withdraw them from the cylinder head. Recover the manifold gasket **(see illustration)**.

Refitting

19 Refitting is a reversal of removal, bearing in mind the following points.

a) Ensure that the cylinder head and manifold mating surfaces are clean and use a new gasket.
b) Where applicable, refit the EGR recirculation valve and pipe with reference to Chapter 4C.
c) Refit the turbocharger as described in Section 21.
d) Ensure that the breather hoses are correctly reconnected as noted before removal.
e) Ensure that any wiring or hose brackets/clips are positioned as noted before removal.

19 Idle speed and anti-stall speed - check and adjustment

1 The usual type of tachometer (rev counter), which works from ignition system pulses, cannot be used on diesel engines. A diagnostic socket is provided for the use of Renault test equipment, but this will not normally be available to the home mechanic. If it is not felt that adjusting the idle speed "by ear" is satisfactory, one of the following alternatives may be used.

a) Purchase or hire of an appropriate tachometer.
b) Delegation of the job to a Renault dealer or other specialist.
c) Timing light (strobe) operated by a petrol engine running at the desired speed. If the timing light is pointed at a mark on the camshaft or injection pump sprocket, the mark will appear stationary when the two engines are running at the same speed (or multiples of that speed). The sprocket will be rotating at half the crankshaft speed but this will not affect the adjustment. (In practice it was found impossible to use this method on the crankshaft pulley due to the acute viewing angle.)

2 Before making adjustments warm up the engine to normal operating temperature. Make sure that the accelerator cable is correctly adjusted (see Section 5).

Idle speed checking and adjustment

3 With the accelerator lever resting against the idle stop, check that the engine idles at the specified speed. If necessary adjust as follows.

Lucas injection pump

4 Loosen the locknut on the idle speed adjustment screw. Turn the screw as required and retighten the locknut **(see illustration)**.

5 Check the anti-stall adjustment as described later in this Section.

6 Stop the engine and disconnect the tachometer, where applicable.

Bosch injection pump

7 Loosen the locknut and unscrew the anti-stall adjustment screw until it is clear of the pump accelerator lever **(see illustration)**.

8 Loosen the locknut and turn the idle speed adjustment screw as required, then retighten the locknut.

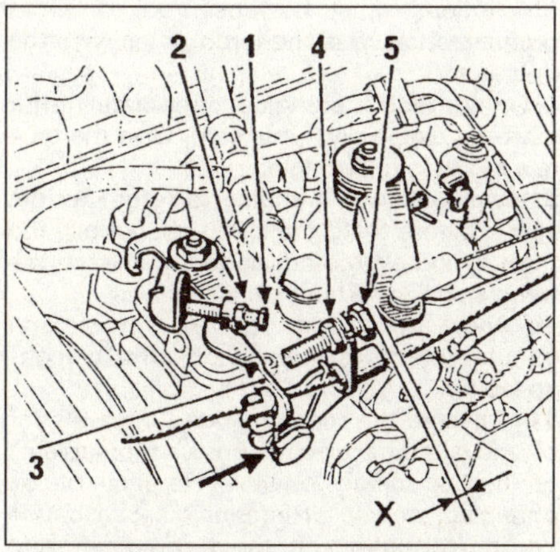

19.4 Idle speed and anti-stall speed adjustment points - Lucas injection pump

1 Idle speed adjustment screw
2 Locknut
3 Fast idle lever
4 Locknut
5 Anti-stall speed adjustment screw
X = Clearance between accelerator lever and anti-stall speed adjustment screw (see Specifications)

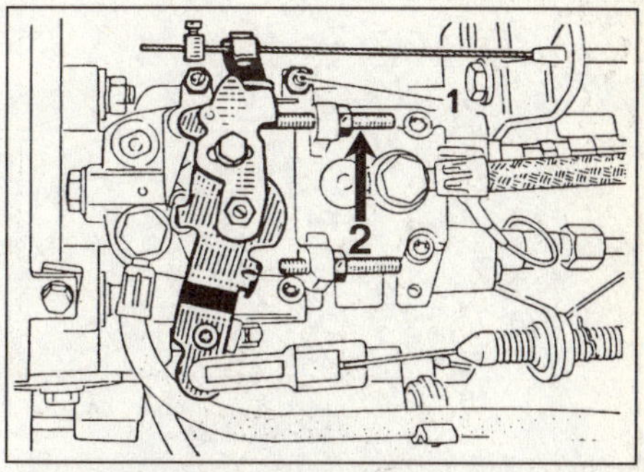

19.7 Idle speed adjustment screw (1) and anti-stall speed adjustment screw (2) - Bosch injection pump

9 Make the anti-stall adjustment as described later in this Section.

10 Stop the engine and disconnect the tachometer, where applicable.

Anti-stall checking and adjustment

11 Make sure that the engine is at normal operating temperature, and idling at the specified speed, as described previously.

Lucas injection pump

12 Insert a shim or feeler blade, of the correct thickness (see Specifications), between the pump accelerator lever and the anti-stall adjustment screw.

13 The engine speed should increase to the specified anti-stall speed (see Specifications).

14 If adjustment is necessary, loosen the locknut, turn the anti-stall adjustment screw as required, then tighten the locknut.

15 Remove the shim or feeler blade and check the idle speed as described previously.

16 Move the pump accelerator lever to increase the engine speed to approximately 3000 rpm, then quickly release the lever. The deceleration period should be approximately 2.5 to 3.5 seconds, and the engine speed should drop to approximately 50 rpm below idle.

17 If the deceleration is too fast and the engine stalls, unscrew the anti-stall adjustment screw 1/4 turn towards the accelerator lever. If the deceleration is too slow, resulting in poor engine braking, turn the screw 1/4 turn away from the lever.

18 Retighten the locknut after making an adjustment, then recheck the idle speed and adjust if necessary as described previously.

19 With the engine idling check the operation of the manual stop control by turning the stop lever clockwise. The engine must stop instantly.

20 Where applicable, disconnect the tachometer on completion.

Bosch injection pump

21 Insert a shim or feeler blade of the correct thickness (see Specifications) between the pump accelerator lever and the anti-stall adjustment screw **(see illustration)**.

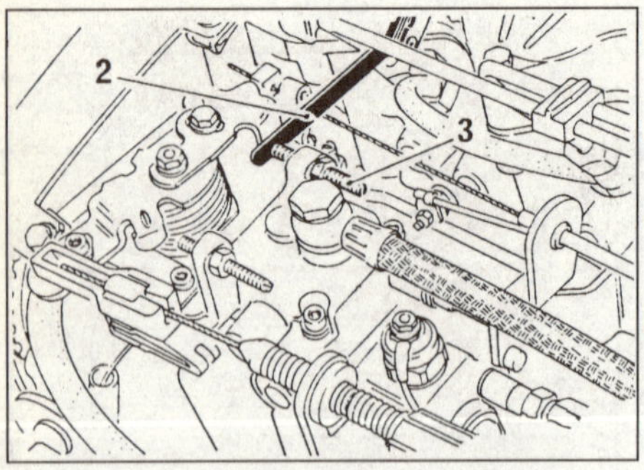

19.21 Checking the anti-stall speed adjustment - Bosch injection pump

2 *Feeler blade*
3 *Anti-stall speed adjustment screw*

22 The engine speed should be as specified for the anti-stall speed (see Specifications).
23 If adjustment is necessary, loosen the locknut and turn the anti-stall adjustment screw as required. Retighten the locknut.
24 Remove the shim or feeler blade and allow the engine to idle.
25 Move the fast idle lever fully towards the flywheel end of the engine and check that the engine speed increases to the specified fast idle speed. If necessary loosen the locknut and turn the fast idle adjusting screw as required, then retighten the locknut.
26 With the engine idling, check the operation of the manual stop control by turning the stop lever. The engine must stop instantly.
27 Where applicable, disconnect the tachometer on completion and stop the engine.
28 After making the adjustment, the deceleration damper must be adjusted as follows.
29 Loosen the two screws securing the damper operating lever to the load lever.
30 Insert a 1.0 mm thick shim or feeler blade between the operating lever and the end of the damper, then push the lever onto the stop. Make sure that the load lever is in contact with the residual flow screw, then tighten the two securing screws.
31 Remove the shim and check that the load lever is still in contact with the residual flow screw.

20 Turbocharger - description and precautions

Description

1 A turbocharger is fitted to the F8Q 784 engine. It increases engine efficiency by raising the pressure in the inlet manifold above atmospheric pressure. Instead of the air simply being sucked into the cylinders, it is forced in. Additional fuel is supplied by the injection pump in proportion to the increased air intake.
2 Energy for the operation of the turbocharger comes from the exhaust gas. The gas flows through a specially-shaped housing (the turbine housing) and in so doing, spins the turbine wheel. The turbine wheel is attached to a shaft, at the end of which is another vaned wheel known as the compressor wheel. The compressor wheel spins in its own housing and compresses the inducted air on the way to the inlet manifold.
3 Between the turbocharger and the inlet manifold, the compressed air passes through an intercooler. This is an air-to-air heat exchanger, mounted on the left-hand side of the car behind the front bumper, and supplied with cooling air ducted through the front spoiler. The purpose of the intercooler is to remove, from the inducted air, some of the heat gained in being compressed. Because cooler air is denser, removal of this heat further increases engine efficiency.
4 Boost pressure (the pressure in the inlet manifold) is limited by a wastegate, which diverts the exhaust gas away from the turbine wheel in response to a pressure-sensitive actuator.
5 The turbo shaft is pressure-lubricated by an oil feed pipe from the main oil gallery. The shaft "floats" on a cushion of oil. A drain pipe returns the oil to the sump.
6 The turbocharger bearings are cooled by circulating engine coolant. After switching off the ignition, an electric pump continues to circulate coolant for several minutes to prevent overheating of the bearings and to prolong their life.

Precautions

7 The turbocharger operates at extremely high speeds and temperatures. Certain precautions must be observed to avoid premature failure of the turbo or injury to the operator.
8 Do not race the engine immediately after start-up, especially if it is cold. Give the oil a few seconds to circulate.
9 Always allow the engine to return to idle speed before switching it off - do not blip the throttle and switch off, as this will leave the turbo spinning without lubrication.
10 Allow the engine to idle for several minutes before switching off after a high-speed run.
11 Observe the recommended intervals for oil and filter changing, and use a reputable oil of the specified quality. Neglect of oil changing, or use of inferior oil, can cause carbon formation on the turbo shaft and subsequent failure.

Warning: Do not operate the turbo with any parts exposed. Foreign objects falling onto the rotating vanes could cause excessive damage and (if ejected) personal injury.

21 Turbocharger - removal and refitting

Note: *New turbocharger-to-manifold nuts must be used on refitting. If a new turbocharger is to be fitted, new exhaust elbow-to-turbocharger nuts and new coolant pipe seals will be required.*

Removal

1 Whilst the engine is still warm, spray the turbocharger mounting bolts with a penetrating oil to ease removal.
2 Apply the handbrake, then jack up the front of the vehicle, and support securely on axle stands (see *"Jacking and vehicle support"*). Where fitted, remove the engine compartment undershield.
3 Ensure that the engine has cooled sufficiently to avoid scalding, then drain the cooling system as described in Chapter 1B.
4 Where applicable, remove the engine undershield.
5 Remove the exhaust front section and downpipe with reference to Section 25.
6 Working under the vehicle, unscrew the upper securing nut and bolt, and the lower bolt, and remove the turbocharger lower support bracket **(see illustrations)**.
7 Unscrew the turbocharger oil return pipe union from the outlet on the underside of the turbocharger **(see illustration)**. Be prepared for oil spillage.
8 Cut or remove the hose clamp, and disconnect the lower coolant hose from the turbocharger pipe **(see illustration)**. Be prepared for coolant spillage.
9 Working in the engine compartment, unscrew the securing bolts, and remove the strengthening bar from between the front suspension strut turrets.
10 Remove the air cleaner assembly as described in Section 2.
11 Loosen the clamps, and remove the air trunking connecting the inlet manifold to the intercooler tube.
12 Loosen the clamps, and remove the air trunking connecting the air cleaner to the turbocharger. Also disconnect the breather hose connecting the air trunking to the inlet manifold. Note the locations of any clips attached to the air trunking.
13 Loosen the clamp and disconnect the air outlet trunking (running to the intercooler) from the turbocharger.
14 Remove the nut securing the preheating system control unit to the bulkhead, and move the unit to one side, leaving the wiring connected.
15 Unscrew the union nut, and disconnect the vacuum pipe from the brake vacuum pump **(see illustration)**.
16 Disconnect the coolant hose from the rear of the cylinder head, and move the hose to one side, clear of the working area **(see illustration)**. Be prepared for coolant spillage.

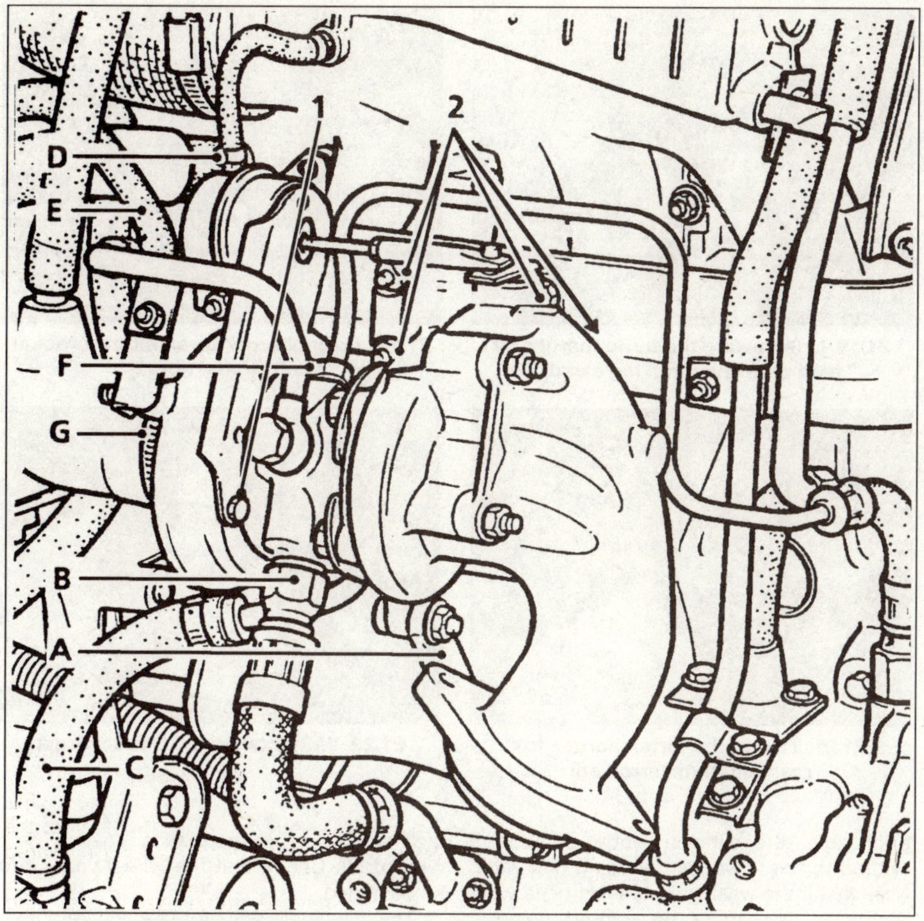

21.6a Turbocharger mountings and hose/pipe connections

1 Wastegate bracket bolts
2 Turbocharger-to-manifold nuts
A Lower support bracket
B Oil return pipe union
C Lower turbocharger coolant hose
D Cylinder head-to-T-piece coolant hose
E Upper turbocharger coolant hose
F Oil feed pipe union
G Air trunking (air cleaner-to-turbocharger)

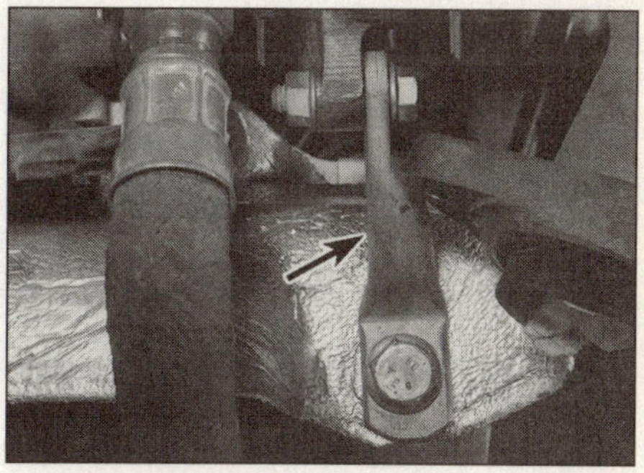

21.6b Turbocharger lower support bracket (arrowed) - viewed with engine removed for clarity

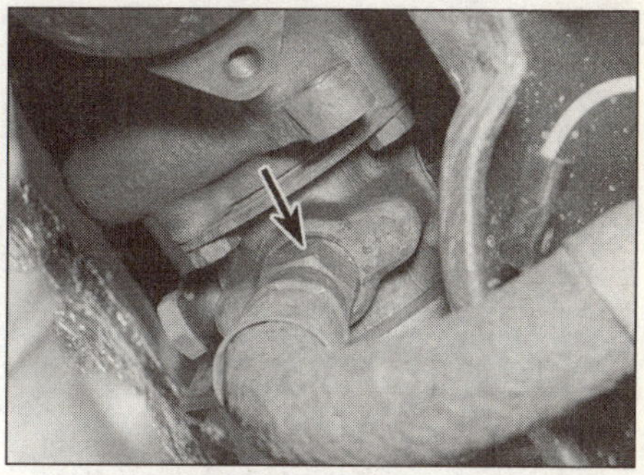

21.7 Turbocharger oil return pipe union (arrowed) - viewed from underneath vehicle

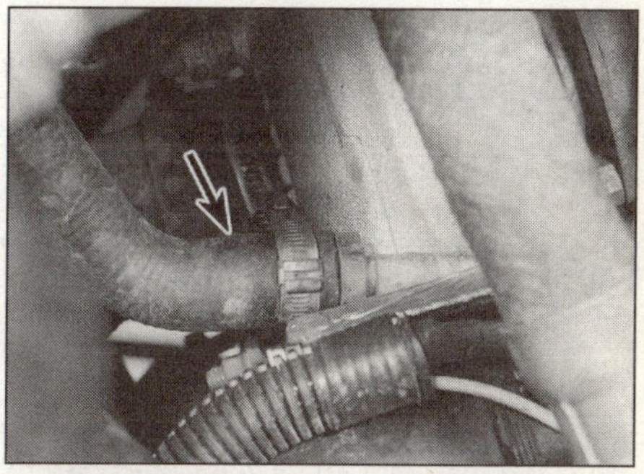

21.8 Disconnect the lower coolant hose (arrowed) from the turbocharger pipe - viewed from underneath vehicle

21.15 Unscrew the brake vacuum pump pipe union

21.16 Disconnect coolant hose (arrowed) from the rear of the cylinder head

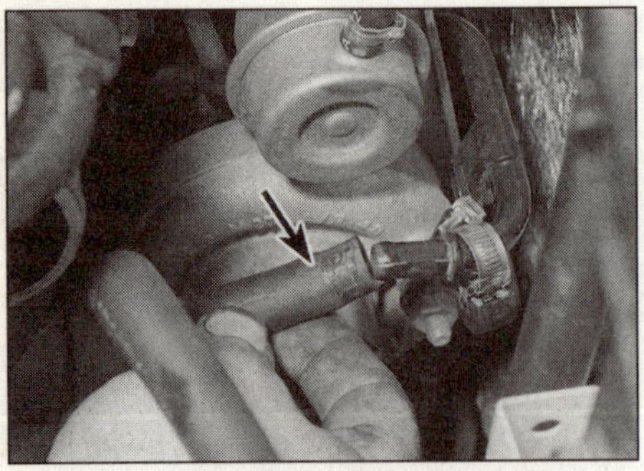

21.17 Disconnecting the upper turbocharger coolant hose (arrowed)

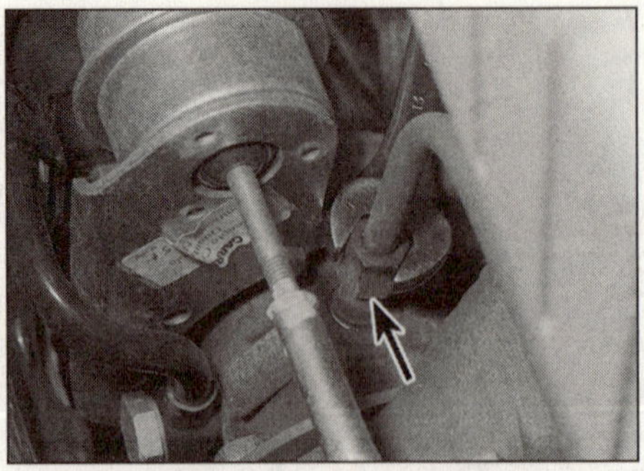

21.18 Unscrewing the turbocharger oil feed pipe union nut (arrowed)

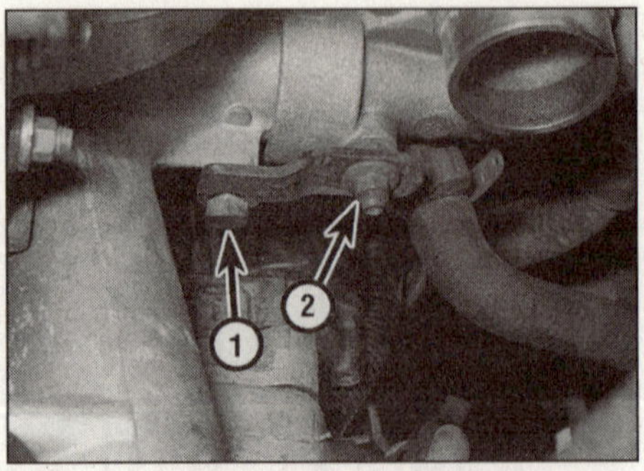

21.19 Turbocharger-to-air elbow bracket bolt (1) and nut (2)

21.20 Disconnecting the boost pressure pipe (arrowed) from the inlet manifold

21.21 Two of the turbocharger-to-manifold nuts (arrowed)

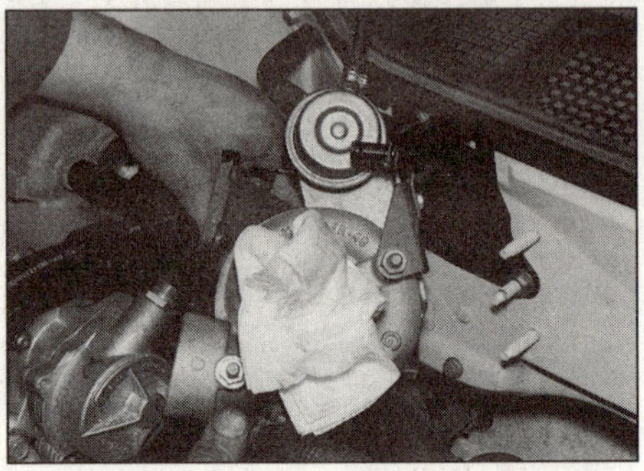

21.22 Withdrawing the turbocharger

17 Similarly, disconnect the upper turbocharger coolant hose **(see illustration)**.

18 Unscrew the union nut, and disconnect the oil feed pipe from the top of the turbocharger **(see illustration)**. It may be necessary to remove the wastegate bracket bolts to enable access to the union nut.

19 Remove the bolt and the nut securing the turbocharger-to-air elbow bracket, and withdraw the bracket **(see illustration)**.

20 Disconnect the boost pressure pipe from the inlet manifold **(see illustration)**.

21 Unscrew the four turbocharger-to-manifold nuts, and slide the turbocharger from the manifold studs **(see illustration)**. Access to the lower right-hand nut is easiest from underneath the vehicle. For improved access to the nuts, unbolt the exhaust elbow from the turbocharger.

22 Withdraw the turbocharger from above the engine compartment, by passing it below the brake vacuum pump, and then manipulating it between the vacuum pump and the coolant reservoir **(see illustration)**.

23 If a new turbocharger is to be fitted, remove the coolant pipes and the exhaust elbow (if not already done) from the old unit, and transfer them to the new unit. Use new nuts when fitting the exhaust elbow, and new seals when fitting the coolant pipes.

Refitting

24 Refitting is a reversal of removal, but renew any damaged hose clamps, and use new turbocharger-to-manifold nuts. Where applicable, the exhaust elbow should be refitted to the turbocharger using new nuts. Note that if the wastegate bracket bolts were removed, they must be refitted using a suitable thread-locking compound. Refill the cooling system as described in Chapter 1B.

25 On completion, the following procedure must be observed before starting the engine.

a) Disconnect the wiring from the stop solenoid on the injection pump, and insulate the connector.

b) Crank the engine on the starter motor until the instrument panel oil pressure warning light goes out (this may take several seconds).

c) Reconnect the wiring to the stop solenoid, then start the engine using the normal procedure.

d) Run the engine at idle speed, and check the turbocharger oil and coolant unions for leakage. Rectify any problems without delay.

26 After the engine has been run, check the engine oil level, and top up if necessary.

22 Turbocharger - examination and renovation

1 With the turbocharger removed, inspect the housing for cracks or other visible damage.

2 Spin the turbine or the compressor wheel to verify that the shaft is intact and to feel for excessive shake or roughness. Some play is normal since in use the shaft is "floating" on a film of oil. Check that the wheel vanes are undamaged.

3 The wastegate actuator is a separate unit, and can be renewed independently of the turbocharger. Consult a Renault dealer or other specialist if it is thought that testing or renewal is necessary.

4 If the exhaust or inlet passages are oil-contaminated, the turbo shaft oil seals have probably failed. (On the inlet side, this will also have contaminated the intercooler, which if necessary should be flushed with a suitable solvent.)

5 Check the oil feed and return pipes for contamination or blockage and clean if necessary.

6 No DIY repair of the turbocharger is possible. A new unit may be available on an exchange basis.

23 Boost pressure fuel delivery corrector (Turbo models) - general information

1 This device is mounted on the side of the injection pump, and its purpose is to adjust the injection pump fuel metering in relation to the turbocharger boost pressure **(see illustration)**. Effectively, the quantity of fuel injected is increased as the boost pressure increases.

2 An adjustment screw is provided, but this is sealed at factory, and no attempt should be

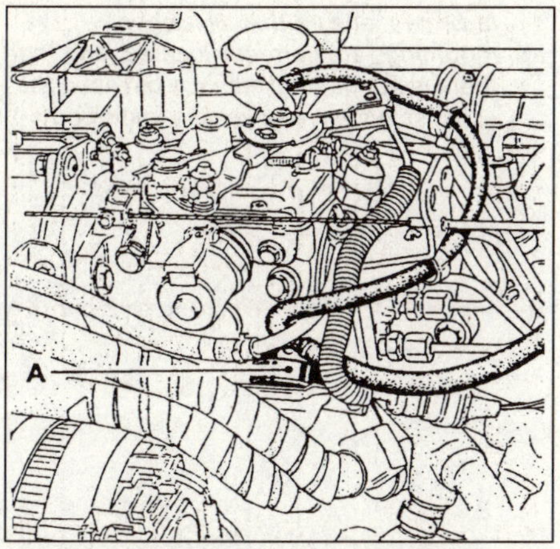

23.1 Boost pressure fuel delivery corrector location (A) - Turbo models

made to carry out adjustments without the use of specialist test equipment.

3 If a fault with the device is suspected, consult a Renault dealer or a suitably qualified specialist.

24 Intercooler - removal and refitting

Removal

1 The intercooler is located at the front left-hand side of the engine compartment, behind the front bumper.

2 Remove the front bumper as described in Chapter 11.

3 Loosen the securing clamps, and disconnect the air trunking from the intercooler.

4 Working from outside the wing panel, unscrew the two bolts securing the intercooler mounting bracket to the body panel. Where applicable, also remove the bracket mounting nut.

5 Working at the inner edge of the intercooler, remove the bolt securing the clamp plate to the body panel, and withdraw the clamp plate.

6 Push the intercooler up to release the upper securing clip from the lip of the body panel (the clip is visible from the engine compartment), then withdraw the intercooler and bracket assembly from under the vehicle.

7 The intercooler can be removed from its bracket after unscrewing the two securing bolts, and removing the lower section of the bracket.

Refitting

8 Refitting is a reversal of removal, bearing in mind the following points.

a) When positioning the assembly, take care not to damage the wiring harness located beneath the wheel arch.

b) Ensure that the upper securing clip is correctly located on the body panel (push the intercooler upwards until the clip engages over the lip of the body panel).

c) Ensure that the air trunking securing clamps are securely tightened.

25 Exhaust system - general information and component renewal

General information

1 On new vehicles, the exhaust system consists of just two sections; the front downpipe and the remaining system consisting of a catalytic converter or resonator (according to model), and the tailpipe and silencer. The downpipe is attached to the rear section by a flanged joint with gasket.

2 The rear section of the exhaust is located above the rear suspension, however the intermediate and tailpipe/silencer may be renewed separately by cutting the intermediate pipe with a hacksaw. Where a catalytic converter is fitted, this can also be renewed separately by cutting from the intermediate pipe. The system is suspended throughout its entire length by rubber mountings.

Removal

3 To remove a part of the system, first jack up the front or rear of the car, and support it on axle stands (see *"Jacking and vehicle support"*). Alternatively, position the car over an inspection pit, or on car ramps. Where fitted, remove the engine compartment undershield.

Front pipe

4 Where applicable, unbolt the front pipe from the mounting bracket.

5 Unscrew and remove the nuts/bolts securing the front pipe flange joint to the manifold or Turbocharger (as applicable), and recover the gasket.

6 Unscrew and remove the flange bolts, and disconnect the front pipe from the catalytic converter or intermediate section. Withdraw the pipe from under the vehicle.

Catalytic converter

7 Unscrew and remove the flange bolts attaching the front pipe to the catalytic converter and recover the gasket.

8 If the original rear section is fitted, it must be cut just behind the catalytic converter using either a hacksaw or pipe cutter. The cutting point is marked with two circular punch marks on the side of the pipe. The punch marks are 90 mm apart and the exhaust section should be cut at the mid-point between the two punch marks. **Note:** *Ensure that the exhaust pipe is cut squarely else it will be difficult to obtain a gas-tight seal when the exhaust is refitted.*

9 With the intermediate pipe cut, withdraw the catalytic converter from under the vehicle.

Intermediate pipe (and resonator where applicable)

10 On models without a catalytic converter, unscrew and remove the flange bolts attaching the front pipe to the intermediate pipe and recover the gasket.

11 On models with a catalytic converter, remove the catalytic converter as described in paragraphs 7 to 9.

12 If the original rear section is fitted, it must be cut in half using either a hacksaw or pipe cutter. The cutting point is marked with two circular punch marks on the side of the pipe. The punch marks are 90 mm apart and the exhaust section should be cut at the mid-point between the two punch marks. **Note:** *Ensure that the exhaust pipe is cut squarely else it will be difficult to obtain a gas-tight seal when the exhaust is refitted.*

13 With the intermediate pipe cut, withdraw the intermediate exhaust section from under the vehicle.

14 If the rear section is already in two halves, unscrew the bolt and slide the clamp sleeve onto the rear section then release the rubber mountings and withdraw the intermediate section from under the vehicle.

Rear tailpipe and silencer

15 If the original rear section is fitted, follow the instructions given in paragraph 8.

16 If the rear section is in two halves, unscrew the bolt and slide the clamp sleeve onto the intermediate section then release the rubber mountings and withdraw the tailpipe and silencer from under the vehicle.

Heat shield(s)

17 The heat shields are secured to the underside of the body by various nuts and bolts. Each shield can be removed separately but note that they overlap making it necessary to loosen another section first. If a shield is being removed to gain access to a component located behind it, it may prove sufficient in some cases to remove the retaining nuts and/or bolts, and simply lower the shield, without disturbing the exhaust system. Otherwise remove the exhaust section as described earlier.

Refitting

18 Each section is refitted by reversing the removal sequence, noting the following points:

a) Ensure that all traces of corrosion have been removed from the flanges, and renew all necessary gaskets.

b) Inspect the rubber mountings for signs of damage or deterioration, and renew as necessary.

c) When reconnecting sections which have been cut, apply a smear of exhaust system jointing paste (Renault recommend the use of Sodicam) to the sleeve inner surface, to ensure a gas-tight seal. Make sure both inner ends of the cut pipe are positioned squarely against the stop of the clamp sleeve. Position the sleeve bolt vertically on the left-hand side of the pipe and securely tighten the nut until it is heard to click; the clamp bolt has a groove in it to ensure that the nut is correctly tightened (equivalent to a tightening torque of approximately 25 Nm).

d) *Prior to tightening the exhaust system fasteners, ensure that all rubber mountings are correctly located, and that there is adequate clearance between the exhaust system and vehicle underbody.*

Chapter 4 Part C:
Emission control systems

Contents

Degrees of difficulty

Easy, suitable for novice with little experience	**Fairly easy,** suitable for beginner with some experience	**Fairly difficult,** suitable for competent DIY mechanic	**Difficult,** suitable for experienced DIY mechanic	**Very difficult,** suitable for expert DIY or professional

Specifications

EGR solenoid valve resistance:	
K7M 703 petrol engine	30 ± 3 ohms
F8Q diesel engine	10 ± 5 ohms
EVAP solenoid valve resistance:	
Single-point injection	34.5 ohms
Multi-point injection	35 ± 5 ohms
Oxygen (Lambda) sensor voltage at 850°C (petrol engines):	
Rich mixture	More than 625 mvolt
Lean mixture	0 to 80 mvolt

Torque wrench settings

Torque wrench settings	Nm	lbf ft
Oxygen (Lambda) sensor (petrol engines)	45	33
EGR solenoid valve bolt (K7M engine)	10	7
Steel pipe for EGR solenoid valve (K7M engine)	10	7

1 General information and precautions

1 All petrol engines have the ability to use unleaded petrol and also have various other features built into the fuel system to help minimise harmful emissions. In addition, all models are equipped with the crankcase emission-control system described below. All models are also equipped with a catalytic converter and an evaporative emission control system. On the K7M 703 engine, an exhaust gas recirculation (EGR) system is fitted.
2 All diesel engine models are also designed to meet strict emission requirements and are also equipped with a crankcase emission control system. In addition to this, certain models are fitted with a catalytic converter to reduce harmful exhaust emissions. To further reduce emissions, an exhaust gas recirculation (EGR) system is fitted.
3 The emission control systems function as follows.

Petrol models

Crankcase emission control

4 To reduce the emission of unburned hydrocarbons from the crankcase into the atmosphere, the engine is sealed and the blow-by gases and oil vapour are drawn from inside the crankcase, through two restrictors, and into the inlet manifold to be burned by the engine during normal combustion. A 6.0 mm (E7J engine) or 5.00 mm (K7M engine) diameter restrictor is located in a hose connected directly between the valve cover and the air filter housing. A smaller restrictor is located in a hose between the valve cover and the inlet manifold, and the evaporative canister solenoid valve is also connected into this hose **(see illustrations)**.
5 Under conditions of high manifold depression (idling, deceleration) the gases will be sucked positively out of the crankcase. Under conditions of low manifold depression (acceleration, full-throttle running), the gases are forced out of the crankcase by the (relatively) higher crankcase pressure; if the engine is worn, the raised crankcase pressure (due to increased blow-by) will cause some of the flow to return under all manifold conditions.
6 The crankcase ventilation hoses and restrictors should be periodically cleaned to ensure correct operation of the system.

Exhaust emission control

7 To minimise the amount of pollutants which escape into the atmosphere, all models are fitted with a catalytic converter in the exhaust system. The system is of the closed-loop type, in which an oxygen (Lambda) sensor in the exhaust system downpipe provides the fuel-injection/ignition system ECU with constant feedback, enabling the ECU to adjust the mixture to provide the best possible conditions for the converter to operate.
8 The oxygen (Lambda) sensor has a heating element built-in that is controlled by the ECU through the sensor relay to quickly bring the sensor's tip to an efficient operating temperature. The sensor's tip is sensitive to oxygen and sends the ECU a varying voltage

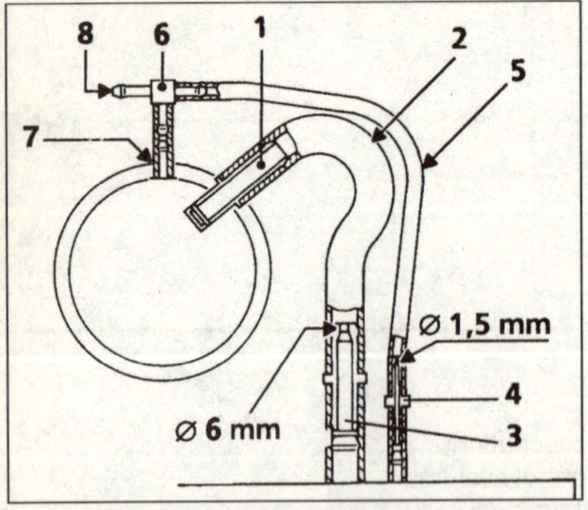

1.4a Crankcase emission control system fitted to the E7J petrol engine

1 *Air cleaner inlet*
2 *Valve cover-to-air cleaner hose*
3 *Large restrictor*
4 *Small restrictor*
5 *Valve cover-to-inlet manifold hose*
6 *Tee-piece*
7 *Inlet manifold inlet*
8 *To canister*

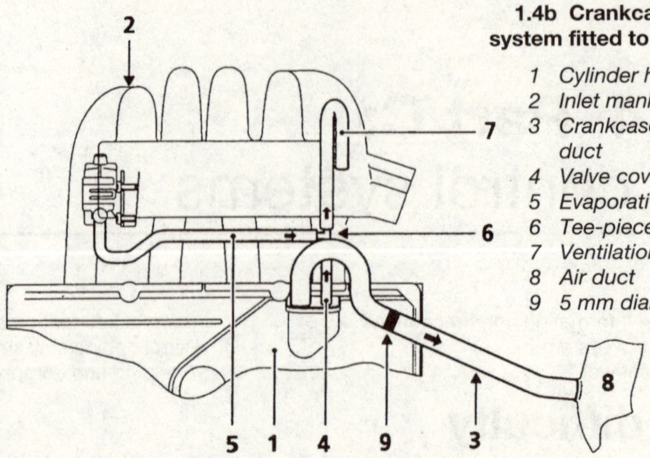

1.4b Crankcase emission control system fitted to the K7M petrol engine

1 *Cylinder head*
2 *Inlet manifold*
3 *Crankcase ventilation hose to air duct*
4 *Valve cover-to-inlet manifold hose*
5 *Evaporative valve hose*
6 *Tee-piece*
7 *Ventilation hose on inlet manifold*
8 *Air duct*
9 *5 mm diameter restrictor*

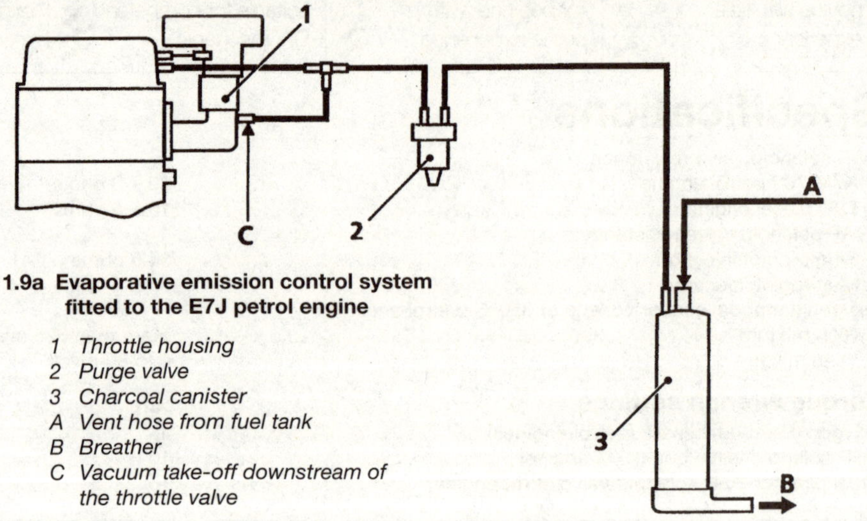

1.9a Evaporative emission control system fitted to the E7J petrol engine

1 *Throttle housing*
2 *Purge valve*
3 *Charcoal canister*
A *Vent hose from fuel tank*
B *Breather*
C *Vacuum take-off downstream of the throttle valve*

depending on the amount of oxygen in the exhaust gases; if the inlet air/fuel mixture is too rich, the exhaust gases are low in oxygen so the sensor sends a low-voltage signal, the voltage rising as the mixture weakens and the amount of oxygen rises in the exhaust gases. Peak conversion efficiency of all major pollutants occurs if the inlet air/fuel mixture is maintained at the chemically-correct ratio for the complete combustion of petrol of 14.7 parts (by weight) of air to 1 part of fuel (the 'stoichiometric' ratio). The sensor output voltage alters in a large step at this point, the ECU using the signal change as a reference point and correcting the inlet air/fuel mixture accordingly by altering the fuel injector pulse width.

Evaporative emission control

9 To minimise the escape into the atmosphere of unburned hydrocarbons, an evaporative emissions control system is also fitted to all models **(see illustrations)**. The fuel tank filler cap is sealed and a charcoal canister is mounted on the front right-hand side of the engine compartment. The canister collects the petrol vapours generated in the tank when the car is parked and stores them until they can be cleared from the canister (under the control of the fuel-injection/ignition system ECU) via the purge valve into the inlet manifold to be burned by the engine during normal combustion.

10 To ensure that the engine runs correctly when it is cold and/or idling and to protect the catalytic converter from the effects of an over-rich mixture, the purge control valve is not opened by the ECU until the engine has warmed up, and the engine is under load; the valve solenoid is then modulated on and off to allow the stored vapour to pass into the inlet manifold.

1.9b Evaporative emission control system fitted to the K7M petrol engine

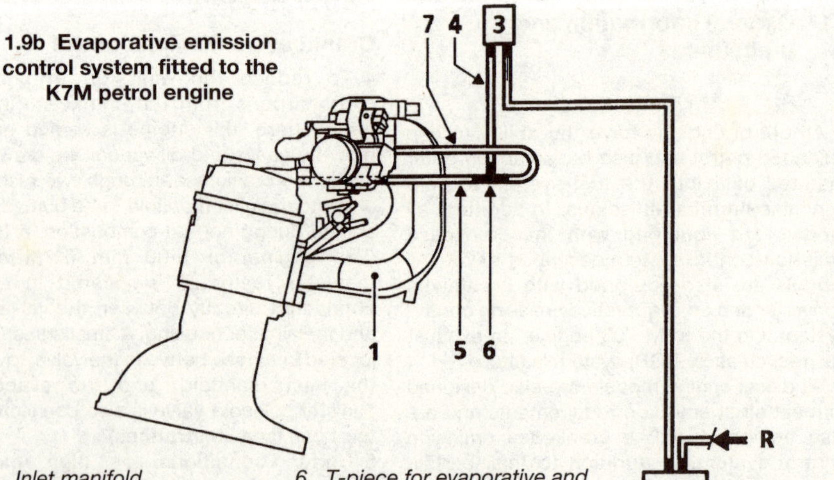

1 *Inlet manifold*
2 *Charcoal canister*
3 *Purge valve*
4 *Recirculation pipe*
5 *Crankcase ventilation pipe*
6 *T-piece for evaporative and crankcase ventilation pipes*
7 *Pipe for crankcase and evaporative ventilation*
R *Pipe from fuel tank*
MA *Breather*

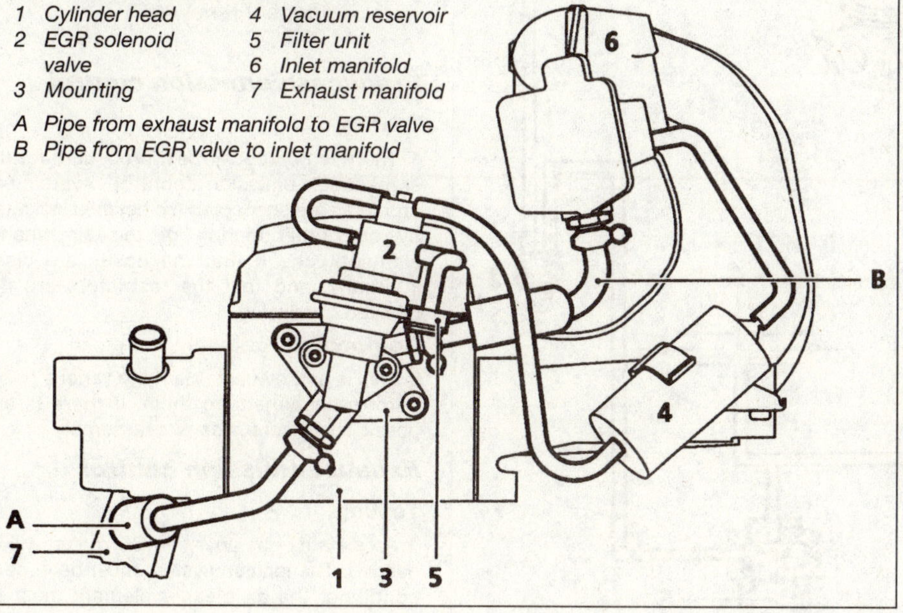

1.11 Exhaust gas recirculation system fitted to the K7M 703 engine

1 Cylinder head
2 EGR solenoid valve
3 Mounting
4 Vacuum reservoir
5 Filter unit
6 Inlet manifold
7 Exhaust manifold
A Pipe from exhaust manifold to EGR valve
B Pipe from EGR valve to inlet manifold

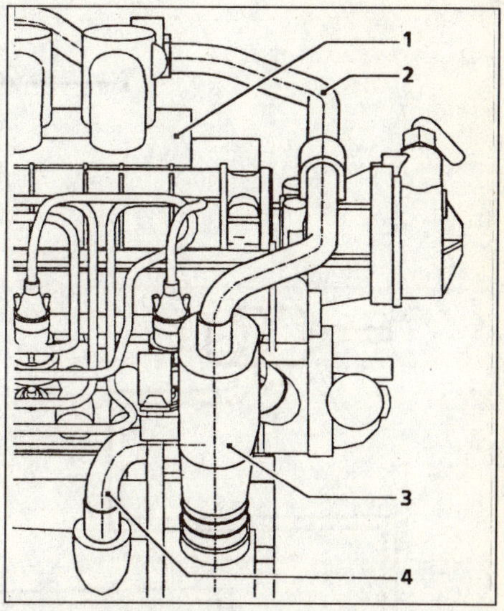

1.12a Typical crankcase emission control system - non-Turbo models

1 Inlet manifold
2 Hose to inlet manifold
3 Oil separator
4 Return pipe to sump

Exhaust gas recirculation (EGR) system

11 This system is fitted to the K7M 703 engine and is designed to reduce oxides of nitrogen in the exhaust gases **(see illustration)**. This is achieved by taking some of the exhaust gases from the cylinder head exhaust ports and recirculating them back into the inlet manifold, where they are burned again during normal combustion. The exhaust gas recirculation (EGR) valve is fitted to the left-hand side of the cylinder head.

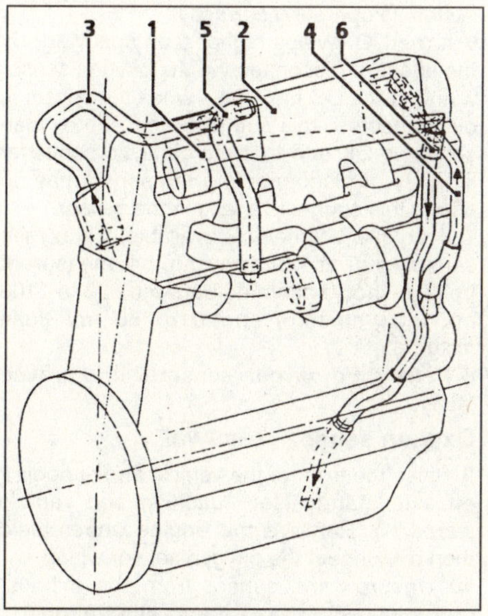

1.12b Typical crankcase emission control system - Turbo models

1 Inlet manifold
2 Vapour separator
3 Breather hose (from crankcase)
4 Breather hose (from crankcase)
5 Breather hose (to turbocharger inlet)
6 Return hose to sump

Diesel models

Crankcase emission control

12 To reduce the emission of unburned hydrocarbons from the crankcase into the atmosphere, the engine is sealed and the blow-by gases and oil vapour are drawn from inside the crankcase, through an oil separator located on the front left-hand side of the cylinder block, and into the inlet manifold to be burned by the engine during normal combustion **(see illustrations)**.

13 There are no restrictors in the system hoses, since the minimal depression in the inlet manifold remains constant during all engine operating conditions.

Exhaust emission control

14 To minimise the amount of pollutants which escape into the atmosphere, an unregulated catalytic converter is fitted in the exhaust system. The catalytic converter operates remotely in the exhaust system, and there is no oxygen sensor as fitted to the petrol engines.

Exhaust gas recirculation (EGR) system

15 The system is designed to recirculate small quantities of exhaust gas into the inlet tract, and therefore into the combustion process **(see illustrations)**. This process reduces the level of oxides of nitrogen present

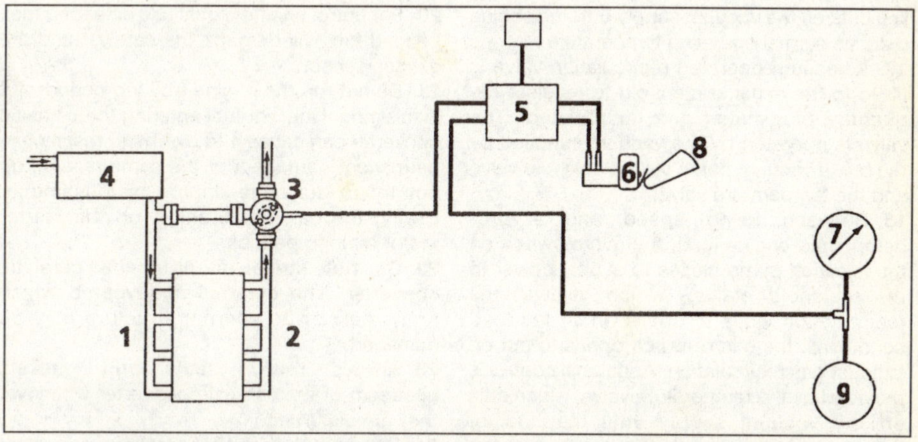

1.15a Exhaust gas recirculation system fitted to the F8Q diesel engine

1 Inlet manifold
2 Exhaust manifold
3 EGR valve
4 Air filter
5 EGR solenoid
6 Injection pump microswitch
7 Vacuum pump
8 Injection pump load lever
9 Brake vacuum servo

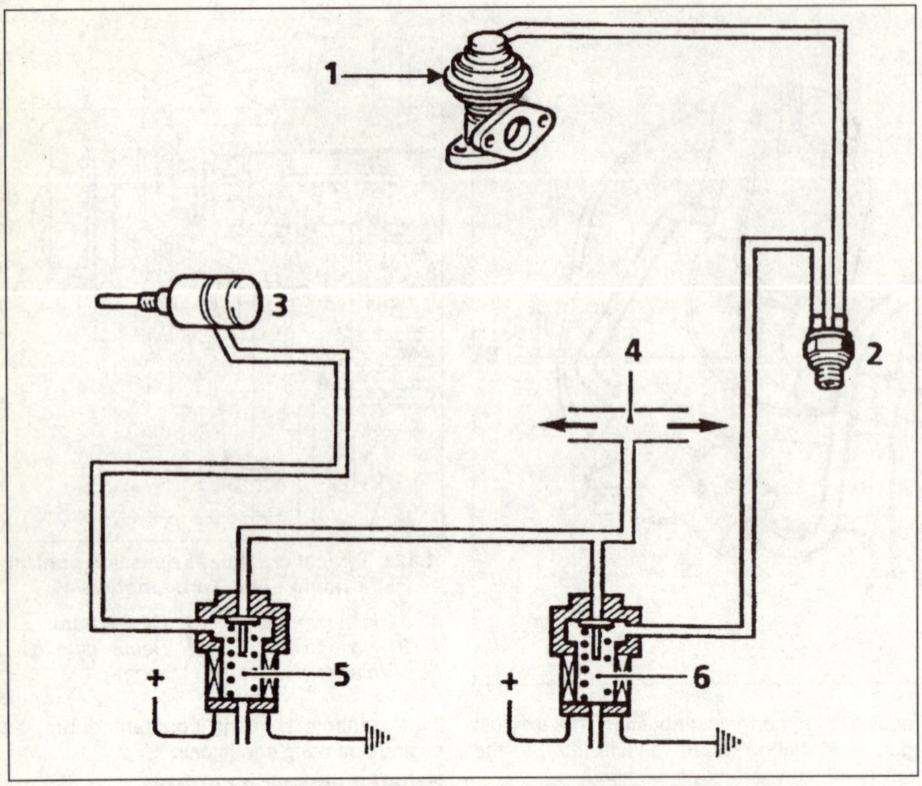

1.15b Schematic view of exhaust gas recirculation (EGR) system

1 *Recirculation valve*
2 *Temperature valve*
3 *Fast idle vacuum actuator*
4 *Vacuum take-off between vacuum pump and servo*
5 *Fast idle solenoid valve*
6 *EGR solenoid valve*

in the final exhaust gas which is released into the atmosphere, and also lowers the combustion temperature.

16 The volume of exhaust gas recirculated is controlled by vacuum supplied from the brake servo vacuum pump, via a solenoid valve controlled by a micro-switch mounted on the injection pump (see Chapter 4B). A temperature valve fitted in the vacuum supply line cuts off the vacuum supply until the engine reaches normal operating temperature.

17 A vacuum-operated recirculation valve is fitted to the exhaust manifold to regulate the quantity of exhaust gas recirculated. The valve is operated by the vacuum supplied by the brake vacuum pump via the solenoid valve and the temperature valve.

18 Between idling speed and a pre-determined engine load, the micro-switch on the injection pump closes to supply power to the solenoid valve, which allows the recirculation valve to open. Under full-load conditions, the micro-switch opens to cut off exhaust gas recirculation. Additional control is provided by the temperature valve, which cuts off the vacuum supply until the coolant temperature reaches 40°C, preventing the recirculation valve from opening during the engine warm-up phase. An "altimetric" capsule is also fitted, which cuts off the electrical supply to the solenoid valve (overriding the injection pump micro-switch) if atmospheric pressure drops below 900 mbars due to altitude.

Catalytic converter - precautions

19 For long life and satisfactory operation of the catalytic converter, certain precautions must be observed. These are as follows.

20 For petrol engines, only use unleaded fuel. Leaded fuel will damage the catalyst and the oxygen sensor.

21 Do not run the engine for long periods if it is misfiring. Unburnt fuel entering the catalytic converter can cause it to overheat, resulting in permanent damage. For the same reason, do not try to start the engine by pushing or towing the car, nor crank it on the starter motor for long periods.

22 Do not strike or drop the catalytic converter. The ceramic honeycomb which forms part of its internal structure may be damaged.

23 Always renew seals and gaskets upstream of the catalytic converter whenever they are disturbed.

24 Remember that the catalytic converter gets hotter than the rest of the exhaust system. Avoid parking on long grass, undergrowth or similar material which could catch fire or be damaged.

2 **Petrol engine emission control systems** - testing and component renewal

Crankcase emission control

Testing

1 There is no specific test procedure for the crankcase emission control system. If problems are suspected (sometimes indicated by oil contamination of the air cleaner element), check that the hoses are clean internally, and that the restrictors are not blocked or missing.

Component renewal

2 This is self-evident. Mark the various hoses before disconnecting them, if there is any possibility of confusion on reassembly.

Exhaust emission control

Testing

3 An exhaust gas analyser (CO meter) will be needed. The ignition system must be in good condition, the air cleaner element must be clean, and the engine must be in good mechanical condition.

4 Bring the engine to normal operating temperature, then connect the exhaust gas analyser in accordance with the equipment maker's instructions.

5 Run the engine at 2500 rpm for about 30 seconds, then allow it to idle and check the CO level (Chapter 1A Specifications). If the CO level is within the specified limits, the system is operating correctly.

6 If the CO level is higher than specified, try the effect of disconnecting the oxygen sensor wiring. If the CO level rises when the sensor is disconnected, this suggests that the oxygen sensor is OK and that the catalytic converter is faulty. If disconnecting the sensor has no effect, this suggests a fault in the sensor.

7 If a digital voltmeter is available, the oxygen sensor output voltage can be measured. Voltage should alternate between 625 to 1100 mV (rich mixture) and 0 to 80 mV (lean mixture).

8 Renew the oxygen sensor if it is proved faulty.

Oxygen sensor - renewal

9 Raise the front of the vehicle and support it on axle stands (see *"Jacking and vehicle support"*). Remove the engine undershield then disconnect the oxygen sensor wiring.

10 Unscrew the sensor from the exhaust downpipe, and remove it **(see illustrations)**.

11 Clean the threads in the exhaust pipe, and the threads of the sensor (if it is to be refitted).

12 Note that if the sensor wires are broken, the sensor must be renewed. No attempt should be made to repair them.

13 Apply high-temperature anti-seize compound to the sensor threads. Screw the sensor in by hand, then tighten it fully.

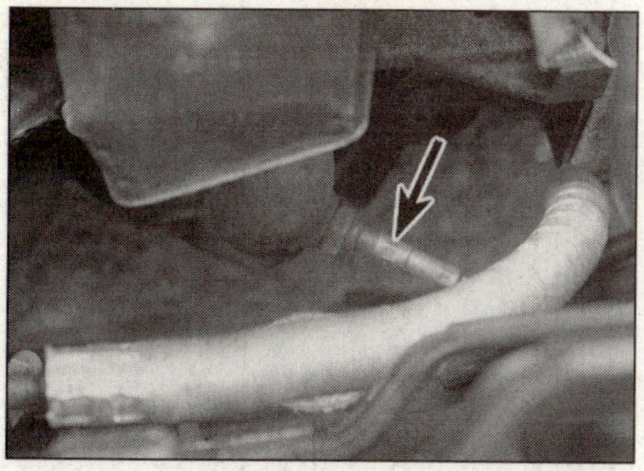

2.10a Oxygen sensor viewed from inside the engine compartment (E7J engine)

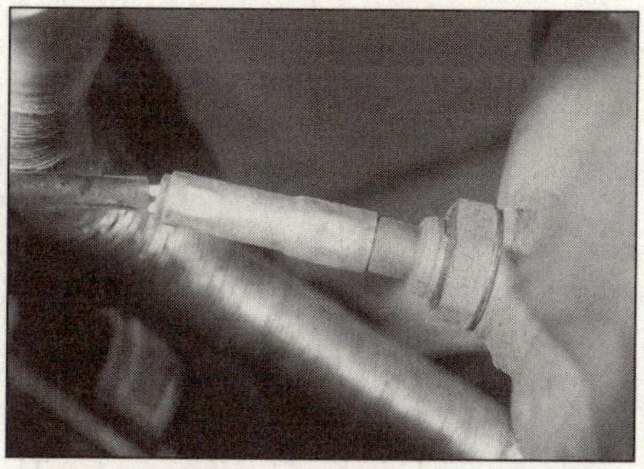

2.10b Oxygen sensor viewed from under the engine (K7M engine)

2.25 Evaporative emission canister located in the right-hand front corner of the engine compartment

2.30 Evaporative emission solenoid valve (E7J engine)

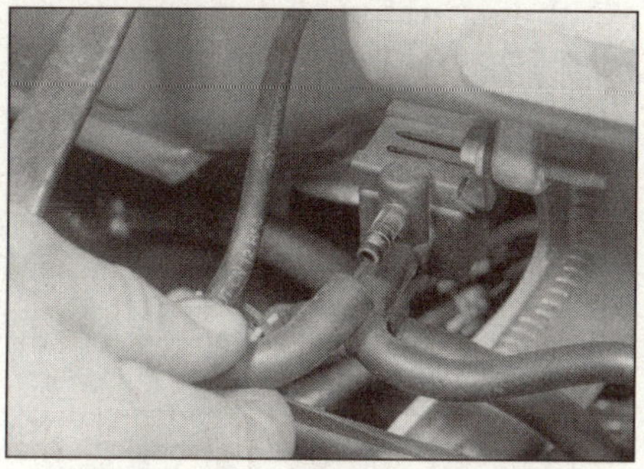

2.31a Disconnecting the hoses . . .

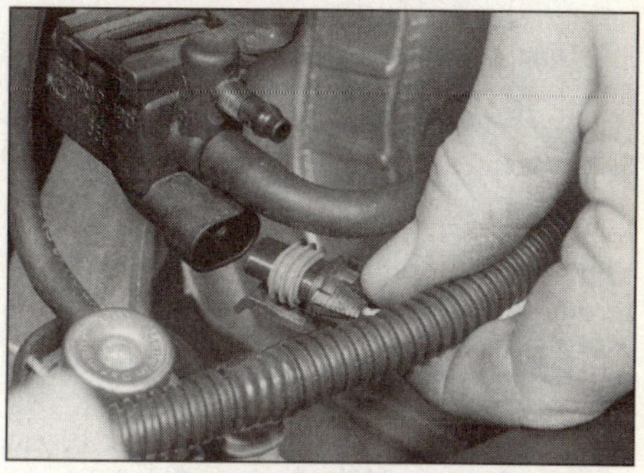

2.31b . . . and wiring multi-plug from the evaporative emission solenoid valve (K7M engine)

14 Reconnect the sensor wiring, refit the undershield and lower the vehicle to the ground.

Catalytic converter - renewal

15 The catalytic converter is renewed as part of the exhaust system. Refer to Part A of this Chapter.

Evaporative emission control

Testing

16 The operating principle of the system is that the solenoid valve is open only when the engine is warm with the throttle at least at the part-throttle position.

17 Bring the engine to normal operating temperature, then switch it off. Connect a vacuum gauge (range 0 to 1000 mbars) into the hose between the canister and the solenoid valve. Connect a voltmeter to the solenoid valve terminals.

18 Start the engine and allow it to idle. There should be no vacuum shown on the gauge, and no voltage present at the solenoid.

19 If manifold vacuum is indicated although no voltage is present, the solenoid valve may be stuck open. Temporarily disconnect the hoses from the solenoid valve and blow through the outlets to dislodge any particles of carbon.

20 If voltage is present at idle, there is a fault in the wiring or the computer.

21 Depress the accelerator slightly. Voltage should appear momentarily at the solenoid terminals, and manifold vacuum be indicated on the gauge.

22 If vacuum is not indicated even though voltage is present, either there is a leak in the hoses, or the valve is not opening.

23 If no voltage appears, there is a fault in the wiring or the computer.

Canister renewal

24 Temporarily remove the fuel injection ECU from the right-hand front of the engine compartment, leaving the wiring connected.

25 Disconnect and plug the fuel tank vent hose from the top of the canister **(see illustration)**. Refer to the fuel pump removal in Part A or B of this Chapter for details of removing the quick release clips.

26 Apply the handbrake, then jack up the front of the vehicle and support it on axle stands (see *"Jacking and vehicle support"*).

27 Using a spanner between the front bumper and mudguard skirt, unscrew and remove the mounting bolts, then remove the canister from the engine compartment.

28 Dispose of the old canister safely, bearing in mind that it may contain liquid fuel and/or fuel vapour.

29 Fit the new canister using a reversal of the removal procedure. Make sure that the hoses are connected correctly.

Solenoid valve renewal

30 On the E7J engine, the solenoid valve is located on top of the inlet manifold, to the right-hand side of the single-point throttle housing **(see illustration)**. On the K7M engine, it is located on the right-hand end of the inlet manifold. First identify, then disconnect the hoses from the valve.

31 Disconnect the hoses and multi-plug from the valve. Release the valve from its mounting bracket (E7J engine) or unbolt it from the inlet manifold (K7M engine), and remove it **(see illustrations)**.

32 Fit the new valve using a reversal of the removal procedure. Make sure that the hoses are connected correctly.

Exhaust gas recirculation

Testing

33 Run the engine to normal operating temperature (cooling fan must operate at least once), then switch it off.

34 Disconnect the wiring from the EGR solenoid valve located on the left-hand end of the cylinder head.

35 Using suitable bridging wires, supply 12 volts to the EGR valve.

36 Disconnect the vacuum reservoir hose from the EGR valve, then connect a vacuum pump in its place and apply vacuum of 300 mbars. Check that the vacuum does not drop by more that 2 mbars per second.

37 Disconnect the vacuum pump, then start the engine and allow it to run at idle speed. Reconnect the vacuum pump and apply vacuum of 300 mbars; the engine should run hesitantly at idle speed. If not, the valve should be renewed, however note that Renault recommend that the valve be cleaned every 50,000 miles (refer to Chapter 1A).

38 After making the check, erase the ECU memory with reference to Chapter 4.

EGR valve - renewal

39 Unbolt and remove the tie-bar from between the front suspension turrets.

40 Remove the air inlet duct from between the air cleaner and throttle housing with reference to Part A or B of this Chapter.

41 Disconnect the crankcase ventilation hose from the valve cover and position it to one side.

42 Disconnect the accelerator cable from the throttle housing and position it to one side.

43 Disconnect the vacuum hose from the EGR valve, then disconnect the wiring.

44 Unscrew the mounting bolts and withdraw the EGR valve from the cylinder head. Recover the seal.

45 Fit the new valve using a reversal of the removal procedure, but use a new seal and tighten the mounting bolts to the specified torque.

3 Diesel engine emission control systems - testing and component renewal

Crankcase emission control

Testing

1 There is no specific test procedure for the crankcase emission control system. If problems are suspected, check that the hoses are clean internally and are not blocked.

Component renewal

2 This is self-evident. Mark the various hoses before disconnecting them, if there is any possibility of confusion on reassembly.

Exhaust emission control

Testing

3 The system can only be tested accurately using a suitable exhaust gas analyser (suitable for use with diesel engines).

Catalytic converter - renewal

4 The catalytic converter is renewed as part of the exhaust system. Refer to Part B of this Chapter.

Exhaust gas recirculation

Testing

5 Start the engine, and run it until it reaches normal operating temperature (the cooling fan should have cut in and out at least once).

6 With the engine idling, disconnect the vacuum hose from the recirculation valve. As the hose is disconnected, it should be possible to hear the valve click shut. If no click is heard, proceed as follows.

3.18 EGR solenoid valve location on diesel engine models

7 Check that vacuum is present at the recirculation valve end of the vacuum hose. If a vacuum gauge is available, check that the vacuum is at least 500 mbars. If vacuum is present, it is likely that the recirculation valve is faulty (jammed or pierced diaphragm). If no vacuum is present, carry out the following checks.

8 Check the security of all vacuum hose connections.

9 Check the electrical feed to the solenoid valve.

10 Check the operation of the temperature valve. This can be done by checking that vacuum will pass through the valve with the engine at normal operating temperature. Stop the engine and disconnect the temperature valve vacuum hoses at the recirculation valve and the solenoid valve, and check that it is possible to blow through the hoses. If not, it is likely that the temperature valve is faulty, or the hoses are obstructed.

11 Check the operation of the micro-switch on the injection pump, using a continuity tester or an ohmmeter. With the pump accelerator lever in the idle position, the switch should be open (no continuity/infinite resistance). As the accelerator lever is moved from the idle position, the switch should close (continuity/zero resistance), until the lever reaches a predetermined position towards the end of its travel, when the switch should open again. Note that adjustment of the switch should **not** be attempted. If it is suspected that adjustment is required, or if a new switch is fitted, consult a Renault dealer who will have access to the specialist calibration equipment required to carry out accurate adjustment.

Component renewal

Recirculation valve

Note: *Where applicable, new gaskets should be used on refitting.*

12 The valve is located at the rear of the engine, and is bolted to the exhaust manifold on F8Q 620 engines, or on the inlet manifold on F8Q 784 engines. The valve is connected to the remaining manifold via a metal pipe bolted between the valve and the manifold.

13 Disconnect the vacuum hose from the valve.

3.25 Altimetric capsule location on diesel engine models

14 Remove the securing bolts, and disconnect the valve pipe from the inlet or exhaust manifold. Recover the gasket.

15 Unbolt the valve body from the relevant manifold, and recover the gasket. Remove the valve complete with the pipe.

16 If a new valve is to be fitted, unscrew the securing bolts, and transfer the pipe to the new valve, using a new gasket.

17 Refitting is a reversal of removal, using new gaskets where applicable.

Solenoid valve

18 The valve is located on a bracket attached to the engine compartment bulkhead **(see illustration)**.

19 Disconnect the wiring plug and the vacuum hoses from the valve.

20 Unscrew the securing nut(s), and withdraw the valve complete with its bracket.

21 Refitting is a reversal of removal, ensuring that the vacuum hoses are securely reconnected.

Temperature valve

22 The temperature valve is located in the coolant hose at the left-hand front corner of the cylinder block.

23 To remove the valve, simply disconnect the wiring plug and unscrew the valve from the hose. Be prepared for coolant spillage.

24 Refitting is reversal of removal, but check for coolant leaks on completion, and if necessary, top up the coolant level.

"Altimetric" capsule

25 The capsule is located on the crossmember at the front of the engine compartment **(see illustration)**.

26 Disconnect the wiring plug from the capsule, then unscrew the securing nut(s) and withdraw the capsule.

27 Refitting is a reversal of removal.

Injection pump micro-switch (EGR cut-off switch)

28 Removal and refitting of the switch is self-explanatory, but note that on completion the switch must be adjusted by a Renault dealer using specialist calibration equipment. Refer to Chapter 5C for details of the switch location.

4 Catalytic converter - general information and precautions

The catalytic converter is a reliable and simple device which needs no maintenance in itself, but there are some facts of which an owner should be aware if the converter is to function properly for its full service life.

Petrol models

a) *DO NOT use leaded petrol in a car equipped with a catalytic converter - the lead will coat the precious metals, reducing their converting efficiency and will eventually destroy the converter.*

b) *Always keep the ignition and fuel systems well-maintained in accordance with the manufacturer's schedule.*

c) *If the engine develops a misfire, do not drive the car at all (or at least as little as possible) until the fault is cured.*

d) *DO NOT push- or tow-start the car - this will soak the catalytic converter in unburned fuel, causing it to overheat when the engine does start.*

e) *DO NOT switch off the ignition at high engine speeds.*

f) *DO NOT use fuel or engine oil additives - these may contain substances harmful to the catalytic converter.*

g) *DO NOT continue to use the car if the engine burns oil to the extent of leaving a visible trail of blue smoke.*

h) *Remember that the catalytic converter operates at very high temperatures. DO NOT, therefore, park the car on dry undergrowth, over long grass or piles of dead leaves after a long run.*

i) *Remember that the catalytic converter is FRAGILE - do not strike it with tools during servicing work.*

j) *In some cases a sulphurous smell (like that of rotten eggs) may be noticed from the exhaust. This is common to many catalytic converter-equipped cars and once the car has covered a few thousand miles the problem should disappear.*

k) *The catalytic converter, used on a well-maintained and well-driven car, should last for between 50 000 and 100 000 miles - if the converter is no longer effective it must be renewed.*

Diesel models

Refer to the information given in parts f, g, h and i of the petrol models information given above.

Chapter 5 Part A:
Starting and charging systems

Contents

Degrees of difficulty

Easy, suitable for novice with little experience	**Fairly easy,** suitable for beginner with some experience	**Fairly difficult,** suitable for competent DIY mechanic	**Difficult,** suitable for experienced DIY mechanic	**Very difficult,** suitable for expert DIY or professional

Specifications

System type	12-volt, negative earth
Battery	
Type	Lead-acid, low-maintenance or "maintenance free"
Charge condition:	
Poor	12.5 volts
Normal	12.6 volts
Good	12.7 volts
Alternator	
Type	Valeo
Output:	
E7J engine	80 or 110 amps (depending on model)
K7M engine	75 or 110 amps (depending on model)
F8Q engine	75 or 110 amps (depending on model)
Regulated voltage	13.5 to 14.8 volts
Starter motor	
Type	Valeo, Bosch or Mitsubishi (depending on model)

1 General information and precautions

General information

The engine electrical system consists mainly of the charging and starting systems. Because of their engine-related functions, these components are covered separately from the body electrical devices such as the lights, instruments, etc (which are covered in Chapter 12). On petrol engine models, refer to Part B for information on the ignition system, and on diesel models, refer to Part C for information on the preheating system.

The electrical system is of the 12-volt negative earth type.

The battery is of the low maintenance or "maintenance-free" (sealed for life) type, and is charged by the alternator, which is belt-driven from the crankshaft pulley.

The starter motor is of the pre-engaged type, incorporating an integral solenoid. On starting, the solenoid moves the drive pinion into engagement with the flywheel ring gear before the starter motor is energised. Once the engine has started, a one-way clutch prevents the motor armature being driven by the engine until the pinion disengages from the flywheel.

Precautions

Further details of the various systems are given in the relevant Sections of this Chapter. While some repair procedures are given, the usual course of action is to renew the component concerned. The owner whose interest extends beyond mere component renewal should obtain a copy of the *"Automobile Electrical & Electronic Systems Manual"*, available from the publishers of this manual.

It is necessary to take extra care when working on the electrical system, to avoid damage to semi-conductor devices (diodes and transistors), and to avoid the risk of personal injury. In addition to the precautions given in *"Safety first!"* at the beginning of this manual, observe the following when working on the system:

Always remove rings, watches, etc before working on the electrical system. Even with the battery disconnected, capacitive discharge could occur if a component's live terminal is earthed through a metal object. This could cause a shock or nasty burn.

⚠ ***Do not reverse the battery connections. Components such as the alternator, electronic control units, or any other components having semi-conductor circuitry could be irreparably damaged.***

If the engine is being started using jump leads and a slave battery, connect the batteries positive-to-positive and negative-to-negative (see "Jump starting" at the start of this Manual). This also applies when connecting a battery charger.

Never disconnect the battery terminals, the alternator, any electrical wiring or any test instruments with the engine running.

Do not allow the engine to turn the alternator when the alternator is not connected.

Never "test" for alternator output by "flashing" the output lead to earth.

Never use an ohmmeter of the type incorporating a hand-cranked generator for circuit or continuity testing.

Always ensure that the battery negative lead is disconnected when working on the electrical system.

Before using electric-arc welding equipment on the car, disconnect the battery, alternator and components such as the fuel injection/ignition electronic control unit to protect them from the risk of damage.

The radio/cassette unit fitted as standard equipment by Renault is equipped with a built-in security code, to deter thieves. If the power source to the unit is cut, the anti-theft system will activate. Refer to "Radio/cassette unit anti-theft system - precaution" in the Reference Section.

2 Electrical fault finding - general information

Refer to Chapter 12.

3 Battery - testing and charging

Standard and low maintenance battery - testing

1 If the vehicle covers a small annual mileage, it is worthwhile checking the specific gravity of the electrolyte every three months to determine the state of charge of the battery. Use a hydrometer to make the check and compare the results with the following table (the temperatures quoted are ambient temperatures). Note that the specific gravity readings assume an electrolyte temperature of 15°C (60°F); for every 10°C (48°F) below 15°C (60°F) subtract 0.007. For every 10°C (48°F) above 15°C (60°F) add 0.007.

	Above 25°C	Below 25°C
Fully-charged	*1.210 to 1.230*	*1.270 to 1.290*
70% charged	*1.170 to 1.190*	*1.230 to 1.250*
Discharged	*1.050 to 1.070*	*1.110 to 1.130*

2 If the battery condition is suspect, first check the specific gravity of electrolyte in each cell. A variation of 0.040 or more between any cells indicates loss of electrolyte or deterioration of the internal plates.

3 If the specific gravity variation is 0.040 or more, the battery should be renewed. If the cell variation is satisfactory but the battery is discharged, it should be charged as described later in this Section.

Maintenance-free battery - testing

4 In cases where a "sealed for life" maintenance-free battery is fitted, topping-up and testing of the electrolyte in each cell is not possible. The condition of the battery can therefore only be tested using a battery condition indicator or a voltmeter.

5 Certain models may be fitted with a "Delco" type maintenance-free battery, with a built-in charge condition indicator. The indicator is located in the top of the battery casing, and indicates the condition of the battery from its colour. If the indicator shows green, then the battery is in a good state of charge. If the indicator turns darker, eventually to black, then the battery requires charging, as described later in this Section. If the indicator shows clear/yellow, then the electrolyte level in the battery is too low to allow further use, and the battery should be renewed. **Do not** attempt to charge, load or jump start a battery when the indicator shows clear/yellow.

6 If testing the battery using a voltmeter, connect the voltmeter across the battery and compare the result with those given in the *Specifications* under "charge condition". The test is only accurate if the battery has not been subjected to any kind of charge for the previous six hours. If this is not the case, switch on the headlights for 30 seconds, then wait four to five minutes before testing the battery after switching off the headlights. All other electrical circuits must be switched off, so check that the doors and tailgate/bootlid are fully shut when making the test.

7 If the voltage reading is less than 12.2 volts, then the battery is discharged, whilst a reading of 12.2 to 12.4 volts indicates a partially discharged condition.

8 If the battery is to be charged, remove it from the vehicle (Section 4) and charge it as described later in this Section.

Standard and low maintenance battery - charging

Note: *The following is intended as a guide only. Always follow the maker's recommendations (often printed on a label attached to the battery) before charging a battery.*

9 Charge the battery at a rate of 3.5 to 4 amps, and continue to charge the battery at this rate until no further rise in specific gravity is noted over a four-hour period.

10 Alternatively, a trickle charger charging at the rate of 1.5 amps can safely be used overnight.

11 Specially rapid `boost' charges which are claimed to restore the power of the battery in 1 to 2 hours are not recommended, as they can cause serious damage to the battery plates through overheating.

12 While charging the battery, note that the temperature of the electrolyte should never exceed 37.8°C (100°F).

Maintenance-free battery - charging

Note: *The following is intended as a guide only. Always follow the maker's recommendations (often printed on a label attached to the battery) before charging a battery.*

13 This battery type takes considerably longer to fully recharge than the standard type, the time taken being dependent on the extent of discharge, but it can take anything up to three days.

14 A constant voltage type charger is required, to be set, when connected, to 13.9 to 14.9 volts with a charger current below 25 amps. Using this method, the battery should be usable within three hours, giving a voltage reading of 12.5 volts, but this is for a partially discharged battery and, as mentioned, full charging can take considerably longer.

15 If the battery is to be charged from a fully discharged state (condition reading less than 12.2 volts), have it recharged by your Renault dealer or local automotive electrician, as the charge rate is higher and constant supervision during charging is necessary.

4 Battery - removal and refitting

Note: *If a Renault radio/cassette unit is fitted, refer to "Radio/cassette unit anti-theft system".*

Removal

1 The battery is located in the right-hand rear bulkhead area of the engine compartment **(see illustration)**.

4.1 The battery is located at the right-hand rear of the engine compartment

4.2a Unscrew the mounting screws . . .

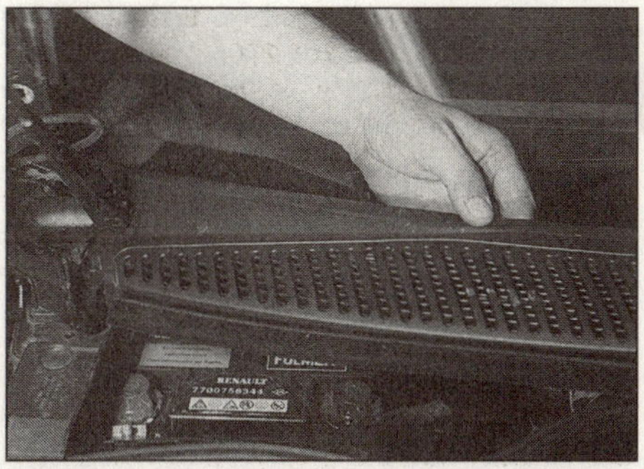

4.2b . . . and remove the scuttle grille from the right-hand side of the bulkhead

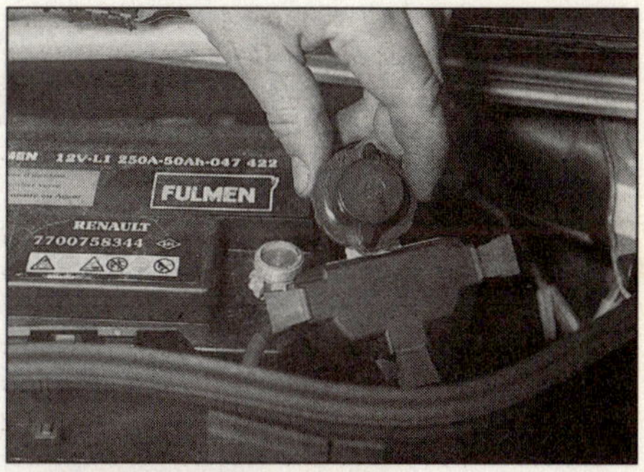

4.4 Removing the positive terminal lead from the battery

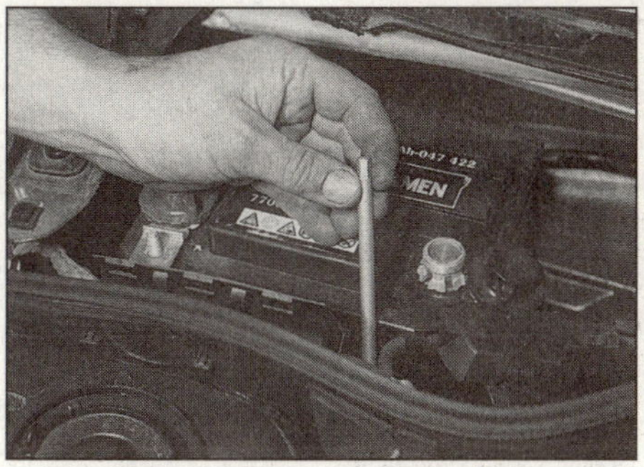

4.5a Unscrew the special bolt . . .

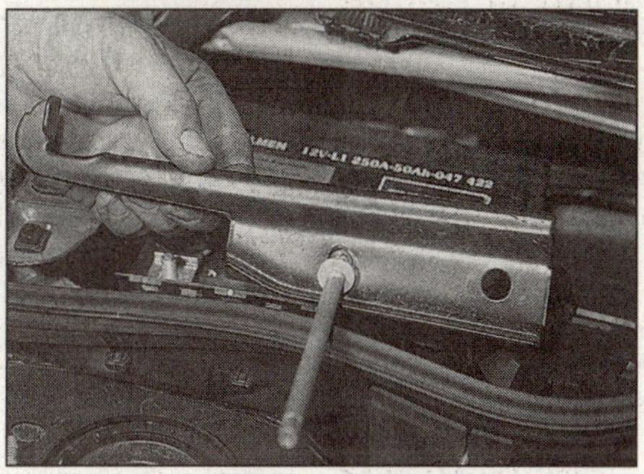

4.5b . . . and remove the battery retaining clamp

2 Unscrew the mounting screws, and remove the scuttle grille from the right-hand side of the bulkhead compartment in front of the windscreen. The inner end of the grille must be unclipped from the left-hand grille **(see illustrations)**.

3 Slacken the plastic terminal nut and disconnect the clamp from the battery negative (earth) terminal.

4 Slacken the plastic terminal nut and disconnect the positive terminal lead in the same way **(see illustration)**.

5 Unscrew the special bolt and remove the battery retaining clamp **(see illustrations)**.

6 Lift the battery out of the bulkhead compartment.

7 If necessary, remove the battery tray.

Refitting

8 Refitting is a reversal of removal, but smear petroleum jelly on the terminals when reconnecting the leads, and always reconnect the positive lead first, and the negative lead last.

5 Charging system - testing

Note: *Refer to the warnings given in "Safety first!" and in Section 1 of this Chapter before starting work.*

1 If the ignition warning light fails to illuminate when the ignition is switched on, first check the alternator wiring connections for security. If satisfactory, check that the warning light bulb has not blown, and that the bulbholder is secure in its location in the instrument panel. If the light still fails to illuminate, check the continuity of the warning light feed wire from the alternator to the bulbholder. If all is satisfactory, the alternator is at fault, and should be renewed or taken to an auto-electrician for testing and repair.

2 If the ignition warning light illuminates when the engine is running, stop the engine and check that the drivebelt is correctly tensioned (see Chapter 1) and that the alternator connections are secure. If all is so far satisfactory, have the alternator checked by an auto-electrician for testing and repair.

3 If the alternator output is suspect even though the warning light functions correctly, the regulated voltage may be checked as follows.

4 Connect a voltmeter across the battery terminals and start the engine.

5 Increase the engine speed until the voltmeter reading remains steady; the reading should be approximately 12 to 13 volts, and no more than 14 volts.

6 Switch on as many electrical accessories (eg, the headlights, heated rear window and heater blower) as possible, and check that the alternator maintains the regulated voltage at around 13 to 14 volts.

7 If the regulated voltage is not as stated, the fault may be due to worn brushes, weak brush springs, a faulty voltage regulator, a faulty diode, a severed phase winding, or worn or damaged slip rings. The alternator should be renewed or taken to an auto-electrician for testing and repair.

6 Alternator drivebelt - removal, refitting and tensioning

Refer to the procedure given for the auxiliary drivebelt(s) in Chapter 1.

7 Alternator - removal and refitting

Removal

1 Disconnect the battery negative lead. Release the auxiliary drivebelt tension as described in Chapter 1 and disengage the drivebelt from the alternator pulley. **Note:** *Renault recommend that the drivebelt be renewed whenever it is disturbed.*

Caution: If the radio/cassette in your vehicle is equipped with an anti-theft system, make sure you have the correct activation code before disconnecting the battery.

Petrol engine models

2 Apply the handbrake, then jack up the front of the vehicle and support on axle stands (see *"Jacking and vehicle support"*). Remove the right-hand roadwheel, the engine undershield and the wheelarch liner. Access to the alternator is best achieved from underneath the vehicle.

3 If the drivebelt is to be renewed on models without air conditioning, remove the power steering drivebelt first as described in

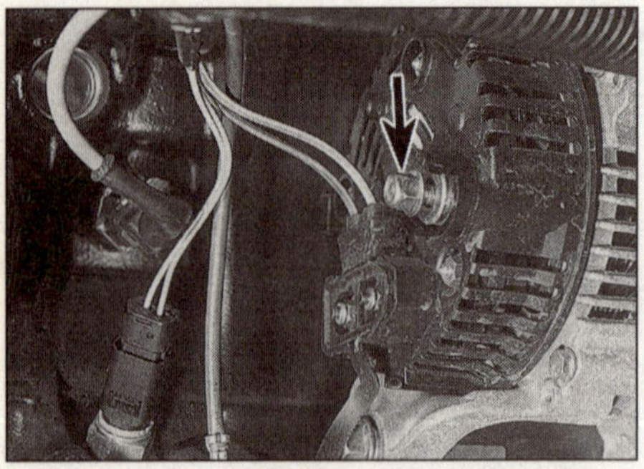

7.4a Main battery cable retaining nut on the alternator B+ terminal

7.4b Disconnecting the wiring from the rear of the alternator

Chapter 1. If it is not being renewed, there is no need to remove the power steering drivebelt.

4 Remove the rubber covers (where fitted) from the alternator terminals, then unscrew the retaining nuts and disconnect the wiring from the rear of the alternator **(see illustrations)**.

5 Loosen the pivot and adjusting bolt and swivel the alternator towards the engine. Slip the drivebelt from the pulley.

6 Unscrew and remove the alternator upper adjusting bolt.

7 Unscrew and remove the alternator lower pivot bolt, and temporarily support the alternator.

8 Disconnect the steering track rod end from the right-hand front hub carrier with reference to Chapter 10.

9 Unscrew and remove the right-hand suspension strut upper mounting bolts from the top of the suspension turret with reference to Chapter 10.

10 Under the vehicle, have an assistant pull the right-hand suspension downwards as far as possible, while the alternator is being removed through the wheelarch inner panel.

Diesel engine models

11 Just behind the alternator, unscrew the bolt from the fuel pipe support bracket, and position the bracket to one side.

12 If the drivebelt is to be renewed, raise the front of the vehicle and support on axle stands (see *"Jacking and vehicle support"*), then unbolt and remove the lower timing cover.

13 Unscrew and remove the mounting bolts, slip the drivebelt from the pulley, then withdraw the alternator from the engine.

Refitting

14 Refitting is a reversal of removal, but refer to Chapter 1 for details of fitting the drivebelt.

8 Alternator - testing and overhaul

If the alternator is thought to be suspect, it should be removed from the vehicle and taken to an automotive electrician for testing. Most auto-electricians will be able to supply and fit brushes at a reasonable cost. However, check on the cost of repairs before proceeding, as it may prove more economical to obtain a new or exchange alternator.

9 Starting system - testing

Note: *Refer to the precautions given in "Safety first!" and in Section 1 of this Chapter before starting work.*

1 If the starter motor fails to operate when the ignition key is turned to the appropriate position, the following possible causes may be to blame.

a) *The battery is faulty.*
b) *The electrical connections between the switch, solenoid, battery and starter motor are somewhere failing to pass the necessary current from the battery through the starter to earth.*
c) *The solenoid is faulty.*
d) *The starter motor is mechanically or electrically defective.*

2 To check the battery, switch on the headlights. If they dim after a few seconds, this indicates that the battery is discharged - recharge (see Section 3) or renew the battery. If the headlights glow brightly, operate the ignition switch and observe the lights. If they dim, then this indicates that current is reaching the starter motor, therefore the fault must lie in the starter motor. If the lights continue to glow brightly (and no clicking sound can be heard from the starter motor solenoid), this indicates that there is a fault in the circuit or solenoid - see following paragraphs. If the starter motor turns slowly when operated, but the battery is in good condition, then this indicates that either the starter motor is faulty, or there is considerable resistance somewhere in the circuit.

3 If a fault in the circuit is suspected, disconnect the battery leads (including the earth connection to the body), the starter/solenoid wiring and the engine/transmission earth strap. Thoroughly clean the connections, and reconnect the leads and wiring, then use a voltmeter or test lamp to check that full battery voltage is available at the battery positive lead connection to the solenoid, and that the earth is sound. Smear petroleum jelly around the battery terminals to prevent corrosion - corroded connections are amongst the most frequent causes of electrical system faults.

4 If the battery and all connections are in good condition, check the circuit by disconnecting the wire from the solenoid blade terminal. Connect a voltmeter or test lamp between the wire end and a good earth (such as the battery negative terminal), and check that the wire is live when the ignition switch is turned to the 'start' position. If it is, then the circuit is sound - if not the circuit wiring can be checked as described in Chapter 12.

5 The solenoid contacts can be checked by connecting a voltmeter or test lamp between the battery positive feed connection on the starter side of the solenoid, and earth. When the ignition switch is turned to the 'start' position, there should be a reading or lighted bulb, as applicable. If there is no reading or lighted bulb, the solenoid is faulty and should be renewed.

6 If the circuit and solenoid are proved sound, the fault must lie in the starter motor. In this event, it may be possible to have the starter motor overhauled by a specialist, but check on the cost of spares before proceeding, as it may prove more economical to obtain a new or exchange motor.

10 Starter motor - removal and refitting

Removal

1 Disconnect the battery negative lead.

Caution: If the radio/cassette in your vehicle is equipped with an anti-theft system, make sure you have the correct activation code before disconnecting the battery.

Petrol engine models

2 Remove the air cleaner housing as described in Chapter 4A.

3 Firmly apply the handbrake, then jack up the front of the vehicle and support it on axle stands (see "*Jacking and vehicle support*"). Remove the engine undershield where fitted.

4 Where fitted, unbolt and remove the support bracket from the rear of the starter motor.

5 Unscrew the nut and disconnect the main cable from the starter motor solenoid. Recover the washers under the nuts.

6 Disconnect the solenoid trigger wire from the terminal **(see illustration)**.

7 Unscrew and remove the three bolts securing the starter motor to the transmission housing, noting the location of the support bracket, then remove the motor downwards and out of position, noting the locating dowel

10.6 Disconnecting the starter motor solenoid trigger wire

10.7a Unscrew the starter motor mounting bolts . . .

10.7b . . . noting the location of the support bracket . . .

10.7c . . . then withdraw the starter motor from the transmission

which is fitted to the rear mounting bolt hole **(see illustrations)**.

Non-turbo diesel engine models

8 Remove the air cleaner housing as described in Chapter 4B.
9 Firmly apply the handbrake, then jack up the front of the vehicle and support it on axle stands (see *"Jacking and vehicle support"*). Remove the engine undershield where fitted.
10 Undo the retaining screws and nut, and remove the protective cover from the starter motor.
11 Disconnect the wiring connector from the starter solenoid, then undo the retaining nut and disconnect the main feed cable from the solenoid. Recover the washer under the nut.
12 Undo the retaining nuts and bolts, and remove the rear mounting bracket from the starter motor.
13 Slacken and remove the three bolts securing the starter motor to the transmission housing, and remove the motor from the engine compartment. Recover the locating dowel which is fitted to the rear mounting bolt hole.

Turbo diesel engine models

14 Firmly apply the handbrake, then jack up the front of the vehicle and support it on axle stands (see *"Jacking and vehicle support"*). Remove the right-hand front roadwheel and the engine undershield.
15 Loosen the securing clips, and remove the air ducting connecting the inlet manifold to the intercooler.
16 Similarly, remove the air ducting connecting the turbocharger to the intercooler.
17 Unscrew the three bolts securing the starter motor to the gearbox bellhousing. Where applicable, note the location of any wiring brackets secured by the bolts.
18 Remove the exhaust front downpipe as described in Chapter 4.
19 Working beneath the right-hand side of the vehicle, use a pin punch to drive out the roll pin securing the right-hand driveshaft to the gearbox differential sun wheel stub shaft.
20 Loosen the two nuts and bolts securing the right-hand stub axle carrier to the suspension strut. Remove the upper nut and bolt, but leave the lower nut and bolt in position. Note that the nuts are on the rear side of the strut.
21 Pull the top of the stub axle carrier outwards until the inner end of the driveshaft is released from the stub shaft. Support the driveshaft using string or wire - do not allow it to hang under its own weight.
22 Move the driveshaft as necessary to allow access to the two nuts securing the turbocharger oil return pipe to the cylinder block.
23 Remove the two securing nuts, and disconnect the turbocharger oil return pipe from the cylinder block. Recover the gasket. Plug the hole in the cylinder block to prevent dirt ingress.
24 Remove the nut securing the wiring bracket to the bottom of the rear starter motor mounting bracket, then disconnect the wiring from the starter motor.
25 Remove the bolt securing the turbocharger support bracket to the rear starter motor support bracket (note that the bolt also secures the heat shield). Loosen the nut and bolt securing the support bracket to the turbocharger, and pivot the support bracket clear of the starter motor.
26 Remove the nuts and bolts securing the rear starter motor mounting bracket to the engine and the starter motor.
27 Remove the remaining bolt securing the heat shield.
28 Withdraw the starter motor, rear mounting bracket and heat shield from underneath the vehicle. Take care not to drop the starter motor.

Refitting

29 Refitting is a reversal of removal, but tighten the mounting bolts to the specified torque.

11 Starter motor - testing and overhaul

If the starter motor is thought to be suspect, it should be removed from the vehicle and taken to an auto-electrician for testing. Most auto-electricians will be able to supply and fit brushes at a reasonable cost. However, check on the cost of repairs before proceeding, as it may prove more economical to obtain a new or exchange motor.

12 Ignition switch - removal and refitting

Removal

1 Disconnect the battery negative lead.
Caution: If the radio/cassette in your vehicle is equipped with an anti-theft system, make sure you have the correct activation code before disconnecting the battery.
2 Remove the steering wheel and steering column shrouds as described in Chapter 10.
3 Release the ignition switch wiring connector from its location beneath the steering column by pivoting it, then disconnect it. Note the routing of the wiring for correct refitting.
4 Unscrew and remove the mounting bolt from the ignition switch.
5 Insert the ignition key and turn it to position '3'.
6 Depress the side peg and withdraw the ignition switch together with the wiring.

Refitting

7 Refitting is a reversal of removal, but make sure the wiring is correctly routed.

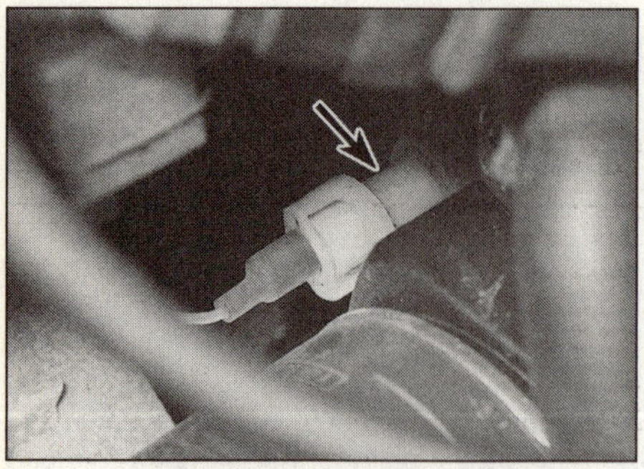

13.1 On E7J and F8Q engines, the oil pressure switch is located next to the oil filter

13 Oil pressure warning light switch - removal and refitting

Removal

1 On the E7J petrol and F8Q diesel engines, the switch is located at the lower front right-hand side of the cylinder block, next to the oil filter **(see illustration)**. On the K7M petrol engine, it is located at the left-hand front of the cylinder block.

2 Apply the handbrake, then jack up the front of the vehicle and support on axle stands (see *"Jacking and vehicle support"*).

13.4 Removing the oil pressure switch on the K7M engine

3 Disconnect the wiring from the switch.

4 Unscrew the switch from the cylinder block, and recover the sealing washer **(see illustration)**. Be prepared for oil spillage, and if the switch is to be left removed from the engine, plug the hole in the cylinder block.

Refitting

5 Examine the sealing washer for signs of damage or deterioration, and if necessary renew it.

6 Refit the switch, complete with washer, and tighten it securely. Reconnect the wiring connector.

7 If necessary, top-up the engine oil as described in *"Weekly checks"*.

14 Oil level sensor - removal and refitting

Removal

1 On E7J petrol and F8Q diesel engines, the oil level sensor is located on the front of the cylinder block, below the oil filter **(see illustration)**. On the K7M petrol engine it is located on the rear of the cylinder block below the manifolds.

2 Apply the handbrake, then jack up the front of the vehicle and support on axle stands (see *"Jacking and vehicle support"*).

3 Disconnect the wiring from the sensor.

4 Unscrew the sensor from the cylinder block, and recover the sealing washer **(see illustrations)**. If the sensor is to be left removed from the engine, plug the hole in the cylinder block.

Refitting

5 Examine the sealing washer for signs of damage or deterioration, and if necessary renew it.

6 Refit the sensor, complete with washer, and tighten it securely. Reconnect the wiring connector.

14.1 On E7J petrol and F8Q diesel engines, the oil level sensor is located below the oil filter

14.4a Unscrew the oil level sensor . . .

14.4b . . . and remove it from the cylinder block (K7M engine)

Chapter 5 Part B:
Ignition system - petrol engine models

Contents

Degrees of difficulty

Easy, suitable for novice with little experience	**Fairly easy,** suitable for beginner with some experience	**Fairly difficult,** suitable for competent DIY mechanic	**Difficult,** suitable for experienced DIY mechanic	**Very difficult,** suitable for expert DIY or professional

Specifications

System type

E7J engine	Electronic ignition system with distributor, controlled by engine management ECU. Ignition power module and single coil
K7M engine	Static (distributorless) ignition system, controlled by engine management ECU. Double wasted-spark coils
Firing order	1-3-4-2 (No 1 cylinder at transmission end)

Ignition system data

Ignition timing	Controlled by the ECU - see text
Ignition HT coil resistances (E7J engine):*	
Primary windings	1.5 ohms
Secondary windings	6.5 k ohms

**These are suggested figures, typical for this type of coil - no exact values are quoted by Renault*

Ignition HT coil resistances (K7M engine only):	
Terminals 1 to 2	0.5 ohms
Terminals 1 to 3	1.0 ohms
Terminals 2 to 3	1.0 ohms
HT terminal to HT terminal	10 k ohms

1 Ignition system - general information and precautions

The ignition system is integrated with the fuel injection system to form a combined engine management system under the control of one ECU (See Chapter 4A for further information). There are two types of ignition system.

Models with a distributor (E7J engine with single-point injection)

The E7J engine is fitted with a distributor, and the ignition system incorporates an amplifier unit and single HT ignition coil.

Depending on the information received from its various sensors, the ECU calculates the relevant ignition advance/coil charge time required for the engine speed and load. The ECU switches the amplifier unit to operate the ignition coil primary (LT) circuit. This causes a high voltage to be induced in the coil secondary (HT) windings which then travels down the HT lead to the distributor and on to the relevant spark plug. The distributor is driven off the left-hand end of the camshaft.

Models without a distributor (K7M engine with multi-point injection)

The K7M engine is fitted with a static (distributorless) ignition system, the ignition system, with a pair of two output ignition coils. Each coil supplies two cylinders each (one coil supplies cylinders 1 and 4, and the other cylinders 2 and 3). Under the control of the ECU, the ignition coil operates on the "wasted spark" principle, ie. each spark plug sparks twice for every cycle of the engine, once on the compression stroke and once on the exhaust stroke. The ECU uses its inputs from the various sensors to calculate the required ignition advance setting and coil charging time.

The following precautions must be observed, to prevent damage to the ignition system components and to reduce risk of personal injury.

a) Ensure the ignition is switched off before disconnecting any of the ignition wiring.
b) Ensure that the ignition is switched off before connecting or disconnecting any ignition test equipment, such as a timing light.
c) Do not earth the coil primary or secondary circuits.

Warning: Voltages produced by an electronic ignition system are considerably higher than those produced by conventional ignition systems. Extreme care must be taken when working on the system with the ignition switched on. Persons with surgically-implanted cardiac pacemaker devices should keep well clear of the ignition circuits, components and test equipment.

2 Ignition system - testing

Models with a distributor (E7J engine with single-point injection)

1 The components of electronic ignition systems are normally very reliable; most faults are far more likely to be due to loose or dirty connections, or to "tracking" of HT voltage due to dirt, dampness or damaged insulation than to the failure of any of the system's components. Always check all wiring thoroughly before condemning an electrical component, and work methodically to eliminate all other possibilities before deciding that a particular component is faulty.

2 The old practice of checking for a spark by holding the live end of an HT lead a short distance away from the engine is not recommended; not only is there a high risk of a powerful electric shock, but the HT coil or ECU will be damaged. Similarly, never try to "diagnose" misfires by pulling off one HT lead at a time.

Engine will not start

3 If the engine either will not turn over at all, or only turns very slowly, check the battery and starter motor as described in Chapter 5A.

4 If the engine turns over at normal speed but will not start, check the HT circuit by connecting a timing light (following the manufacturer's instructions) and turning the engine over on the starter motor; if the light flashes, voltage is reaching the spark plugs, so these should be checked first. If the light does not flash, check the HT leads themselves followed by the distributor cap, carbon brush and rotor arm using the information given in Chapter 1.

5 If there is still no spark, check the coil's primary and secondary winding resistance as described later in this Section; renew the coil if faulty, but be careful to check carefully the condition of the LT connections themselves before doing so, to ensure that the fault is not due to dirty or poorly-fastened connectors.

6 If these checks fail to reveal the cause of the problem, the vehicle should be taken to a suitably-equipped Renault dealer for testing. A wiring block connector is incorporated in the engine management circuit, into which a special electronic diagnostic tester can be plugged. The tester will locate the fault quickly and simply, alleviating the need to test all the system components individually, which is a time-consuming operation that carries a high risk of damaging the ECU. If necessary, the system wiring and wiring connectors can be checked as described in Chapter 12, ensuring that the ECU wiring connector is first disconnected with the ignition switched off.

Engine misfires

7 An irregular misfire suggests either a loose connection or intermittent fault in the primary circuit, or an HT fault on the coil side of the rotor arm.

8 With the ignition switched off, check carefully through the system ensuring that all connections are clean and securely fastened.

9 Check that the HT coil, the distributor cap and the HT leads are clean and dry. Check the leads themselves and the spark plugs (by substitution, if necessary), then check the distributor cap, carbon brush and rotor arm as described in Chapter 1.

10 Regular misfiring is probably due to a fault in the distributor cap, HT leads or spark plugs. Use a timing light (as described above) to check whether HT voltage is present at all leads.

11 If HT voltage is not present on any particular lead, the fault will be in that lead or in the distributor cap. If HT is present on all leads, the fault will be in the spark plugs; check and renew them if there is any doubt about their condition.

12 If no HT is present, check the HT coil; its secondary windings may be breaking down under load.

Models without a distributor (K7M engine with multi-point injection)

13 Check the ignition system as described in paragraphs 3 to 6, ignoring the references to the distributor cap and rotor arm.

3 Ignition HT coil(s) - removal, testing and refitting

Removal

1 Disconnect the battery negative lead.

Caution: If the radio/cassette in your vehicle is equipped with an anti-theft system, make sure you have the correct activation code before disconnecting the battery.

Models with a distributor (E7J engine with single-point injection)

2 The ignition HT coil is bolted onto the front of the amplifier unit, which is mounted onto the right hand side of the engine compartment bulkhead. Note that Renault do not supply the coil separate to the amplifier unit, but it may be possible to obtain one from other sources.

3 Slacken and remove the coil retaining bolts, and carefully remove the coil from the amplifier unit.

Models without a distributor (K7M engine with multi-point injection)

4 The ignition coils are located on the right-hand end of the engine valve cover. Unclip and remove the plastic cover from the front of the ignition coil assembly **(see illustrations)**.

5 Disconnect the HT leads from the spark plugs.

6 Undo the screws securing the HT lead retainer to the coil assembly, and move the retainer to one side **(see illustrations)**.

3.4a Double wasted-spark ignition coils fitted to the K7M engine

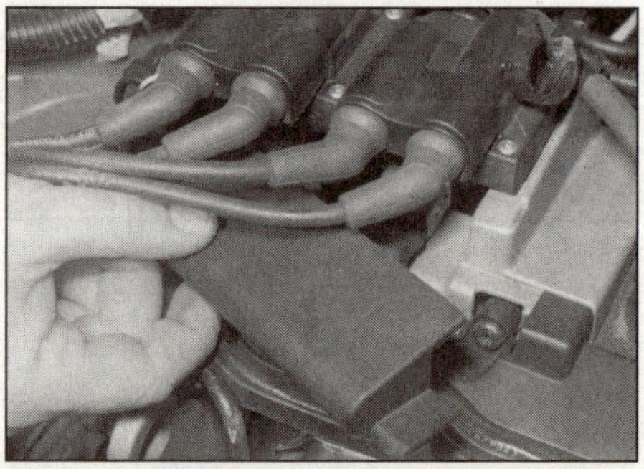

3.4b Removing the plastic cover from the ignition coils (K7M engine)

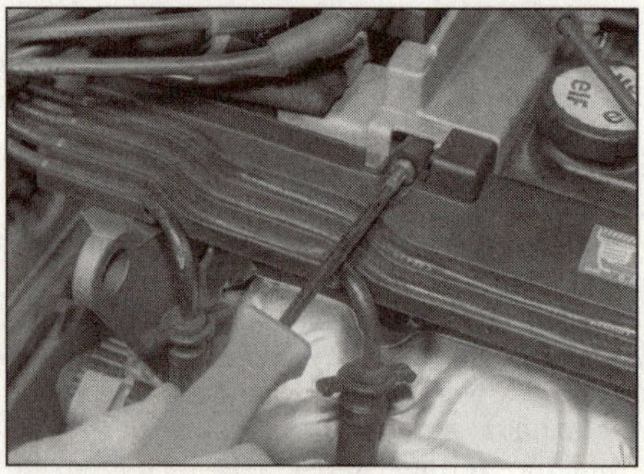

3.6a Remove the screws . . .

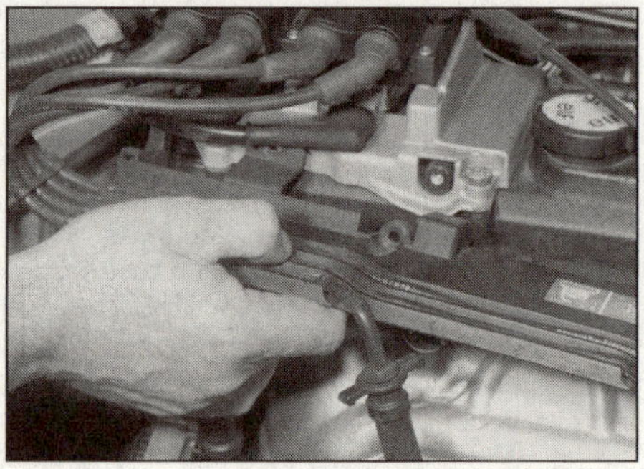

3.6b . . . and move the HT lead retainer to one side (K7M engine)

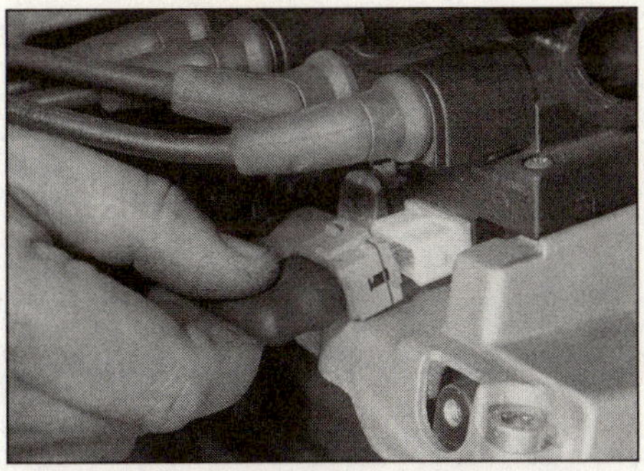

3.7 Disconnecting the wiring plugs from the coils (K7M engine)

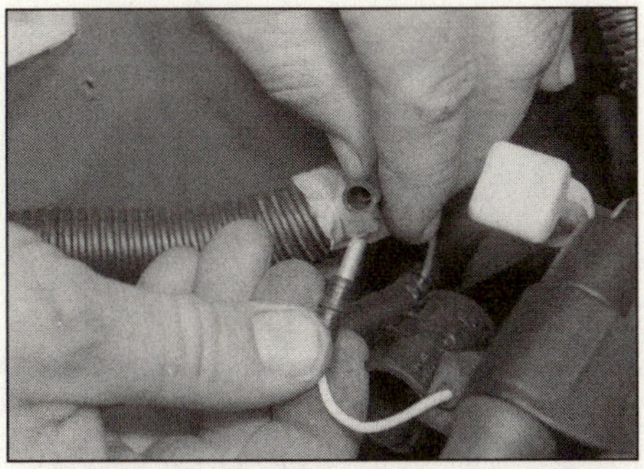

3.9 Disconnect the single wire (K7M engine)

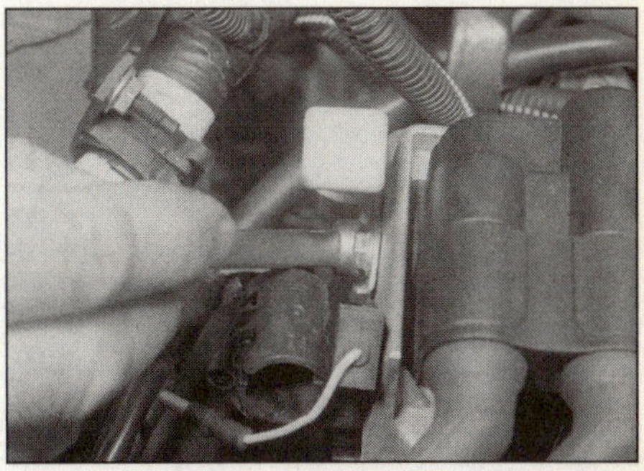

3.10 Unscrew the dipstick mounting bolt from the coil mounting plate (K7M engine)

3.11a Unscrew the mounting screws . . .

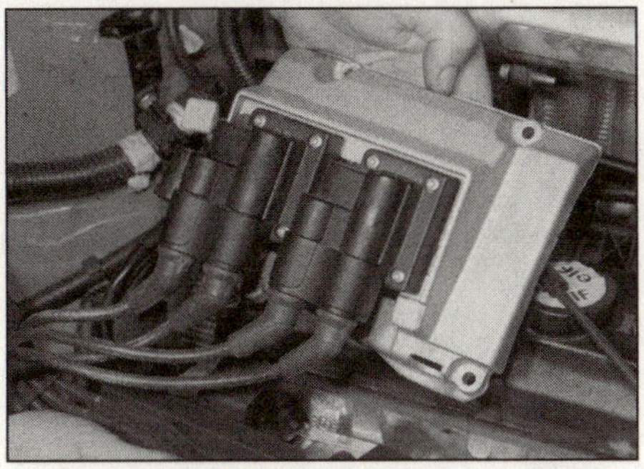

3.11b . . . and withdraw the coil assembly from the valve cover (K7M engine)

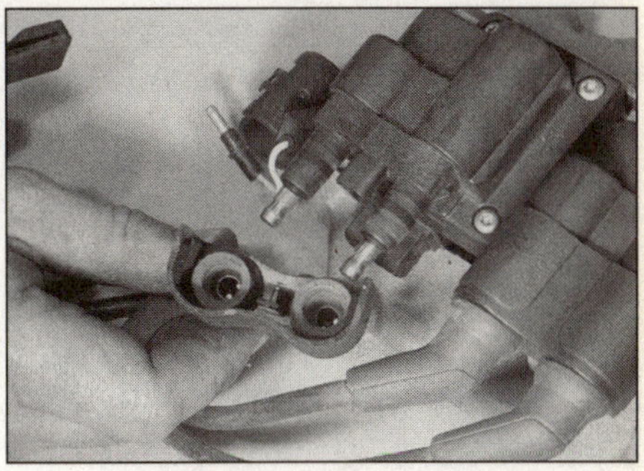

3.12 Disconnecting the HT lead connectors from the coils (K7M engine)

7 Disconnect the wiring **(see illustration)**.

8 Release the wiring from the support on the right-hand end of the assembly.

9 Disconnect the single wire **(see illustration)**.

10 Unscrew the mounting bolt for the engine oil level dipstick **(see illustration)**.

11 Unscrew the four mounting screws, and lift the coil assembly together with the HT leads from the engine valve cover. Note that the screws are located in the tops of the main valve cover screws **(see illustrations)**.

12 Carefully pull the HT lead connectors from each coil **(see illustration)**.

13 Unscrew the mounting screws and remove the coils from the mounting plate **(see illustrations). Note:** *The base of each coil is different to ensure that they only locate in one position.*

Testing

Models with a distributor (E7J engine with single-point injection)

14 Testing of the coil consists of using a multimeter set to its resistance function, to check the primary (LT '+' to '-' terminals) and secondary (LT '+' to HT lead terminal) windings for continuity. Compare the results obtained to those given in the Specifications at the start of this Chapter. Note the resistance of the coil windings will vary slightly according to the coil temperature, the results in the Specifications are approximate values only since no exact figures are quoted by Renault.

15 Check that there is no continuity between the HT lead terminal and the mounting bracket.

16 If the coil is faulty, renew it.

Models without a distributor (K7M engine with multi-point injection)

17 Each coil can be tested using a multimeter set to its resistance function. Connect the multimeter to the terminals, and compare with the information given in the Specifications. The three low tension terminals are marked with their track numbers on the socket. Connect the multimeter between the two HT terminals for the final test **(see illustrations)**.

18 Renew the coil if faulty, noting that the coils are different.

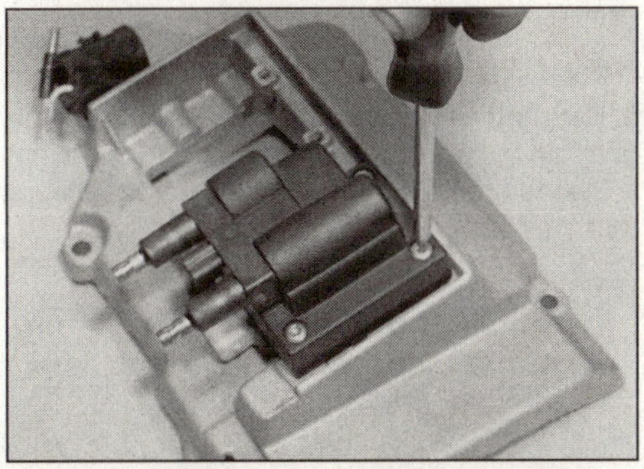

3.13a Unscrew the mounting screws . . .

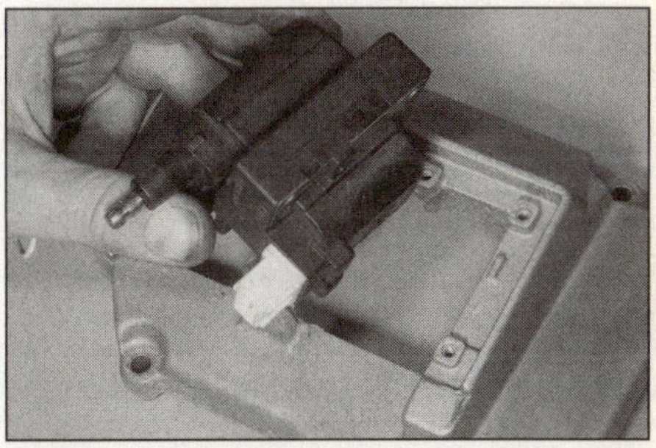

3.13b . . . and remove the coils from the mounting plate (K7M engine)

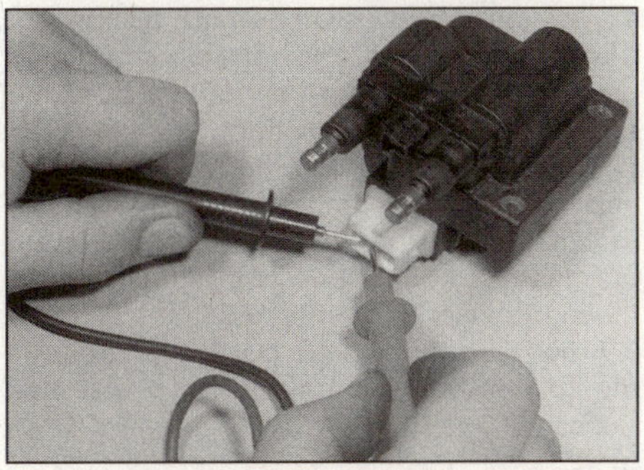

3.17a Testing the low tension circuits of the ignition coil (K7M engine)

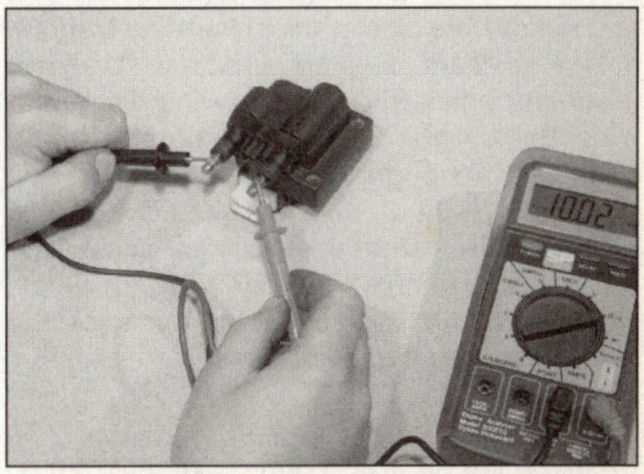

3.17b Testing the resistance between the coil HT terminals (K7M engine)

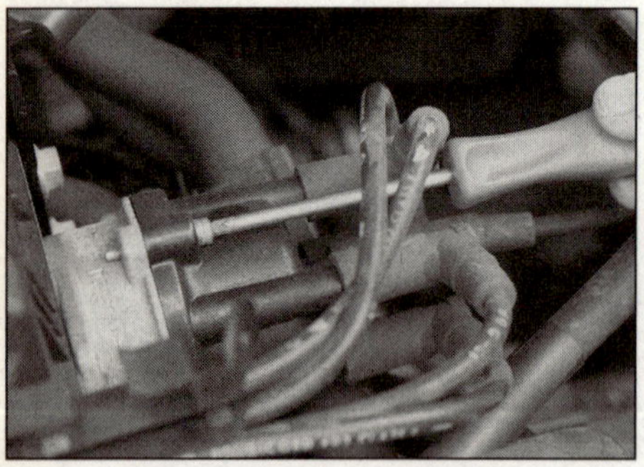

4.2a Undo the screws . . .

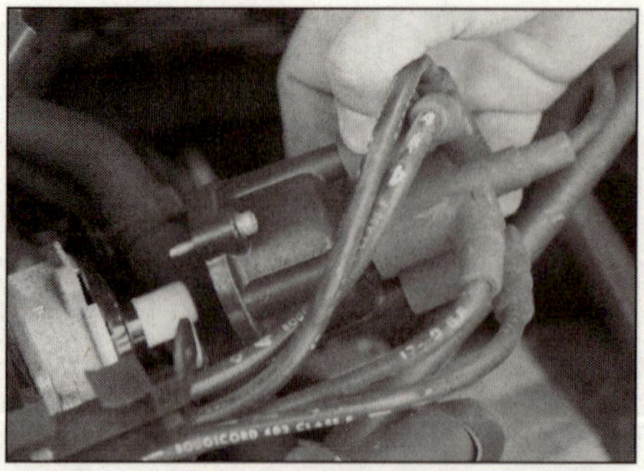

4.2b . . . and remove the distributor cap (E7J engine)

4.3 Removing the rotor arm (E7J engine)

4.4a Distributor mounting bolts (E7J engine)

4.4b Removing the distributor (E7J engine)

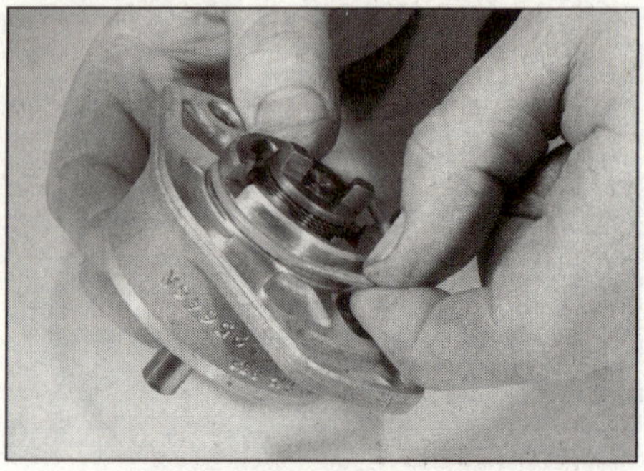

4.5 Removing the seal from the base of the distributor (E7J engine)

Refitting

19 Refitting is a reversal of the relevant removal procedure, ensuring that the wiring connectors and HT leads are correctly and securely connected.

4 Distributor - removal, inspection and refitting

Removal

1 Remove the air cleaner ducting from the left-hand side of the engine (see Chapter 4A).
2 Undo the retaining screws and remove the cap from the distributor body. Position the cap and HT leads to one side, or if required identify and remove the leads from the cap, spark plugs and holder **(see illustrations)**.
3 Pull the rotor arm off the distributor shaft, and remove the plastic shield **(see illustration)**.
4 Use a Torx key to unscrew the two mounting bolts, and withdraw the distributor from the cylinder head **(see illustrations)**. Note that although the upper mounting bolt hole in the distributor body is slotted, the lower bolt hole is not slotted, so it is not possible to adjust the position of the distributor.
5 Remove the seal from the base of the distributor **(see illustration)**.

Inspection

6 Check the rotor arm and distributor cap as described in Chapter 1, and renew them if necessary.
7 Check the amount of play between the distributor shaft and body. This should not be excessive unless the engine has covered a high mileage. If it is excessive, the complete distributor should be renewed.
8 Check the condition of the distributor base seal, and renew it if necessary.

Refitting

9 Refitting is the reverse of removal. If the distributor cap is to be renewed, transfer the HT leads one by one to the new cap to ensure that they are correctly positioned.

5 Amplifier unit (models with a distributor) - removal and refitting

Removal

1 With the ignition switched off, disconnect the HT lead from the ignition coil, then disconnect the wiring connectors from the amplifier unit **(see illustration)**.
2 Unscrew and remove the mounting nuts, and remove the amplifier unit from its mounting studs on the engine compartment bulkhead. Note that one of the nuts secures the map sensor hose support.
3 If necessary undo the bolts and separate the ignition coil from the amplifier unit (see Section 3).

Refitting

4 Refitting is the reverse of removal.

6 Ignition timing - checking and adjustment

Due to the nature of the ignition system, the ignition timing is constantly being monitored and adjusted by the engine management ECU, and nominal values cannot be given. Therefore, it is not possible for the home mechanic to check the ignition timing.

The only way in which the ignition timing can be checked is using special electronic test equipment, connected to the engine management system diagnostic connector (refer to Chapter 4A). No adjustment of the ignition timing is possible. Should the ignition timing be incorrect, then a fault must be present in the engine management system.

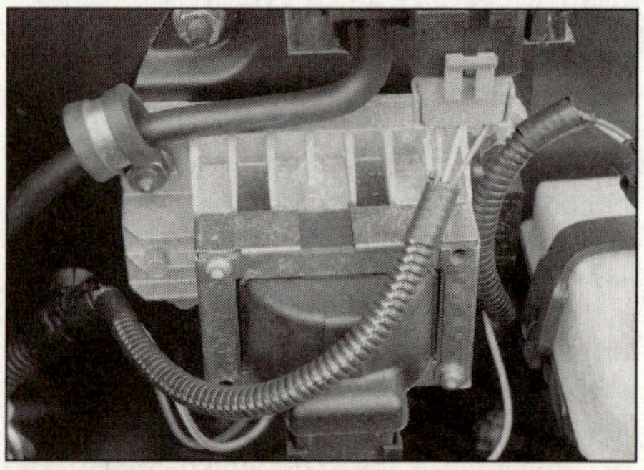

5.1 Amplifier and HT coil mounted on the bulkhead (E7J engine)

Chapter 5 Part C:
Pre/post-heating system - diesel engine models

Contents

Degrees of difficulty

Easy, suitable for novice with little experience	**Fairly easy,** suitable for beginner with some experience	**Fairly difficult,** suitable for competent DIY mechanic	**Difficult,** suitable for experienced DIY mechanic	**Very difficult,** suitable for expert DIY or professional

Specifications

Torque wrench settings	**Nm**	**lbf ft**
Glow plugs	20	15
Glow plug terminal nut	4	3

1 Pre/post-heating system - description and testing

Description

1 Each swirl chamber has a heater plug (commonly called a glow plug) screwed into it. The plugs are electrically operated before, during, and a short time after start-up when the engine is cold. Electrical feed to the glow plugs is controlled by the pre/post-heating control unit.

2 The glow plugs also have a "post-heating" function, whereby they remain switched on for a period after the engine has started. Once the starter is switched off, the glow plugs are supplied with full current for 10 seconds, then the plugs are supplied with current for up to 3 minutes depending on the engine coolant temperature and throttle position. Current to the glow plugs is controlled by the pre/post-heating unit which incorporates an air temperature sensor, and additionally controls the air conditioning control solenoid valve and EGR control solenoid valve **(see illustration)**. The supply to the plugs will be interrupted by:

a) Opening of the "post-heating cut-out" microswitch (also cuts out EGR where applicable). The supply is cut off 3 seconds after the switch opens (ie 3 seconds after the accelerator pedal is depressed). The current supply is restored as soon as the switch closes.

b) A coolant temperature of more than 65°C.

3 A warning light in the instrument panel tells the driver that pre-heating is taking place. When the light goes out, the engine is ready to be started. The voltage supply to the glow plugs continues for up to 8 seconds after the light goes out. If no attempt is made to start, the timer then cuts off the supply in order to avoid draining the battery and overheating of the glow plugs.

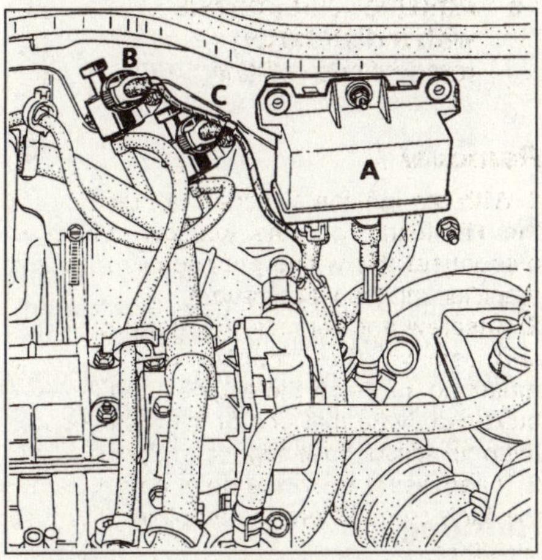

1.2 Pre/post-heating control unit (A), air conditioning control solenoid valve (B) and EGR control solenoid valve (C)

Testing

4 If the system malfunctions, testing is ultimately by substitution of known good units, but some preliminary checks may be made as follows.

5 Connect a voltmeter or 12 volt test lamp between the glow plug supply cable and earth (engine or vehicle metal). Make sure that the live connection is kept clear of the engine and bodywork.

6 Have an assistant switch on the ignition and check that voltage is applied to the glow plugs. Note the time for which the warning light is lit and the total time for which voltage is applied before the system cuts out. Switch off the ignition.

7 At an under-bonnet temperature of 20°C typical times noted should be 5 or 6 seconds for warning light operation, followed by a further 4 to 5 seconds supply after the light goes out (provided the starter motor is not operated). Warning light time will increase with lower temperatures and decrease with higher temperatures.

8 If there is no supply at all, the relay or associated wiring is at fault.

9 To locate a defective glow plug, disconnect the main supply cable and the interconnecting wire or strap from the top of the glow plugs. Be careful not to drop the nuts and washers.

10 Use a continuity tester, or a 12 volt test lamp connected to the battery positive terminal, to check for continuity between each

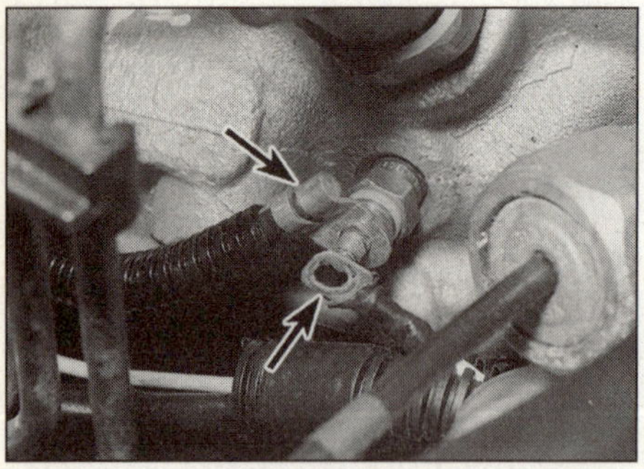

2.2 Disconnecting the wiring from the glow plug terminal

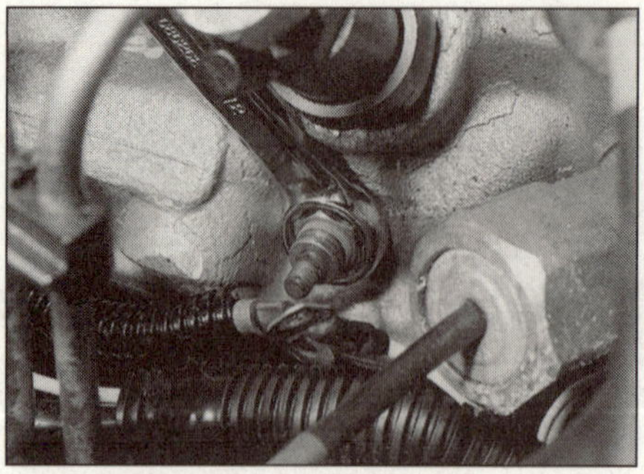

2.4a Unscrew the glow plug . . .

2.4b . . . and remove it from the cylinder head

glow plug terminal and earth. The resistance of a glow plug in good condition is very low (less than 1 ohm), so if the test lamp does not light or the continuity tester shows a high resistance the glow plug is certainly defective.

11 If an ammeter is available, the current draw of each glow plug can be checked. After an initial surge of around 15 to 20 amps, each plug should draw around 10 amps. Any plug which draws much more or less than 10 amps is probably defective.

12 As a final check, the glow plugs can be removed and inspected as described in Section 2.

2 Glow plugs - removal, inspection and refitting

Removal

Caution: If the pre-heating system has just been energised, or if the engine has been running, the glow plugs may be very hot.

1 Disconnect the battery negative lead.

Caution: If the radio/cassette in your vehicle is equipped with an anti-theft system, make sure you have the correct activation code before disconnecting the battery.

2 Unscrew the nuts from the glow plug terminals, and recover the washers. Disconnect the wiring **(see illustration)**. Note that the main electrical feed wiring is connected to two of the plugs.

3 Where applicable, carefully move any obstructing pipes or wires to one side to enable access to the glow plugs.

4 Unscrew the glow plugs and remove them from the cylinder head **(see illustrations)**.

Inspection

5 Inspect the glow plugs for physical damage. Burnt or eroded glow plug tips can be caused by a bad injector spray pattern. Have the injectors checked if this sort of damage is found.

6 If the glow plugs are in good physical condition, check them electrically using a 12 volt test lamp or continuity tester as described in the previous Section.

7 The glow plugs can be energised by applying 12 volts to them to verify that they heat up evenly and in the required time. Observe the following precautions:

a) *Support the glow plug by clamping it carefully in a vice or self-locking pliers. Remember it will become red-hot.*
b) *Make sure that the power supply or test lead incorporates a fuse or overload trip to protect against damage from a short-circuit.*
c) *After testing, allow the glow plug to cool for several minutes before attempting to handle it.*

8 A glow plug in good condition will start to glow red at the tip after drawing current for 5 seconds or so. Any plug which takes much longer to start glowing, or which starts glowing in the middle instead of at the tip, is defective.

Refitting

9 Refit by reversing the removal operations. Apply a smear of copper-based anti-seize compound to the plug threads and tighten the glow plugs to the specified torque. Do not overtighten, as this can damage the glow plug element.

3 Pre/post-heating system control unit - removal and refitting

Removal

1 The unit is located on the engine compartment bulkhead **(see illustration)**. Disconnect the battery negative lead.

2 Unscrew the securing nut, and withdraw the unit from the bracket on the bulkhead.

3 Disconnect the wiring from the base of the unit, noting the connector locations.

Refitting

4 Refitting is a reversal of removal, ensuring that the wiring connectors are correctly connected.

4 Post-heating cut-out (no-load) microswitch - testing, adjustment, removal and refitting

Testing and adjustment

Note: *A continuity tester or an ohmmeter will be required for testing.*

1 Ensure that the idle speed and anti-stall settings are correctly adjusted as described in Chapter 1B.

2 To test and adjust the switch, first disconnect the switch wiring connector.

Bosch injection pump

3 Connect a continuity tester or an ohmmeter across the switch wiring connector terminals B1 and C1.

4 Insert a 4.5 mm feeler blade between the accelerator lever and the maximum engine speed stop screw, and check that the microswitch contacts are open (infinity resistance) **(see illustration)**.

5 Now insert a 5.5 mm feeler blade and check that the contacts are closed (zero resistance).

6 If necessary, loosen the mounting bolts, adjust the position of the switch, then tighten the mounting bolts.

7 Repeat the procedure described in paragraphs 4 and 5.

8 If the switch is permanently open or closed, renew it.

3.1 Pre/post-heating control unit located on the bulkhead

4.4 Post-heating cut-out (no-load) microswitch adjustment

1 Microswitch for cutting out post-heating and EGR
2 Microswitch for cutting out air conditioning (where applicable)
A Accelerator lever
B Maximum engine speed stop screw
X Adjustment shim location

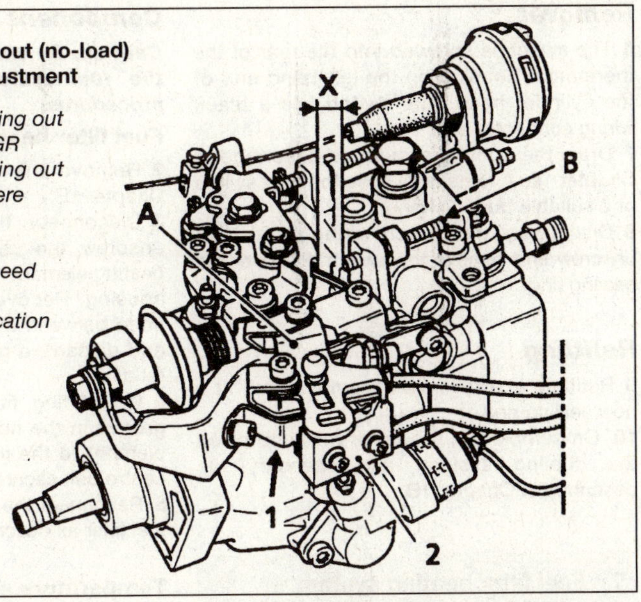

Lucas injection pump

9 Connect a continuity tester or an ohmmeter across wiring connectors "B" and "C".

10 Insert feeler blades between the injection pump accelerator lever and the anti-stall adjustment screw, and note the readings on the continuity tester or ohmmeter **(see illustrations)**. The readings obtained should be as follows.

Spacer thickness	Test reading
Up to 9.0 mm	*Continuity/zero resistance*
More than 11.0 mm	*No continuity/infinite resistance*

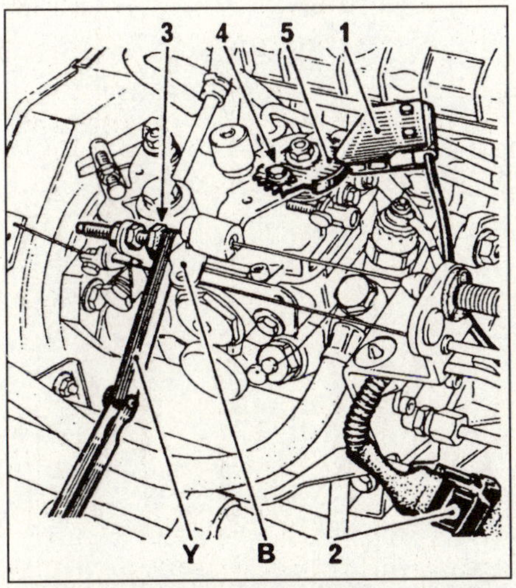

4.10a Adjusting the post-heating cut-out (no-load) microswitch on the Lucas injection pump

B Accelerator lever
Y Feeler blades
1 Microswitch
2 Switch wiring connector
3 Anti-stall adjustment screw
4 Bolt
5 Cam

11 If necessary, loosen the mounting bolts, adjust the position of the switch, then tighten the mounting bolts. Note that on certain injection pumps, the adjustment is made by loosening the screw and adjusting the position of the cam in relation to the accelerator lever.

12 Repeat the procedure in paragraph 10.

13 If the switch is permanently open or closed, renew it.

Removal

14 Disconnect the switch wiring connector, then remove the securing screws, and withdraw the switch from the bracket on the injection pump.

Refitting

15 Refitting is a reversal of removal. Adjust the switch as described above before tightening the retaining screws.

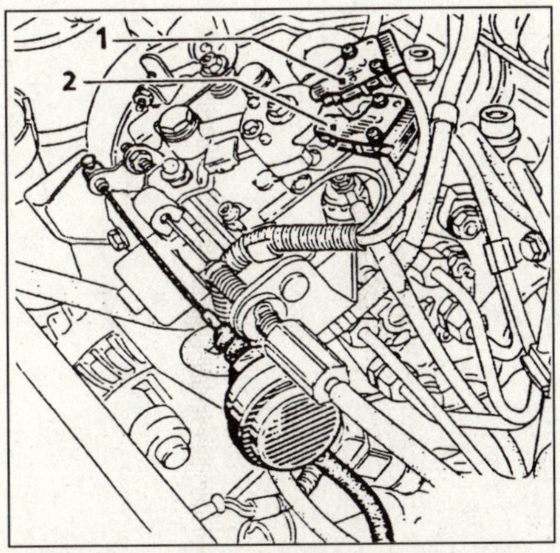

4.10b Post-heating (no-load) microswitch location on Turbo models fitted with EGR
1 Microswitch *2 EGR cut-off switch*

5 Air conditioning cut-off microswitch - testing, adjustment, removal and refitting

Testing and adjustment

Note: *A continuity tester or an ohmmeter will be required for testing.*

1 The cut-off switch is fitted in order to switch off the air conditioning compressor during starting and warm-up. The switch is located on the front of the injection pump.

2 To test and adjust the switch, first disconnect the switch wiring connector.

Bosch injection pump

3 Connect a continuity tester or an ohmmeter across the switch wiring connector terminals B2 and C2.

4 Insert a 4.5 mm feeler blade between the accelerator lever and the maximum engine speed stop screw, and check that the microswitch contacts are open (infinity resistance).

5 Now insert a 5.5 mm feeler blade and check that the contacts are closed (zero resistance).

6 If necessary, loosen the mounting bolts, adjust the position of the switch, then tighten the mounting bolts.

7 Repeat the procedure described in paragraphs 4 and 5.

8 If the switch is permanently open or closed, renew it.

Lucas injection pump

9 Connect a continuity tester or an ohmmeter across wiring connectors "B" and "C".

10 Insert feeler blades between the injection pump accelerator lever and the anti-stall adjustment screw, and note the readings on the continuity tester or ohmmeter. The readings obtained should be as follows.

Spacer thickness	Test reading
Up to 27.0 mm	*Continuity/zero resistance*
More than 29.0 mm	*No continuity/infinite resistance*

11 If necessary, loosen the mounting bolts, adjust the position of the switch, then tighten the mounting bolts. Note that on certain injection pumps, the adjustment is made by loosening the screw and adjusting the position of the cam in relation to the accelerator lever.

12 Repeat the procedure in paragraph 10.

13 If the switch is permanently open or closed, renew it.

Removal

14 Disconnect the switch wiring plug, then remove the screws, and withdraw the switch from the bracket on the injection pump.

Refitting

15 Refitting is a reversal of removal. Adjust the switch as described above before tightening the retaining screws.

6 Coolant temperature switch (post-heating cut-off) - testing, removal and refitting

Testing

Note: *A continuity tester or an ohmmeter will be required for testing.*

1 The switch has two functions. The thermistor part of the switch varies in resistance according to the temperature of the coolant, and is connected to the pre/post-heating control unit. The temperature switch part is used to operate the EGR valve. The switch must be removed as described in paragraphs 7 to 9 for testing.

2 Connect a continuity tester or an ohmmeter across the switch terminals 1 and 4. There should be infinity resistance between the terminals at room temperature proving that the contacts are open.

3 Now suspend the switch in a container of water (using wire or string), along with a thermometer. Heat the water and check that the contacts remain open up to 20°C but close at temperatures above 30°C. With the contacts closed, the ohmmeter must show zero resistance or the tester show continuity.

4 Connect an ohmmeter across the switch terminals 2 and 3. Heat the water and check that the resistance of the thermistor is 2610 to 2990 ohms at 25°C, and 310 to 350 ohms at 80°C.

5 If the results obtained are not as specified, the switch is proved faulty and should be renewed.

Removal

6 The switch is screwed into the rear of the thermostat housing on the left-hand end of the cylinder head. The switch has a black wiring connector.

7 Drain the cooling system as described in Chapter 1B. Alternatively have the new switch or a suitable bung to hand.

8 Disconnect the switch wiring plug, then unscrew and remove the switch. Recover the sealing ring.

Refitting

9 Refitting is a reversal of removal, but fit a new sealing ring.

10 On completion, top-up or refill and bleed the cooling system, as necessary, as described in Chapter 1B.

7 Fuel filter heating system - general information and component renewal

General information

1 An electrically-operated heating element is fitted in the fuel filter housing, to prevent the fuel "waxing" at low temperatures. The heater is controlled by an internal thermostat. When the fuel temperature is below 0°C, current to the heater warms the fuel in the filter housing and filter. When the fuel temperature reaches 8°C, current supply to the heater ceases.

Component renewal

Caution: Be careful not to allow dirt into the fuel system during the following procedures.

Fuel filter heating element

2 Remove the fuel filter as described in Chapter 1B.

3 Disconnect the wiring connector then unscrew the centre bolt and remove the heating element from the base of the fuel filter housing. Recover the sealing ring which is fitted between the element and filter housing and discard; a new one should be used on refitting.

4 On refitting, fit the new sealing ring to the groove in the heating element and refit the element to the filter housing, tightening the centre bolt securely.

5 Reconnect the wiring connector and fit the fuel filter as described in Chapter 1B.

Temperature switch

6 Disconnect the wiring connector from the temperature switch.

7 Wipe clean the area around the temperature switch and fuel filter housing. Position a container to catch any spilt fuel then slacken and remove the switch from the fuel filter housing. Recover the sealing washers which are fitted on each side of the fuel hose union and discard them; new ones must be used on refitting.

8 Refitting is the reverse of removal, positioning a new sealing washer on each side of the fuel hose union.

Chapter 6
Clutch

Contents

Degrees of difficulty

Easy, suitable for novice with little experience	**Fairly easy,** suitable for beginner with some experience	**Fairly difficult,** suitable for competent DIY mechanic	**Difficult,** suitable for experienced DIY mechanic	**Very difficult,** suitable for expert DIY or professional

Specifications

General

Clutch type	Single dry plate, diaphragm spring, cable-operated with automatic adjustment

Clutch disc

Diameter:	
Petrol models:	
1390 cc engine	181.5 mm
1598 cc engine	200.0 mm
Diesel models	200.0 mm
Friction material thickness (new)	8.3 mm (approximate)

Torque wrench settings

Torque wrench settings	Nm	lbf ft
Pressure plate bolts:		
7 mm diameter	18	13
8 mm diameter	22	17

1 General information

The clutch consists of a friction plate, a pressure plate assembly, a release bearing and the release fork; all of these components are contained in the large cast-aluminium alloy bellhousing, sandwiched between the engine and the transmission.

The friction plate is fitted between the engine flywheel and the clutch pressure plate, and is allowed to slide on the transmission input shaft splines.

The pressure plate assembly is bolted to the engine flywheel. When the engine is running, drive is transmitted from the crankshaft, via the flywheel, to the friction plate (these components being clamped securely together by the pressure plate assembly) and from the friction plate to the transmission input shaft.

To interrupt the drive, the spring pressure must be relaxed by the cable-operated release mechanism. Depressing the clutch pedal pulls the control cable inner wire, and this pivots the release fork. The release fork then presses the release bearing against the pressure plate spring fingers. This causes the springs to deform, and releases the clamping force on the pressure plate.

When the pedal is released, the diaphragm spring forces the pressure plate into contact with the friction linings on the clutch disc. The disc is now firmly sandwiched between the pressure plate and the flywheel, thus transmitting engine power to the transmission.

Wear of the friction material on the clutch disc is automatically compensated for by an adjustment mechanism attached to the clutch pedal. The mechanism consists of a serrated quadrant, a notched cam and a tension spring. One end of the clutch cable is attached to the quadrant, which is free to pivot on the pedal, but is kept in tension by a spring. When the pedal is depressed, the notched cam contacts the quadrant, thus locking it and allowing the pedal to pull the cable and operate the clutch. When the pedal is released, the tension spring causes the notched cam to move free of the quadrant; at the same time tension is maintained on the cable, keeping the release bearing in contact with the diaphragm spring. As the friction material on the disc wears, the self-adjusting quadrant will rotate when the pedal is released, and the pedal free play will be maintained between the notched cam and the quadrant.

Some later models may be fitted with an automatically-operated clutch which operates when the gear lever is depressed. At the time of writing, no information was available on this type of clutch.

2 Clutch cable - removal and refitting

Removal

1 Remove the clutch pedal as described in Section 3 and tie a length of string to the end of the clutch cable.

2 On all engines except E7J, from within the engine compartment, remove the air cleaner assembly and air ducting as described in Chapter 4.

3 With the inner cable already disconnected, release the outer cable from its bracket on the transmission, then work along its length, freeing it from the retaining clips and ties whilst noting its correct routing.

4 Release the clutch cable end fitting from the bulkhead and remove the cable from the vehicle. When the end of the cable appears, untie the string and leave it in position; it can then be used to draw the cable back through into position on refitting.

5 Examine the cable, looking for worn end fittings or a damaged outer casing, and for signs of fraying of the inner wire. Check the cable's operation; the inner wire should move smoothly and easily through the outer casing. Remember that a cable that appears serviceable when tested off the car may well be much heavier in operation when in its working position. Renew the cable if it shows signs of excessive wear or any damage.

Refitting

6 Manoeuvre the cable into position, making sure it is correctly routed, and use the string to draw the cable end through into the passenger compartment. Untie the string, and seat the cable end fitting securely in the bulkhead.

7 Secure the cable in position with all the relevant clips and ties, and locate the transmission end of the outer cable in its mounting bracket.

8 Refit the clutch pedal and check the operation of the automatic adjustment mechanism as described in Section 3.

9 Refit the air cleaner assembly and air ducting.

3 Clutch pedal - checking, removal and refitting

Checking

1 The clutch pedal assembly incorporates the automatic adjustment mechanism, and the operation of the mechanism can be checked as follows.

2 First measure the clutch cable freeplay. To check this, with the clutch pedal in the at-rest position, pull out the transmission end of the clutch inner cable and measure the length of exposed cable. If the mechanism is operating correctly, there should be at least 20 mm freeplay in the cable **(see illustration)**. If there is less than this, the automatic adjustment quadrant should be checked for seizure or possible restricted movement.

3 Have an assistant depress and release the clutch pedal fully whilst you measure the total movement of the release fork end; if the mechanism is operating correctly, this should be between 27.4 and 30.7 mm **(see illustration)**. This movement ensures that the clutch pedal stroke is correct. If not, make sure that the automatic adjustment mechanism teeth are undamaged, and that the spring has not lost its tension. Also check that the inner cable is not seizing in the outer cable.

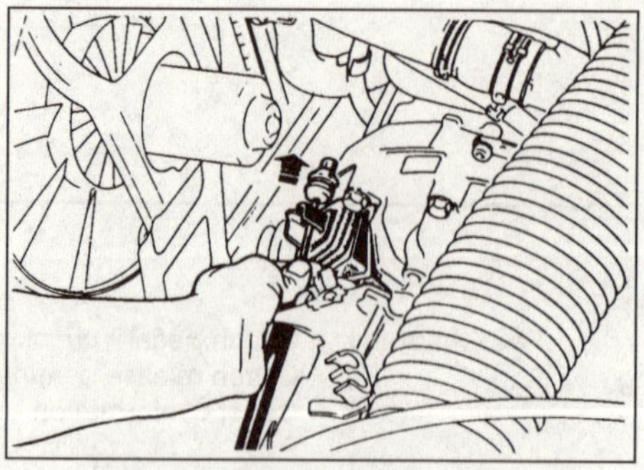

3.2 Checking the clutch inner cable freeplay at the release fork end

Removal

4 Disconnect the battery negative lead.

Caution: If the radio/cassette in your vehicle is equipped with an anti-theft system, make sure you have the correct activation code before disconnecting the battery.

5 On all engines except E7J, from within the engine compartment, remove the air cleaner assembly and air ducting as described in Chapter 4.

6 Working in the engine compartment, unhook the end of the clutch cable from the release fork on the transmission **(see illustration)**.

7 From inside the vehicle, where applicable remove the facia lower cover to gain access to the clutch pedal assembly.

8 Unscrew and remove the nut and washer from the pedal pivot bolt **(see illustration)**. Where applicable, undo the retaining nut/bolt and remove the support bracket from the end of the pivot bolt.

9 Partially withdraw the clutch/brake pedal pivot shaft until it is possible to remove the clutch pedal along with its pivot bushes and spacer. Note the positions of the pivot bushes to ensure correct refitting.

10 Detach the clutch inner cable from the top of the pedal, then free it from the quadrant and remove the pedal assembly from the vehicle. With the clutch pedal removed, slide the pivot bolt back into position to securely retain the brake pedal.

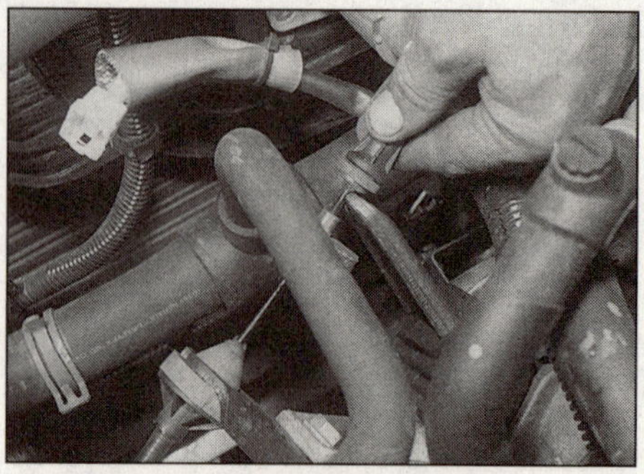

3.6 Releasing the end of the clutch cable from the release fork

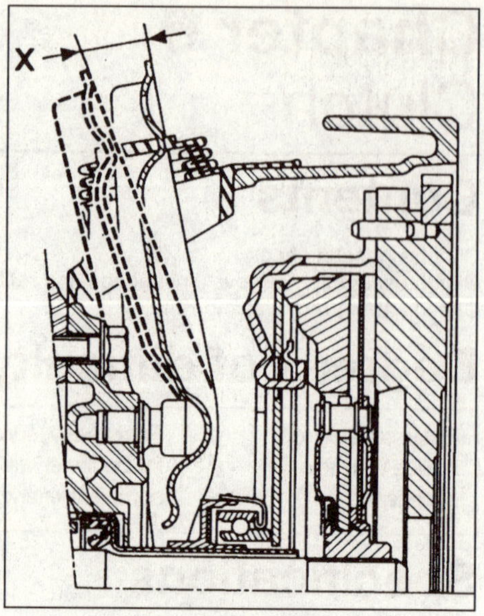

3.3 Checking the clutch release fork movement (X)
For dimension X, see text

11 Examine the pedal assembly for signs of wear or damage, paying particular attention to the automatic adjustment mechanism. Do not attempt to dismantle the assembly as no components are available separately; the pedal assembly must be renewed as a unit if damaged. Renew the pivot bushes if they show signs of wear or damage.

Refitting

12 Apply some multi-purpose grease to the bearing surfaces of the pedal spacer and bushes, and also to the surface of the adjustment mechanism teeth.

13 Fit the bushes to the pedal, making sure they are located in the previously noted positions. Ensure that the locating lug of each bush is correctly engaged with the pedal cut-out, then slide the spacer into position.

14 Manoeuvre the pedal assembly into position under the facia. Engage the clutch cable end fitting with the self-adjust mechanism, then seat the pedal in position.

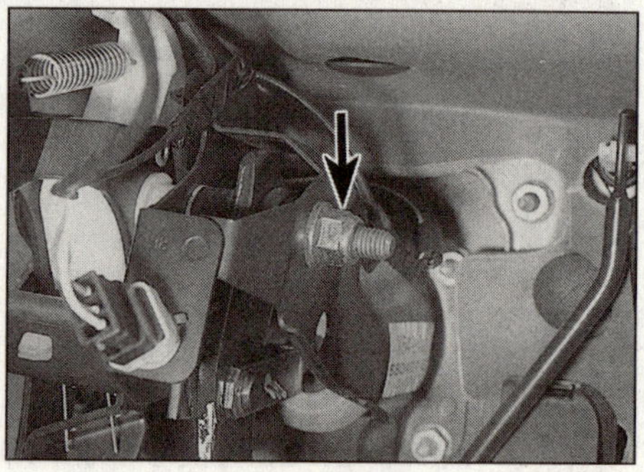

3.8 Nut on the clutch/brake pedal pivot bolt

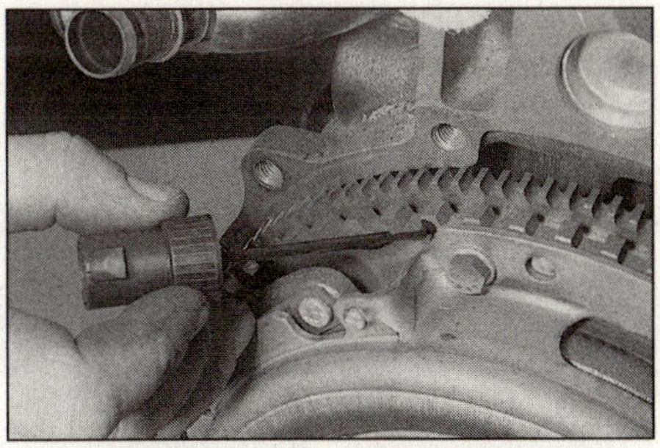

4.2 Mark the relationship of the clutch pressure plate and flywheel before removing the pressure plate

4.3a Slacken the pressure plate bolts progressively . . .

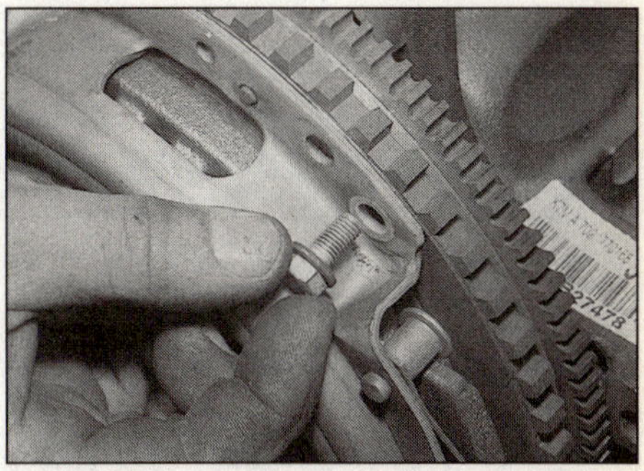

4.3b . . . before removing them

15 Ensure the clutch cable is correctly seated in the adjustment mechanism quadrant, then refit the support bracket (where fitted), tightening its nut/bolt lightly only at this stage.
16 Slide the pivot bolt fully into position, then refit the washer and nut. Tighten both the pivot bolt and support bracket nut/bolt securely.
17 Hook the transmission end of the cable back onto the release fork.
18 Depress the clutch pedal several times, whilst ensuring that the cable remains correctly seated in the quadrant. This will allow the automatic adjustment mechanism to set the correct free play/pedal stroke.
19 Check the operation of the adjustment mechanism as described in paragraphs 1 to 3.
20 Refit the facia lower cover, the air cleaner assembly and air ducting (as applicable), and reconnect the battery negative lead.

4 Clutch assembly - removal, inspection and refitting

Warning: Dust created by clutch wear and deposited on the clutch components may contain asbestos, which is a health hazard. DO NOT blow it out with compressed air, nor inhale any of it. DO NOT use petrol or petroleum-based solvents to clean off the dust. Brake system cleaner or methylated spirit should be used to flush the dust into a suitable receptacle. After the clutch components are wiped clean with rags, dispose of the contaminated rags and cleaner in a sealed, marked container.

Removal

1 The complete engine/transmission assembly must be removed from the car, and the transmission separated on the bench as described in Chapter 2A or 2B (as applicable).
2 Before disturbing the clutch, use paint or a marker pen to mark the relationship of the pressure plate assembly to the flywheel **(see illustration)**.
3 Working in a diagonal sequence, slacken the pressure plate bolts by half a turn at a time, until spring pressure is released and the bolts can be unscrewed by hand **(see illustrations)**.
4 Prise the pressure plate assembly off its locating dowels, and collect the friction plate, noting which way round the friction plate is fitted.

Inspection

Note: *Due to the amount of work necessary to remove and refit clutch components, it is usually considered good practice to renew the clutch friction plate, pressure plate assembly and release bearing as a matched set, even if only one of these is actually worn enough to require renewal.*
5 When cleaning clutch components, read first the warning at the beginning of this Section; remove any dust using a clean, dry cloth, and working in a well-ventilated atmosphere.
6 Check the friction plate facings for signs of wear, damage or oil contamination. If the friction material is cracked, burnt, scored or damaged, or if it is contaminated with oil or grease (shown by shiny black patches), the friction plate must be renewed. Check the depth of the rivets below the friction material surface; if the friction material is worn down to, or close to (less than approximately 0.5 mm) the rivet heads, then the friction plate must be renewed.
7 If the friction material is still serviceable, check that the centre boss splines are unworn, that the torsion springs are in good condition and securely fastened, and that all the rivets are tight. If any wear or damage is found, the friction plate must be renewed.
8 If the friction material is fouled with oil, this must be due to an oil leak from the crankshaft left-hand oil seal, from the sump-to-cylinder block joint, or from the transmission input shaft. Renew the seal or repair the joint, as appropriate, as described in Chapter 2 or 7, before installing the new friction plate.
9 Check the pressure plate assembly for obvious signs of wear or damage; shake it to check for loose rivets or worn or damaged fulcrum rings, and check that the drive straps securing the pressure plate to the cover do not show signs (such as a deep yellow or blue discoloration) of overheating. If the diaphragm spring is worn or damaged, or if its pressure is in any way suspect, the pressure plate assembly should be renewed.
10 Examine the machined bearing surfaces of the pressure plate and of the flywheel; they should be clean, completely flat, and free from scratches or scoring. If either is discoloured from excessive heat, or shows signs of cracks, it should be renewed - although minor damage of this nature can sometimes be polished away using emery paper.
11 Check that the release bearing contact surface rotates smoothly and easily, with no sign of noise or roughness. Also check that the surface itself is smooth and unworn, with no signs of cracks, pitting or scoring. If there is any doubt about its condition, the bearing must be renewed.

Refitting

12 On reassembly, ensure that the bearing surfaces of the flywheel and pressure plate are completely clean, smooth, and free from oil or grease. Use solvent to remove any protective grease from new components.
13 Fit the friction plate so that its spring hub assembly faces away from the flywheel; there may also be a marking showing which way round the plate is to be refitted. If required at this stage, the friction plate may be held in position using a clutch aligning tool **(see illustration)**.
14 Refit the pressure plate assembly, aligning the marks made on dismantling (if the original pressure plate is re-used), and locating the pressure plate on its three

4.13 Using the clutch aligning tool to hold the friction plate in position

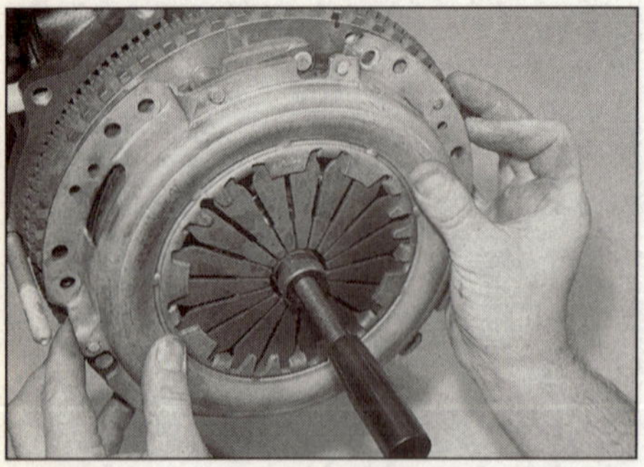

4.14 Refitting the pressure plate assembly

locating dowels **(see illustration)**. Fit the pressure plate bolts, but tighten them only finger-tight, so that the friction plate can still be moved.

15 The friction plate must now be centralised, so that when the transmission is refitted, its input shaft will pass through the splines at the centre of the friction plate.

16 Centralisation can be achieved by passing a screwdriver or other long bar through the friction plate and into the hole in the crankshaft; the friction plate can then be moved around until it is centred on the crankshaft hole. Alternatively, a clutch-aligning -tool can be used to eliminate the guesswork; these can be obtained from most accessory shops. A home-made aligning tool can be fabricated from a length of metal rod or wooden dowel which fits closely inside the crankshaft hole, and has insulating tape wound around it to match the diameter of the friction plate splined hole. **Note:** *If a genuine Renault replacement clutch kit is being fitted a centring tool will be included.*

17 When the friction plate is centralised, tighten the pressure plate bolts evenly and in a diagonal sequence to the specified torque setting. Remove the centring tool.

18 Apply a thin smear of molybdenum disulphide grease (Renault recommend the use of Molykote BR2 - available from your Renault dealer) to the contact areas of the release bearing, guide tube and fork.

Caution: Do not apply too much grease, as there is a risk that it will contaminate the friction plate material.

19 Refit the transmission to the engine, then refit the assembly to the car as described in Chapter 2A or 2B (as applicable).

5 Clutch release bearing - removal, inspection and refitting

Removal

1 For access to the clutch release bearing components, the complete engine/transmission assembly must be removed from the car, and the transmission separated on the bench as described in Chapter 2A or 2B (as applicable).

2 Tilt the release fork, and at the same time slide the bearing assembly off the transmission input shaft guide tube **(see illustration)**.

3 Lift the fork off its pivot stud, and withdraw it from inside the transmission bellhousing. If necessary, remove the rubber gaiters from the bellhousing aperture and from the pivot stud.

Inspection

4 Check the release fork, bearing and rubber gaiters, renewing any component which is worn or damaged. Carefully check all bearing surfaces and points of contact.

5 When checking the release bearing itself, note that it is often considered worthwhile to renew it as a matter of course. Check that the contact surface rotates smoothly and easily, with no sign of noise or roughness, and that the surface itself is smooth and unworn, with no signs of cracks, pitting or scoring. If there is any doubt about its condition, the bearing must be renewed. **Note:** *Due to the amount of work necessary to remove and refit clutch components, it is usually considered good practice to renew the clutch friction plate, pressure plate assembly and release bearing as a matched set, even if only one of these is actually worn enough to require renewal.*

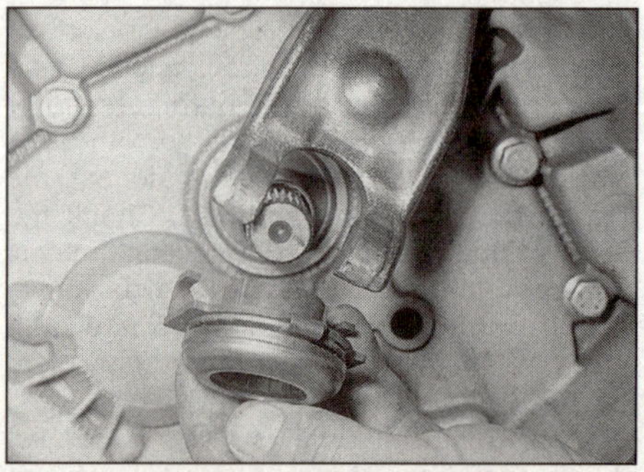

5.2 Removing the release bearing from the guide tube and fork

Refitting

6 Prior to refitting, smear molybdenum disulphide grease (Renault recommend the use of Molykote BR2 - available from your Renault dealer) to the contact surfaces of the release bearing, release fork and pivot stud and the transmission input shaft guide tube **(see illustration)**.

Caution: Do not apply too much grease, as there is a risk that it will contaminate the friction plate material.

7 Ensure the rubber gaiters are correctly located in the bellhousing aperture and in the groove on the stud, then locate the fork through the bellhousing gaiter and onto the stud.

8 Locate the bearing assembly onto the guide tube, and at the same time, engage the release fork ends with the lugs on the bearing. Slide the bearing fully onto the guide tube and make sure the dimple on the release fork locates fully on the stud.

9 Refit the transmission to the engine, then refit the assembly to the car as described in Chapter 2A or 2B (as applicable). Check the operation of the release mechanism as soon as the transmission is attached to the engine and before refitting the assembly.

5.6 Smear molybdenum disulphide grease on the pivot stud before refitting the release fork

Chapter 7 Part A: Manual transmission

Contents

Degrees of difficulty

Easy, suitable for novice with little experience	**Fairly easy,** suitable for beginner with some experience	**Fairly difficult,** suitable for competent DIY mechanic	**Difficult,** suitable for experienced DIY mechanic	**Very difficult,** suitable for expert DIY or professional

Specifications

General

Type	Manual, five forward speeds and reverse. Synchromesh on all forward speeds
Designation:	
Petrol engine models	JB1
Diesel engine models	JB1 or JC5 (depending on model)

Note: *Transmission code is stamped on a plate attached to the transmission (see Section 1).*

Lubrication

Capacity (approximate):	
JB1 transmission	3.4 litres
JC5 transmission	3.1 litres
Recommended oil	See *"Lubricants, fluids and tyre pressures"*

Torque wrench settings

Torque wrench settings	Nm	lbf ft
Clutch release bearing guide tube bolts - JC5 transmission	24	18
Engine-to-transmission bolts/nuts	35	26
Engine/gearbox mountings for petrol engines:		
Rear mounting bracket to transmission	65	48
Rear mounting to subframe	75	55
RH and LH front engine mountings to subframe	45	33
Engine/gearbox mountings for diesel engines:		
Engine mounting bolt on transmission	55	41
Engine mounting bolt on front left-hand side member	25	19
Front right-hand engine mounting cover bolt on engine	40	30
Front right-hand engine mounting movement limiter bolt	60	44
Gearchange lever assembly to floor	15	11
Gearchange link rod:		
Link rod clamp bolt	30	22
Link rod to transmission bolt	30	22
Roadwheel bolts	90	66

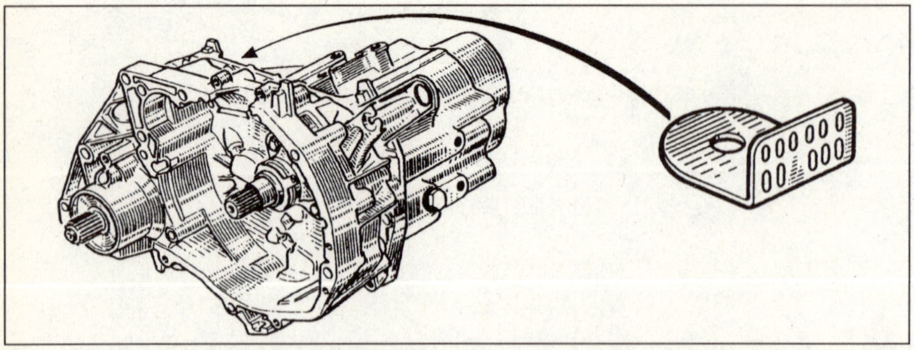

1.1a Transmission identification plate location on the transmission

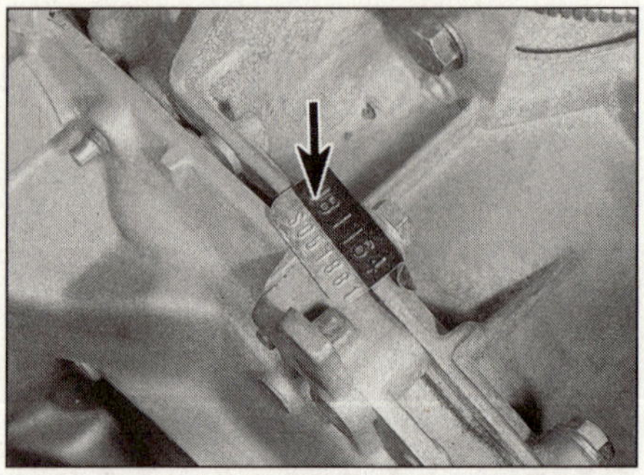

1.1b The identification number starts with the transmission code

1 General information

The transmission is contained in a cast-aluminium alloy casing bolted to the engine's left-hand end, and consists of the gearbox and final drive differential - often called a "transaxle". The transmission type is stamped on an identification plate attached to the transmission **(see illustrations)**.

Drive is transmitted from the crankshaft via the clutch to the input shaft, which has a splined extension to accept the clutch friction plate, and rotates in sealed ball-bearings. From the input shaft, drive is transmitted to the output shaft, which rotates in a roller bearing at its right-hand end, and a sealed ball-bearing at its left-hand end. From the output shaft, the drive is transmitted to the differential crownwheel, which rotates with the differential case and planetary gears, thus driving the sun gears and driveshafts. The rotation of the planetary gears on their shaft allows the inner roadwheel to rotate at a slower speed than the outer roadwheel when the car is cornering.

The input and output shafts are arranged side by side, parallel to the crankshaft and driveshafts, so that their gear pinion teeth are in constant mesh. In the neutral position, the output shaft gear pinions rotate freely, so that drive cannot be transmitted to the crownwheel.

Gear selection is via a floor-mounted lever and selector rod mechanism.

The transmission selector rod causes the appropriate selector fork to move its respective synchro-sleeve along the shaft, to lock the gear pinion to the synchro-hub. Since the synchro-hubs are splined to the output shaft, this locks the pinion to the shaft, so that drive can be transmitted. To ensure that gear-changing can be made quickly and quietly, a synchro-mesh system is fitted to all forward gears, consisting of baulk rings and spring-loaded fingers, as well as the gear pinions and synchro-hubs. The synchro-mesh cones are formed on the mating faces of the baulk rings and gear pinions.

2 Manual transmission - draining and refilling

Note: *The filler plug is also used as the level plug.*

1 This operation is much quicker and more efficient if the car is first taken on a journey of sufficient length to warm the engine/transmission up to operating temperature.

2 Park the car on level ground, switch off the ignition and apply the handbrake firmly. For improved access, jack up the front of the car and support it securely on axle stands (see *"Jacking and vehicle support"*). Where fitted, remove the engine undershield. Note that the car must be lowered to the ground and level, to ensure accuracy, when refilling and checking the oil level.

3 Remove all traces of dirt from around the plug, then unscrew the filler/level plug from the front face of the transmission. Recover the sealing washer.

4 Position a suitable container under the base of the transmission housing. Undo the screws and remove the protective cover from the transmission to gain access to the drain plug.

5 Unscrew the drain plug and allow the oil to drain completely into the container. If the oil is hot, take precautions against scalding. Clean both the filler/level and the drain plugs, being especially careful to wipe any metallic particles off the magnetic inserts. Discard the original sealing washers; they should be renewed whenever they are disturbed.

6 When the oil has finished draining, clean the drain plug threads and those of the transmission casing, fit a new sealing washer and refit the drain plug, tightening it securely. Refit the protective cover to the transmission. It the car was raised for the draining operation, now lower it to the ground.

7 Refilling the transmission is an extremely awkward operation. Above all, allow plenty of time for the oil level to settle properly before checking it. Note that the car must be parked on flat level ground when checking the oil level.

8 Refill the transmission with the exact amount of the specified type of oil, then check the oil level as described in Chapter 1. When the level is correct, refit the filler or filler level plug with a new sealing washer and tighten securely. **Note:** *If the correct amount was poured into the transmission, and a large amount flows out on checking the level, refit the filler/level plug and take the car on a short journey so that the new oil is distributed fully around the transmission components, then check the level again on your return.*

9 Where necessary, refit the engine undershield and lower the car to the ground.

3 Gearchange mechanism - adjustment

Note: *A special Renault service tool (B.Vi. 1133) will be required to accurately adjust the gearchange linkage.*

1 Firmly apply the handbrake, then jack up the front of the vehicle and support it securely on axle stands (see *"Jacking and vehicle support"*). Where fitted, remove the engine undershield.

2 Undo the retaining screws and remove the protective cover from the base of the transmission. Unbolt the heat shields in order to gain access to the bottom of the gear lever.

3 Select 1st gear on the transmission by moving the lever to the appropriate position. To hold the lever securely in position, fit the Renault service tool (B.Vi. 1133) as shown **(see illustration)**. In the absence of the

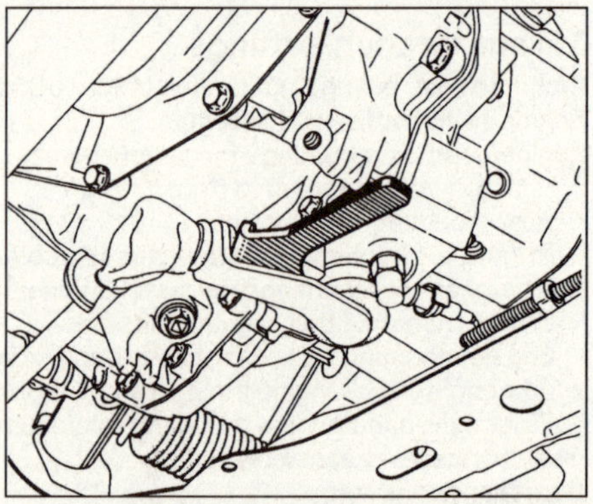

3.3 Using the special Renault tool (B.Vi. 1133) to hold the gearchange lever in position

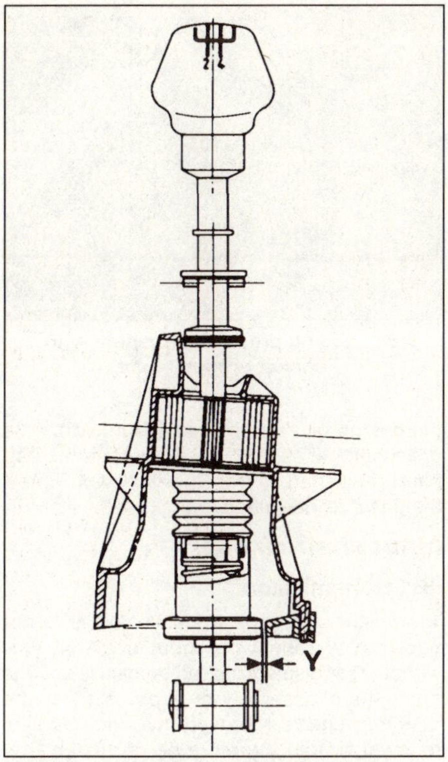

3.4 The clearance (Y) between the gear lever and housing should be 3 mm

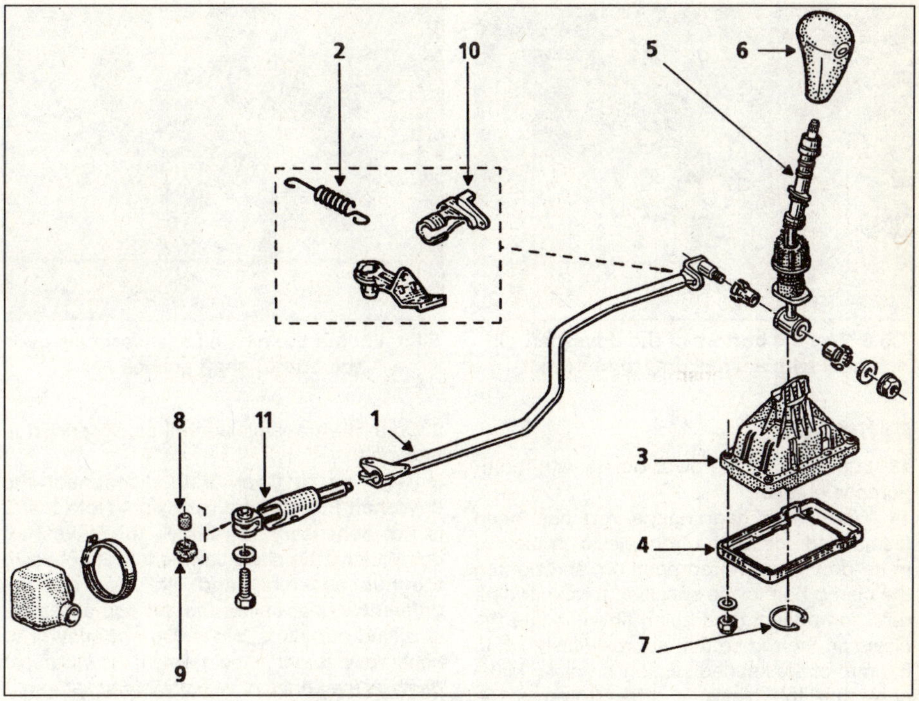

4.1 Gearchange linkage components

1 *Gearchange rod*
2 *Return spring*
3 *Base assembly*
4 *Retaining frame*
5 *Gear lever assembly*
6 *Gear lever knob*
7 *Circlip*
8 *Bush*
9 *Spacer*
10 *Retaining clip*
11 *Gearchange clevis*

special tool, a suitable alternative tool can be made from flat metal bar or a piece of wood.

4 Using a feeler blade, check that the clearance between the reverse stop-ring on the gear lever and the inclined plane on the right-hand side of the gear lever housing is as shown **(see illustration)**.

5 If adjustment is necessary, unhook the return spring from the gear lever end of the link rod, then loosen the clamp bolt at the transmission end of the link rod so that the rod can be moved.

6 Move the gear lever so that the reverse stop-ring is against the inclined plane on the housing, then insert a 3 mm feeler blade between the ring and plane. Hold the lever in this position, then tighten the clamp bolt to the specified torque setting.

7 Remove the holding tool, then refit the return spring and recheck the clearance as described in paragraph 4.

8 Check that all gears can be selected, then refit the heat shields, protective cover and/or undershield before lowering the vehicle to the ground.

4 Gearchange mechanism - removal and refitting

Removal

1 Working inside the vehicle, carefully prise the gearchange lever gaiter out from the centre console. Pull the knob off from the top of the gearchange lever and remove it along with the gaiter **(see illustration)**.

2 Firmly apply the handbrake, then jack up the front of the vehicle and support it on axle stands (see *"Jacking and vehicle support"*).

3 Unbolt and remove the heat shields from under the gear lever position. If necessary for improved access, remove the exhaust system as described in Chapter 4A or 4B.

4 If the exhaust is still in position, unscrew and remove the downpipe mounting bolts and remove the rubber mounting brackets.

5 Unhook the return spring from the gearchange rod or floor tunnel **(see illustration)**.

6 Unscrew the nut, then remove the washer and disconnect the gearchange rod from the bottom of the gear lever **(see illustration)**.

7 Unscrew the mounting nuts securing the gear lever base assembly to the floor, then lower the assembly and remove it from under the car. If the exhaust is still in position, it will be necessary to push it to one side while removing the gear lever assembly.

8 Remove the frame from the base.

9 Mount the gear lever in a soft-jawed vice, then extract the circlip and separate the base and gaiter.

10 To remove the gearchange rod, release the rubber gaiter and unscrew the bolt at the front of the rod. Note the location of the bush and spacer. Remove the rod from the transmission.

11 To separate the front clevis, first mark it in relation to the main rod, then loosen the clamp bolt and separate the two sections.

12 Examine all components for signs of wear or damage, and renew as necessary.

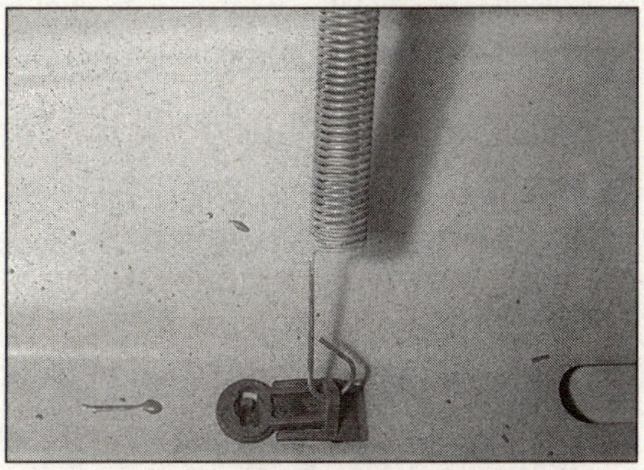

4.5 Return spring attachment to the floor tunnel

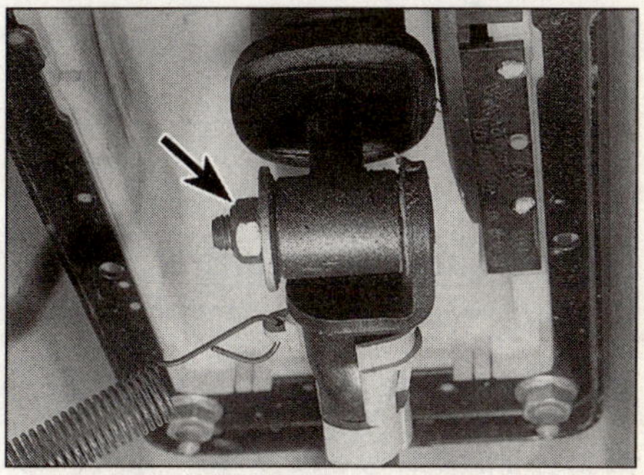

4.6 Nut securing the gearchange rod to the bottom of the gear lever

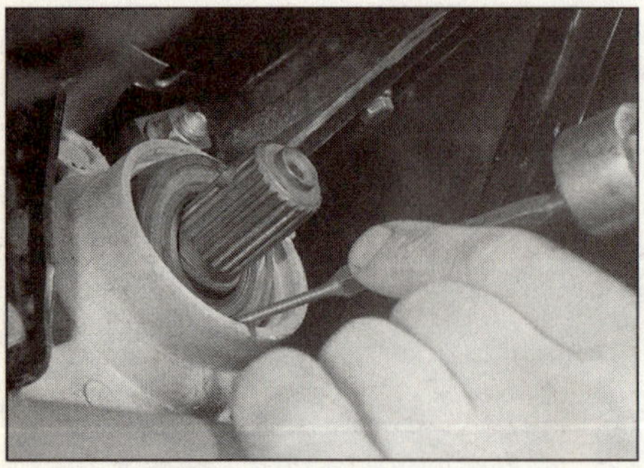
5.6 Tap the bottom of the driveshaft oil seal with a small drift to remove it

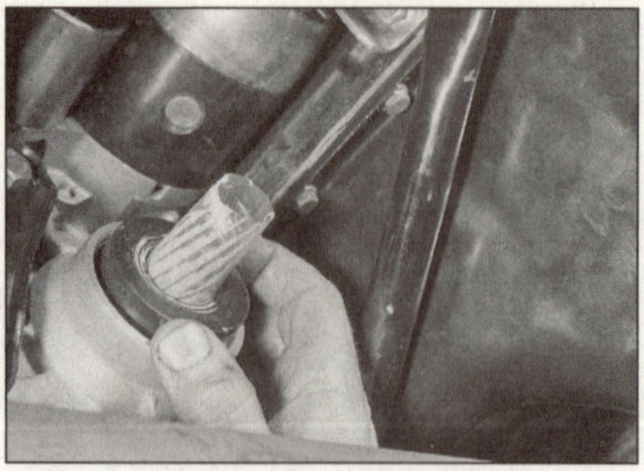
5.8a Locate the new oil seal carefully over the output shaft splines . . .

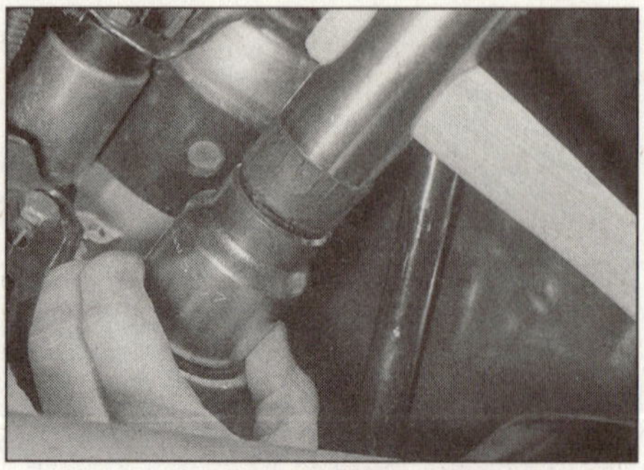
5.8b . . . and press into position with a socket or metal tube

Refitting

13 Lubricate all the pivot points with multi-purpose grease.
14 Where the gearchange rod has been dismantled, locate the front clevis on the rod in its previously noted position, and tighten the clamp bolt to the specified torque. Where new components are being fitted, locate the clevis on the rod so that approximately 7.0 to 8.0 mm of the knurled section is visible. Make sure that the clevis is offset towards the transmission.
15 Attach the clevis to the transmission lever, and refit the bolt making sure that the bush and spacer are in their correct positions. Tighten the bolt to the specified torque.
16 Reassemble the gear lever components using a reversal of the removal procedure. Make sure the circlip is secure.
17 Lift the gear lever assembly into position in the floor aperture, then locate the frame and refit the mounting nuts. Tighten the nuts to the specified torque.
18 Connect the gearchange rod to the bottom of the gear lever, making sure that the bushes are correctly located and that the nut is positioned on the left-hand side. Refit the washer, then tighten the nut to the specified torque.
19 Hook the return spring onto the gearchange rod.
20 Check and adjust the gearchange with reference to Section 3.
21 Refit the heat shields and exhaust system components as applicable, then lower the car to the ground.

5 Oil seals - renewal

Right-hand driveshaft oil seal

Note: *New driveshaft-to-differential side gear roll pins will be required on refitting (see Chapter 8).*
1 Apply the handbrake, then jack up the front of the car and support it on axle stands (see *"Jacking and vehicle support"*). Remove the right-hand wheel.
2 Drain the transmission oil as described in Section 2.
3 Referring to Chapter 8, disconnect the driveshaft from the transmission. Note that it is not necessary to remove the driveshaft completely, the shaft can be left attached to the hub assembly and slid off from the differential gear splines as the hub assembly is pulled outwards. **Note:** *Do not allow it to hang down under its own weight, as this could damage the constant velocity joints/gaiters.*
4 Remove the O-ring from the side gear shaft.
5 Wipe clean the old oil seal, and measure its fitted depth below the casing edge. This is necessary to determine the correct fitted position of the new oil seal, if the special Renault fitting tool is not being used.
6 Free the old oil seal, using a small drift to tap the outer edge of the seal inwards so that the opposite edge of the seal tilts out of the casing **(see illustration)**. A pair of pliers or grips can then be used to pull out the oil seal. Take care not to damage the splines of the differential side gear.
7 Wipe clean the oil seal seating in the casing. Wrap tape around the end of the differential gear splines to prevent the new seal being damaged.
8 Apply a smear of grease to the sealing lip of the new oil seal and, making sure its sealing lip is facing inwards, carefully slide it onto the differential gear shaft. Press the seal squarely into the transmission until it is positioned at the same depth as the original was prior to removal. If necessary, the seal can be tapped into position using a piece of metal tube or a socket which bears only on the hard outer edge of the seal **(see illustrations)**.
9 Remove the tape from the end of the differential shaft, and slide a new O-ring into position.
10 Reconnect the driveshaft to the transmission as described in Chapter 8.
11 Refill the transmission with oil as described in Section 2.
12 Refit the roadwheel and lower the car to the ground. Tighten the wheel bolts to the specified torque.

Left-hand driveshaft oil seal

13 On the left-hand side of the transmission, there is no oil seal; the seal is formed by the driveshaft gaiter. If oil is leaking from the left-hand driveshaft to transmission joint, renew the gaiter as described in Chapter 8.

Input shaft oil seal

JB1 transmission

14 On the JB1 type transmission, it is not possible to renew the input shaft oil seal without first dismantling the transmission. The guide tube assembly is a press fit in the housing, and is removed inwards. Oil seal renewal should therefore be entrusted to a Renault dealer or transmission overhaul specialist.

JC5 transmission

Note: *The release bearing guide tube is a press fit in the transmission housing, and it is likely that the special Renault service tools (Emb.880 and Emb.1163) will be require to remove and refit it safely. Removal and refitting without the special tools is likely to lead to the guide tube/transmission casing being damaged.*
15 Remove the transmission as described in Section 8, and remove the clutch release mechanism as described in Chapter 6.
16 Slacken and remove the two bolts securing the release bearing guide tube to the transmission. Withdraw the guide tube from the transmission housing (see Note) and slide it off the input shaft; the Renault service tool is in the form of a slide hammer which clamps to the outside of the guide tube.
17 Where necessary, carefully lever out the oil seal, taking care not to mark the shaft or casing, and slide it off the end of the shaft. Discard the seal and guide tube, they must be renewed as an assembly.
18 Wrap tape around the end of the input shaft, and slide the new seal and guide tube assembly into position.
19 Remove the tape from the input shaft, and press the guide tube squarely into the transmission casing.
20 Refit the guide tube retaining bolts and tighten them to the specified torque.
21 Refit the clutch release mechanism as described in Chapter 6, then refit the transmission with reference to Section 8.

6 Reversing light switch - testing, removal and refitting

Testing

1 The reversing light circuit is controlled by a plunger-type switch that is screwed into the left-hand side of the transmission casing, next to the driveshaft inner joint. If a fault develops in the circuit, first ensure that the circuit fuse has not blown.

2 To test the switch, disconnect the wiring connector, and use a multimeter (set to the resistance function) or a battery-and-bulb test circuit to check that there is continuity between the switch terminals only when reverse gear is selected. If this is not the case, and there are no obvious breaks or other damage to the wires, the switch is faulty, and must be renewed.

Removal

3 Firmly apply the handbrake, then jack up the front of the vehicle and support it on axle stands (see *"Jacking and vehicle support"*).

4 Undo the retaining screws and remove the engine undershield.

5 Disconnect the wiring connector, then unscrew it from the transmission casing along with its sealing washer.

Refitting

6 Fit a new sealing washer to the switch, then screw it back into position in the top of the transmission housing and tighten it securely. Reconnect the wiring connector, and test the operation of the circuit. Refit the engine undershield and lower the vehicle to the ground. If any oil was lost when the switch was removed, check the oil level as described in Chapter 1.

7 Speedometer drive - removal and refitting

Note: *On some later models, the speedometer drive is taken from the right-hand side of the transmission, just above the driveshaft. It is not possible to remove the drive on this type. The later type can be identified by the plastic cable connection to the transmission instead of the clip type connection on earlier models.*

Removal

1 Disconnect the left-hand driveshaft at the transmission end - refer to Chapter 8. Note that it is not necessary to remove the driveshaft completely; the shaft can be left attached to the hub assembly and slid off from the differential gear splines as the hub assembly is pulled outwards. **Note:** *Do not allow it to hang down under its own weight, as this could damage the constant velocity joints/gaiters.*

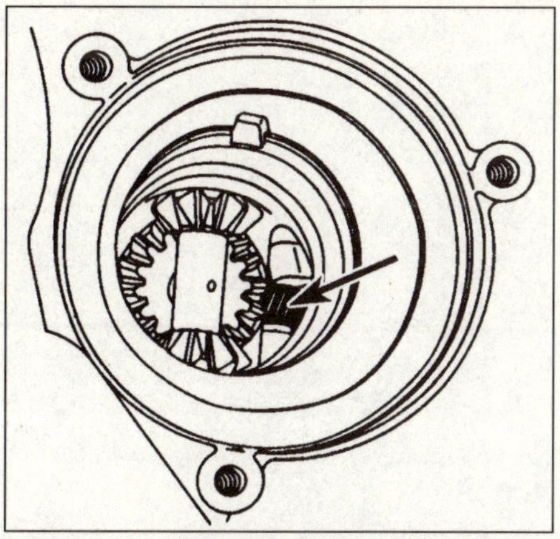

7.3 View of the speedometer drivegear (arrowed) with the differential sun wheel removed

2 Extract the circlip and thrustwasher, then withdraw the left-hand sun wheel from the differential. The sun wheel also acts as the driveshaft spider housing.

3 Turn the differential until the planet wheels are in a vertical plane so that the speedometer drivegear is visible **(see illustration)**.

4 Pull out the clip and disconnect the speedometer transmitter from the outside of the transmission **(see illustration)**.

5 Using long-nosed pliers, extract the speedometer drivegear shaft vertically from the outside of the transmission.

6 Using the same pliers, extract the speedometer drivegear from inside the differential housing, being very careful not to drop it.

7 Examine the drivegear teeth for wear and damage. Renew it if necessary. Note that if the drivegear teeth on the differential are worn or damaged, it will be necessary to dismantle the transmission - this work should be entrusted to a Renault dealer.

Refitting

8 Using long-nosed pliers, insert the speedometer drivegear into its location.

9 From outside the transmission, refit the drivegear shaft. Make sure that it engages with the gear location notches correctly **(see illustration)**.

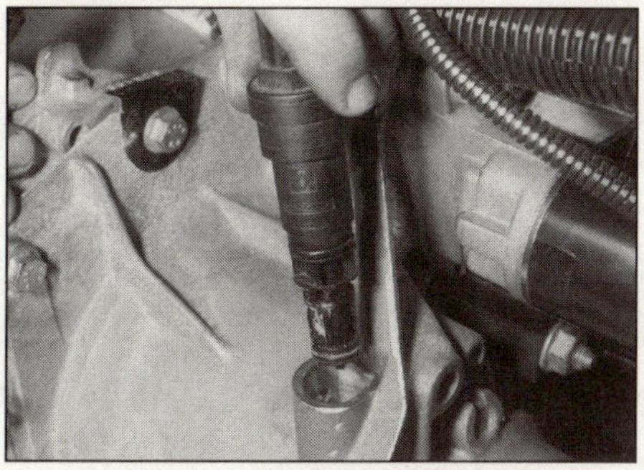

7.4 Removing the speedometer transmitter from the transmission

7.9 Notches in the speedometer drivegear engage with the shaft as shown
Left - Correct Right - Incorrect

10 Refit the speedometer transmitter, and secure with the clip.

11 Insert the differential sun wheel, then refit the thrustwasher and circlip.

12 Reconnect the left-hand driveshaft with reference to Chapter 8.

8 Manual transmission - removal and refitting

On all models, the engine/transmission assembly must be removed from the car and the manual transmission separated from the engine on the bench. Refer to Chapter 2C for details.

9 Manual transmission overhaul - general information

1 Overhauling a manual transmission is a difficult and involved job for the DIY home mechanic. In addition to dismantling and reassembling many small parts, clearances must be precisely measured and, if necessary, changed by selecting shims and spacers. Internal transmission components are also often difficult to obtain, and in many instances, extremely expensive. Because of this, if the transmission develops a fault or becomes noisy, the best course of action is to have the unit overhauled by a specialist repairer, or to obtain an exchange reconditioned unit.

2 Nevertheless, it is not impossible for the more experienced mechanic to overhaul the transmission, provided the special tools are available, and the job is done in a deliberate step-by-step manner, so that nothing is overlooked.

3 The tools necessary for an overhaul include internal and external circlip pliers, bearing pullers, a slide hammer, a set of pin punches, a dial test indicator, and possibly a hydraulic press. In addition, a large, sturdy workbench and a vice will be required.

4 During dismantling of the transmission, make careful notes of how each component is fitted, to make reassembly easier and more accurate.

5 Before dismantling the transmission, it will help if you have some idea what area is malfunctioning. Certain problems can be closely related to specific areas in the transmission, which can make component examination and replacement easier. Refer to the *"Fault finding"* Section at the rear of this manual for more information.

Chapter 7 Part B: Automatic transmission

Contents

Degrees of difficulty

Easy, suitable for novice with little experience	**Fairly easy,** suitable for beginner with some experience	**Fairly difficult,** suitable for competent DIY mechanic	**Difficult,** suitable for experienced DIY mechanic	**Very difficult,** suitable for expert DIY or professional

Specifications

General

Type	Electronically controlled with four forward speeds and reverse
Designation	AD4

Torque wrench settings

	Nm	lbf ft
Driveplate cover plate bolts	24	18
Driveplate-to-torque converter nuts	25	18
Engine-to-transmission bolts/nuts	35	26
Fluid cooler bolts	25	18
Fluid filter mounting bolts	5	4
Left-hand engine/transmission mounting to body	20	15
Rear mounting bracket to transmission	65	48
Roadwheel bolts	90	66
Transmission sump bolts	10	7

1 General information

Models with the K7M 1598 cc petrol engine are available with a four-speed fully-automatic transmission, consisting of a torque converter, an epicyclic geartrain, and hydraulically-operated clutches and brakes **(see illustration)**.

The transmission consists of a torque converter, an epicyclic geartrain, hydraulically-operated clutches and brakes, and a computer control unit.

The torque converter provides a fluid coupling between engine and transmission, which acts as an automatic clutch, and also provides a degree of torque multiplication when accelerating.

The epicyclic geartrain provides the forward gears or reverse gear, depending on which of its component parts are held stationary or allowed to turn. The components of the geartrain are held or released by brakes and clutches which are activated by a hydraulic control unit. An fluid pump within the transmission provides the necessary hydraulic pressure to operate the brakes and clutches.

Impulses from switches and sensors connected to the transmission throttle and selector linkages are directed to a computer module, which determines the ratio to be selected from the information received. The computer activates solenoid valves, which in turn open or close ducts within the hydraulic control unit. This causes the clutches and brakes to hold or release the various components of the geartrain, and provide the correct ratio for the particular engine speed or load. The information from the computer module can be overridden by use of the selector lever, and a particular gear can be held if required, regardless of engine speed.

The automatic transmission fluid is cooled by passing it through a cooler located on the top of the transmission. Coolant from the cooling system passes through the cooler.

Due to the complexity of the automatic transmission, any repair or overhaul work must be left to a Renault dealer with the necessary special equipment for fault diagnosis and repair. The contents of the following Sections are therefore confined to supplying general information, and any service information and instructions that can be used by the owner.

7B

2 Transmission fluid filter - renewal

1 The fluid filter should be changed whenever the transmission fluid has become contaminated.

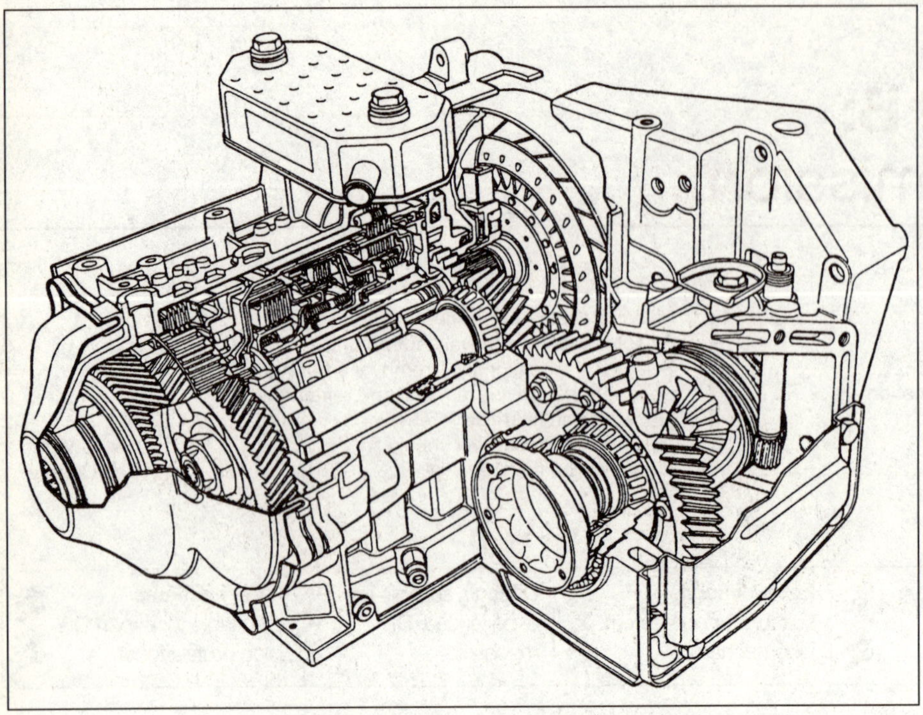

1.1 Cutaway view of the AD4 transmission unit

2 Drain the transmission fluid as described in Chapter 1.
3 Unbolt the coolant expansion tank from the left-hand side of the engine compartment, and position it to one side.
4 Remove the air cleaner assembly as described in Chapter 4A.
5 Apply the handbrake, then jack up the left-hand front of the car and support on axle stands (see *"Jacking and vehicle support"*). Remove the roadwheel.
6 Remove the left-hand front wheelarch liner which is retained by plastic clips and screws.
7 Using a trolley jack and block of wood, support the weight of the transmission.
8 Unscrew and remove the bolts securing the left-hand engine/transmission mounting to the body.

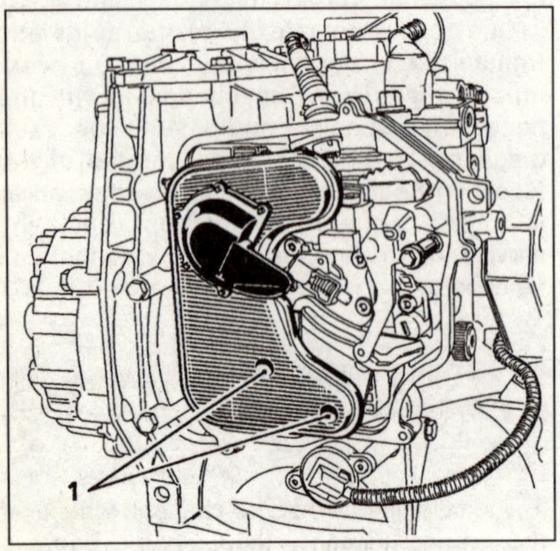

2.12 Fluid filter retaining bolts (1)

9 On models with ABS, unscrew and ABS pump mounting nut on the left-hand engine mounting.
10 Raise the transmission on the trolley jack to provide room to remove the transmission sump.
11 Slacken and remove the sump retaining bolts, and lower the sump away from the transmission. Recover the sump seal.
12 Unbolt the filter from the base of the transmission, and recover the filter gasket **(see illustration)**.
13 Remove the magnet from inside the sump, noting its correct fitted position, and clean away all traces of metal filings. The filings (if any) should be very fine; any sizeable chips of metal indicate a worn component in the transmission. Refit the magnet in the correct position.
14 Fit a new gasket on top of the filter element, and offer the filter to the transmission. Refit the retaining bolts, ensuring that the gasket is still correctly positioned, and tighten them to the specified torque.
15 Ensure that the seal is correctly located on the sump and refit the sump to the transmission, tightening its retaining bolts to the specified torque.
16 Lower the transmission, then refit the left-hand engine/transmission mounting bolts and tighten to the specified torque.
17 On models with ABS, refit and tighten the pump mounting nut on the left-hand engine mounting.
18 Refit the wheelarch liner and roadwheel, and lower the vehicle to the ground.
19 Refit the air cleaner and coolant expansion tank.
20 Fill the transmission with the specified type and amount of fluid as described in Chapter 1.

3 Selector cable - adjustment

1 Apply the handbrake, then jack up the front of the car and support it on axle stands (see *"Jacking and vehicle support"*). Remove the heatshield(s) for access to the underside of the selector assembly.
2 Remove the centre console as described in Chapter 11.
3 Remove the air cleaner assembly as described in Chapter 4A.
4 Disconnect the selector cable from the transmission lever by removing the clip(s) and pulling the end fitting from the balljoint.
5 Make sure the cable is attached to the lever assembly correctly, then turn the cable lock ring to ensure that the adjustment sleeve is free.
6 Position both the lever on the transmission and the selector lever inside the vehicle in 1st gear hold.
7 Offer the cable end fitting over the balljoint (but do not refit it at this stage) on the transmission, and check that it is aligned correctly. If not, adjust the cable by turning the sleeve. When correctly aligned, push the end fitting onto the balljoint, and lock the adjustment by turning the ring on the sleeve.
8 Refit the cable retaining clip(s).
9 Refit the centre console as described in Chapter 11.
10 Check that the selector lever moves freely, and that the starter motor will only operate with "P" or "N" selected. Also check that the Park function operates correctly.
11 Refit the air cleaner assembly as described in Chapter 4A.
12 Refit the heatshield(s) then lower the vehicle to the ground.

4 Selector lever assembly - removal and refitting

Removal

1 Apply the handbrake, then jack up the front of the car and support it on axle stands (see *"Jacking and vehicle support"*). Remove the heatshield(s) for access to the underside of the selector assembly.
2 Inside the vehicle, unscrew the small screw and remove the knob from the top of the gear lever, then remove the centre console as described in Chapter 11 and disconnect the wiring **(see illustration)**.
3 Remove the air cleaner assembly as described in Chapter 4A.

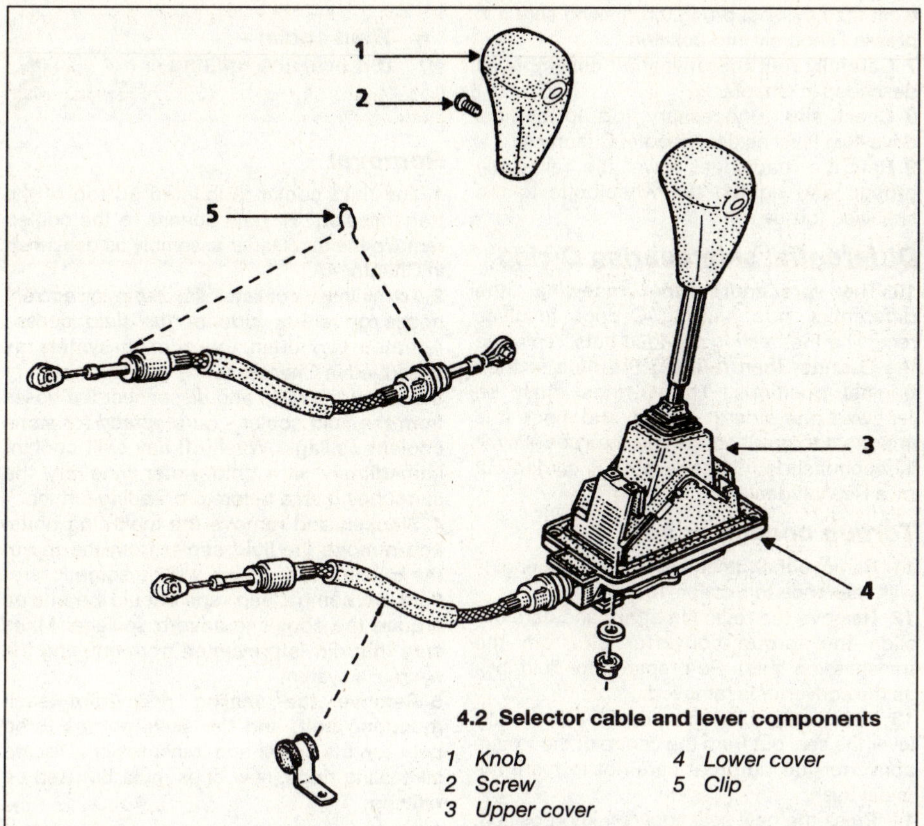

4.2 Selector cable and lever components

1 *Knob*
2 *Screw*
3 *Upper cover*
4 *Lower cover*
5 *Clip*

4 Disconnect the selector cable from the transmission lever by removing the clip(s) and pulling the end fitting from the balljoint.

5 Unscrew the two bolts, and lift off the upper cover from the lever assembly **(see illustration)**.

6 Under the vehicle, unscrew the nuts securing the lever assembly to the floor.

7 Unscrew the bolt securing the selector cable to the underbody, and remove the assembly from under the vehicle.

8 Remove the cable from the lever assembly with reference to Section 5.

Refitting

9 Refitting is the reverse of removal, noting the following points.

a) *Prior to refitting, apply a smear of multi-purpose grease to the sliding surfaces of the selector lever mechanism.*

b) *On completion, check and adjust the selector lever as described in Section 3.*

5 Selector cable - removal and refitting

Removal

1 Remove the selector lever assembly as described in Section 4.

2 Unclip the cable from the bottom of the gear lever.

3 Using a screwdriver, release the clip and pull the cable from the lever assembly **(see illustration)**.

Refitting

4 Refitting is the reverse of removal. Refit the selector lever as described in Section 4, and adjust the cable as described in Section 3.

6 Speedometer drive - removal and refitting

Removal

1 The speedometer drivegear can be removed once the final drive cover plate has been unbolted from the rear of the transmission.

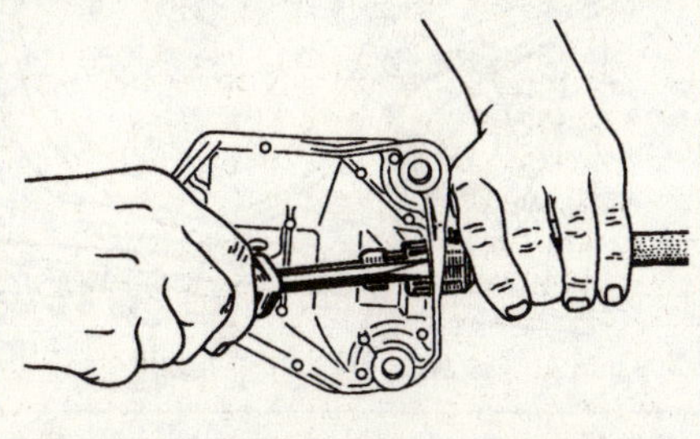

5.3 Using a screwdriver to release the cable from the lever assembly

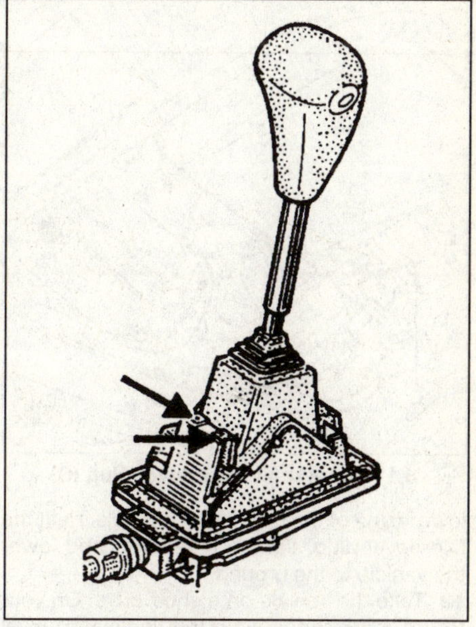

4.5 Upper cover retaining bolts

2 Firmly apply the handbrake, then jack up the front of the vehicle and support it on axle stands (see *"Jacking and vehicle support"*). Where fitted, remove the engine/transmission undershield.

3 Remove the air cleaner assembly as described in Chapter 4A.

4 Disconnect the wiring and unscrew the speedometer transmitter from the top of the transmission.

5 Place a suitable container under the final drive housing, to catch the fluid which will be released as the cover plate is removed.

6 Unscrew the securing bolts, and remove the final drive housing cover plate. Recover the gasket.

7 Unscrew the speedometer gear housing from the top of the transmission, then unclip the drivegear and withdraw the gear. Recover the sealing ring.

Refitting

8 Fit a new sealing ring to the speedometer gear housing, and insert the housing into the transmission. Clip the drivegear into the housing and seat the housing in position, making sure the gear is correctly engaged, and tighten securely.

9 Ensure that the cover plate and transmission surfaces are clean and dry, and refit the cover using a new gasket. Fit the cover bolts and tighten them securely.

10 Refit the speedometer transmitter and reconnect the wiring.

11 It is now necessary to refill the final drive unit. The unit is refilled via the filler/level plug on the right-hand end of the transmission **(see illustration)**. Slacken and remove the filler/level plug, and refill the final drive with the exact amount of the specified type of oil (*see the end of "Weekly checks" and Chapter 1 Specifications*) until the fluid level is up to the

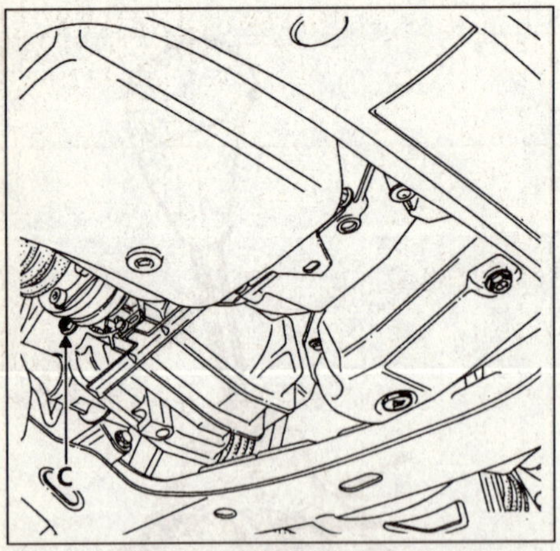

6.11 Final drive filler/level plug (C)

lower edge of the filler/level plug hole. Refit the filler/level plug, tighten it securely and lower the vehicle to the ground.

12 Take the vehicle on a short drive. On your return, park the vehicle the vehicle on level ground and check the final drive unit oil level as described in Chapter 1.

7 Oil seals - renewal

Differential oil seal

1 Apply the handbrake, then jack up the front of the car and support it on axle stands (see *"Jacking and vehicle support"*). Remove the relevant roadwheel.

2 Referring to Chapter 8, disconnect the complete driveshaft assembly from the transmission. Note that it is not necessary to remove the driveshaft completely, the shaft can be left attached to the hub assembly, and freed from the transmission as the hub assembly is pulled outwards. **Note:** *Do not allow it to hang down under its own weight, as this could damage the constant velocity joints/gaiters.*

3 Carefully lever off the oil seal protector from the outside of the transmission housing **(see illustration)**.

4 Wipe clean the oil seal and housing, and note its fitted depth as a guide for fitting the new seal. Carefully lever out the old seal using a small screwdriver **(see illustration)**. Some loss of fluid will occur, so position a container beneath the oil seal position.

5 Wipe clean the oil seal seating in the casing, and apply a smear of oil to the seal lip. Making sure the seal lip is facing inwards, carefully ease the new seal into position over the differential shaft. Press the seal squarely into the transmission until it is positioned at the same depth as the original was prior to removal. If necessary, the seal can be tapped into position using a piece of metal tube, or a socket which bears only on the hard outer edge of the seal.

6 Fit the new seal protector, making sure it is pressed securely into position.

7 Carefully refit the driveshaft assembly as described in Chapter 8.

8 Check and, if necessary, top-up the final drive fluid level as described in Chapter 1.

9 Refit the roadwheel, lower the car to the ground and tighten the wheelbolts to the specified torque.

Differential side bearing O-ring

10 The procedure for renewing the differential side bearing O-rings involves removing the bearing pre-load nuts, replacing the O-rings, then refitting the nuts in their original positions. The O-rings must be renewed one side at a time, and since it is important to pre-load the bearings correctly, it is recommended that this work is carried out by a Renault dealer.

Torque converter seal

11 Remove the transmission from the engine with reference to Section 10.

12 Remove the retaining strap, and carefully slide the torque converter off from the transmission shaft. Be prepared for fluid loss as the converter is removed.

13 Using a flat-bladed screwdriver, carefully lever the seal out from the centre of the torque converter, taking great care not to mark the metal bush.

14 Press the new seal squarely into position, making sure its sealing lip is facing inwards.

15 Lubricate the lip of the seal with clean transmission fluid, and carefully slide the converter onto the transmission shaft.

16 Make sure the torque converter is correctly engaged with the transmission shaft splines, then refit the transmission as described in Section 10.

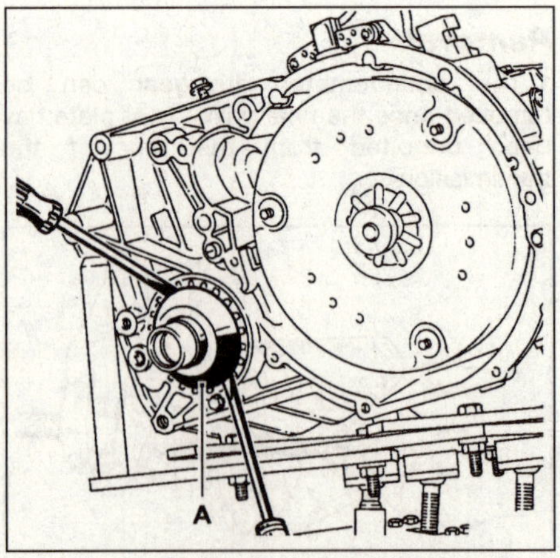

7.3 Using two screwdrivers to prise the oil seal protector (A) from the transmission housing

8 Fluid cooler - removal and refitting

Removal

1 The fluid cooler is located on top of the transmission. To gain access to the cooler, remove the air cleaner assembly as described in Chapter 4A.

2 To minimise coolant loss, clamp the coolant hoses on either side of the fluid cooler. Alternatively, drain the cooling system as described in Chapter 1.

3 Loosen the clips and disconnect the hoses from the fluid cooler - be prepared for some coolant spillage. Wash off any spilt coolant immediately with cold water, and dry the surrounding area before proceeding further.

4 Slacken and remove the mounting bolts, and remove the fluid cooler from the top of the transmission. There will be some loss of fluid, so some clean rags should be placed around the cooler to absorb spillage. Make sure that dirt is prevented from entering the hydraulic system.

5 Remove the sealing ring from each mounting bolt, and the sealing rings fitted between the cooler and transmission. Discard all sealing rings; new ones must be used on refitting.

Refitting

6 Lubricate the new seals with clean automatic transmission fluid, then fit the two new seals to the base of the fluid cooler, and a new seal to each mounting bolt.

7 Locate the fluid cooler on the top of the transmission housing, ensuring its lower seals remain in position. Refit the mounting bolts and tighten them to the specified torque setting.

8 Reconnect the coolant hoses to the fluid cooler, and securely tighten their retaining clips. Remove the hose clamps.

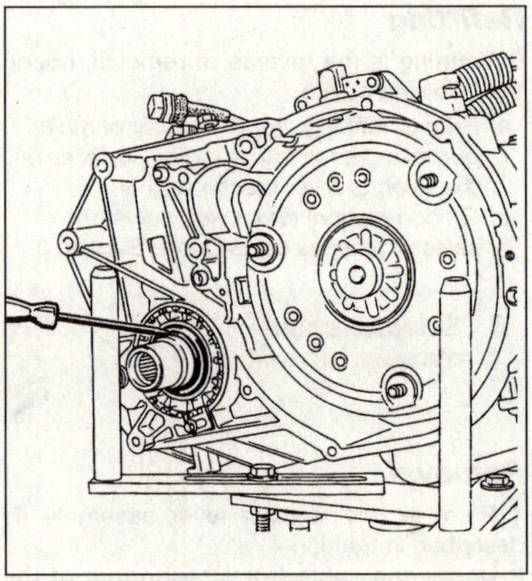

7.4 Levering out the differential oil seal

9 Refit the air cleaner assembly as described in Chapter 4A.
10 On completion, top-up the cooling system and check the automatic transmission fluid level as described in *"Weekly checks"* and Chapter 1.

9 Automatic transmission electronic components - removal and refitting

Note: *Whenever any of the transmission electronic control components are renewed, on completion it is necessary to validate the no-load/full-load position of the transmission. To do this, the Renault (XR25) diagnostic equipment is needed. Therefore, it will be necessary to entrust the following work to a Renault dealer if access to the necessary equipment cannot be gained. Failure to validate the no-load/full-load position could lead to incorrect gearchange thresholds, and the instrument panel warning light illuminating randomly.*

Removal

1 Disconnect the battery negative lead, and proceed as described under the relevant sub-heading.
Caution: If the radio/cassette in your vehicle is equipped with an anti-theft system, make sure you have the correct activation code before disconnecting the battery.

Multi-function switch

2 The multi-function switch informs the electronic control unit of the selector lever position, prevents the starter motor operating when the transmission is in gear, and also controls the reversing lights. The switch is located on the rear of the transmission, above the left-hand driveshaft.
3 To remove the switch, trace the wiring back from the switch and disconnect it at the connector.
4 Unscrew the mounting bolt and remove the clamp plate. Unscrew the earth wire bolt, then pull the switch out of the transmission along with its sealing ring.

Throttle potentiometer

5 Refer to Chapter 4A.

Vehicle speed sensor

6 The speed sensor is mounted onto the top of the transmission, on its front left-hand end.
7 If necessary, to improve access, remove the air cleaner assembly and mounting bracket, as described in Chapter 4A.
8 Trace the wiring back from the sensor, and disconnect at the wiring connector.
9 Unscrew the mounting bolt and remove the clamp plate, then pull the switch out of the transmission along with its sealing ring.

Line pressure sensor

10 The line pressure sensor is fitted to the base of the transmission, on the front end.
11 To gain access to the sensor, firmly apply the handbrake then jack up the front of the vehicle and support it on axle stands. Undo the retaining screws and remove the plastic undercover.
12 Trace the wiring back from the sensor and disconnect it at the wiring connector.
13 Be prepared for some fluid loss as the sensor is removed, then slacken the retaining bolts and remove the sensor retaining plate. Remove the sensor and recover the sealing ring.

Electronic control unit

14 The electronic control unit (ECU) is located in the front left-hand corner of the engine compartment.
15 Release the rubber retaining strap, then disconnect the wiring connectors and remove the ECU from the engine compartment.

Refitting

16 Refitting is a reversal of the removal procedure. On completion, reconnect the battery and validate the no-load/full-load position of the transmission.

10 Automatic transmission - removal and refitting

The engine and automatic transmission assembly must be removed from the vehicle, and the transmission separated from the engine on the bench. Refer to Chapter 2C for details.

11 Automatic transmission overhaul - general information

In the event of a fault occurring with the transmission, it is first necessary to determine whether it is of an electrical, mechanical or hydraulic nature, and to do this, special test equipment is required. It is therefore essential to have the work carried out by a Renault dealer if a transmission fault is suspected.

Do not remove the transmission from the car for possible repair before professional fault diagnosis has been carried out, since most tests require the transmission to be in the vehicle.

Chapter 8
Driveshafts

Contents

Degrees of difficulty

Easy, suitable for novice with little experience	**Fairly easy,** suitable for beginner with some experience	**Fairly difficult,** suitable for competent DIY mechanic	**Difficult,** suitable for experienced DIY mechanic	**Very difficult,** suitable for expert DIY or professional

Specifications

General

Driveshaft type	Equal-length solid steel shafts, splined to inner and outer constant velocity joints, vibration damper fitted on some shafts
Lubricant:	
Type/specification	Special grease supplied in sachets with gaiter kits - joints are otherwise pre-packed with grease and sealed
Quantity - manual transmission models (per joint):	
GE86 joint	320 g
AC1700 joint	140 g
GI62 joint	130 g
RC490 joint	160 g

Torque wrench settings

	Nm	lbf ft
Driveshaft retaining nut	250	185
Left-hand driveshaft inner gaiter retaining plate bolts - manual transmission models	25	18
Roadwheel bolts	90	66

1 General information

Drive is transmitted from the differential to the front wheels by means of two, equal-length, open driveshafts.

Both driveshafts are fitted with a constant velocity (CV) joint at their outer ends, which may be of the spider-and-yoke type or of the ball-and-cage type. Each joint has an outer member, which is splined at its outer end to accept the wheel hub and is threaded so that it can be fastened to the hub by a large nut. The joint contains either a spring-loaded plunger or six balls within a cage, which engage with the inner member. The complete assembly is protected by a flexible gaiter secured to the driveshaft and joint outer member.

On vehicles with a manual gearbox, a different inner constant velocity joint arrangement is fitted to each driveshaft. On the right-hand side, the driveshaft is splined to engage with a tripod joint, containing needle roller bearings and cups. The tripod joint is free to slide within the yoke of the joint outer member, which is splined and retained by a roll pin to the differential sun wheel stub shaft. As on the outer joints, a flexible gaiter secured to the driveshaft and outer member protects the complete assembly. On the left-hand side, the driveshaft also engages with a tripod joint, but the yoke in which the tripod joint is free to slide is an integral part of the differential sun wheel. On this side, the gaiter is secured to the transmission casing with a retaining plate, and to a ball-bearing on the driveshaft with a retaining clip. The bearing allows the driveshaft to turn within the gaiter, which does not revolve.

On vehicles with automatic transmission, both the left- and right-hand inner joints are the same. Each joint is a splined-fit in the differential sun wheel stub shaft. As with the outer joints, the complete assembly is protected by a flexible gaiter which is secured to the driveshaft and joint outer member.

2 Driveshaft - removal and refitting

Removal

Note: *A new driveshaft retaining nut will be required on refitting. On manual transmission models, new roll pins will be required on refitting if the right-hand driveshaft is to be removed.*

Note: *The driveshaft outer joint splines are coated with locking compound prior to refitting. Therefore it is likely that a puller/extractor will be required to draw the hub assembly off the driveshaft end on removal.*

1 Remove the wheel trim/hub cap (as applicable), then slacken the driveshaft nut with the vehicle resting on its wheels. Also slacken the wheel bolts.

TOOL TiP

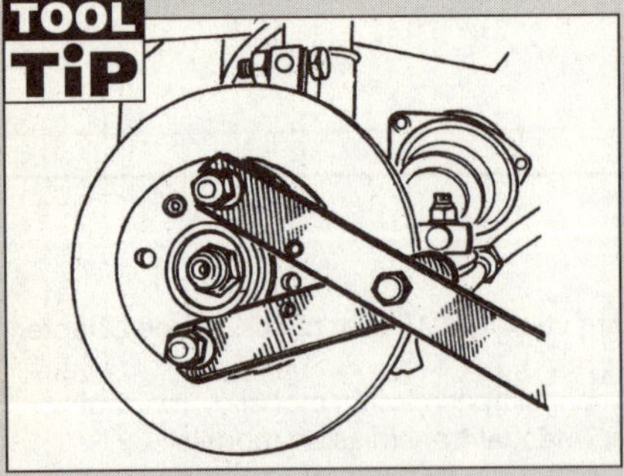

Using a fabricated tool to hold the front hub stationary whilst the driveshaft nut is slackened

2 Chock the rear wheels of the car, firmly apply the handbrake, then jack up the front of the car and support it on axle stands. Remove the appropriate front roadwheel.

3 On models equipped with ABS, remove the wheel sensor as described in Chapter 9.

4 Slacken and remove the driveshaft retaining nut. If the nut was not slackened with the wheels on the ground (see paragraph 1), refit at least two roadwheel bolts to the front hub, tightening them securely, then have an assistant firmly depress the brake pedal to prevent the front hub from rotating, whilst you slacken and remove the driveshaft retaining nut. Alternatively, a tool can be fabricated from two lengths of steel strip (one long, one short) and a nut and bolt; the nut and bolt forming the pivot of a forked tool **(see Tool tip)**.

5 Unscrew the two bolts securing the brake caliper assembly to the swivel hub, and slide the caliper assembly off the disc. Using a piece of wire or string, tie the caliper to the front suspension coil spring, to avoid placing any strain on the hydraulic brake hose.

6 Slacken and remove the nut securing the steering gear track rod end balljoint to the swivel hub. Release the balljoint tapered shank using a universal balljoint separator **(see illustration)**.

7 Slacken and remove the two nuts and washers from the bolts securing the swivel hub to the suspension strut, noting that the nuts are positioned on the rear side of the strut. Withdraw the upper bolt, but leave the lower bolt in position at this stage **(see illustration)**. Now proceed as described under the relevant sub-heading.

Left-hand driveshaft - manual transmission models

8 Drain the transmission oil as described in Chapter 7.

9 Slacken and remove the three bolts securing the flexible gaiter retaining plate to the side of the gearbox/transmission **(see illustration)**.

10 Pull the top of the swivel hub outwards until the driveshaft tripod joint is released from its yoke; be prepared for some oil spillage as the joint is withdrawn **(see illustration)**. Be careful that the rollers on the end of the tripod do not fall off.

11 Remove the lower bolt securing the swivel hub to the suspension strut. Taking care not to damage the driveshaft gaiters, release the outer constant velocity joint from the hub and remove the driveshaft. Note that it is likely the joint will be a tight fit in the hub splines (see Note at the start of this Section). Try tapping the joint out of position using a hammer and a soft metal drift, whilst an assistant supports the hub assembly. If this fails to move the joint, a suitable puller/extractor will be required to draw the hub assembly off the driveshaft end **(see illustration)**. Whilst the driveshaft is removed, support the hub assembly by refitting the bolts to the base of the strut.

Right-hand driveshaft - manual transmission models

12 Rotate the driveshaft until the double roll pin, securing the inner constant velocity joint to the sun wheel shaft, is visible. Using a hammer and a 5 mm diameter pin punch, drive out the double roll pin **(see illustration)**. New roll pins must be used on refitting.

13 Pull the top of the swivel hub outwards until the inner constant velocity joint splines are released from the sun wheel shaft. Remove the sealing ring (where fitted) from the sun wheel shaft splines.

14 Remove the driveshaft as described in paragraph 11.

2.6 Using a balljoint separator to release the track rod end balljoint from the swivel hub

2.7 Withdraw the upper swivel hub-to-strut bolt, noting which way around it is fitted

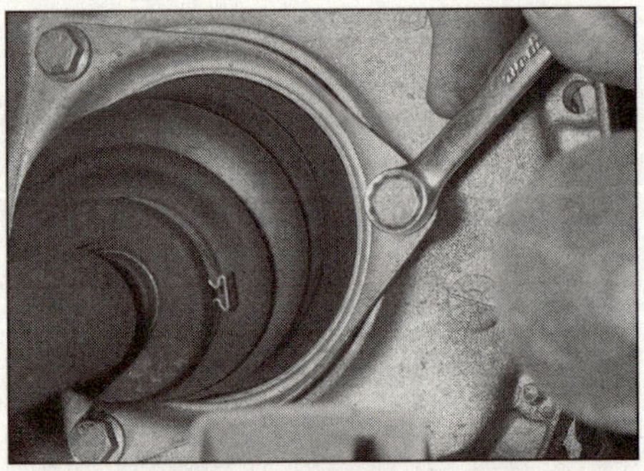

2.9 On manual transmission models, on the left-hand driveshaft, slacken and remove the flexible gaiter retaining plate bolts . . .

2.10 . . . and release the tripod joint from the transmission

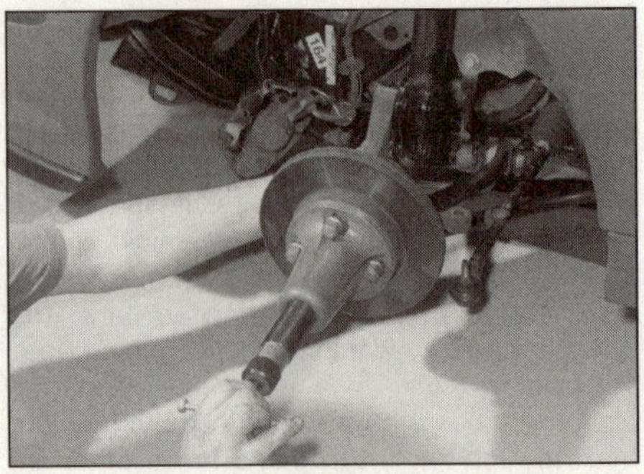

2.11 Using an extractor to press the driveshaft out of the front hub

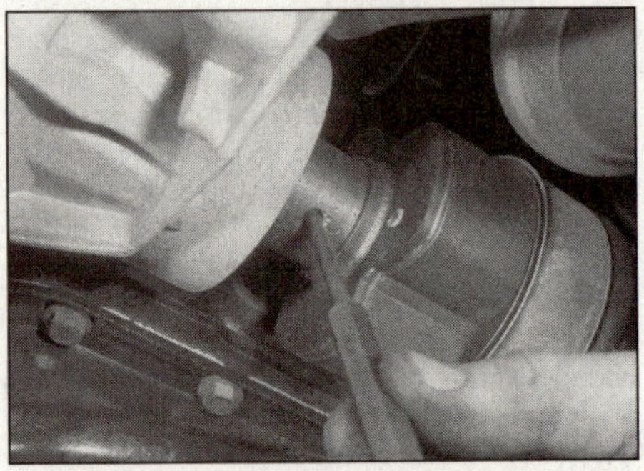

2.12 On manual transmission models, on the right-hand driveshaft, tap out the roll pins with a suitable pin punch

Both driveshafts - automatic transmission models

15 Pull the top of the swivel hub outwards, and disengage the inner constant velocity joint from the drive flange.

16 Remove the driveshaft as described in paragraph 11.

Refitting

17 All new driveshafts supplied by Renault are equipped with cardboard or plastic protectors, to prevent damage to the gaiters. Even the slightest knock to the gaiter can puncture it, allowing the entry of water or dirt at a later date, which may lead to the premature failure of the joint. If the original driveshaft is being refitted, it is worthwhile making up some cardboard protectors as a precaution. They can be held in position with elastic bands. The protectors should be left on the driveshafts until the end of the refitting procedure.

Left-hand driveshaft - manual transmission models

18 Wipe clean the side of the transmission and the outer constant velocity joint splines.

19 Insert the tripod joint into the sun wheel yoke, keeping the driveshaft horizontal as far as possible. Align the gaiter retaining plate with its bolt holes. Refit the retaining bolts, and tighten them to the specified torque. Ensure that the gaiter is not twisted.

20 Ensure both the hub and driveshaft outer constant velocity joint splines are clean and dry, and apply a coat of locking fluid (Renault recommend Loctite Scelbloc - available from your Renault dealer) to the splines of the driveshaft.

21 Move the top of the swivel hub inwards, at the same time engaging the driveshaft with the hub.

22 Slide the hub fully onto the driveshaft splines, then insert the two suspension strut mounting bolts from the front side of the strut. Refit the washers and nuts to the rear of the bolts, and tighten them to the specified torque (see Chapter 10 Specifications).

23 Slide on the washer, then fit the new driveshaft retaining nut, tightening it by hand only at this stage.

24 Reconnect the steering track rod balljoint to the swivel hub, and tighten its retaining nut to the specified torque (see Chapter 10 Specifications).

25 Clean the threads of the caliper bracket mounting bolts, and coat them with thread locking compound (Renault recommend Loctite Frenbloc - available from your Renault dealer). Slide the caliper into position, making sure the pads pass either side of the disc, and tighten the caliper bracket bolts to the specified torque setting (see Chapter 9 Specifications).

26 Using the method employed during removal to prevent the hub from rotating, tighten a new driveshaft retaining nut to the specified torque. Alternatively, lightly tighten

2.30 Fit the new O-ring (where fitted) onto the sun wheel shaft . . .

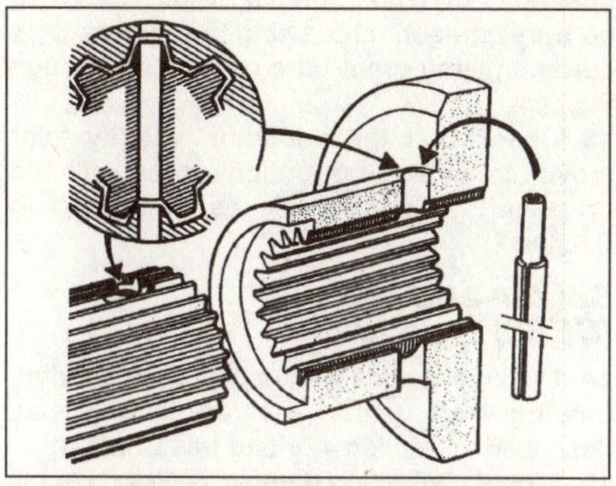

2.32a Right-hand driveshaft inner constant velocity joint roll pin arrangement - manual transmission models

the nut at this stage, and tighten it to the specified torque once the vehicle is resting on its wheels again.

27 Check that the hub rotates freely, then remove the protectors (where fitted) from the driveshaft, taking great care not to damage the flexible gaiters.

28 Refit the roadwheel. Lower the car to the ground and tighten the roadwheel bolts to the specified torque. If not already done, also tighten the driveshaft retaining nut to the specified torque.

29 Refill the transmission with the specified type and amount of oil, and check the level using the information given in Chapter 1.

Right-hand driveshaft - manual transmission models

30 Ensure that the inner constant velocity joint and sun wheel shaft splines are clean and dry. Apply a smear of molybdenum disulphide grease to the splines (Renault recommend the use of Molykote BR2 - available from your Renault dealer). Where necessary, fit a new sealing ring over the end of the sun wheel shaft, and slide the O-ring along the shaft until it abuts the transmission oil seal **(see illustration)**.

31 Engage the driveshaft splines with those of the sun wheel shaft, making sure that the roll pin holes are in alignment **(see illustration)**. Slide the driveshaft onto the sun wheel shaft until the roll pin holes are aligned.

32 Drive in new roll pins with their slots 180°

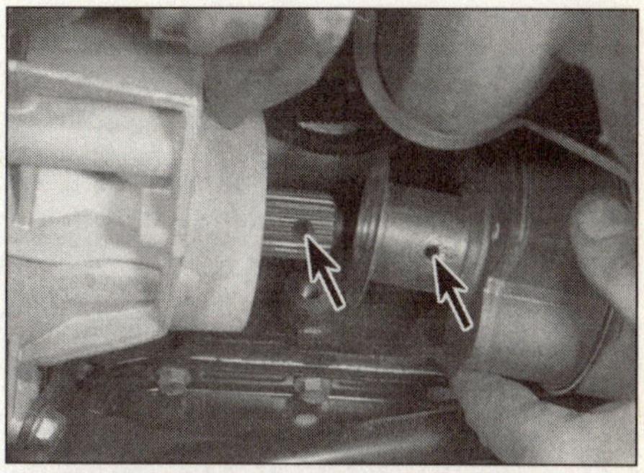

2.31 . . . and engage the driveshaft, ensuring that the roll pin holes (arrowed) are correctly aligned

2.32b On the right-hand driveshaft on manual transmission models, tap the roll pins securely into position, and seal their ends with a suitable sealing compound

apart, then seal the ends of the pins with sealing compound (Renault recommend the use of CAF 4/60 THIXO or Rhodoseal 5661 - available from your Renault dealer) **(see illustrations)**.

33 Carry out the procedures described in paragraphs 20 to 28.

Both driveshafts - models with AD4 automatic transmission

34 Ensure that the inner constant velocity joint and sun wheel shaft splines are clean and dry. Apply a smear of molybdenum disulphide grease to the splines (Renault recommend the use of Molykote BR2 - available from your Renault dealer).

35 Engage the driveshaft inner constant velocity joint with the sun wheel shaft splines.

36 Carry out the procedures described in paragraphs 20 to 28.

3 Driveshaft outer joint gaiter (manual transmission models) - renewal

1 Remove the driveshaft (refer to Section 2).

2 Cut through the gaiter retaining clips, and slide the gaiter down the shaft to expose the outer constant velocity joint.

3 Scoop out as much grease as possible from the joint, and determine which type of constant velocity joint is fitted. Proceed as described under the relevant sub-heading.

Ball-and-cage (AC1700) type joint

4 Using circlip pliers, expand the joint internal circlip. At the same time, tap the exposed face of the ball hub with a mallet to separate the joint from the driveshaft. Slide off the gaiter and rubber collar.

5 With the constant velocity joint removed from the driveshaft, clean the joint using paraffin, or a suitable solvent, and dry it thoroughly. Carry out a visual inspection of the joint.

6 Move the inner splined driving member from side to side, to expose each ball in turn at the top of its track. Examine the balls for cracks, flat spots or signs of surface pitting.

7 Inspect the ball tracks on the inner and outer members. If the tracks have widened, the balls will no longer be a tight fit. At the same time, check the ball cage windows for wear or cracking between the windows.

8 If on inspection any of the constant velocity joint components are found to be worn or damaged, it will be necessary to renew the complete driveshaft assembly, since no components are available separately. If the joint is in satisfactory condition, obtain a repair kit from your Renault dealer consisting of a new gaiter, rubber collar, retaining spring, and the correct type and quantity of grease.

9 Tape over the splines on the end of the driveshaft, then slide the rubber collar and gaiter onto the shaft. Locate the inner end of the gaiter on the driveshaft, and secure it in position with the rubber collar.

10 Remove the tape, then slide the constant velocity joint coupling onto the driveshaft until the internal circlip locates in the driveshaft groove.

11 Check that the circlip holds the joint securely on the driveshaft, then pack the joint with the grease supplied. Work the grease well into the ball tracks, and fill the gaiter with any excess.

12 Locate the outer lip of the gaiter in the groove on the joint outer member. With the coupling aligned with the driveshaft, lift the lip of the gaiter to equalise the air pressure. Fit both the inner and outer retaining clips to the gaiter, and secure each one in position by compressing its raised section. In the absence of the special tool, carefully compress each clip using a pair of side cutters, taking great care not to cut through the clip

13 Check that the constant velocity joint moves freely in all directions, then refit the driveshaft to the vehicle as described in Section 2.

Spider-and-yoke (GE86) type joint

14 Remove the inner constant velocity joint, bearing and gaiter (as applicable), as described in Section 4 or 5 of this Chapter.

15 Where a vibration damper is fitted, clearly mark the position of the damper on the driveshaft, then use a puller or press to remove it from the inner end of the driveshaft, noting which way around it is fitted **(see illustration)**. Ensure that the legs of the puller or support plate rest only on the damper inner rubber bush, otherwise the damper will distort and break away from the outer metal housing as it is removed.

16 Slide the outer constant velocity joint gaiter off the inner end of the driveshaft.

17 Clean the outer constant velocity joint using paraffin or a suitable solvent, and dry it thoroughly. Carry out a visual inspection of the joint.

18 Check the driveshaft spider and outer member yoke for signs of wear, pitting or scuffing on their bearing surfaces. Also check that the outer member pivots smoothly and easily, with no traces of roughness.

19 If inspection reveals signs of wear or damage, it will be necessary to renew the driveshaft complete, since no components are available separately. If the joint components are in satisfactory condition, obtain a repair kit consisting of a new gaiter, retaining clips, and the correct type and quantity of grease **(see illustration)**.

20 Tape over the splines on the inner end of the driveshaft, then carefully slide the outer gaiter onto the shaft.

21 Pack the joint with the grease supplied in the repair kit **(see illustration)**. Work the grease well into the joint, and fill the gaiter with any excess.

22 Ease the gaiter over the joint, and ensure that the gaiter lips are correctly located in the grooves on the driveshaft and on the joint **(see illustration)**. With the coupling aligned with the driveshaft, lift the lip of the gaiter to equalise the air pressure.

3.15 Removing the vibration damper from the driveshaft

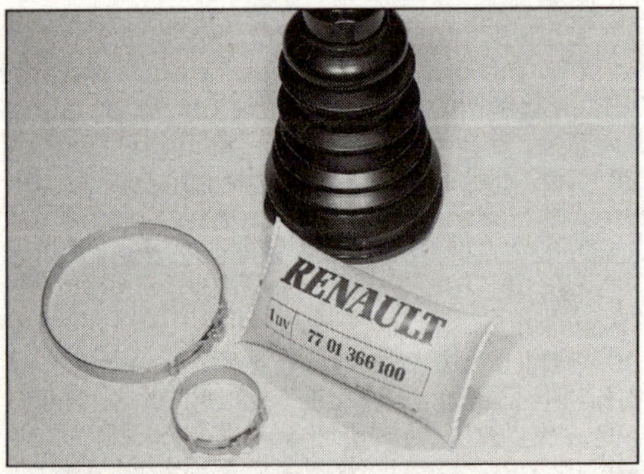

3.19 Renault driveshaft gaiter repair kit

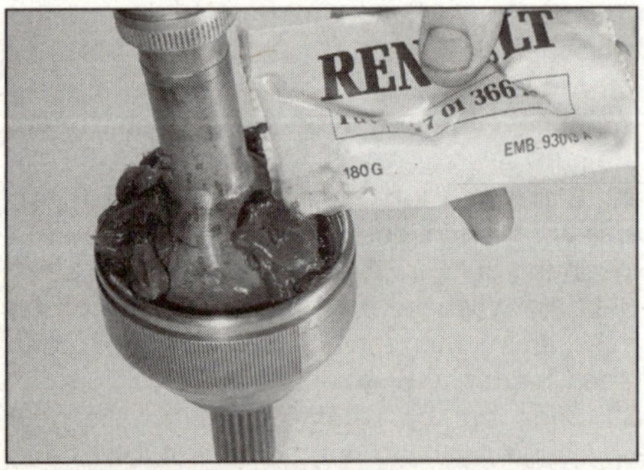

3.21 Pack the joint with the grease supplied in the repair kit . . .

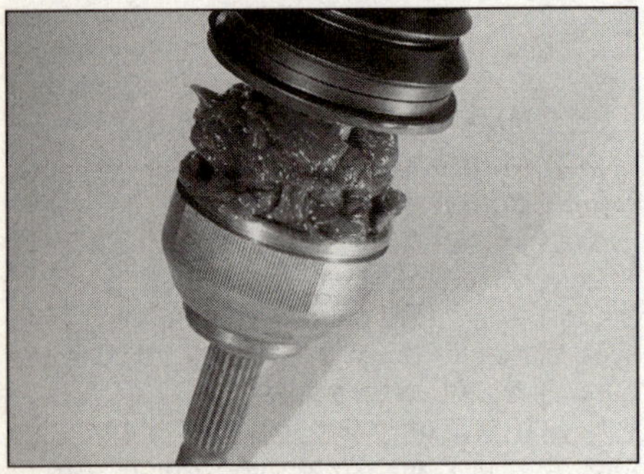

3.22 . . . then slide the gaiter into position over the joint

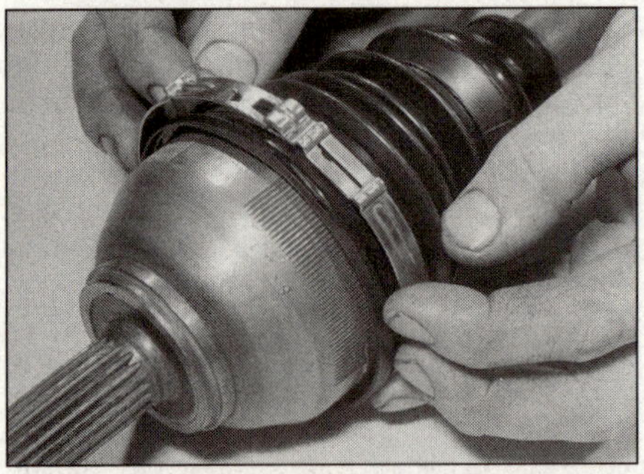

3.23a Fit the large gaiter retaining clip . . .

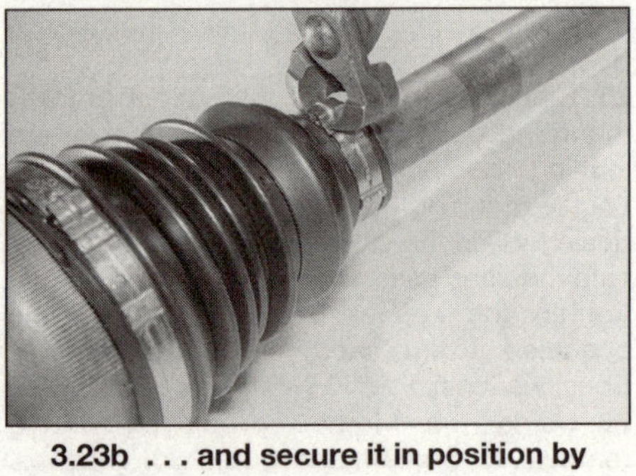

3.23b . . . and secure it in position by carefully compressing the raised section of the clip

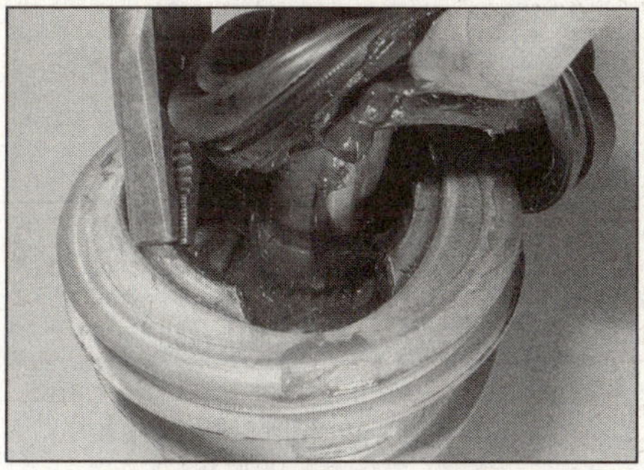

4.4 On the Gl62-type joint, bend up the anti-separation tangs with pliers to release the joint outer member

4.5 Remove the circlip . . .

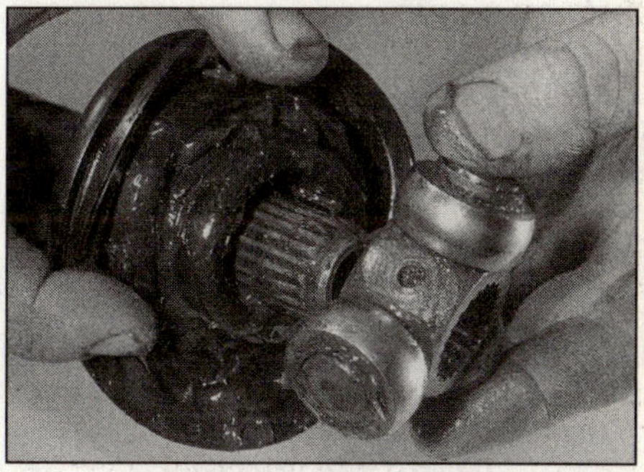

4.6 . . . and withdraw the tripod joint from the driveshaft end

23 Fit the large metal retaining clip to the gaiter. Remove any slack in the gaiter retaining clip by carefully compressing the raised section of the clip. In the absence of the special tool, a pair of side cutters may be used, so long as great care is taken not to cut the clip **(see illustrations)**. Secure the small retaining clip using the same procedure. Check that the constant velocity joint moves freely in all directions before proceeding further.

24 To refit the vibration damper (when applicable), lubricate the driveshaft with a solution of soapy water. Press or drive the vibration damper along the shaft, using a tubular spacer which bears only on the damper inner bush, until it is aligned with the mark made prior to removal.

25 Refit the inner constant velocity joint components as described in Section 4 or 5 (as applicable), then refit the driveshaft to the vehicle as described in Section 2.

4 Right-hand inner gaiter (manual transmission models) - renewal

1 Remove the driveshaft as described in Section 2.

2 On these models, two different types of inner constant velocity joint are used on the right-hand driveshaft: type Gl62 and type RC490. The joints can be identified by the shape of their outer members. The Gl62 joint has a smooth, circular outer member; the RC490 joint has a recessed outer member which appears clover-shaped when viewed end-on. Identify the type of joint fitted, then proceed as described under the relevant sub-heading.

Gl62-type joint

3 Cut the retaining clips, then slide the gaiter down the shaft to expose the joint.

4 Using pliers, carefully bend up the anti-separation plate tangs at their corners **(see illustration)**. Slide the outer member off the tripod joint. Be prepared to hold the rollers in place, otherwise they may fall off the tripod ends as the outer member is withdrawn. If necessary, secure the rollers in place using tape after removal of the outer member. **Note:** *The rollers are matched to the tripod joint stems, and it is important that they are not interchanged.*

5 Using circlip pliers, extract the circlip securing the tripod joint to the driveshaft **(see illustration)**. Note that on some models, the joint may be staked in position; if so, relieve the staking using a file. Mark the position of the tripod in relation to the driveshaft, using a dab of paint or a punch.

6 The tripod joint can now be removed **(see illustration)**. If it is tight, draw the joint off the driveshaft end using a puller. Ensure that the legs of the puller are located behind the joint inner member and do not contact the joint rollers. Alternatively, support the inner member of the tripod joint, and press the shaft out using a hydraulic press, again ensuring that no load is applied to the joint rollers.

7 With the tripod joint removed, slide the gaiter and inner retaining collar off the end of the driveshaft.

8 Wipe clean the joint components, taking care not to remove the alignment marks made on dismantling. **Do not** use paraffin or other solvents to clean this type of joint.

9 Examine the tripod joint, rollers and outer member for any signs of scoring or wear. Check that the rollers move smoothly on the tripod stems. If wear is evident, the tripod joint and roller assembly can be renewed, but it is not possible to obtain a replacement outer member. Obtain a new gaiter, retaining spring/collar, and a quantity of the special lubricating grease. These parts are available in the form of a repair kit from your Renault dealer.

10 Tape over the splines on the end of the driveshaft, then carefully slide the inner retaining collar and gaiter onto the shaft.

11 Remove the tape, then, aligning the marks made on dismantling, engage the tripod joint with the driveshaft splines. Use a hammer and soft metal drift to tap the joint onto the shaft, taking great care not to damage the driveshaft splines or joint rollers. Alternatively, support the driveshaft, and press the joint into position using a hydraulic press and suitable tubular spacer which bears only on the joint inner member.

12 Secure the tripod joint in position with the circlip, ensuring that it is correctly located in the driveshaft groove. Where no circlip is fitted, secure the joint in position by staking the end of the driveshaft in three places, at intervals of 120°, using a hammer and punch.

13 Evenly distribute the grease contained in the repair kit around the tripod joint and inside the outer member. Pack the gaiter with the remainder of the grease.

14 Slide the outer member into position over the tripod joint.

15 Using a piece of 2.5 mm thick steel or similar material, make up a support plate to the dimensions shown **(see illustration)**.

16 Position the support plate under each anti-separation plate tang in the outer member in turn, and tap the tang down onto the support plate. Remove the plate when all the tangs have been returned to their original shape.

17 Slide the gaiter up the driveshaft. Locate the gaiter in the grooves on the driveshaft and outer member.

18 Slide the inner retaining collar into place over the inner end of the gaiter.

19 Using a blunt rod, carefully lift the outer lip of the gaiter to equalise the air pressure. With the rod in position, compress the joint until the dimension from the inner end of the gaiter to

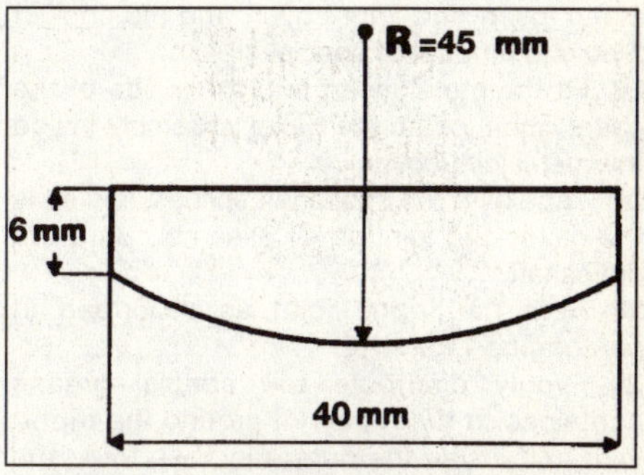

4.15 Support plate dimensions for re-forming the anti-separation plate tangs - Gl62-type joint

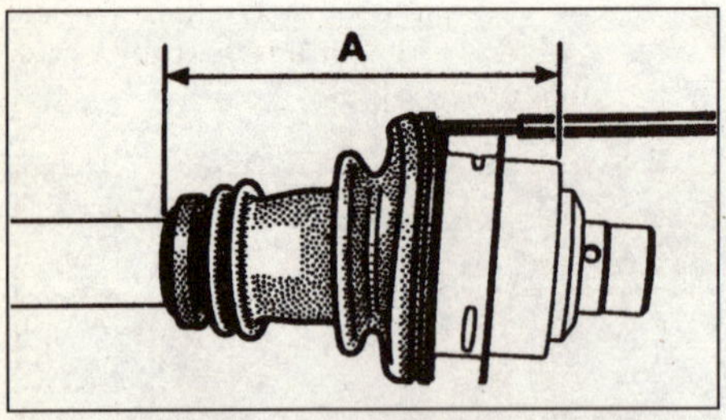

4.19 Fitting dimension for the right-hand driveshaft inner constant velocity joint gaiter - GI62-type joint
A = 153 ± 1 mm

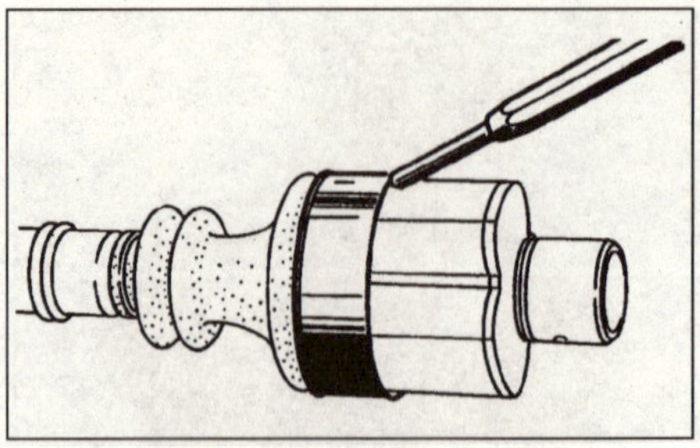

4.24 Removing the metal cover from the right-hand driveshaft inner constant velocity joint - RC490 type joint

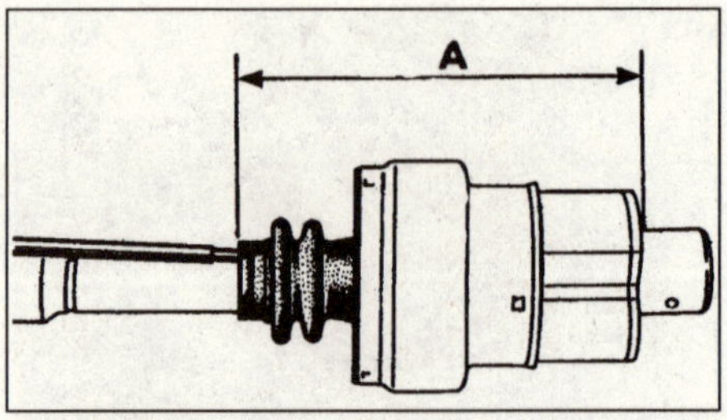

4.32 Fitting dimension for the right-hand driveshaft inner constant velocity joint gaiter - RC490 type joint
A = 156 ± 1 mm

the flat end face of the outer member is as shown **(see illustration)**. Hold the outer member in this position and withdraw the rod.

20 Fit both the inner and outer retaining clips to the gaiter, and secure each one in position by compressing its raised section. In the absence of the special tool, carefully compress each clip using a pair of side cutters, taking great care not to cut through the clip

21 Check that the constant velocity joint moves freely in all directions, then refit the driveshaft as described in Section 2.

RC490-type joint

22 Using a pair of grips, bend up the metal joint cover at the points where it has been staked into the outer member recesses.

23 Using a pair of snips, cut the gaiter inner retaining clip.

24 Using a soft metal drift, tap the metal joint cover off the outer member **(see illustration)**. Slide the outer member off the end of the tripod joint. Be prepared to hold the rollers in place, otherwise they may fall off the tripod ends as the outer member is withdrawn. If necessary, secure the rollers in place using tape after removal of the outer member. **Note:** *The rollers are matched to the tripod joint stems, and it is important that they are not interchanged.*

25 Remove the tripod joint and gaiter assembly, and examine the joint components for wear, using the information given in paragraphs 6 to 9 of this Section. Obtain a repair kit consisting of a gaiter, retaining clip, metal insert and joint cover, and the correct type and amount of special grease.

26 Fit the metal insert into the inside of the gaiter, then locate the gaiter assembly inside the metal joint cover.

27 Tape over the driveshaft splines, and slide the gaiter and joint cover assembly onto the driveshaft.

28 Refit the tripod joint as described in paragraphs 11 and 12.

29 Evenly distribute the special grease contained in the repair kit around the tripod joint and inside the outer member. Pack the gaiter with the remainder of the grease.

30 Slide the outer member into position over the tripod joint.

31 Slide the metal joint cover onto the outer member until it is flush with the outer member guide panel. Secure the joint cover in position by staking it into the recesses in the outer member, using a hammer and a round-ended punch.

32 Using a blunt rod, carefully lift the inner lip of the gaiter to equalise the air pressure. With the rod in position, compress the joint until the dimension from the inner end of the gaiter to the flat end face of the outer member is as shown **(see illustration)**. Hold the outer member in this position and withdraw the rod.

33 Fit the retaining clip to the inner end of the gaiter. Remove any slack in the gaiter retaining clip by carefully compressing the raised section of the clip. In the absence of the special tool, a pair of side cutters may be used, taking great care not to cut the clip

34 Check that the constant velocity joint moves freely in all directions, then refit the driveshaft as described in Section 2.

5 Left-hand driveshaft inner gaiter (manual transmission models) - renewal

1 Remove the driveshaft (refer to Section 2).

2 Using circlip pliers, extract the circlip securing the tripod joint to the driveshaft. Note that on some models, the joint may be staked in position; if so, relieve the staking using a file. Using a dab of paint or a hammer and punch, mark the position of the tripod joint in relation to the driveshaft, to use as a guide to refitting.

3 The tripod joint can now be removed. If it is tight, draw the joint off the driveshaft end using a puller. Ensure that the legs of the puller are located behind the joint inner member and do not contact the joint rollers. Alternatively, support the inner member of the tripod joint and press the shaft out of the joint, again ensuring that no load is applied to the joint rollers.

4 The gaiter and bearing assembly is removed in the same way, either by drawing the bearing off the driveshaft, or by pressing the driveshaft out of the bearing. Remove the retaining plate, noting which way round it is fitted.

5 Obtain a new gaiter, which is supplied complete with the small bearing.

6 Owing to the lip-type seal used in the bearing, the bearing and gaiter must be pressed into position. If a hammer and tubular drift are used to drive the assembly onto the driveshaft, there is a risk of distorting the seal.

7 Refit the retaining plate to the driveshaft, ensuring that it is fitted the correct way around.

8 Support the driveshaft, and press the gaiter bearing onto the shaft, using a tubular spacer which bears only on the bearing inner race. Position the bearing so that the distance from the end of the driveshaft to the inner face of the bearing is as shown **(see illustrations)**.

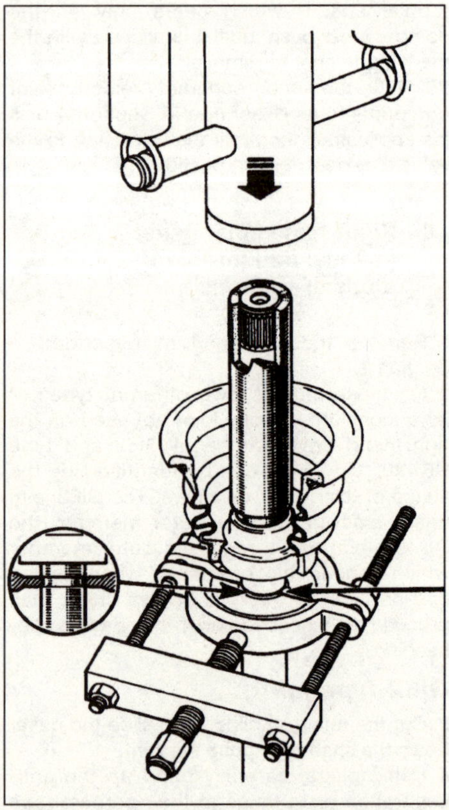

5.8a Pressing the inner bearing/gaiter onto the end of the left-hand driveshaft - manual transmission models

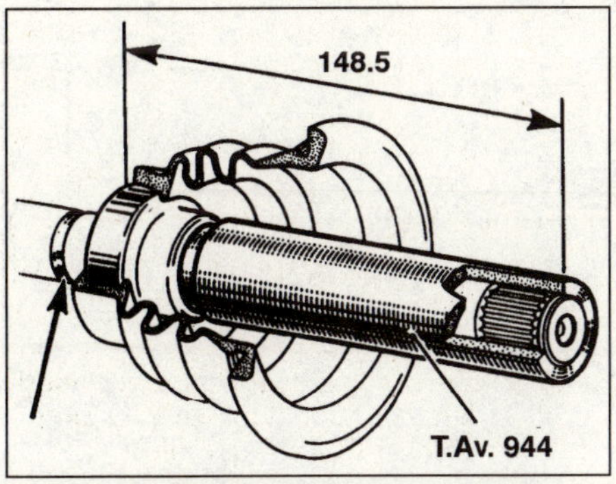

5.8b Fitting dimension (in mm) for the left-hand driveshaft inner bearing/gaiter - manual transmission models

9 Align the marks made on dismantling, and engage the tripod joint with the driveshaft splines. Use a hammer and soft metal drift to tap the joint onto the shaft, taking care not to damage the driveshaft splines or joint rollers. Alternatively, support the driveshaft, and press the joint into position using a tubular spacer which bears only on the joint inner member.

10 Secure the tripod joint in position with the circlip, ensuring that it is correctly located in the driveshaft groove. Where no circlip is fitted, secure the joint in position by staking the end of the driveshaft in three places, at intervals of 120°, using a hammer and punch.

11 Refit the driveshaft to the vehicle as described in Section 2.

6 Driveshaft gaiter renewal (automatic transmission models) - general information

At the time of writing, no information on driveshaft dismantling was available for these models. If gaiter renewal is necessary, the driveshaft should be removed from the vehicle, as described in Section 2, and taken to a Renault dealer.

7 Driveshaft overhaul - general information

If any of the checks described in Chapter 1 reveal wear in a driveshaft joint, first remove the roadwheel trim or centre cap (as appropriate) and check that the driveshaft retaining nut is still correctly tightened; if in doubt, use a torque wrench to check it. Refit the centre cap or trim, and repeat the check on the other driveshaft.

Road test the vehicle, and listen for a metallic clicking from the front as the vehicle is driven slowly in a circle on full-lock. If a clicking noise is heard, this indicates wear in the outer constant velocity joint.

If vibration, consistent with road speed, is felt through the vehicle when accelerating, there is a possibility of wear in the inner constant velocity joints.

Constant velocity joints can be dismantled and inspected for wear as described in Sections 3, 4 and 5.

On models with manual transmission, wear in the outer constant velocity joint can only be rectified by renewing the driveshaft. This is necessary since no outer joint components are available separately. For the inner joint, the tripod joint and roller assembly is available separately, but wear in any of the other components will also necessitate driveshaft renewal.

On models with automatic transmission, wear in either constant velocity joint will necessitate driveshaft renewal; no components for either joint are available separately.

Chapter 9
Braking system

Contents

Degrees of difficulty

Easy, suitable for novice with little experience	**Fairly easy,** suitable for beginner with some experience	**Fairly difficult,** suitable for competent DIY mechanic	**Difficult,** suitable for experienced DIY mechanic	**Very difficult,** suitable for expert DIY or professional

Specifications

Front brakes

Disc thickness:	
Models with solid discs:	
New	12.0 mm
Minimum	10.3 mm
Models with ventilated discs:	
1.6 litre (K7M 702 and 703 engine) models:	
New	20.0 mm
Minimum	18.3 mm
All other models:	
New	20.6 mm
Minimum	19.0 mm
Maximum disc run-out	0.07 mm
Brake pad minimum thickness (backing plate and friction material)	7.0 mm

Rear drum brakes

Drum internal diameter:	
New	203.2 mm
Maximum diameter after machining	204.4 mm
Brake shoe minimum thickness (shoe and friction material)	2.5 mm

Torque wrench settings

Torque wrench settings	Nm	lbf ft
Bleed screw	6	5
Front disc to hub bolts	14	10
Front brake caliper:		
Guide pin bolts	32	24
Mounting bolts	100	74
Hydraulic hose union	15	11
Hydraulic pipe union nut	13	10
Master cylinder retaining nuts	23	17
Rear hub nut	175	127
Roadwheel bolts	90	66
Vacuum servo unit mounting nuts	20	15

1 General information

The braking system is of the servo-assisted, dual-circuit hydraulic type. The arrangement of the hydraulic system is such that each circuit operates one front and one rear brake from a tandem master cylinder. Under normal circumstances, both circuits operate in unison. However, in the event of hydraulic failure in one circuit, full braking force will still be available at two wheels.

All models are fitted with front disc brakes and rear drum brakes. The front disc brakes are actuated by single-piston sliding type calipers, which ensure that equal pressure is applied to each disc pad. The rear drum brakes, the rear brakes incorporate leading and trailing shoes, which are actuated by twin-piston wheel cylinders. A self-adjust mechanism is incorporated, to automatically compensate for brake shoe wear. As the brake shoe linings wear, the footbrake operation automatically operates the adjuster mechanism, which effectively lengthens the shoe strut and repositions the brake shoes, to remove the lining-to-drum clearance.

A load-sensitive pressure-regulating valve is situated in the hydraulic circuit to each rear brake. The valve is linked to the rear axle and regulates the hydraulic pressure applied to the rear brakes, according to the loading on the axle, and so helps prevent rear wheel lock-up during emergency braking.

The handbrake provides an independent mechanical means of rear brake application.

Warning: When servicing any part of the system, work carefully and methodically; also observe scrupulous cleanliness when overhauling any part of the hydraulic system. Always renew components (in axle sets, where applicable) if in doubt about their condition, and use only genuine Renault replacement parts, or at least those of known good quality. Note the warnings given in "Safety first" and at relevant points in this Chapter concerning the dangers of asbestos dust and hydraulic fluid.

2 Hydraulic system - bleeding

Warning: Hydraulic fluid is poisonous; wash off immediately and thoroughly in the case of skin contact, and seek immediate medical advice if any fluid is swallowed or gets into the eyes. Certain types of hydraulic fluid are inflammable, and may ignite when allowed into contact with hot components; when servicing any hydraulic system, it is safest to assume that the fluid is inflammable, and to take precautions against the risk of fire as though it is petrol that is being handled. Hydraulic fluid is also an effective paint stripper, and will attack plastics; if any is spilt, it should be washed off immediately, using copious quantities of fresh water. Finally, it is hygroscopic (it absorbs moisture from the air) - old fluid may be contaminated and unfit for further use. When topping-up or renewing the fluid, always use the recommended type, and ensure that it comes from a freshly-opened sealed container.

Models not equipped with ABS

General

1 The correct operation of any hydraulic system is only possible after removing all air from the components and circuit; this is achieved by bleeding the system.

2 During the bleeding procedure, add only clean, unused hydraulic fluid of the recommended type; never re-use fluid that has already been bled from the system. Ensure that sufficient fluid is available before starting work.

3 If there is any possibility of incorrect fluid being already in the system, the brake components and circuit must be flushed completely with uncontaminated, correct fluid, and new seals should be fitted to the various components.

4 If hydraulic fluid has been lost from the system, or air has entered because of a leak, ensure that the fault is cured before proceeding further.

5 Park the vehicle on level ground, switch off the engine and select first or reverse gear then chock the wheels and release the handbrake.

6 Check that all pipes and hoses are secure, unions tight and bleed screws closed. Clean any dirt from around the bleed screws.

7 Unscrew the master cylinder reservoir cap, and top the master cylinder reservoir up to the "MAXI" level line; refit the cap loosely, and remember to maintain the fluid level at least above the "MINI" level line throughout the procedure, or there is a risk of further air entering the system.

8 There are a number of one-man, do-it-yourself brake bleeding kits currently available from motor accessory shops. It is recommended that one of these kits is used whenever possible, as they greatly simplify the bleeding operation, and also reduce the risk of expelled air and fluid being drawn back into the system. If such a kit is not available, the basic (two-man) method must be used, which is described in detail below.

9 If a kit is to be used, prepare the vehicle as described previously, and follow the kit manufacturer's instructions, as the procedure may vary slightly according to the type being used; generally, they are as outlined below in the relevant sub-section.

10 Whichever method is used, the same sequence must be followed (paragraphs 11 and 12) to ensure the removal of all air from the system.

Bleeding sequence

11 If the system has been only partially disconnected, and suitable precautions were taken to minimise fluid loss, it should be necessary only to bleed that part of the system (ie the primary or secondary circuit).

12 If the complete system is to be bled, then it should be done working in the following sequence:

a) Right-hand rear brake.
b) Left-hand front brake.
c) Left-hand rear brake.
d) Right-hand front brake.

Bleeding - basic (two-man) method

13 Collect a clean glass jar, a suitable length of plastic or rubber tubing which is a tight fit over the bleed screw, and a ring spanner to fit the screw. The help of an assistant will also be required.

14 Remove the dust cap from the first screw in the sequence. Fit the spanner and tube to the screw, place the other end of the tube in the jar, and pour in sufficient fluid to cover the end of the tube.

15 Ensure that the master cylinder reservoir fluid level is maintained at least above the "MINI" level line throughout the procedure.

16 Have the assistant fully depress the brake pedal several times to build up pressure, then maintain it on the final downstroke.

17 While pedal pressure is maintained, unscrew the bleed screw (approximately one turn) and allow the compressed fluid and air to flow into the jar **(see illustration)**. The assistant should maintain pedal pressure, following it down to the floor if necessary, and should not release it until instructed to do so. When the flow stops, tighten the bleed screw again, have the assistant release the pedal slowly, and recheck the reservoir fluid level.

18 Repeat the steps given in paragraphs 16 and 17 until the fluid emerging from the bleed screw is free from air bubbles. If the master cylinder has been drained and refilled, and air is being bled from the first screw in the sequence, allow approximately five seconds between cycles for the master cylinder passages to refill.

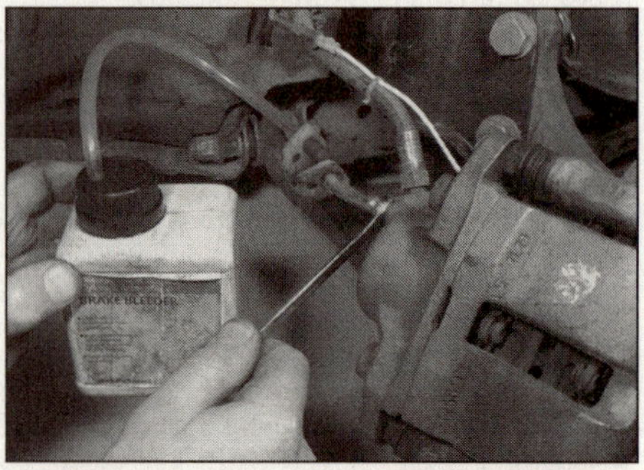

2.17 Bleeding a front brake caliper

19 When no more air bubbles appear, tighten the bleed screw to the specified torque, remove the tube and spanner, and refit the dust cap. Do not overtighten the bleed screw.
20 Repeat the procedure on the remaining screws in the sequence, until all air is removed from the system and the brake pedal feels firm again.

Bleeding - using a one-way valve kit

21 As their name implies, these kits consist of a length of tubing with a one-way valve fitted, to prevent expelled air and fluid being drawn back into the system; some kits include a translucent container, which can be positioned so that the air bubbles can be more easily seen flowing from the end of the tube.
22 The kit is connected to the bleed screw, which is then opened. The user returns to the driver's seat, depresses the brake pedal with a smooth, steady stroke, and slowly releases it; this is repeated until the expelled fluid is clear of air bubbles.
23 Note that these kits simplify work so much that it is easy to forget the master cylinder reservoir fluid level; ensure that this is maintained at least above the "MINI" level line at all times.

Bleeding - using a pressure-bleeding kit

24 These kits are usually operated by the reservoir of pressurised air contained in the spare tyre. However, note that it will probably be necessary to reduce the pressure to a lower level than normal; refer to the instructions supplied with the kit.
25 By connecting a pressurised, fluid-filled container to the master cylinder reservoir, bleeding can be carried out simply by opening each screw in turn (in the specified sequence), and allowing the fluid to flow out until no more air bubbles can be seen in the expelled fluid.
26 This method has the advantage that the large reservoir of fluid provides an additional safeguard against air being drawn into the system during bleeding.
27 Pressure-bleeding is particularly effective when bleeding "difficult" systems, or when bleeding the complete system at the time of routine fluid renewal.

All methods

28 When bleeding is complete, and firm pedal feel is restored, wash off any spilt fluid, tighten the bleed screws to the specified torque, and refit their dust caps.
29 Check the hydraulic fluid level in the master cylinder reservoir, and top-up if necessary (see "*Weekly checks*").
30 Discard any hydraulic fluid that has been bled from the system; it will not be fit for re-use.
31 Check the feel of the brake pedal. If it feels at all spongy, air must still be present in the system, and further bleeding is required. Failure to bleed satisfactorily after a reasonable repetition of the bleeding procedure may be due to worn master cylinder seals.

Models equipped with ABS

Caution: On models equipped with ABS, disconnect the battery before disconnecting any braking system hydraulic union and do not reconnect the battery until after the hydraulic system has been bled. Failure to do this could lead to air entering the hydraulic unit. If air enters the hydraulic unit pump, it will prove very difficult (in some cases impossible) to bleed the unit.
Caution: Do not use a pressure-bleeding kit on models equipped with ABS.
32 On models with ABS due to the complexity of the hydraulic system, special precautions/procedures are necessary to bleed the hydraulic system of air. These procedures differ depending on which part of the system has been disconnected.
33 The basic information given for the conventional system (ignoring the information about pressure-bleeding) also applies to models equipped with ABS. One notable change is the sequence that the complete system is bled in. On models with ABS the order is as follows.

a) Left-hand front brake.
b) Right-hand rear brake.
c) Right-hand front brake.
d) Left-hand rear brake.

34 In addition to this, the following bleeding procedures should be used to ensure that all air is removed from the system.

Bleeding procedure after disconnecting a brake caliper/ wheel cylinder union

Note: *If more than one caliper/wheel cylinder has been removed, carry out this procedure on each brake, working in the order given in paragraph 33.*
35 With the container connected to the bleed screw and the master cylinder topped-up, continue as follows, ensuring that the brake fluid never fails below the "MINI" level in the reservoir.
36 Open up the bleed screw then have your assistant fully depress the brake pedal and hold it down. Close the bleed screw securely then have your assistant release the brake pedal slowly and wait for approximately 3 seconds. Repeat this procedure at least ten times until the fluid flowing from bleed screw is free of air bubbles.
37 Next build up pressure in the braking system by having your assistant pump the brake pedal at least 3 times, keep the pedal depressed on the last stroke. Open the bleed screw again and allow the pedal to reach the floor. Securely close the screw then have your assistant release the pedal slowly and wait at least 3 seconds. Repeat this procedure again and check that the fluid flowing from the bleed screw is free of air bubbles.
38 Check the feel of the brake pedal then remove the container and top-up the fluid level (see "*Weekly checks*"). Ensure that the bleed screw is tightened to the specified torque then refit the dust cap.

Bleeding procedure after disconnecting the pressure-regulating unions

39 Bleed both rear brakes using the information in paragraphs 34 to 37.

Bleeding procedure after disconnecting a master cylinder union

40 If the master cylinder is disconnected for any reason, to minimise the risk of trapping air in the hydraulic unit it is necessary to bleed the cylinder of air prior to reconnecting the pipes. In order to do this two blanking plugs which screw into the master cylinder ports will be required.
41 With both pipes disconnected, screw the plugs into the master cylinder ports and tighten securely.
42 Fill the master cylinder reservoir then open up the rear port (primary circuit) plug/bleed screw and have an assistant depress and hold the brake pedal. Securely close the plug then have your assistant release the brake pedal slowly and wait for approximately 3 seconds. Repeat this procedure 5 or 6 times.
43 Repeat the procedure on the master cylinder front port (secondary circuit) to remove all traces of air from the master cylinder.
44 To ensure that the fluid does not escape (allowing air to enter) when the plugs are disconnected, it is necessary to depress the brake pedal slightly (approximately 30 mm) and hold it in position. Special tools are available from accessory shops to hold pedals in position or alternately have an assistant hold the pedal.
45 Remove the plugs from the brake pipes and check that both pipes are full of fluid; if not top them up. With the brake pedal held in position, quickly remove one of the plugs from the master cylinder port and reconnect the brake pipe tightening its union nut to the specified torque. Repeat the operation on the second master cylinder port then wash off any spilt fluid.
46 Following the information given in paragraphs 34 to 37, the complete braking system should then be bled in the specified sequence.

Bleeding procedure after disconnecting a hydraulic unit union

47 Prior to reconnecting the pipes to the hydraulic unit ensure that all pipes and the hydraulic unit ports are full of fluid. This will minimise the amount of air present in the circuit and lessen the risk of getting air trapped in the hydraulic unit. Tighten all union nuts to the specified torque and wash off any spilt fluid.
48 Following the information given in paragraphs 34 to 37, the complete braking system should then be bled in the specified sequence.

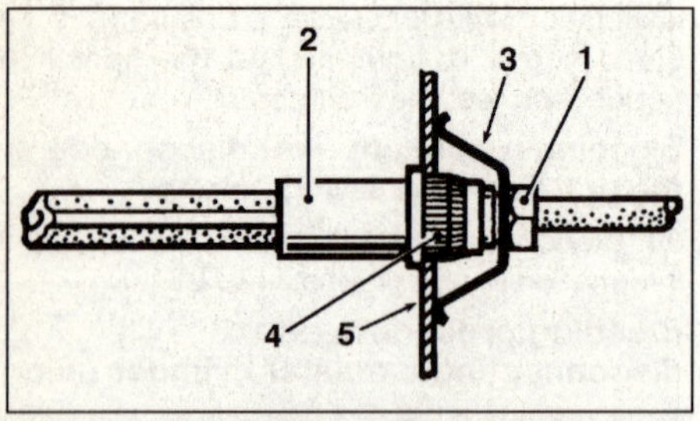

3.2 Typical hydraulic pipe connection to a flexible hose

1 Union nut
2 Flexible hose
3 Spring clip
4 End fitting
5 Bodywork

3 Hydraulic pipes and hoses - renewal

Note: *Before starting work, refer to the note at the beginning of Section 2 concerning the dangers of hydraulic fluid.*

Models not equipped with ABS

1 If any pipe or hose is to be renewed, minimise fluid loss by first removing the master cylinder reservoir cap, then tightening it down onto a piece of polythene to obtain an airtight seal. Alternatively, flexible hoses can be sealed, if required, using a proprietary brake hose clamp; metal brake pipe unions can be plugged (if care is taken not to allow dirt into the system) or capped immediately they are disconnected. Place a wad of rag under any union that is to be disconnected, to catch any spilt fluid.

2 If a flexible hose is to be disconnected, unscrew the brake pipe union nut before removing the spring clip which secures the hose to its mounting bracket **(see illustration)**.

3 To unscrew the union nuts, it is preferable to obtain a brake pipe spanner of the correct size; these are available from most large motor accessory shops **(see illustration)**. Failing this, a close-fitting open-ended spanner will be required, though if the nuts are tight or corroded, their flats may be rounded-off if the spanner slips. In such a case, a self-locking wrench is often the only way to unscrew a stubborn union, but it follows that the pipe and the damaged nuts must be renewed on reassembly. Always clean a union and surrounding area before disconnecting it. If disconnecting a component with more than one union, make a careful note of the connections before disturbing any of them.

4 If a brake pipe is to be renewed, it can be obtained, cut to length and with the union nuts and end flares in place, from Renault dealers. All that is then necessary is to bend it to shape, following the line of the original, before fitting it to the car. Alternatively, most motor accessory shops can make up brake pipes from kits, but this requires very careful measurement of the original, to ensure that the replacement is of the correct length. The safest answer is usually to take the original to the shop as a pattern.

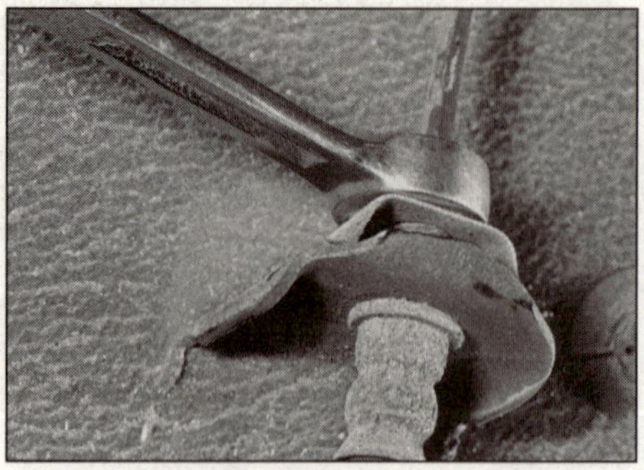

3.3 Using a brake pipe spanner to slacken a hydraulic pipe union nut

5 On refitting, do not overtighten the union nuts. It is not necessary to exercise brute force to obtain a sound joint.

6 Ensure that the pipes and hoses are correctly routed, with no kinks, and that they are secured in the clips or brackets provided. After fitting, remove the polythene from the reservoir, and bleed the hydraulic system as described in Section 2. Wash off any spilt fluid, and check carefully for fluid leaks.

Models equipped with ABS

Caution: On models equipped with ABS, disconnect the battery before disconnecting any braking system hydraulic union and do not reconnect the battery until after the hydraulic system has been bled. Failure to do this could lead to air entering the hydraulic unit. If air enters the hydraulic unit pump, it will prove very difficult (in some cases impossible) to bleed the unit.

7 Refer to the information given in paragraphs 1 to 6, noting the following.

a) *If a brake pipe(s) is to be disconnected from the hydraulic unit, it is essential that the unit port(s) are sealed up immediately to prevent fluid loss and allow the entry of air. This will minimise the risk of air getting trapped in the unit (see Caution). Prior to reconnecting the pipe(s), ensure both the port(s) and pipe(s) are full of fluid (see Section 2).*

b) *If a brake pipe(s) is disconnected from the master cylinder, it will be necessary to bleed the master cylinder prior to reconnecting the pipe (see Section 2).*

4 Front brake pads - renewal

Warning: Renew both sets of front brake pads at the same time - never renew the pads on only one wheel, as uneven braking may result. Note that the dust created by wear of the pads may contain asbestos, which is a health hazard. Never blow it out with compressed air, and don't inhale any of it. An approved filtering mask should be worn when working on the brakes. DO NOT use petrol or petroleum-based solvents to clean brake parts; use brake cleaner or methylated spirit only.

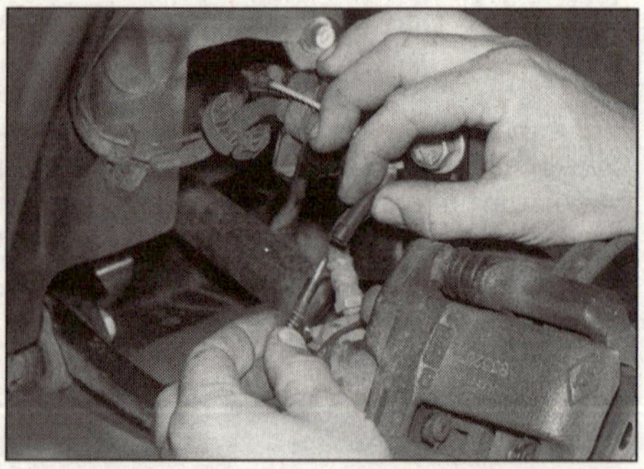

4.2 Disconnect the brake pad wear sensor wiring connector and free the wiring

Note: *New caliper guide pin bolts will be required on refitting.*

1 Chock the rear wheels, apply the handbrake, then jack up the front of the vehicle and support it on axle stands. Remove the front roadwheels.

2 Trace the brake pad wear sensor wiring back from the pads, and disconnect it from the wiring connector **(see illustration)**. Note the routing of the wiring, and free it from any relevant retaining clips.

3 Push the piston into its bore by pulling the caliper outwards.

4 Slacken and remove the caliper lower guide pin bolt, using a slim open-ended spanner to prevent the guide pin themselves from rotating **(see illustration)**. Discard the guide pin bolt - a new one must be used on refitting.

5 Pivot the caliper away from the brake pads and mounting bracket, and tie it to the suspension strut using a suitable piece of wire **(see illustration)**. Do not allow the caliper to hang down by the hose.

6 Withdraw the two brake pads from the caliper mounting bracket **(see illustration)**.

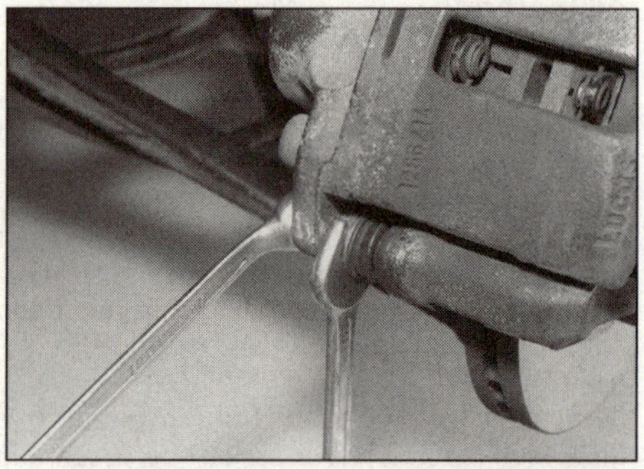

4.4 Slacken and remove the lower guide pin bolt whilst retaining the guide pin with an open-ended spanner

4.5 Pivot the caliper upwards and away from the disc . . .

4.6 . . . and lift out the pads from the caliper mounting bracket

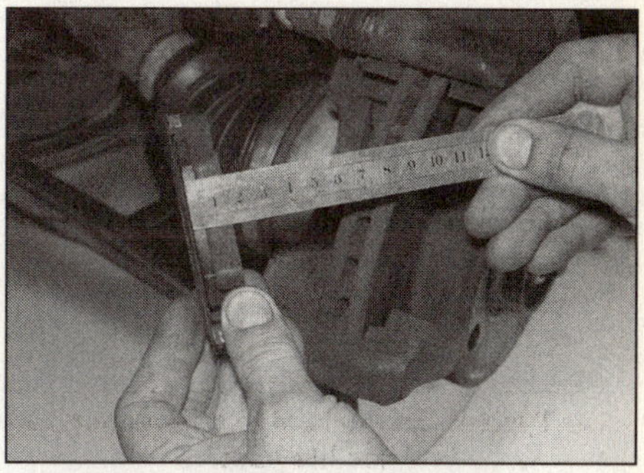

4.7 Measuring brake pad thickness

7 First measure the thickness of each brake pad (friction material and backing plate) **(see illustration).** If either pad is worn at any point to the specified minimum thickness or less, all four pads must be renewed. The pads should also be renewed if any are fouled with oil or grease; note that there is no satisfactory way of degreasing friction material, once contaminated. If any of the brake pads are worn unevenly, or are fouled with oil or grease, trace and rectify the cause before reassembly.

8 If the brake pads are still serviceable, carefully clean them using a clean, fine wire brush or similar, paying particular attention to the sides and back of the metal backing. Clean out the grooves in the friction material, and pick out any large embedded particles of dirt or debris. Carefully clean the pad locations in the caliper mounting bracket.

9 Prior to fitting the pads, check that the guide pins are free to slide easily in the caliper mounting bracket, and check that the rubber guide pin gaiters are undamaged **(see illustration).** Brush the dust and dirt from the caliper and piston, but **do not** inhale it, as it is a health hazard. Inspect the dust seal around the piston for damage, and the piston for evidence of fluid leaks, corrosion or damage. If attention to any of these components is necessary, refer to Section 8.

10 If new brake pads are to be fitted, the caliper piston must be pushed back into the cylinder to make room for them. Either use a G-clamp or similar tool, or use suitable pieces of wood as levers; take care not to damage the piston seals. Provided that the master cylinder reservoir has not been overfilled with hydraulic fluid, there should be no spillage, but keep a careful watch on the fluid level while retracting the piston. If the fluid level rises above the "MAXI" level line at any time, the surplus should be siphoned off or ejected via a plastic tube connected to the bleed screw (see Section 2). **Note:** *Do not syphon the fluid by mouth, as it is poisonous; use a syringe or an old poultry baster.*

11 Ensuring that the friction material of each pad is against the brake disc, fit the pads to the caliper mounting bracket noting that the pad with the wear sensor wiring should be fitted as the inner pad.

4.9 Prior to fitting the pads check the guide pin gaiters for signs of wear or damage

12 Pivot the caliper down into position over the pads, ensuring that the pad warning sensor wiring is correctly routed. Ensure that the pad anti-rattle springs are correctly positioned against the caliper then press down on the caliper and install the new guide pin bolt **(see illustration).** Tighten the guide pin bolt to the specified torque setting, while retaining the guide pin with an open-ended spanner.

13 Reconnect the brake pad wear sensor wiring connector, ensuring that the wiring is securely clipped in position.

14 Depress the brake pedal repeatedly, until the pads are pressed into firm contact with the brake disc, and normal (non-assisted) pedal pressure is restored.

15 Repeat the above procedure on the remaining front brake caliper.

16 Refit the roadwheels, then lower the vehicle to the ground and tighten the roadwheel bolts to the specified torque setting.

17 Check the hydraulic fluid level as described in "*Weekly checks*".

HAYNES HiNT ***New pads will not give full braking efficiency until they have bedded in. Be prepared for this, and avoid hard braking as far as possible for the first hundred miles or so after pad renewal.***

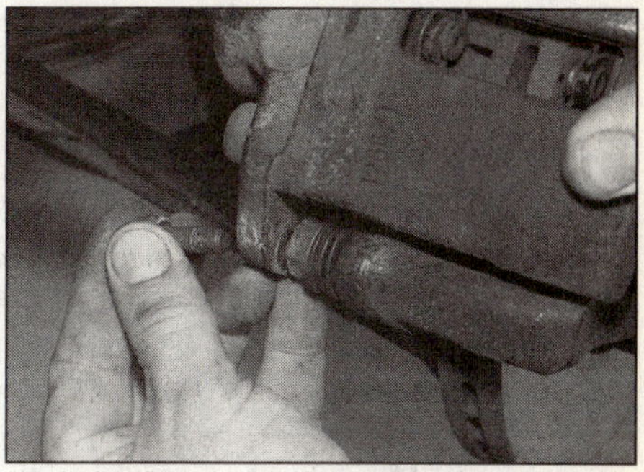

4.12 Pivot the caliper down into position and fit the new guide pin bolt

5 Rear brake shoes - renewal

Warning: Brake shoes must be renewed on both rear wheels at the same time - never renew the shoes on only one wheel, as uneven braking may result. Also, the dust created by wear of the shoes may contain asbestos, which is a health hazard. Never blow it out with compressed air, and don't inhale any of it. An approved filtering mask should be worn when working on the brakes. DO NOT use petrol or petroleum-based solvents to clean brake parts; use brake cleaner or methylated spirit only.

1 Remove the brake drum as described in Section 7.

2 Working carefully, and taking the necessary precautions, remove all traces of brake dust from the brake drum, backplate and shoes.

3 Measure the thickness of each brake shoe at several points (friction material and shoe); if either shoe is worn at any point to the specified minimum thickness or less, all four shoes must be renewed as a set. The shoes should also be renewed if any are fouled with oil or grease; there is no satisfactory way of degreasing friction material, once contaminated.

4 If any of the brake shoes are worn unevenly, or fouled with oil or grease, trace and rectify the cause before reassembly.

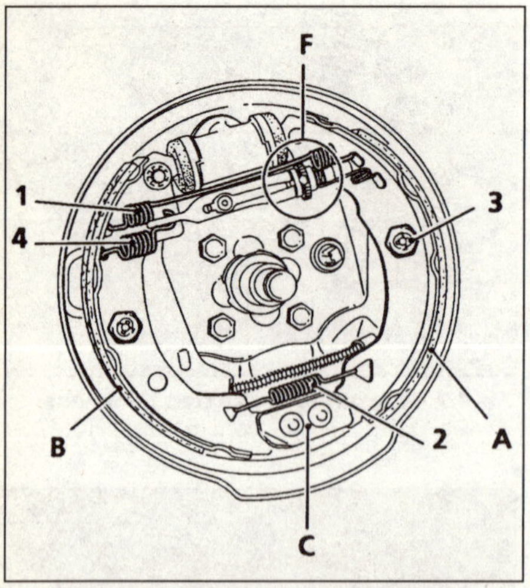

5.6 Rear drum brake assembly

1 *Upper return spring*
2 *Lower return spring*
3 *Retainer spring cup*
4 *Adjuster strut spring*
A *Leading shoe*
B *Trailing shoe*
C *Lower pivot point*
F *Adjuster strut knurled wheel*

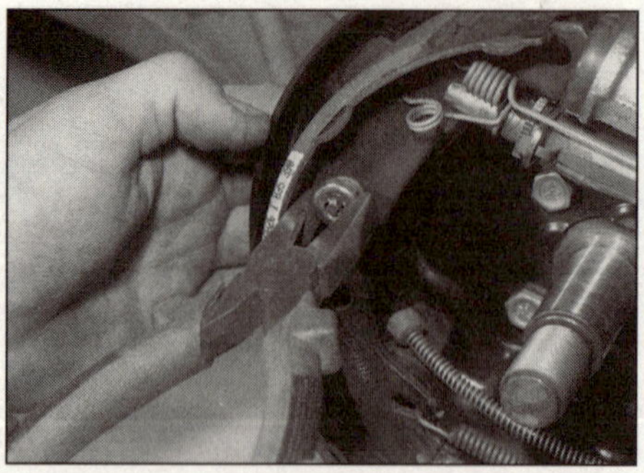

5.7a Using pliers, depress the spring cup and rotate through 90° . . .

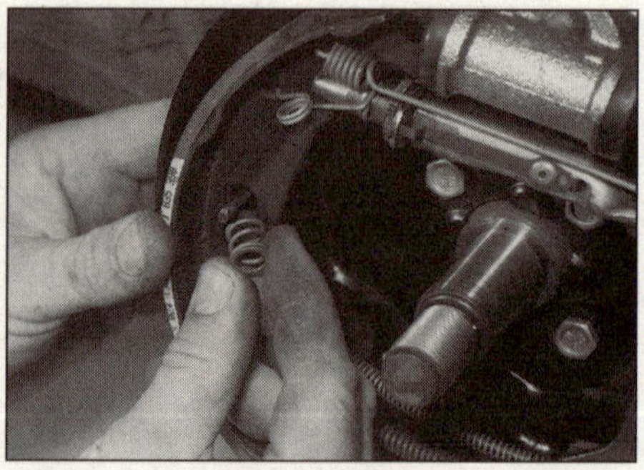

5.7b . . . then lift off the spring and remove the retainer pin

5 To renew the brake shoes, proceed as follows. If the all components are in good condition, refit the brake drum as described in Section 7.

6 Note the position of each shoe, and the location of each of the springs **(see illustration)**. Also make a note of the self-adjuster component locations, to aid refitting later.

7 Using a pair of pliers, remove the shoe retainer spring cups by depressing and turning them through 90°. With the cups removed, lift off the springs and withdraw the retainer pins **(see illustrations)**.

8 Ease the shoes out one at a time from the lower pivot point, to release the tension of the return spring, then disconnect the lower return spring from both shoes **(see illustration)**.

9 Ease the upper end of both shoes out from their wheel cylinder locations, taking care not to damage the wheel cylinder seals, and disconnect the handbrake cable from the trailing shoe. The brake shoe and adjuster strut assembly can then be manoeuvred out of position and away from the backplate. Do not depress the brake pedal until the brakes are reassembled; wrap a strong elastic band around the wheel cylinder pistons to retain them **(see illustrations)**.

10 With the shoe and adjuster strut assembly on the bench, make a note of the correct fitted positions of the springs and adjuster strut, to use as a guide on reassembly. Release the handbrake lever stop-peg (if not already done), then detach the adjuster strut bolt retaining spring from the leading shoe. Disconnect the upper return spring, then detach the leading shoe and return spring from the trailing shoe and strut assembly. Unhook the adjuster strut from the trailing shoe, and remove its spring noting which way around it is fitted **(see illustrations)**.

5.8 Ease the shoes out from the lower pivot point and remove the lower return spring

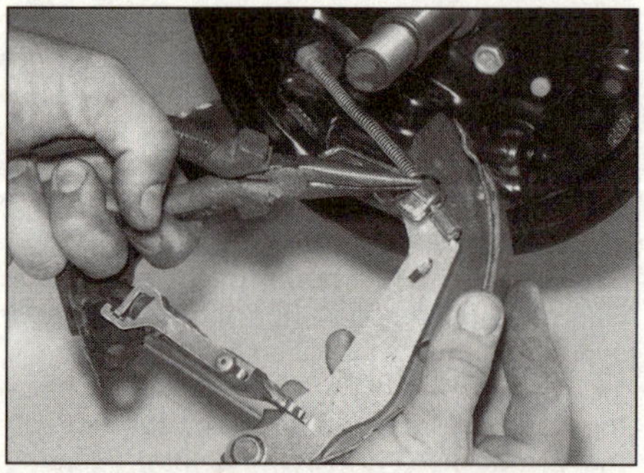

5.9a Remove the shoe assembly from the backplate and detach it from the handbrake cable

5.9b Whilst the shoes are removed wrap a strong elastic band around the wheel cylinder to retain the pistons

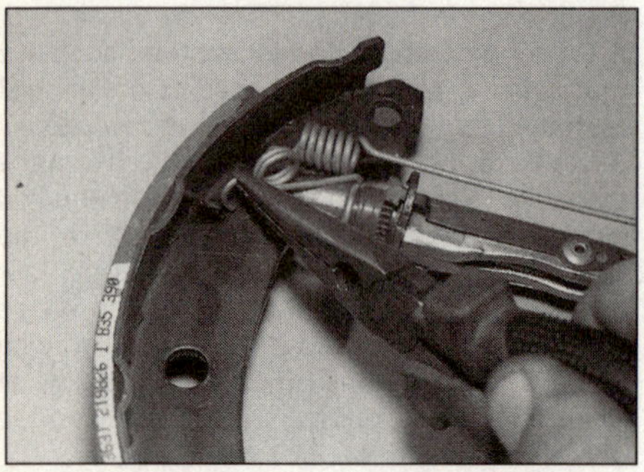

5.10a Unhook the adjuster strut bolt spring from the leading shoe . . .

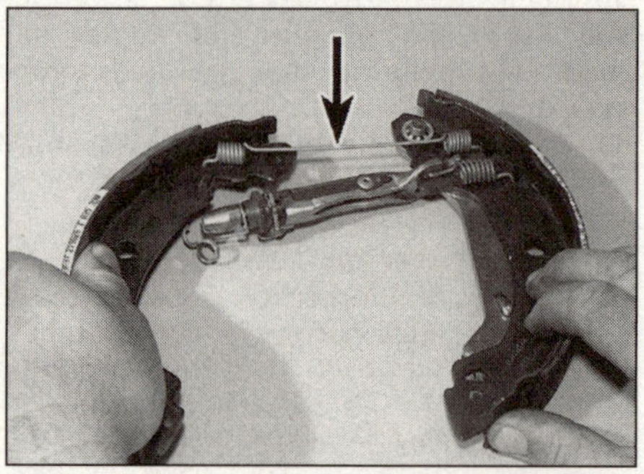

5.10b . . . then separate the shoe from the strut and remove the upper return spring (arrowed)

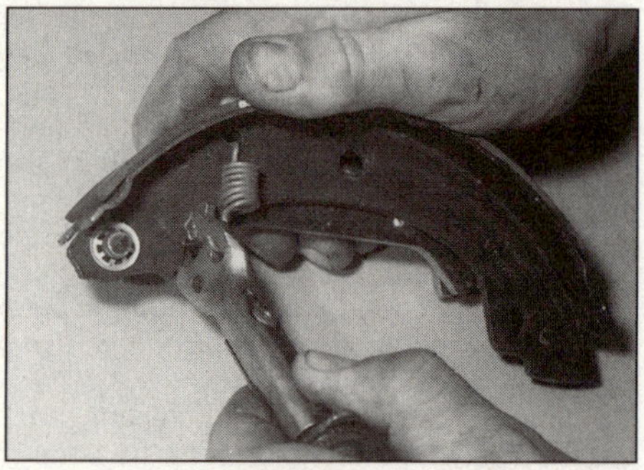

5.10c Unhook the adjuster strut from the trailing shoe . . .

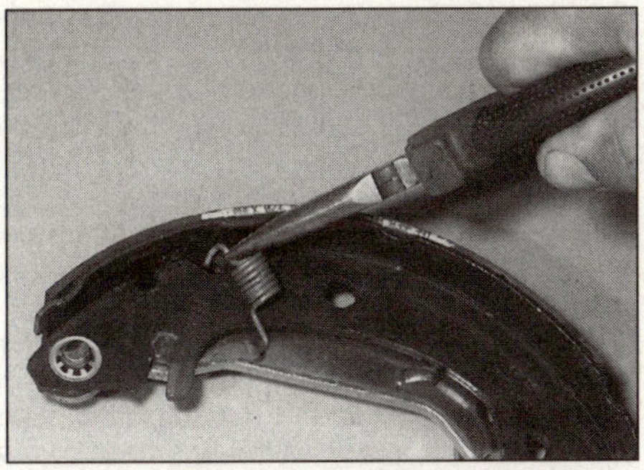

5.10d . . . and remove the spring noting that it is fitted with its shorter hook engaged with the shoe

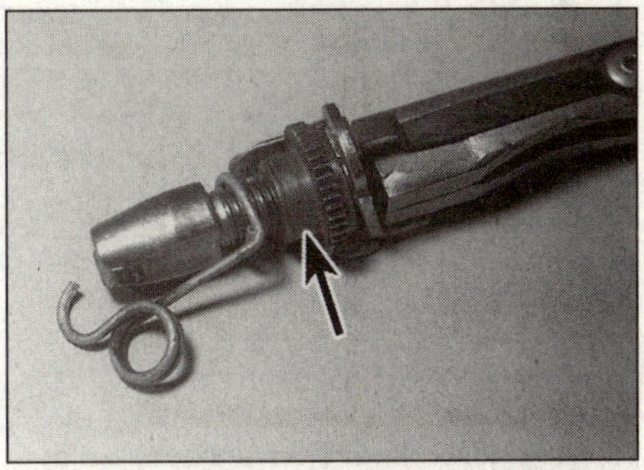

5.11 The adjuster struts can be identified by the grooves (arrowed) on the knurled wheel - left-hand strut assembly shown

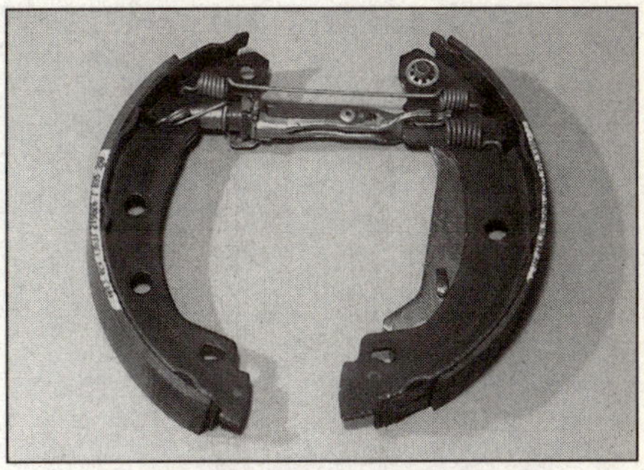

5.15 Ensure that all components are correctly reassembled before refitting

11 Withdraw the adjuster bolt from the strut, and carefully examine the assembly for signs of wear or damage, paying particular attention to the threads of the adjuster bolt and the knurled adjuster wheel, and renew if necessary. Note that left-hand and right-hand struts are not interchangeable; the strut bolts can also be identified by the grooves on the adjuster wheel collars; the left-hand strut bolt has two grooves whereas the right-hand bolt has only a single groove **(see illustration)**.

12 Depending on the type of brake shoes being installed, it may be necessary to remove the handbrake lever from the original trailing shoe, and install it on the new shoe. Secure the lever in position with a new retaining clip. All return springs should be renewed, regardless of their apparent condition; spring kits are available from Renault dealers.

13 Ensure that the components on the end of the strut are correctly positioned, then apply a little high melting-point grease to the threads of the adjuster bolt. Screw the adjuster wheel onto the bolt until only a small gap exists between the wheel and the head of the bolt, then install the bolt in the strut.

14 Fit the adjuster strut retaining spring to the trailing shoe, ensuring that the shorter hook of the spring is engaged with the shoe. Attach the adjuster strut to the spring end, then ease the strut into position in its slot in the trailing shoe.

15 Engage the upper return spring with the trailing shoe. Hook the leading shoe onto the other end of the spring, and lever the leading shoe down until the adjuster bolt head is correctly located in its groove. Once the bolt is correctly located, hook its retaining spring into the slot on the leading shoe **(see illustration)**.

16 Remove the elastic band fitted to the wheel cylinder. Peel back the rubber protective caps, and check the wheel cylinder for fluid leaks or other damage. Also check that both cylinder pistons are free to move easily. Refer to Section 9, if necessary, for information on wheel cylinder renewal.

17 Prior to installation, clean the backplate and apply a thin smear of high-temperature brake grease or anti-seize compound to all those surfaces of the backplate which bear on the shoes, particularly the wheel cylinder pistons and lower pivot point. Do not use too much, and don't allow the lubricant to foul the friction material.

18 Ensure that the handbrake lever stop-peg is correctly located against the edge of the trailing shoe.

19 Manoeuvre the shoe and strut assembly into position on the vehicle and attach the handbrake cable to the trailing shoe lever. Engage the upper ends of both shoes with the wheel cylinder pistons then fit the lower return spring to both shoes, and ease the shoes into position on the lower pivot point.

20 Centralise the shoes relative to the backplate by tapping them. Refit the shoe retainer pins and springs, and secure them in position with the spring cups.

21 Using a screwdriver, turn the strut adjuster wheel until the diameter of the shoes is between 202.5 and 202.7 mm. This should allow the brake drum to just pass over the shoes.

22 Slide the drum into position over the linings, but do not refit the hub nut yet.

23 Repeat the above procedure on the remaining rear brake.

24 Once both sets of rear shoes have been renewed, adjust the lining-to-drum clearance by alternately depressing the brake pedal then applying the handbrake. Whilst doing this, have an assistant listen to the rear drums, to check that the adjuster strut is functioning correctly; if this is so, a clicking sound will be emitted by the strut as the pedal/lever is operated.

25 Remove both the rear drums and adjust the handbrake as described in Section 14.

26 Refit the brake drums as described in Section 7.

27 On completion, check the hydraulic fluid level as described in "*Weekly checks*".

> **HAYNES HiNT** ***New shoes will not give full braking efficiency until they have bedded in. Be prepared for this, and avoid hard braking as far as possible for the first hundred miles or so after shoe renewal.***

6 Front brake disc - inspection, removal and refitting

Note: *Before starting work, refer to the note at the beginning of Section 4 concerning the dangers of asbestos dust.*

Inspection

Note: *If either disc requires renewal, BOTH should be renewed at the same time, to ensure even and consistent braking. New brake pads should also be fitted.*

1 Apply the handbrake, then jack up the front of the car and support it on axle stands. Remove the appropriate front roadwheel.

2 Slowly rotate the brake disc so that the full area of both sides can be checked; remove the brake pads if better access is required to the inboard surface. Light scoring is normal in the area swept by the brake pads, but if heavy scoring or cracks are found, the disc must be renewed.

3 It is normal to find a lip of rust and brake dust around the disc's perimeter; this can be scraped off if required. If, however, a lip has formed due to excessive wear of the brake pad swept area, then the disc's thickness must be measured using a micrometer **(see illustration)**. Take measurements at several places around the disc, at the inside and outside of the pad swept area; if the disc has worn at any point to the specified minimum thickness or less, the disc must be renewed.

6.3 Using a micrometer to measure brake disc thickness

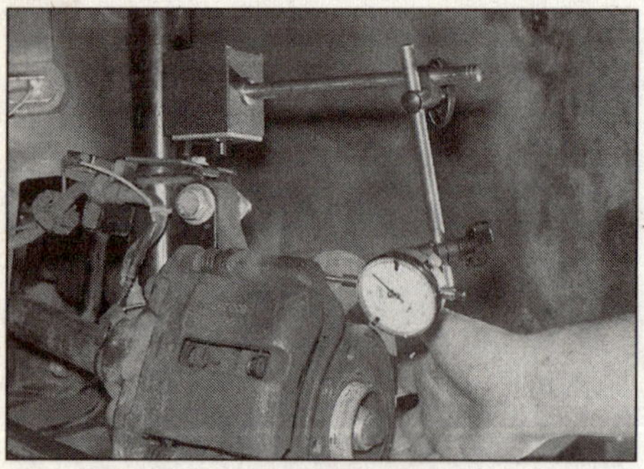

6.4 Using a dial gauge to measure brake disc runout

4 If the disc is thought to be warped, it can be checked for run-out. Either use a dial gauge mounted on any convenient fixed point, while the disc is slowly rotated, or use feeler blades to measure (at several points all around the disc) the clearance between the disc and a fixed point, such as the caliper mounting bracket **(see illustration)**. If the measurements obtained are at the specified maximum or beyond, the disc is excessively warped, and must be renewed; however, it is worth checking first that the hub bearing is in good condition (Chapters 1 and/or 10).

5 Check the disc for cracks, especially around the wheel bolt holes, and any other wear or damage, and renew if necessary.

Removal

6 Slacken and remove the two bolts securing the brake caliper mounting bracket to the swivel hub **(see illustration)**. Slide the caliper assembly off the disc and tie the assembly to the front coil spring, using a piece of wire or string, to avoid placing any strain on the hydraulic brake hose.

7 Use chalk or paint to mark the relationship of the disc to the hub, then remove the screws securing the brake disc to the hub, and remove the disc. If it is tight, lightly tap its rear face with a hide or plastic mallet.

Refitting

8 Refitting is the reverse of the removal procedure, noting the following points:

a) *Ensure that the mating surfaces of the disc and hub are clean and flat.*

b) *Align (if applicable) the marks made on removal, and tighten the disc retaining screws to the specified torque setting.*

c) *If a new disc has been fitted, use a suitable solvent to wipe any preservative coating from the disc, before refitting the caliper.*

d) *Prior to installation, clean the threads of the caliper bracket mounting bolts and coat them with thread locking compound (Renault recommend Loctite Frenbloc - available from your Renault dealer). Slide the caliper into position, making sure the pads pass either side of the disc, and tighten the caliper bracket bolts to the specified torque setting.*

e) *Refit the roadwheel, then lower the vehicle to the ground and tighten the wheel bolts to the specified torque. Apply the footbrake several times to force the pads back into contact with the disc before driving the vehicle.*

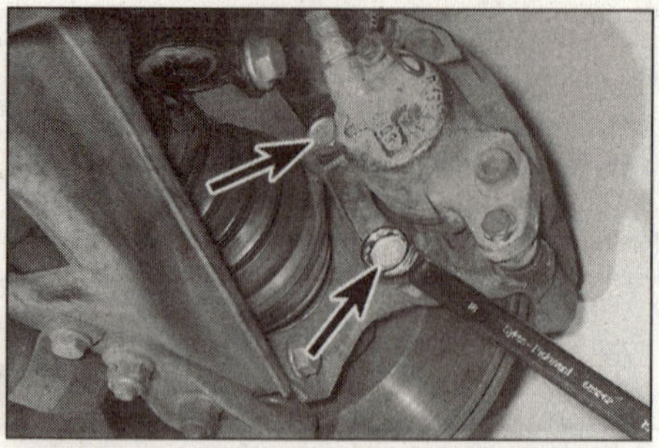

6.6 Slacken and remove the mounting bracket bolts (arrowed) and slide the brake caliper off from the disc

7 Rear brake drum - removal, inspection and refitting

Note: *Before starting work, refer to the note at the beginning of Section 5 concerning the dangers of asbestos dust.*

Note: *A new hub nut will be required on refitting.*

Removal

1 Chock the front wheels, then jack up the rear of the vehicle and support it on axle stands. Remove the appropriate rear wheel.

2 Using a hammer and suitable large flat-bladed screwdriver, carefully tap and prise the cap out of the centre of the brake drum **(see illustration)**. If the cap is damaged, it must be renewed.

3 Using a socket and long bar, slacken and remove the rear hub nut and (where fitted) the washer. Discard the hub nut; a new nut must used on refitting.

4 It should now be possible to withdraw the brake drum and hub bearing assembly from the stub axle by hand. It may be difficult to remove the drum due to the tightness of the hub bearing on the stub axle, or due to the brake shoes binding on the inner circumference of the drum. If the bearing is tight, tap the periphery of the drum using a hide or plastic mallet, or use a universal puller, secured to the drum with the wheel bolts, to pull it off. If the brake shoes are binding, proceed as follows.

5 First ensure that the handbrake is fully off. From underneath the vehicle, slacken the handbrake cable adjuster locknut, then back off the adjuster nut on the handbrake lever rod. Note that on some models, it will first be necessary to remove the mounting nut(s) and lower the exhaust heat shield to gain access to the adjuster nut.

6 Insert a screwdriver through one of the wheel bolt holes in the brake drum, so that it contacts the handbrake operating lever on the trailing brake shoe. Push the lever until the stop-peg slips behind the brake shoe web, allowing the brake shoes to retract fully. Withdraw the brake drum, and slide the spacer off the stub axle **(see illustrations)**.

7.2 Levering out the hub cap from the rear brake drum

7.6a If necessary to increase clearance, insert a screwdriver through one of the brake drum holes . . .

7.6b . . . push the handbrake lever inwards so that the stop-peg (arrowed) slips behind the web of the shoe

7.6c Remove the brake drum and slide the spacer off the stub axle

7.15a On refitting, fit a new hub nut . . .

7.15b . . . and tighten it to the specified torque setting

Inspection

7 Working carefully, remove all traces of brake dust from the drum, but *avoid inhaling the dust, as it is injurious to health.*

8 Scrub clean the outside of the drum, and check it for obvious signs of wear or damage such as cracks around the roadwheel bolt holes; renew the drum if necessary.

9 Examine carefully the inside of the drum. Light scoring of the friction surface is normal, but if heavy scoring is found, the drum must be renewed. It is usual to find a lip on the drum's inboard edge which consists of a mixture of rust and brake dust; this should be scraped away to leave a smooth surface which can be polished with fine (120 to 150 grade) emery paper. If the lip is due to the friction surface being recessed by wear, then the drum must be refinished (within the specified limits) or renewed.

10 If the drum is thought to be excessively worn or oval, its internal diameter must be measured at several points using an internal micrometer. Take measurements in pairs, the second at right-angles to the first, and compare the two to check for signs of ovality. Minor ovality can be corrected by machining; otherwise, renew the drum.

Refitting

11 If a new brake drum is to be installed, use a suitable solvent to remove any preservative coating that may have been applied to its interior.

12 Ensure that the handbrake lever stop-peg is correctly repositioned against the edge of the brake shoe web. Apply a smear of gear oil to the stub axle, and slide on the spacer and brake drum, being careful not to get oil onto the brake shoes or the friction surface of the drum. Depress the footbrake several times to operate the self-adjusting mechanism until normal, non-assisted pedal action returns.

13 Repeat the above procedure on the remaining rear brake assembly (as necessary).

14 Remove the brake drum(s) and adjust the handbrake as described in Section 14.

15 Once the handbrake is correctly adjusted, refit the drum then fit the washer (where necessary) and a new hub nut. Tighten the nut to the specified torque and tap the hub cap into place in the centre of the brake drum **(see illustrations)**.

16 On completion, refit the roadwheel(s), lower the vehicle to the ground and tighten the wheel bolts to the specified torque.

8 Front brake caliper - removal, overhaul and refitting

Caution: On models equipped with ABS, disconnect the battery before disconnecting any braking system hydraulic union and do not reconnect the battery until after the hydraulic system has been bled. Failure to do this could lead to air entering the hydraulic unit. If air enters the hydraulic unit pump, it will prove very difficult (in some cases impossible) to bleed the unit (see Section 2).

Note: *Before starting work, refer to the note at the beginning of Section 2 concerning the dangers of hydraulic fluid, and to the warning at the beginning of Section 4 concerning the dangers of asbestos dust.*

Removal

Note: *New guide pin bolts will be required on refitting.*

1 Apply the handbrake, then jack up the front of the vehicle and support it on axle stands. Remove the appropriate roadwheel.

2 Minimise fluid loss by first removing the master cylinder reservoir cap, and then tightening it down onto a piece of polythene, to obtain an airtight seal. Alternatively, use a brake hose clamp, a G-clamp or a similar tool to clamp the flexible hose.

3 Clean the area around the caliper hose union, then loosen the union. Disconnect the pad wear warning sensor wiring connector, and free it from any relevant retaining clips.

4 Slacken and remove the upper and lower caliper guide pin bolts, using a slim open-ended spanner to prevent the guide pin itself from rotating. Discard the bolts, new ones must be used on refitting. Lift the caliper away from the brake disc, then unscrew the caliper from the end of the brake hose; plug the hose end to minimise fluid loss and prevent dirt entry. Note that the brake pads need not be disturbed, and can be left in position in the caliper mounting bracket.

Overhaul

5 With the caliper on the bench, wipe away all traces of dust and dirt, but *avoid inhaling the dust, as it is injurious to health.*

6 Carefully remove the retaining ring from the outside of the caliper dust seal. Withdraw the partially ejected piston from the caliper body, and remove the dust seal. If the piston cannot be withdrawn by hand, it can be pushed out by applying compressed air to the brake hose union hole. Only low pressure should be required, such as is generated by a foot pump. As the piston is expelled take great care not to trap your fingers between the piston and caliper.

7 Using a small screwdriver, extract the piston hydraulic seal, taking great care not to damage the caliper bore.

8 Thoroughly clean all components, using only methylated spirit, isopropyl alcohol or clean hydraulic fluid as a cleaning medium. Never use mineral-based solvents such as petrol or paraffin, as they will attack the hydraulic system's rubber components. Dry the components immediately, using compressed air or a clean, lint-free cloth. Use compressed air to blow clear the fluid passages.

9 Check all components, and renew any that are worn or damaged. Check particularly the cylinder bore and piston; these should be renewed (note that this means the renewal of the complete body assembly) if they are scratched, worn or corroded in any way. Similarly check the condition of the guide pins and their gaiters; both pins should be undamaged and (when cleaned) a reasonably tight sliding fit in the caliper bracket. If there is any doubt about the condition of any component, renew it.

10 If the assembly is fit for further use, obtain the appropriate repair kit; the components are available from Renault dealers in various combinations. All rubber seals should be renewed as a matter of course; these should never be re-used.

11 On reassembly, ensure that all components are clean and dry.

12 Soak the piston and the new piston (fluid) seal in clean hydraulic fluid. Smear clean fluid on the cylinder bore surface.

13 Fit the new piston (fluid) seal, using only your fingers (no tools) to manipulate it into the cylinder bore groove.

14 Fit the new dust seal to the piston groove then carefully ease the piston squarely into the cylinder bore using a twisting motion. Press the piston fully into position then seat the outer lip of the dust seal on the caliper body and secure it in position with the retaining ring.

15 If the guide pins are being renewed, lubricate the pin shafts with the special grease supplied in the repair kit and fit the gaiters to the pin grooves. Insert the pins into the caliper bracket and seat the gaiters correctly in the bracket grooves.

Refitting

16 Screw the caliper body fully onto the flexible hose union.

17 Ensure that the brake pads are still correctly fitted in the caliper mounting bracket and refit the caliper, making sure the pad wear sensor wiring is correctly routed.

18 Make sure the pad anti-rattle springs are correctly engaged with the caliper body then press the caliper into position and fit the new guide pin bolts. Tighten both guide pin bolts to the specified torque, starting with the lower bolt, while retaining the guide pins with an open-ended spanner.

19 Tighten the brake hose union nut to the specified torque then remove the brake hose clamp or polythene (where fitted).

20 Bleed the hydraulic system as described in Section 2. Note that, providing the precautions described were taken to minimise brake fluid loss, it should only be necessary to bleed the relevant front brake.

21 Refit the roadwheel, then lower the vehicle to the ground and tighten the roadwheel bolts to the specified torque.

9 Rear wheel cylinder - removal and refitting

Caution: On models equipped with ABS, disconnect the battery before disconnecting any braking system hydraulic union and do not reconnect the battery until after the hydraulic system has been bled. Failure to do this could lead to air entering the hydraulic unit. If air enters the hydraulic unit pump, it will prove very difficult (in some cases impossible) to bleed the unit (see Section 2).

Note: *Before starting work, refer to the note at the beginning of Section 2 concerning the dangers of hydraulic fluid, and to the warning at the beginning of Section 5 concerning the dangers of asbestos dust.*

Removal

1 Remove the brake drum as described in Section 7.

2 Minimise fluid loss by first removing the master cylinder reservoir cap, and then tightening it down onto a piece of polythene, to obtain an airtight seal. Alternatively, use a brake hose clamp, a G-clamp or a similar tool to clamp the flexible hose at the nearest convenient point to the wheel cylinder.

3 Carefully unhook the brake shoe upper return spring, and remove it from both brake shoes. Pull the upper ends of the shoes away from the wheel cylinder to disengage them from the pistons.

4 Wipe away all traces of dirt around the brake pipe union at the rear of the wheel cylinder, and unscrew the union nut **(see illustration)**. Carefully ease the pipe out of the wheel cylinder, and plug or tape over its end to prevent dirt entry. Wipe off any spilt fluid immediately.

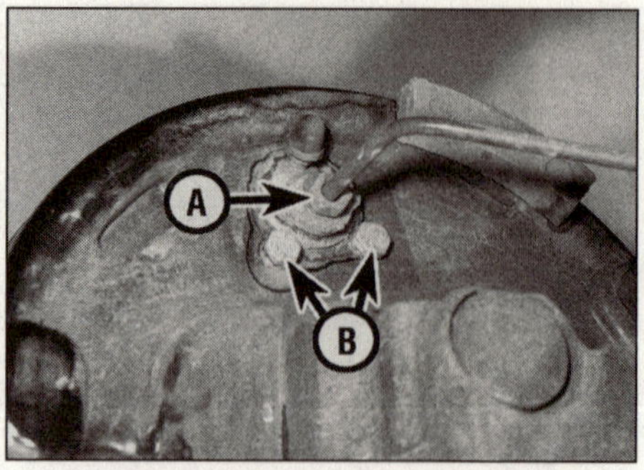

9.4 Rear wheel cylinder brake pipe union nut (A) and retaining bolts (B)

5 Unscrew the two wheel cylinder retaining bolts from the rear of the backplate, and remove the cylinder, along with its sealing ring (where fitted), taking great care not to allow surplus hydraulic fluid to contaminate the brake shoe linings.

6 It is not possible to overhaul the cylinder, since no components are available separately. If faulty, the complete wheel cylinder assembly must be renewed.

Refitting

7 Ensure that the backplate and wheel cylinder mating surfaces are clean and dry. On models where a sealing ring is fitted to the cylinder, use a new sealing ring. Where no sealing ring is fitted, apply a smear of sealing compound to the wheel cylinder mating surface.

8 Spread the brake shoes and manoeuvre the wheel cylinder into position. Engage the brake pipe, and screw in the union nut two or three turns to ensure that the thread has started.

9 Insert the two wheel cylinder retaining bolts, tightening them securely, then tighten the brake pipe union nut to the specified torque.

10 Remove the clamp from the flexible brake hose, or the polythene from the master cylinder reservoir (as applicable).

11 Ensure that the brake shoes are correctly located in the cylinder pistons, then carefully refit the brake shoe return spring, ensuring it is correctly located in both shoes.

12 Refit the brake drum as described in Section 7.

13 Bleed the brake hydraulic system as described in Section 2. Providing suitable precautions were taken to minimise loss of fluid, it should only be necessary to bleed the relevant rear brake.

10 Master cylinder - removal, overhaul and refitting

Caution: On models equipped with ABS, disconnect the battery before disconnecting any braking system hydraulic union and do not reconnect the battery until after the hydraulic system has been bled. Failure to do this could lead to air entering the hydraulic unit. If air enters the hydraulic unit pump, it will prove very difficult (in some cases impossible) to bleed the unit (see Section 2).

Note: *Before starting work, refer to the warning at the beginning of Section 2 concerning the dangers of hydraulic fluid.*

Removal

1 Position a container beneath the master cylinder. Disconnect both fluid hoses from the top of the master cylinder and allow the contents of the reservoir to drain into the container.

2 On left-hand drive models, if necessary, to improve access to the master cylinder undo the retaining nuts and remove the fluid reservoir assembly from the engine compartment bulkhead. On right-hand drive models access to the master cylinder is very poor but can only be significantly improved by removing the inlet manifold.

3 Wipe clean the area around the brake pipe unions on the side of the master cylinder, and place absorbent rags beneath the pipe unions to catch any surplus fluid. Make a note of the correct fitted positions of the unions, then unscrew the union nuts and carefully withdraw the pipes. Plug or tape over the pipe ends and master cylinder orifices, to minimise the loss of brake fluid, and to prevent the entry of dirt into the system. Wash off any spilt fluid immediately with cold water.

4 Slacken and remove the nuts and washers securing the master cylinder to the vacuum servo unit. Remove the master cylinder from the engine compartment along with its sealing ring; discard the sealing ring a new one must be used on refitting.

5 It is not possible to overhaul the cylinder, since no internal components are available separately. If faulty, the complete master cylinder assembly must be renewed. The only components which can be renewed separately are the fluid supply hose union and their seals; both are a push-fit and can simply be pulled out of position.

Refitting

6 Prior to refitting, check that the pushrod end-to-servo unit mating surface distance is correctly set (see Section 12) and adjust as necessary.

7 Ensure that the mating surfaces are clean and dry then fit the new sealing ring to the rear of the master cylinder.

8 Carefully fit the master cylinder to the servo unit, ensuring that the servo unit pushrod enters the master cylinder bore centrally. fit the retaining nuts and tighten them to the specified torque setting.

9 Securely reconnect the fluid hoses to the master cylinder.

10 On models not equipped with ABS, wipe clean the brake pipe unions and refit them to the master cylinder ports, tightening them to the specified torque. Refill the master cylinder reservoir with new fluid. Bleed the complete hydraulic system as described in Section 2.

11 On models equipped with ABS, referring to Section 2, refill the reservoir then bleed the master cylinder of air. Reconnect the brake pipes, tightening the union nuts to the specified torque, then bleed the complete braking system.

11 Brake pedal - removal and refitting

Removal

Manual transmission models with standard clutch

1 Remove the clutch pedal as described in Chapter 6.

2 Slide off the retaining clip and remove the clevis pin securing the vacuum servo unit pushrod to the pedal.

3 Slide the pedal pivot bolt out of position and manoeuvre the brake pedal out from underneath the facia along with its mounting bushes and spacer.

4 Inspect the pedal for signs of wear or damage, paying particular attention to the pivot bushes, and renew worn components as necessary.

Automatic transmission/clutch models

5 From inside the vehicle, remove the facia undercover (where fitted) to gain access to the clutch pedal assembly. Slacken and remove the nut from the pedal pivot bolt then remove the pedal as described in paragraphs 2 to 4.

Refitting

Manual transmission models with standard clutch

6 Apply some multi-purpose grease to the bearing surfaces of the pedal, spacer and bushes. Insert the spacer into the pedal and fit the bushes with their collars outermost. Also apply a smear of grease to the servo unit pushrod clevis and pin.

7 Manoeuvre the pedal assembly into position under the facia, engaging it with the servo unit pushrod, then slide the pivot bolt into position.

8 Align the pedal with the pushrod and insert the clevis pin. Secure the pin in position with the retaining clip making sure it is securely located in the pin groove.

9 Check the operation of the brake pedal then refit the clutch pedal as described in Chapter 6.

Automatic transmission/clutch models

10 Fit the pedal as described above in paragraphs 6 to 8 then securely tighten the pivot bolt nut. Check the operation of the brake pedal before using the vehicle on the road.

12.7 Slide off the retaining clip (arrowed) and withdraw the clevis pin

12.8 Vacuum servo unit retaining nuts (arrowed)

12 Vacuum servo unit - testing, removal and refitting

Testing

1 To test the operation of the servo unit, depress the footbrake several times to exhaust the vacuum, then start the engine whilst keeping the pedal firmly depressed. As the engine starts, there should be a noticeable "give" in the brake pedal as the vacuum builds up. Allow the engine to run for at least two minutes, then switch it off. If the brake pedal is now depressed it should feel normal, but further applications should result in the pedal feeling firmer, with the pedal stroke decreasing with each application.

2 If the servo does not operate as described, first inspect the servo unit check valve as described in Section 13. If the check valve is functioning correctly, renew the servo unit air filter (see paragraph 9).

3 If the servo unit still fails to operate satisfactorily, the fault lies within the unit itself. Repairs to the unit are not possible - if faulty, the servo unit must be renewed.

Removal

4 Remove the master cylinder as described in Section 10.

5 Disconnect the vacuum hose from the servo unit check valve.

6 From inside the vehicle, remove the retaining clips and remove the undercover (where fitted) from the driver's side of the facia.

7 Slide off the retaining clip and remove the clevis pin securing the vacuum servo unit pushrod to the pedal **(see illustration)**.

8 Slacken and remove the four retaining nuts then return to the engine compartment and manoeuvre the servo unit out of position, along with its gasket **(see illustration)**. The gasket must be renewed if it shows signs of wear or damage.

9 If the servo unit is faulty it must be renewed; overhaul of the unit is not possible. The only item which is available separately is the servo unit air filter. To renew the filter, ease the rubber gaiter off the rear of the servo unit, slide it along the pushrod then carefully hook out the old air filter from the rear of the servo. Make a cut in the new filter as shown **(see illustration)**. Place the filter over the pushrod, slide it securely into position in the servo then refit the rubber gaiter.

Refitting

10 Prior to refitting, ensure that the servo unit pushrod is correctly adjusted. Measure the length of the brake pedal end of the pushrod (from the clevis pin centre to the servo unit rear surface) and the distance from the master end of the pushrod to the servo unit front mating surface and check that they are as specified **(see illustration)**. To adjust the length of the brake pedal end of the pushrod, slacken the clevis locknut and rotate the clevis as necessary; once the clevis is correctly positioned hold it stationary and securely tighten the locknut. To adjust the master cylinder end-to-pushrod distance, simply rotate the nut on the end of the pushrod until the distance is as specified.

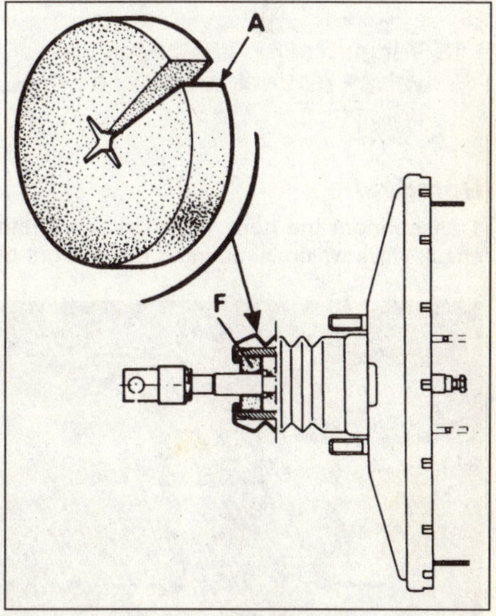

12.9 Vacuum servo unit air filter renewal

A Cut the filter as shown
F Correct position of filter in servo unit

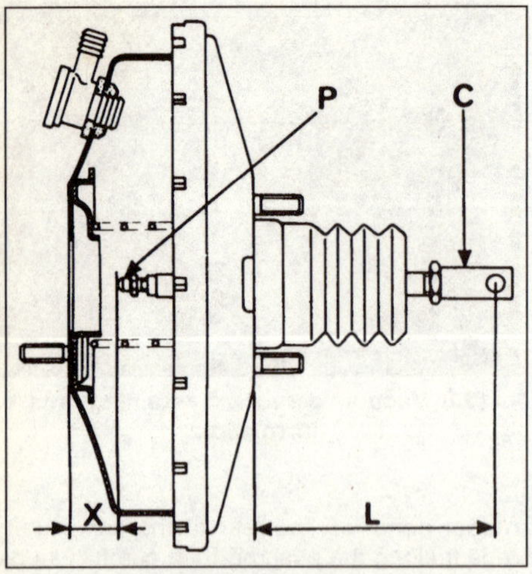

12.10 Vacuum servo unit adjustment dimensions

C Pushrod clevis L 133 mm
P Pushrod nut X 22.3 mm

11 Ensure that all mating surfaces are clean and dry and fit the gasket to the rear of the servo unit. Manoeuvre the servo unit into position and locate it in the bulkhead.

12 From inside the vehicle, ensure that the servo unit is correctly engaged with the brake pedal then refit the retaining nuts and tighten them to the specified torque setting.

13 Align the pushrod clevis with the brake pedal and slide in the clevis pin. Secure the pin in position with the retaining clip, ensuring it is correctly located in the pin groove. Refit the undercover to the facia.

14 From within the engine compartment, reconnect the vacuum hose to the servo unit.

15 Refit the master cylinder as described in Section 10 and bleed the complete hydraulic system as described in Section 2.

13 Vacuum servo unit check valve - removal, testing and refitting

Removal

1 Disconnect the hose from the servo unit check valve which is mounted on the front of the servo unit. On right-hand drive models, access to the valve is poor but can only be significantly improved by removing the inlet manifold (see Chapter 4).

2 Withdraw the valve from its rubber sealing grommet, using a pulling and twisting motion. Remove the grommet from the servo.

Testing

3 Examine the check valve for signs of damage, and renew if necessary. The valve may be tested by blowing through it in both directions. Air should flow through the valve in one direction only - when blown through from the servo unit end of the valve. Renew the valve if this is not the case.

4 Examine the rubber sealing grommet and flexible vacuum hose for signs of damage or deterioration, and renew as necessary.

Refitting

5 Fit the sealing grommet into position in the servo unit.

6 Ease the check valve into position, taking care not to displace or damage the grommet, then reconnect the vacuum hose.

7 On completion, start the engine and check that there are no air leaks.

14 Handbrake - adjustment

Caution: If the handbrake is incorrectly adjusted, the rear brake automatic adjustment mechanism will not be able to function correctly. This will lead to the brake shoe-to-drum clearance becoming excessive as the shoe linings wear, resulting in excessive brake pedal travel.

1 The handbrake will normally be kept in adjustment by the action of the rear drum brake automatic adjusters. Occasionally, the handbrake mechanism may require adjustment to compensate for cable stretch but adjustment should only be needed if the brake shoes, drums, cables or handbrake lever are disturbed. Adjustment is carried out as follows.

2 Remove both rear brake drums as described in Section 7.

3 Check that both the left- and right-hand brake shoe adjuster strut knurled wheels are free to rotate easily then back each wheel off by 5 to 6 teeth.

4 Have an assistant operate the handbrake lever and check that the left- and right-hand brake shoe handbrake levers move smoothly and easily. If not, check the cables/brake shoes for signs of wear or damage before proceeding.

5 With the handbrake lever fully released, check that the stop pegs on both the left- and right-hand brake shoe handbrake levers are resting against the edge of each rear brake shoe. If not, adjust the handbrake cable using the adjuster which is situated on the underside of the handbrake lever. To gain access to the adjuster; it will be necessary to unbolt and remove the exhaust system heatshield which is situated just to the rear of the handbrake lever. Slacken the locknut (where fitted) and back the adjuster nut off until a slight amount of freeplay is present in the cable **(see illustrations)**.

6 With the stop pegs in contact with the shoes, have an assistant slowly operate the handbrake lever whilst you observe the movement of the brake shoe levers. The lever on each rear brake assembly should start to move as the handbrake lever is moved between the first and second notch (click) of its ratchet mechanism, ie so that the stop-pegs are still in contact with the shoes when the handbrake is on the first notch of the ratchet, but no longer contact the shoes when the handbrake is on the second notch. If adjustment is necessary, slacken the locknut (where fitted) and rotate the adjuster nut.

7 Once the handbrake adjustment is correctly set and both rear brake assembly handbrake levers are operating correctly, where necessary, hold the adjuster nut stationary and securely tighten the locknut. Refit the exhaust system heatshield to the vehicle underbody and securely tighten its retaining bolts/nuts.

8 Refit the brake drums as described in Section 7. Check the operation of the rear brake automatic adjustment mechanism by depressing the brake pedal whilst having an assistant listen to the rear drums. If the adjustment mechanism is functioning

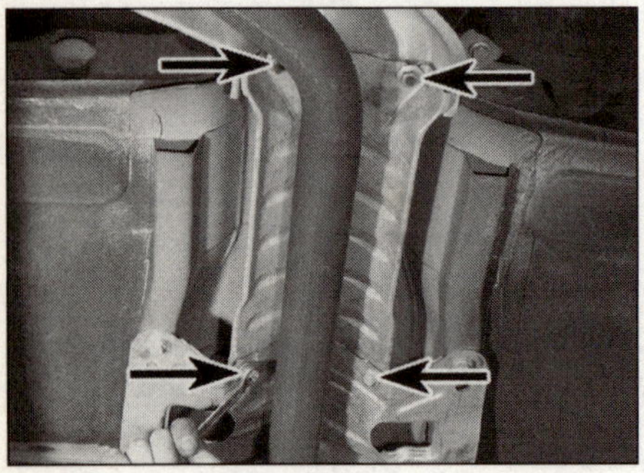

14.5a Undo the retaining nuts and bolts (arrowed) . . .

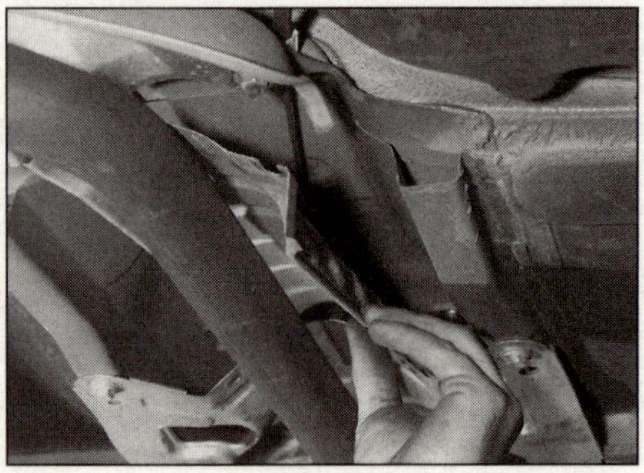

14.5b . . . and remove the heatshield from underneath the handbrake lever . . .

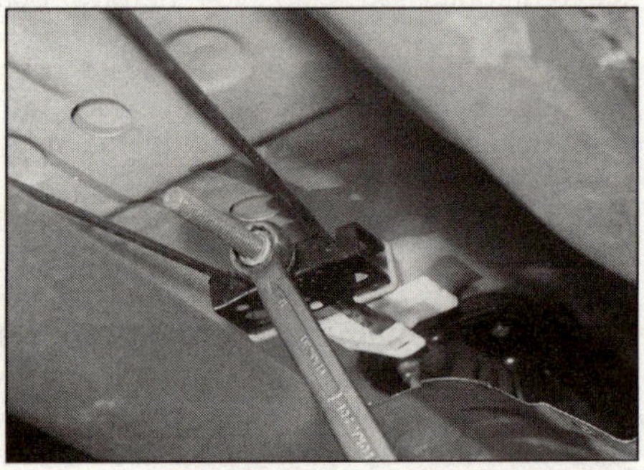

14.5c . . . to gain access to the handbrake cable adjuster nut

correctly, a clicking sound will be emitted by the strut as the pedal is depressed.

9 Once the shoe-to-drum clearance is correctly adjusted, refit the roadwheels then lower the vehicle to the ground and tighten the wheel bolts to the specified torque.

15 Handbrake lever - removal and refitting

Removal

1 Remove the rear section of the centre console as described in Chapter 11 to gain access to the handbrake lever. Disconnect the wiring connector from the handbrake warning light switch and fully release the lever.

2 Chock the front wheels then jack up the rear of the vehicle and support it on axle stands.

3 Slacken and remove the retaining nuts/bolts and remove the exhaust system heatshield(s) from underneath the handbrake lever. **Note:** *On some models it may be necessary to separate the exhaust system intermediate pipe front joint to enable the heatshield(s) to be removed (see Chapter 4).*

4 Slacken the locknut (where fitted) then unscrew the adjuster nut to obtain maximum freeplay in the handbrake cables **(see illustration)**.

5 Unhook the cables from the handbrake lever equalizer plate then unclip the lever rod from its guide.

6 Slacken and remove the lever retaining nuts then manoeuvre the assembly out from underneath the vehicle **(see illustration)**.

Refitting

7 Refitting is a reversal of removal tightening the lever retaining nuts securely. Prior to refitting the heatshield(s), adjust the handbrake cable as described in Section 14.

16 Handbrake cables - removal and refitting

Removal

1 The handbrake cable consists of two sections, a right- and left-hand section which connect the rear brakes to the handbrake lever. Each section can be removed individually as follows.

2 Firmly chock the front wheels, then jack up the rear of the vehicle and support it on axle stands.

3 To gain access to the underside of the handbrake lever and the cable, undo the retaining nuts/bolts (as applicable) and remove the exhaust system heatshield(s).

4 Remove the relevant set of rear brake shoes as described in Section 5.

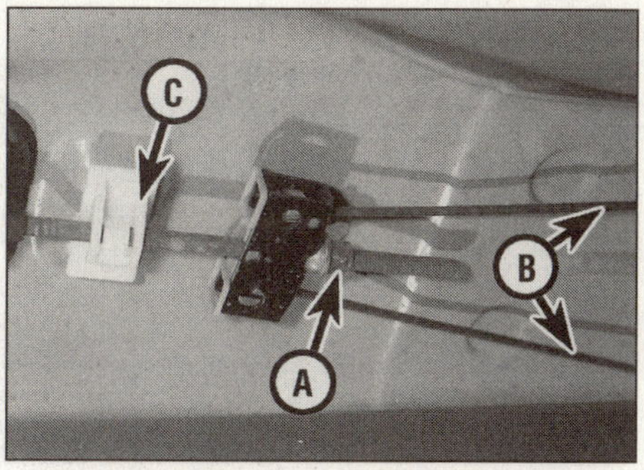

15.4 Slacken the adjuster nut (A) then detach the handbrake cables (B) and unclip the lever rod (C) from its guide

5 Unclip the handbrake cable end fitting and free the cable from the brake backplate **(see illustration)**.

6 Work back along the cable, noting its correct routing whilst freeing it from any clips or ties **(see illustration)**. Detach the front end of the inner cable from the handbrake lever equalizer plate then free the outer cable end fitting from the vehicle body and remove the cable from the vehicle.

Refitting

7 Refitting is a reversal of the removal procedure ensuring that the cable is correctly routed and retained by all the relevant clips and ties. Prior to refitting the heatshields, adjust the handbrake as described in Section 14.

17 Rear brake pressure-regulating valve - testing, removal and refitting

Testing

1 A load-sensitive pressure regulating valve is fitted into the hydraulic circuit to each rear brake. The valve is mounted onto the underside of the rear of the vehicle and is attached to the rear axle. The valve measures the load on the rear axle, via the movement of the axle, and regulates the hydraulic pressure being applied to the rear brakes to help prevent rear wheels locking up under hard braking.

16.5 Unclip the handbrake cable from the backplate . . .

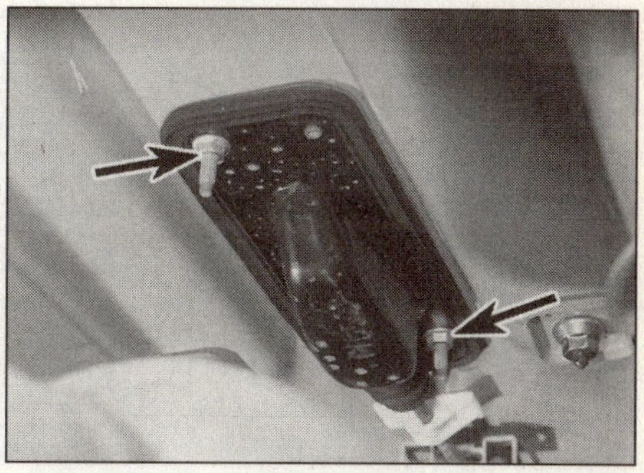

15.6 Undo the retaining nuts (arrowed) and remove the handbrake lever from underneath the vehicle

2 Specialist equipment is required to check the performance of the valve(s), therefore if the valve is thought to be faulty the car should be taken to a suitably equipped Renault dealer for testing. Repairs are not possible and, if faulty, the valve must be renewed.

Removal

Caution: On models equipped with ABS, disconnect the battery before disconnecting any braking system hydraulic union and do not reconnect the battery until after the hydraulic system has been bled. Failure to do this could lead to air entering the hydraulic unit. If air enters the hydraulic unit pump, it will prove very difficult (in some cases impossible) to bleed the unit (see Section 2).

Note: *Before starting work, refer to the warning at the beginning of Section 2 concerning the dangers of hydraulic fluid.*

3 Minimise fluid loss by first removing the master cylinder reservoir cap, and then tightening it down onto a piece of polythene, to obtain an airtight seal.

4 Check the front wheels then jack up the rear of the vehicle and support it on axle stands.

5 Wipe clean the area around the brake pipe unions on the valve, and place absorbent rags beneath the pipe unions to catch any surplus fluid. To avoid confusion on refitting, make alignment marks between the pipes and valve assembly.

6 Slacken the union nuts and disconnect the brake pipes from the valve. Plug or tape over

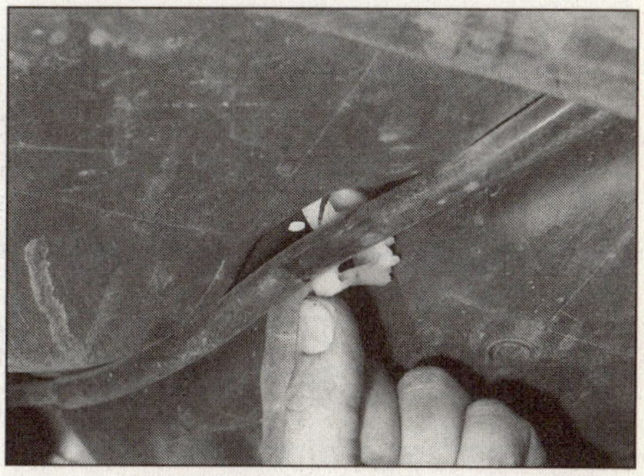

16.6 . . . and release it from all the relevant retaining clips

17.7a Slacken and remove the rear brake pressure-regulating valve retaining bolts (arrowed) . . .

17.7b . . . then unclip the link rod from the axle and remove the assembly from underneath the vehicle

the pipe ends and valve orifices, to minimise the loss of brake fluid, and to prevent the entry of dirt into the system. Wash off any spilt fluid immediately with cold water.

7 Slacken and remove the retaining bolts then unclip the valve link rod and remove the valve assembly from underneath the vehicle **(see illustrations)**. **Note:** *Do not slacken the link rod clamp bolt. If the bolt is slackened and the link rod length altered the valve will need to be adjusted on refitting.*

Refitting

8 Manoeuvre the valve assembly into position and tighten its retaining bolts securely.

9 Refit the brake pipes to their specific unions on the valve and tighten the union nuts to the specified torque setting.

10 Clip the valve link rod back into position in its retaining clip and lower the vehicle to the ground.

11 If a new valve assembly is being fitted, it will be noted that a spacer is fitted to the link rod; this is to adjust the link rod length. With the vehicle resting on its wheels, with a full tank of fuel and one person in the drivers seat, slacken the link rod clamp bolt and allow the valve spring to set the link rod length. Securely tighten the clamp bolt and remove the spacer from the link rod. Although not strictly necessary, it is recommended that the valve operation is tested by a Renault dealer

12 Remove the polythene from the master cylinder reservoir and bleed the complete hydraulic system as described in Section 2.

18 Stop-light switch - removal, refitting and adjustment

Removal

1 The stop-light switch is located on the pedal bracket behind the facia.

2 To remove the switch, reach up behind the facia, disconnect the wiring connector and unscrew the switch from the bracket.

Refitting and adjustment

3 Screw the switch back into position in the mounting bracket.

4 Connect a continuity tester (ohmmeter or self-powered test light) across the switch terminals. Screw the switch in until an open-circuit is present between the switch terminals (infinite resistance, or light goes out). Gently depress the pedal and check that continuity exists between the switch terminals (zero resistance, or light comes on) after the pedal has travelled approximately 5 mm. If necessary, reposition the switch until it operates as specified.

5 In the absence of a continuity tester, the same adjustment can be made by reconnecting the switch and having an assistant observe the stop-lights (ignition on).

6 Once the stop-light switch is correctly adjusted, remake the original wiring connections, and recheck the operation of the stop-lights.

19 Anti-lock braking system (ABS) - general information

ABS was available as an option on most models. The system comprises of the hydraulic unit and the four roadwheel sensors. The hydraulic unit contains the electronic control unit (ECU), the eight hydraulic solenoid valves (two for each brake - one inlet and one outlet) and the electrically driven pump. The purpose of the system is to prevent the wheel(s) locking during heavy braking. This is achieved by automatic release of the brake on the relevant wheel, followed by re-application of the brake. In the case of the rear wheels both brakes are applied at the same time.

The solenoid valves are controlled by the ECU, which itself receives signals from the four wheel sensors (which are fitted to the wheel hubs), which monitor the speed of rotation of each wheel. By comparing these signals, the ECU can determine the speed at which the vehicle is travelling. It can then use this speed to determine when a wheel is decelerating at an abnormal rate, compared to the speed of the vehicle, and therefore predicts when a wheel is about to lock. During normal operation, the system functions in the same way as a non-ABS braking system.

If the ECU senses that a wheel is about to lock, it closes the relevant outlet solenoid valves in the hydraulic unit, which then isolates the relevant brake(s) on the wheel(s) which is/are about to lock from the master cylinder, effectively sealing-in the hydraulic pressure.

If the speed of rotation of the wheel continues to decrease at an abnormal rate, the ECU opens the inlet solenoid valves on the relevant brake(s) and operates the electrically-driven return pump which pumps the hydraulic fluid back into the master cylinder, releasing the brake. Once the speed of rotation of the wheel returns to an acceptable rate, the pump stops; the solenoid valves switch again, allowing the hydraulic master cylinder pressure to return to the caliper/wheel cylinder (as applicable), which then re-applies the brake. This cycle can be carried out many times a second.

The action of the solenoid valves and return pump creates pulses in the hydraulic circuit. When the ABS system is functioning, these pulses can be felt through the brake pedal.

The operation of the ABS system is entirely dependent on electrical signals. To prevent the system responding to any inaccurate signals, a built-in safety circuit monitors all signals received by the ECU. If an inaccurate signal or low battery voltage is detected, the ABS system is automatically shut down, and the warning light on the instrument panel is illuminated, to inform the driver that the ABS system is not operational. Normal braking should still be available, however.

If a fault does develop in the ABS system, the vehicle must be taken to a Renault dealer for fault diagnosis and repair.

20 Anti-lock braking system (ABS) components - removal and refitting

Hydraulic unit

Caution: Disconnect the battery before disconnecting any braking system hydraulic union and do not reconnect the battery until after the hydraulic system has been bled. Failure to do this could lead to air entering the hydraulic unit. If air enters the hydraulic unit pump, it will prove very difficult (in some cases impossible) to bleed the unit (see Section 2).

Note: *Before starting work, refer to the warning at the beginning of Section 2 concerning the dangers of hydraulic fluid. Blanking plugs will be needed to seal the hydraulic unit unions once the pipes have been disconnected.*

Removal

1 Disconnect the battery negative lead.

2 Firmly apply the handbrake then jack up the front of the vehicle and support it on axle stands. Remove the right-hand front roadwheel.

3 Unscrew the fasteners and remove the wheelarch liner from underneath the right-hand wing.
4 Release the retaining clip and disconnect the wiring connector from the base of the hydraulic unit.
5 Minimise fluid loss by first removing the master cylinder reservoir cap, and then tightening it down onto a piece of polythene, to obtain an airtight seal.
6 Mark the locations of the hydraulic fluid pipes to ensure correct refitting. Unscrew the union nuts, and disconnect the pipes from the regulator assembly, whilst being prepared for fluid spillage. Seal the hydraulic unit ports with the blanking plugs and plug the pipes to prevent dirt ingress and further fluid loss. **Note:** *If hydraulic unit unions are not securely plugged there is a risk of air entering the hydraulic unit pump (see Caution at the start of this Section).*
7 Slacken and remove the hydraulic unit mounting bolts and remove the assembly from the engine compartment.

Refitting

8 Refitting is the reverse of removal, noting the following points.

a) *Prior to reconnection, ensure that the pipes and the hydraulic unit unions are full of fluid (see Section 2). Reconnect the pipes to their original unions and tighten the union nuts to the specified torque.*
b) *Ensure that the wiring connector is securely held in position with its retaining clip.*
c) *Bleed the hydraulic system as described in Section 2 then reconnect the battery.*
d) *On completion, it is recommended that the operation of the ABS system is checked at the earliest opportunity by a Renault dealer using special electronic test equipment.*

Electronic control unit (ECU)

Removal

9 Carry out the operations described in paragraphs 1 to 4.
10 Slacken and remove the hydraulic unit and mounting bracket bolts then carefully manoeuvre the bracket away from the unit. Securely support the hydraulic unit to avoid placing any strain on the hydraulic pipes.
11 Undo the retaining screws then pull the ECU squarely away from the hydraulic unit and remove it from the vehicle.

Refitting

12 Refitting is the reverse of removal, taking great care to ensure that the ECU is correctly engaged with the hydraulic unit connectors.

Front wheel sensor

Removal

13 Disconnect the battery negative lead.
14 Apply the handbrake, then jack up the front of the vehicle and support securely on axle stands. To improve access, remove the roadwheel.
15 Trace the wiring back from the sensor, releasing it from all the relevant clips and ties whilst noting its correct routing, and disconnect the wiring connector.
16 Slacken and remove the retaining bolt and withdraw the sensor from the swivel hub.

Refitting

17 Ensure that the mating faces of the sensor and the swivel hub are clean, and apply a little multi-purpose grease to the swivel hub bore before refitting.
18 Make sure the sensor tip is clean and ease it into position in the swivel hub. Refit the retaining bolt and tighten it to the specified torque.
19 Work along the sensor wiring, making sure it is correctly routed, securing it in position with all the relevant clips and ties. Reconnect the wiring connector then lower the vehicle and (where necessary) tighten the wheel bolts to the specified torque.

Rear wheel sensor

Removal

20 Chock the front wheels, then jack up the rear of the vehicle and support it on axle stands. To improve access, remove the appropriate roadwheel.
21 Trace the wiring back from the sensor, releasing it from all the relevant clips and ties whilst noting its correct routing, and disconnect the wiring connector.
22 Slacken and remove the retaining bolt and withdraw the sensor.

Refitting

23 Ensure that the mating faces of the sensor and the hub are clean, and apply a little multi-purpose grease to the hub bore before refitting.
24 Make sure the sensor tip is clean and ease it into position in the swivel hub. Refit the retaining bolt and tighten it to the specified torque.
25 Work along the sensor wiring, making sure it is correctly routed, securing it in position with all the relevant clips and ties. Reconnect the wiring connector then lower the vehicle and (where necessary) tighten the wheel bolts to the specified torque.

21 Vacuum pump (diesel engine models) - removal and refitting

Removal

1 If necessary, to improve access to the vacuum pump, remove the air cleaner duct (see Chapter 4).
2 Release the retaining clip and disconnect the vacuum hose from the pump.
3 Slacken and remove the mounting bolts/nuts securing the pump to the left-hand end of the cylinder head, then remove the pump. Recover the pump gasket and discard it; a new one should be used on refitting.

Refitting

4 Ensure that the pump and cylinder head mating surfaces are clean and dry and fit the new gasket to the head.
5 Manoeuvre the pump into position, aligning its drive dog with the camshaft slot, then refit the pump mounting bolts/nuts and tighten them securely.
6 Reconnect the vacuum hose to the pump, tightening its retaining clip securely, and (where necessary) refit the air cleaner duct.

22 Vacuum pump (diesel engine models) - testing and overhaul

1 The operation of the braking system vacuum pump can be checked using a vacuum gauge.
2 Disconnect the vacuum pipe from the pump, and connect the gauge to the pump union using a suitable length of hose.
3 Start the engine and allow it to idle, then measure the vacuum created by the pump. As a guide, after one minute, a minimum of approximately 500 mm Hg should be recorded. If the vacuum registered is significantly less than this, it is likely that the pump is faulty. However, seek the advice of a Renault dealer before condemning the pump.
4 Overhaul of the vacuum pump is not possible, since no components are available separately for it. If faulty, the complete pump assembly must be renewed.

Chapter 10
Suspension and steering

Contents

Degrees of difficulty

Easy, suitable for novice with little experience	**Fairly easy,** suitable for beginner with some experience	**Fairly difficult,** suitable for competent DIY mechanic	**Difficult,** suitable for experienced DIY mechanic	**Very difficult,** suitable for expert DIY or professional

Specifications

Front suspension

Front hub assembly endplay	0 to 0.05 mm
Vehicle front ride height (H1 minus H2 - see Section 14):	
Hatchback models:	
Petrol models	105 ± 5 mm
Diesel models:	
Normally aspirated models	105 ± 5 mm
Turbo models	108 ± 5 mm
Saloon models:	
Petrol models:	
1.4 litre engine	105 ± 5 mm
1.6 litre engine	108 ± 5 mm
Diesel models:	
Normally aspirated models	105 ± 5 mm
Turbo models	108 ± 5 mm
Coupe models	113 ± 5 mm

Rear suspension

Rear hub assembly endplay	0 to 0.03 mm
Vehicle rear ride height (H4 minus H5 - see Section 14):	
Hatchback models:	
Petrol models	14 ± 7.5 mm
Diesel models:	
Normally aspirated models	14 ± 7.5 mm
Turbo models	22 ± 7.5 mm
Saloon models:	
Petrol models:	
1.4 litre engine	18.2 ± 7.5 mm
1.6 litre engine	19.2 ± 7.5 mm
Diesel models:	
Normally aspirated models	18.2 ± 7.5 mm
Turbo models	19.2 ± 7.5 mm
Coupe models	34 ± 7.5 mm

Roadwheel rim runout (max)	1.2 mm

Wheel alignment and steering angles

Front wheel camber angle at underbody height (H1 minus H2) stated:	
Models with manual steering:	
33 mm	1° 14' ± 30'
90 mm	-0° 8' ± 30'
109 mm	0° 25' ± 30'
188 mm	0° 30' ± 30'
Models with power steering:	
22 mm	1° 35' ± 30'
102 mm	-0° 20' ± 30'
122 mm	-0° 35' ± 30'
188 mm	-0° 30' ± 30'
Maximum difference between left- and right-hand sides	1°
Castor angle at underbody height (H5 minus H2) stated:	
Models with manual steering:	
76.5 mm	2° 30' ± 30'
96.5 mm	2° ± 30'
116.5 mm	1° 30' ± 30'
136.5 mm	1° ± 30'
156.5 mm	0° 30' ± 30'
Models with power steering:	
Turbo diesel models:	
71.5 mm	4° 30' ± 30'
91.5 mm	4° ± 30'
111.5 mm	3° 30' ± 30'
131.5 mm	3° ± 30'
151.5 mm	2° 30' ± 30'
All other models:	
76.5 mm	4° 30' ± 30'
96.5 mm	4° ± 30'
116.5 mm	3° 30' ± 30'
136.5 mm	3° ± 30'
156.5 mm	2° 30' ± 30'
Maximum difference between left- and right-hand sides	1°
Steering axis inclination/kingpin inclination at underbody height (H1 minus H2) stated:	
Models with manual steering:	
33 mm	11° 10' ± 30'
90 mm	13° 1' ± 30'
109 mm	13° 29' ± 30'
188 mm	14° 29' ± 30'
Models with power steering:	
22 mm	10° 45' ± 30'
102 mm	13° 20' ± 30'
122 mm	13° 53' ± 30'
188 mm	14° 32' ± 30'
Maximum difference between left- and right-hand sides	1°
Front wheel toe setting (vehicle unladen)	0° ± 10' (1 ± 1 mm) toe-out
Rear wheel camber setting (vehicle unladen):	
Enclosed-bar rear axle	-1° ± 15'
Open-bar rear axle	-0° 50' ± 15'
Rear wheel toe setting (vehicle unladen):	
Enclosed-bar rear axle	-0° 50' ± 20' (-5 ± 2 mm) toe-in
Open-bar rear axle	-0° 30' ± 20' (-3 ± 2 mm) toe-in

Tyre pressures - see end of "*Weekly checks*"

Torque wrench settings

Torque wrench settings	Nm	lbf ft
Front suspension		
Anti-roll bar mounting clamp bolts	32	23
Driveshaft nut	250	185
Lower arm:		
Pivot bolt nuts	90	67
Balljoint clamp bolt	60	44
Balljoint retaining bolt nuts	60	44
Suspension strut:		
Strut-to-swivel hub bolts	170	125
Strut upper mounting bolts	30	22
Strut upper mounting plate nut	60	44

Torque wrench settings

	Nm	lbf ft
Rear suspension		
Anti-roll bar bolts - enclosed bar axle	60	44
Axle mounting bolt nuts	100	74
Shock absorber:		
Upper mounting bolt	100	74
Shock absorber lower mounting bolt	135	100
Hub nut	175	129
Steering		
Hydraulic pipe union nuts:		
Feed pipe (pump to steering rack)	24	17
Return pipe (steering rack to pump):		
Aluminium casting	22	16
Steel casting	28	20
Steering gear pipe (pinion to rack housing)	15	11
Steering column:		
Mounting nuts and bolts	20	15
Universal joint clamp bolt nut	25	18
Steering gear mounting bolts	50	37
Steering wheel retaining bolt	45	33
Track rod balljoint:		
Retaining nut	35	26
Clamp bolt	20	15
Track rod inner balljoint-to steering rack:		
SMI steering gear assembly	50	37
TRW steering gear assembly	80	58
Roadwheels		
Wheel bolts	90	66

1 General information

The independent front suspension is of the MacPherson strut type, incorporating coil springs and integral telescopic shock absorbers. The MacPherson struts are located by transverse lower suspension arms, which utilise rubber inner mounting bushes and incorporate a balljoint at the outer ends. The front swivel hubs, which carry the wheel bearings, brake calipers and the hub/disc assemblies, are bolted to the MacPherson struts and connected to the lower arms via the balljoints. The anti-roll bar is rubber-mounted onto the subframe, and connects both the lower suspension arms.

The rear suspension differs according to specification. As a rule, most smaller-engine vehicles are fitted with an enclosed-bar type rear axle and larger-engine models are fitted with an open-bar rear axle. The enclosed-bar type axle consists of two torsion bars within a tubular crossmember which connects both the rear trailing arms. An anti-roll bar is situated just to the rear of the crossmember, and is also connected to both trailing arms. The open-bar type axle consists of two torsion bars, two anti-roll bars and an L-section metal crossmember which is connected to both the trailing arms. The two torsion bars and two anti-roll bars are connected at the centre with a link, and at their outer ends to the trailing arms.

The steering column has a universal joint approximately halfway along its length and a second universal joint at its lower end. The lower universal joint is attached to the steering gear pinion by means of a clamp bolt.

The steering gear is mounted onto the front subframe. It is connected by two track rods and balljoints to steering arms projecting rearwards from the swivel hubs. The track rod ends are threaded to enable wheel alignment adjustment.

Power-assisted steering was available on most models. The hydraulic system is powered a pump, which is driven off the crankshaft pulley via a drivebelt.

2 Front swivel hub assembly - removal and refitting

Note: *A new driveshaft nut will be required on refitting.*

Note: *The driveshaft outer joint splines are coated with locking compound prior to refitting. Therefore it is likely that a puller/extractor will be required to draw the hub assembly off of the driveshaft end on removal.*

Removal

1 Remove the wheel trim/hub cap (as applicable) then slacken the driveshaft nut with the vehicle resting on its wheels. Also slacken the wheel bolts.

2 Chock the rear wheels of the car, firmly apply the handbrake, then jack up the front of the car and support it on axle stands. Remove the appropriate front roadwheel.

3 On models equipped with ABS, unbolt the wheel sensor and position it clear of the hub assembly (see Chapter 9). Note that there is no need to disconnect the wiring.

4 Slacken and remove the driveshaft retaining nut. If the nut was not slackened with the wheels on the ground (see paragraph 1), refit at least two roadwheel bolts to the front hub, tightening them securely, then have an assistant firmly depress the brake pedal to prevent the front hub from rotating, whilst you slacken and remove the driveshaft retaining nut. Alternatively, a tool can be fabricated to hold the hub stationary (see Chapter 8, Section 2).

5 Slacken and remove the nut securing the steering gear track rod to the swivel hub then free the balljoint from the hub. If the balljoint is tight, use a universal balljoint separator to free it.

6 If the hub bearings are to be disturbed, remove the brake disc as described in Chapter 9. If not, unscrew the two bolts securing the brake caliper mounting bracket assembly to the swivel hub, and slide the caliper assembly off the disc **(see illustration)**. Using a piece of wire or string, tie the caliper to the front suspension coil spring, to avoid placing any strain on the hydraulic brake hose. If necessary, unbolt the brake hose/wiring bracket from the strut.

7 Remove the nut and clamp bolt securing the lower suspension arm to the swivel hub. Carefully lever the balljoint out of the swivel

2.6 Undo the mounting bracket bolts and slide the brake caliper off the disc

hub, taking care not to damage the balljoint or driveshaft gaiters. Note the plastic protector plate which is fitted to the balljoint shank.

8 Remove the two nuts and washers from the bolts securing the swivel hub to the suspension strut, noting that the nuts are positioned on the rear side of the strut **(see illustration)**. Withdraw the bolts and support the swivel hub assembly.

9 Release the driveshaft joint from the hub, and remove the swivel hub assembly from the vehicle. Note that locking fluid is applied to the joint splines during assembly, so it is likely that they will be a tight fit in the hub. Use a hammer and soft metal drift to tap the joint out of the hub, or use a puller to draw the swivel hub assembly off the joint splines **(see illustration)**. Whilst the hub assembly is removed, support the driveshaft to avoid damage to the constant velocity joints/gaiters.

Refitting

10 Ensure that the driveshaft joint and hub splines are clean and dry, then apply a coat of locking fluid (Renault recommend Loctite Scelbloc - available from your Renault dealer) to the joint splines.

11 Engage the joint splines with the hub, and slide the hub fully onto the driveshaft. Insert the two swivel hub-to-suspension strut mounting bolts from the front side of the strut, then refit the washers and nuts to the rear of the bolts and tighten them to the specified torque **(see illustration)**.

12 Slide on the washer and fit the new driveshaft nut, tightening it by hand only at this stage.

13 Ensure that the plastic protector is still fitted to the lower arm balljoint, then locate the balljoint shank in the swivel hub. Refit the balljoint clamp bolt, and tighten its retaining nut to the specified torque.

14 Reconnect the track rod end balljoint to the swivel hub, and tighten its retaining nut to the specified torque.

15 On models equipped with ABS, refit the sensor to the hub, and tighten its retaining bolt to the specified torque (see Chapter 9).

16 Refit the brake disc (if removed), aligning the marks made on removal, and tighten its retaining screws to the specified torque. Slide the brake caliper assembly into position over

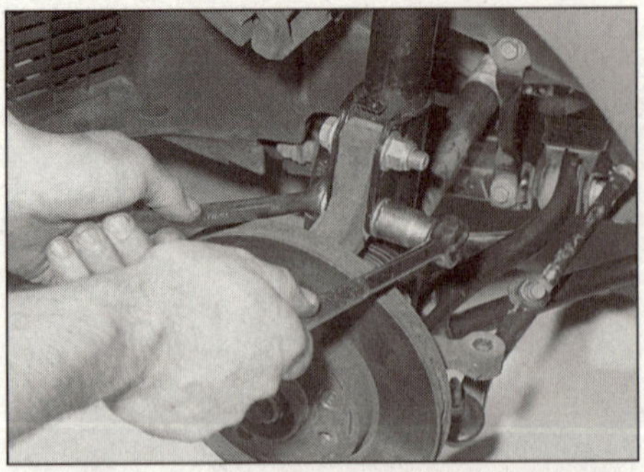

2.8 Undo the nuts and remove the suspension strut-to-swivel hub bolts

2.11 Ensure the strut-to-swivel hub bolts are fitted from the front of strut

the brake disc. Apply a few drops of locking fluid to the caliper bolt threads (Renault recommend Loctite Frenbloc - available from your Renault dealer) then refit the bolts and tighten them to the specified torque (Chapter 9) **(see illustration)**.

17 Using the method employed during removal to prevent the hub from rotating, tighten a new driveshaft retaining nut to the specified torque. Alternatively, lightly tighten the nut at this stage and tighten it to the specified torque once the vehicle is resting on its wheels again.

18 Check that the hub rotates freely, then refit the roadwheel and lower the vehicle to the ground. Tighten the roadwheel bolts to the specified torque. If not already having done so, also tighten the driveshaft retaining nut to the specified torque.

3 Front hub bearings - checking and renewal

Note: *The bearing is a sealed, pre-adjusted and pre-lubricated, double-row roller type, and is intended to last the car's entire service life without maintenance or attention. Do not attempt to remove the bearing unless absolutely necessary, as it will be damaged during the removal operation. Never overtighten the driveshaft nut in an attempt to 'adjust' the bearing.*

Note: *A press will be required to dismantle and rebuild the assembly; if such a tool is not*

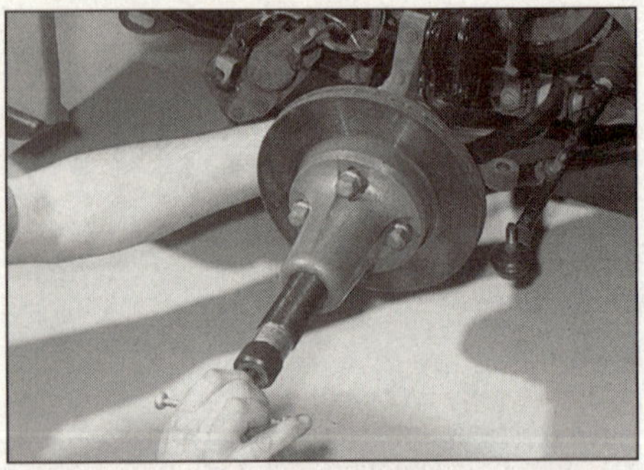

2.9 Using a puller to draw the hub assembly off the driveshaft joint splines

2.16 Apply thread locking compound to the threads of the brake caliper mounting bolts prior to installation

available, a large bench vice and suitable spacers (such as large sockets) will serve as an adequate substitute. The bearing's inner races are an interference fit on the hub; if the inner race remains on the hub when it is pressed out of the hub carrier, a suitable knife-edged bearing puller will be required to remove it.

Checking

1 Wear in the front hub bearings can be checked by measuring the amount of side play present. To do this, a dial gauge should be fixed so that its probe is in contact with the disc face of the hub. The play should be between 0 and 0.05 mm. If it is greater than this, the bearings are worn excessively, and should be renewed.

Renewal

2 Remove the swivel hub assembly as described in Section 2, then undo the brake disc shield retaining screws and remove the shield from the hub.

3 Support the swivel hub securely on blocks or in a vice. Using a suitable tubular spacer which bears only on the inner end of the hub flange, press the hub flange out of the bearing. If the bearing outboard inner race remains on the hub, remove it using a suitable bearing puller (see note above).

4 Extract the bearing retaining circlip from the inner end of the swivel hub assembly.

5 Where necessary, refit the inner race in position over the ball cage, and support the inner face of the swivel hub. Using a suitable

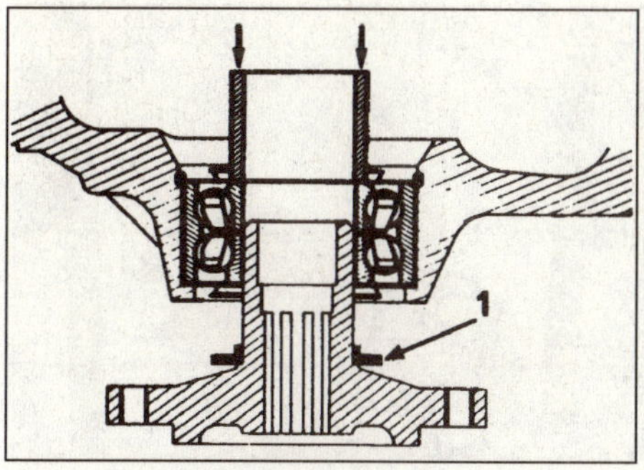

3.11 Pressing the hub assembly onto the hub flange
1 Thrustwasher

tubular spacer which bears only on the inner race, press the complete bearing assembly out of the swivel hub.

6 Thoroughly clean the hub and swivel hub, removing all traces of dirt and grease. Polish away any burrs or raised edges which might hinder reassembly. Check for cracks or any other signs of wear or damage, and renew the components if necessary. As noted above, the bearing and its circlip must be renewed whenever they are disturbed. A replacement bearing kit, which consists of the bearing, circlip, driveshaft nut and washer is available from Renault dealers.

7 On reassembly, check (if possible) that the new bearing is packed with grease. Apply a light film of oil to the bearing outer race and to the hub flange shaft.

8 Before fitting the new bearing, remove the plastic covers protecting the seals at each end, but leave the inner plastic sleeve in position to hold the inner races together.

9 Securely support the swivel hub, and locate the bearing in the hub. Press the bearing into position, ensuring that it enters the hub squarely, using a suitable tubular spacer which bears only on the outer race.

10 Once the bearing is correctly seated, secure it with the new circlip and remove the plastic sleeve. Apply a smear of grease to the oil seal lips.

11 Securely support the outer face of the hub flange then locate the bearing inner race over the end of the hub flange. Press the bearing onto the hub, using a tubular spacer which bears only on the inner race, until it seats against the shoulder **(see illustration)**. Check that the hub flange rotates freely. Wipe off any excess oil or grease.

12 Refit the brake disc shield to the swivel hub, and tighten its retaining screws.

13 Refit the swivel hub assembly (Section 2).

4 Front strut - removal and refitting

Removal

1 Chock the rear wheels, firmly apply the handbrake, then jack up the front of the vehicle and support it on axle stands. Remove the appropriate roadwheel.

2 Unscrew the retaining bolt and free the brake hose/wiring bracket from the base of the strut **(see illustration)**.

3 Remove the two nuts and washers from the bolts securing the swivel hub to the suspension strut, noting that the nuts are positioned on the rear side of the strut. Withdraw the bolts, and support the swivel hub assembly **(see illustration)**.

4 From within the engine compartment, note which bolt holes the strut upper mounting bolts are fitted to the, whilst an assistant supports the weight of the strut, slacken and remove both bolts **(see illustration)**.

5 Release the strut from the swivel hub, and withdraw it from under the wheel arch, taking care to prevent damage to the driveshaft gaiter **(see illustration)**.

Refitting

6 Manoeuvre the strut assembly into position, taking care not damage the driveshaft gaiter. Position the upper mounting plate correctly to align with the relevant set of the mounting bolt holes then refit the mounting bolts and tighten them to the specified torque setting. The rear bolt holes are for use if the vehicle is fitted with power steering and front bolt holes for use when the vehicle is fitted with manual steering; rotate the mounting plate as necessary **(see illustration)**.

7 Insert the two swivel hub-to-suspension strut mounting bolts from the front side of the strut. Refit the washers and nuts to the rear of the bolts, and tighten them to the specified torque.

8 Refit the brake hose/wiring mounting bracket to the strut and securely tighten its retaining bolt.

9 Refit the roadwheel, lower the vehicle to the ground and tighten the roadwheel bolts to the specified torque.

5 Front strut - overhaul

Warning: Before attempting to dismantle the front suspension strut, a suitable tool to hold the coil spring in compression must be obtained. Adjustable coil spring compressors are readily available, and are

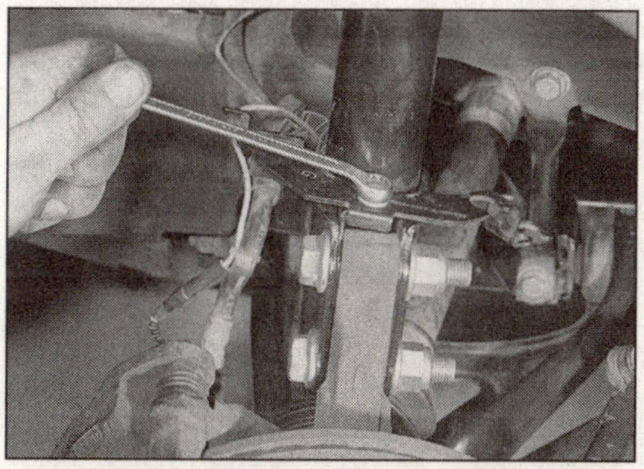

4.2 Undo the retaining bolt and free the hose/wiring bracket from the base of the strut

4.3 Undo the nuts and withdraw the bolts securing the strut to the swivel hub, noting which way around they are fitted

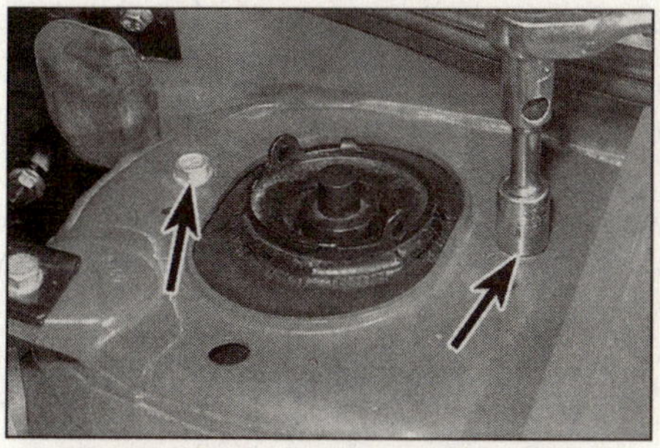

4.4 Slacken and remove the upper mounting bolts (arrowed), noting which holes the bolts are fitted to . . .

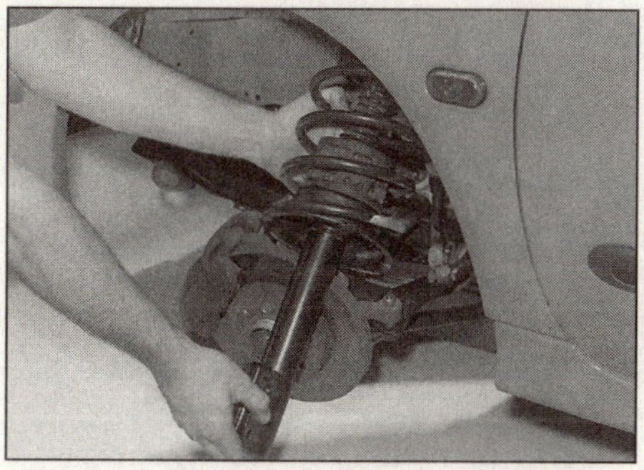

4.5 . . . and manoeuvre the strut out from underneath the wheel arch

4.6 Ensure the strut mounting is correctly positioned so that the upper mounting bolts are fitted to the correct holes (see text) - power steering model shown

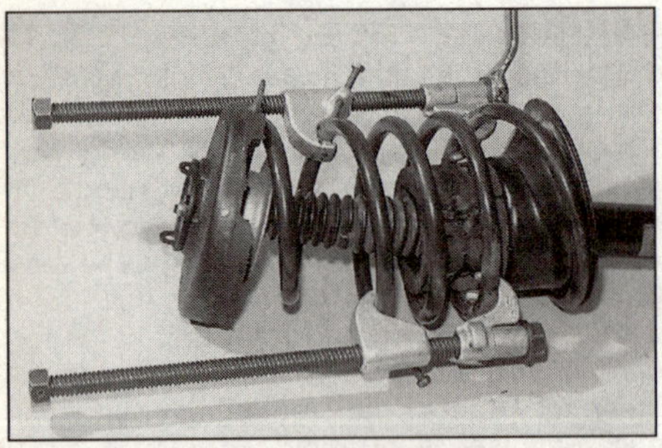

5.1 Using spring compressors, compress the strut coil spring until tension is relieved from the spring seats

5.2 Remove the cover from the upper mounting plate

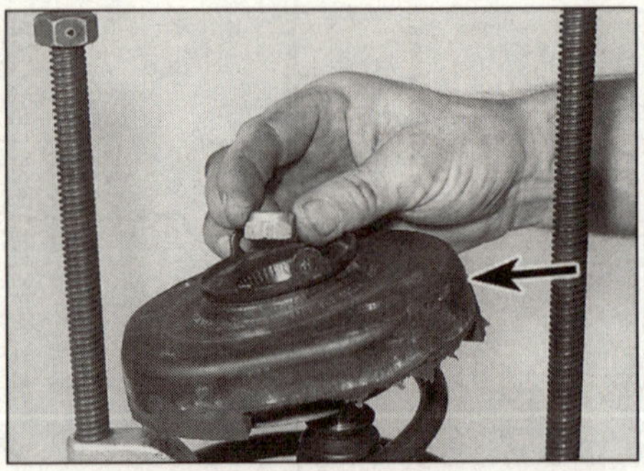

5.3 Slacken and remove the nut then lift off the upper mounting plate (arrowed)

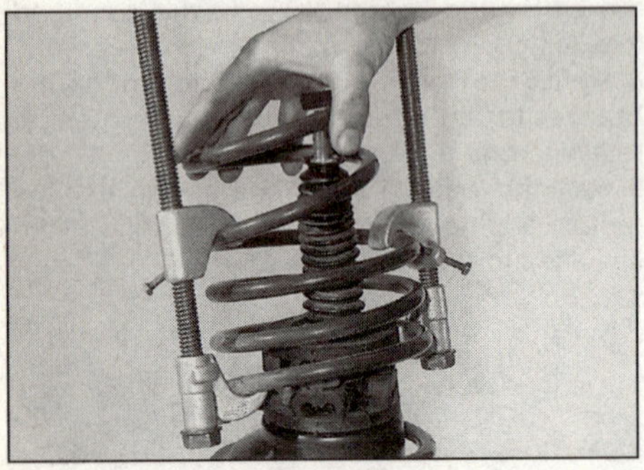

5.4a Lift off the coil spring . . .

5.4b . . . and the spring stop/lower seat . . .

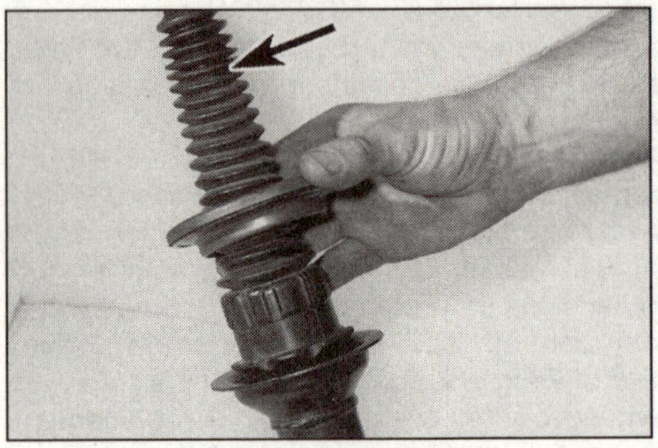

5.4c . . . then remove the lower seat bearing and the dust gaiter (arrowed)

recommended for this operation. Any attempt to dismantle the strut without such a tool is likely to result in damage or personal injury.

Note: *A new upper mounting plate nut will be required.*

1 With the strut removed from the car, clean away all external dirt, then mount it upright in a vice. Fit the spring compressor and compress the coil spring until tension is relieved from the spring seats **(see illustration)**.

2 Remove the cover from the upper mounting plate nut then slacken the nut whilst retaining the shock absorber piston with a suitable Allen key **(see illustration)**.

3 Remove the nut and lift off the mounting plate **(see illustration)**.

4 Lift off the spring, followed by the spring stop/lower seat and the lower seat bearing then slide the dust gaiter off of the piston **(see illustrations)**.

5 Examine the shock absorber for signs of fluid leakage. Check the piston for signs of pitting along its entire length, and check the shock body for signs of damage. While holding it in an upright position, test the operation of the shock absorber by moving the piston through a full stroke, and then through short strokes of 50 to 100 mm. In both cases, the resistance felt should be smooth and continuous. If the resistance is jerky, or uneven, or if there is any visible sign of wear or damage to the shock absorber, renewal is necessary.

6 Inspect all other components for signs of damage or deterioration, paying particular attention to the bearing, and renew any that are suspect.

7 Slide the dust gaiter onto the piston and seat its lower end in the strut.

8 Fit the bearing, ensuring it is the correct way up, followed by the spring stop/lower seat assembly.

9 Refit the coil spring, making sure its lower end is correctly seated against the spring seat stop.

10 Fit the upper mounting plate, aligning its stop with the spring end, then screw on the new nut **(see illustration)**. Retain the shock absorber piston and tighten the mounting plate nut to the specified torque. Refit the cover to the nut.

11 Slowly release the spring compressor, ensuring the spring ends remain correctly seated, and remove it from the strut assembly.

12 Refit the strut to the vehicle as described in Section 4.

5.10 On refitting ensure the upper mounting plate stop is correctly located against the spring end

6 Front anti-roll bar - removal and refitting

Removal

1 Chock the rear wheels, apply the handbrake, then jack up the front of the vehicle and support it on axle stands. Remove both front roadwheels.

2 Remove the exhaust downpipe as described in Chapter 4.

3 On models fitted with manual transmission, disconnect the gearchange mechanism from the transmission as described in Chapter 7A.

4 On models where the anti-roll bar is clamped directly to the lower arms, slacken and remove the retaining nut and bolt securing each mounting clamp in position **(see illustration)**. Unhook both clamps and remove them from the undersides of the arms.

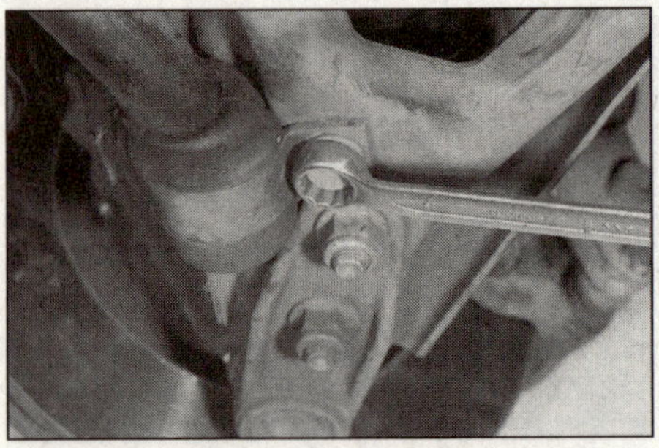

6.4 On models where the anti-roll bar is attached to the lower arms, slacken and remove the mounting clamp nut and bolt

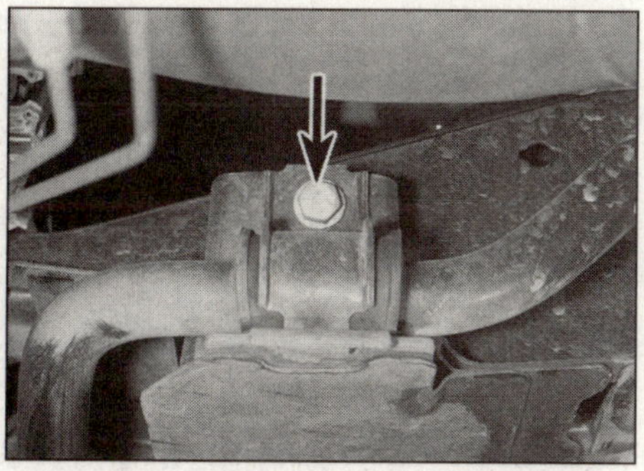

6.6 Slacken and remove the retaining bolt (arrowed) and remove the anti-roll bar mounting clamps

5 On models where the anti-roll bar is attached to the lower arms via link bolts, unscrew the link bolt nuts and remove the upper spacers. Withdraw the link bolts and remove the spacers which are fitted between the anti-roll bar and the lower arms.

6 Slacken and remove the retaining nuts and bolts and remove the anti-roll bar mounting clamps from the subframe **(see illustration)**.

7 Manoeuvre the anti-roll bar out from underneath the vehicle, and remove the rubber mounting bushes.

8 Carefully examine the anti-roll bar components for signs of wear, damage or deterioration, paying particular attention to the mounting bushes. Renew worn components as necessary.

Refitting

9 Fit the mounting bushes onto the anti-roll bar then manoeuvre the anti-roll bar into position underneath the vehicle. Position the bushes so that their splits are against the subframe, then refit the mounting clamps, tightening the bolts lightly only at this stage.

10 On models where the anti-roll bar is clamped to the lower arms, lubricate the bar end mounting rubbers with a smear of multi-purpose grease (Renault recommend the use of Molykote 33 medium grease - available from your Renault dealer). Fit the mounting rubbers to the end of the bar then hook the mounting clamps into position. Refit the mounting clamp bolts and nuts, tightening them lightly only at this stage.

11 On models with link bolts, position the spacers between the anti-roll bar and the lower arms. Insert the link bolts then fit the upper spacers and nuts, tightening them securely.

12 On manual transmission models, reconnect the selector mechanism as described in Chapter 7A.

13 Refit the exhaust downpipe as described in Chapter 4.

14 Refit the roadwheels, lower the vehicle to the ground and tighten the wheel bolts to the specified torque.

15 With the vehicle resting on its wheels, rock the vehicle to settle the anti-roll bar in position then tighten the anti-roll bar mounting clamp nuts and bolts to the specified torque.

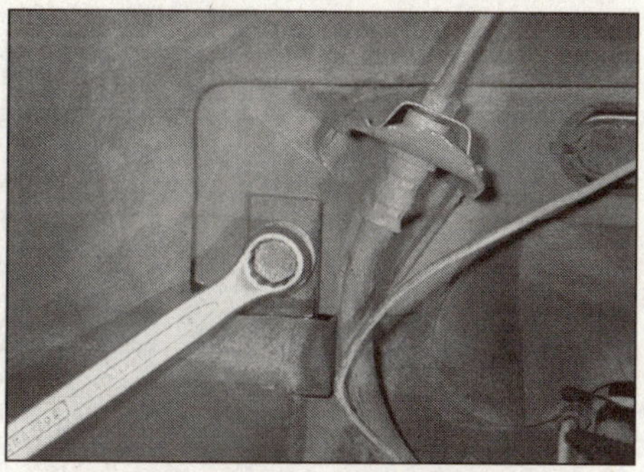

7.3a Undo the upper mounting bolt . . .

7 Front lower arm - removal, overhaul and refitting

Removal

1 Chock the rear wheels, apply the handbrake, jack up the front of the vehicle and support it on axle stands. Remove the appropriate front roadwheel.

2 Working as described in Section 6, disconnect the ends of the anti-roll bar from the lower suspension arms then slacken the mounting clamps and pivot the bar clear of the lower arm.

3 Remove the bolt securing the support bar (where fitted) to the front wing valance, and the nut securing it to the lower arm front pivot bolt and remove the support bar from the vehicle **(see illustrations)**.

4 Remove the nut and clamp bolt securing the lower suspension arm balljoint to the swivel hub **(see illustration)**. Slacken the lower arm pivot bolts then carefully lever the balljoint out of the swivel hub, taking care not to damage the balljoint or driveshaft gaiters. Remove the plastic protector plate which is fitted to the balljoint shank.

5 Remove the two pivot nuts and bolts, and remove the lower suspension arm from the vehicle.

Overhaul

6 Clean the lower arm and the area around the arm mountings, then check for cracks, distortion or any other signs of damage. On some models, a brake disc cooling shield is clipped to the arm; this should also be removed. Check that the lower arm balljoint moves freely, without any sign of roughness, and that the balljoint gaiter is free from cracks and splits. Examine the shank of the pivot bolts for signs of wear or scoring. Renew worn components as necessary.

7.4 Remove the clamp bolt and free the lower arm balljoint from the swivel hub

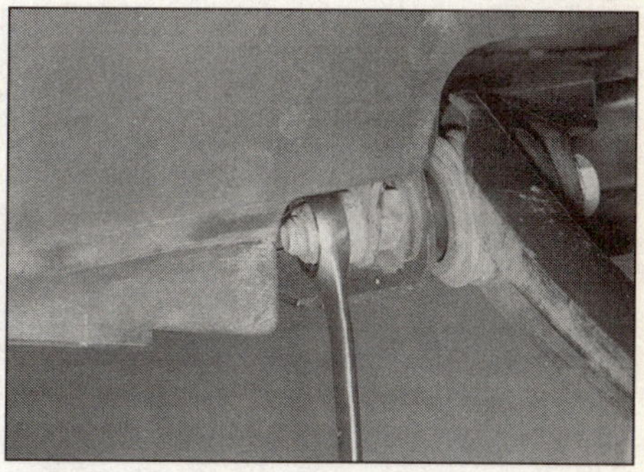

7.3b . . . then unscrew the lower nut from the pivot bolt and remove the support bar

7 Inspect the lower arm pivot bushes. If they are worn, cracked, split or perished, they must be renewed. To renew the bushes, support the lower arm, and press the first bush out using a tubular spacer, such as a socket, which bears only on the hard, outer edge of the bush. **Note:** *Remove only one bush at a time from the arm, to ensure that each new bush is correctly positioned on installation.* Thoroughly clean the lower arm bore, removing all traces of dirt and grease, and polish away any burrs or raised edges which might hinder reassembly. Apply a smear of a suitable grease to the outer edge of the new bush. Press the bush into position until the distance 'A' between the inner edges of the lower arm bushes is as shown **(see illustration)**. Wipe away surplus grease and repeat the procedure on the remaining bush.

Refitting

8 Offer up the lower suspension arm, and insert the two pivot bolts from the rear of the

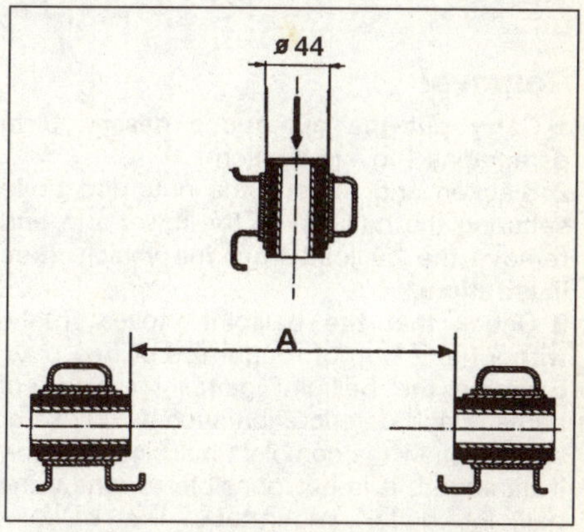

7.7 Front lower arm pivot bush fitting dimension
A = 147 ± 0.5 mm

10

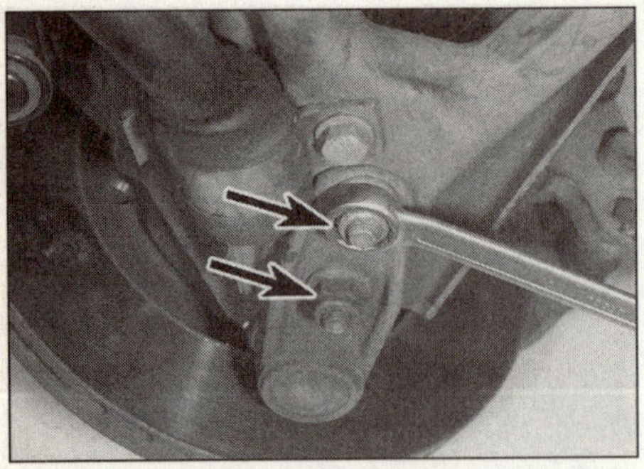

8.2 Slacken and remove the nuts and bolts (arrowed) securing the balljoint to the lower arm

suspension arm. Refit the nuts, but tighten them by hand only at this stage.

9 Refit the plastic protector to the lower arm balljoint, then locate the balljoint shank in the swivel hub. Refit the balljoint clamp bolt, and tighten its retaining nut to the specified torque.

10 When applicable, refit the support bar on the front pivot bolt shank. Fit the retaining bolt and nut, tightening both by hand only at this stage.

11 Working as described in Section 6, reconnect the ends of the anti-roll bar to the lower suspension arms, tightening the mounting bolts only lightly at this stage.

12 Refit the roadwheel, lower the vehicle and tighten the roadwheel bolts to the specified torque.

13 With the vehicle standing on its wheels, rock the vehicle to settle the lower arm and anti-roll bar in position. Tighten the lower arm pivot bolts to the specified torque setting then tighten the anti-roll bar mounting clamp nuts and bolts to their specified torque settings. Once the pivot bolts and anti-roll bar fixings have been tightened, secure the support bar (where fitted) in position by securely tightening its retaining nut and bolt.

8 Front lower arm balljoint - removal and refitting

Removal

1 Carry out the operations described in paragraphs 1 to 4 of Section 7.

2 Slacken and remove the nuts and bolts securing the balljoint to the lower arm and remove the balljoint from the vehicle **(see illustration).**

3 Check that the balljoint moves freely, without any sign of roughness or free play. Examine the balljoint gaiter for signs of damage and deterioration such as cracks or splits. Renew the complete balljoint assembly if damaged; it is not possible to renew the balljoint gaiter separately. The balljoint renewal kit obtainable from Renault dealers contains the balljoint, the plastic protector plate and all fixings.

Refitting

4 Offer up the balljoint then insert the retaining bolts and tighten the retaining nuts to the specified torque.

5 Carry out the operations described in paragraphs 9 to 13 of Section 7.

9 Rear hub bearings - checking and renewal

Note: *The bearing is a sealed, pre-adjusted and pre-lubricated, double-row tapered-roller type, and is intended to last the car's entire service life without maintenance or attention. Never overtighten the hub nut in an attempt to 'adjust' the bearings.*

Checking

1 Chock the front wheels and engage reverse gear (or 'P'). Jack up the rear of the vehicle and support it on axle stands. Remove the appropriate rear roadwheel, and fully release the handbrake.

2 Wear in the rear hub bearings can be checked by measuring the amount of side play present. To do this, a dial test indicator should be fixed so that its probe is in contact with the hub outer face. The play should be between 0 and 0.03 mm. If it is greater than this, the bearings are worn excessively and should be renewed.

Renewal

3 Remove the rear brake drum as described in Chapter 9.

4 Using circlip pliers, extract the bearing retaining circlip from the front of the brake drum **(see illustration).**

5 Securely support the front of the drum hub and press or drive the bearing out of the hub using a tubular drift which bears on the bearing outer race.

6 Thoroughly clean the hub, removing all traces of dirt and grease. Polish away any burrs or raised edges which might hinder reassembly. Check the hub for cracks or any other signs of damage, and renew if necessary. The bearing and its circlip must be renewed whenever they are disturbed. A replacement bearing kit is available from Renault dealers, consisting of the bearing, circlip, spacer, thrustwasher, hub nut and grease cap.

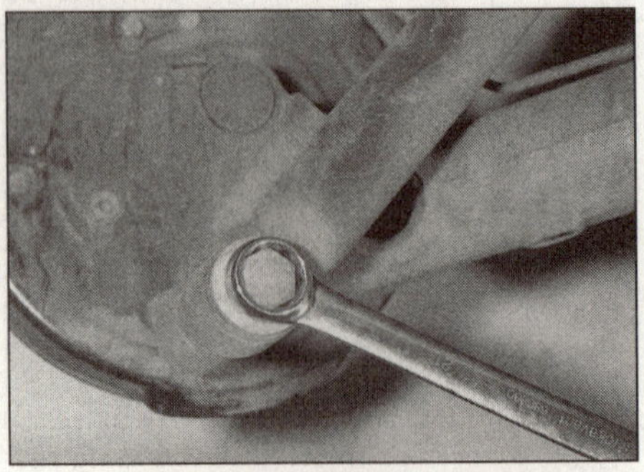

10.2 Slacken and remove the lower mounting bolt . . .

9.4 Rear hub bearing circlip (arrowed)

7 On reassembly, check (if possible) that the new bearing is packed with grease. Apply a light film of gear oil to the bearing outer race and stub axle.

8 Securely support the rear of the drum hub then press/tap the bearing into position using a suitable tubular spacer which bears only on the bearing outer race. Ensure that the bearing enters the hub squarely until it is in contact with the hub shoulder.

9 Secure the bearing in position with the new circlip, ensuring the circlip is correctly seated in its groove.

10 Refit the brake drum as described in Chapter 9.

10 Rear shock absorber - removal, testing and refitting

Removal

1 Chock the front wheels and engage reverse gear (or 'P'). Jack up the rear of the vehicle and support it on axle stands. Remove the appropriate rear roadwheel.

2 Using a jack, raise the trailing arm slightly until the shock absorber is slightly compressed. Remove the lower mounting bolt and washer **(see illustration).**

3 Slacken and remove the upper mounting bolt and remove the shock absorber from the vehicle **(see illustration).**

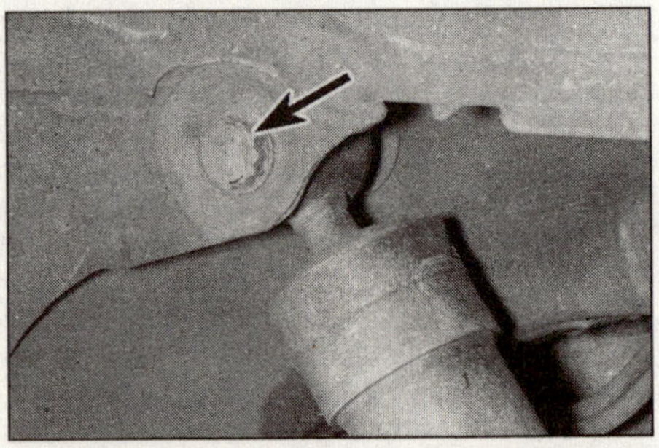

10.3 . . . and the upper mounting bolt (arrowed) then remove the rear shock absorber from the vehicle

Testing

4 Examine the shock absorber for signs of fluid leakage or damage. Test the operation of the shock absorber, while holding it in an upright position, by moving the piston through a full stroke and then through short strokes of 50 to 100 mm. In both cases, the resistance felt should be smooth and continuous. If the resistance is jerky, or uneven, or if there is any visible sign of wear or damage, renewal is necessary. Also check the rubber mountings for damage and deterioration. Renew the shock absorber complete if any damage or wear is evident; the mounting bushes are not available separately. Inspect the mounting bolts for signs of wear or damage, and renew as necessary.

Refitting

5 Prior to refitting the shock absorber, mount it upright in the vice, and operate it fully through several strokes in order to prime it. Apply a smear of multi-purpose grease (Renault recommend the use of Molykote BR2 - available from your Renault dealer) to the shock absorber mounting bolts.

6 Offer up the shock absorber, then install its upper mounting bolt, tightening it by hand only at this stage.

7 Refit the lower shock absorber mounting bolt and washer, again tightening it by hand only. Remove the jack from under the trailing arm.

8 Refit the roadwheel, lower the vehicle to the ground and tighten the roadwheel bolts to the specified torque.

9 Rock the vehicle to settle the shock absorber mounting bushes in position, then tighten both the mounting bolts to the specified torque.

11 Rear anti-roll bar (enclosed-bar rear axle) - removal and refitting

Removal

1 Chock the front wheels and engage reverse gear (or 'P'). Jack up the rear of the vehicle and support it on axle stands.

2 Remove the bolts securing the ends of the anti-roll bar to the rear suspension trailing arms **(see illustration)**. Recover the anti-roll bar retaining nut plates from the top of the trailing arms.

3 Withdraw the anti-roll bar from under the vehicle.

Refitting

4 Refit the anti-roll bar to the vehicle, noting that the cutaway ends of the anti-roll bar retaining bolt brackets must face towards the front of the vehicle **(see illustration)**.

5 Position the retaining nut plates on the top of the trailing arms. Refit the retaining bolts, ensuring that the handbrake cable retaining clips are correctly positioned. Tighten the

11.2 Undo the bolts (arrowed) securing the anti-roll bar to each trailing arm and remove the bar - enclosed-bar rear axle

anti-roll bar retaining bolts to the specified torque.

6 Lower the vehicle to the ground.

12 Rear axle - removal and refitting

Removal

Note: *New axle mounting bolt nuts will be required on refitting.*

1 Chock the front wheels and engage reverse gear (or 'P'). Jack up the rear of the vehicle and support it on axle stands. Remove both rear roadwheels.

2 Working as described in Chapter 9, carry out the following.

a) *Detach both handbrake cables from the lever and release the cables from the vehicle underbody so that they are free to be removed with the axle.*

b) *Trace the brake pipes/hoses back from the wheel cylinders to their unions just in front of the axle* **(see illustration)**. *Disconnect both unions and plug the hose/pipe ends to minimise fluid loss and prevent dirt entry then slide out the retaining clip and free the hose from its mounting bracket. On models with ABS, disconnect the battery before disconnect the hose unions (see Warnings in Chapter 9).*

c) *Unhook the pressure regulating valve link rod from the rear axle.*

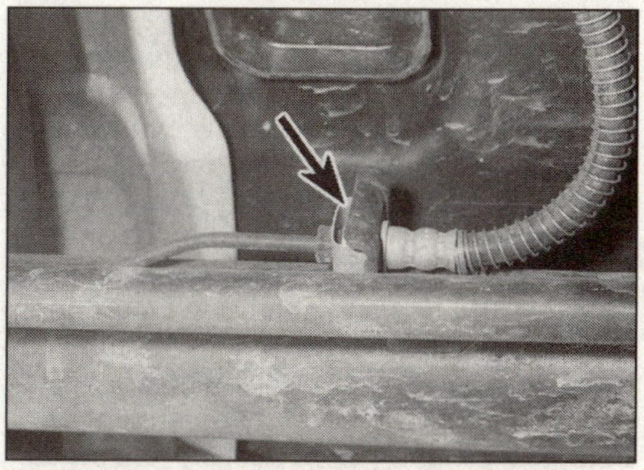

12.2 Disconnect both the left and right-hand rear brake pipes/hoses at their unions (arrowed) in front of the axle

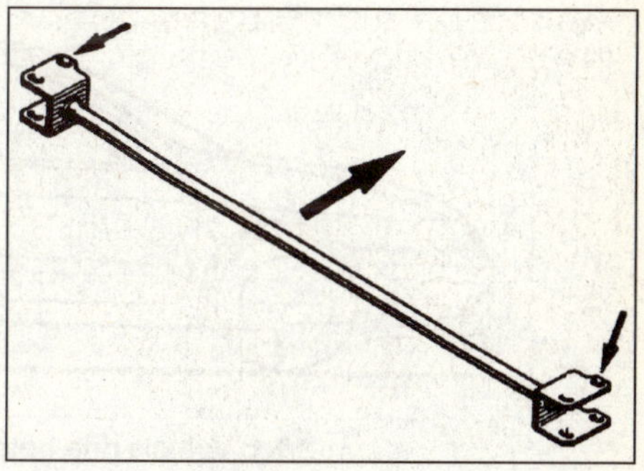

11.4 On refitting ensure the cutaway ends of the anti-roll bar brackets (small arrows) face towards the front of the vehicle

d) *On models with ABS, trace the wiring back from the rear wheel sensors and disconnect the wiring connectors. Free the wiring from all the relevant clips so it is free to be removed with the axle.*

3 Support the weight of the rear axle on a trolley jack.

4 Slacken and remove the left- and right-hand shock absorber lower mounting bolts.

5 Slacken and remove the nuts securing the axle in position and remove the bracing bar which is fitted to the axle mounting bolts **(see illustration)**. Discard the nuts new ones should be used on refitting.

6 Ensure that the axle cables/wiring are freed from all the necessary clips and ties then carefully lower the axle assembly out of position and remove it from underneath the vehicle.

7 To gain access to the axle mounting bolts, remove the rear seat cushion and the sill/side trim panel (see Chapter 11). Undo the retaining bolt then remove the retaining plate and mounting bolts from the vehicle.

Refitting

8 Where necessary, refit the axle mounting bolts and retaining plate in position, tightening the retaining plate bolt securely.

9 Manoeuvre the axle assembly into position, aligning it with the mounting bolts. Refit the bracing bar to the axle bolts then fit the new mounting nuts and tighten them to the specified torque. Where necessary, refit the trim panel and rear seat cushion.

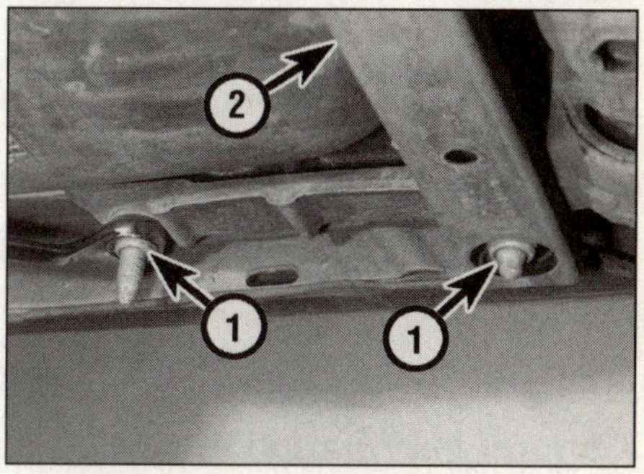

12.5 Slacken and remove the axle retaining nuts (1) and remove the bracing bar (2)

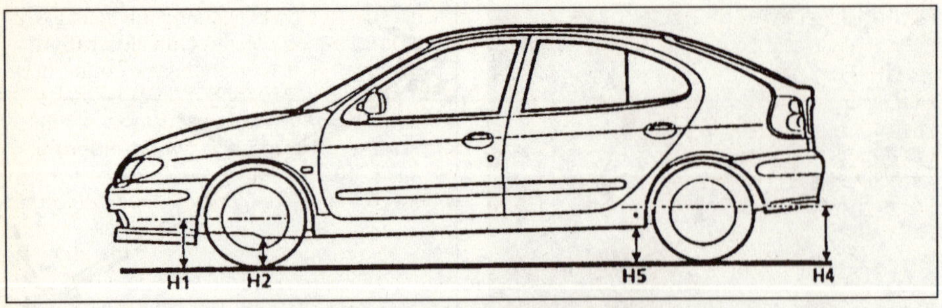

14.1 Vehicle ride height measurement points

10 Apply a smear of multi-purpose grease (Renault recommend the use of Molykote BR2 - available from your Renault dealer) to the shock absorber lower mounting bolts. Align the shock absorbers with the trailing arms and refit the mounting bolts, tightening them by hand only at this stage.

11 Working as described in Chapter 9, carry out the following.

a) Reconnect the rear brake lines, tightening the union nuts to the specified torque, and secure them in position with the retaining clips.
b) On models with ABS, reconnect the wheel sensor wiring connectors ensuring the wiring is correctly routed and retained by all the necessary clips.
c) Reconnect and adjust the handbrake cables.
d) Clip the pressure regulating valve link rod back into position.

12 Refit the roadwheels then lower the vehicle to the ground and tighten the wheel bolts to the specified torque.

13 Rock the vehicle to settle the rear axle in position then tighten the shock absorber lower mounting bolts to the specified torque.

14 Bleed the complete braking system as described in Chapter 9.

13 Rear axle overhaul - general information

Overhaul of the rear axle assembly is a complex task, requiring the use of several special tools. It is also critical that the axle is reassembled correctly, otherwise the vehicle ride height will be incorrect and/or the rear suspension performance will be adversely affected. For these reasons, axle overhaul should be entrusted to a Renault dealer with access to all the necessary equipment.

14 Vehicle ride height - general information and checking

General information

1 The vehicle ride height measurements are used to ensure accuracy when checking the front suspension and steering angles (see Section 23). This is since the angles will very slightly according to the ride height of the vehicle (see Specifications). The ride height measuring points are as shown **(see illustration)**. The front and rear ride heights can also be calculated as follows.

Checking

2 To accurately check the ride height, position the unladen vehicle on a level surface, with the tyres correctly inflated and the fuel tank full.

Front ride height

3 To check the front ride height, measure and record the dimensions H1 (centre of the wheel axis to the ground) and H2 (subframe to the ground) on both sides of the vehicle as shown. Subtract H5 from H4 to find the underbody height checking dimension. Check that this dimension is within the range given in the Specifications at the start of this Chapter. Note that the difference between the heights on each side must not exceed 5 mm, with the driver's side slightly higher than the passenger side.

4 Note that no adjustment of the front ride height is possible although changes of the rear suspension will obviously have some effect. If the ride height differs greatly from that specified, examine the front suspension components for signs of wear or damage.

Rear ride height

5 Measure and record the dimensions H4 (centre of the wheel axis to the ground) and H5 (centre of the rear axle torsion bar to the ground) on both sides of the vehicle as shown. Subtract H5 from H4 to find the underbody height checking dimension. Check that this dimension is within the range given in the Specifications at the start of this Chapter. Note that the difference between the heights on each side must not exceed 5 mm, with the driver's side slightly higher than the passenger side.

6 Adjustment of the rear ride height is possible by altering the position of the rear axle torsion bars (and anti-roll bars on open-bar axle). However the task should be entrusted to a Renault dealer who will access to the necessary tools. Adjustment is a complex procedure, especially on models with an open-bar rear axle, and if done incorrectly can adversely affect the handling of the vehicle.

15 Steering wheel - removal and refitting

Note: *Models equipped with a driver's side airbag have the word AIRBAG stamped on the steering wheel pad. A new steering wheel retaining bolt will be required on refitting.*

Models without airbag

Removal

1 Set the front wheels in the straight-ahead position, and engage the steering lock.

2 Taking care not to damage the steering wheel, carefully prise the horn button out from the centre of the wheel. Disconnect the wiring connector and remove the horn button.

3 Slacken and remove the steering wheel retaining bolt and discard it; a new one must be used on refitting.

4 Mark the steering wheel and steering column shaft in relation to each other, then lift the steering wheel off the column splines.If the wheel is tight, tap it up near the centre, using the palm of your hand, or twist it from side to side, whilst pulling upwards to release it from the shaft splines.

Refitting

5 Refitting is a reversal of removal, noting the following points:

a) Prior to refitting, ensure that the indicator switch stem is in its central position. Failure to do this could lead to the steering wheel lug breaking the switch tab as the steering wheel is refitted.
b) Refit the wheel, aligning the marks made on removal, then fit the new retaining bolt and tighten to the specified torque.

Models with an airbag

Warning: Refer to the precautions in Chapter 12 before proceeding.

Removal

6 Remove the airbag unit (see Chapter 12).

7 Position the front wheels in the straight-ahead position and engage the steering lock. Disconnect the horn button wiring connector **(see illustration)**.

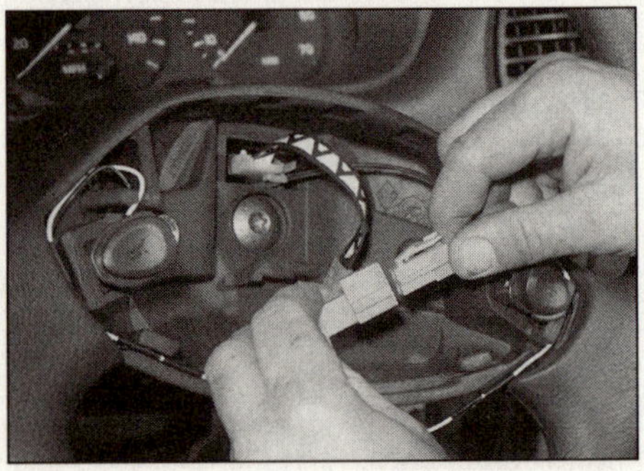

15.7 On models with an airbag, disconnect the horn wiring connector . . .

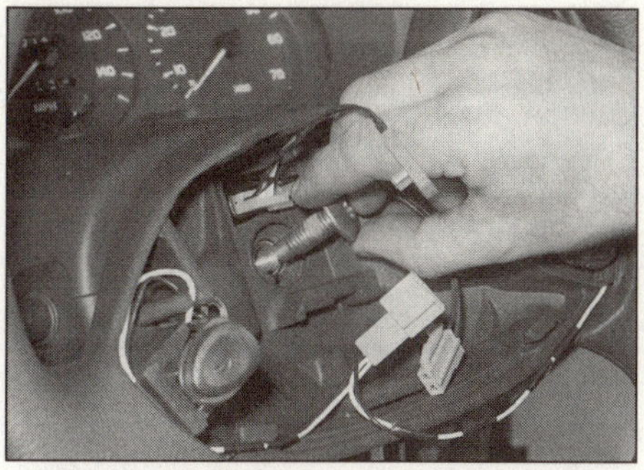

15.8 . . . then slacken and remove the retaining bolt and make alignment marks between the wheel and column

8 Slacken and remove the steering wheel retaining bolt then mark the steering wheel and steering column shaft in relation to each other **(see illustration)**. Discard the bolt; a new one must be used on refitting.

9 Disconnect the horn button wiring connector(s) then lift the steering wheel off the column splines, taking care not to damage the wiring **(see illustration)**.

HAYNES HiNT *If the wheel is tight, tap it up near the centre, using the palm of your hand, or twist it from side to side, whilst pulling upwards to release it from the shaft splines.*

Refitting

10 Refitting is a reversal of removal, noting the points listed in paragraph 5. Prior to refitting the steering wheel, ensure the front wheels are still in the straight-ahead position and check that the airbag contact unit is correctly centralised (Chapter 12, Section 25). On completion, refit the airbag unit as described in Chapter 12.

16 Steering column - removal, inspection and refitting

Warning: On models equipped with a driver's side airbag, refer to the precautions given in Chapter 12 before proceeding.

Removal

1 Disconnect the battery negative lead.

2 Remove the steering wheel as described in Section 15.

3 Remove the complete combination switch assembly as described in Chapter 12.

4 Slacken and remove the retaining screws and remove the steering column trim panel from the facia **(see illustration)**.

5 Trace the wiring back from the ignition switch, freeing it from any relevant clips and ties, and disconnect the wiring connectors **(see illustration)**.

6 Undo the two bolts securing the facia to the top of the steering column **(see illustration)**.

7 Apply the handbrake then jack up the front of the vehicle and support it on axle stands.

8 Working from underneath the vehicle, release the retaining clip (where fitted) and fold back the protective rubber cover to gain access to the steering column lower universal joint. Using paint or a suitable marker pen, make alignment marks between the steering column universal joint and the steering gear pinion. Slacken and remove the nut and clamp bolt securing the joint to the pinion.

9 From inside the vehicle, slacken and remove the four steering column mounting nuts and bolts **(see illustration)**. Unclip the steering column from the facia then release it from its mountings. On models equipped with power steering, also release the steering column lower gaiter from the bulkhead.

10 Disengage the universal joint from the steering gear pinion, and remove the steering column assembly from the vehicle **(see illustration)**.

Inspection

11 The lower section of the steering column incorporates a telescopic safety feature. In the event of a front-end crash, the shaft collapses and prevents the steering wheel injuring the driver. Before refitting the steering column, the length of the intermediate shaft must be checked **(see illustration)**. If the length is

15.9 Lift the steering wheel off the column feeding the wiring carefully through the wheel aperture

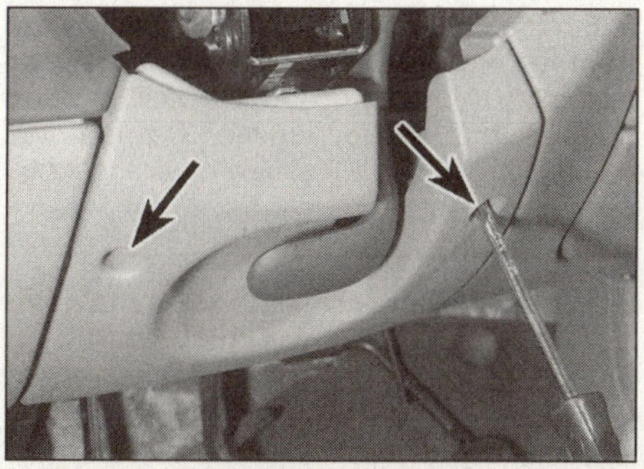

16.4 Undo the retaining screws (arrowed) and remove the steering column trim panel from the facia

16.5 Disconnect the ignition switch wiring and free it from any relevant clips and ties

16.6 Slacken and remove the screws (arrowed) securing the facia to the top of the steering column

16.9 Slacken and remove the mounting bolts and nuts . . .

16.10 . . . then free the universal joint and remove the column assembly from the vehicle

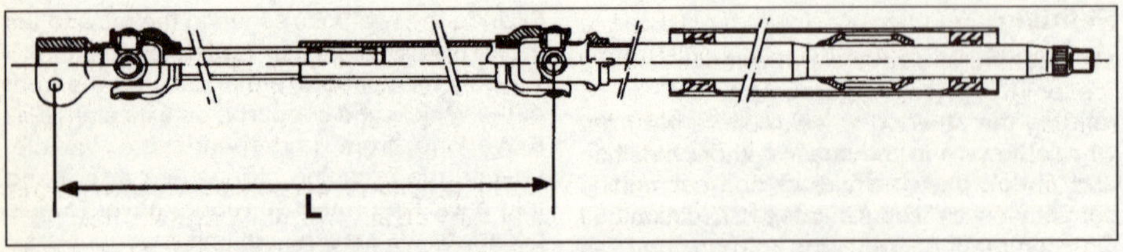

16.11 Steering column lower section checking dimension

Right-hand drive models - 373.5 ± 0.5 mm
Left-hand drive models with manual steering - 426.6 ± 0.5 mm
Left-hand drive models with power steering - 370.4 ± 0.5 mm

shorter than specified, the complete steering column must be renewed. Damage to the intermediate shaft is also implied if it is found that the clamp bolt at its base cannot be inserted freely when refitting the column.

12 Check the steering shaft for signs of free play in the column bushes, and check the universal joints for signs of damage or roughness in the joint bearings. If damage or wear is found on the steering shaft universal joints or shaft bushes, the column must be renewed as an assembly.

13 Inspect the column rubber gaiter/ protective cover for signs of damage or deterioration and renew if necessary.

Refitting

14 Manoeuvre the steering column assembly into position. Engage the universal joint with the steering gear pinion, aligning the marks made prior to removal.

15 Refit and tighten the steering column mounting nuts and bolts. On models with power steering, refit the column gaiter to the bulkhead **(see illustration)**.

16 From underneath the vehicle, refit the universal joint clamp bolt and nut, and tighten it to the specified torque. Relocate the protective cover securely in position and (where necessary) secure it in position with a new clip. Lower the vehicle to the ground.

17 Refit the two bolts securing the facia to the top of the steering column, and tighten them securely. Refit the steering column trim panel to the facia.

18 Ensure the ignition switch wiring is correctly routed then reconnect it to the main wiring harness.

19 Refit the combination switch assembly as described in Chapter 12. On models with an airbag, also refit the contact unit to the steering column.

20 Refit the steering wheel as described in Section 15 and reconnect the battery.

17 Steering gear - removal, inspection and refitting

Removal

Note: *New steering gear mounting nuts and bolts will be required on refitting.*

1 Chock the rear wheels, firmly apply the handbrake then jack up the front of the vehicle and support it on axle stands. Remove both front roadwheels and continue as described under the relevant sub-heading.

Manual steering gear

2 Remove the nuts securing the track rod balljoints to the swivel hubs. Release the balljoint tapered shanks using a universal balljoint separator **(see illustration)**.

3 Working from underneath the vehicle, release the retaining clip (where fitted) and fold back the protective rubber cover to gain access to the steering column lower universal joint. Using paint or a suitable marker pen, make alignment marks between the steering column universal joint and the steering gear pinion. Slacken and remove the nut and clamp bolt securing the joint to the pinion.

4 Remove the two nuts and bolts securing the steering gear assembly to the front subframe. Release the steering gear pinion from the universal joint, and manoeuvre the assembly sideways out of position and recover the rubber cover. Discard the mounting nuts and bolts; new ones should be used on refitting.

Power steering gear

5 Remove the exhaust downpipe as described in Chapter 4. On left-hand drive models, if necessary, remove the air cleaner housing to improve access to the steering gear assembly.

6 Carry out the operations described in paragraphs 2 and 3.

7 Using brake hose clamps, clamp both the supply and return hoses near the power

16.15 On refitting, tighten the column mounting nuts and bolts to the specified torque

steering fluid reservoir. This will minimise fluid loss during subsequent operations.

8 Mark the unions to ensure that they are correctly positioned on reassembly, then unscrew the feed and return pipe union nuts from the steering gear assembly; be prepared for fluid spillage, and position a suitable container beneath the pipes whilst unscrewing the union nuts. Disconnect both pipes, and plug the pipe ends and steering gear orifices, to prevent fluid leakage and to keep dirt out of the hydraulic system. Remove the sealing rings from the union nuts and discard them; new ones should be used on refitting.

9 Unscrew the nuts/bolts securing the fluid feed and return pipes to the subframe and position both pipes clear of the steering gear.

10 Slacken and remove the steering gear mounting bolts and nuts then remove the left- and right-hand mounting clamp halves and the mounting rubbers **(see illustration)**. Discard the mounting nuts and bolts; new ones should be used on refitting.

11 Release the steering gear pinion from the universal joint then manoeuvre the assembly out from the driver's side of the vehicle.

Inspection

12 Examine the steering gear assembly for signs of wear or damage, and check that the rack moves freely throughout the full length of its travel, with no signs of roughness or excessive free play between the steering gear pinion and rack. On models with power-

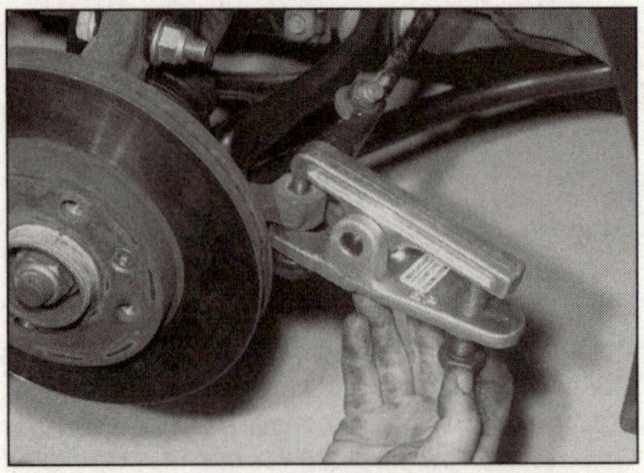

17.2 Using a universal balljoint separator to free the track rod balljoint from the swivel hub

17.10 Removing a steering gear mounting clamp retaining bolt nut

assisted steering, inspect all the steering gear fluid unions for signs of leakage, and check that all union nuts are securely tightened.
13 Overhaul of the steering gear rack and pinion assembly is not possible. The only components which can be renewed are the steering gear gaiters, the track rod balljoints and the track rods. Track rod balljoint, steering gear gaiter and track rod renewal procedures are covered elsewhere in this Chapter.

Refitting

Manual steering gear

14 Manoeuvre the steering gear assembly and rubber cover into position. Engage the universal joint with the steering gear pinion splines, aligning the marks made prior to removal.
15 Insert the new mounting bolts then fit the nuts and tighten them to the specified torque.
16 Refit the universal joint clamp bolt and nut, and tighten it to the specified torque. Relocate the protective cover and secure it in position with a new cable tie.
17 Reconnect the track rod balljoints to the swivel hubs, and tighten their retaining nuts to the specified torque.
18 Refit the roadwheels, lower the vehicle to the ground and tighten the wheel bolts to the specified torque.
19 On completion check and, if necessary, adjust the front wheel alignment as described in Section 23.

Power steering gear

20 Manoeuvre the steering gear assembly and rubber cover into position from the driver's side of the vehicle. Engage the universal joint with the steering gear pinion splines, aligning the marks made prior to removal.
21 Fit the mounting rubbers to the steering gear and manoeuvre the mounting clamp halves into position. Fit the new mounting bolts and nuts tightening them by hand only at this stage.
22 Fit new sealing rings to the steering pipe unions nuts and reconnect the pipes to the steering gear. Screw each union nut in by a few turns then refit the pipe retaining clip nuts/bolts (as applicable) but do not tighten yet.
23 Tighten the steering gear mounting bolts and nuts to the specified torque setting then tighten the feed and return pipe union nuts to their specified torque settings. The pipe retaining clip nuts/bolts should then be securely tightened.
24 Refit the universal joint clamp bolt and nut, tightening it to the specified torque, then relocate the protective cover securely in position.
25 Reconnect the track rod balljoints to the swivel hubs, and tighten their retaining nuts to the specified torque.
26 Refit the exhaust system downpipe and (where removed) the air cleaner housing as described in Chapter 4.
27 Refit the roadwheels, lower the vehicle to the ground and tighten the wheel bolts to the specified torque.
28 Remove the hose clamps and bleed the hydraulic system as described in Section 19.
29 On completion check and, if necessary, adjust the front wheel alignment as described in Section 23.

18 Power steering pump - removal and refitting

Removal

1 On models were a remote fluid reservoir is fitted, to minimise fluid loss, using brake hose clamps, clamp both the supply and return hoses near the power steering fluid reservoir. This will minimise fluid loss during subsequent operations.
2 Working as described in Chapter 1, release the drivebelt tension and disengage the drivebelt from the pump pulley, noting that the power steering pump pulley retaining bolts should be slackened before the belt is removed.
3 Remove the pulley retaining bolts and washers. Remove the pulley from the pump, noting which way round it is fitted.
4 Slacken the retaining clip, and disconnect the fluid supply hose from the pump then slacken the union nut, and disconnect the feed pipe; on some models it will be necessary to undo the feed pipe mounting clamp bolt to allow the pipe to be disconnected. Be prepared for some fluid spillage as the pipe and hose are disconnected, and plug the hose/pipe end and pump unions, to minimise fluid loss and prevent the entry of dirt into the system. Discard the feed pipe union nut sealing ring, a new one should be used on refitting.
5 Slacken and remove the bolts securing the power steering pump to its mounting bracket and remove the pump from the engine.

Refitting

6 Manoeuvre the pump assembly back into position and securely tighten the mounting bolts.
7 Fit a new sealing ring to the feed pipe union and refit the pipe to the pump, tightening the union nut to the specified torque. Reconnect the supply pipe and securely tighten its retaining clip. Where necessary, refit the feed pipe clamp retaining bolt and tighten securely.
8 Refit the pulley, ensuring it is the correct way around, and refit the retaining bolts.
9 Refit the drivebelt and tension as described in Chapter 1. The pump pulley retaining bolts can then be securely tightened.
10 Remove the hose clamps (where fitted) and bleed the hydraulic system as described in Section 19.

19 Power steering hydraulic system - bleeding

1 This procedure will only be necessary when any part of the hydraulic system has been disconnected, or if air has entered because of leakage.
2 Remove the fluid reservoir filler cap, and top-up the fluid level to the maximum mark, using only the specified fluid as described in "*Weekly checks*".
3 With the engine stopped, slowly move the steering from lock-to-lock several times to expel trapped air, then top-up the level in the fluid reservoir. Repeat this procedure until the fluid level in the reservoir does not drop any further.
4 Start the engine. Slowly move the steering from lock-to-lock several times to expel any air remaining in the system. Repeat this procedure until bubbles cease to appear in the fluid reservoir.
5 If, when turning the steering, an abnormal noise is heard from the fluid pipes, it indicates that there is still air in the system. Check this by turning the wheels to the straight-ahead position and switching off the engine. If the fluid level in the reservoir rises, air is still present in the system, and further bleeding is necessary.
6 Once all traces of air have been removed, stop the engine and allow the system to cool. Once cool, check that the fluid level is up to the maximum mark on the power steering fluid reservoir; top-up if necessary.

20 Steering gear rubber gaiter - renewal

1 Remove the track rod end balljoint as described in Section 21.
2 Mark the correct fitted position of the gaiter on the track rod. Release the retaining clips, and slide the gaiter off the steering gear housing and track rod end.
3 Thoroughly clean the track rod and the steering gear housing, using fine abrasive paper to polish off any corrosion, burrs or sharp edges which might damage the sealing lips of the new gaiter on installation.
4 Recover the grease from inside the old gaiter. If it is uncontaminated with dirt or grit, apply it to the track rod inner balljoint. If the old grease is contaminated, or it is suspected that some has been lost, apply some new molybdenum disulphide grease.
5 Grease the inside of the new gaiter. Carefully slide the gaiter onto the track rod, and locate it on the steering gear housing. Align the outer edge of the gaiter with the mark made on the track rod prior to removal, then secure it in position with new clips.
6 Refit the track rod balljoint as described in Section 21.

21 Track rod balljoint - removal and refitting

Removal

1 Apply the handbrake, then jack up the front of the vehicle and support it on axle stands. Remove the appropriate front roadwheel.
2 If the balljoint is to be re-used, use a straight-edge and a scriber, or similar, to mark its relationship to the track rod.
3 Slacken the balljoint clamp bolt.
4 Remove the nut securing the track rod balljoint to the swivel hub. Release the balljoint tapered shank using a universal balljoint separator. If the balljoint is to be re-used, protect the threaded end of the shank by screwing the nut back on a few turns before using the separator.
5 Counting the **exact** number of turns necessary to do so, unscrew the balljoint from the track rod end.
6 Carefully clean the balljoint and the threads. Renew the balljoint if its movement is sloppy or if it is too stiff, if it is excessively worn, or if it is damaged in any way. Carefully check the shank taper and threads. If the balljoint gaiter is damaged, the complete balljoint must be renewed; it is not possible to obtain the gaiter separately.

Refitting

7 Screw the balljoint into the track rod by the number of turns noted on removal. This align the marks that were made (if applicable) on removal.
8 Refit the balljoint shank to the swivel hub, and tighten the retaining nut to the specified torque. If difficulty is experienced due to the balljoint shank rotating, jam it by exerting pressure on the underside of the balljoint, using a tyre lever or a jack.
9 Refit the roadwheel, lower the vehicle to the ground and tighten the roadwheel bolts to the specified torque.
10 Check the front wheel toe setting as described in Section 23 then tighten the balljoint clamp bolt to the specified torque.

22 Track rod - removal and refitting

Note: *In order to safely remove the track rod, without the risk of damaging the steering rack, a special track rod wrench (Renault number Dir.1305) and rack retaining clamp (Renault number Dir.1306 for SMI steering rack or 1306-01 for TRW steering rack) will be required. The special wrench engages with the track balljoint housing allowing the track rod to be easily slackened/tightened and the retaining clamp secures the rack to the steering gear housing to prevent any stress being exerted on the steering gear pinion assembly. Note that without access to the special tools, track rod removal will be difficult, especially without causing damage.*
Note: *There are two different types of steering gear assemblies fitted to these vehicles; the steering racks are manufactured by either SMI or TRW; the type of rack fitted can be identified from the manufacturing markings cast onto the steering gear housing. If work is being carried out on a SMI steering rack note that a new track rod lockwasher assembly must be used on refitting.*

Removal

1 Remove the track rod end balljoint as described in Section 21.
2 Cut the retaining clips, and slide the steering gear gaiter off the track rod. It is recommended that the gaiter is renewed, regardless of its apparent condition.
3 In the absence of the special tools, using a suitable pair of grips, unscrew the track rod inner balljoint from the steering rack end. Prevent the steering rack from turning by holding the balljoint lockwasher/steering rack with a second pair of grips.
Caution: Take care not to mark the surfaces of the rack and balljoint.
4 Remove the track rod assembly from the steering rack. On SMI steering gear assemblies, discard the lockwasher assembly; a new one must be used on refitting.
5 Examine the inner balljoint for signs of slackness or tight spots. Check that the track rod itself is straight and free from damage. If necessary, renew the track rod.

Refitting

6 Prior to refitting the track rod remove all traces of locking compound from the rack and track rod threads.
7 On models fitted with a SMI steering gear assembly, ensure the new lockwasher and locking ring are correctly assembled and fit them to the end of the track rod.
8 Apply a coat of locking compound (Renault recommend the use of Loctite Frenbloc - available from your Renault dealer) to the threads of the track rod then fit the track rod to the end of the steering rack. Using the method employed on removal, tighten the track rod inner balljoint to the specified torque, taking great care not to marks either the rack or the balljoint. On the SMI steering gear, whilst tightening the track rod ensure the tabs on the lock washer engage with the flats on the steering end.
9 Slide the new gaiter onto the track rod end, and locate it on the steering gear housing. Turn the steering from lock-to-lock to check that the gaiter is correctly positioned, then secure it with new retaining clips.
10 Refit the track rod end balljoint as described in Section 21.

23 Wheel alignment and steering angles - general information

Definitions

1 A car's steering and suspension geometry is defined in four basic settings - all angles are expressed in degrees (toe settings are also expressed as a measurement); the steering axis is defined as an imaginary line drawn through the axis of the suspension strut, extended where necessary to contact the ground.
2 Camber is the angle between each roadwheel and a vertical line drawn through its centre and tyre contact patch, when viewed from the front or rear of the car. Positive camber is when the roadwheels are tilted outwards from the vertical at the top; negative camber is when they are tilted inwards. The camber angle is not adjustable
3 Castor is the angle between the steering axis and a vertical line drawn through each roadwheel's centre and tyre contact patch, when viewed from the side of the car. Positive castor is when the steering axis is tilted so that it contacts the ground ahead of the vertical; negative castor is when it contacts the ground behind the vertical. The castor angle is not adjustable.
4 Toe is the difference, viewed from above, between lines drawn through the roadwheel centres and the car's centre-line. "Toe-in" is when the roadwheels point inwards, towards each other at the front, while "toe-out" is when they splay outwards from each other at the front.
5 The front wheel alignment (toe-setting) is adjusted by screwing the track rod in or out of its balljoints, to alter the effective length of the track rod assembly. The rear wheel toe setting is not adjustable.

Checking and adjustment

6 Due to the special measuring equipment necessary to check the wheel alignment and steering angles, and the skill required to use it properly, the checking and adjustment of these settings is best left to a Renault dealer or similar expert. Note that most tyre-fitting shops now possess sophisticated checking equipment. The following is provided as a guide, should the owner decide to carry out a DIY check.

Front wheel alignment (toe-setting)

7 The front wheel alignment (toe-setting) is checked by measuring the distance between the front and rear inside edges of the roadwheel rims. Proprietary toe measurement gauges are available from motor accessory shops. Adjustment is made by screwing the balljoints in or out of their track rods, to alter the effective length of the track rod assemblies.

8 For **accurate** checking, the vehicle **must** be at the kerb weight, ie unladen and with a full tank of fuel.

9 Before starting work, check first that the tyre sizes and types are as specified, then check the tyre pressures and tread wear, the roadwheel run-out, the condition of the hub bearings, the steering wheel free play, and the condition of the front suspension components (see "*Weekly checks*" and Chapter 1). Correct any faults found.

10 Park the vehicle on level ground, check that the front roadwheels are in the straight-ahead position, then rock the rear and front ends to settle the suspension. Release the handbrake, and roll the vehicle backwards 1 metre, then forwards again, to relieve any stresses in the steering and suspension components.

11 Measure the distance between the front edges of the wheel rims and the rear edges of the rims. Subtract the rear measurement from the front measurement, and check that the result is within the specified range.

12 If adjustment is necessary, apply the handbrake, then jack up the front of the vehicle and support it securely on axle stands. Turn the steering wheel onto full-left lock, and record the number of exposed threads on the right-hand track rod end. Now turn the steering onto full-right lock, and record the number of threads on the left-hand side. If there are the same number of threads visible on both sides, then subsequent adjustment should be made equally on both sides. If there are more threads visible on one side than the other, it will be necessary to compensate for this during adjustment. **Note:** *It is most important that after adjustment, the same number of threads are visible on each track rod end.*

13 First clean the track rod threads; if they are corroded, apply penetrating fluid before starting adjustment. Release the rubber gaiter outboard clips, and peel back the gaiters; apply a smear of grease to the inside of the gaiters, so that both are free, and will not be twisted or strained as their respective track rods are rotated.

14 Use a straight-edge and a scriber or similar to mark the relationship of each track rod to its balljoint then slacken the balljoint clamp bolt.

15 Alter the length of the track rods, bearing in mind the note made in paragraph 12. **Note:** *One complete rotation of the track rod equals approximately 3 mm of adjustment*. Screw them into or out off the balljoints, rotating the track rod using an open-ended spanner fitted to the flats provided on the track rod **(see illustration)**. Shortening the track rods (screwing them into their balljoints) will reduce toe-in/increase toe-out.

16 When the setting is correct, tighten the balljoint clamp bolts to the specified torque. Check that the balljoints are seated correctly in their sockets, and count the exposed threads to check the length of both track rods. If they are not the same, then the adjustment has not been made equally, and problems will be encountered with tyre scrubbing in turns; also, the steering wheel spokes will no longer be horizontal when the wheels are in the straight-ahead position.

23.15 Adjusting the front wheel alignment (toe setting)

17 If the track rod lengths are the same, lower the vehicle to the ground and re-check the toe setting; re-adjust if necessary. When the setting is correct, tighten the track rod balljoint clamp bolts to the specified torque. Ensure that the rubber gaiters are seated correctly, and are not twisted or strained, and secure them in position with new retaining clips.

Chapter 11
Bodywork and fittings

Contents

Degrees of difficulty

Easy, suitable for novice with little experience	**Fairly easy,** suitable for beginner with some experience	**Fairly difficult,** suitable for competent DIY mechanic	**Difficult,** suitable for experienced DIY mechanic	**Very difficult,** suitable for expert DIY or professional

1 General information

The bodyshell is made of pressed-steel sections. Most components are welded together, but some use is made of structural adhesives.

The bonnet, door, and some other vulnerable panels are made of zinc-coated metal, and are further protected by being coated with an anti-chip primer before being sprayed.

Extensive use is made of plastic materials, mainly in the interior, but also in exterior components. The front and rear bumpers and front grille are injection-moulded from a synthetic material that is very strong and yet light. Plastic components such as wheel arch liners are fitted to the underside of the vehicle, to improve the body's resistance to corrosion.

2 Maintenance - bodywork and underframe

The general condition of a vehicle's bodywork is the one thing that significantly affects its value. Maintenance is easy, but needs to be regular. Neglect, particularly after minor damage, can lead quickly to further deterioration and costly repair bills. It is important also to keep watch on those parts of the vehicle not immediately visible, for instance the underside, inside all the wheel arches, and the lower part of the engine compartment.

The basic maintenance routine for the bodywork is washing - preferably with a lot of water, from a hose. This will remove all the loose solids which may have stuck to the vehicle. It is important to flush these off in such a way as to prevent grit from scratching the finish. The wheel arches and underframe need washing in the same way, to remove any accumulated mud which will retain moisture and tend to encourage rust. Paradoxically enough, the best time to clean the underframe and wheel arches is in wet weather, when the mud is thoroughly wet and soft. In very wet weather, the underframe is usually cleaned of large accumulations automatically, and this is a good time for inspection.

Periodically, except on vehicles with a wax-based underbody protective coating, it is a good idea to have the whole of the underframe of the vehicle steam-cleaned, engine compartment included, so that a thorough inspection can be carried out to see what minor repairs and renovations are necessary. Steam cleaning is available at many garages, and is necessary for the removal of the accumulation of oily grime, which sometimes is allowed to become thick in certain areas. If steam-cleaning facilities are not available, there are some excellent grease solvents available which can be brush-applied; the dirt can then be simply hosed off. Note that these methods should not be used on vehicles with wax-based underbody protective coating, or the coating will be removed. Such vehicles should be inspected annually, preferably just before Winter, when the underbody should be washed down, and repair any damage to the wax coating. Ideally, a completely fresh coat should be applied. It would also be worth considering the use of such wax-based protection for injection into door panels, sills, box sections, etc, as an additional safeguard against rust damage, where such protection is not provided by the vehicle manufacturer.

After washing paintwork, wipe off with a chamois leather to give an unspotted clear finish. A coat of clear protective wax polish will give added protection against chemical pollutants in the air. If the paintwork sheen has dulled or oxidised, use a cleaner/polisher combination to restore the brilliance of the shine. This requires a little effort, but such dulling is usually caused because regular washing has been neglected. Care needs to be taken with metallic paintwork, as special non-abrasive cleaner/polisher is required to avoid damage to the finish. Always check that the door and ventilator opening drain holes and pipes are completely clear, so that water can be drained out. Brightwork should be treated in the same way as paintwork. Windscreens and windows can be kept clear of the smeary film which often appears, by proprietary glass cleaner. Never use any form of wax or other body or chromium polish on glass.

3 Maintenance - upholstery and carpets

Mats and carpets should be brushed or vacuum-cleaned regularly, to keep them free of grit. If they are badly stained, remove them from the vehicle for scrubbing or sponging, and make quite sure they are dry before refitting. Seats and interior trim panels can be kept clean by wiping with a damp cloth. If they do become stained (which can be more apparent on light-coloured upholstery), use a little liquid detergent and a soft nail brush to scour the grime out of the grain of the material. Do not forget to keep the headlining clean in the same way as the upholstery. When using liquid cleaners inside the vehicle, do not over-wet the surfaces being cleaned. Excessive damp could get into the seams and padded interior, causing stains, offensive odours or even rot. If the inside of the vehicle gets wet accidentally, it is worthwhile taking some trouble to dry it out properly, particularly where carpets are involved. *Do not leave oil or electric heaters inside the vehicle for this purpose.*

4 Minor body damage - repair

Repairs of minor scratches in bodywork

If the scratch is very superficial, and does not penetrate to the metal of the bodywork, repair is very simple. Lightly rub the area of the scratch with a paintwork renovator or a very fine cutting paste to remove loose paint from the scratch, and to clear the surrounding bodywork of wax polish. Rinse the area with clean water.

Apply touch-up paint to the scratch using a fine paint brush; continue to apply fine layers of paint until the surface of the paint in the scratch is level with the surrounding paintwork. Allow the new paint at least two weeks to harden, then blend it into the surrounding paintwork by rubbing the scratch area with a paintwork renovator or a very fine cutting paste. Finally, apply wax polish.

Where the scratch has penetrated right through to the metal of the bodywork, causing the metal to rust, a different repair technique is required. Remove any loose rust from the bottom of the scratch with a penknife, then apply rust-inhibiting paint to prevent the formation of rust in the future. Using a rubber or nylon applicator, fill the scratch with bodystopper paste. If required, this paste can be mixed with cellulose thinners to provide a very thin paste which is ideal for filling narrow scratches. Before the stopper-paste in the scratch hardens, wrap a piece of smooth cotton rag around the top of a finger. Dip the finger in cellulose thinners, and quickly sweep it across the surface of the stopper-paste in the scratch; this will ensure that the surface of the stopper-paste is slightly hollowed. The scratch can now be painted over as described earlier in this Section.

Repairs of dents in bodywork

When deep denting of the vehicle's bodywork has taken place, the first task is to pull the dent out, until the affected bodywork almost attains its original shape. There is little point in trying to restore the original shape completely, as the metal in the damaged area will have stretched on impact, and cannot be reshaped fully to its original contour. It is better to bring the level of the dent up to a point which is about 3 mm below the level of the surrounding bodywork. In cases where the dent is very shallow anyway, it is not worth trying to pull it out at all. If the underside of the dent is accessible, it can be hammered out gently from behind, using a mallet with a wooden or plastic head. Whilst doing this, hold a suitable block of wood firmly against the outside of the panel, to absorb the impact from the hammer blows and thus prevent a large area of the bodywork from being "belled-out".

Should the dent be in a section of the bodywork which has a double skin, or some other factor making it inaccessible from behind, a different technique is called for. Drill several small holes through the metal inside the area - particularly in the deeper section. Then screw long self-tapping screws into the holes, just sufficiently for them to gain a good purchase in the metal. Now the dent can be pulled out by pulling on the protruding heads of the screws with a pair of pliers.

The next stage of the repair is the removal of the paint from the damaged area, and from an inch or so of the surrounding "sound" bodywork. This is accomplished most easily by using a wire brush or abrasive pad on a power drill, although it can be done just as effectively by hand, using sheets of abrasive paper. To complete the preparation for filling, score the surface of the bare metal with a screwdriver or the tang of a file, or alternatively, drill small holes in the affected area. This will provide a good "key" for the filler paste.

To complete the repair, see the Section on filling and respraying.

Repairs of rust holes or gashes in bodywork

Remove all paint from the affected area, and from an inch or so of the surrounding "sound" bodywork, using an abrasive pad or a wire brush on a power drill. If these are not available, a few sheets of abrasive paper will do the job most effectively. With the paint removed, you will be able to judge the severity of the corrosion, and therefore decide whether to renew the whole panel (if this is possible) or to repair the affected area. New body panels are not as expensive as most people think, and it is often quicker and more satisfactory to fit a new panel than to attempt to repair large areas of corrosion.

Remove all fittings from the affected area, except those which will act as a guide to the original shape of the damaged bodywork (eg headlamp shells etc). Then, using tin snips or a hacksaw blade, remove all loose metal and any other metal badly affected by corrosion. Hammer the edges of the hole inwards, to create a slight depression for the filler paste.

Wire-brush the affected area to remove the powdery rust from the surface of the remaining metal. Paint the affected area with rust-inhibiting paint; if the back of the rusted area is accessible, treat this also.

Before filling can take place, it will be necessary to block the hole in some way. This can be achieved with aluminium or plastic mesh, or aluminium tape.

Aluminium or plastic mesh, or glass-fibre matting, is probably the best material to use for a large hole. Cut a piece to the approximate size and shape of the hole to be filled, then position it in the hole so that its edges are below the level of the surrounding bodywork. It can be retained in position by several blobs of filler paste around its periphery.

Aluminium tape should be used for small or very narrow holes. Pull a piece off the roll, trim it to the approximate size and shape required, then pull off the backing paper (if used) and stick the tape over the hole; it can be overlapped if the thickness of one piece is insufficient. Burnish down the edges of the tape with the handle of a screwdriver or similar, to ensure that the tape is securely attached to the metal underneath.

Bodywork repairs - filling and respraying

Before using this Section, see the Sections on dent, deep scratch, rust holes and gash repairs.

Many types of bodyfiller are available, but generally speaking, those proprietary kits which contain a tin of filler paste and a tube of resin hardener are best for this type of repair which can be used directly from the tube. A wide, flexible plastic or nylon applicator will be found invaluable for imparting a smooth and well-contoured finish to the surface of the filler.

Mix up a little filler on a clean piece of card or board - measure the hardener carefully (follow the maker's instructions on the pack), otherwise the filler will set too rapidly or too slowly. Using the applicator, apply the filler paste to the prepared area; draw the applicator across the surface of the filler to achieve the correct contour and to level the surface. When a contour that approximates to the correct one is achieved, stop working the paste - if you carry on too long, the paste will become sticky and begin to "pick-up" on the applicator. Continue to add thin layers of filler paste at 20-minute intervals, until the level of the filler is just proud of the surrounding bodywork.

Once the filler has hardened, the excess can be removed using a metal plane or file. From then on, progressively-finer grades of abrasive paper should be used, starting with a 40-grade production paper, and finishing with a 400-grade wet-and-dry paper. Always wrap the abrasive paper around a flat rubber, cork, or wooden block - otherwise the surface of the filler will not be completely flat. During the smoothing of the filler surface, the wet-and-dry paper should be periodically rinsed in water. This will ensure that a very smooth finish is imparted to the filler at the final stage.

At this stage, the "dent" should be surrounded by a ring of bare metal, which in turn should be encircled by the finely "feathered" edge of the good paintwork. Rinse the repair area with clean water, until all the dust produced by the rubbing-down operation has gone.

Spray the whole area with a light coat of primer - this will show up any imperfections in the surface of the filler. Repair these imperfections with fresh filler paste or bodystopper, and again smooth the surface with abrasive paper. If bodystopper is used, it can be mixed with cellulose thinners, to form a thin paste which is ideal for filling small holes. Repeat this spray-and-repair procedure until you are satisfied that the surface of the filler, and the feathered edge of the paintwork, are perfect. Clean the repair area with clean water, and allow to dry fully.

The repair area is now ready for final spraying. Paint spraying must be carried out in a warm, dry, windless and dust-free atmosphere. This condition can be created artificially if you have access to a large indoor working area, but if you are forced to work in the open, you will have to pick your day very carefully. If you are working indoors, dousing the floor in the work area with water will help to settle the dust which would otherwise be in the atmosphere. If the repair area is confined to one body panel, mask off the surrounding panels; this will help to minimise the effects of a slight mis-match in paint colours. Bodywork fittings (eg chrome strips, door handles etc) will also need to be masked off. Use genuine masking tape, and several thickness of newspaper, for the masking operations.

Before starting to spray, agitate the aerosol can thoroughly, then spray a test area (an old tin, or similar) until the technique is mastered. Cover the repair area with a thick coat of primer; the thickness should be built up using several thin layers of paint, rather than one thick one. Using 400 grade wet-and-dry paper, rub down the surface of the primer until it is smooth. While doing this, the work area should be thoroughly doused with water, and the wet-and-dry paper periodically rinsed in water. Allow to dry before spraying on more paint.

Spray on the top coat, again building up the thickness by using several thin layers of paint. Start spraying in the centre of the repair area, and then, using a circular motion, work outwards until the whole repair area and about 2 inches of the surrounding original paintwork is covered. Remove all masking material 10 to 15 minutes after spraying on the final coat of paint.

Allow the new paint at least two weeks to harden, then, using a paintwork renovator or a very fine cutting paste, blend the edges of the paint into the existing paintwork. Finally, apply wax polish.

Plastic components

With the use of more and more plastic body components by the vehicle manufacturers (eg bumpers. spoilers, and in some cases major body panels), rectification of more serious damage to such items has become a matter of either entrusting repair work to a specialist in this field, or renewing complete components. Repair of such damage by the DIY owner is not feasible, owing to the cost of the equipment and materials required for effecting such repairs. The basic technique involves making a groove along the line of the crack in the plastic, using a rotary burr in a power drill. The damaged part is then welded back together, using a hot air gun to heat up and fuse a plastic filler rod into the groove. Any excess plastic is then removed, and the area rubbed down to a smooth finish. It is important that a filler rod of the correct plastic is used, as body components can be made of a variety of different types (eg polycarbonate, ABS, polypropylene).

Damage of a less serious nature (abrasions, minor cracks etc) can be repaired by the DIY owner using a two-part epoxy filler repair material which can be used directly from the tube. Once mixed in equal proportions, this is used in similar fashion to the bodywork filler used on metal panels. The filler is usually cured in twenty to thirty minutes, ready for sanding and painting.

If the owner is renewing a complete component himself, or if he has repaired it with epoxy filler, he will be left with the problem of finding a suitable paint for finishing which is compatible with the type of plastic used. At one time, the use of a universal paint was not possible, owing to the complex range of plastics met with in body component applications. Standard paints, generally speaking, will not bond to plastic or rubber satisfactorily, but professional matched paints, to match any plastic or rubber finish, can be obtained from some dealers. However, it is now possible to obtain a plastic body parts finishing kit which consists of a pre-primer treatment, a primer and coloured top coat. Full instructions are normally supplied with a kit, but basically the method of use is to first apply the pre-primer to the component concerned, and allow it to dry for up to 30 minutes. Then the primer is applied, and left to dry for about an hour before finally applying the special-coloured top coat. The result is a correctly coloured component, where the paint will flex with the plastic or rubber, a property that standard paint does not normally posses.

5 Major body damage - repair

Where serious damage has occurred, or large areas need renewal due to neglect, it means that complete new panels will need welding-in, and this is best left to professionals. If the damage is due to impact, it will also be necessary to check completely the alignment of the bodyshell, and this can only be carried out accurately by a Renault dealer using special jigs. If the body is left misaligned, it is primarily dangerous, as the car will not handle properly, and secondly, uneven stresses will be imposed on the steering, suspension and possibly transmission, causing abnormal wear, or complete failure, particularly to such items as the tyres.

6 Front bumper - removal and refitting

Removal

1 Firmly apply the handbrake then jack up the front of the vehicle and support it on axle stands (see *"Jacking and vehicle support"*). Remove both front roadwheels.

2 Slacken and remove the screws securing the front wheelarch liners to each end of the bumper. Remove the fasteners securing the front section of each wheelarch liner to the body and remove both the left- and right-hand sections.

3 On models with front foglights, trace the foglight wiring back to the connector and disconnect it from the main wiring harness.

4 On models with headlight washers, release the retaining clip(s) and disconnect the washer supply hose(s) (as applicable).

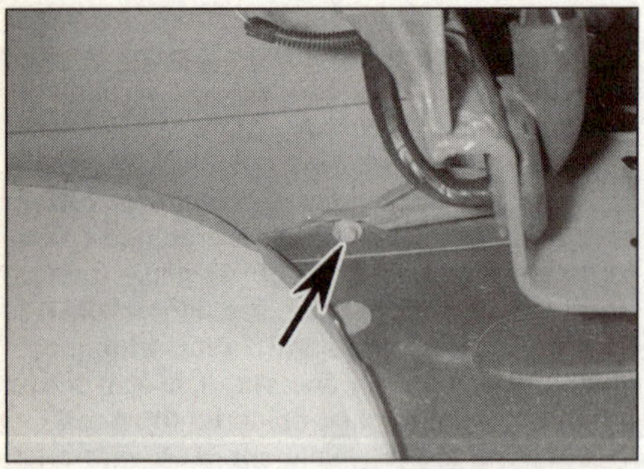

6.5a Slacken and remove the bumper end bolts (arrowed) . . .

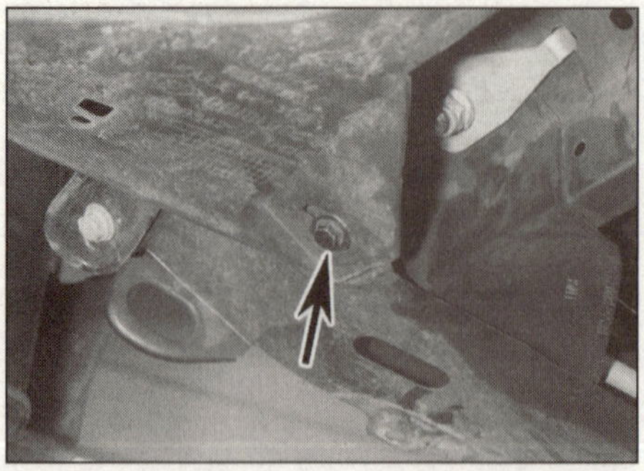

6.5b . . . and the bolts (arrowed) securing the undercover to the subframe

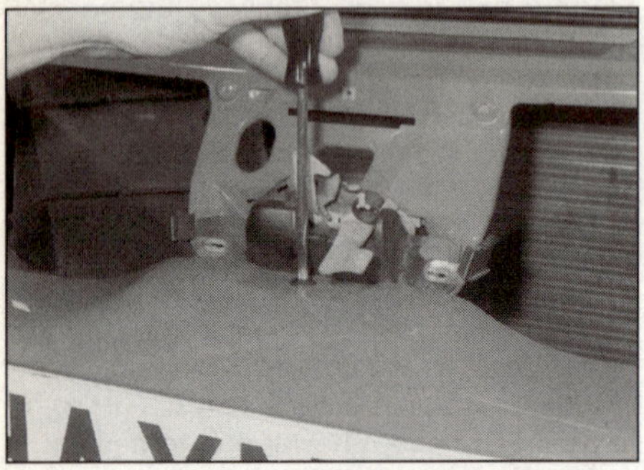

6.6a Undo the retaining bolt from the centre of the bumper . . .

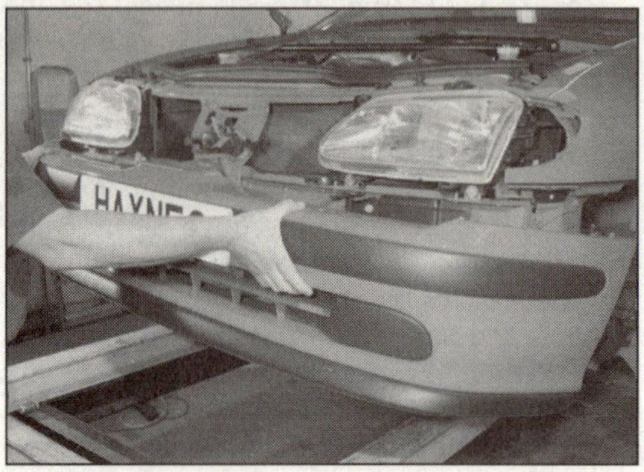

6.6b . . . then release the bumper ends and manoeuvre the bumper forwards and away from the vehicle

5 From underneath, slacken and remove the bolts securing the left- and right-hand ends of the bumper in position then undo the two bolts securing the rear of the bumper under cover to the subframe **(see illustrations)**.

6 Unscrew the centre retaining bolt from the top of the bumper then, with the aid of an assistant, release the bumper ends then manoeuvre the bumper assembly forwards and away from the vehicle **(see illustrations)**.

7 Inspect the bumper mountings for signs of damage and renew if necessary.

Refitting

8 Refitting is a reverse of the removal procedure, ensuring that the bumper mounting bolts are securely tightened.

7 Rear bumper - removal and refitting

Removal

1 Open up the tailgate/boot lid then slacken and remove the bumper upper retaining screws **(see illustration)**.

2 Slacken and remove the screws securing the rear wheelarch liner rear sections to the left- and right-hand ends of the bumper **(see illustration)**.

3 From underneath, slacken and remove the five retaining bolts (one at each end and three in the centre) securing the bumper in position, noting the correct fitted location of the clips on the three centre bolts **(see illustrations)**.

4 With the aid of an assistant, carefully release the bumper ends and remove the bumper from the vehicle **(see illustration)**.

Refitting

5 Refitting is a reverse of the removal procedure, ensuring that the retaining bolts are securely tightened.

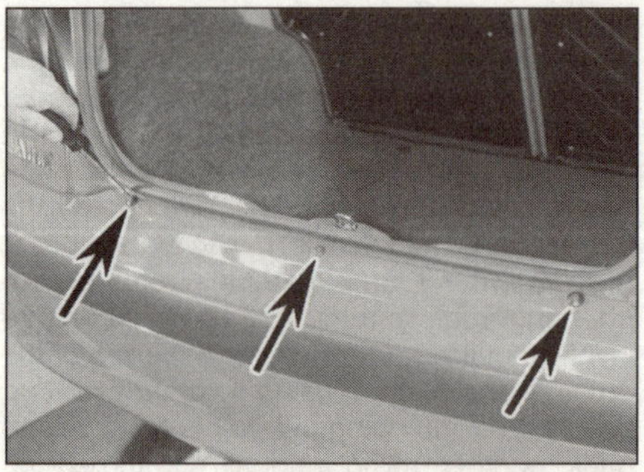

7.1 Slacken and remove the three rear bumper upper retaining screws (arrowed)

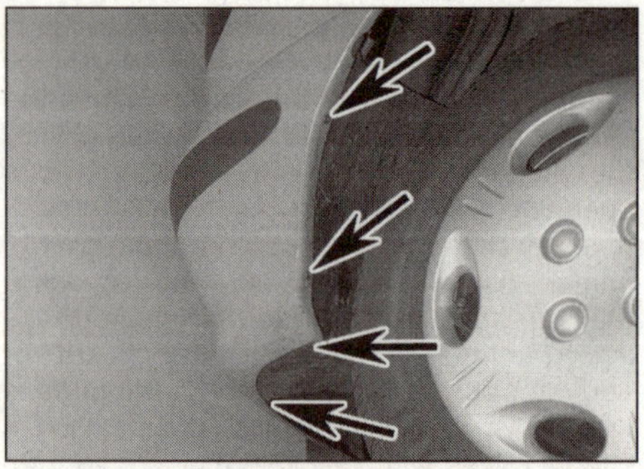

7.2 Undo the retaining screws (arrowed) securing the wheelarch liners to the rear bumper

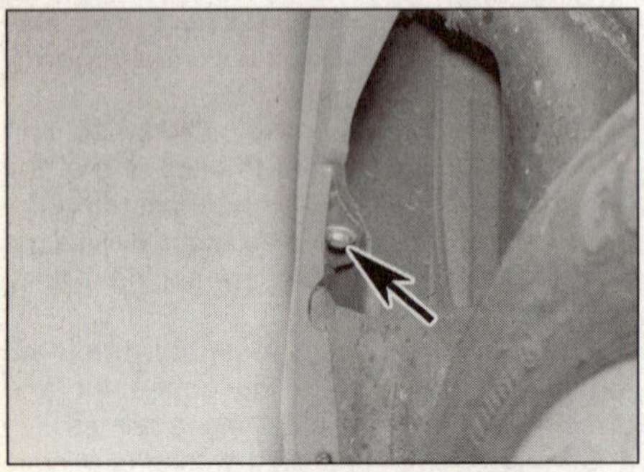

7.3a Slacken and remove the end retaining bolts (arrowed) . . .

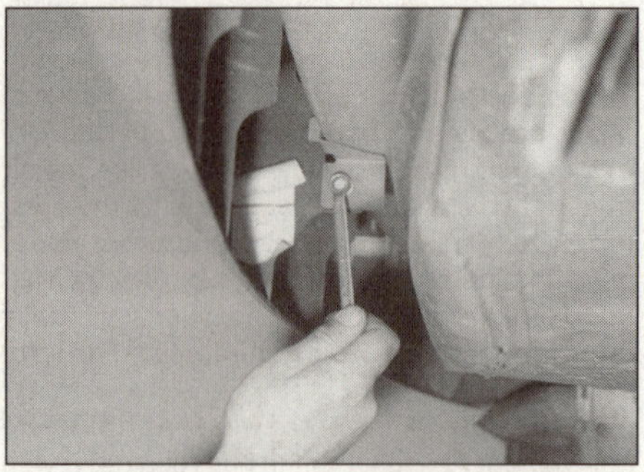

7.3b . . . and the centre retaining bolts . . .

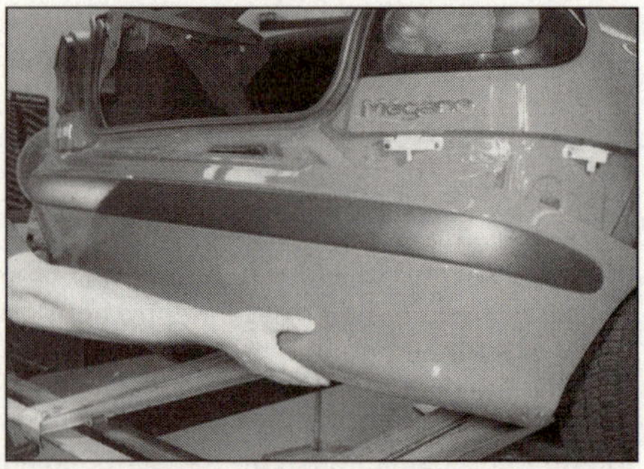

7.4 . . . then release the bumper ends and remove the rear bumper from the vehicle

8 Bonnet - removal, refitting and adjustment

Removal

1 Open the bonnet and place a wad of rag underneath each corner of the bonnet to protect against possible damage should the bonnet slip.
2 Disconnect the washer hose from the right-hand side of the bonnet.
3 Using a pencil or felt tip pen, mark the outline of each retaining bolt relative to the bonnet, to use as a guide on refitting.
4 With the aid of an assistant, slacken and remove the left- and right-hand hinge to bonnet bolts and carefully remove the bonnet from the vehicle.
5 Inspect the bonnet hinges for signs of wear and free play at the pivots, and if necessary renew; the hinges are secured in position by two nuts and can be removed once the vent panel has been removed from the base of the windscreen.

Refitting and adjustment

6 With the aid of an assistant, engage the bonnet with the hinges. Refit the retaining bolts and tighten them by hand only. Align the bolts with the marks made on removal, then tighten them securely. Where necessary, clip the support struts securely onto the bonnet.
7 Close the bonnet, and check for alignment with the adjacent panels. If necessary, slacken the hinge bolts and re-align the bonnet to suit; the height of the bonnet is altered by moving the rubber stops on the bonnet crossmember. Once the bonnet is correctly aligned, check that the bonnet fastens and releases satisfactorily.

9 Bonnet release cable - removal and refitting

Removal

1 Remove the bonnet lock as described in Section 10.
2 Tie a long piece of string to the end of the release cable and free the cable from all the necessary retaining clips and ties.
3 On right-hand drive models, carefully prise the left- and right-hand front speaker grilles out from the top of the facia panel. Slacken and remove the retaining screws from the left- and right-hand end and unclip the top panel from the main facia and remove it from the vehicle. Unclip the release cable from its bulkhead clips so that it is free to be withdrawn from the vehicle **(see illustration)**.
4 On all models, peel back the door sealing strip then carefully unclip the front section of the driver's side sill trim panel until access can be gained to the bonnet release lever retaining screw.

9.3 On right-hand drive models, unclip the bonnet release cable retaining clips situated at the top of the bulkhead

5 Slacken and remove the bonnet release lever screw then remove the lever from underneath the facia, withdrawing the cable into the passenger engine compartment. When the end of the string appears from the bulkhead, untie it and leave it in position; the string can then be used on refitting to draw the cable back into position.

Refitting

6 Refitting is the reverse of removal ensuring that the cable is correctly routed, and secured to all the relevant retaining clips. Prior to closing the bonnet, operate the release handle and have an assistant check that the lock hook moves easily and smoothly to its stop.

10 Bonnet lock - removal and refitting

Removal

1 Remove the direction indicator light units as described in Chapter 12.
2 Slacken and remove the retaining screw from the centre of the radiator grille and remove the joining piece **(see illustration)**.
3 Undo the retaining screws securing the left-hand section of the radiator grille in position. Using a flat-bladed screwdriver, release the lower retaining clips then slide the left-hand radiator grille section forwards and away from the vehicle **(see illustrations)**. Repeat the procedure on the right-hand section of the grille.
4 Unclip the protective covers from the front and rear of the lock assembly then, using a suitable marker pen, make alignment marks between the lock and crossmember **(see illustration)**.
5 Slacken and remove the retaining nuts and remove the lock mounting plate from the rear of the crossmember, freeing it from the release outer cable. Detach the lock assembly from the inner cable and remove it from the vehicle **(see illustrations)**.

Refitting

6 Refitting is the reverse of removal, aligning the lock with the marks made prior to

10.2 Undo the retaining screw and remove the joining piece from the centre of the radiator grille

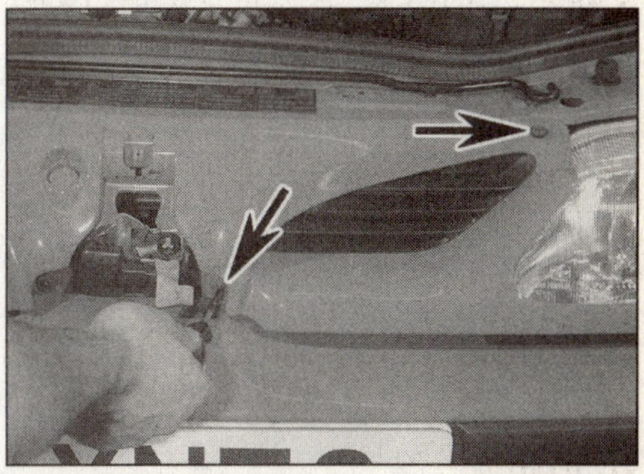

10.3a Slacken and remove the retaining screws (arrowed) . . .

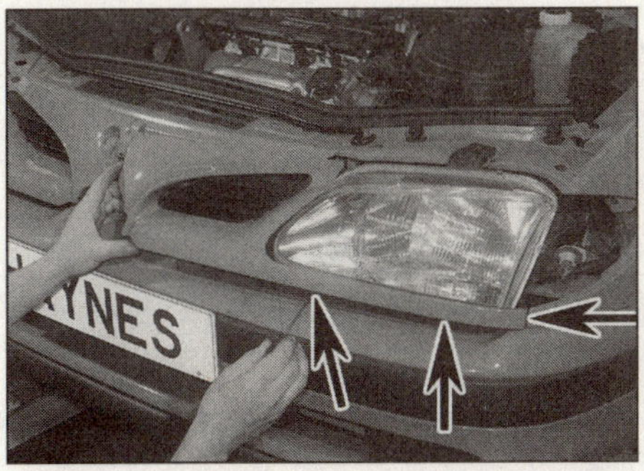

10.3b . . . then release the retaining clips (locations arrowed) and remove the grille panel from the vehicle

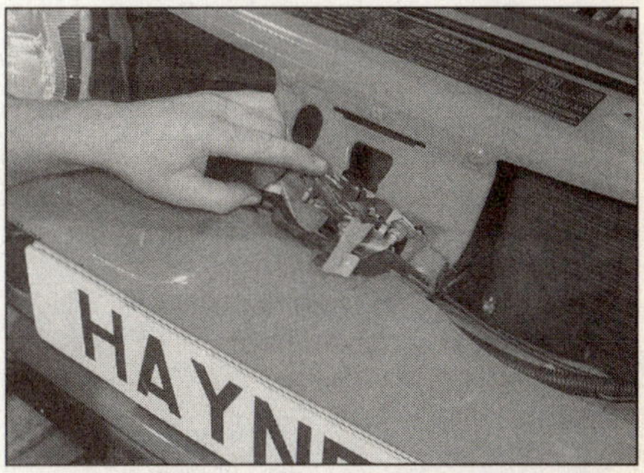

10.4 Unclip the protective covers and remove them from the front and rear of the bonnet lock assembly

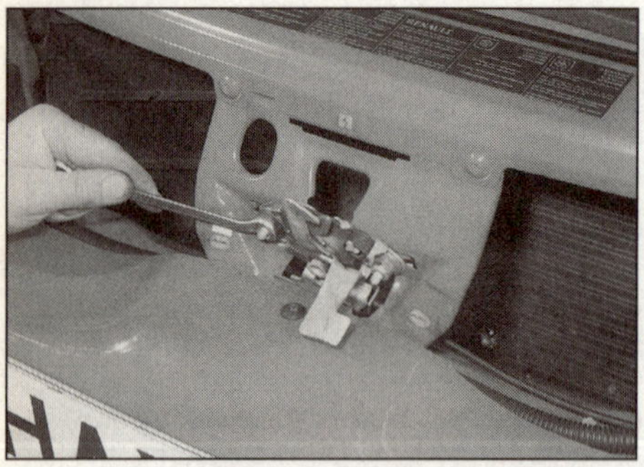

10.5a Slacken and remove the retaining nuts . . .

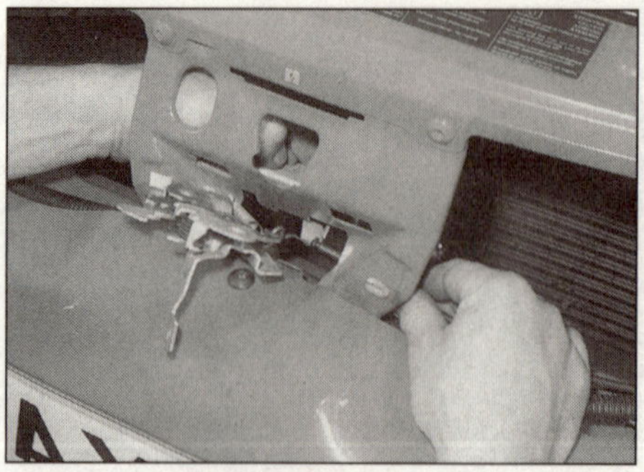

10.5b . . . and remove the mounting plate, freeing it from the outer cable . . .

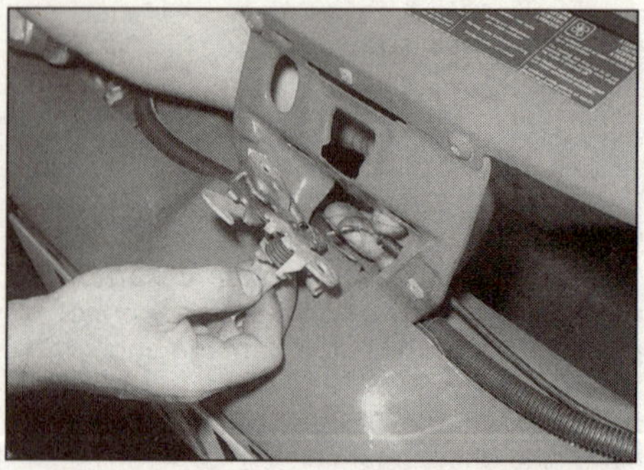

10.5c . . . then detach the lock from the inner cable and remove it from the vehicle

removal. Prior to refitting the grille panel check the lock operation and lubricate it with multi-purpose grease.

11 Door - removal and refitting

Note: *If the hinge pins are a tight fit, a special tool will be required to pull them out of position. The Renault tool is in the form of a hooked adapter (Car.1055-02) which engages with the head of the pin; a slide hammer is then attached to the adapter and used to force each pin out of position.*

Removal

1 Disconnect the battery negative terminal.

2 Undo the retaining bolt securing the check link to the pillar **(see illustration)**.

3 Lift up the retaining clip and disconnect the door wiring connector from the pillar **(see illustration)**.

4 Remove the retaining clips from the upper and lower hinge pins **(see illustration)**.

5 Have an assistant support the weight of the door then carefully tap out the upper and lower hinge pins using a hammer and punch; the upper pin is removed in a downwards direction and the lower pin is removed in an upwards direction **(see illustrations)**. Since it is not possible to hit the pin squarely with a punch, great care must be taken not to bend the pin (see Note at the start of this Section).

6 Once both pins have been removed, move the door away from the vehicle.

7 Examine the hinge pins and bushes for signs of wear or damage and renew if necessary; the bushes are a press-fit in the hinge. It is recommended that the pins are renewed every time they are renewed. If the hinges themselves are damaged, it will be necessary to consult a Renault dealer on the best course of action; the hinges are welded in position and cannot be easily renewed.

Refitting

8 Ensure that the bushes are pressed securely into the hinges and lubricate them with a smear of multi-purpose grease.

9 Manoeuvre the door into position and fit the hinge pins; the upper pin is fitted from below and the lower pin from above. Tap the pins fully into position and secure them in position with the retaining clips.

10 Reconnect the wiring connector to the pillar and secure it in position with the retaining clip.

11 Align the check link with the pillar, securely tighten its retaining bolt then reconnect the battery negative terminal.

12 Door inner trim panel - removal and refitting

Removal

Note: *Door trim panel design varies according to the equipment level of the vehicle therefore some trim panel fastener locations on your vehicle might be different to those shown in the accompanying illustrations.*

1 Disconnect the battery negative terminal and proceed as described under the relevant sub-heading.

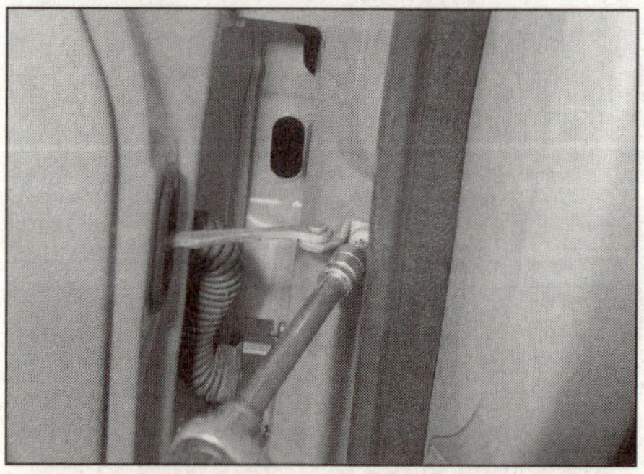

11.2 Slacken and remove the check link retaining bolt

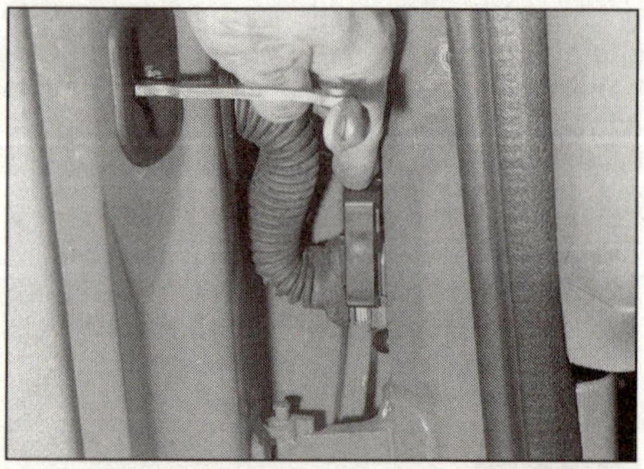

11.3 Lift up the retaining clip and disconnect the door wiring connector

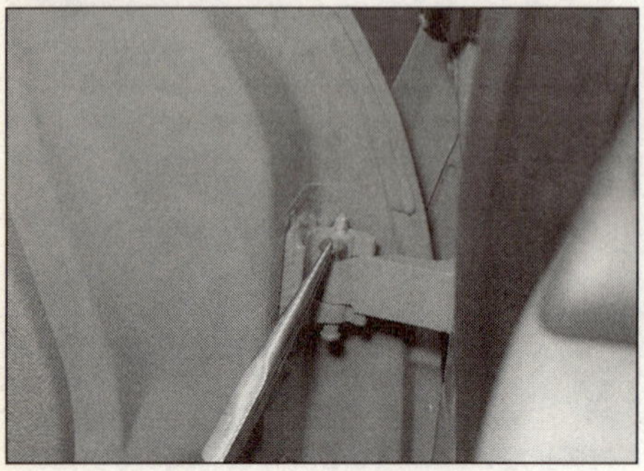

11.4 Remove the retaining clips from the hinge pins

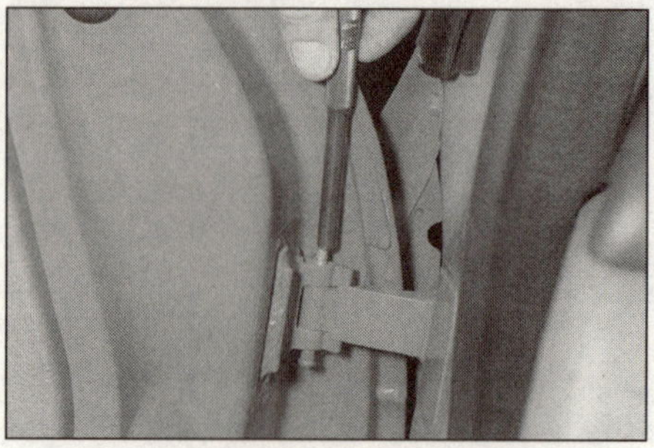

11.5a Tap the hinge pins out of position noting that the upper pin is removed in a downwards direction . . .

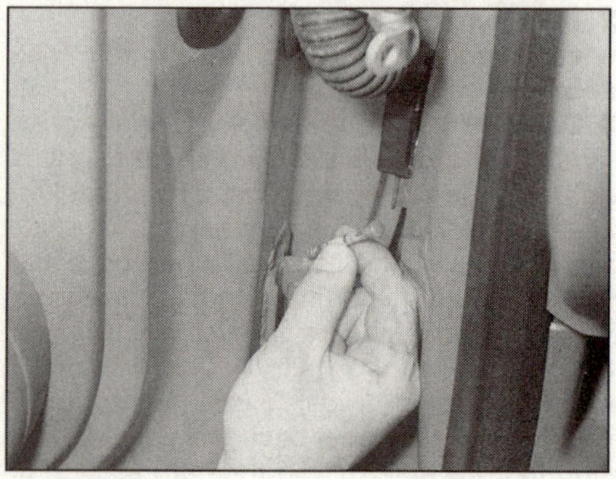

11.5b . . . and the lower pin upwards

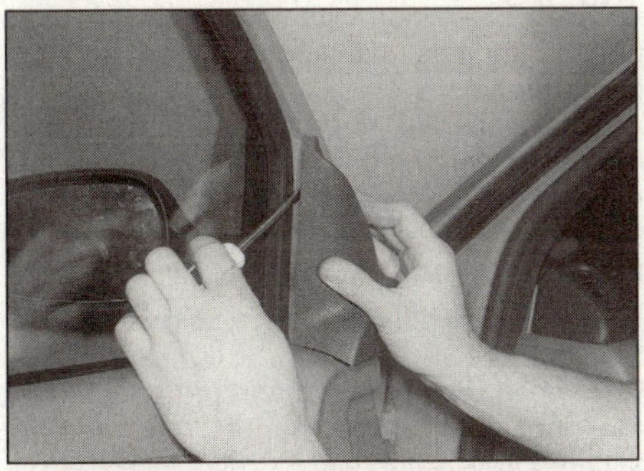

12.2a Unclip the mirror inner trim panel and remove it from the door

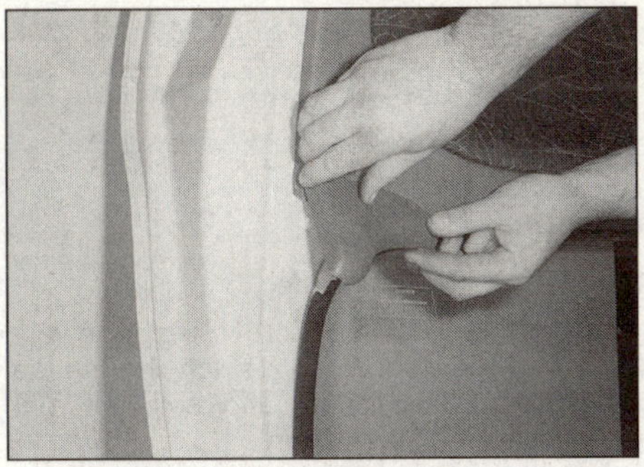

12.2b On Coupe models also unclip the small trim panel from the rear of the door

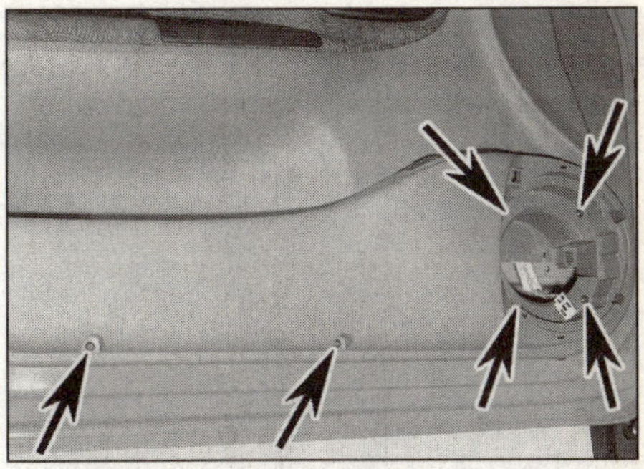

12.4a Undo the retaining screws (arrowed) . . .

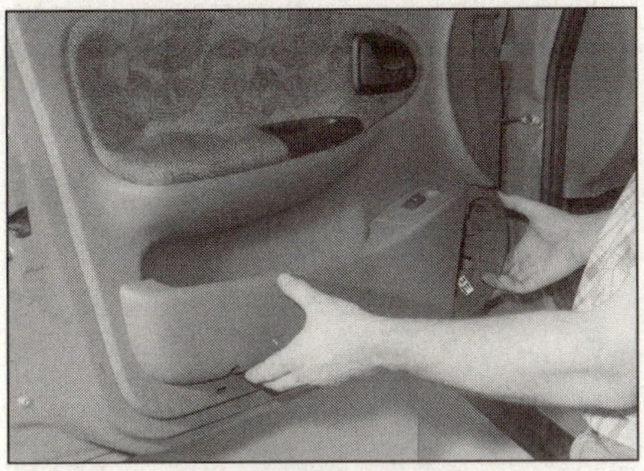

12.4b . . . then unclip the storage compartment and remove it from the door . . .

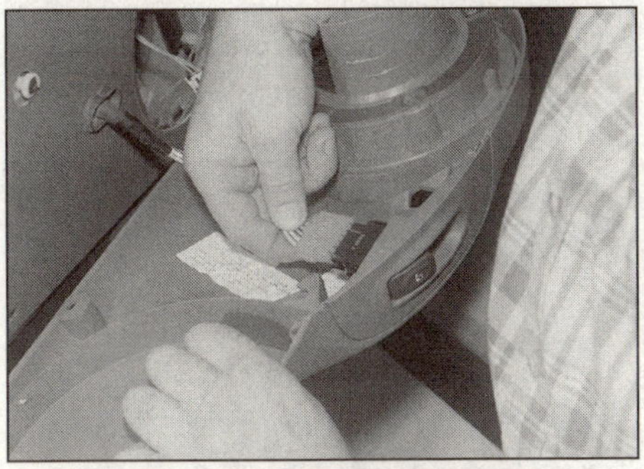

12.4c . . . disconnect the wiring connectors as they become accessible

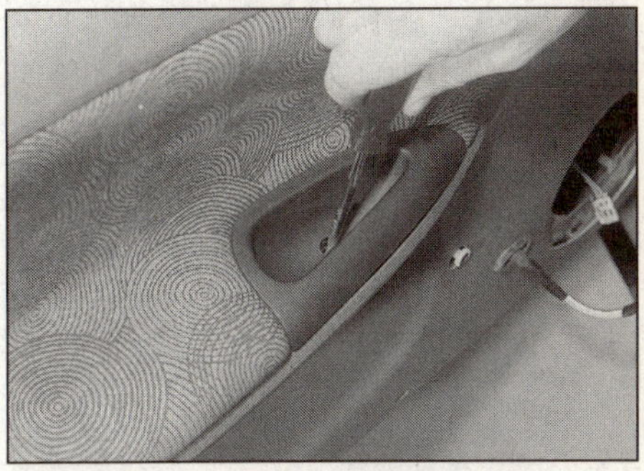

12.5a Undo the retaining screw . . .

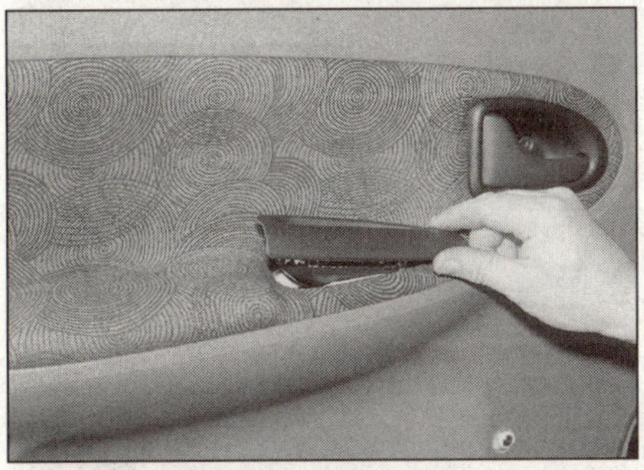

12.5b . . . and remove the armrest pocket from the trim panel

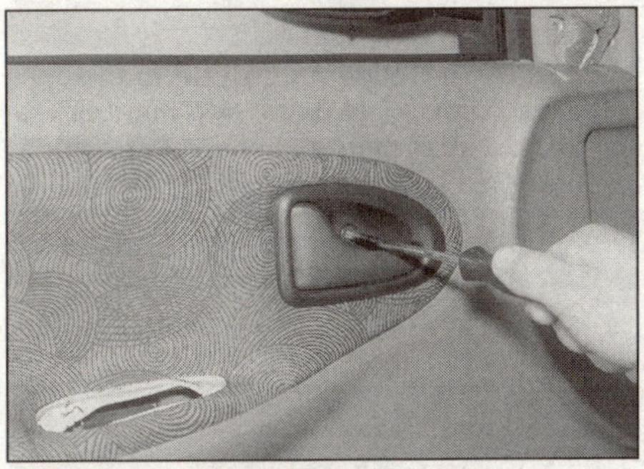

12.6a Undo the retaining screw . . .

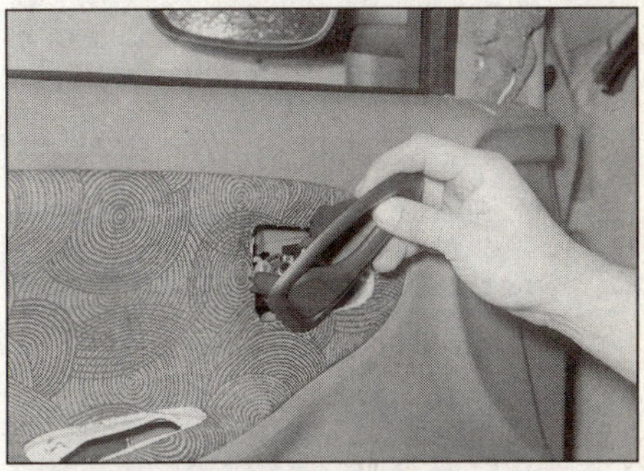

12.6b . . . then unhook the lock inner handle from the link rod

Front door

2 Unclip the exterior mirror inner trim panel from the front of the door. On Coupe models also unclip the small trim panel from the upper rear corner of the panel **(see illustrations)**.

3 Remove the loudspeaker from the door as described in Chapter 12.

4 Undo the retaining screws securing the storage compartment to the bottom of the trim panel. Move the storage compartment upwards to release it from its retaining clips then remove it from the door, disconnecting the wiring connector(s) (where necessary) as they become accessible **(see illustrations)**.

5 Remove the trim cover from the base of the armrest pocket then slacken and remove the retaining screw and lift the pocket out of position **(see illustrations)**.

6 Undo the retaining screw and remove the lock inner handle, disconnect it from its link rod **(see illustrations)**.

7 On models with manual windows, pull/lever the window regulator handle off its spindle and remove the spacer **(see illustration)**.

8 Make a final check that all the screws have been removed then carefully unclip the base of the trim panel from the door then manoeuvre the panel upwards and out of position **(see illustration)**. **Note:** *The panel is stuck in position with sealant as well as being clipped in position and some force will be required to unclip it.* As the panel is removed, free the wiring harness noting its correct routing.

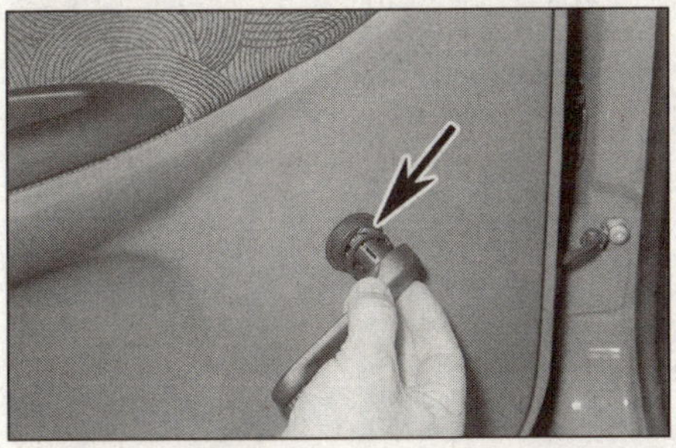

12.7 On models with manual windows, unclip the regulator handle and remove the spacer (arrowed)

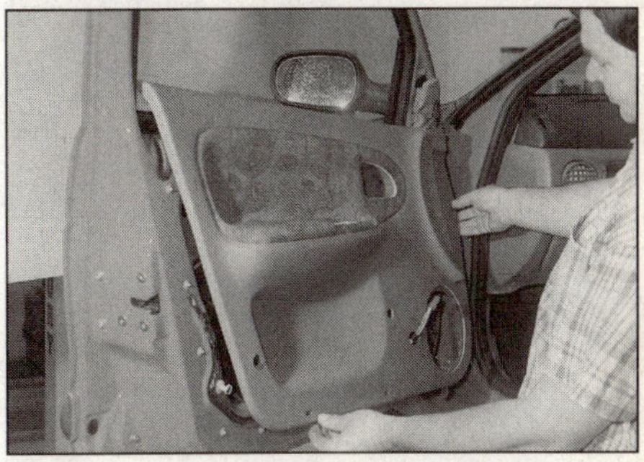

12.8 Unclip the base of the trim panel and remove it from the door

Rear door

9 Undo the trim panel retaining screws from the base of the storage compartment (where fitted).

10 Remove the trim panel as described in paragraphs 5 to 8, ignoring the remark about the wiring.

Refitting

11 Refitting of the trim panel is the reverse of removal. Prior to clipping the panel in position, ensure that all the wiring (where necessary) is correctly routed and passed through the relevant apertures.

13 Door handle and lock components - removal and refitting

Removal

Door lock inner handle

1 Undo the retaining screw securing the handle to the door, then detach the handle assembly from the link rod and remove it from the vehicle **(see illustration)**.

Front door lock exterior handle

2 Remove the door inner trim panel as described in Section 12.

3 Remove the impact absorbing panel from the door. On Coupe models the panel is secured in position with screws whereas on all other models it is clipped in position **(see illustrations)**.

4 Detach the link rod from the handle then undo the retaining nuts and remove the handle from the door **(see illustrations)**.

Front door lock cylinder

5 Remove the door trim panel as described in Section 12.

6 Remove the impact absorbing panel from the door. On Coupe models the panel is secured in position with screws whereas on all other models it is clipped in position **(see illustrations 13.3a and 13.3b)**.

7 Slacken and remove the retaining screws and remove the anti-theft cover/bracket from the lock assembly **(see illustrations)**.

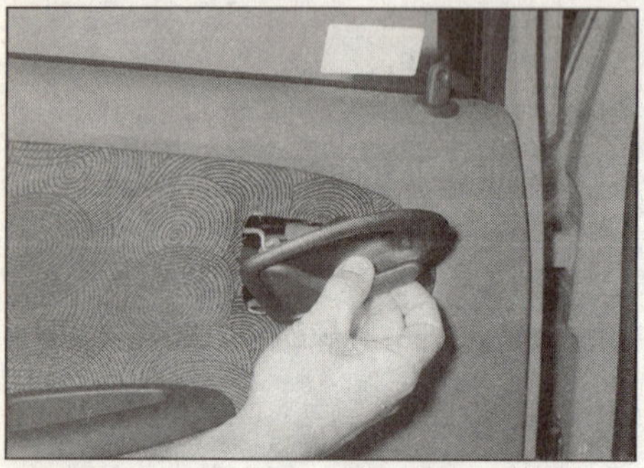

13.1 Removing the lock inner handle

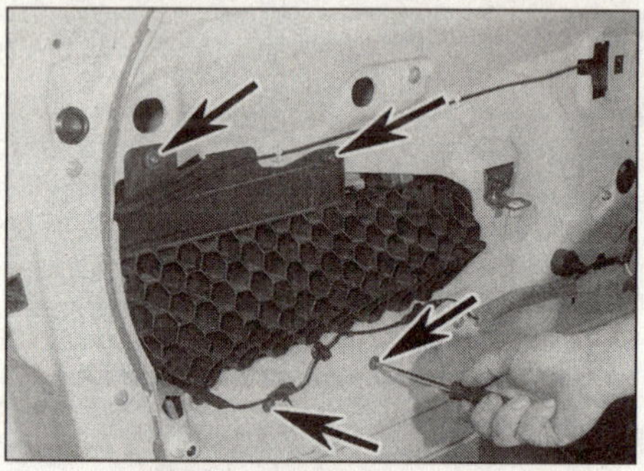

13.3a On Coupe models undo the retaining screws (arrowed) and manoeuvre the impact absorbing panel out of position

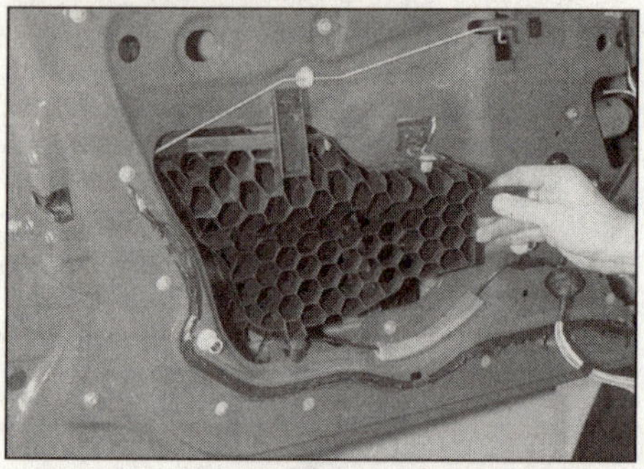

13.3b On Hatchback and Saloon models unclip the impact absorbing panel and remove it from the door

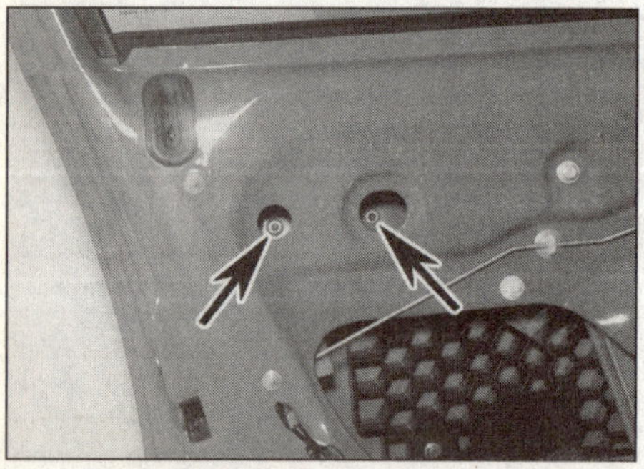

13.4a Undo the retaining nuts (arrowed) . . .

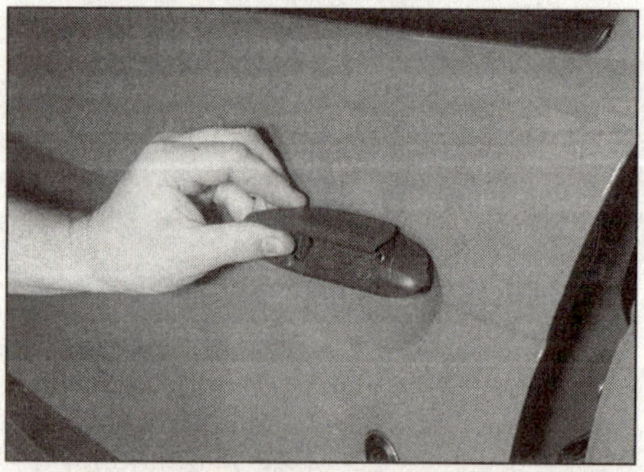

13.4b . . . and remove the exterior handle from the door

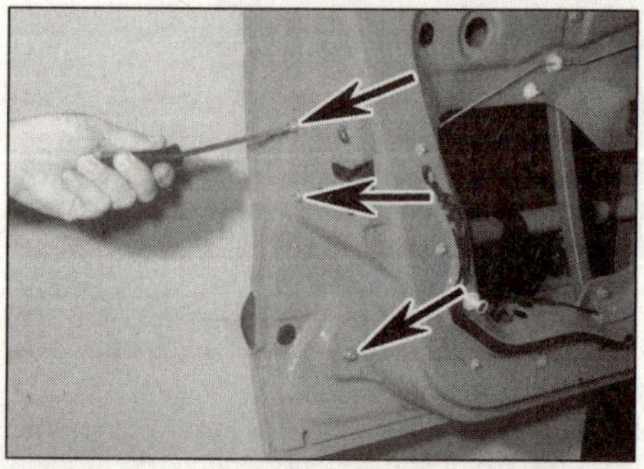

13.7a On Hatchback and Saloon models, undo the retaining screws (arrowed) . . .

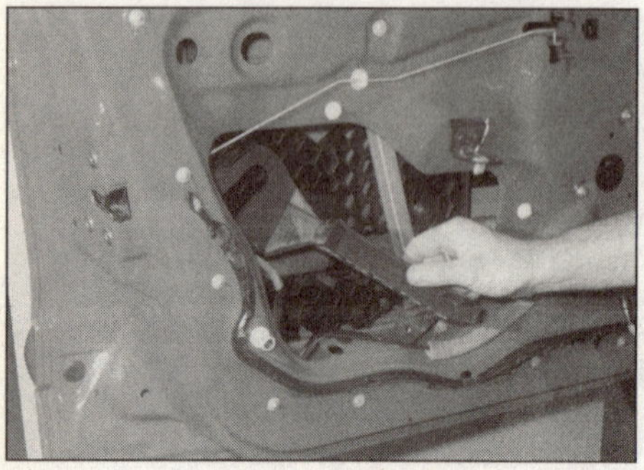

13.7b . . . and remove the anti-theft cover from the door

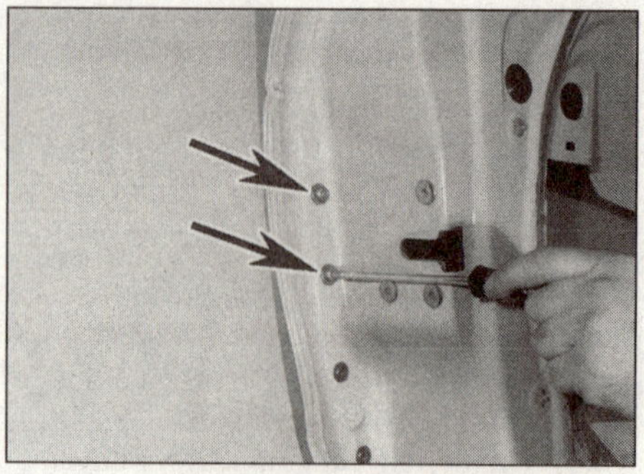

13.7c On Coupe models undo the two retaining screws (arrowed) . . .

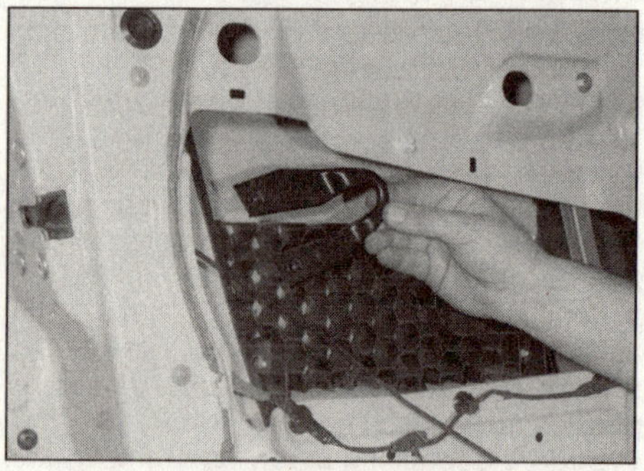

13.7d . . . and remove the anti-theft bracket from the door

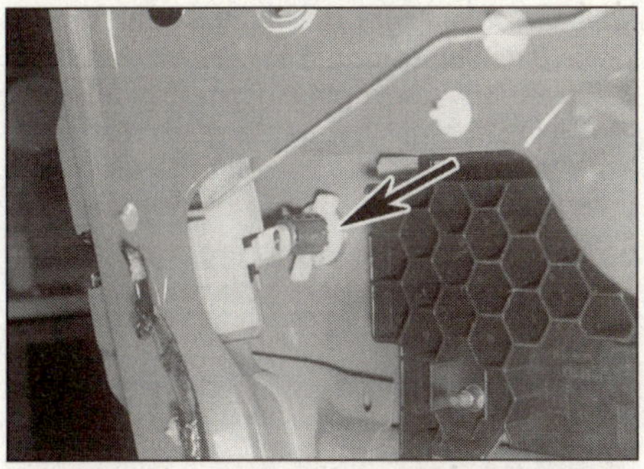

13.8a Rotate the retaining clip (arrowed) through 90° . . .

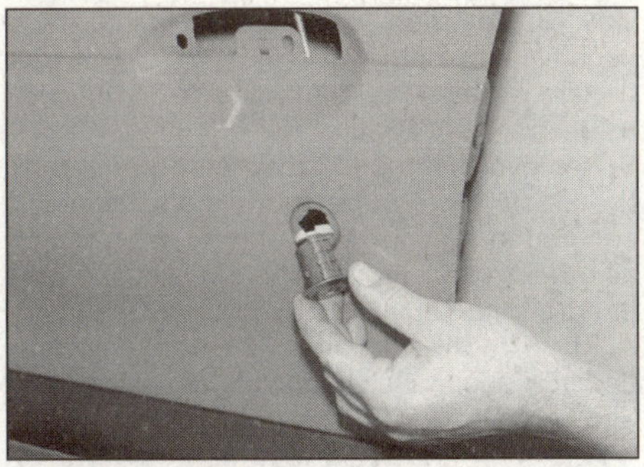

13.8b . . . and remove the lock cylinder from the door

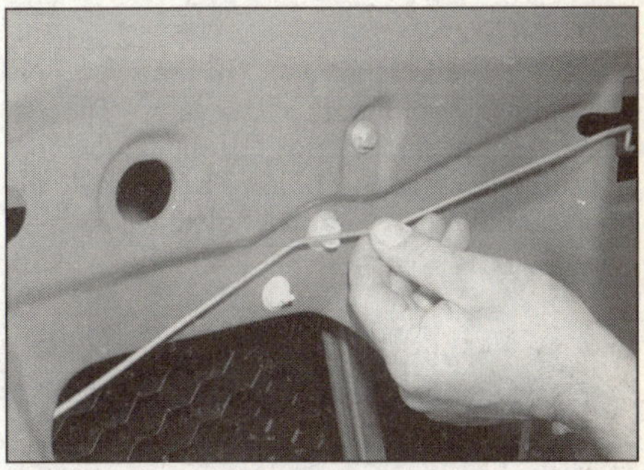

13.10 Unclip the inner handle link rod from its guide

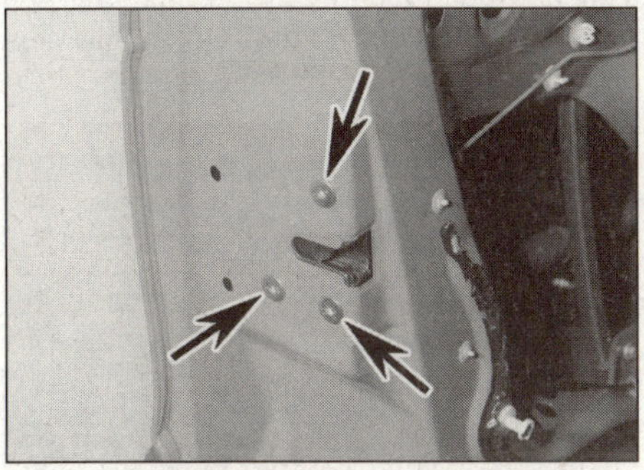

13.11a Undo the retaining screws (arrowed) . . .

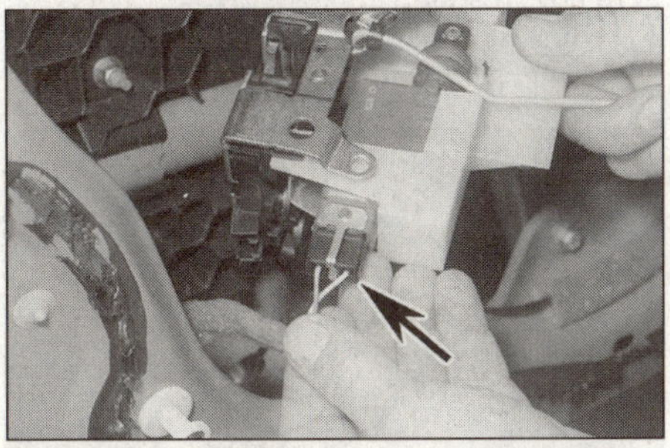

13.11b . . . then remove the lock, where necessary, release the retaining clip and disconnect the wiring connector (arrowed) from the central locking motor

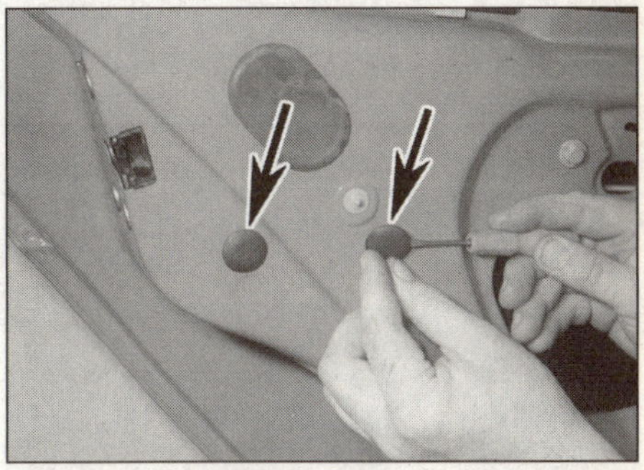

13.13 Prise out the rubber covers (arrowed) to gain access to the rear door exterior handle nuts

8 Release the retaining clip by rotating it through 90° then remove the lock cylinder and clip from the door **(see illustrations)**.

Front door lock assembly

9 Remove the lock cylinder as described earlier in this Section.

10 Detach the lock assembly link rod from the exterior handle then release the interior handle link rod from its guide on the door **(see illustration)**.

11 Slacken and remove the lock retaining screws and remove the lock assembly from the door, disconnecting the wiring connector from the central locking motor (where fitted) **(see illustrations)**.

Rear door exterior handle

12 Remove the door inner trim panel as described in Section 12.

13 Prise out the rubber covers from the door to gain access to the handle retaining nuts **(see illustration)**.

14 Detach the link rod from the handle then undo the retaining nuts and remove the handle from the door.

Rear door lock assembly

15 Remove the exterior handle as described in paragraphs 12 to 14.

16 Unclip the lock inner button pivot then free the link cable from the button and remove the button assembly from the door **(see illustration)**.

17 Remove the rear window glass as described in Section 14.

18 Release the inner handle link rod from its guides and detach the link cable clips from the door **(see illustration)**.

19 Undo the retaining screws and remove the lock assembly from the door, complete with link rods and cable, disconnecting the central locking motor wiring connector (where fitted) **(see illustrations)**.

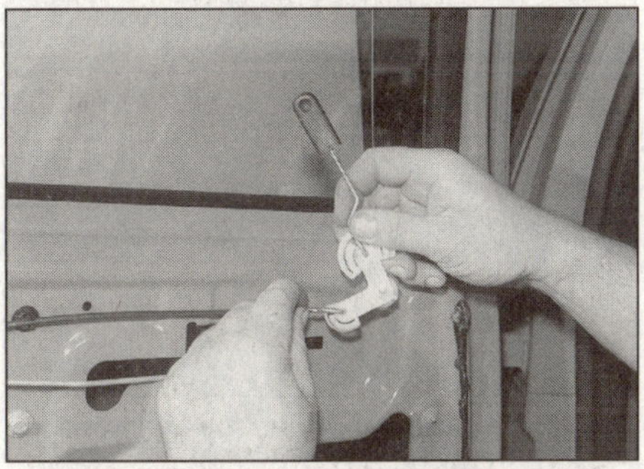

13.16 Unclip the lock inner button from the door and detach it from the link rod

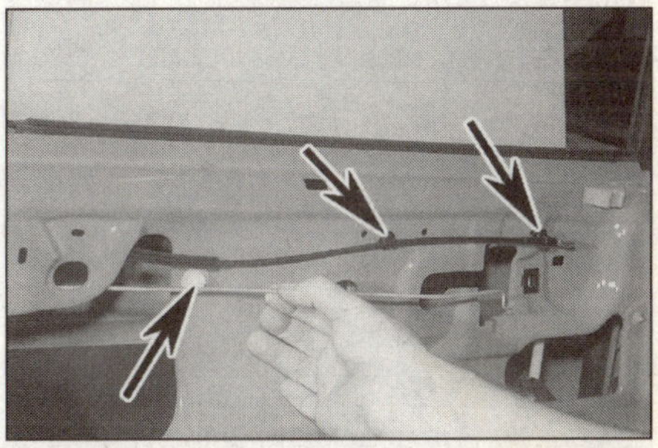

13.18 Unclip the link rod from its guides and release the cable from the door (arrowed)

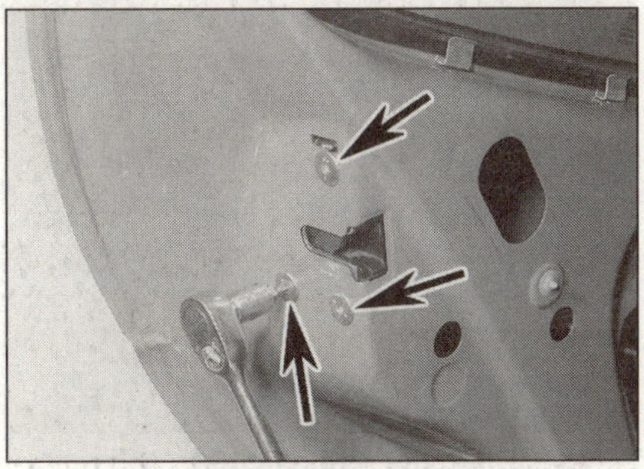

13.19a Undo the retaining screws (arrowed) . . .

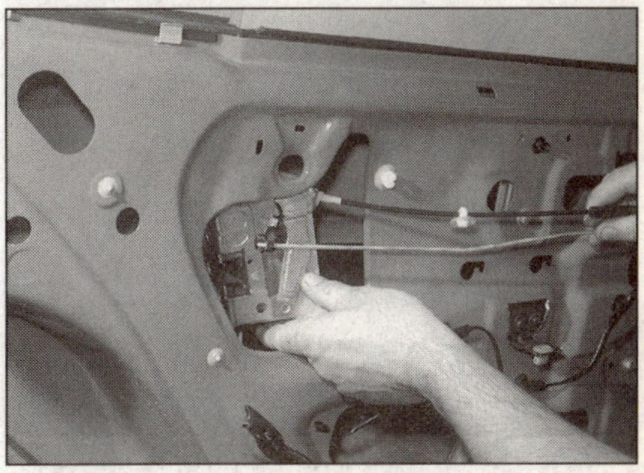

13.19b . . . and manoeuvre the lock assembly out from the rear door

Refitting

20 Refitting is the reverse of removal, ensuring that all link rods/cable are clipped securely in position. Prior to refitting the trim panel, reconnect the battery and check the operation of the lock and handles. If all is well, refit the trim panel as described in Section 12.

14 Door window glass and regulator - removal and refitting

Removal

Front door window

1 Position the window glass about 3/4 of the way down then disconnect the battery negative terminal.

2 Remove the inner trim panel from the door as described in Section 12.

3 Remove the impact absorbing panel from the door. On Coupe models the panel is secured in position with screws whereas on all other models it is clipped in position **(see illustrations 13.3a and 13.3b)**.

4 On Hatchback and Saloon models, carefully unclip the window inner sealing strip and remove it from the door, noting which way around it is fitted **(see illustration)**. On Coupe models unclip the lower and rear sections of the sealing strip; there is no need to remove the strip completely.

5 Remove the window guide retaining clip by sliding it off the regulator; a 13 mm ring spanner is ideal for releasing the retaining clip tangs **(see illustrations)**.

6 Free the window glass from the regulator then lower it slightly. Free the glass rear guide from the sealing strip then tilt the glass and manoeuvre it out of the top of the door **(see illustrations)**.

Front door window regulator

7 Remove the window as described in paragraphs 1 to 6.

8 On models with electric windows disconnect the wiring connector from the window motor **(see illustration)**.

9 Slacken and remove the regulator mounting nuts and bolts then free the assembly from the door and manoeuvre it out through the door aperture **(see illustrations)**.

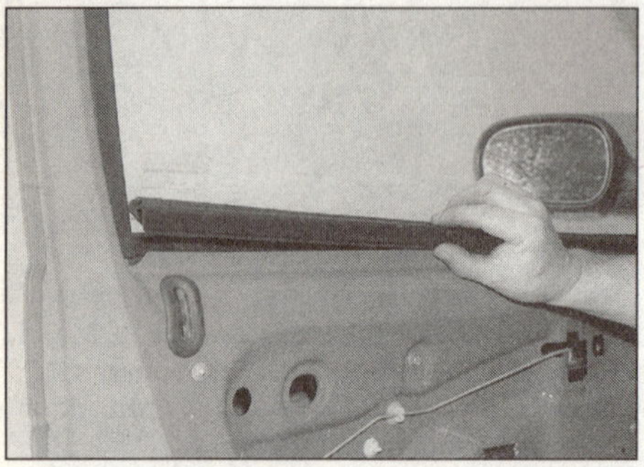

14.4 On Hatchback and Saloon models unclip the window inner sealing strip and remove it from the door

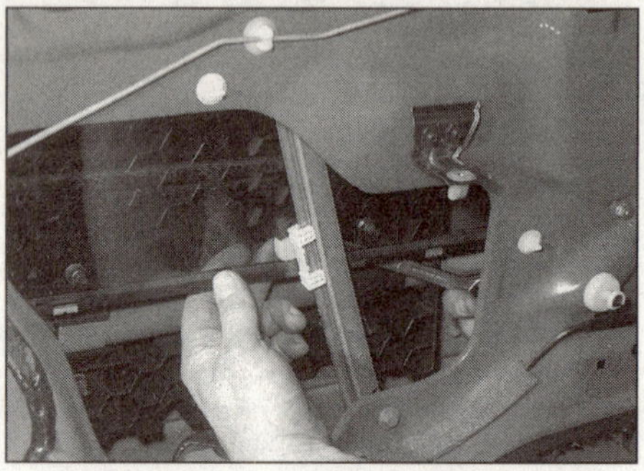

14.5a Using a 13 mm ring spanner, carefully release the retaining clip tangs . . .

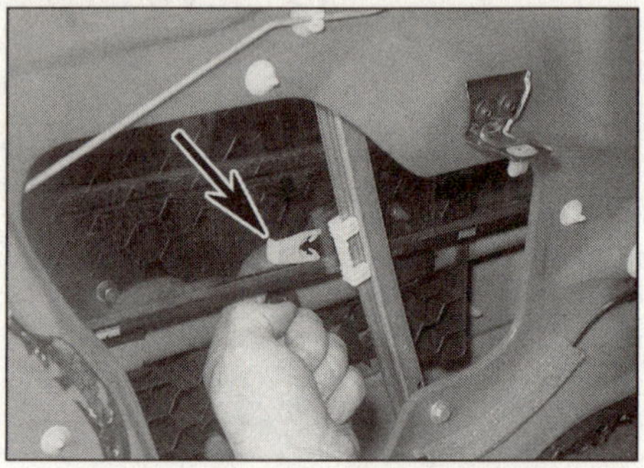

14.5b . . . and pull the retaining clip (arrowed) off from the regulator

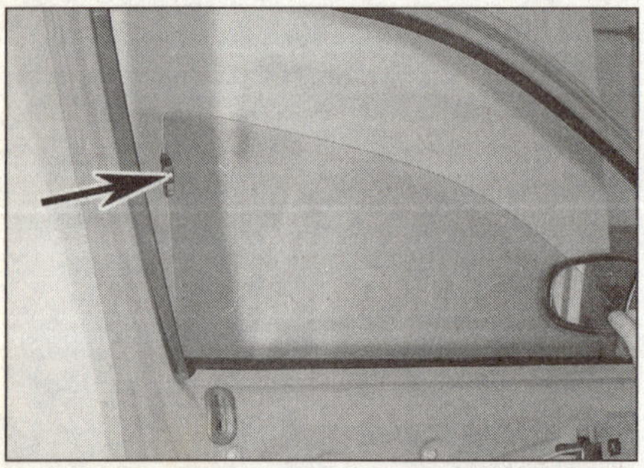

14.6a On Hatchback and Saloon models, tilt the glass to free its guide (arrowed) then lift the glass through the top of the door

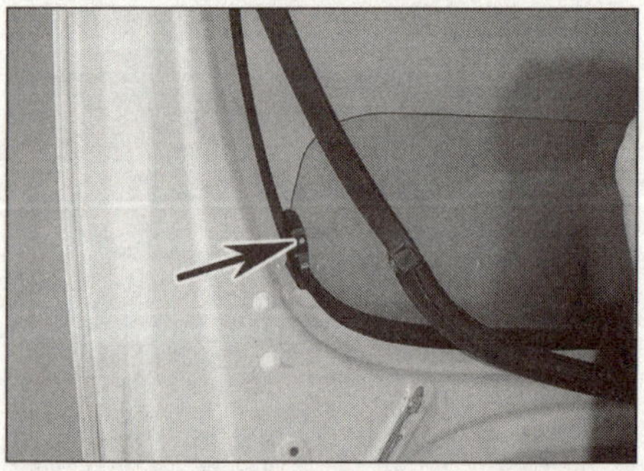

14.6b On Coupe models free the window rear guide (arrowed) . . .

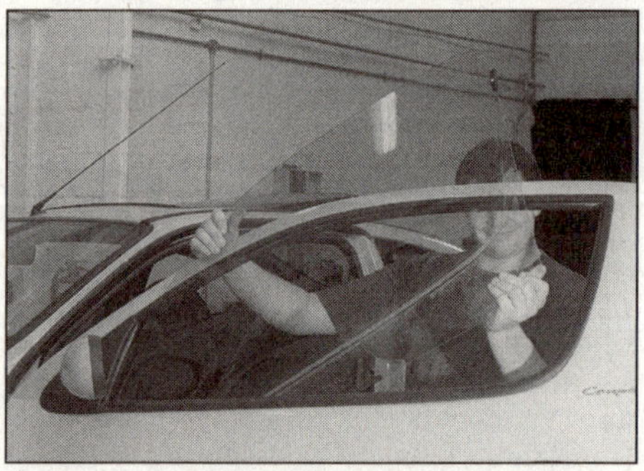

14.6c . . . then lift the glass out from the top of the door

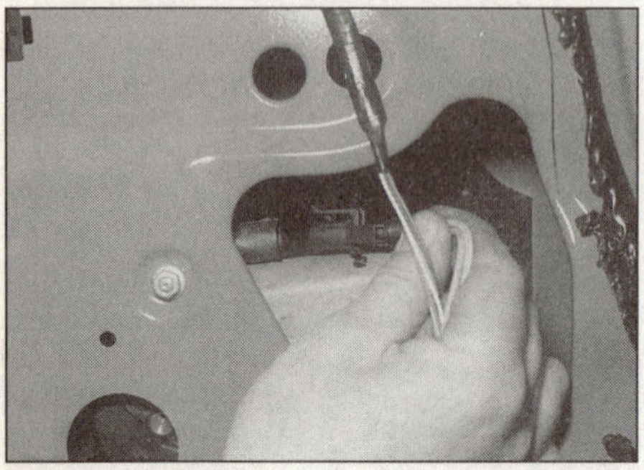

14.8 On models with electric windows disconnect the wiring connector from the regulator motor

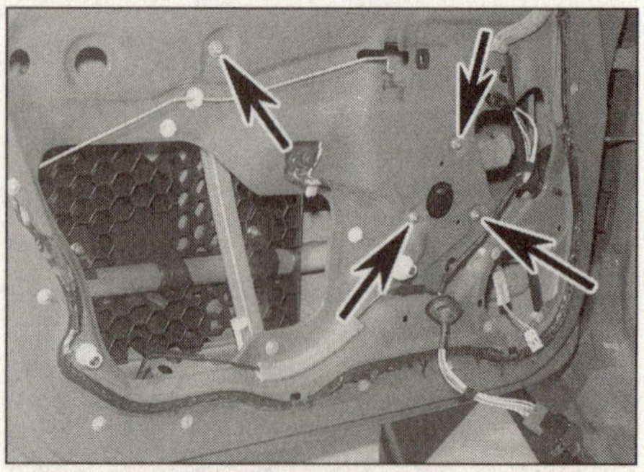

14.9a Undo the retaining nuts and bolts (arrowed) . . .

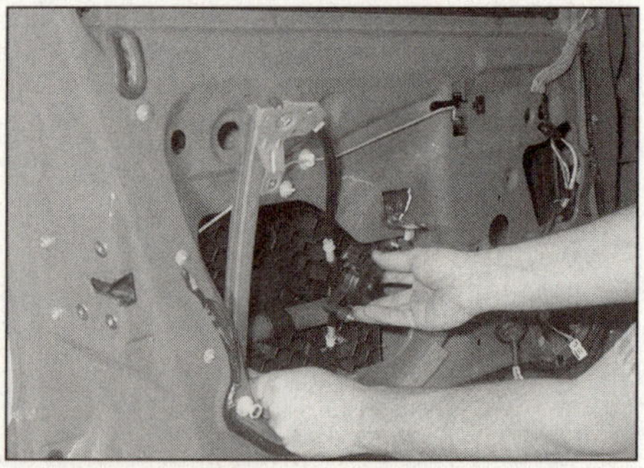

14.9b . . . and manoeuvre the regulator assembly out from the door (Hatchback shown)

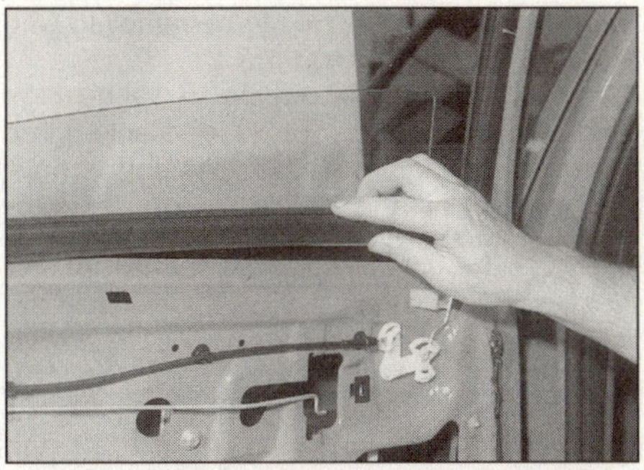

14.12 Unclip the inner sealing strip and remove it from the top of the door

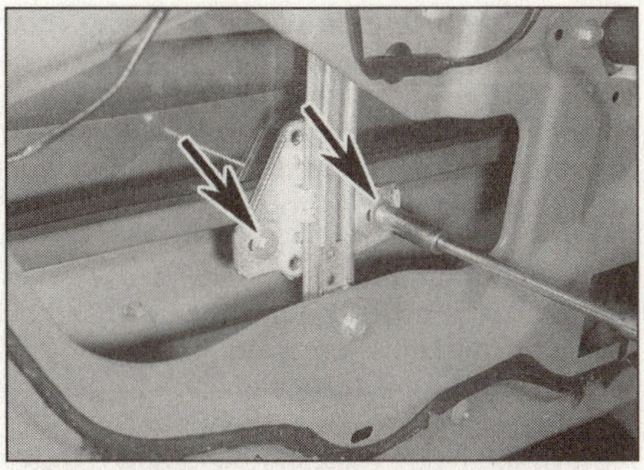

14.13 Undo the retaining bolts (arrowed) and free the glass from the rear door regulator

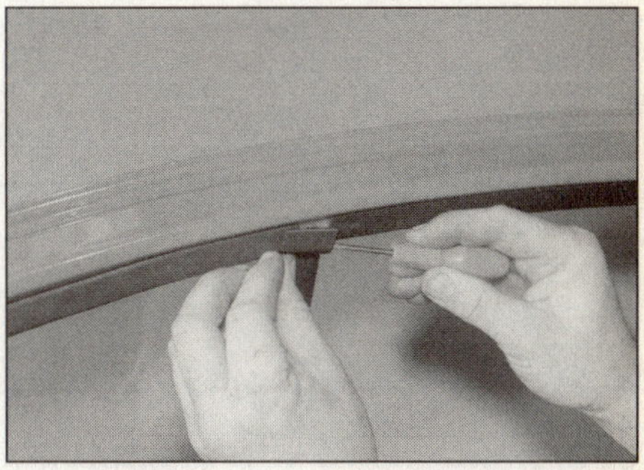

14.14a Prise off the trim cover . . .

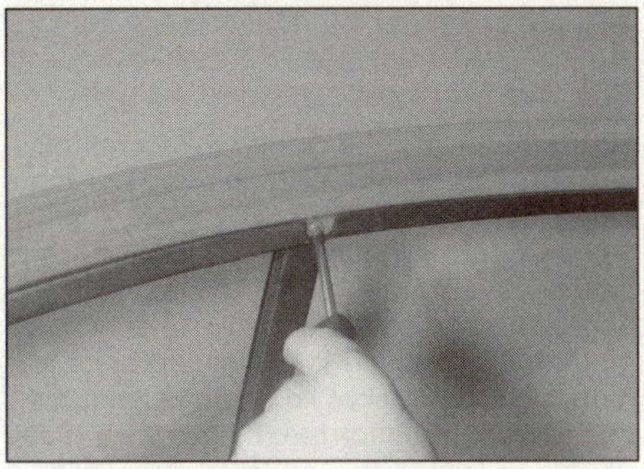

14.14b . . . then slacken and remove the upper retaining screw . . .

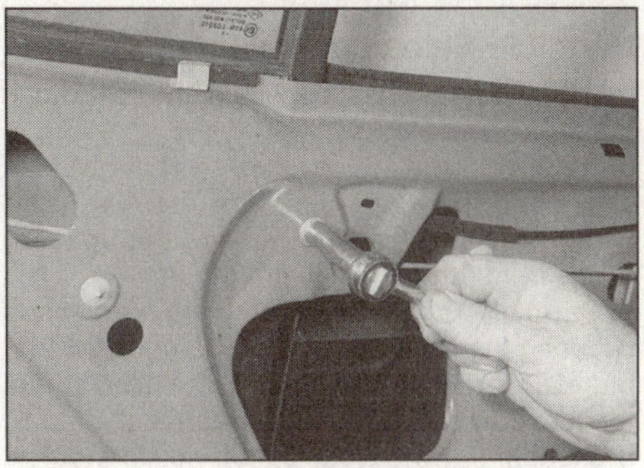

14.14c . . . and the lower screw . . .

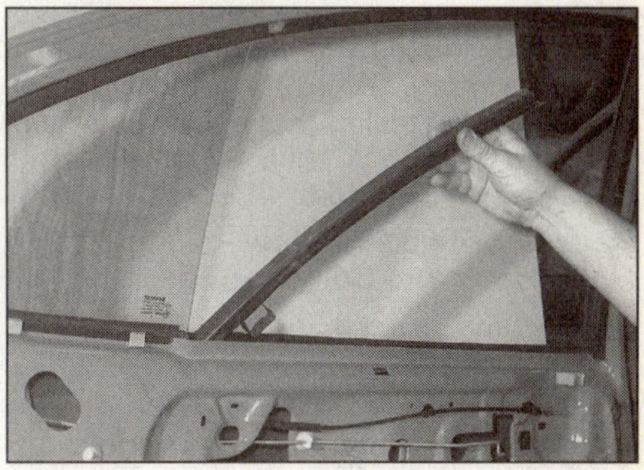

14.14d . . . and manoeuvre the window guide rail out from the top of the door

Rear door window - Hatchback and Saloon models

10 Position the window glass about 3/4 of the way down then disconnect the battery.

11 Remove the inner trim panel from the door as described in Section 12.

12 Carefully unclip the window inner sealing strip and remove it from the door, noting which way around it is fitted **(see illustration)**.

13 Slacken and remove the bolts securing the window glass to the regulator then free the glass and fully lower the glass in the door **(see illustration)**.

14 Remove the trim cover to gain access to the window rear guide rail upper retaining screw, then slacken and remove both the upper and lower screws. Free the guide rail from the fixed window then rotate it through 90° and remove it from the door **(see illustrations)**.

15 Carefully lift the window glass out of the top of the door **(see illustration)**.

Rear door fixed window - Hatchback and Saloon models

16 Remove the window as described in paragraphs 10 to 15.

17 Slide the fixed window glass forwards and remove it from the vehicle.

Rear side window - Coupe models

18 Open up the window then undo the retaining screw and free the opening mechanism from the motor spindle.

19 Undo the screw securing the opening

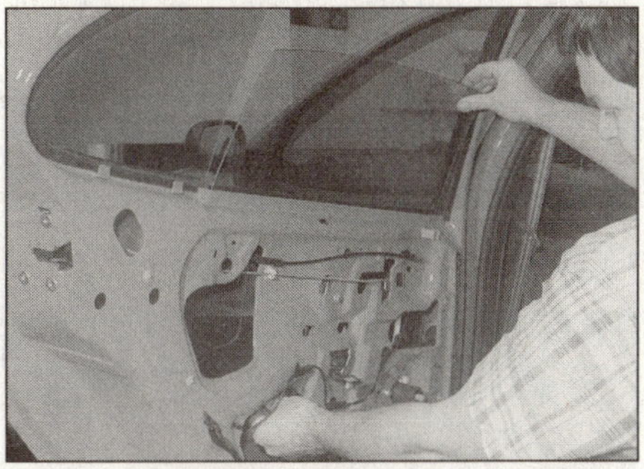

14.15 Removing the rear door window (Hatchback shown)

mechanism to the window and recover the washers, rubbers and end cap from the window, noting the correct fitted location and orientation of each component.

20 Have an assistant support the window then slacken and remove the screws securing the front of the window to its hinges. Recover the spacers, mounting rubbers and the end caps, noting each components correct fitted position and orientation, and remove the window from the vehicle.

Rear door window regulator - Hatchback and Saloon models

21 Remove the window as described in paragraphs 10 to 15.

22 On models with electric windows disconnect the wiring connector from the window motor.

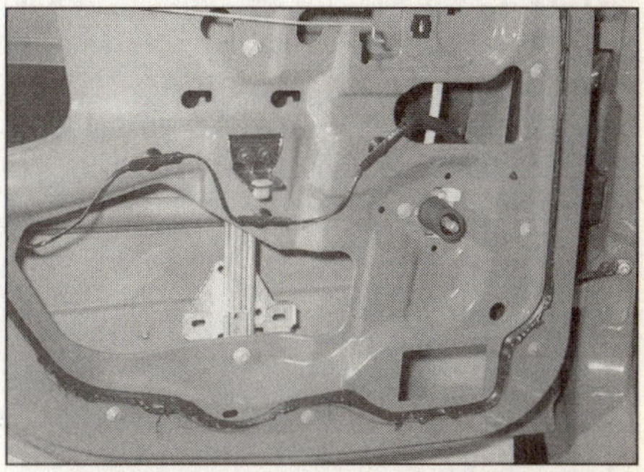

14.23 Rear door window regulator retaining bolts (Hatchback shown)

23 Slacken and remove the regulator mounting bolts then free the assembly from the door and manoeuvre it out through the door aperture **(see illustration)**.

Refitting

Front door window

24 Refitting is the reverse of removal, checking the operation of the window prior to refitting the trim panel.

Front door window regulator

25 Refitting is the reverse of removal, checking the operation of the window prior to refitting the trim panel.

Rear door window - Hatchback and Saloon models

26 Refitting is the reverse of removal. Prior to

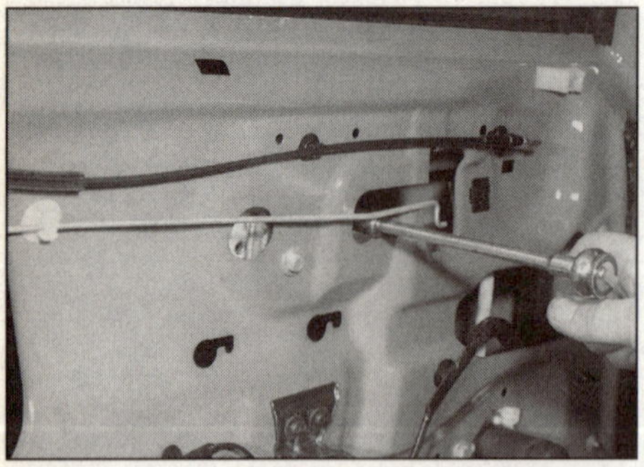

14.26 On the rear door, fully raise the rear window before tightening its retaining bolts securely

refitting the trim panel, adjust the position of the window glass by slackening the retaining bolts then fully raising the window before securely retightening them; this will ensure that the window is squarely seated in its guides **(see illustration)**.

Rear door fixed window - Hatchback and Saloon models

27 Refitting is the reverse of removal ensuring that the glass is correctly seated in its sealing strip.

Rear side window - Coupe models

28 Refitting is the reverse of removal making sure the window fixing components are correctly positioned. Prior to refitting the window screws, apply a few drops of locking compound to their threads.

Rear door window regulator

29 Refitting is the reverse of removal, checking the operation of the window prior to refitting the trim panel.

15 Boot lid - removal and refitting

Removal

Boot lid - Coupe models

1 Open up the boot lid and disconnect the battery negative terminal.

2 Disconnect the wiring connectors from the boot lid electrical components, noting their correct fitted locations, and tie a piece of string to the end of the wiring. Noting the correct routing of the wiring harness, then release the grommet from the left-hand side of the boot lid and withdraw the wiring **(see illustration)**. When the end of the wiring appears, untie the string and leave it in position in the boot lid; it can then be used on refitting to draw the wiring into position.

3 Draw around the outline of each hinge with a suitable marker pen then slacken and remove the hinge retaining bolts and remove the boot lid from the vehicle **(see illustration)**.

4 Inspect the hinges for signs of wear or damage and renew if necessary; each hinge is secured to the body by two nuts.

Boot lid - Saloon models

5 Lift up the rear seat cushions and fold down the seat backs.

6 Remove the high-level stoplight as described in Chapter 12. On models fitted with rear speakers also remove the speaker grilles from the parcel shelf.

7 Carefully prise out the retaining clips securing the front edge of the parcel shelf trim panel in position then slide the trim panel forwards and position it clear of the shelf.

8 Disconnect the boot lid wiring connector, which is clipped to the left-hand side of the shelf, and free the wiring from its retaining clips **(see illustration)**.

9 Open up the boot lid and free the wiring harness grommet from the body. Withdraw the wiring so that it is free to be removed with the boot lid **(see illustration)**.

10 Draw around the outline of each hinge with a suitable marker pen then slacken and remove the hinge retaining bolts and remove the boot lid from the vehicle.

11 Inspect the hinges for signs of wear or damage and renew if necessary; each hinge is secured to the body by two nuts.

Support strut

12 To remove a support strut, using a small flat-bladed screwdriver, carefully lift the retaining clip(s) and unhook the strut from its balljoint(s); on some models one end of the strut may be secured in position with a retaining nut **(see illustration)**.

Refitting

Boot lid

13 Refitting is the reverse of removal, aligning the hinges with the marks made before removal.

14 On completion, close the boot lid and check its alignment with the surrounding panels. If necessary slight adjustment can be made by slackening the retaining bolts and repositioning the boot lid on its hinges. If the paint work around the hinges has been damaged, paint the affected area with a suitable touch-in brush to prevent corrosion.

Support strut

15 Refitting is the reverse of removal ensuring that the strut is securely clipped in position.

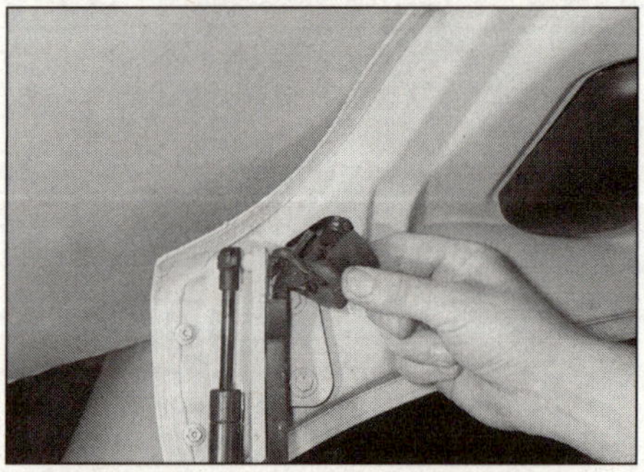

15.2 On Coupe models disconnect the wiring plugs from the boot lid then withdraw the wiring harness from the boot lid

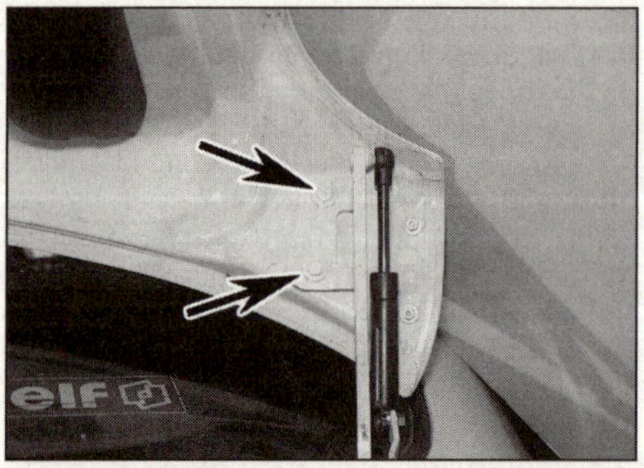

15.3 Undo the retaining bolts (arrowed) and remove the boot lid from the vehicle

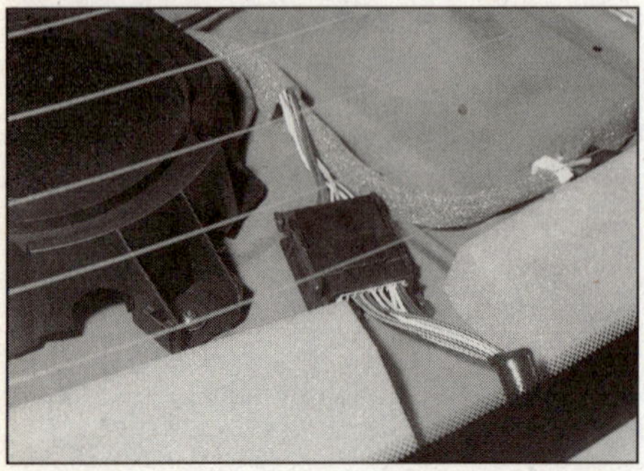

15.8 On Saloon models disconnect the wiring connector located behind the parcel shelf trim . . .

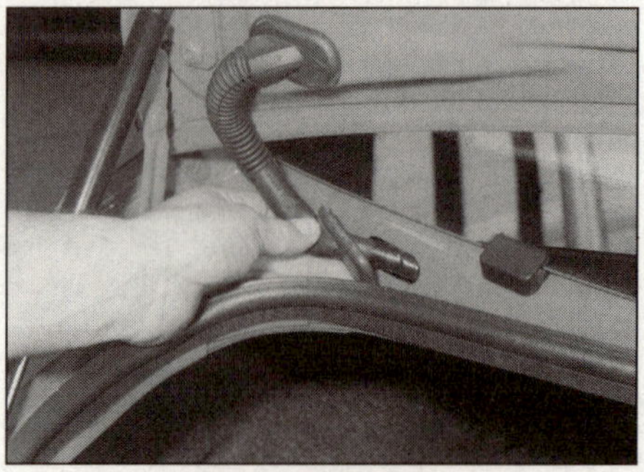

15.9 . . . then release the grommet and withdraw the wiring from the body

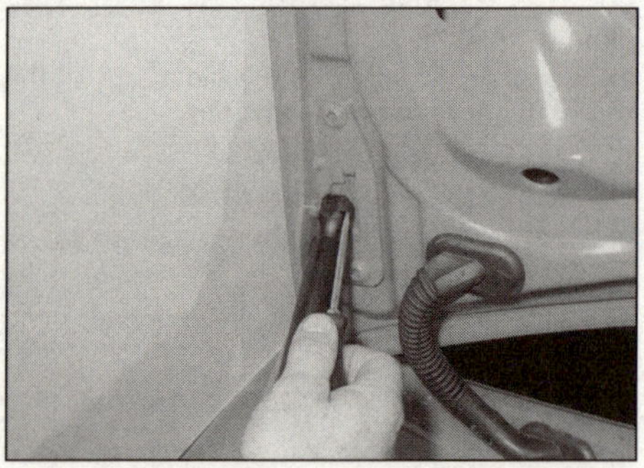

15.12 Release the boot lid support strut from its balljoints by carefully lifting its retaining clips

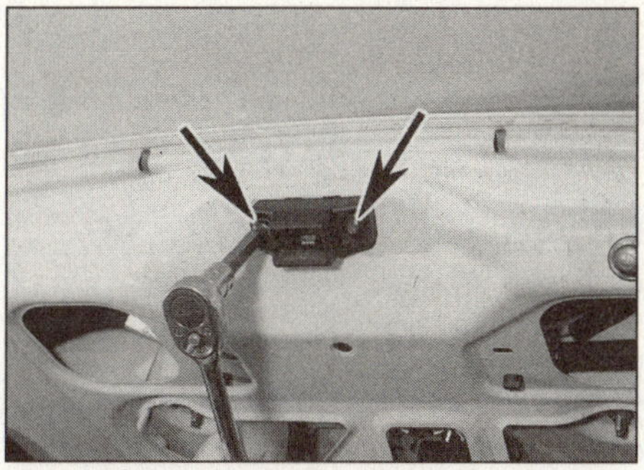

16.2a On Coupe models, undo the two retaining bolts (arrowed) . . .

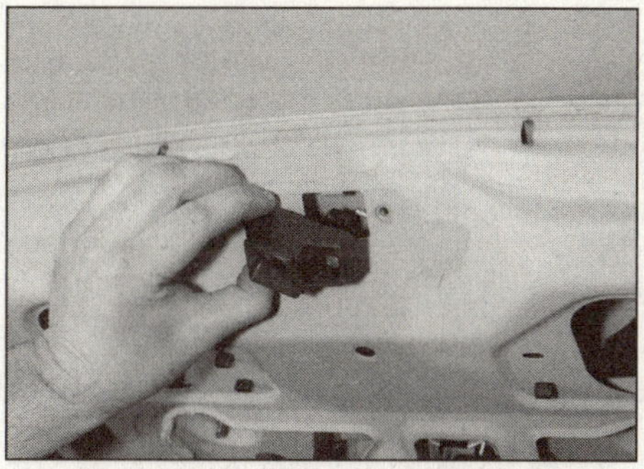

16.2b . . . then detach the lock from the link rod and remove it from the boot lid

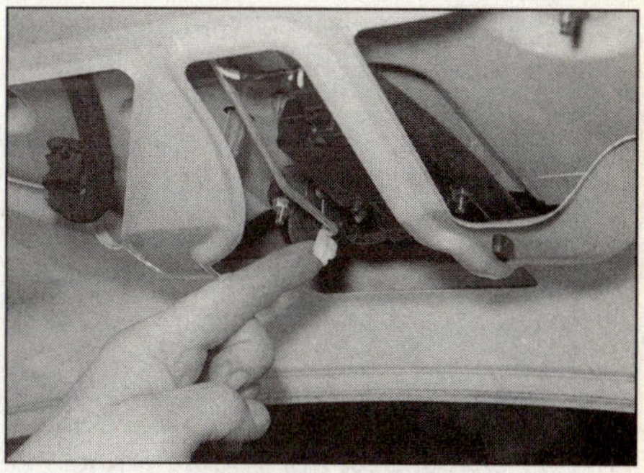

16.3 Unclip the link rod . . .

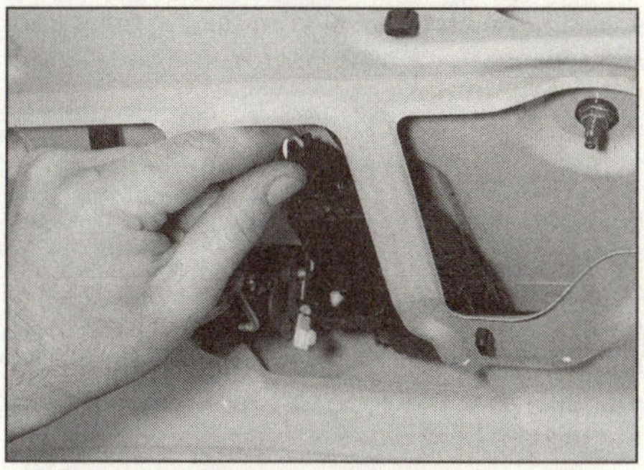

16.4 . . . then disconnect the central locking motor wiring connector

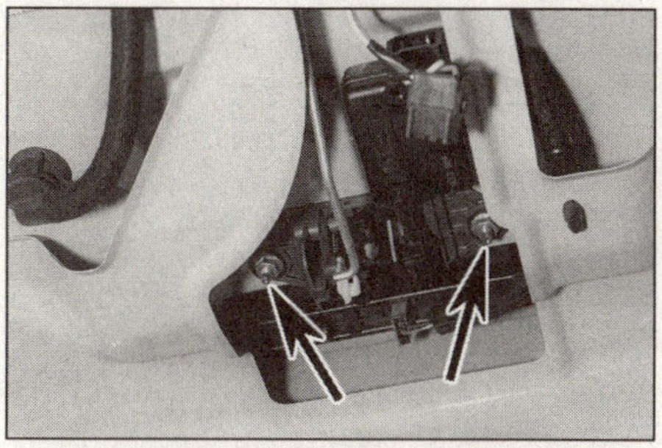

16.5 Undo the retaining nuts (arrowed) and remove the boot lid lock button assembly from the vehicle

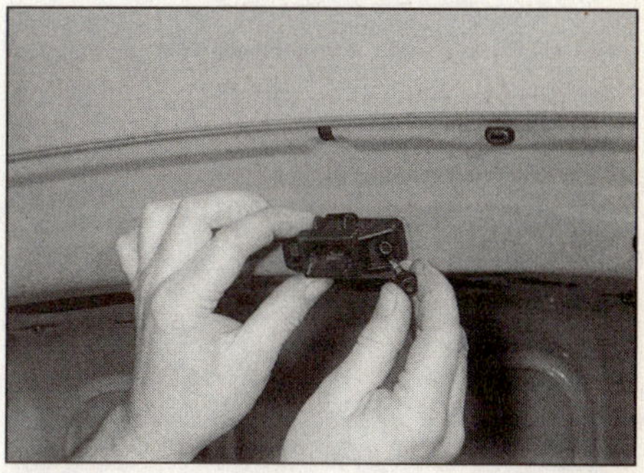

16.6 On Saloon models, undo the retaining bolts and remove the lock from the boot lid

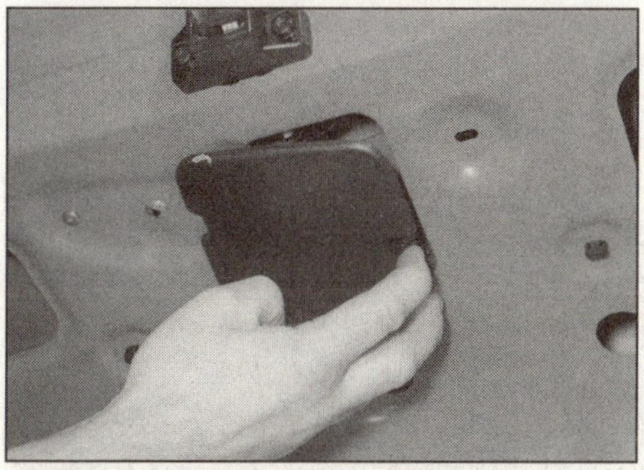

16.7a Unclip the plastic trim cover . . .

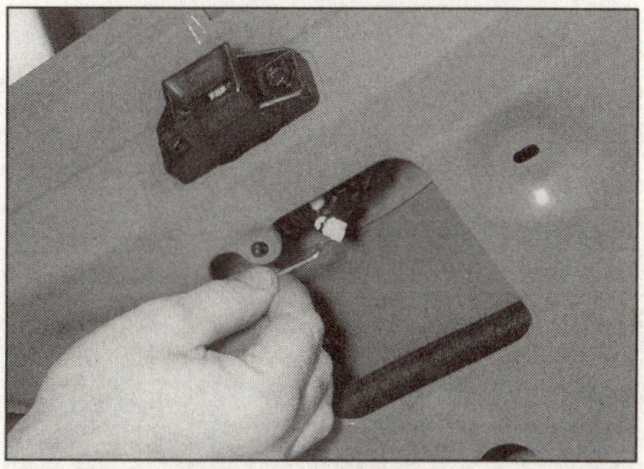

16.7b . . . and detach the link rod from the lock button

16.8 Unscrew the retaining nuts and remove the lock button assembly from the boot lid

16 Boot lid lock components - removal and refitting

Removal

1 Open up the boot lid then disconnect the battery negative terminal.

Lock mechanism - Coupe models

2 Slacken and remove the retaining bolts then remove the lock, unhooking it from the link rod **(see illustrations)**.

Lock button assembly - Coupe models

3 Unclip the link rod and detach it from the lock button assembly **(see illustration)**.

4 Disconnect the wiring connector from the central locking motor (where fitted) **(see illustration)**.

5 Slacken and remove the retaining nuts and manoeuvre the button assembly out of position **(see illustration)**.

Lock mechanism - Saloon models

6 Undo the retaining bolts and remove the lock from the boot lid, noting which way around it is fitted **(see illustration)**.

Lock button assembly - Saloon models

7 Unclip the plastic trim cover then release the retaining clip and detach the link rod from the button assembly **(see illustrations)**.

8 Disconnect the drain pipe from the lock then undo the two retaining nuts and manoeuvre the assembly out from the boot lid **(see illustration)**.

Refitting

9 Refitting is the reverse of removal, checking the operation of the lock assembly and (where necessary) the central locking motor.

17 Tailgate - removal and refitting

Removal

Tailgate

1 Open up the tailgate and disconnect the battery negative terminal.

2 Working as described in paragraphs 25 to 29 of Section 27, remove the upper trim panel from the left-hand rear pillar to gain access to the tailgate wiring connector.

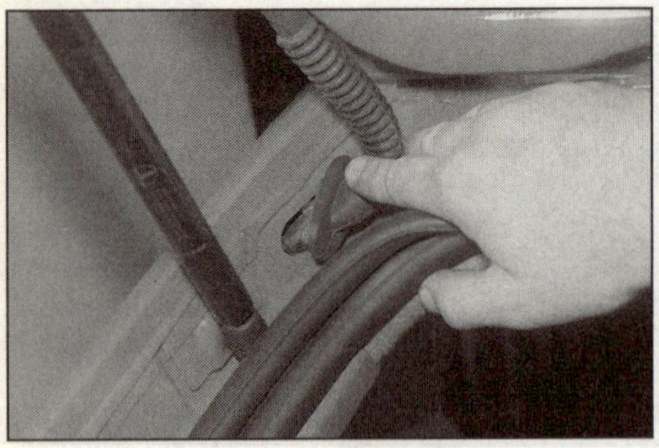

17.3 Disconnect the tailgate wiring connector then withdraw the wiring harness from the body

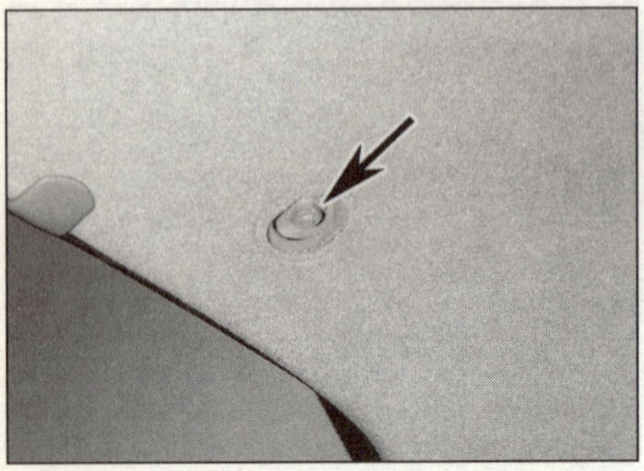

17.4b . . . then undo the hinge retaining nuts and remove the tailgate

3 Disconnect the tailgate wiring connector then free the wiring grommet and withdraw the wiring harness from the body so that it is free to be removed with the tailgate **(see illustration)**.

4 Prise out the trim plugs from the rear of the headlining to gain access to the tailgate hinge retaining nuts **(see illustrations)**.

5 Have an assistant support the tailgate, then carefully lift the retaining clips and detach the support struts.

6 Slacken and remove the nuts securing the tailgate hinges to the body and remove the tailgate from the vehicle.

Support strut

7 Have an assistant support the tailgate then, using a small flat-bladed screwdriver, carefully lift the retaining clips and unhook the strut from its balljoints **(see illustration)**.

Refitting

Tailgate

8 Refitting is the reverse of removal. On completion, ensure that the tailgate is correctly aligned with all its surrounding body components; adjustments can be made by slacken the bolts and repositioning the tailgate adjusters located on either side.

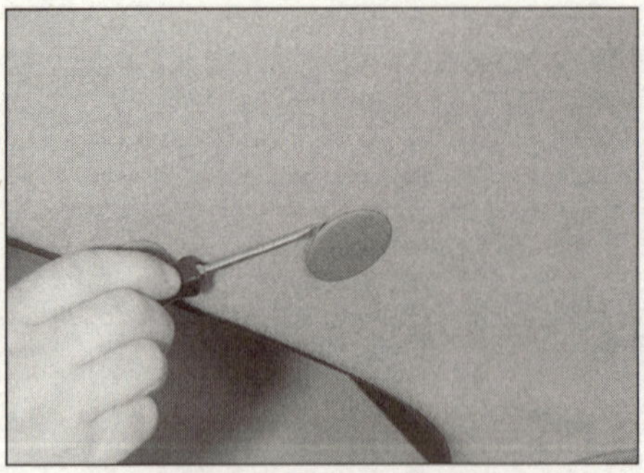

17.4a Prise out the trim plugs . . .

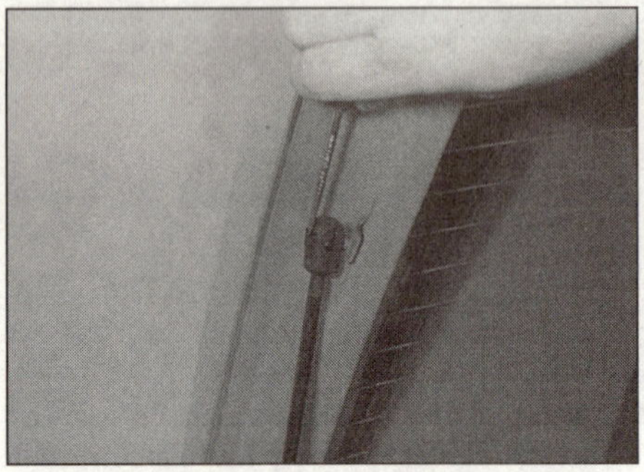

17.7 Release the tailgate support strut from its balljoints by lifting its clips

Support strut

9 Refitting is the reverse of removal making sure the strut is clipped securely in position.

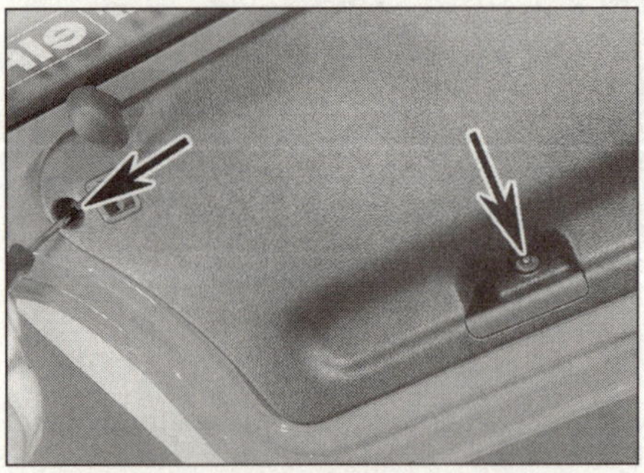

18.2a Undo the retaining screws (arrowed) . . .

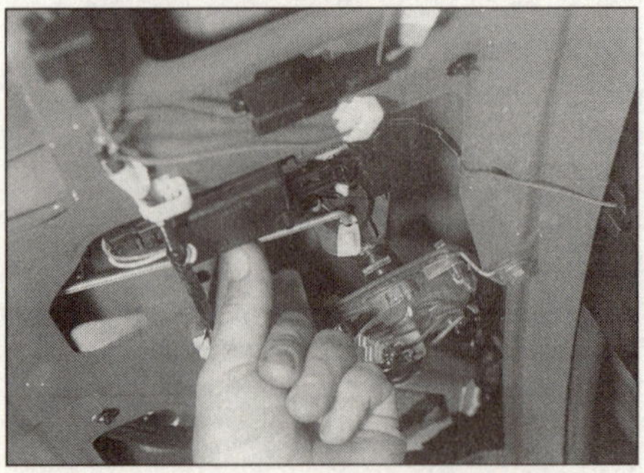

18.3 Disconnect the link rod from the lock button . . .

18 Tailgate lock components - removal and refitting

Removal

Lock assembly

1 Open up the tailgate then disconnect the battery negative terminal.

2 Undo the retaining screws then unclip the inner trim panel and remove it from the tailgate **(see illustrations)**.

3 Detach the link rod from the lock button assembly **(see illustration)**.

4 Slacken and remove the retaining bolts and remove the lock assembly from the tailgate **(see illustration)**.

Lock button assembly

5 Carry out the operations described in paragraphs 1 to 3.

6 Disconnect the wiring connector from the central locking motor (where fitted) then slacken and remove the retaining nuts and manoeuvre the lock button assembly out from the tailgate **(see illustrations)**.

Lock striker

7 Open up the tailgate and unclip the plastic cover from the lock striker.

8 Draw around the outline of each retaining bolt then undo the bolts and remove the striker from the vehicle.

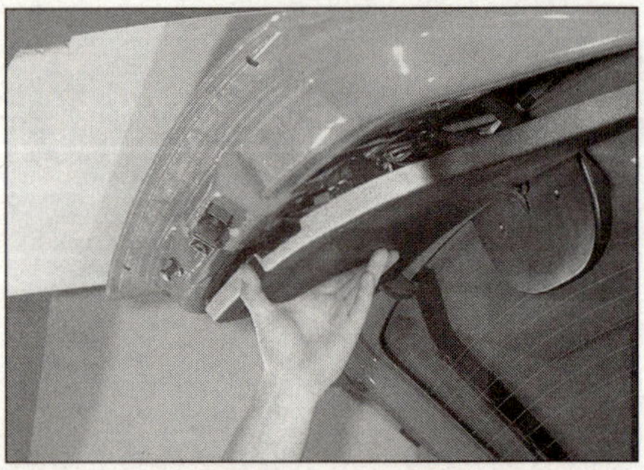

18.2b . . . and remove the inner trim panel from the tailgate

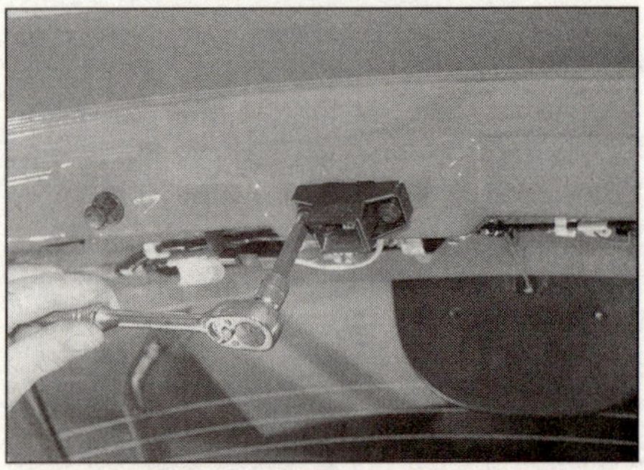

18.4 . . . then undo the retaining bolts and remove the lock from the tailgate

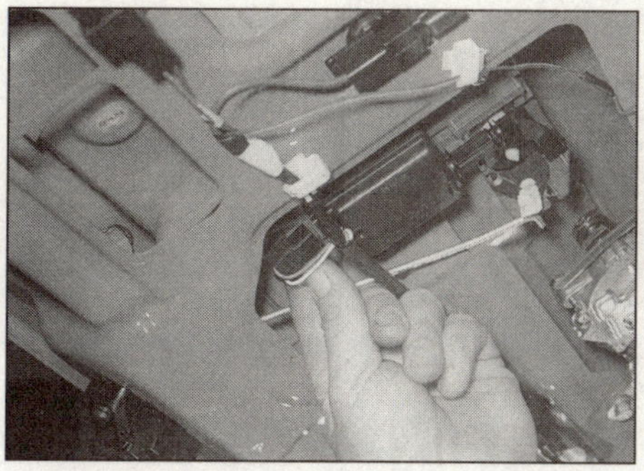

18.6a Disconnect the wiring connector from the central locking motor . . .

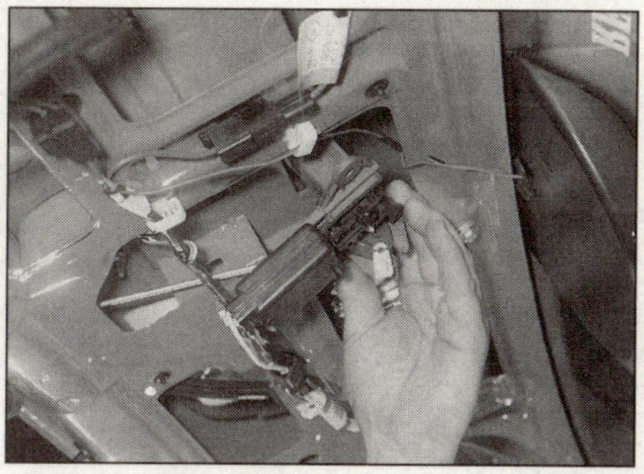

18.6b . . . then undo the retaining nuts and remove the lock button assembly

Refitting

Lock assembly

9 Refitting is the reverse of removal. Check the lock operation before refitting the trim panel.

Lock button assembly

10 Refitting is the reverse of removal, checking the operation of the lock prior to refitting the trim panel.

Lock striker

11 Fit the striker to the vehicle, and refit the bolts. Align the bolts with the marks made before removal and tighten them securely.

12 Check the operation of the lock, if necessary adjustments can be made by slackening the bolts and repositioning the striker, then refit the plastic cover.

19 Central locking components - removal and refitting

Removal

Door lock motor

1 Remove the door lock assembly as described in Section 13.

2 Carefully prise off the retaining clip, taking care not to lose it, then unclip the plastic cover and remove it from the lock **(see illustrations)**.

3 Undo the retaining screw and detach the motor from the lock assembly **(see illustrations)**.

Tailgate lock motor - Hatchback models

4 Remove the lock button assembly as described in Section 18.

5 Remove the retaining screw and clip and detach the motor from the mounting bracket.

Boot lid lock motor - Coupe models

6 Remove the lock button assembly as described in Section 16.

7 Remove the retaining screw and clip and detach the motor from the mounting bracket **(see illustration)**.

Boot lid lock motor - Saloon models

8 Open up the boot lid and unclip the lock access cover.

9 Release the retaining clip and detach the link rod from the motor **(see illustration)**.

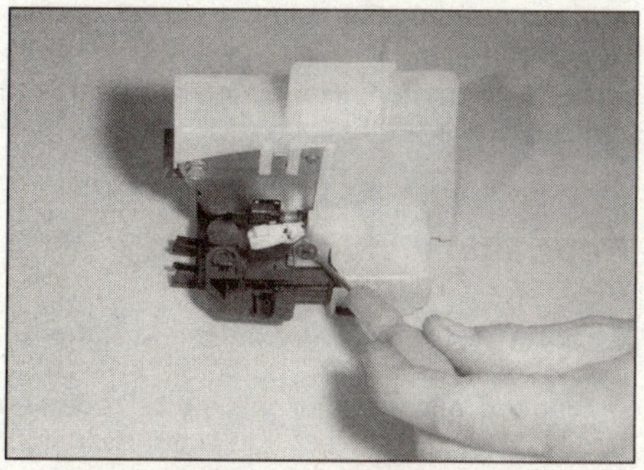

19.2a Prise off the retaining clip . . .

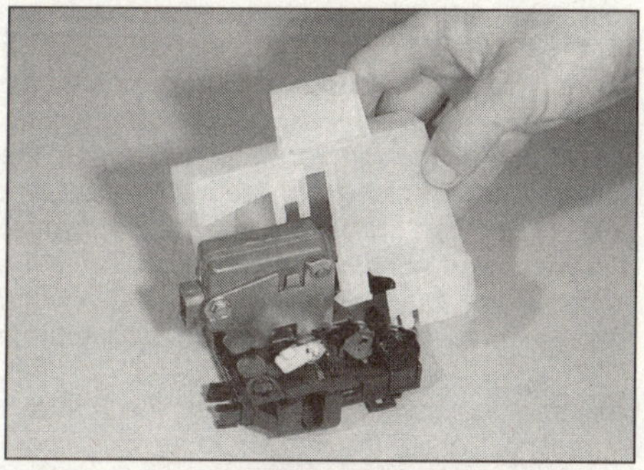

19.2b . . . then unclip the plastic cover and remove it from the door lock

19.3a Undo the retaining screw . . .

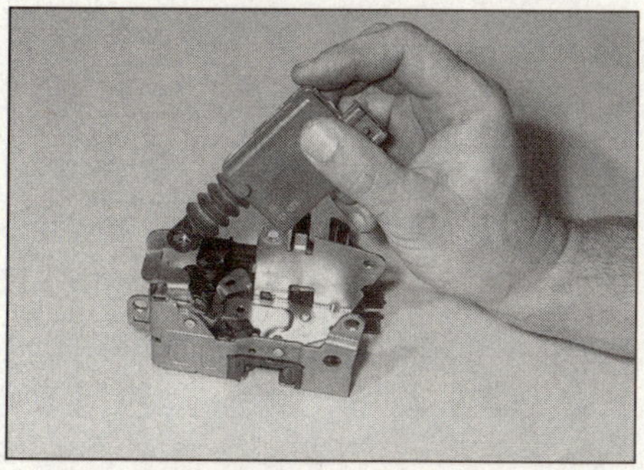

19.3b . . . then pivot the motor away from its mounting bracket and remove it

19.7 Boot lid lock motor retaining screw and clip (arrowed) - Coupe models

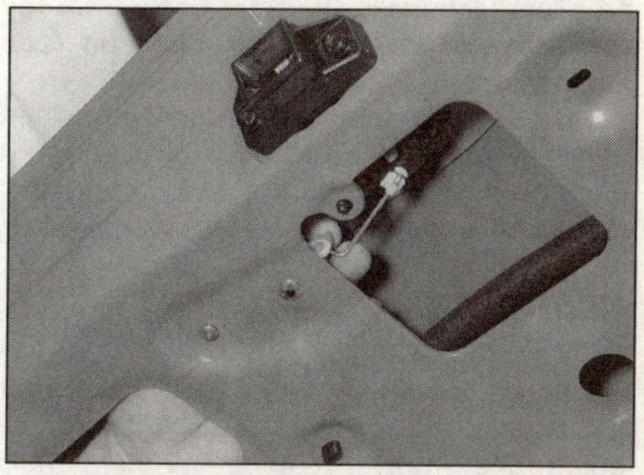

19.9 On Saloon models remove the access cover and detach the link rod . . .

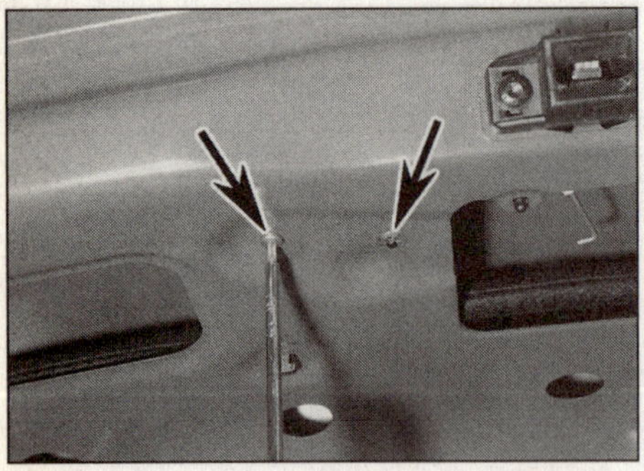
19.10a . . . then remove the retaining screw and clip (arrowed) . . .

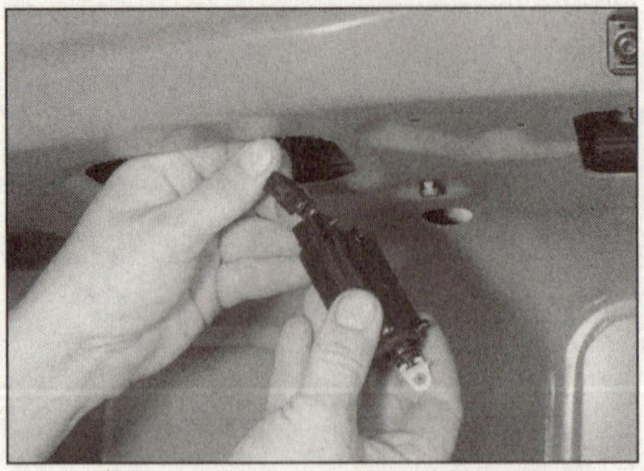
19.10b . . . and remove the central locking motor, disconnecting its wiring connector

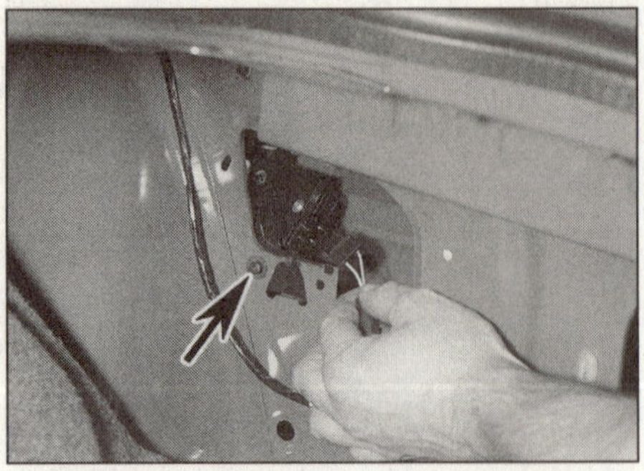
19.12 Disconnect the fuel filler flap wiring plug, then undo the screw (arrowed) . . .

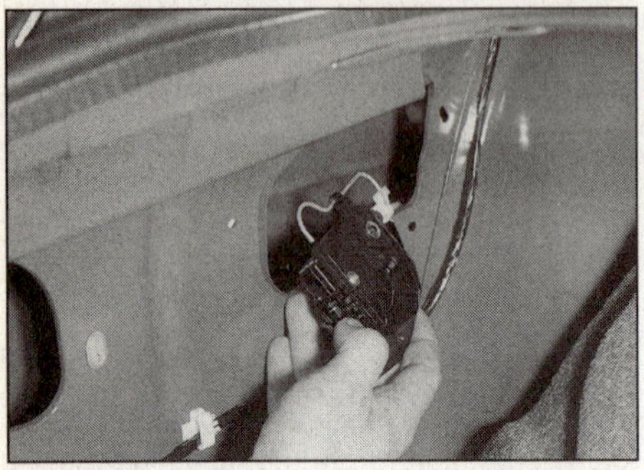
19.13a . . . then unclip and remove the motor and mounting bracket assembly

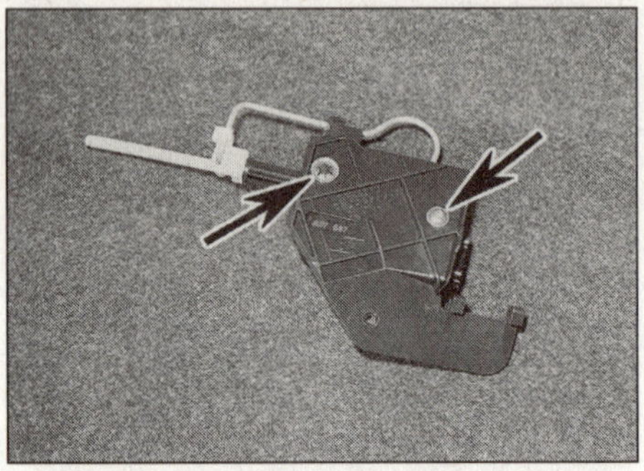
19.13b If necessary, remove the retaining screw and clip and separate the bracket and filler flap motor

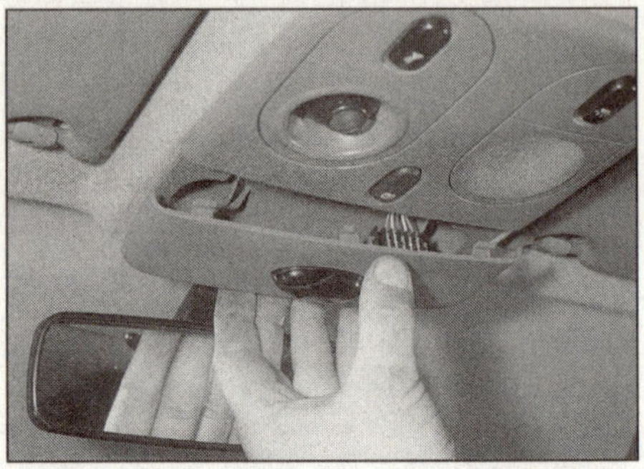
19.14 Carefully unclip the front panel from the roof console and disconnect its wiring connector

10 Remove the retaining screw and clip and withdraw the motor from the boot lid, disconnect its wiring connector as it becomes accessible **(see illustrations).**

Fuel filler flap motor

11 Prise out the retaining clips and peel back the luggage compartment side trim panel to gain access to the motor; on some models it will be necessary to remove the rear light covers to free the panel.

12 Disconnect the wiring connector then undo the screw securing the motor mounting bracket to the body **(see illustration)**.

13 Carefully unclip the mounting bracket and remove the motor assembly from the vehicle. The bracket and motor can be separated once the retaining clip and screw have been removed **(see illustrations)**.

Central locking remote receiver unit

14 Carefully unclip the front panel from the roof console in a forwards direction **(see illustration)**.

15 Disconnect the wiring connector then carefully unclip the central locking printed circuit board and separate it from the panel **(see illustration)**.

Refitting

16 Refitting is the reverse of removal; on refitting the fuel filler flap motor ensure that the flap locking rod is correctly seated in its guide tube. Prior to refitting any trim panels removed for access, thoroughly check the operation of the central locking system.

20 Electric window components - removal and refitting

Note: *If the battery is disconnected with a window open, the window will need reprogramming once the battery is reconnected. To do this, turn on the ignition and fully lower the window to its stop. Hold the switch until the relay clicks then, within two seconds, fully raise the window (it will rise in short steps). Once the window is fully closed, keep the switch depressed for a few seconds; this initialises the ECU so it learns the fully closed position. Release the switch and check the window operation. Repeat the procedure as necessary.*

Window switches

1 Refer to Chapter 12.

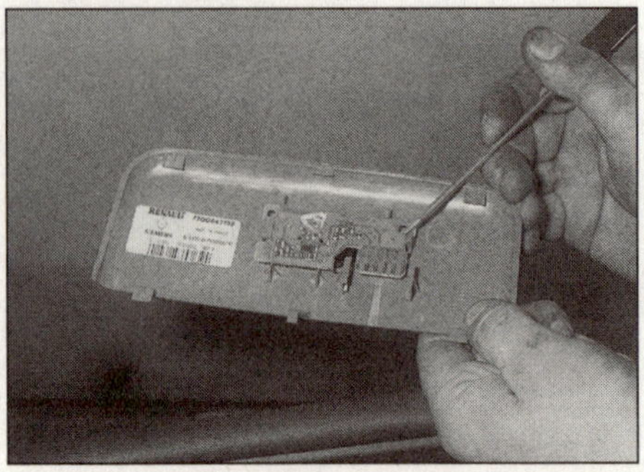
19.15 The central locking remote receiver is clipped into the panel

Door window winder motors

2 At the time of writing, it was unclear whether the window winder motors were available separately or whether the complete regulator assembly would have to be renewed. Refer to your Renault dealer for the latest information. Regulator removal and refitting is given in Section 14.

Rear side window opening motor - Coupe models

3 Remove the rear seat side trim panel as described in Section 27, paragraphs 12 to 23.

4 Slacken and remove the motor retaining screws then disconnect the wiring connector and remove the motor from the vehicle **(see illustration)**.

5 Refitting is the reverse of removal.

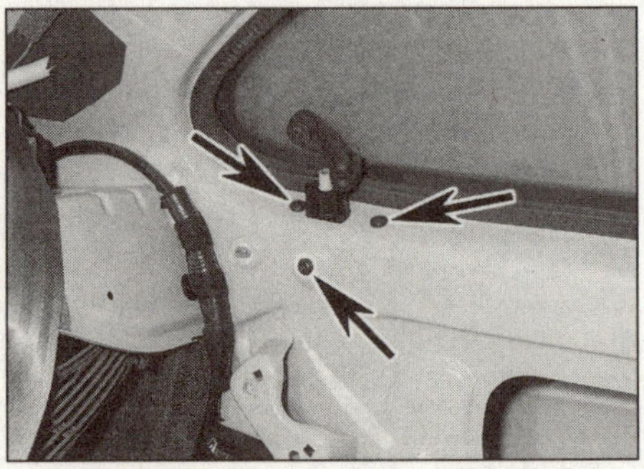
20.4 Rear side window opening motor retaining screws (arrowed) - Coupe models

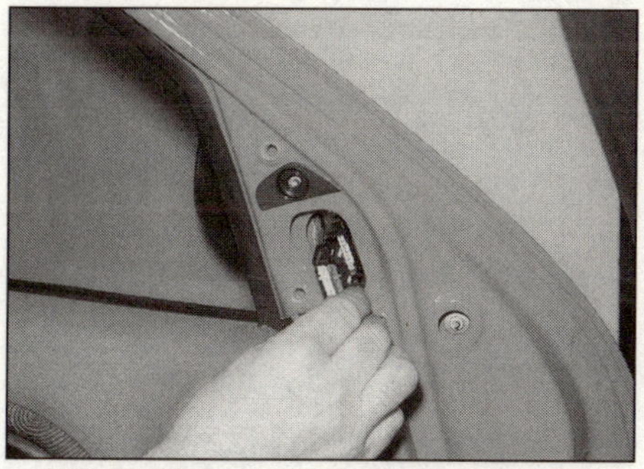

21.6 Disconnect the mirror wiring connector

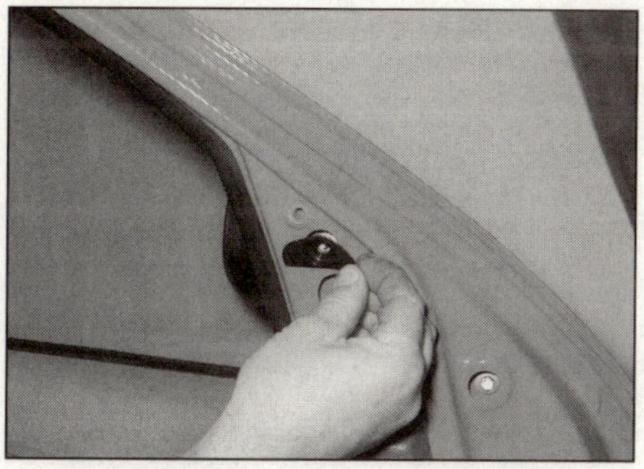

21.7a Remove the covers . . .

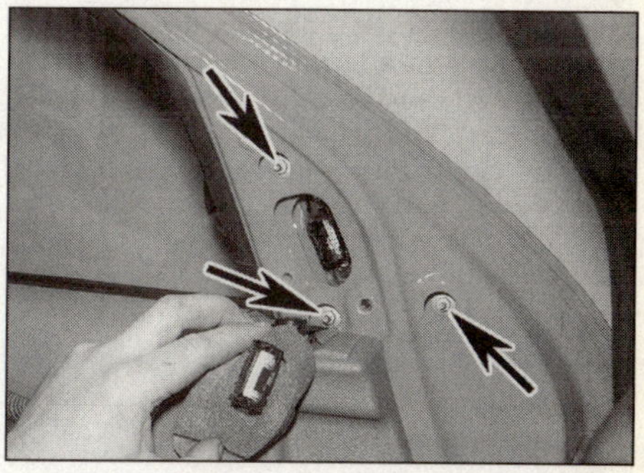

21.7b . . . then slacken and remove the retaining nuts (arrowed) . . .

21.7c . . . and remove the mirror

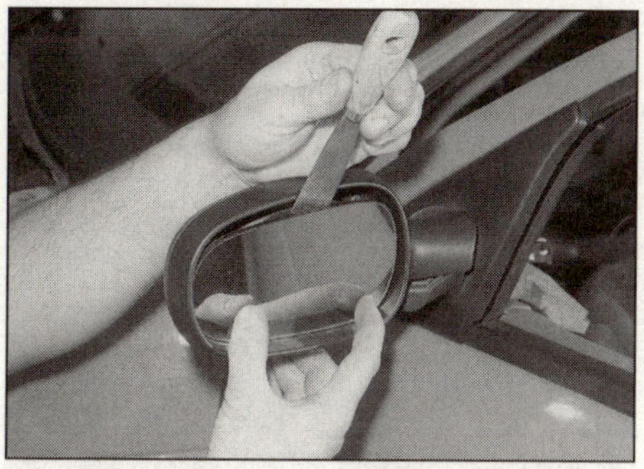

21.9 Carefully prise out the mirror glass . . .

21.10 . . . and disconnect its wiring plugs

21 Exterior mirrors and associated components - removal and refitting

Manually-adjusted mirror

1 Carefully unclip the top of the mirror inner trim panel from the door then free the trim panel from the adjustment lever and remove it from the door.

2 Remove the seal from the adjustment lever and the insulating foam.

3 Prise out the trim covers and insulation to reveal the mirror retaining nuts then undo the nuts and remove the mirror from the door. Recover the rubber seal which is fitted between the door and the mirror; if the seal is damaged it must be renewed.

4 Refitting is the reverse of removal.

Electrically-operated mirror

5 Carefully unclip the top of the mirror inner trim panel and remove the panel from the door.

6 Remove the insulating foam and disconnect the motor wiring connector **(see illustration)**.

7 Prise out the trim covers to reveal the mirror retaining nuts then undo the nuts and remove the mirror from the door **(see illustrations)**. Recover the rubber seal which is fitted between the door and mirror; if the seal is damaged it must be renewed.

8 Refitting is the reverse of removal.

Mirror glass

Note: *The mirror glass is clipped into position. Removal of the glass is likely to result in breakage if carried out carelessly.*

9 Tilt the mirror glass fully downwards and insert a wide plastic or wooden wedge in-between the centre of the mirror glass and the mirror housing **(see illustration)**. Carefully prise the glass from the motor/adjuster; take great care when removing the glass; do not use excessive force as the glass is easily broken.

10 Remove the glass from the mirror, where necessary, disconnect the wiring from the mirror heating element **(see illustration)**.

11 On refitting, reconnect the wiring to the glass and clip the glass onto the motor/adjuster, taking great care not to break it. Ensure that the glass is clipped securely into position and adjust as necessary.

Mirror switch

12 Refer to Chapter 12.

22 Windscreen and rear screen glass - general information

These areas of glass are secured by the tight fit of the weatherstrip in the body aperture, and are bonded in position with a special adhesive. Renewal of such fixed glass is a difficult, messy and time-consuming task, which is beyond the scope of the home mechanic. It is difficult, unless one has plenty of practice, to obtain a secure, waterproof fit. Furthermore, the task carries a high risk of breakage; this applies especially to the laminated glass windscreen. In view of this, owners are strongly advised to have this sort of work carried out by one of the many specialist windscreen fitters.

23 Sunroof - general information

1 Due to the complexity of the sunroof mechanism, considerable expertise is needed to repair, replace or adjust the sunroof components successfully. Removal of the roof first requires the headlining to be removed, which is a complex and tedious operation, and not a task to be undertaken lightly. Therefore, any problems with the sunroof should be referred to a Renault dealer.

2 On models with an electric sunroof, if the sunroof motor fails to operate, first check the relevant fuse. If the fault cannot be traced and rectified, the sunroof can be opened and closed manually using an Allen key to turn the motor spindle **(see illustration)**. To gain access to the motor spindle, on models with an electric sunroof it is necessary to remove

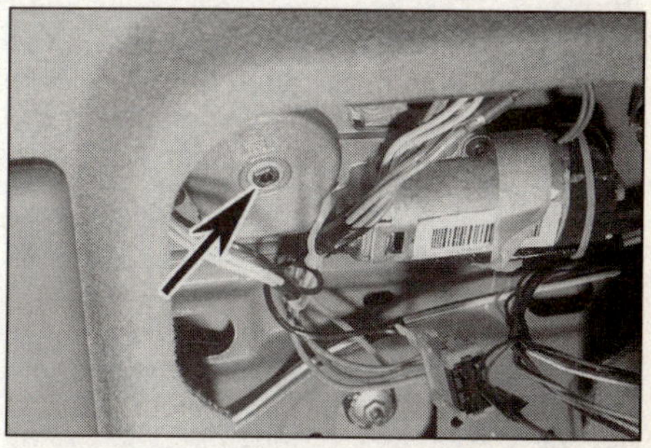

23.2 In an emergency, the electric sunroof can be moved by rotating the motor spindle with a suitable Allen key

the left-hand of the courtesy lights and on models not fitted with a sunroof it is necessary to remove the complete overhead console from the roof (see Chapter 12). Using an Allen key, rotate the motor spindle and move the sunroof to the required position.

24 Body exterior fittings - removal and refitting

Wheel arch liners and body under-panels

1 The various plastic covers fitted to the underside of the vehicle are secured in position by a mixture of screws, nuts and retaining clips and removal will be fairly obvious on inspection. Work methodically around the panel removing its retaining screws and releasing its retaining clips until the panel is free and can be removed from the underside of the vehicle. Most clips used on the vehicle are simply prised out of position. Other clips can be released by unscrewing/ prising out the centre pins and then removing the clip.
2 On refitting, renew any retaining clips that may have been broken on removal, and ensure that the panel is securely retained by all the relevant clips and screws.

Body trim strips and badges

3 The various body trim strips and badges are held in position with a special adhesive tape. Removal requires the trim/badge to be heated, to soften the adhesive, and then cut away from the surface. Due to the high risk of damage to the vehicle's paintwork during this operation, it is recommended that this task should be entrusted to a Renault dealer.

25 Seats - removal and refitting

Removal

Front seat

Warning: ***On models with seat belt tensioners, disable the tensioning mechanism (see Section 26) before removing the seat.***

1 Firmly apply the handbrake then jack up the front of the vehicle and support it on axle stands (see "*Jacking and vehicle support*").
2 From underneath the vehicle, slacken and remove the four seat mounting nuts.
3 From inside the vehicle, lift the front seat out of position, disconnecting its wiring connectors as they become accessible **(see illustration)**.

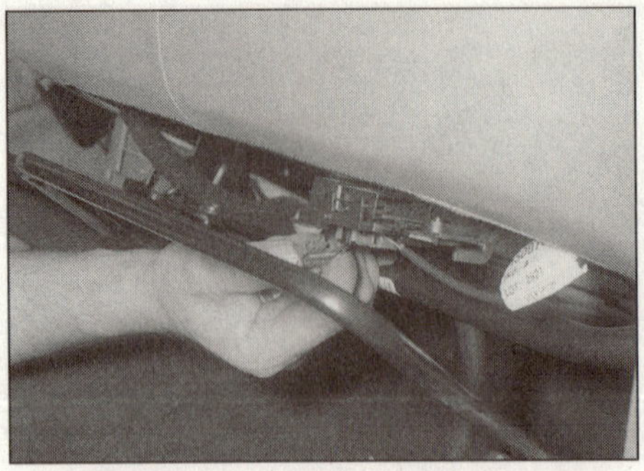

25.3 Disconnect the wiring connectors and remove the front seat from the vehicle

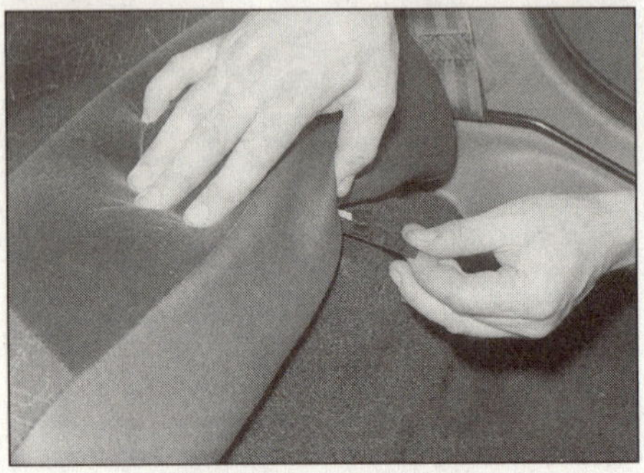

25.7 . . . then slide out the retaining clips to allow the seat cushion to be lifted out of position

Rear seat - Hatchback and Saloon models

4 To remove the rear seat cushions, fold the cushion(s) up and unhook the side front pivots from the floor.
5 To remove the seat back, remove the seat cushion(s) then slacken and remove the retaining nuts from the seat back hinges. Release the seat back from its locking pins and remove it from the vehicle.

Rear seat - Coupe models

6 Open up the rear seat storage compartment then peel back the felt cover and unscrew the retaining nuts **(see illustration)**.
7 Using a pair of pliers, pull out the retaining clips from the left- and right-hand side of the seat cushion then lift the cushion out from the vehicle **(see illustration)**.
8 Slacken and remove the seat back retaining nuts and remove the assembly from the vehicle; if the complete seat assembly is to be removed undo the three hinge main securing nuts, if only one side of the seat is to be removed undo the outer hinge nut and smaller nut securing the seat to the centre hinge assembly **(see illustration)**.

Refitting

9 Refitting is the reverse of removal ensuring all nuts are securely tightened.

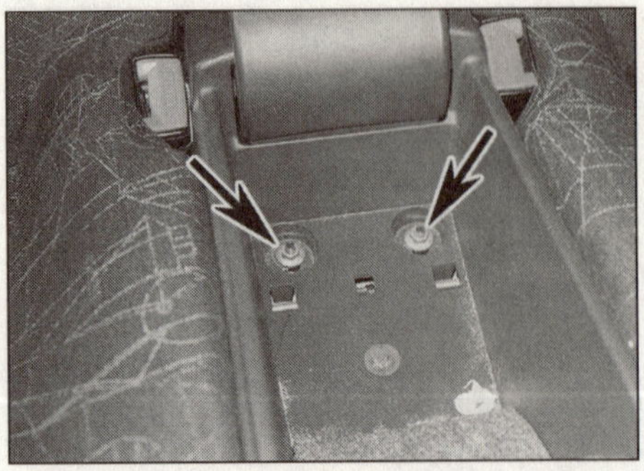

25.6 On Coupe models, undo the retaining nuts (arrowed) which are situated in the storage compartment . . .

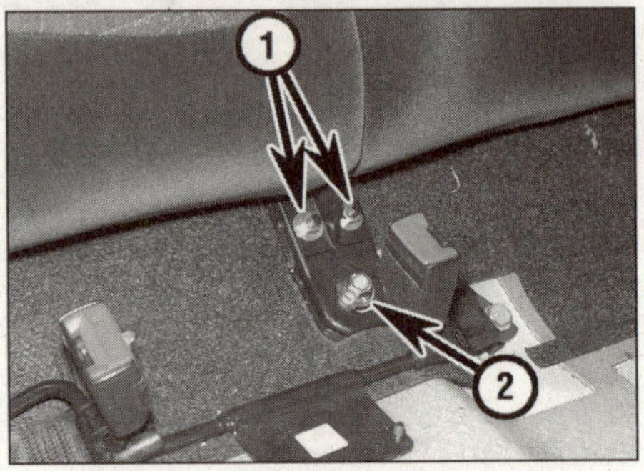

25.8 To remove the complete seat assembly, undo the main centre hinge nut (2); to remove either side of the seat undo the relevant smaller nut (1)

26 Front seat belt tensioning mechanism - general information

Most models covered in this manual are fitted with a front seat belt tensioner system. The system is designed to instantaneously take up any slack in the seat belt in the case of a sudden frontal impact, therefore reducing the possibility of injury to the front seat occupants. Each front seat is fitted with the system, the tensioner being attached directly to the seat belt stalk.

The seat belt tensioner is operated by the same control unit as the airbag (Chapter 12). The tensioner is electrically triggered by a frontal impact above a pre-determined force. Lesser impacts, including impacts from behind, will not trigger the system.

When the system is triggered, the gas generator inside the tensioner cylinder ignites. This forces the tensioner piston forwards which then removes all slack from the seat belt by retracting the seat belt stalk. The strength of the explosion in the tensioner cylinder is calibrated to retract the seat belt sufficiently to securely retain the occupant of the seat without forcing them into the seat. Once the tensioner has been triggered, the seat belt will be permanently locked and the assembly must be renewed.

To prevent the risk of injury if the system is triggered inadvertently when working on the

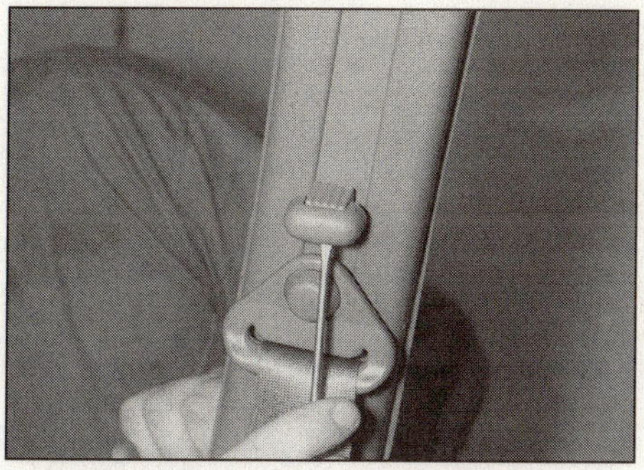

27.3 Lever between the adjustment lever inner and outer sections and remove the lever from the vehicle

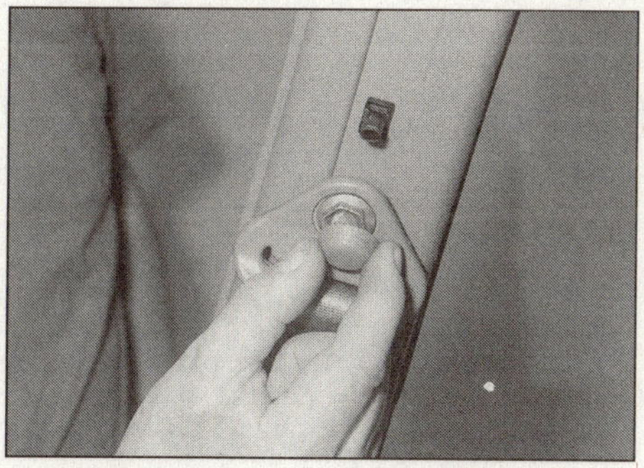

27.4a Remove the trim cover . . .

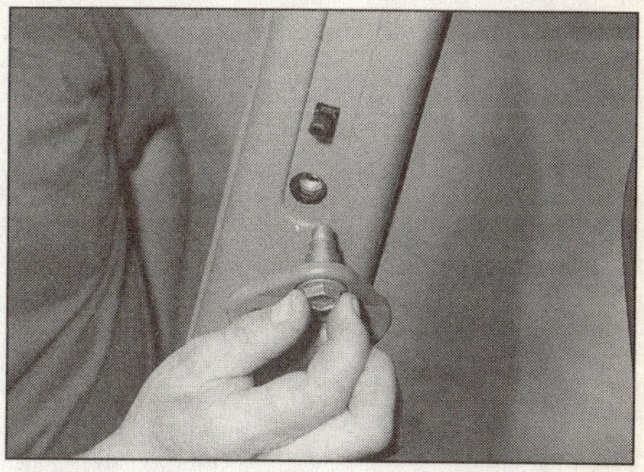

27.4b . . . then unscrew the mounting bolt and detach the seat belt upper mounting from the pillar

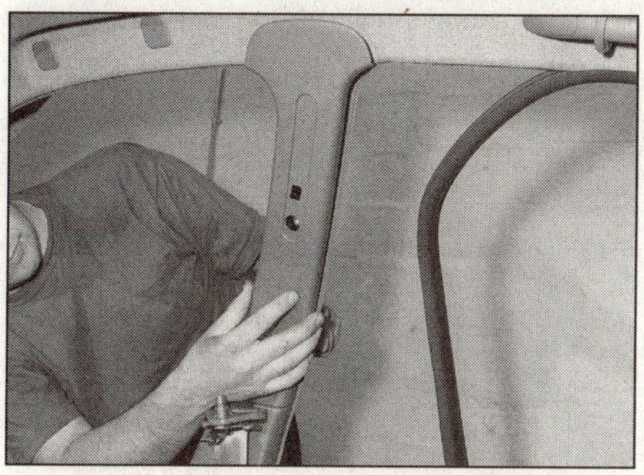

27.6 Peel away the sealing strips then unclip the pillar upper trim panel and remove it from the vehicle

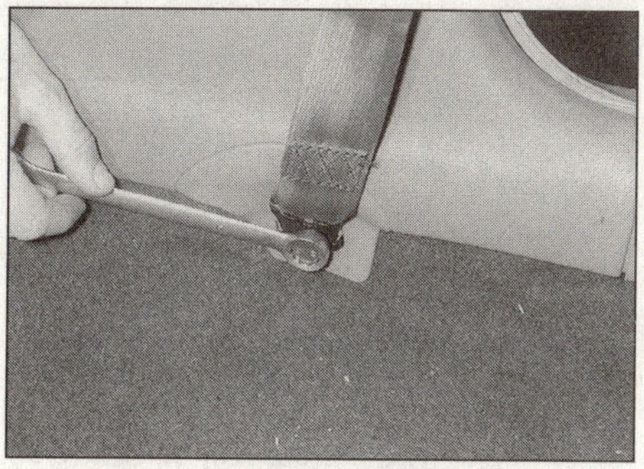

27.7 Unscrew the seat belt lower mounting bolt

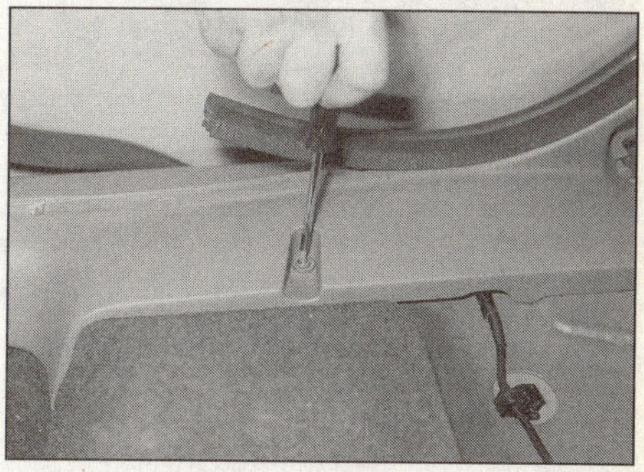

27.8 If necessary, undo the retaining screw to allow the rear door sill trim panel to be unclipped sufficiently to gain access to the front panel rear screw

vehicle, if any work is to be carried out on the front seats, disconnect the battery and disable the tensioner. To do this, remove the seat belt tensioner/airbag system fuse (see "*Wiring diagrams*") from the fusebox and wait for at least five minutes. This will allow the reserve power capacitors in the control unit to discharge.

Also note the following warnings before contemplating any work on the front seat belts.

Warning: If the seat belt tensioners are activated in an accident, both the tensioners and the control unit must be renewed.

Warning: If the tensioner mechanism is dropped, it must be renewed, even it has suffered no apparent damage.

Warning: Do not allow any solvents to come into contact with the tensioner mechanism.

Warning: Do not subject the seat to any form of shock as this could accidentally trigger to the seat belt tensioner.

Warning: Do not subject the tensioner assembly to excessive temperatures.

27 Seat belt components - removal and refitting

Warning: On models equipped with front seat belt tensioners refer to Section 26 before proceeding.

Removal

Front seat belt - Hatchback and Saloon models

1 Disconnect the battery negative terminal.
2 To improve access, remove the relevant front seat as described in Section 25.
3 Using a flat-bladed screwdriver, lever between the inner and outer sections of the height adjustment lever button to release the button from its clip **(see illustration)**.
4 Unclip the trim cover then slacken and remove the mounting bolt and free the seat belt from its upper mounting **(see illustrations)**.
5 Peel away the front and rear sealing strips from the door centre pillar.
6 Unclip the centre pillar trim panel (starting at the top and working down) and remove it from the vehicle **(see illustration)**.
7 Remove the trim cover then slacken and remove the seat belt lower mounting bolt **(see illustration)**.
8 Carefully unclip the front of the rear door sill trim panel to reveal the pillar/front sill trim panel rear retaining screw; if necessary, lift up the rear seat cushion and undo the retaining screw to allow the panel to be sufficiently unclipped **(see illustration)**.
9 Slacken and remove the rear and upper retaining screws then, starting at the rear and working forwards, unclip the trim panel from the door pillar/sill. Free the seat belt from the panel and remove it from the vehicle **(see illustrations)**.

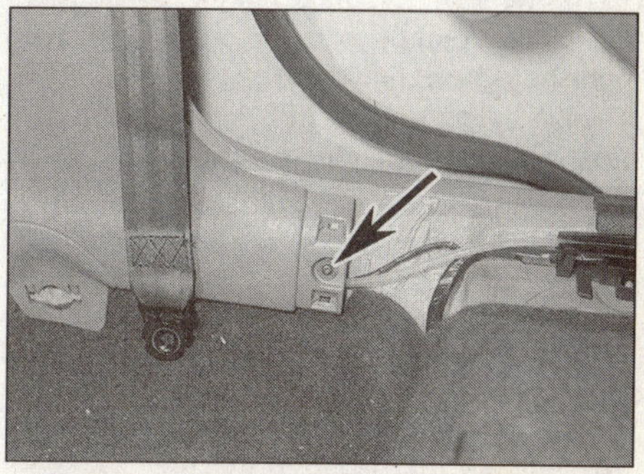

27.9a Slacken and remove the rear screw (arrowed) . . .

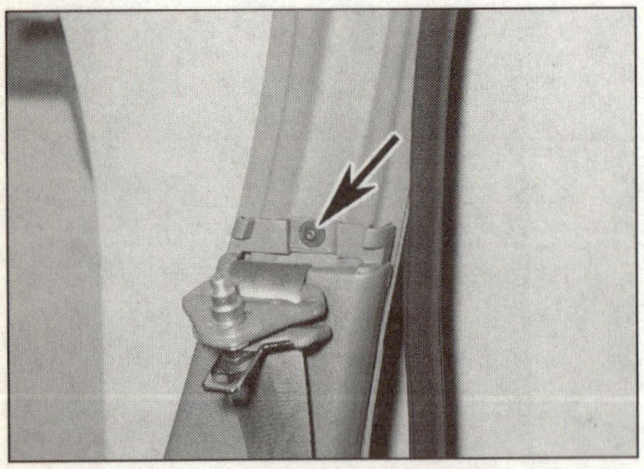

27.9b ... and the upper screw (arrowed) ...

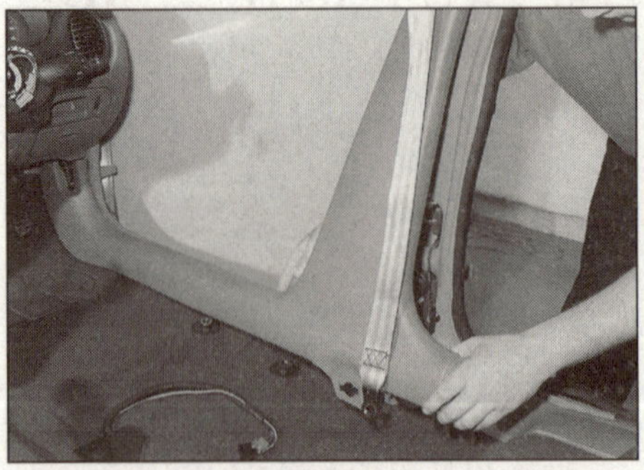

27.9c ... then unclip the front door pillar/sill trim panel and remove it

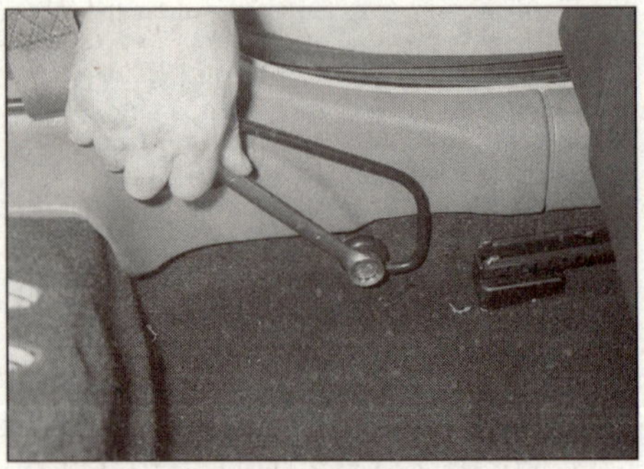

27.13a On Coupe models, unscrew the mounting bolt ...

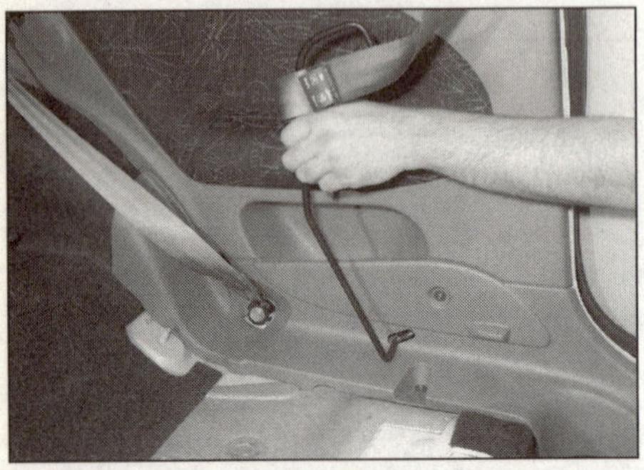

27.13b ... then unhook the lower mounting rail from the body and free it from the end of the seat belt

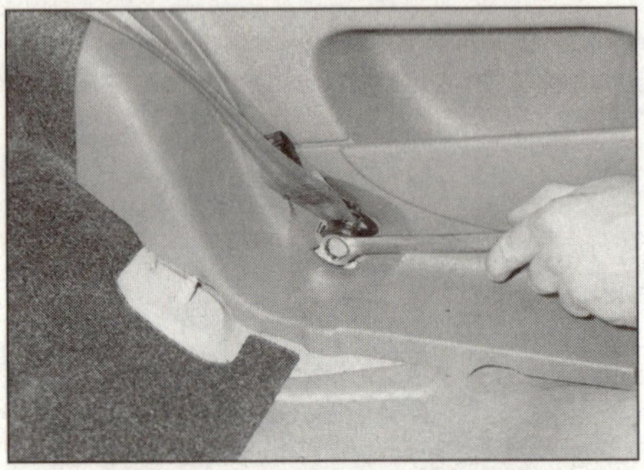

27.14 Unscrew the mounting bolt and free the rear seat belt lower mounting

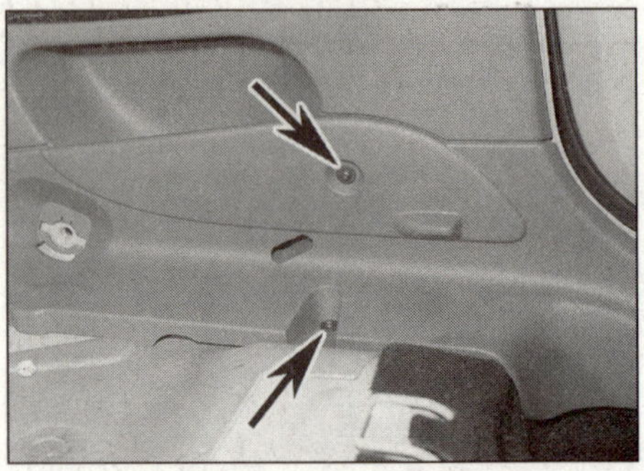

27.16a Undo the retaining screws (arrowed) ...

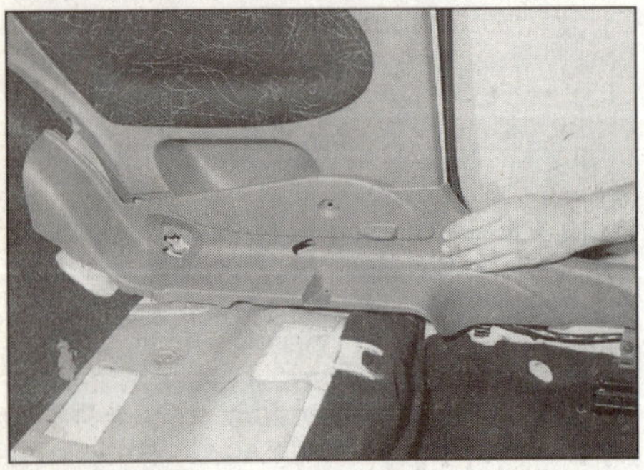

27.16b ... then unclip the lower section of the side trim panel and remove it

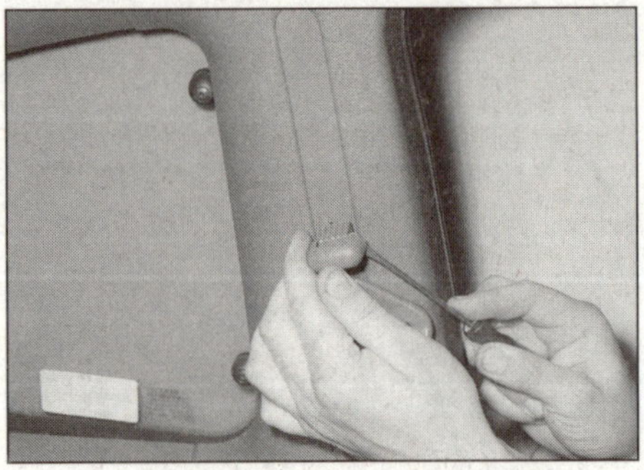

27.17 Remove the height adjustment lever button (see text) ...

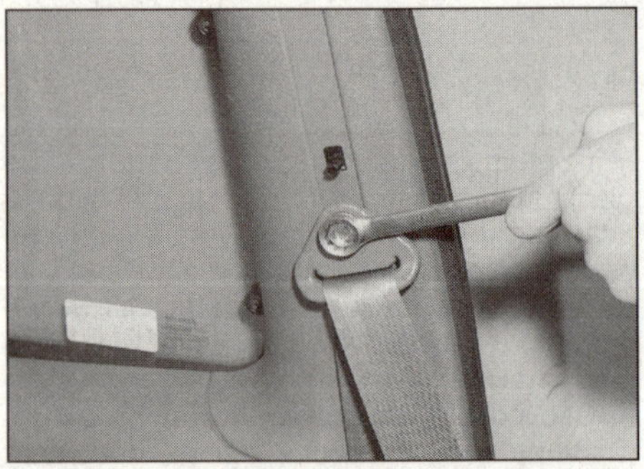

27.18 ... then remove the trim cover and unscrew the seat belt upper mounting bolt

10 Slacken and remove the inertia reel mounting bolt and remove the seat belt assembly from the vehicle.

11 If necessary, the height adjustment mechanism can be removed once its retaining bolts have been undone.

Front seat belt - Coupe models

12 Remove the rear seat as described in Section 25 and slide the front seat fully forwards.

13 Remove the trim cap then unscrew the seat belt lower mounting rail retaining bolt. Unhook the rail from the body and free it from the end of the seat belt **(see illustrations)**.

14 Slacken and remove the rear seat belt lower mounting bolt and free the belt from the trim panel **(see illustration)**.

15 Peel the sealing strip from the rear of the door aperture.

16 Undo the retaining screws then unclip the lower section of the rear seat side trim panel and remove it from the vehicle **(see illustrations)**.

17 Using a flat-bladed screwdriver, lever between the inner and outer sections of the height adjustment lever button to release the button from its clip **(see illustration)**.

18 Unclip the trim cover then slacken and remove the mounting bolt and free the seat belt from its upper mounting **(see illustration)**.

19 Unclip the pillar trim panel (starting at the top and working down) and remove it from the vehicle **(see illustration)**.

20 Remove the rear pillar trim panel as described in paragraphs 34 and 35.

21 Slacken and remove the rear seat back locking pin from the side of the vehicle **(see illustration)**.

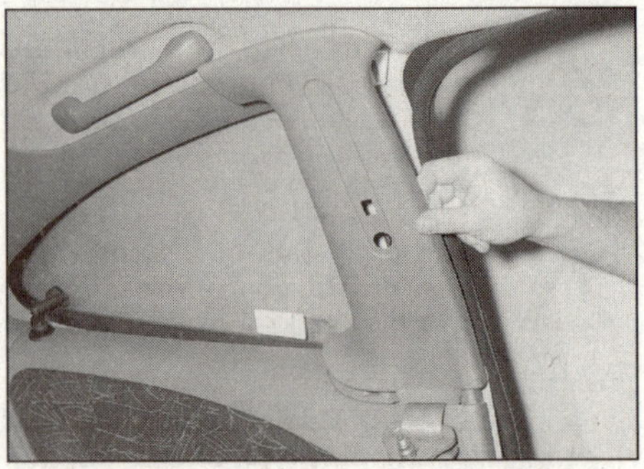

27.19 Unclip the pillar upper trim panel and remove it from the vehicle

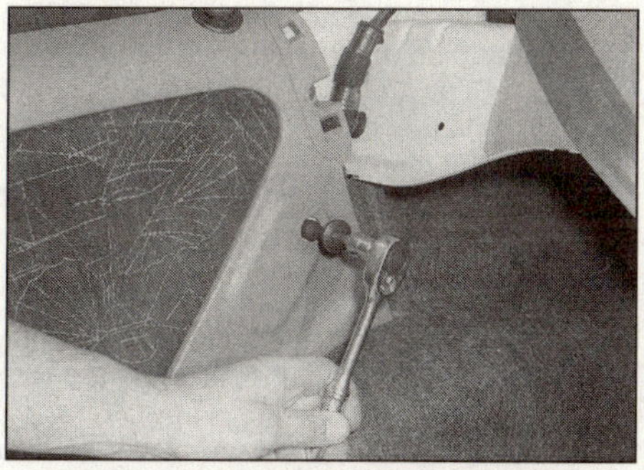

27.21 Unscrew the rear seat back locking pin

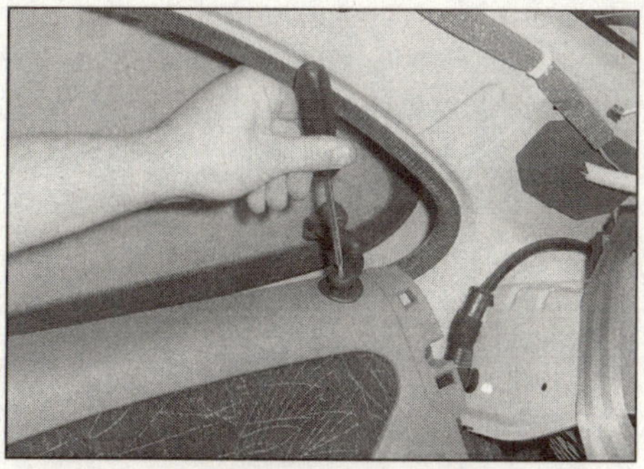

27.22 Undo the retaining screw and free the side window linkage from the spindle

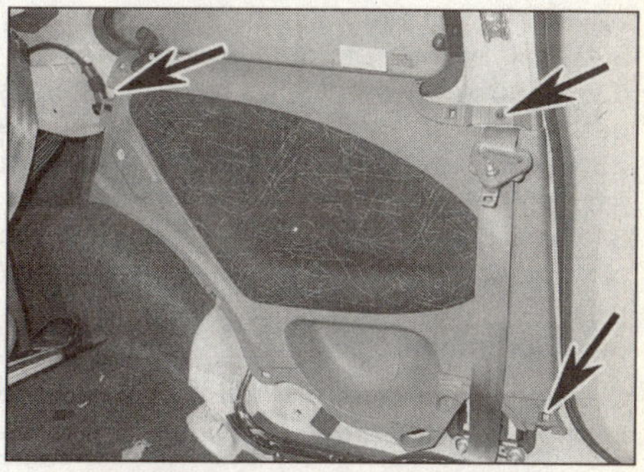

27.23a Undo the retaining screws (arrowed) . . .

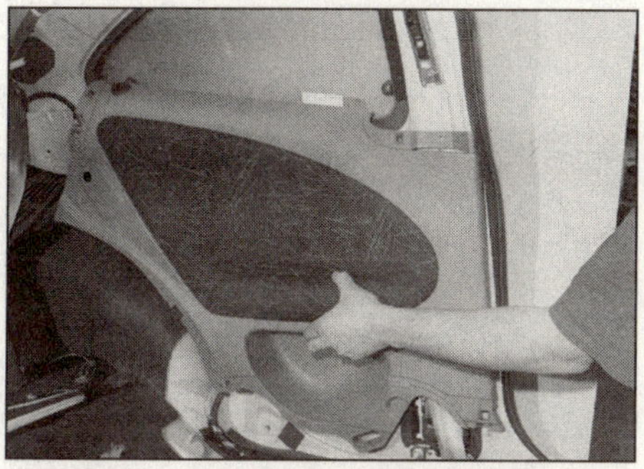

27.23b . . . then unclip the side trim panel and remove it from the vehicle

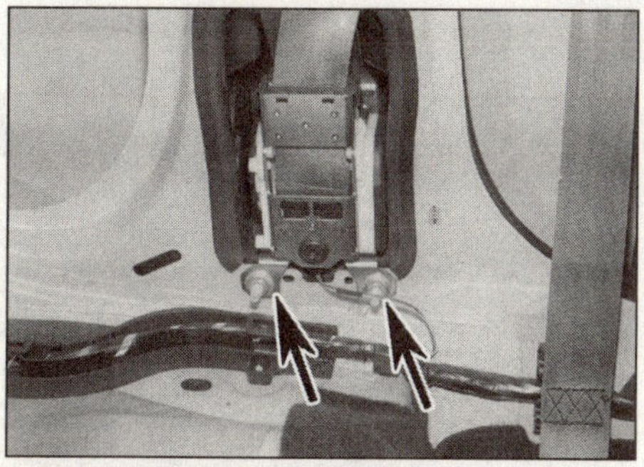

27.24 Front seat belt inertia reel retaining nuts (arrowed) - Coupe models

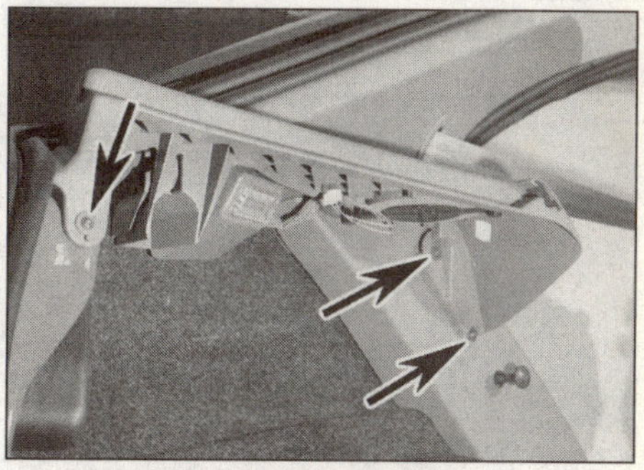

27.26a On Hatchback models undo the retaining screws (arrowed) . . .

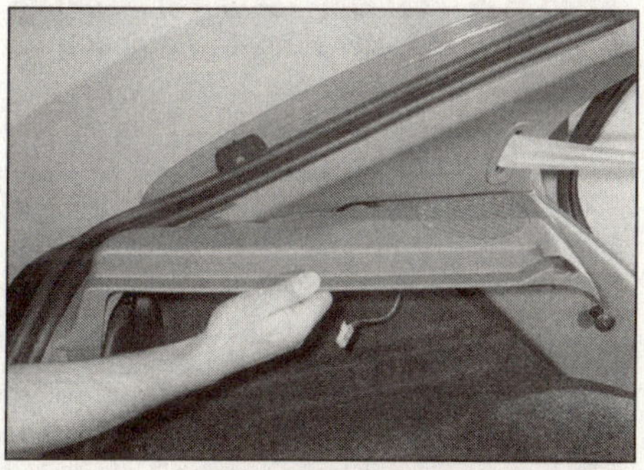

27.26b . . . and remove the parcel shelf side mounting panel

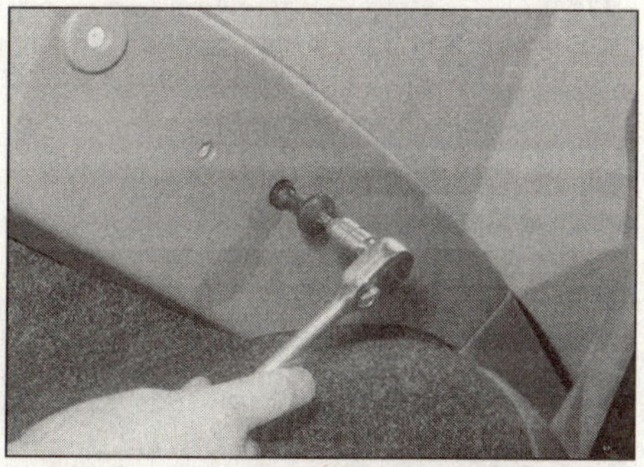

27.27a Unscrew the rear seat back locking pin . . .

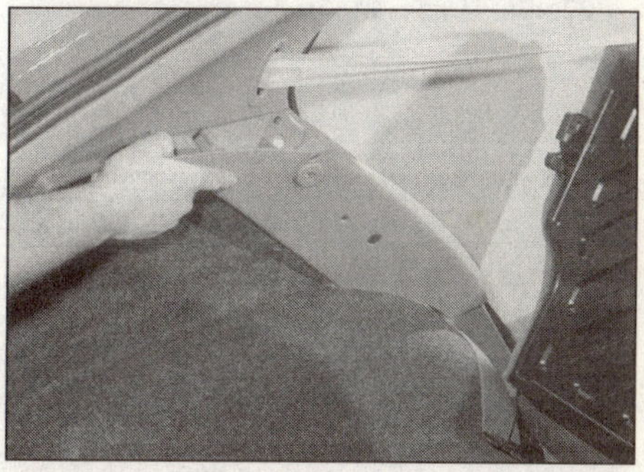

27.27b . . . then unclip the side trim panel and remove it from the vehicle

22 Open up the rear side window then undo the retaining screw securing the opening linkage to the motor spindle **(see illustration)**. Carefully ease the linkage off the spindle.

23 Undo the retaining screws then unclip the side trim panel and remove it from the vehicle, freeing it from the front seat belt **(see illustrations)**.

24 Slacken and remove the inertia reel retaining nuts and remove the front seat belt from the vehicle **(see illustration)**.

Rear seat belt - Hatchback models

25 Remove the parcel shelf from the luggage compartment.

26 Fold the rear seats down and disconnect the wiring connectors from the rear loudspeaker on the side that the seat belt is to be removed. Slacken and remove the retaining screws and remove the relevant parcel shelf side mounting panel from the luggage compartment **(see illustrations)**.

27 Peel the rear door sealing strip away from the rear seat side trim panel. Slacken and remove the seat locking pin then unclip the trim panel and remove it from the vehicle **(see illustrations)**.

28 Slacken and remove the seat belt lower mounting bolt **(see illustration)**.

29 Unclip the upper trim panel from the rear pillar then free the panel from the belt and remove it from the vehicle. If work is being carried out on the left-hand side, it will be necessary to disconnect the luggage compartment light as the panel is removed **(see illustrations)**.

30 Slacken and remove the inertia reel mounting bolt and remove the belt from the vehicle **(see illustration)**.

Rear seat belt - Coupe models

31 Remove the rear seat cushion as described in Section 25 and fold down the seat back.

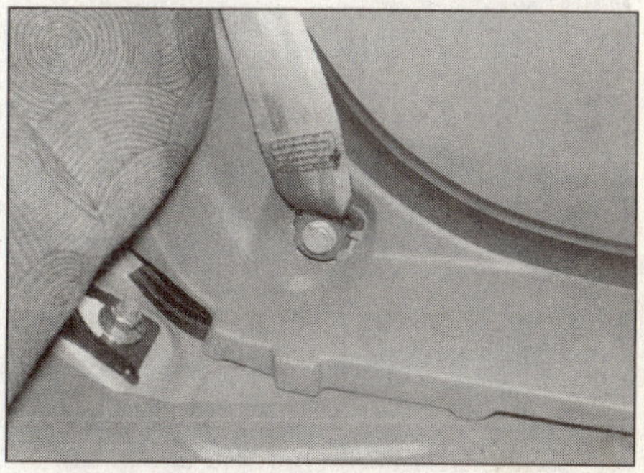

27.28 Unscrew the bolt and free the seat belt lower mounting from the vehicle

32 Slacken and remove the lower mounting bolt and free the belt from the side trim panel.
33 Remove the front pillar upper panel as described in paragraphs 17 to 19, peeling the door sealing strip away from the front edge of the panel.
34 Carefully ease the rear window heated element wiring connector out from the rear pillar trim panel and disconnect the connector **(see illustration)**.

Caution: Take great care not to bend the window wire too much as it is easily broken

35 Undo the retaining screw then unclip the trim panel from the rear pillar **(see illustrations)**. Free the seat belt from the panel and remove the panel from the vehicle.
36 Slacken and remove the inertia reel mounting bolt and remove the belt from the vehicle.

Rear seat belt - Saloon models

37 Remove the high-level stoplight as described in Chapter 12. On models with rear speakers also remove the speaker grilles from the parcel shelf trim.
38 Unclip and remove the rear seat cushions then slacken and remove the seat belt lower mounting bolt (where necessary).
39 Unclip the relevant seat belt trim cover from the parcel shelf trim **(see illustration)**.
40 Fold down the rear seat backs then carefully prise out the retaining clips securing the front edge of the parcel shelf trim panel in position. Slide the trim panel forwards and position it clear of the shelf **(see illustrations)**.
41 Undo the inertia reel mounting bolt then remove the belt from the vehicle, freeing it from the parcel shelf trim panel.

Refitting

42 Refitting is the reverse of removal, making sure the inertia reel is correctly engaged with the body and that all mounting bolts are securely tightened. Also ensure that all trim panels removed are correctly clipped into position and retained by all the relevant clips and screws.

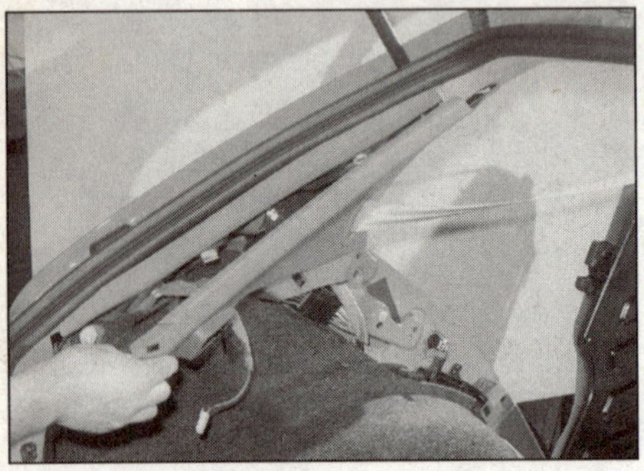

27.29a Unclip the upper trim panel . . .

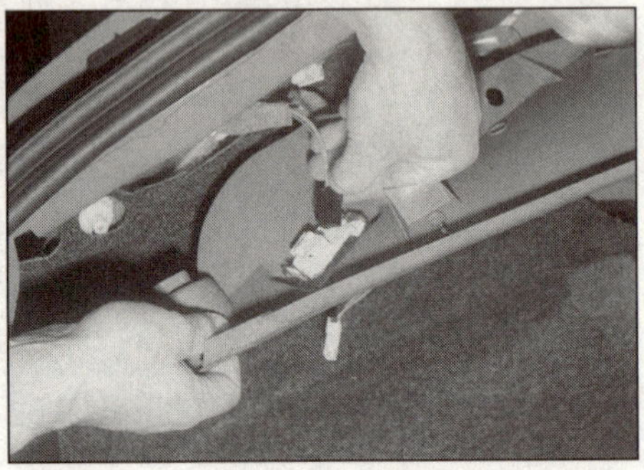

27.29b . . . noting that it will be necessary to disconnect the wiring connector from the light if work is being carried out on the left-hand side

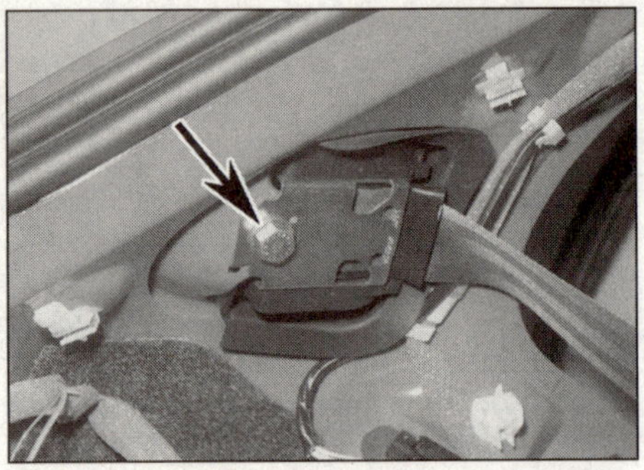

27.30 Rear seat belt inertia reel mounting bolt (arrowed) - Hatchback model

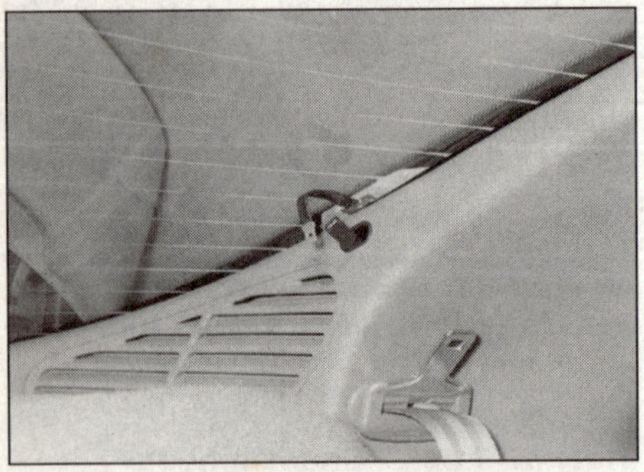

27.34 On Coupe models, disconnect the rear window heating element wiring . . .

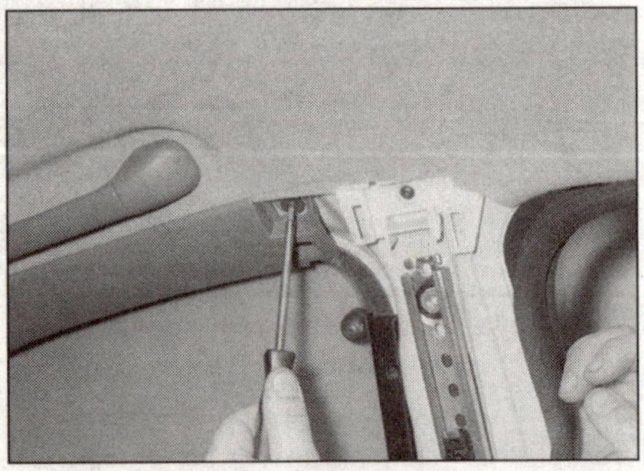

27.35a . . . then undo the retaining screw . . .

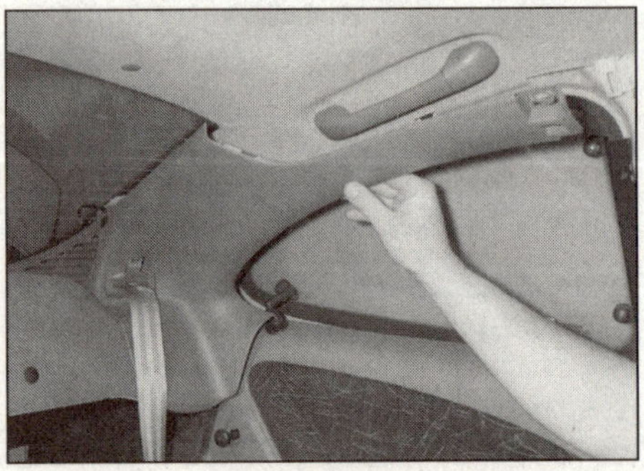

27.35b . . . and unclip the rear pillar trim panel

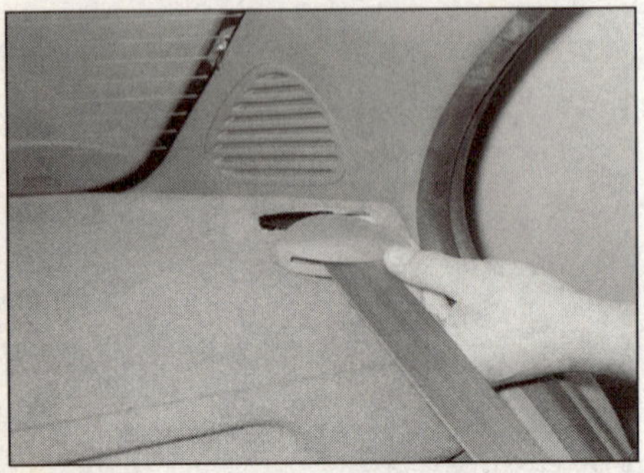

27.39 On Saloon models unclip the relevant seat belt trim cover from the parcel shelf trim

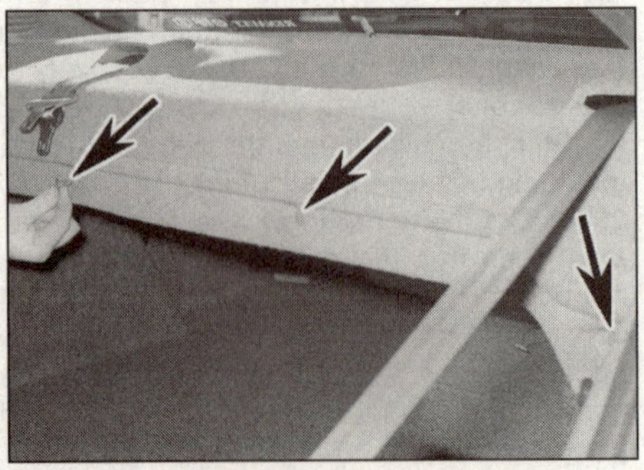

27.40a Prise out the retaining clips (arrowed) . . .

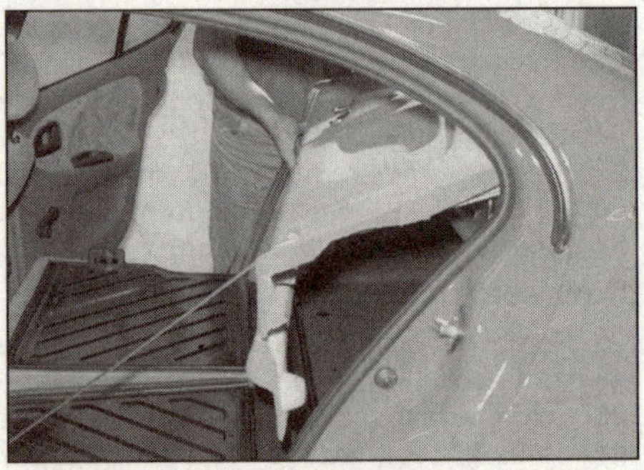

27.40b . . . and release the parcel shelf trim panel in a forwards direction

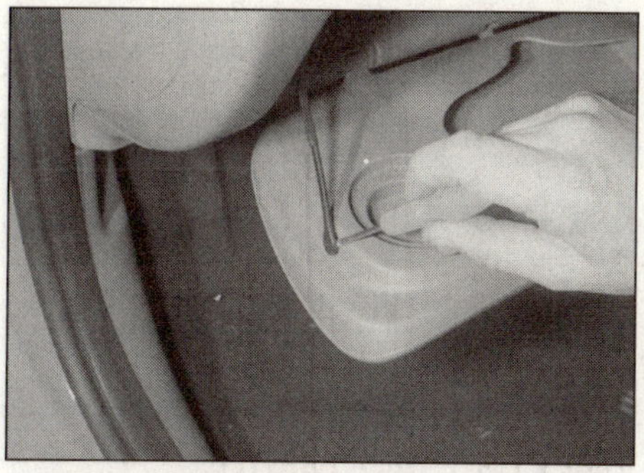

28.6 Carefully unclip the support cables from the sides of the glovebox lid . . .

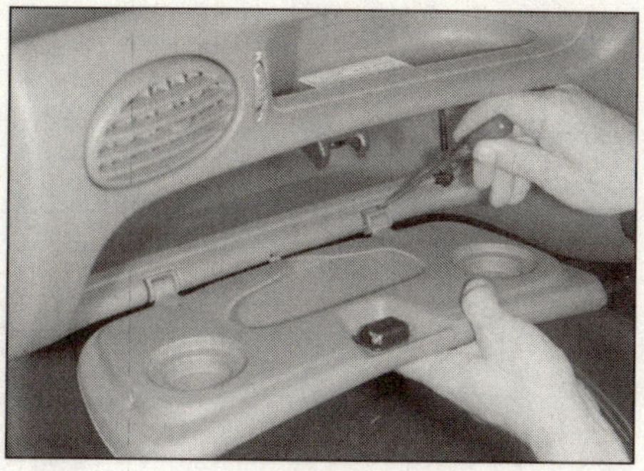

28.7 . . . then carefully unclip the hinges from the facia panel

28 Interior trim - removal and refitting

Interior trim panels

1 The interior trim panels are secured using either screws or various types of trim fasteners, usually studs or clips.

2 Check that there are no other panels overlapping the one to be removed; usually there is a sequence that has to be followed that will become obvious on close inspection.

3 Remove all obvious fasteners, such as screws. If the panel will not come free, it is held by hidden clips or fasteners. These are usually situated around the edge of the panel and can be prised up to release them; note, however that they can break quite easily so replacements should be available. The best way of releasing such clips without the correct type of tool, is to use a large flat-bladed screwdriver. Note in many cases that the adjacent sealing strip must be prised back to release a panel.

4 When removing a panel, **never** use excessive force or the panel may be damaged; always check carefully that all fasteners have been removed or released before attempting to withdraw a panel.

5 Refitting is the reverse of the removal procedure; secure the fasteners by pressing them firmly into place and ensure that all disturbed components are correctly secured to prevent rattles.

Glovebox lid

6 Open up the glovebox lid and unclip the support cables **(see illustration)**.

7 Release the lid by gently pulling it backwards whilst carefully prising its hinge assemblies out from the facia using a screwdriver **(see illustration)**. As the lid is removed, take care not to lose the hinge retaining clips.

8 On refitting ensure that the hinges are securely retained by their clips and the support cables are clipped securely in position.

Carpets

9 The passenger compartment floor carpet is in one piece and is secured at its edges by screws or clips, usually the same fasteners used to secure the various adjoining trim panels.

10 Carpet removal and refitting is reasonably straightforward but very time-consuming because all adjoining trim panels must be removed first, as must components such as the seats, the centre console and seat belt lower anchorages.

Headlining

11 The headlining is clipped to the roof and can be withdrawn only once all fittings such as the grab handles, sun visors, sunroof (if fitted), windscreen and rear quarter windows and related trim panels have been removed and the door, tailgate and sunroof aperture sealing strips have been prised clear.

12 Note that headlining removal requires considerable skill and experience if it is to be carried out without damage and is therefore best entrusted to an expert.

29 Centre console - removal and refitting

Removal

1 Disconnect the battery negative terminal.

2 On models with electric rear windows, remove the window switch panel from the rear of the centre console.

3 On models with manual rear windows, unclip the ashtray and remove it from the rear of the centre console **(see illustration)**.

4 On all models, slacken and remove the rear retaining nut then lift the rear section of the centre console off from the handbrake lever **(see illustrations)**.

5 On manual transmission models, unclip the gearchange lever gaiter from the console **(see illustration)**.

6 On automatic transmission models, undo the retaining screw and remove the knob from the top of the gear selector lever. Carefully prise the selector lever surround out from the top of the centre console and remove it from the vehicle.

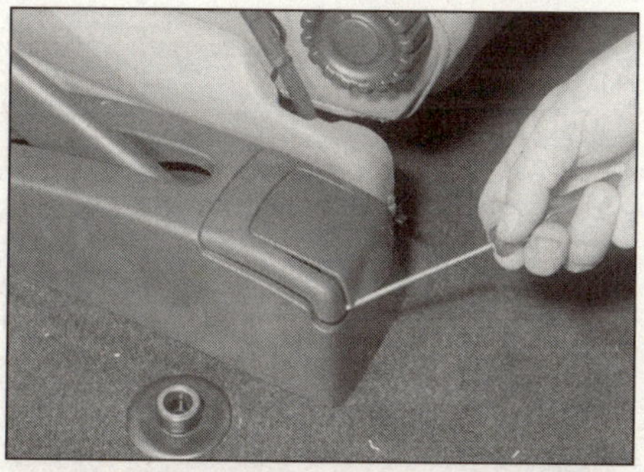

29.3 On models with manual rear windows unclip the ashtray from the rear of the centre console rear section

29.4a Slacken and remove the retaining nut (arrowed) . . .

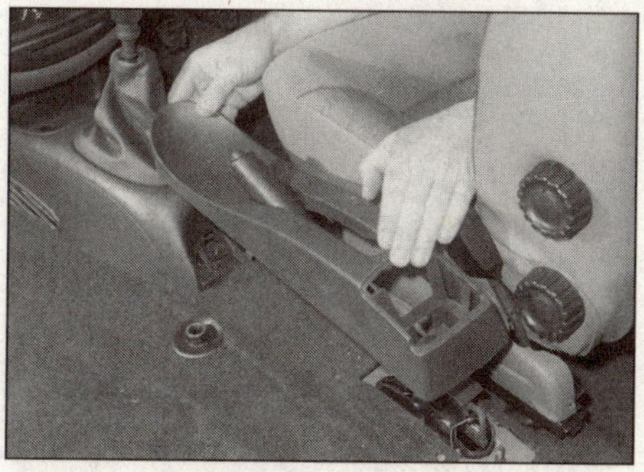

29.4b . . . then lift the centre console rear section off the handbrake lever

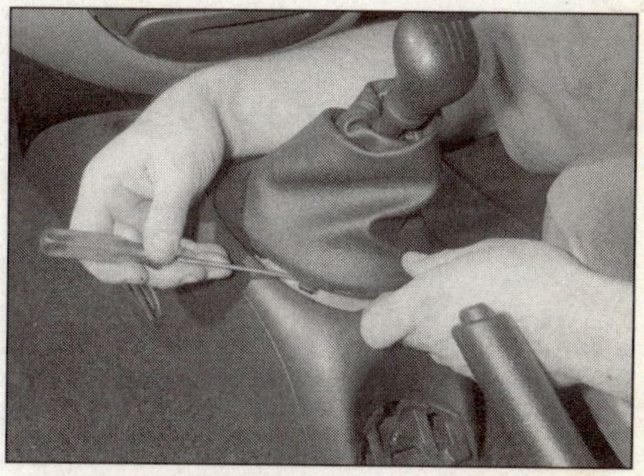

29.5 On manual transmission models unclip the gearchange lever gaiter

7 On all models, slacken and remove the retaining screws then remove the front section of the centre console from the vehicle, disconnecting its wiring connectors (where necessary) as they become accessible **(see illustrations)**.

Refitting

8 Refitting is the reverse of removal, ensuring all wiring is correctly routed and all fasteners are securely tightened.

30 Facia panel assembly - removal and refitting

Removal

HAYNES HiNT ***Attach an identification label to each wiring connector as it is disconnected. The labels can then be used on refitting to help ensure that all wiring is correctly routed through the relevant facia apertures.***

1 Remove the battery (refer to Chapter 5).
2 Remove the front seats as described in Section 25 then prise out the seat mounting spacers from the floor **(see illustration)**.
3 Remove the steering column as described in Chapter 10.
4 Remove the centre console as described in Section 29.

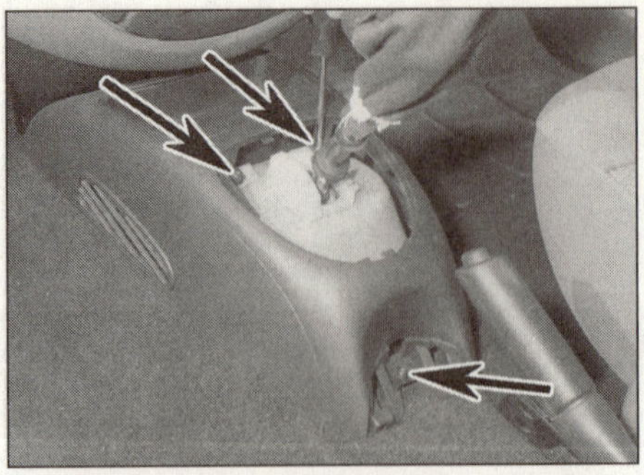

29.7a Slacken and remove the retaining screws (arrowed) . . .

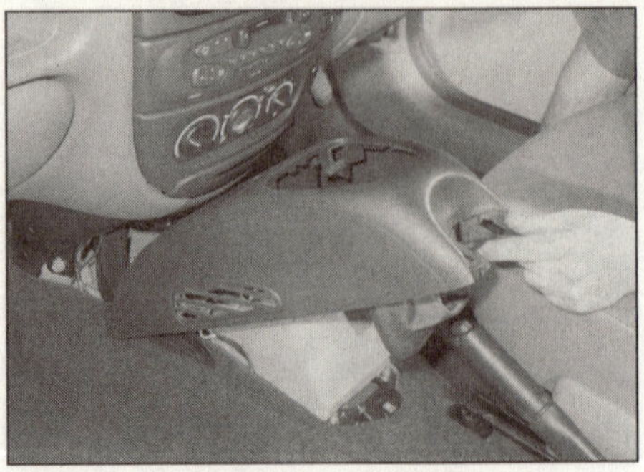

29.7b . . . and remove the front section of the centre console

5 Working as described in Chapter 12, remove the following components.

a) Driver's side switch panel.
b) Instrument panel.
c) Cigarette lighter.
d) Multi-function display.
e) Windscreen wiper arms.

6 Remove the heater control unit as described in Chapter 3.
7 Remove the lower retaining clips then release the upper clips and remove the centre console trim cover from the base of the facia panel. Remove the fastener and unclip the heater duct from the base of the facia **(see illustrations)**.
8 Peel off the rubber sealing strip from the top of the engine compartment bulkhead.
9 Slacken and remove the retaining screws then carefully unclip the left-hand vent panel from the base of the windscreen and remove it from the vehicle.
10 Working in the left-hand side of the engine compartment, remove the cover then slide out the front panel from the fuse/relay box. Unclip the fuse/relay panel then release the retaining clips and disconnect the main (engine wiring harness) wiring connectors. Remove the retaining clips and screws and remove the front and rear sections of the left-hand wheelarch liner. Free the bulbholder from the direction indicator side repeater light then release the wiring harness from all the relevant retaining clips so that it is free to be pulled through into the passenger compartment **(see illustrations)**. If necessary to improve access, jack up the front of the vehicle, support it on axle stands and remove the front wheel.
11 Working on the right-hand side of the engine compartment, release the large fuses

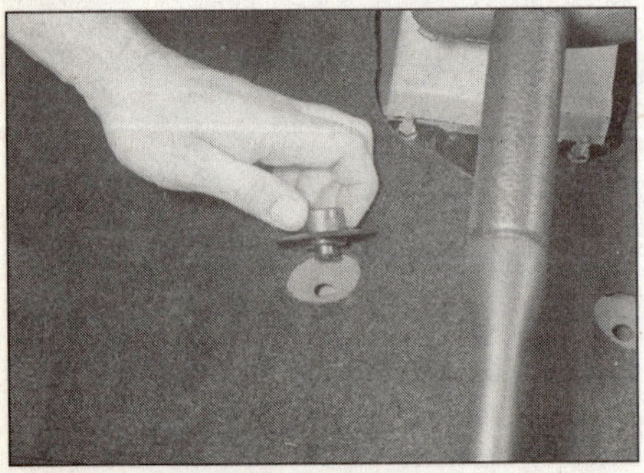

30.2 Remove the front seats and lift out the seat spacers

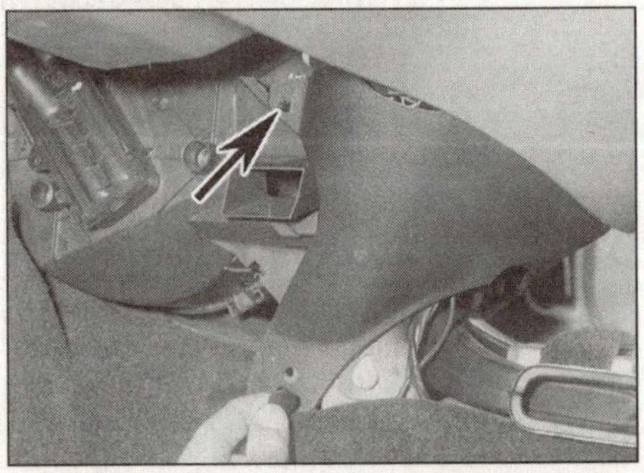

30.7a Remove the lower retaining clips then release the upper clips (arrowed) . . .

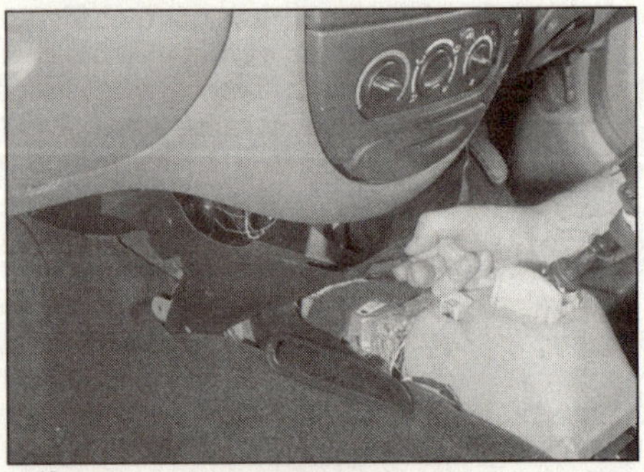

30.7b . . . and remove the centre console trim cover from the facia

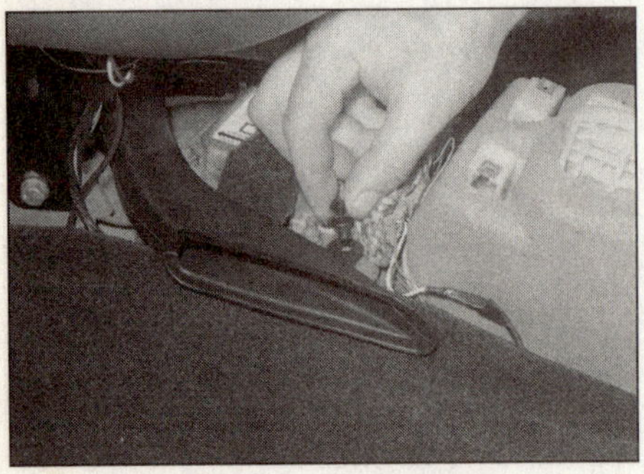

30.7c Remove the fastener . . .

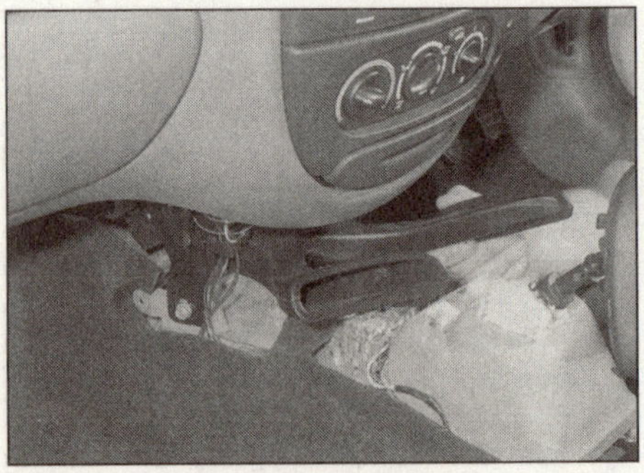

30.7d . . . and unclip the heater duct from the base of the facia

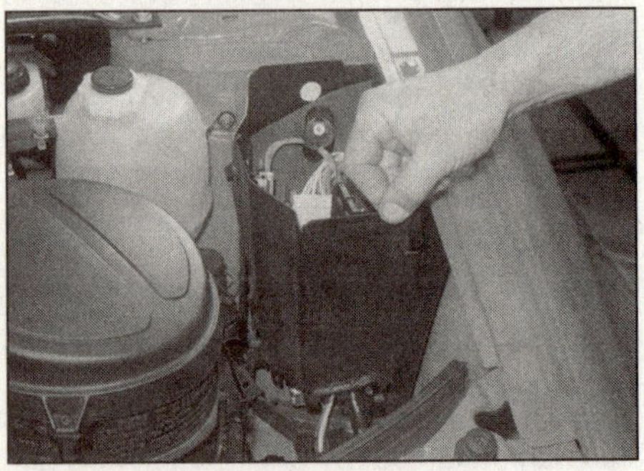

30.10a Remove the fuse/relay box lid then unclip the front panel . . .

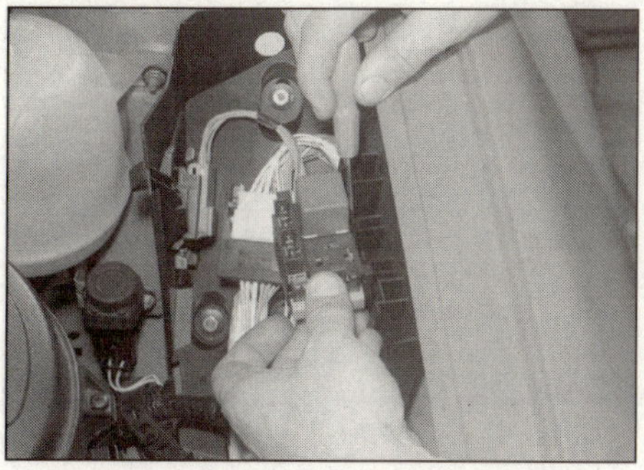

30.10b . . . and release the fuse/relay panel from the box

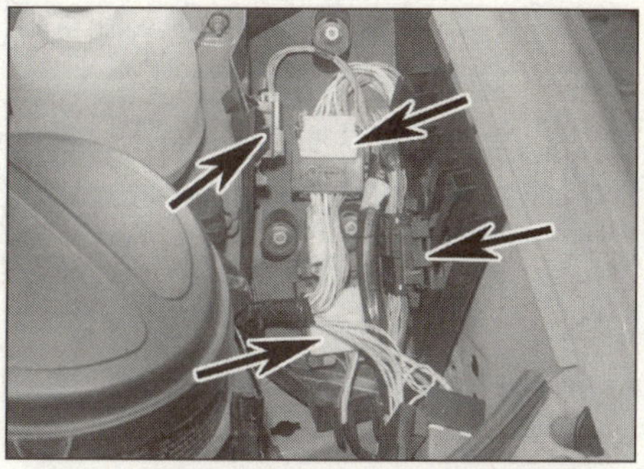

30.10c Disconnect the wiring connectors (arrowed) and free them from their retaining clips

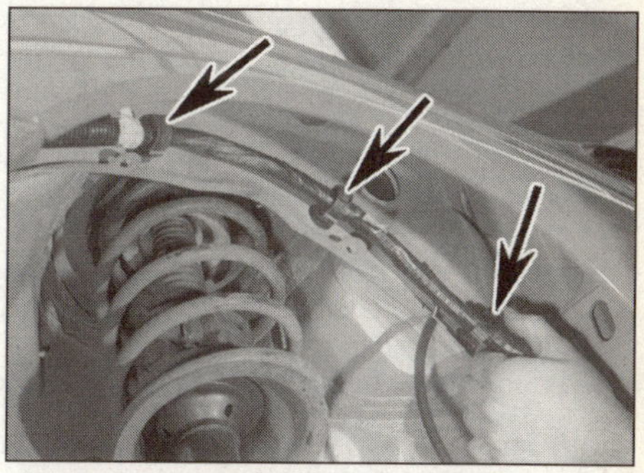

30.10d Free the wiring harness from its retaining clips (arrowed) underneath the wing . . .

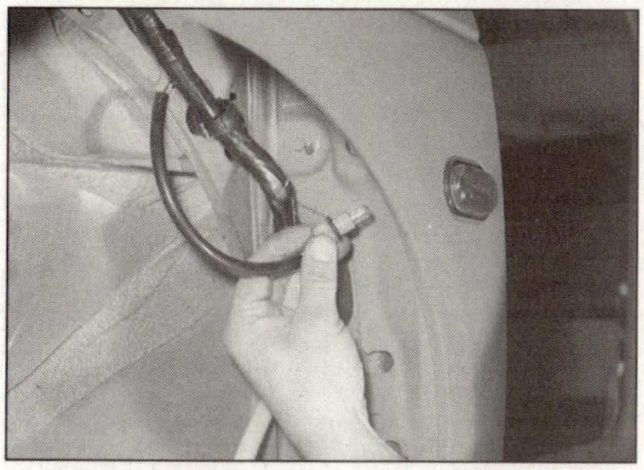

30.10e . . . then detach the bulbholder from the side repeater light . . .

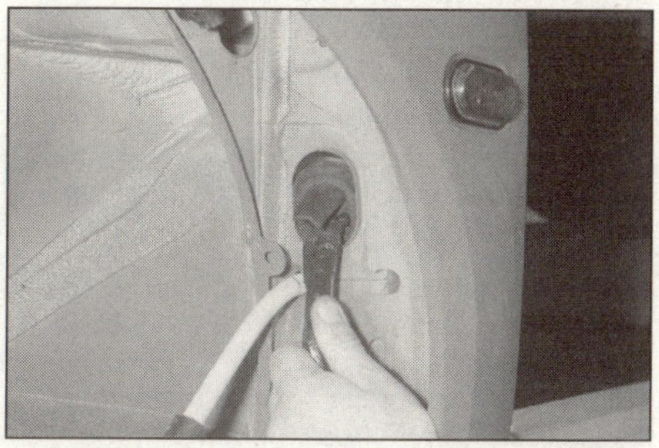

30.10f . . . and free the wiring grommet from the body so the harness is free to be removed with the facia assembly

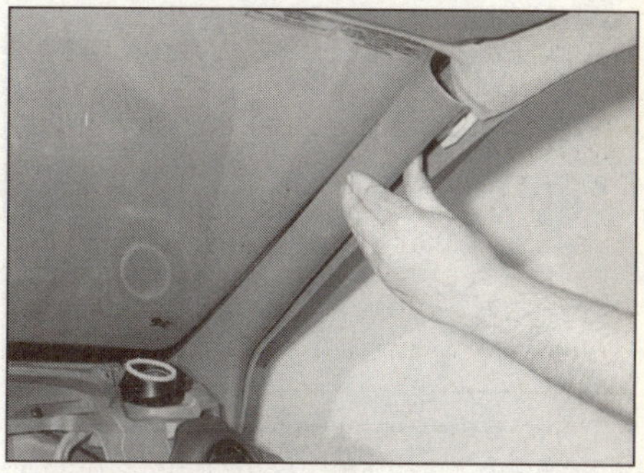

30.12 Unclip the windscreen pillar trim panels and remove them from the vehicle

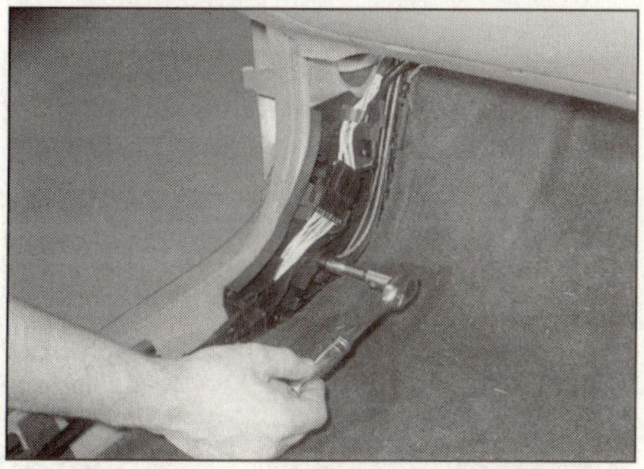

30.14a Unscrew the earth bolts . . .

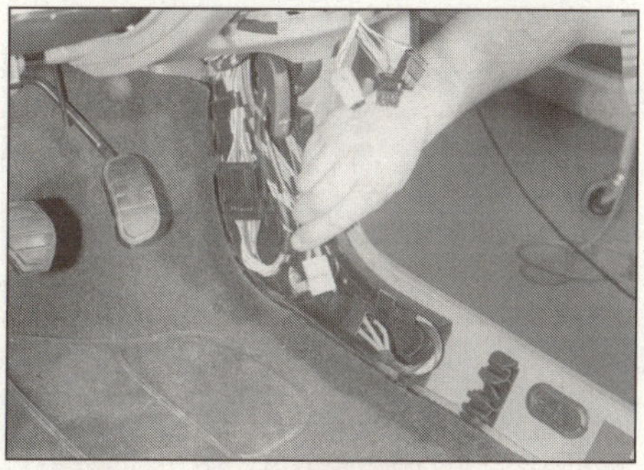

30.14b . . . and disconnect the wiring connectors situated behind the left- and right-hand front door pillar/sill trim panels

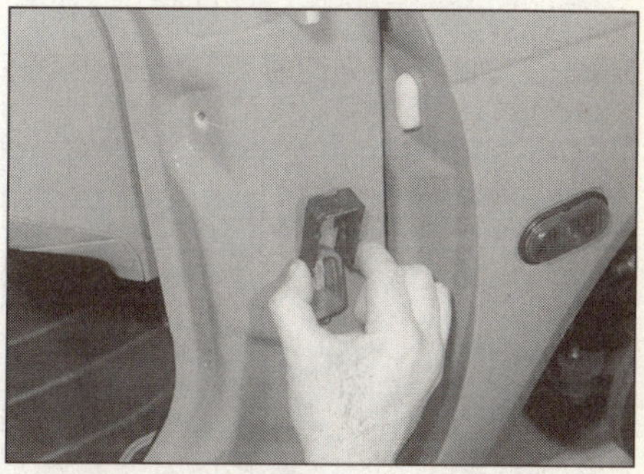

30.15a Slide off the retaining clip . . .

from their retaining clips. Disconnect the wiring connectors from the wiper motor and heater blower motor and release the harness from its retaining clips so that it is free to be pulled through into the passenger compartment. **Note:** *This is not necessary on all models; on some models the wiring harness is not connected to the facia and can be left in position.*

12 Peel the front door sealing strips away from the windscreen pillar trim panels then unclip (starting at the top and working downwards) both panels and remove them from the vehicle **(see illustration)**.

13 On Hatchback models remove the left- and right-hand pillar trim panels as described in paragraphs 3 to 8 of Section 27. On Coupe models remove the rear seat cushion as described in Section 25 then remove the left- and right-hand lower sections of rear seat side trim panels as described in paragraphs 13 to 16 of Section 27.

14 On all models, undo the retaining screws and remove the left- and right-hand door pillar/front sill trim panels (see Section 27, paragraphs 3 to 9). Unscrew the retaining screws and free the earth leads from the left- and right-hand sills then disconnect the various facia wiring connectors and free the wiring harness from all its retaining clips so that it is free to be removed with the facia **(see illustrations)**.

15 Release the retaining clips and disconnect the left- and right-hand front door wiring connectors. Slide the retaining clips fully off the connector then undo the screws and free each door wiring connector from the pillar **(see illustrations)**.

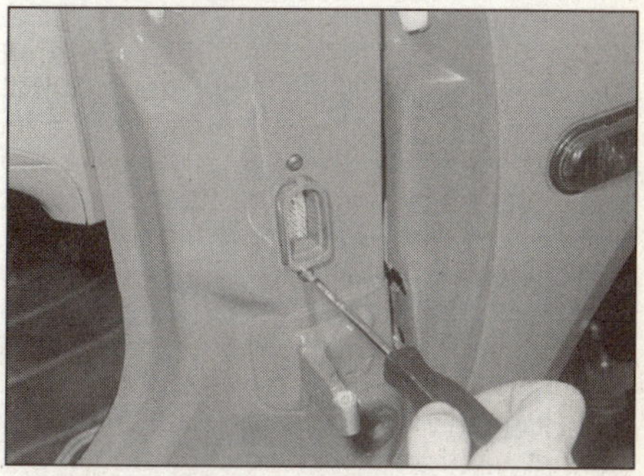

30.15b . . . then undo the screws and free the front door wiring connectors from the pillars (shown with doors removed)

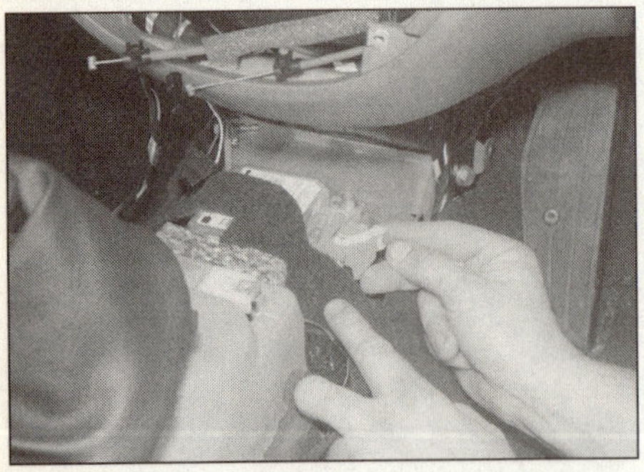

30.16a Disconnect the wiring connector from the airbag/seat belt tensioning control unit . . .

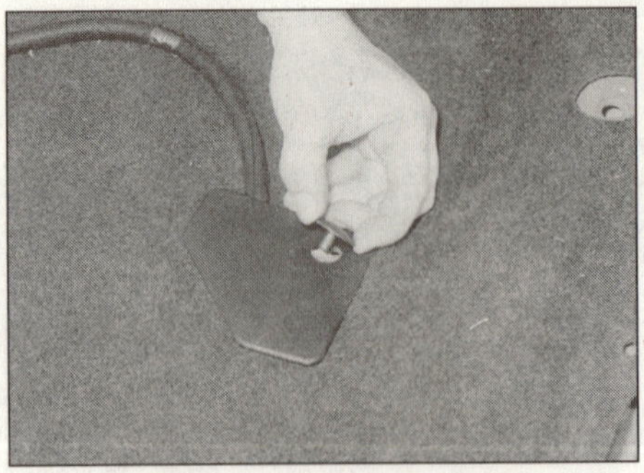

30.16b . . . then remove the retaining clip . . .

30.16c . . . and free the seat wiring from its pivot assembly . . .

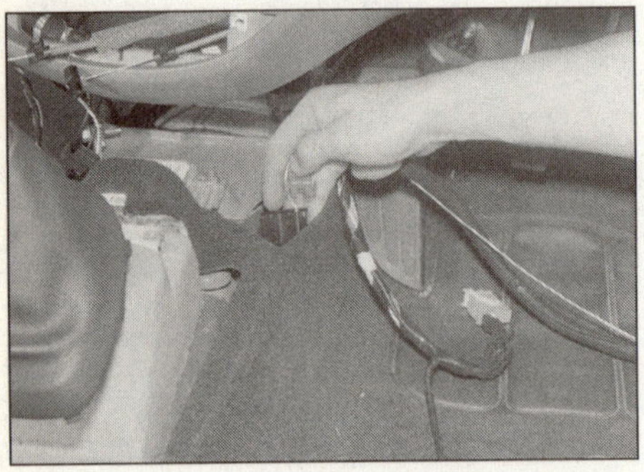

30.16d . . . so that it is free to be withdrawn from under the carpet and be removed with the facia

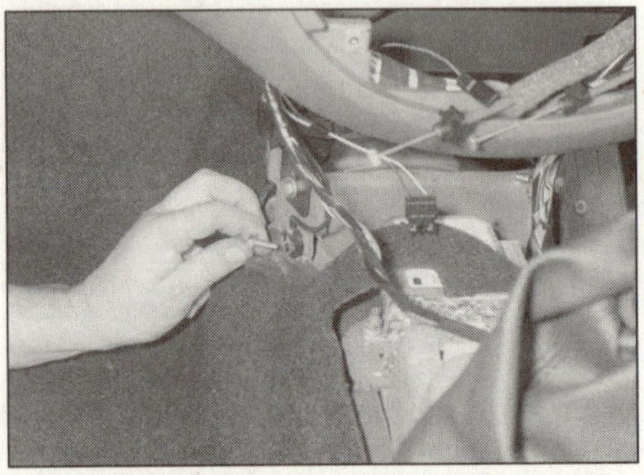

30.17a Undo the earth lead bolt from the left-hand side of the facia . . .

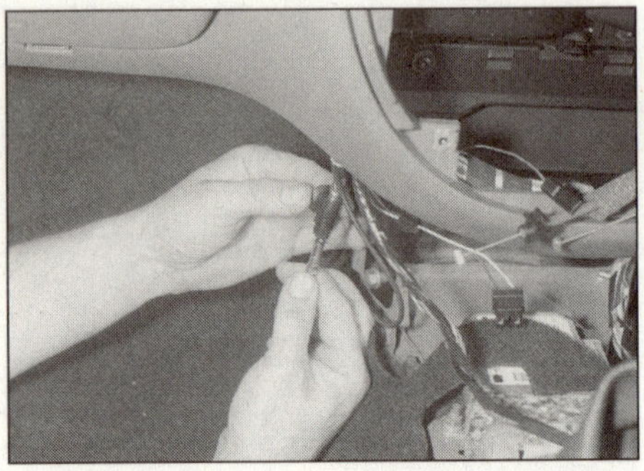

30.17b . . . and disconnect the aerial lead connector

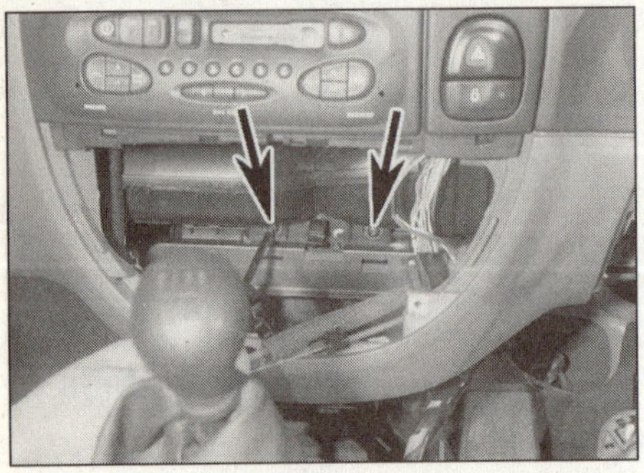

30.18 Slacken and remove the retaining screws (arrowed) which are located in the heater control panel aperture

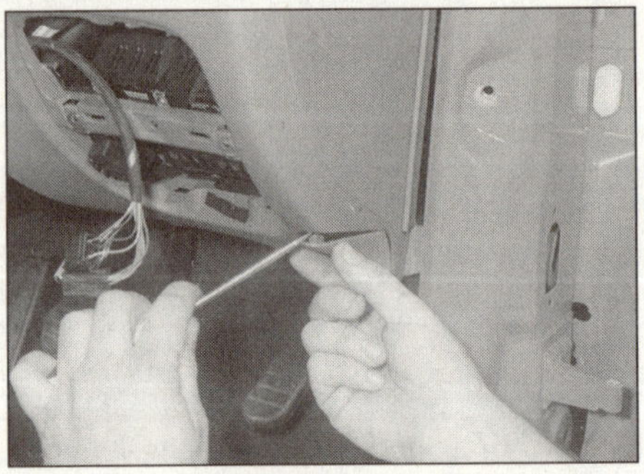

30.19a Unclip the trim covers from the left- and right-hand sides of the facia . . .

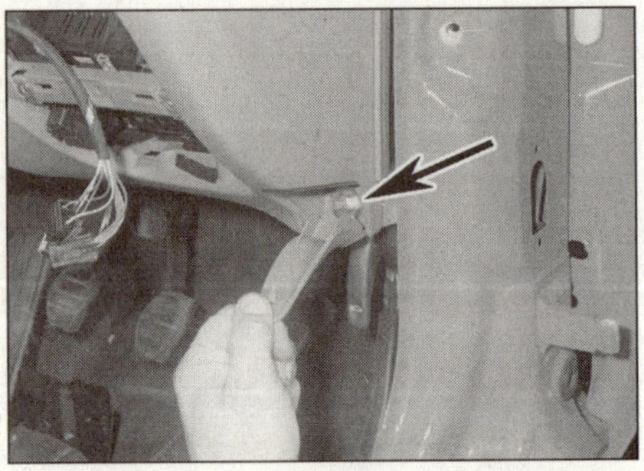

30.19b . . . then slacken and remove the facia lower mounting nuts (arrowed) . . .

16 Refer to the warnings in Section 26, then release the retaining clip and disconnect the wiring connector from the airbag/seat belt tensioning control unit. Remove the fasteners securing the seat wiring loom covers in position then unhook the retaining clips and free both the left- and right-hand front seat wiring from their pivot clips. Lift the carpet and release both left- and right-hand harnesses from their retaining clips so they can be pulled through and are free to be withdrawn with the facia **(see illustrations)**.

17 Slacken and remove the earth bolt from the left-hand side of the centre of the facia and disconnect the aerial lead connector **(see illustrations)**.

18 Slacken and remove the two facia retaining screws which are situated in the heater control panel aperture **(see illustration)**.

19 Carefully prise off the trim covers from the left- and right-hand bottom corners of the facia to gain access to the lower mounting nuts **(see illustrations)**.

20 Slacken and remove the upper and lower mounting nuts from the left- and right-hand ends of the facia **(see illustration)**.

21 With the aid of an assistant, carefully manoeuvre the facia assembly away from the bulkhead **(see illustration)**. As the facia is withdrawn, release the wiring grommets from the bulkhead, noting the correct routing of all the wiring, and release the wiring from all the relevant clips and ties.

30.20 . . . and the upper mounting nuts

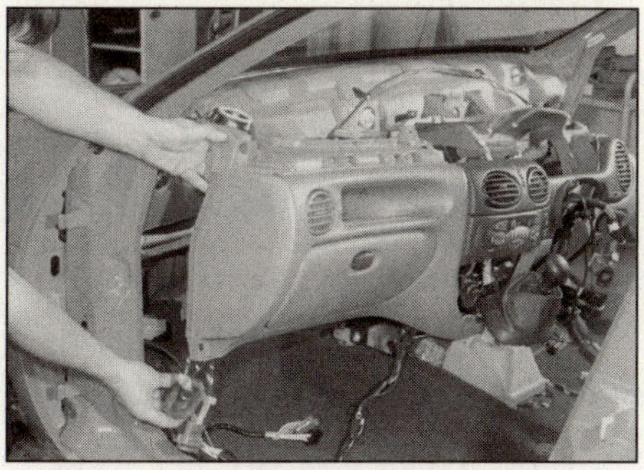

30.21 Removing the facia assembly from the vehicle

Refitting

22 Refitting is a reversal of the removal procedure. Prior to refitting the facia mounting nuts, ensure that all the necessary wiring connectors are fed through the relevant facia apertures and make sure the wiring grommets are correctly seated in the bulkhead. On completion, reconnect the battery and check that all the electrical components and switches function correctly.

Chapter 12
Body electrical system

Contents

Degrees of difficulty

Easy, suitable for novice with little experience	**Fairly easy,** suitable for beginner with some experience	**Fairly difficult,** suitable for competent DIY mechanic	**Difficult,** suitable for experienced DIY mechanic	**Very difficult,** suitable for expert DIY or professional

Specifications

System type	12-volt negative earth	
Bulbs	**Wattage**	
Exterior lights		
Headlight	55/60 (H4 type)	
Front foglight	55 (H1 type)	
Front sidelight	5	
Direction indicator	21	
Direction indicator side repeater	5	
Stop/tail	21/5	
Reversing light	21	
Rear foglight	21	
Rear sidelight	5	
High-level stoplight	5	
Number plate light	5	
Interior light		
Front courtesy lights	5	
Rear courtesy lights	5	
Luggage compartment light	5	
Torque wrench settings	**Nm**	**lbf ft**
Driver's airbag unit screws	5	4

1 General information and precautions

Warning: Before carrying out any work on the electrical system, read through the precautions given in "Safety first!" at the beginning of this manual and Chapter 5.

The electrical system is of the 12-volt negative earth type. Power for the lights and all electrical accessories is supplied by a lead/acid type battery which is charged by the alternator.

This Chapter covers repair and service procedures for the various electrical components not associated with the engine. Information on the battery, alternator and starter motor can be found in Chapter 5A.

It should be noted that prior to working on any component in the electrical system, the battery negative terminal should first be disconnected to prevent the possibility of electrical short circuits and/or fires.

2 Electrical fault finding - general information

Note: *Refer to the precautions given in 'Safety first!' and in Section 1 of this Chapter before starting work. The following tests relate to testing of the main electrical circuits, and should not be used to test delicate electronic circuits (such as anti-lock braking systems), particularly where an electronic control module (ECU) is used.*

General

A typical electrical circuit consists of an electrical component, any switches, relays, motors, fuses, fusible links or circuit breakers related to that component, and the wiring and connectors which link the component to both the battery and the chassis. To help to pinpoint a problem in an electrical circuit, wiring diagrams are included at the end of this Manual.

Before attempting to diagnose an electrical fault, first study the appropriate wiring diagram to obtain a complete understanding of the components included in the particular circuit concerned. The possible sources of a fault can be narrowed down by noting if other components related to the circuit are operating properly. If several components or circuits fail at one time, the problem is likely to be related to a shared fuse or earth connection.

Electrical problems usually stem from simple causes, such as loose or corroded connections, a faulty earth connection, a blown fuse, a melted fusible link, or a faulty relay (refer to Section 3 for details of testing relays). Visually inspect the condition of all fuses, wires and connections in a problem circuit before testing the components. Use the wiring diagrams to determine which terminal connections will need to be checked in order to pinpoint the trouble spot.

The basic tools required for electrical fault-finding include a circuit tester or voltmeter (a 12-volt bulb with a set of test leads can also be used for certain tests); a self-powered test light (sometimes known as a continuity tester); an ohmmeter (to measure resistance); a battery and set of test leads; and a jumper wire, preferably with a circuit breaker or fuse incorporated, which can be used to bypass suspect wires or electrical components. Before attempting to locate a problem with test instruments, use the wiring diagram to determine where to make the connections.

To find the source of an intermittent wiring fault (usually due to a poor or dirty connection, or damaged wiring insulation), a 'wiggle' test can be performed on the wiring. This involves wiggling the wiring by hand to see if the fault occurs as the wiring is moved. It should be possible to narrow down the source of the fault to a particular section of wiring. This method of testing can be used in conjunction with any of the tests described in the following sub-Sections.

Apart from problems due to poor connections, two basic types of fault can occur in an electrical circuit - open circuit, or short circuit.

Open circuit faults are caused by a break somewhere in the circuit, which prevents current from flowing. An open circuit fault will prevent a component from working, but will not cause the relevant circuit fuse to blow.

Short circuit faults are caused by a `short' somewhere in the circuit, which allows the current flowing in the circuit to 'escape' along an alternative route, usually to earth. Short circuit faults are normally caused by a breakdown in wiring insulation, which allows a feed wire to touch either another wire, or an earthed component such as the bodyshell. A short circuit fault will normally cause the relevant circuit fuse to blow.

Finding an open circuit

To check for an open circuit, connect one lead of a circuit tester or voltmeter to either the negative battery terminal or a known good earth.

Connect the other lead to a connector in the circuit being tested, preferably nearest to the battery or fuse.

Switch on the circuit, bearing in mind that some circuits are live only when the ignition switch is moved to a particular position.

If voltage is present (indicated either by the tester bulb lighting or a voltmeter reading, as applicable), this means that the section of the circuit between the relevant connector and the battery is problem-free.

Continue to check the remainder of the circuit in the same fashion.

When a point is reached at which no voltage is present, the problem must lie between that point and the previous test point with voltage. Most problems can be traced to a broken, corroded or loose connection.

Finding a short circuit

To check for a short circuit, first disconnect the load(s) from the circuit (loads are the components which draw current from a circuit, such as bulbs, motors, heating elements, etc).

Remove the relevant fuse from the circuit, and connect a circuit tester or voltmeter to the fuse connections.

Switch on the circuit, bearing in mind that some circuits are live only when the ignition switch is moved to a particular position.

If voltage is present (indicated either by the tester bulb lighting or a voltmeter reading, as applicable), this means that there is a short circuit.

If no voltage is present, but the fuse still blows with the load(s) connected, this indicates an internal fault in the load(s).

Finding an earth fault

The battery negative terminal is connected to 'earth'- the metal of the engine/ transmission and the car body - and most systems are wired so that they only receive a positive feed, the current returning through the metal of the car body. This means that the component mounting and the body form part of that circuit. Loose or corroded mountings can therefore cause a range of electrical faults, ranging from total failure of a circuit, to a puzzling partial fault. In particular, lights may shine dimly (especially when another circuit sharing the same earth point is in operation), motors (eg. wiper motors or the radiator cooling fan motor) may run slowly, and the operation of one circuit may have an apparently unrelated effect on another. Note that on many vehicles, earth straps are used between certain components, such as the engine/transmission and the body, usually where there is no metal-to-metal contact between components due to flexible rubber mountings, etc.

To check whether a component is properly earthed, disconnect the battery and connect one lead of an ohmmeter to a known good earth point. Connect the other lead to the wire or earth connection being tested. The resistance reading should be zero; if not, check the connection as follows.

If an earth connection is thought to be faulty, dismantle the connection and clean back to bare metal both the bodyshell and the wire terminal or the component earth connection mating surface. Be careful to remove all traces of dirt and corrosion, then use a knife to trim away any paint, so that a clean metal-to-metal joint is made. On reassembly, tighten the joint fasteners securely; if a wire terminal is being refitted, use serrated washers between the terminal and the bodyshell to ensure a clean and secure connection. When the connection is remade, prevent the onset of corrosion in the future by applying a coat of petroleum jelly or silicone-based grease or by spraying on (at regular intervals) a proprietary ignition sealer or a water dispersant lubricant.

3 Fuses and relays - general information

Fuses

1 The majority of fuses are located in the fusebox situated on the driver's side of the facia with a few additional fuses being located in the fuse/relay box on the left-hand side of the engine compartment. The larger, higher-rated fuses can be found in the right-hand rear corner of the engine compartment, next to the battery.

2 To gain access to the main fusebox, open up the cover. To gain access to the fuses in the engine compartment fuse/relay box, simply unclip the cover **(see illustrations)**. The larger, higher-rated fuses can be accessed by undoing the retaining screw and removing the battery access cover from the right-hand side of the windscreen vent panel.

3 A list of the circuits each fuse protects is given on the relevant fusebox cover.

4 To remove a fuse, first switch off the circuit concerned (or the ignition), then pull the fuse out of its terminals. The wire within the fuse should be visible; if the fuse is blown it will be broken or melted.

5 Always renew a fuse with one of an identical rating; never use a fuse with a different rating from the original or substitute anything else. Never renew a fuse more than once without tracing the source of the trouble. The fuse rating is stamped on top of the fuse; note that the fuses are also colour-coded for easy recognition.

6 If a new fuse blows immediately, find the cause before renewing it again; a short to earth as a result of faulty insulation is most likely. Where a fuse protects more than one circuit, try to isolate the defect by switching on each circuit in turn (if possible) until the fuse blows again. Always carry a supply of spare fuses of each relevant rating on the vehicle, a spare of each rating should be clipped into the base of the fusebox.

Relays

7 The majority of relays are either located in the relay box which is attached to the fusebox located behind the driver's side of the facia. Other relays can be found in the fuse/relay box in the engine compartment.

8 To gain access to the relay box behind the facia, remove the driver's side facia switch panel as described in Section 4. Slacken and remove the fusebox retaining screw then reach up behind the facia and unclip the plastic protector plate from the base of the fusebox. Release the retaining clips and disconnect the four wiring connectors from the base of the fusebox. Release the fusebox from its side retaining clips by pushing it towards the front of the vehicle then carefully manoeuvre the assembly out from underneath the facia. Release the retaining clips and remove the cover to gain access to the various relays **(see illustrations)**. Refitting is the reverse of removal.

9 If a circuit or system controlled by a relay develops a fault and the relay is suspect, operate the system; if the relay is functioning it should be possible to hear it click as it is energised. If this is the case the fault lies with the components or wiring of the system. If the relay is not being energised then either the relay is not receiving a main supply or a switching voltage or the relay itself is faulty. Testing is by the substitution of a known good unit but be careful; some relays are identical in appearance and in operation, others look similar but perform different functions.

10 To renew a relay first ensure that the ignition switch is off. The relay and then simply be pulled out from the socket and the new relay pressed in.

4 Switches - removal and refitting

Note: *Disconnect the battery negative lead before removing any switch, and reconnect the lead after refitting the switch.*

Ignition switch/ steering column lock

1 Insert the key into the ignition switch so the steering wheel is free to turn.

2 Position the steering wheel so that access can be gained to the radio control switch

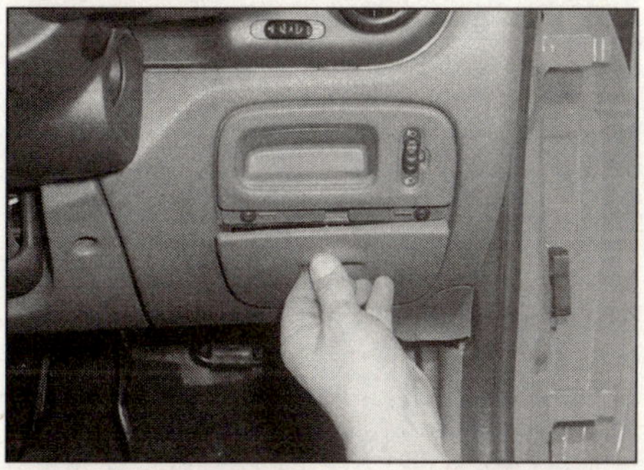

3.2a Most fuses are located behind the cover on the driver's side of the facia . . .

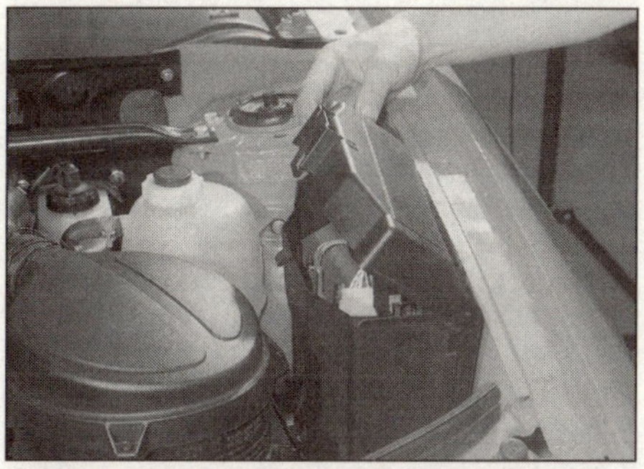

3.2b . . . with the others located in the fuse/relay box in the engine compartment

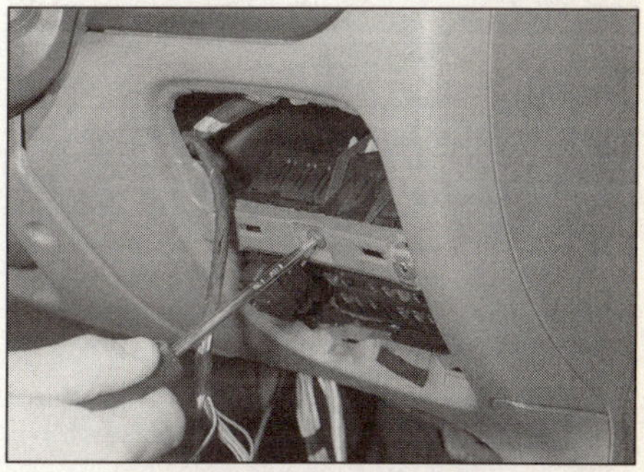

3.8a To gain access to the fusebox relays, undo the retaining screw . . .

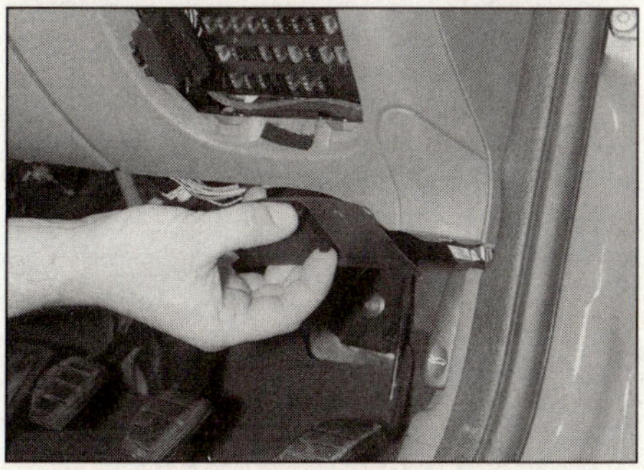

3.8b . . . then unclip the protective cover from the base of the unit

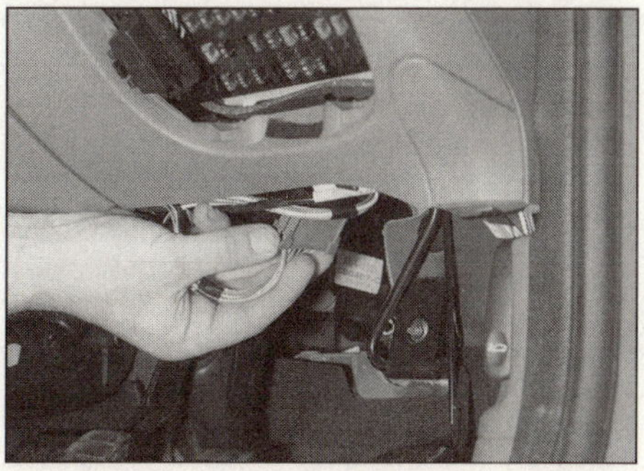

3.8c Disconnect the wiring connectors then unclip the fusebox . . .

3.8d . . . and remove the cover to gain access to the fuses

4.2 Unclip the cover and slacken the radio control switch screw a few turns

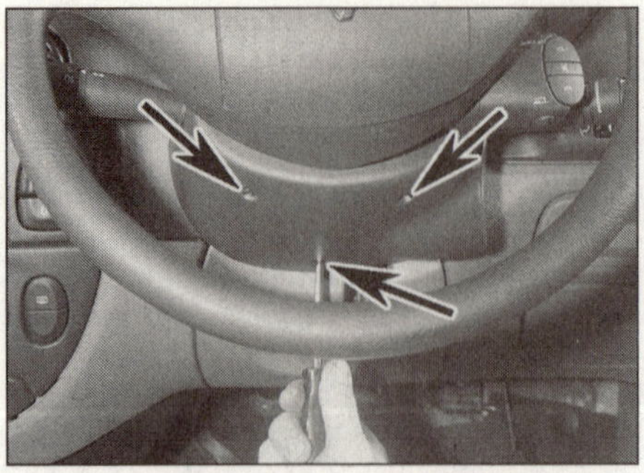
4.3a Undo the retaining screws (arrowed) . . .

4.3b . . . and remove the steering column lower shroud . . .

4.3c . . . then undo the screws . . .

4.3d . . . and lift off the upper shroud

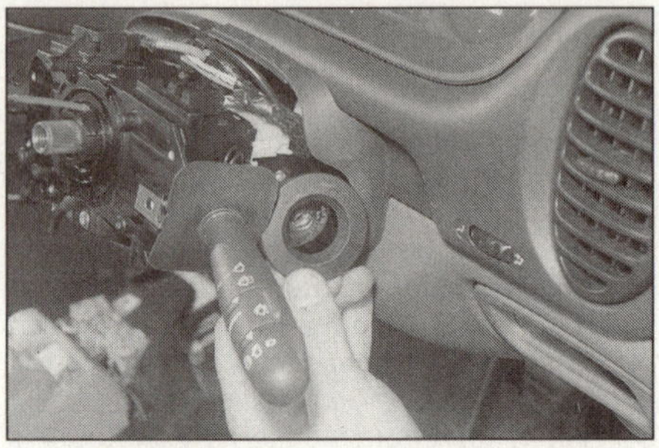
4.6a Remove the trim cover from the switch (shown with steering wheel removed for clarity) . . .

4.6b . . . then remove the switch retaining screw

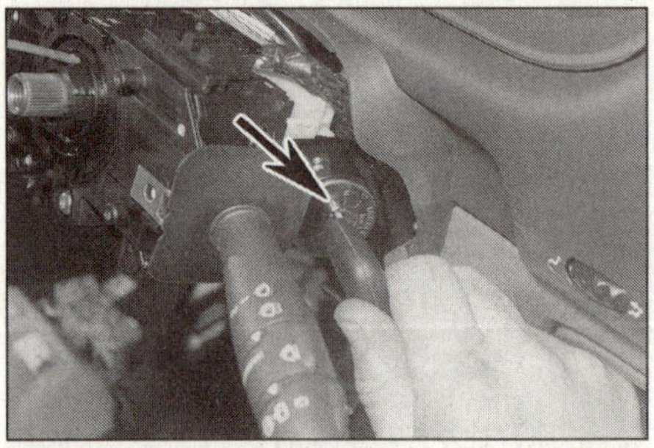
4.7a Insert the key and turn it to align with the mark (arrowed) midway between the "A" and "M" marks . . .

4.7b . . . then depress the retaining clip and slide the switch out of position

assembly. Unclip the cover from the inside of the switch and slacken the switch retaining screw by a few turns **(see illustration)**.

3 Slacken and remove the retaining screws and remove the lower half of the steering column shroud. Undo the two retaining screws, positioning the steering wheel as necessary, and remove the upper half of the steering column shroud **(see illustrations)**.

4 Undo the retaining screws and remove the steering column lower cover from the facia.

5 Trace the wiring back from the switch and disconnect it from the main harness.

6 Remove the plastic trim cover (where fitted) from around the switch and unscrew the switch retaining screw **(see illustrations)**.

7 Insert the ignition key and turn the key until it aligns with the mark halfway between the "A" and "M" (1st and 2nd) switch positions. Hold the key in this position then depress the retaining clip and slide the switch assembly out of position **(see illustrations)**.

8 Refitting is the reverse of removal, making sure the switch is clipped securely into position. Ensure that the wiring is correctly routed and retained by all the necessary clips.

Steering column combination switches

9 Remove the steering column shrouds as described in paragraphs 1 and 3.

10 To remove an individual switch assembly, disconnect the wiring then remove the retaining screws and slide the switch out of its mounting bracket **(see illustration)**.

11 To remove the complete switch and mounting bracket assembly it will first be necessary to remove the steering wheel (see Chapter 10); on models with a driver's airbag also remove the contact unit as described in Section 25. On all models, unscrew the radio control switch stalk mounting screw and position the stalk clear. Disconnect the wiring from the switches then slacken the clamp

4.10 Undo the retaining screws and slide the combination switch out of its mounting bracket

4.11 Removing the complete combination switch assembly

4.12 If the complete assembly has been removed, centralise the stalks with the shrouds and tighten the clamp screw

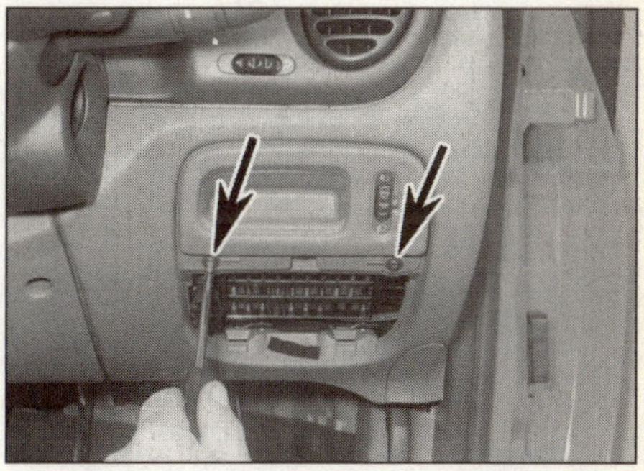

4.13 Remove the fusebox cover then undo the retaining screws (arrowed) . . .

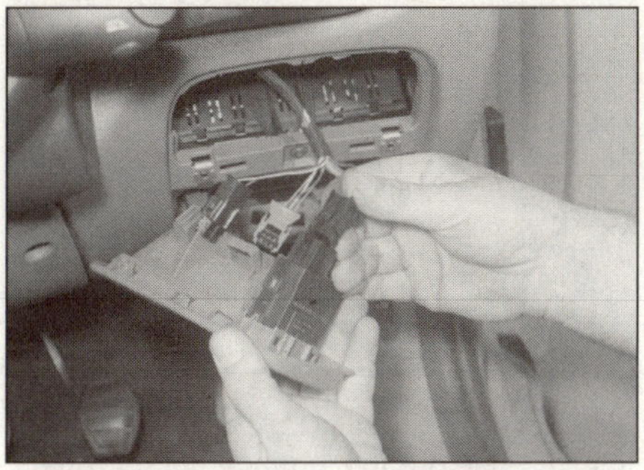

4.14a . . . and remove the switch panel, disconnecting its wiring connectors as they become accessible

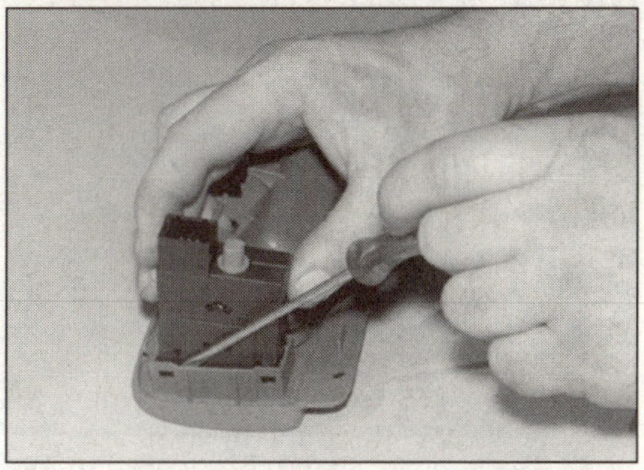

4.14b The switch can then be unclipped and removed from the panel

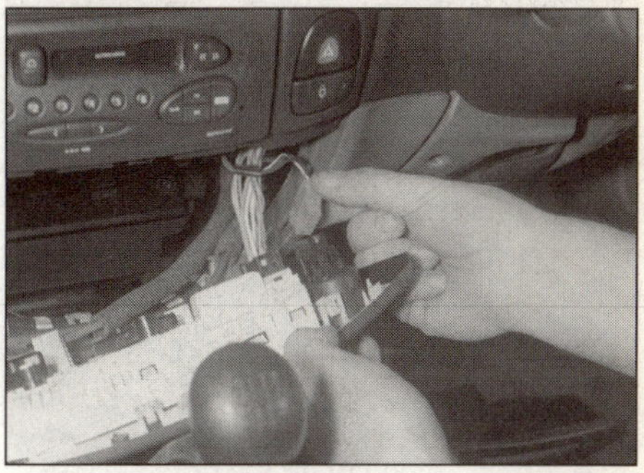

4.17a Withdraw the heater control panel then disconnect the wiring connector . . .

screw a few turns and release the switch clamp by tapping the screw lightly. The complete switch assembly can then be lifted off from the top of the steering column **(see illustration)**.

12 Refitting is the reverse of removal ensuring all wiring connectors are securely reconnected. If the complete switch assembly has been removed, prior to refitting the steering wheel, ensure that the switch stalks are correctly positioned in relation to the shrouds; if necessary slacken the mounting clamp screw and reposition the switch correctly before securely retightening the screw **(see illustration)**.

Driver's side facia switches

13 Remove the fusebox cover from the driver's side of the facia then slacken and remove the switch panel retaining screws **(see illustration)**.

14 Remove the switch panel from the facia, disconnecting its wiring connectors as they become accessible. The relevant switch can then be unclipped and removed from the panel **(see illustrations)**.

15 Refitting is the reverse of removal.

Centre facia switches

16 Referring to Chapter 3, remove the heater control panel from the facia noting that it is not necessary to remove the panel completely, only to slide it sufficiently out of position so

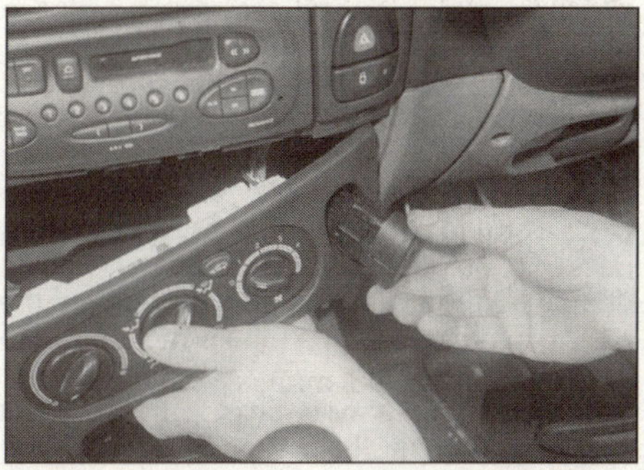

4.17b . . . and unclip the switch assembly

that access can be gained to the relevant switch.

17 Depress the retaining clips then push the relevant switch out of position and disconnect it from the wiring connector **(see illustrations)**.

18 Refitting is the reverse of removal.

Centre console front switches

19 Carefully unclip the switch panel and remove it from the front of the centre console, disconnecting its wiring connectors as they become accessible.

20 Undo the retaining screws/unclip the switch from the panel.

21 Refitting is the reverse of removal ensuring that the switch panel is clipped securely in position.

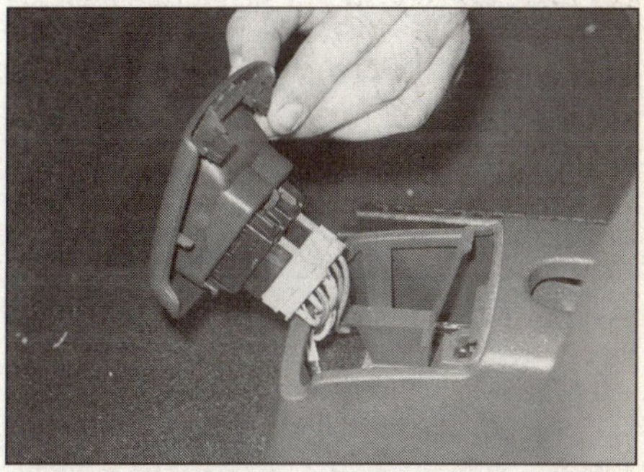

4.22 Unclip the rear window switch panel from the rear of the centre console and disconnect the wiring connectors

Centre console rear switches

22 Carefully unclip the switch panel from the rear section of the centre console and disconnect the wiring connectors **(see illustration)**. The wiring connector are colour-coded for identification; the left-hand one is green and the right-hand one grey.

23 Depress the clips and slide the switch(es) out of position noting their correct locations.

24 Refitting is the reverse of removal.

Door mounted switches

25 Remove the loudspeaker from the door as described in Section 20.

26 Undo the retaining screws and remove the storage compartment from the bottom of the

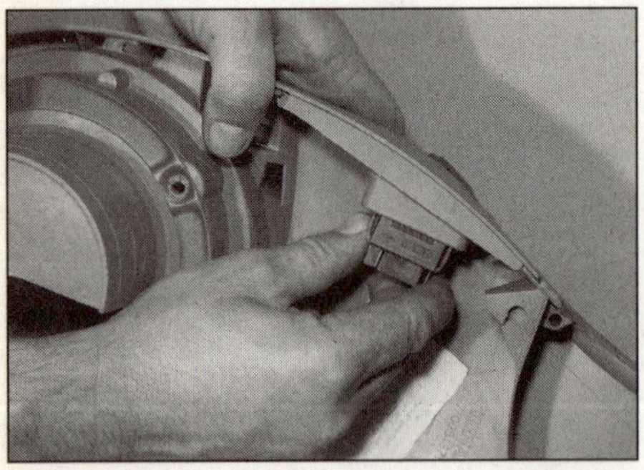

4.27 Depress the retaining clips and slide the window switch out from the storage compartment

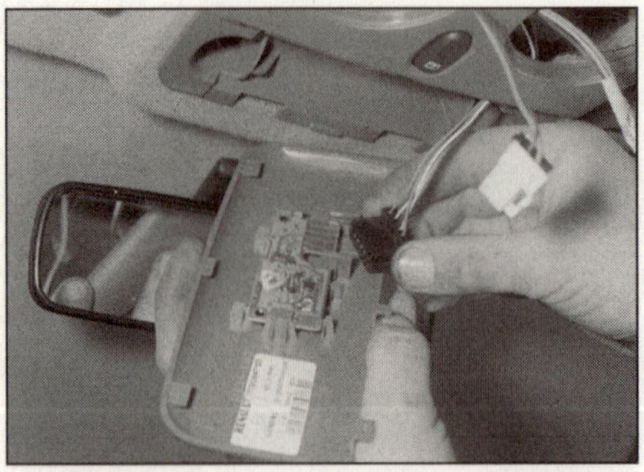

4.30 Slide the plastic cover off from the overhead console and disconnect its wiring connector

4.31a Undo the retaining screws (arrowed) . . .

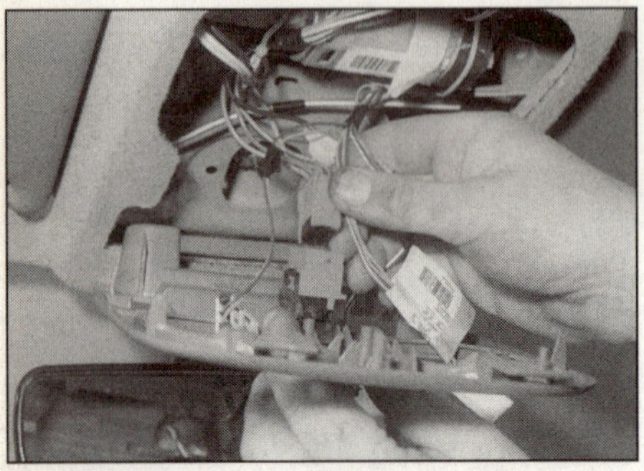

4.31b . . . then unclip the console and disconnect its wiring connectors

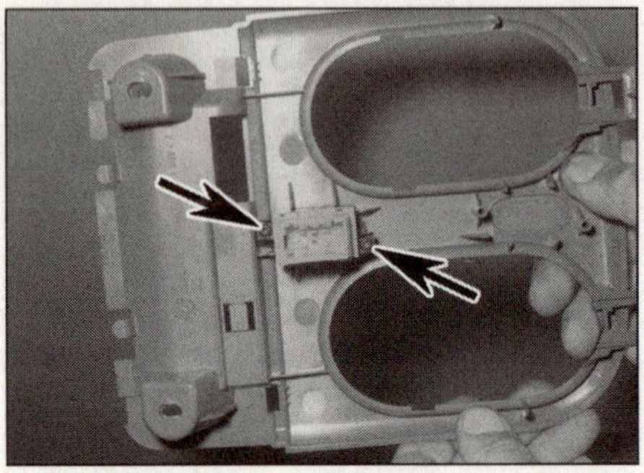

4.32 The sunroof switch is secured to the console by two screws (arrowed)

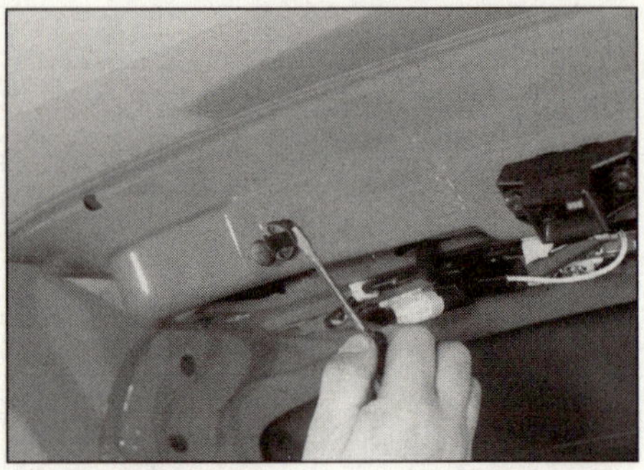

4.40a Prise out the luggage compartment light switch . . .

door in an upwards direction, disconnecting the wiring connectors as they become accessible.

27 Each switch can then be removed by depressing its retaining clips and sliding it out of position **(see illustration)**.

28 Refitting is the reverse of removal.

Overhead console switches

29 The light switches are integral with the courtesy light units (see Section 6). The sunroof switch can be removed as follows.

30 Slide off the plastic cover from the front of the overhead console, in a forwards direction **(see illustration)**.

31 Slacken and remove the retaining screws then remove the console assembly from the roof, disconnecting its wiring connectors as they become accessible **(see illustrations)**.

32 Slacken and remove the retaining screws and remove the switch from the console **(see illustration)**.

33 Refitting is the reverse of removal.

Seat switches

34 Carefully prise the switch panel out from the side of the seat and disconnect it from the wiring connector(s).

35 Reconnect the wiring connector and clip the switch back into position.

Stop-light switch

36 Refer to Chapter 9.

Courtesy light switch

37 Open the door and remove the rubber cover from the switch.

38 Using a small, flat-bladed screwdriver depress the switch retaining clips then carefully ease the switch out from the pillar. Disconnect the wiring connector and tie a piece of string to the wiring to prevent it falling back into the door pillar. If necessary, to avoid damaging the switch/paintwork, remove the pillar trim panel and push the switch out from behind.

39 Refitting is a reverse of the removal procedure. Refit the rubber cover to the switch prior to clipping it into the door pillar.

Luggage compartment light switch

40 Open up the tailgate/boot lid then carefully prise the switch out of position and disconnect its wiring connector. If necessary, to avoid damaging the switch/paintwork, remove the inner trim (where fitted) panel and push the switch out of position **(see illustrations)**.

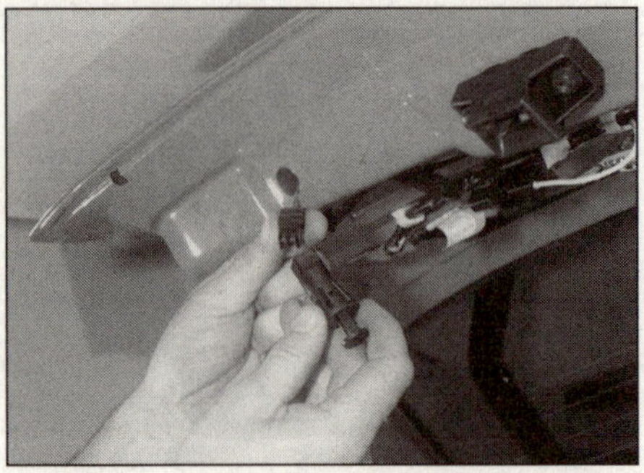

4.40b . . . and disconnect its wiring

41 Refitting is the reverse of removal.

5 Bulbs (exterior lights) - renewal

1 Whenever a bulb is renewed, note the following points.

a) Disconnect the battery negative lead before starting work.

b) Remember that if the light has just been in use the bulb may be extremely hot.

c) Always check the bulb contacts and holder, ensuring that there is clean metal-to-metal contact between the bulb and its live(s) and earth. Clean off any corrosion or dirt before fitting a new bulb.

d) Wherever bayonet-type bulbs are fitted (see Specifications) ensure that the live contact(s) bear firmly against the bulb contact.

e) Always ensure that the new bulb is of the correct rating and that it is completely clean before fitting it; this applies particularly to headlight/foglight bulbs (see below).

Headlight

2 Remove the access cover from the rear of the headlight unit; where a round cover is

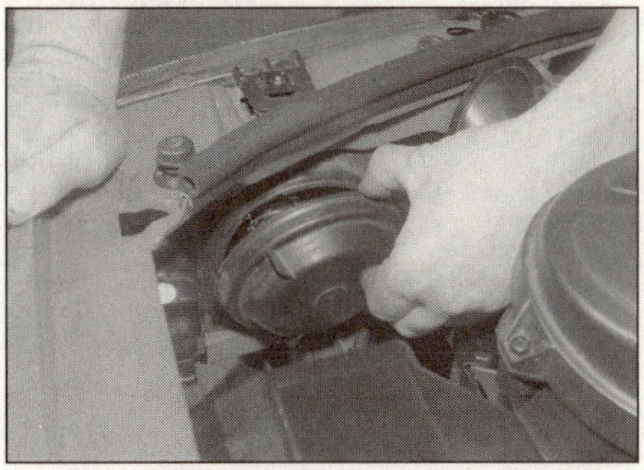

5.2 On models with a round access cover, twist the cover through 90° to release it from the headlight

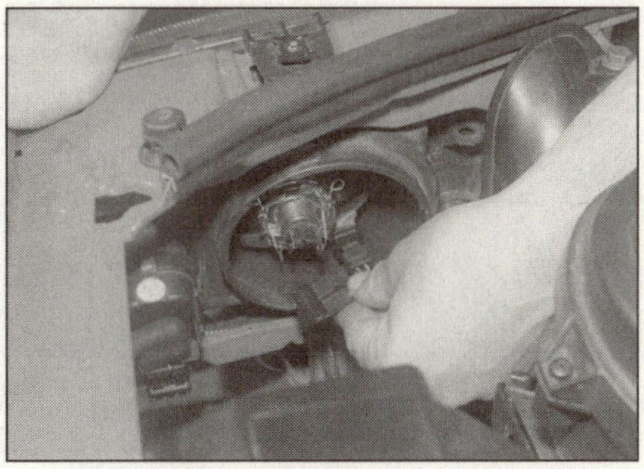

5.3 Disconnect the wiring connector . . .

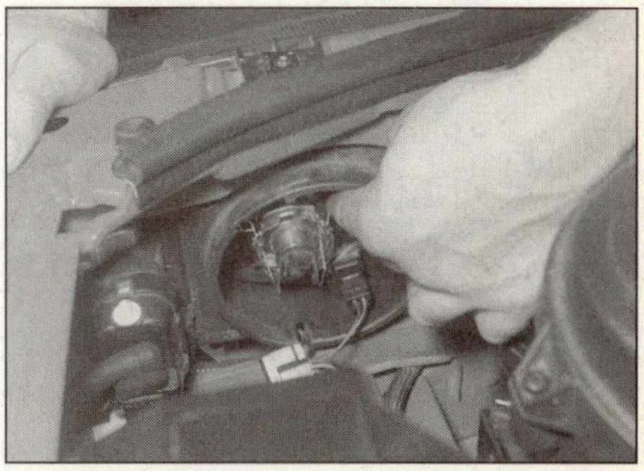

5.4a . . . then release the retaining clip . . .

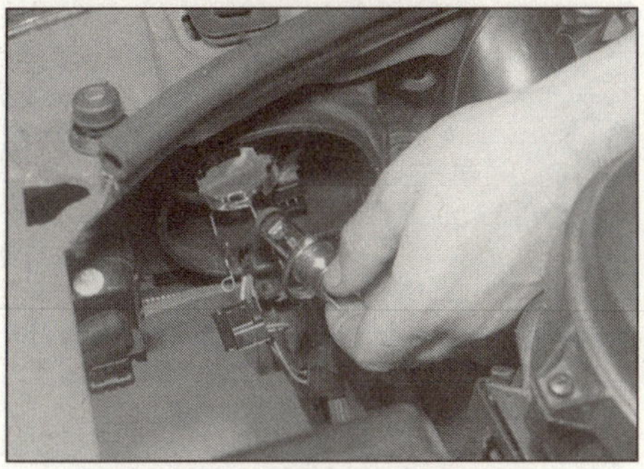

5.4b . . . and withdraw the bulb from the headlight

5.9a Free the sidelight bulbholder from the rear of the headlight . . .

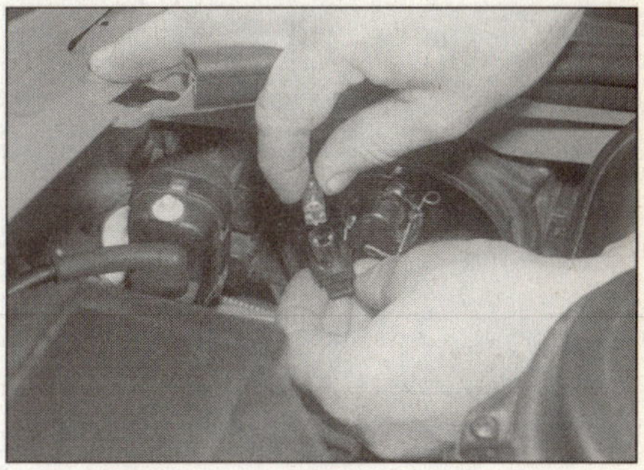

5.9b . . . and pull out the bulb

fitted twist the cover through 90° to release it and where a rectangular cover is fitted release the retaining clip **(see illustration)**.

3 Disconnect the wiring connector from the rear of the bulb **(see illustration)**.

4 Unhook and release the ends of the bulb retaining clip and release it from the rear of the light unit. Withdraw the bulb **(see illustrations)**.

5 When handling the new bulb, use a tissue or clean cloth to avoid touching the glass with the fingers; moisture and grease from the skin can cause blackening and rapid failure of this type of bulb. If the glass is accidentally touched, wipe it clean using methylated spirit.

6 Install the new bulb, ensuring that its locating tabs are correctly located in the light cutouts, and secure it in position with the retaining clip

7 Reconnect the wiring connector and refit the access cover, making sure it is securely refitted.

Front sidelight

8 Remove the access cover from the rear of the headlight unit (see paragraph 2).

9 Rotate the sidelight bulbholder and release it from the headlight unit. The bulb is of the capless (push-fit) type and can be removed by simply pulling it out of the bulbholder **(see illustrations)**.

10 Refitting is the reverse of the removal procedure making sure the access cover is securely refitted.

Front foglight

Hatchback and Saloon models

11 Remove the foglight as described in Section 7.

12 Rotate the access cover and remove it from the rear of the light unit. Unhook and release the ends of the bulb retaining clip and release it from the rear of the light unit. Withdraw the bulb.

13 When handling the new bulb, use a tissue or clean cloth to avoid touching the glass with the fingers; moisture and grease from the skin can cause blackening and rapid failure of this type of bulb. If the glass is accidentally touched, wipe it clean using methylated spirit.

14 Install the new bulb, ensuring that its locating tabs are correctly located in the light cutouts, and secure it in position with the retaining clip

15 Reconnect the wiring connector and refit the access cover, making sure it is securely refitted.

16 Refit the foglight as described in Section 7.

Coupe models

17 If necessary, to improve access, firmly apply the handbrake then jack up the front of the vehicle and support it on axle stands (see "*Jacking and vehicle support*").

18 Twist the cover anti-clockwise and release it from the rear of the light unit **(see illustration)**.

19 Disconnect the wiring connector then release the retaining clip and remove the bulb **(see illustration)**.

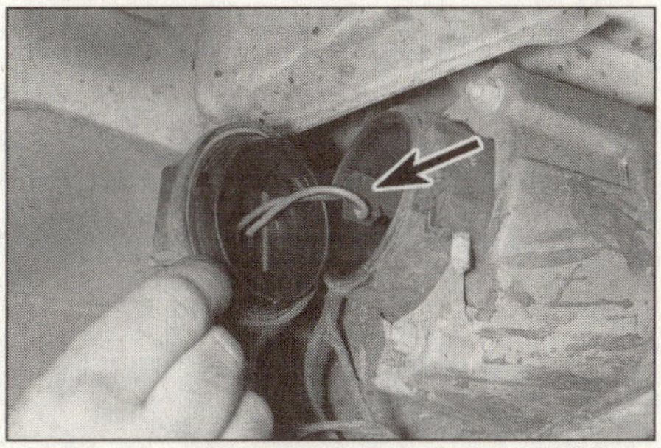

5.18 Release the cover from the rear of the foglight and disconnect the wiring connector (arrowed)

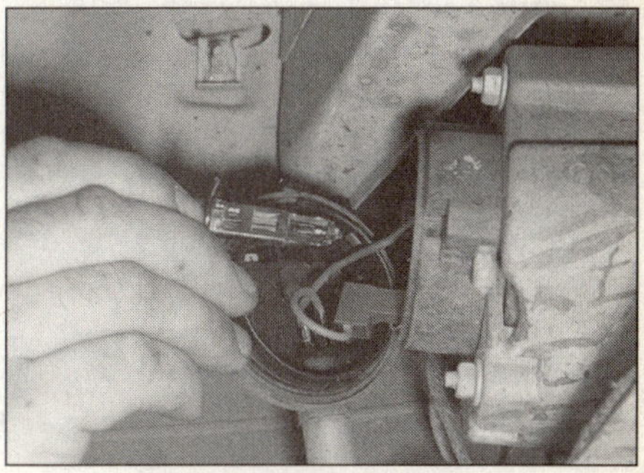

5.19 Release the retaining clip and remove the foglight bulb

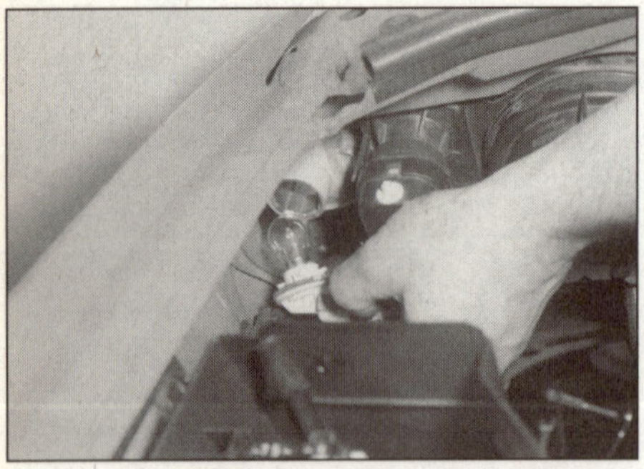

5.21a Rotate the bulbholder anti-clockwise to free it from the direction indicator light . . .

5.21b . . . then push the bulb in and twist it anti-clockwise to free it from the holder

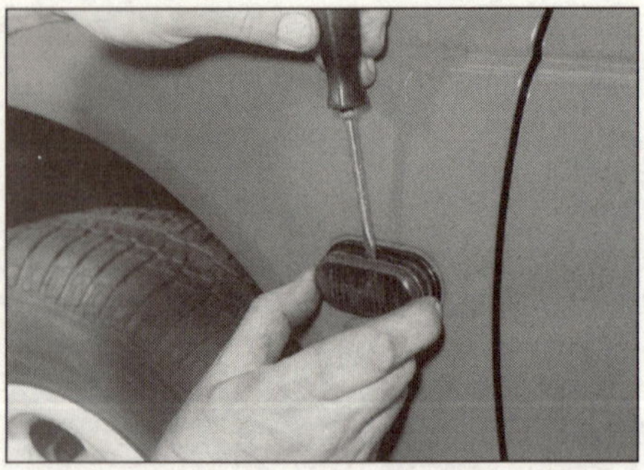

5.23 Carefully prise the side repeater light out from the wing . . .

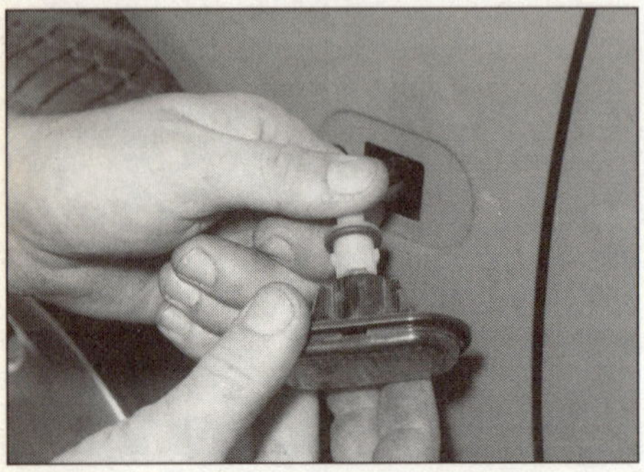

5.24a . . . then rotate the bulbholder to free it from the light . . .

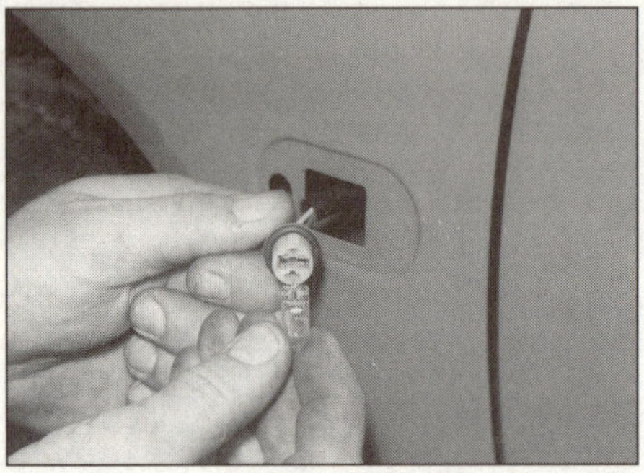

5.24b . . . and pull out the bulb

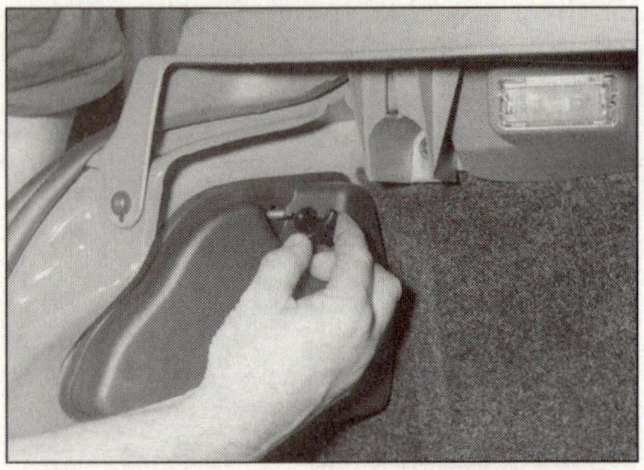

5.26 On Hatchback (shown here) and Coupe models, unscrew the nut and pivot the cover away from the light cluster

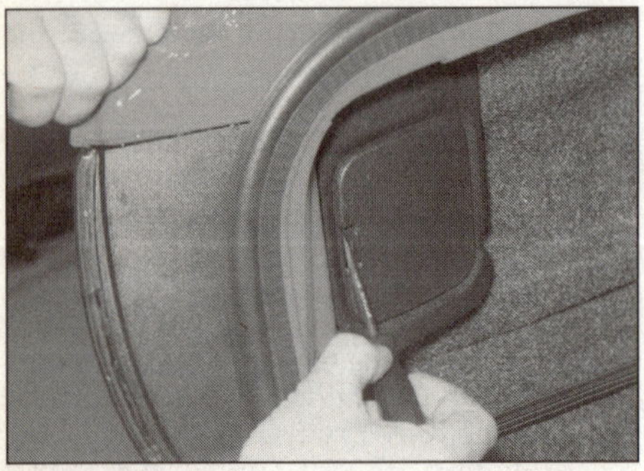

5.27 On Saloons, unclip the access cover to reach the back of the rear light cluster

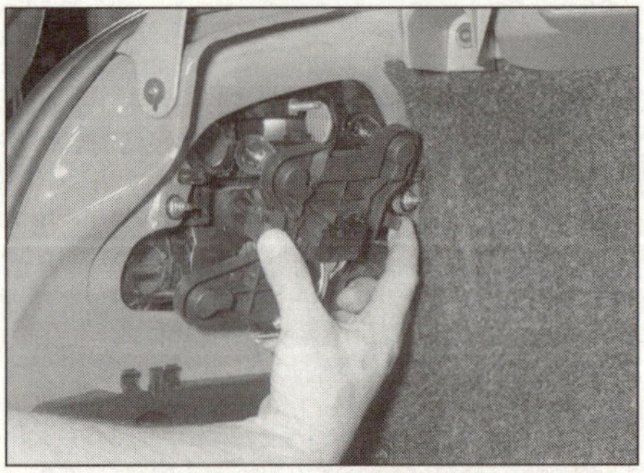

5.28 Depress the retaining clips and remove the bulbholder . . .

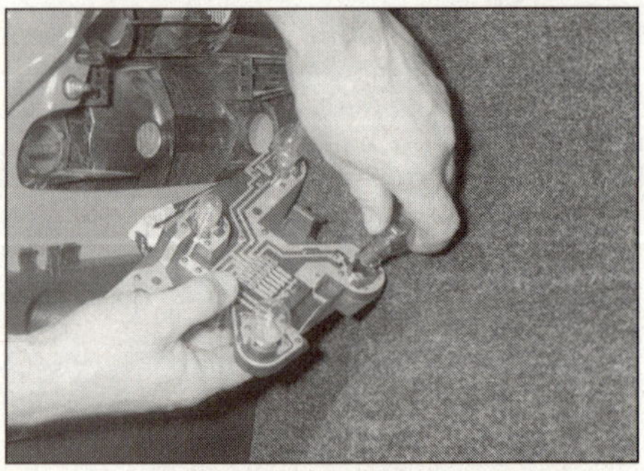

5.29 . . . the bulb can then be removed by pushing it in and turning anti-clockwise

20 Refitting is the reverse of removal noting the information given in paragraph 13.

Front direction indicator

21 Rotate the bulbholder anti-clockwise and remove it from the rear of the light unit. The bulb is a bayonet fit in the holder and can be removed by pressing it and twisting in an anti-clockwise direction **(see illustrations)**.

22 Refitting is a reverse of the removal procedure.

Front direction indicator side repeater

23 Carefully unclip the light unit and withdraw it from the wing **(see illustration)**.

24 Twist the bulbholder anti-clockwise and remove it from the light. The bulb is of the capless (push-fit) type and can be removed by simply pulling it out of the bulbholder **(see illustrations)**.

25 Refitting is a reverse of the removal procedure.

Rear light cluster

26 On Hatchback and Coupe models, from inside the vehicle luggage compartment, unscrew the upper retaining nut and pivot the cover away from the rear of the light cluster; on some models it will also be necessary to unhook the cover rubber retaining band **(see illustration)**.

27 On Saloon models, unclip the access cover from the light unit cover **(see illustration)**.

28 Release the retaining clips and remove the bulbholder assembly from the rear of the light cluster **(see illustration)**.

29 All bulbs have bayonet fittings. The relevant bulb can be removed by pressing in and rotating anti-clockwise **(see illustration)**. Note that the stop/taillight bulb has offset pins to ensure it is fitted the correct way around.

30 Refitting is the reverse of the removal sequence.

Boot lid-mounted rear lights - Coupe models

31 Open up the boot lid then unclip the cover to access to the rear of the light **(see illustration)**.

32 Twist the bulbholder and free it from the rear of the light unit then remove the bulb by

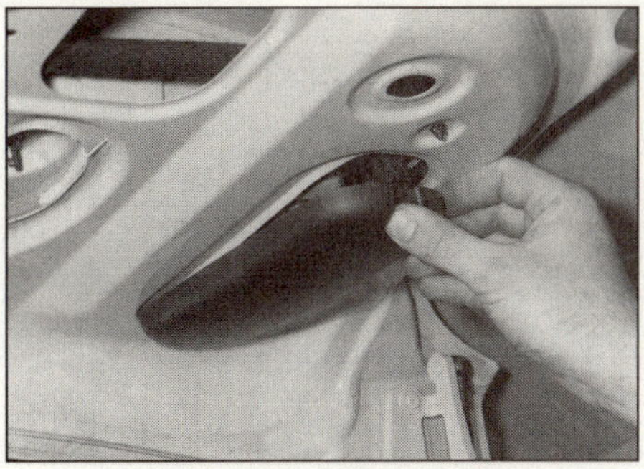

5.31 On Coupe models, unclip the cover . . .

pressing it in and turning it anti-clockwise **(see illustration)**.

33 Refitting is the reverse of the removal sequence.

Boot lid-mounted rear lights - Saloon models

34 Open up the boot lid then unclip the cover to access to the rear of the light.

35 Release the retaining clips and remove the bulbholder assembly from the rear of the light cluster then remove the bulb by pressing it in and turning it anti-clockwise.

36 Refitting is the reverse of the removal sequence.

High-level stoplight

37 On Hatchback models, open up the tailgate then depress the retaining clips and slide off the cover from the underside of the light unit **(see illustrations)**.

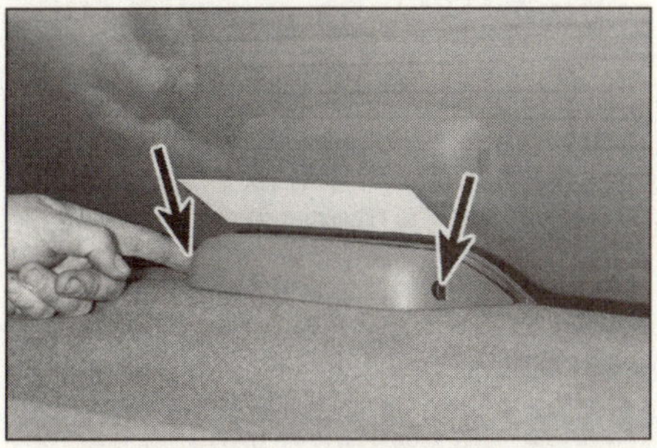

5.39 On Saloon models, depress the retaining clips (arrowed) and slide off the high-level stoplight cover

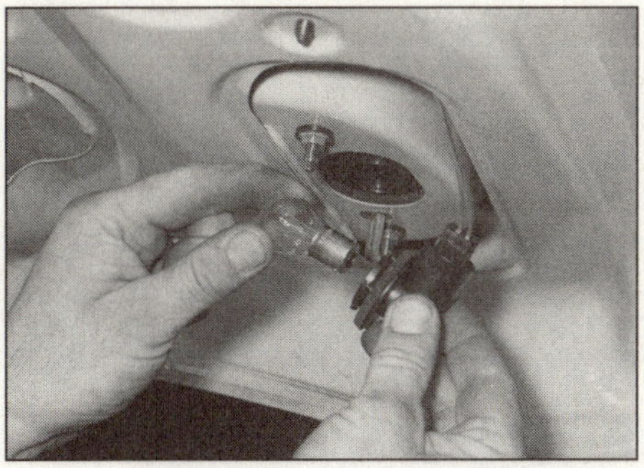

5.32 . . . then rotate the bulbholder to free it from the light unit and remove the bulb by pushing it in and turning anti-clockwise

38 On Coupe models, open up the boot lid to gain access to the rear of the light unit.

39 On Saloon models, from inside the vehicle, depress the retaining clips and slide off the cover from the rear of the light unit **(see illustration)**.

40 On all models, release the retaining clips and remove the bulbholder from the rear of the light unit. Each bulb is of the capless (push-fit) type and can be removed by simply pulling it out of the bulbholder **(see illustrations)**.

41 Refitting is the reverse of the removal sequence.

Number plate light

42 Carefully prise out the light unit out of position then unclip the cover (where fitted) and remove the bulb from the light unit contacts **(see illustrations)**.

5.40a Unclip the bulbholder from the light unit . . .

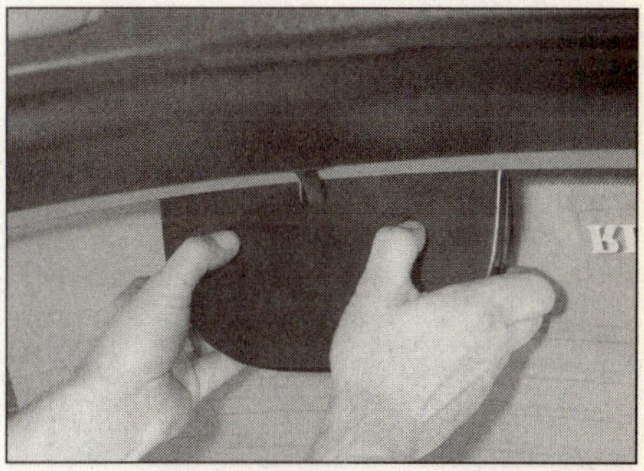

5.37 On Hatchback models, depress the retaining clips and slide the cover off from the high-level stoplight

43 Refitting is the reverse of removal ensuring that the bulb and lens are clipped securely in position.

6 Bulbs (interior lights) - renewal

1 Refer to Section 5, paragraph 1.

Front seat courtesy light and reading light

Models with a combined courtesy/reading light unit

2 Using a flat-bladed screwdriver, carefully

5.40b . . . and pull out the relevant bulb

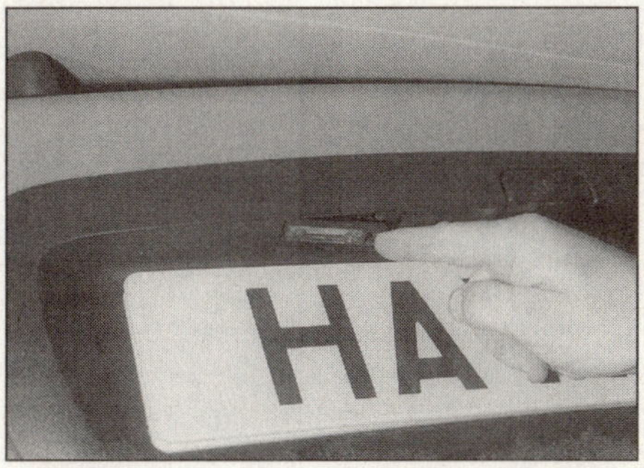

5.42a Unclip the number plate light . . .

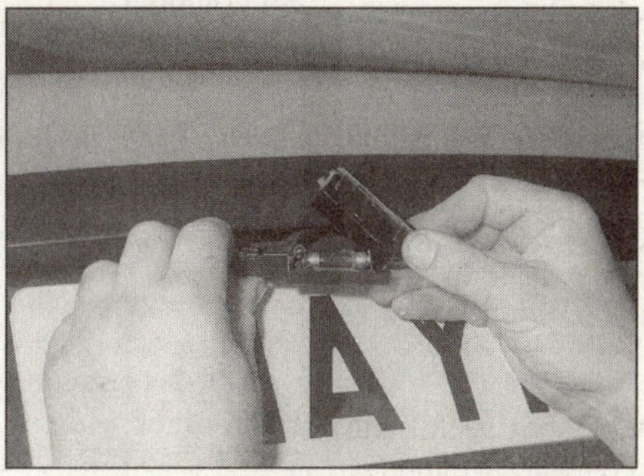

5.42b . . . then unclip the cover and remove the bulb from its contacts

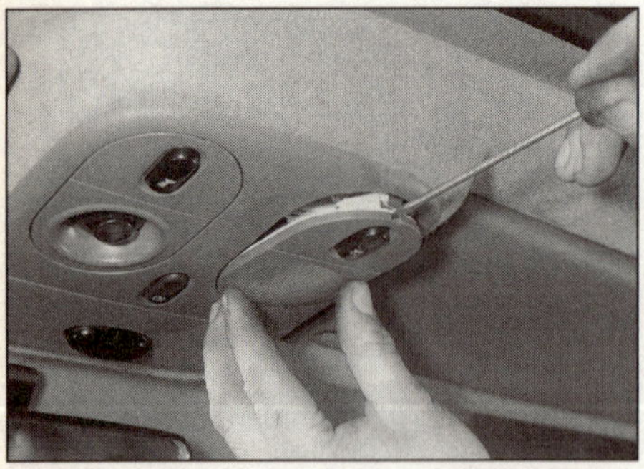

6.5 Carefully prise out the light unit from the overhead console . . .

6.6 . . . then rotate the bulbholder anti-clockwise and pull out the bulb

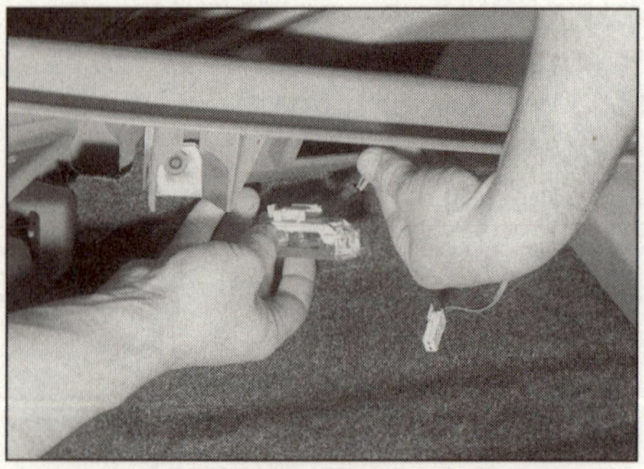

6.10 Prise out the luggage compartment light unit and disconnect its wiring connector (Hatchback model shown) . . .

6.11 . . . then unclip the lens from the bulbholder and remove the bulb

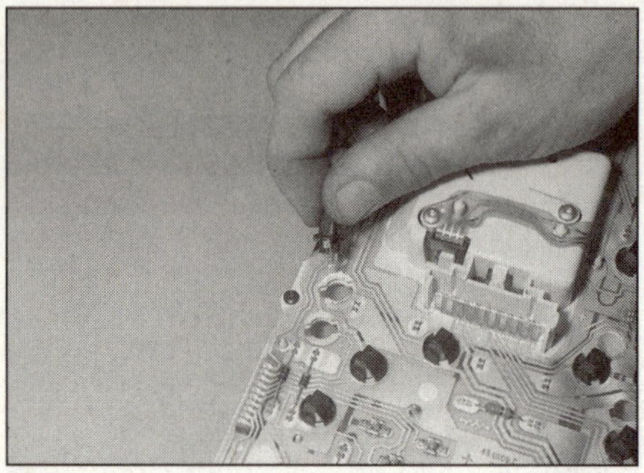

6.14 Removing and instrument panel bulb

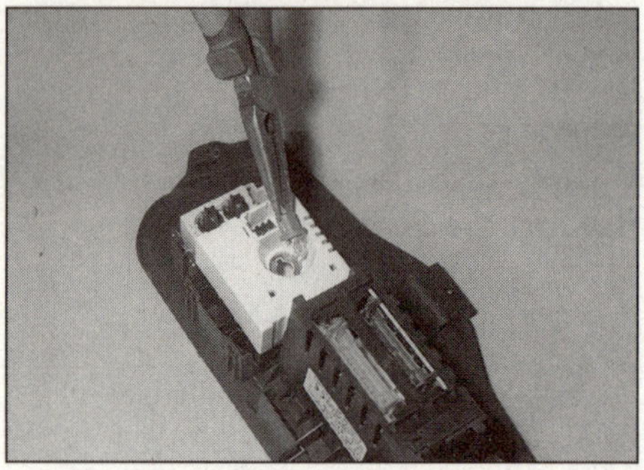

6.19 Removing the clock illumination bulb

unclip the light unit cover from the overhead console taking care not damage either the cover or console.

3 Each individual bulb can then be removed from the console.

4 Refitting is the reverse of removal.

Models with separate courtesy light and reading lights

5 Using a small, flat-bladed screwdriver, carefully prise the light unit assembly out of position **(see illustration)**.

6 Twist the bulbholder anti-clockwise and remove it from the light. The bulb is of the capless (push-fit) type and can be removed by simply pulling it out of the bulbholder **(see illustration)**.

7 Refitting is the reverse of removal.

Rear courtesy light

8 Using a flat-bladed screwdriver, carefully unclip the lens then release the bulb and remove it from the light unit.

9 Refitting is the reverse of removal ensuring that the bulb is securely held by the contacts.

Luggage compartment light

10 Using a flat-bladed screwdriver, carefully unclip the light unit from the trim panel/body and disconnect it from the wiring connector **(see illustration)**.

11 Depress the retaining clip and release the lens from the light unit; the bulb can then be released and removed **(see illustration)**.

12 Refitting is the reverse of removal ensuring that the bulb is securely held by the contacts.

Instrument panel illumination/warning lights

13 Remove the instrument panel as described in Section 9.

14 Twist the relevant bulbholder anti-clockwise and withdraw it from the rear of the panel **(see illustration)**.

15 All bulbs are integral with their holders. Be very careful to ensure that the new bulbs are of the correct rating, the same as those removed; this is especially important in the case of the ignition/battery charging warning light.

16 Refit the bulbholder to the rear of the instrument panel then refit the instrument panel as described in Section 9.

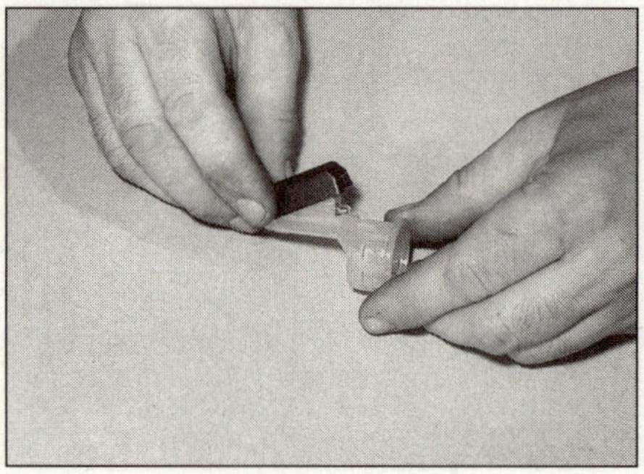

6.22a Unclip the cover from the cigarette lighter surround . . .

Glovebox illumination light bulb

17 Open up the glovebox and renew the bulb as described in paragraphs 10 to 12.

Multi-function display bulb

18 Remove the multi-function display as described in Section 11.

19 Twist the bulbholders through 90° and remove it from the display **(see illustration)**.

20 Refitting is the reverse of removal.

Cigarette lighter illumination bulb

21 Remove the cigarette lighter as described in Section 12.

22 Unclip the cover from around the bulb. The bulb is of the capless (push-fit) type and can be removed by simply pulling it out of the bulbholder **(see illustrations)**.

23 Refitting is the reverse of removal.

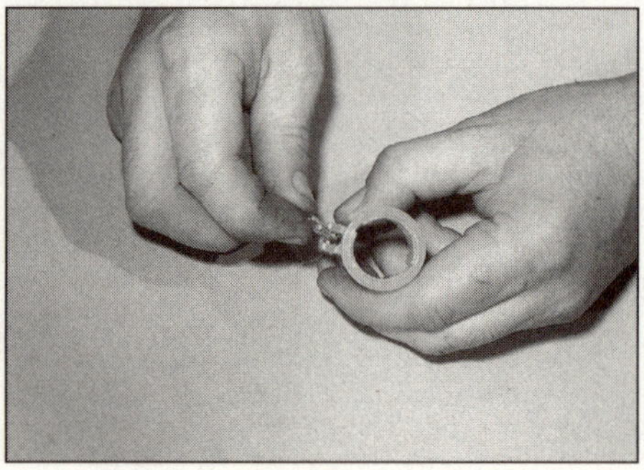

6.22b . . . and pull out the bulb

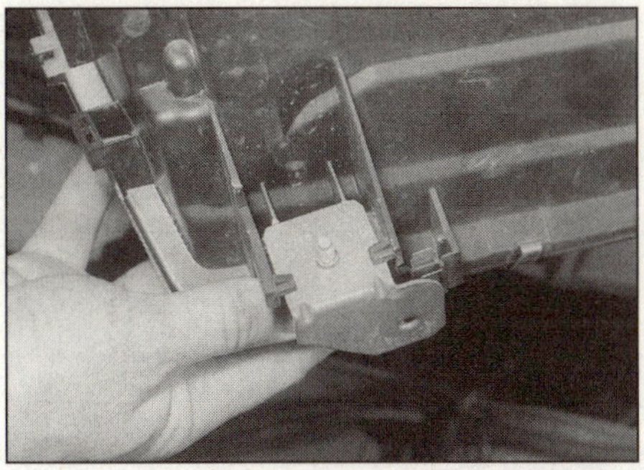

7.1 On later models, modified (slotted) headlight lower mounting brackets were fitted. On models with unslotted brackets (shown here) it will be necessary to remove the front bumper to allow the headlight to be removed

Switch illumination bulbs

24 All of the switches are fitted with illuminating bulbs; some are also fitted with a bulb to show when the circuit concerned is operating. On some switches, these bulbs can be removed by simply twisting the bulbholder through 90° once access has been gained to the rear of the switch. However, on some switches, it will be found that the bulbs are an integral part of the switch assembly and cannot be obtained separately. Bulb replacement will therefore require the renewal of the complete switch assembly.

7 Exterior light units - removal and refitting

Note: *Disconnect the battery negative lead before removing any light unit, and reconnect the lead after refitting the light.*

Headlight

Note: *The following procedure describes headlight removal and refitting on models with original headlight lower mounting brackets* **(see illustration)**. *During production, the headlight lower mounting brackets were modified to allow headlight removal with the front bumper in position. The modification involved slots being fitted to each lower mounting bracket; on models with these brackets the headlight lower mounting bolts can be removed with the bumper in position and the light unit can then be manoeuvred out by lifting it slightly and pulling it forwards.*

1 Remove the bumper as described in Chapter 11 (see Note at the start of this Section).

2 Remove the front direction indicator as described later in this Section.

3 Open up the bonnet then slacken and remove the retaining screw from the centre of the radiator grille and remove the joining section.

4 Undo the retaining screws then, using a flat-bladed screwdriver, release the lower retaining clips and slide the relevant half of the radiator grille forwards and away from the vehicle **(see illustration)**.

7.4 Release the retaining clips (arrowed) and remove the radiator grille panel

5 On headlights with a circular bulb access cover, rotate the cover anti-clockwise and remove it from the rear of the light unit. Disconnect the wiring connectors and free the wiring harness from the light unit **(see illustration)**.

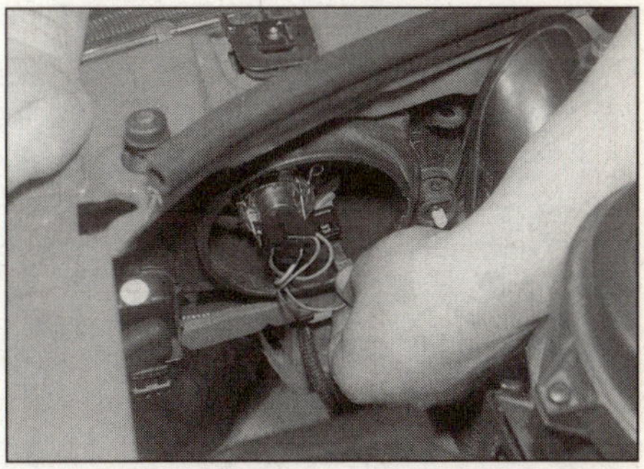

7.5 Disconnect the wiring connectors from the rear of the headlight unit

6 On headlights with a rectangular bulb access cover, depress the retaining clip and disconnect the wiring connector from the rear of the headlight unit.

7 On models equipped with a headlight levelling system, disconnect the wiring connector from the headlight motor **(see illustration)**.

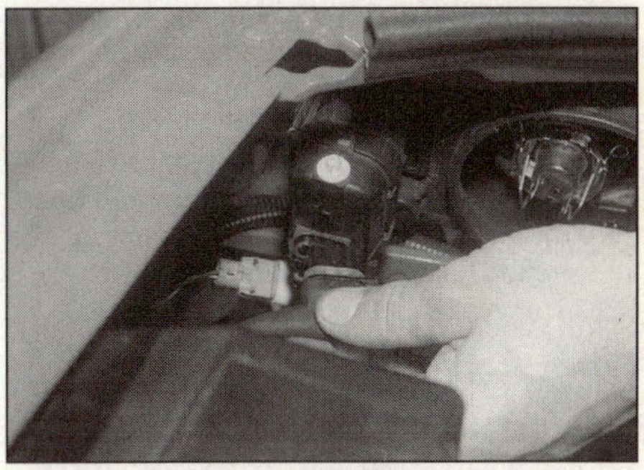

7.7 On models with a headlight levelling system disconnect the wiring connector from the motor

8 Slacken and remove the bolts securing the headlight unit lower mounting brackets to the body then remove the upper retaining screw and clip and manoeuvre the light unit forwards and away from the vehicle **(see illustrations)**.

9 On models equipped with a headlight levelling system, if necessary, remove the

7.8a Slacken and remove the lower mounting bolts (arrowed) . . .

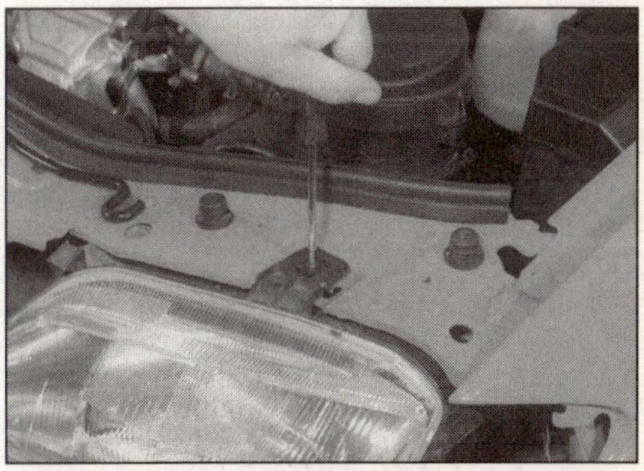

7.8b . . . then slacken and remove the upper retaining screw . . .

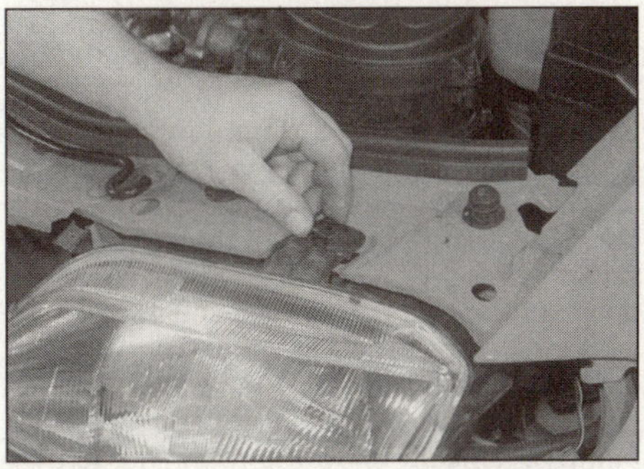

7.8c . . . and the retaining clip . . .

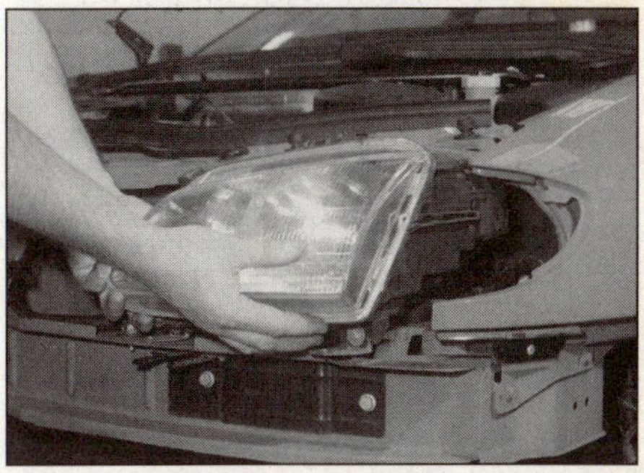

7.8d . . . and manoeuvre the headlight unit out of position

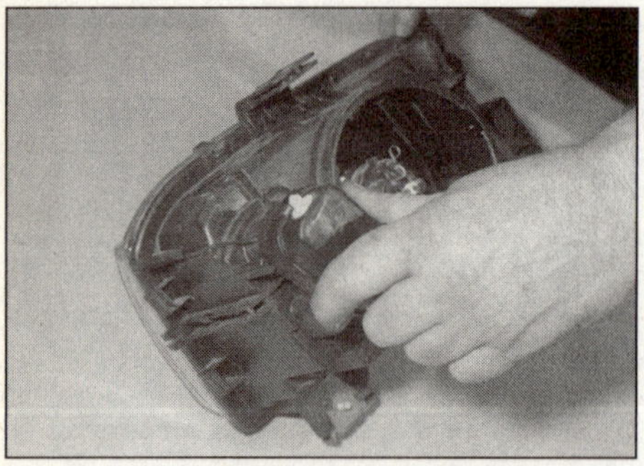

7.9a To remove the headlight levelling motor, rotate it anti-clockwise . . .

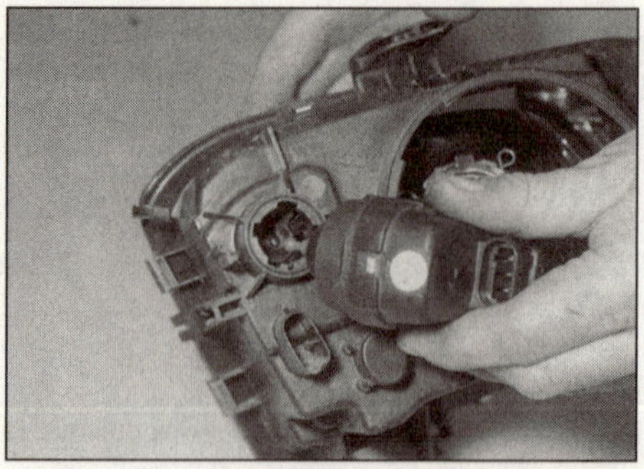

7.9b . . . then carefully unclip its balljoint from the headlight reflector

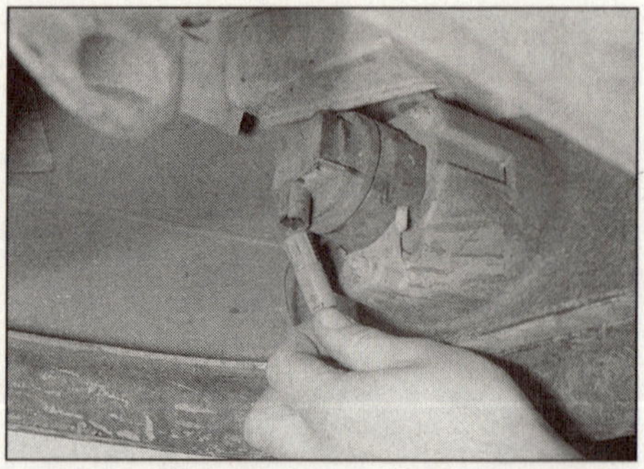

7.15a On Coupe models, disconnect the wiring connector . . .

motor from the rear of the light unit by rotating it anti-clockwise and carefully unclipping its balljoint from the rear of the reflector assembly **(see illustrations)**.

10 Refitting is a direct reversal of the removal procedure. On completion the headlight beam alignment should be checked using the information given in Section 8.

Front foglight

Hatchback and Saloon models

11 Carefully unclip the inner cover from the side of the foglight to gain access to the mounting screws.

12 Slacken and remove the foglight mounting screws and withdraw the light unit from the bumper, disconnecting its wiring connectors as they become accessible.

13 Refitting is the reverse of removal. If necessary, the foglight aim can be adjusted using the screw on the inner edge of the light unit; the screw is accessible once the inner cover has been removed.

Coupe models

14 If necessary, to improve access, firmly apply the handbrake then jack up the front of the vehicle and support it on axle stands (see "*Jacking and vehicle support*").

15 Disconnect the wiring connector then undo the retaining nuts and remove the light unit from the bumper **(see illustrations)**.

16 Refitting is the reverse of removal. If necessary, adjust the foglight aim by rotating the adjuster on the rear of the light unit **(see illustration)**.

Front direction indicator

17 Working through the hole in the bonnet crossmember, slacken the direction indicator retaining screw by a few turns then slide the light unit out of position in a forwards direction and disconnect its wiring connector **(see illustrations)**.

18 Refitting is the reverse of removal.

Front direction indicator side repeater

19 Carefully unclip the light unit and

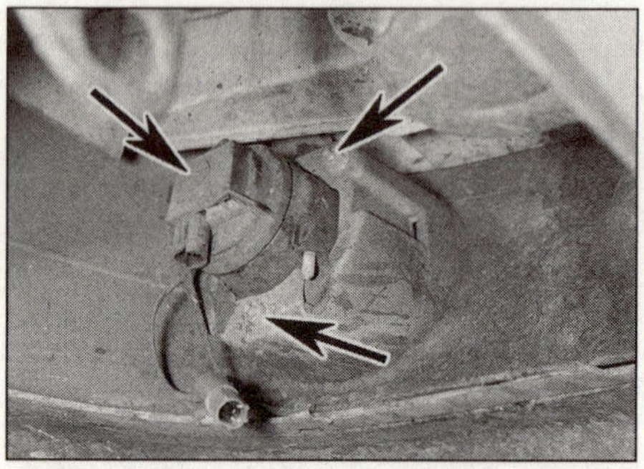

7.15b . . . then undo the retaining nuts and remove the foglight from the bumper

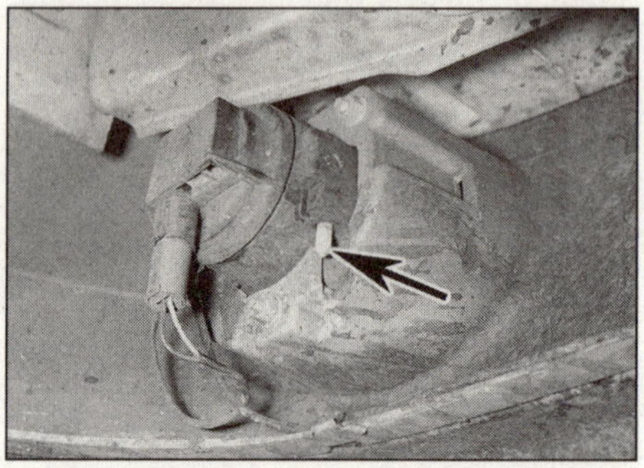

7.16 On refitting, adjust the foglight aim by rotating the adjuster (arrowed)

7.17a Slacken the retaining screw . . .

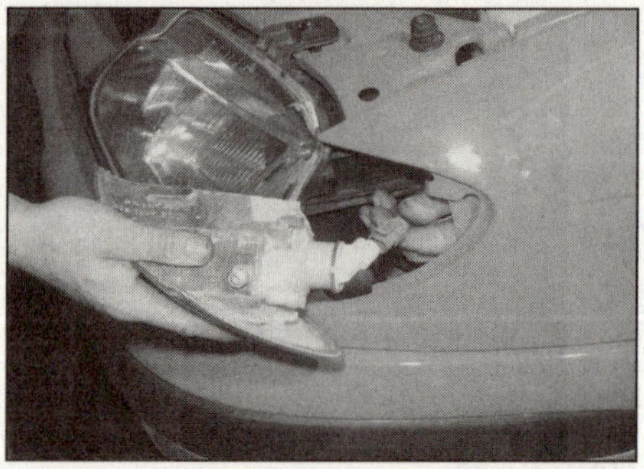

7.17b . . . then slide the direction indicator light unit out of position and disconnect its wiring connector

withdraw it from the wing. Free the bulbholder by rotating it anti-clockwise and remove the light unit from the vehicle.

20 Refitting is a reverse of the removal procedure.

Rear light cluster

21 On Hatchback and Coupe models, from inside the vehicle luggage compartment, unscrew the upper retaining nut and pivot the cover away from the rear of the light cluster; on some models it will also be necessary to unhook the cover rubber retaining band.

22 On Saloon models, from inside the luggage compartment, unclip the access cover from the light unit cover. To further improve access, undo the retaining screws and unclip the light unit cover **(see illustration)**.

23 Disconnect the wiring connector then undo the retaining nuts and remove the light unit from the rear of the vehicle **(see illustrations)**.

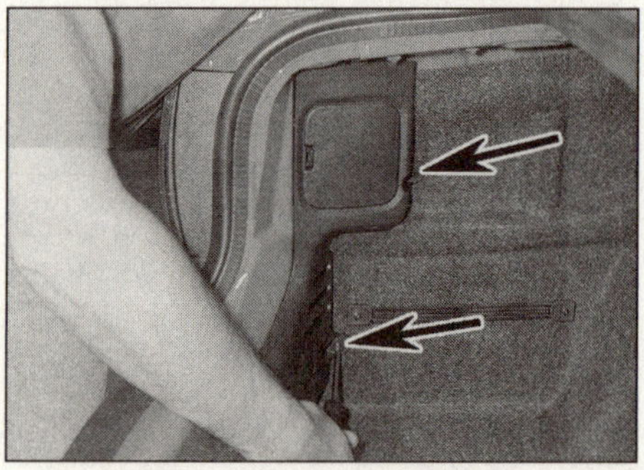

7.22 To improve access to the rear light cluster undo the retaining screws (arrowed) and unclip the complete cover

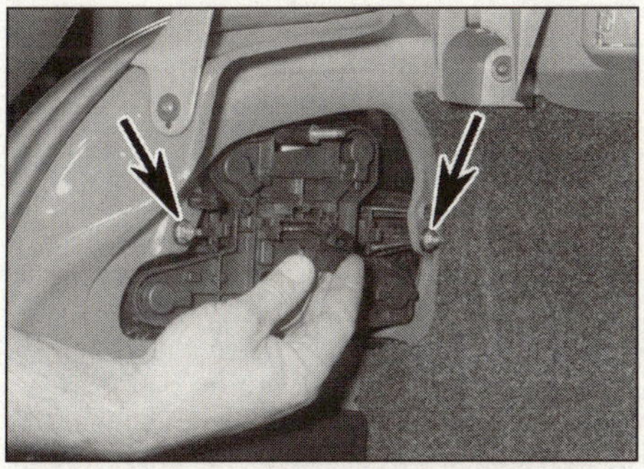

7.23a Disconnect the wiring connector then undo the retaining nuts (arrowed) . . .

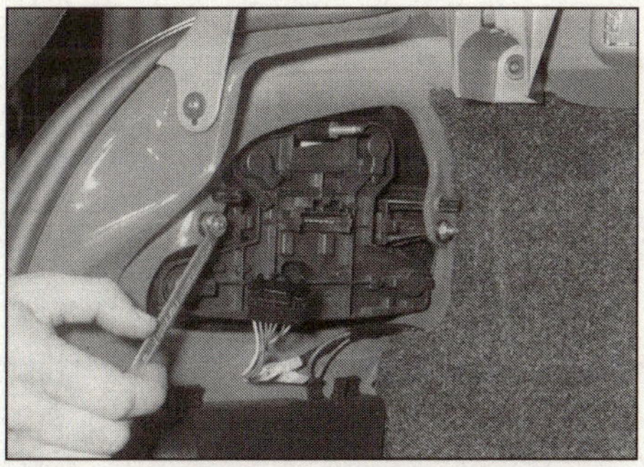

7.23b . . . and remove the rear light cluster unit (Hatchback model shown)

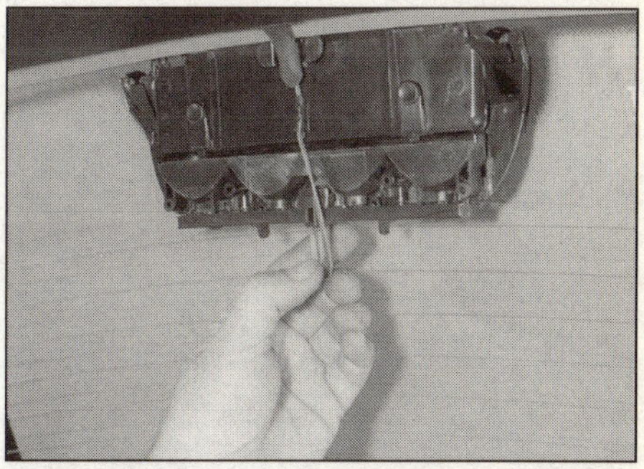

7.29a On Hatchback models, disconnect the wiring connector . . .

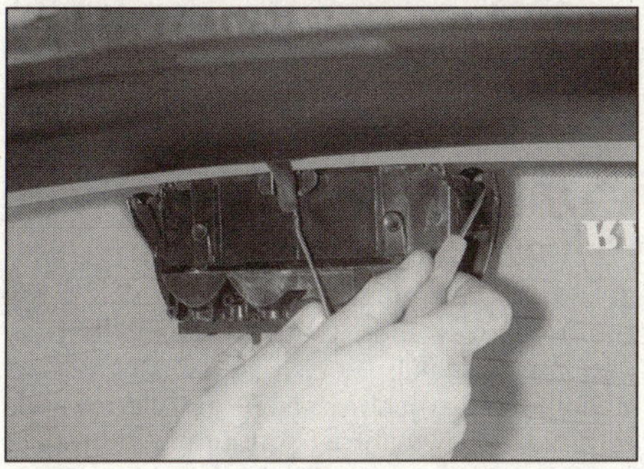

7.29b . . . then release the retaining clips and slide the high-level stoplight off from the tailgate

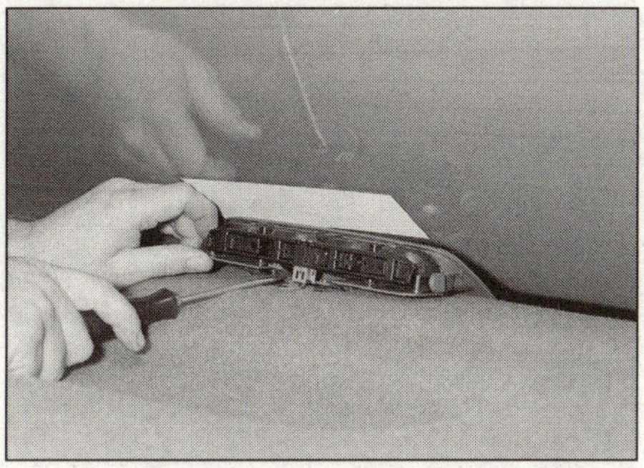

7.35a On Saloon models, release the retaining clips . . .

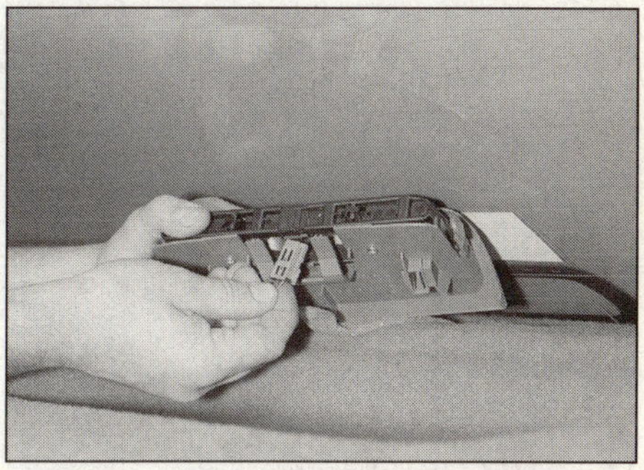

7.35b . . . then slide the high-level stoplight out of position and disconnect its wiring connector

24 Refitting is the reverse of removal, ensuring that the retaining nuts are securely tightened.

Boot lid-mounted rear lights - Coupe and Saloon models

25 Open up the boot lid then unclip the cover to access to the rear of the light.
26 Disconnect the wiring connector then undo the retaining nuts and remove the light unit.
27 Refitting is the reverse of removal.

High-level stoplight

Hatchback models

28 Depress the retaining clips and slide off the cover from the rear of the light unit.
29 Disconnect the wiring connector then release the retaining clips and slide the light assembly off its bracket **(see illustrations)**.
30 Refitting is the reverse of removal ensuring that both the light and cover are clipped securely in position.

Coupe models

31 Open up the boot lid and disconnect the wiring connector from the rear of the light unit.
32 Carefully release the retaining clips and push the light unit out from the boot lid.
33 Refitting is the reverse of removal.

Saloon models

34 Depress the retaining clips and slide off the cover from the rear of the light unit.
35 Using a screwdriver, carefully lift the lower retaining clips then slide the light unit forwards and out of position, disconnecting its wiring connector **(see illustrations)**.
36 Refitting is the reverse of removal ensuring that both the light and cover are clipped securely in position.

Number plate light

37 Carefully prise the light unit out from the tailgate, disconnecting its wiring connector as it becomes accessible.
38 On refitting, reconnect the wiring connector and clip the light unit securely in position.

8 Headlight beam alignment - general information

1 Accurate adjustment of the headlight beam is only possible using optical beam setting equipment and this work should therefore be carried out by a Renault dealer or suitably equipped workshop.
2 For reference the headlights can be adjusted by rotating the adjuster screws on the rear of the headlight unit. The upper screw adjusts the headlight beam vertical aim and the lower screw the headlight beam horizontal aim; on models equipped with headlight levelling the upper screw is built into the levelling motor **(see illustration)**.
3 On models equipped with a headlight levelling system, ensure that the headlight beam adjuster switch is set to position "0" before the headlights are adjusted. On models not equipped with a headlight levelling system, ensure that the manual adjuster on the rear of each light unit is set to position "A" before adjustment.

9 Instrument panel - removal and refitting

Removal

1 Undo the retaining screw and remove the battery cover then disconnect the battery negative terminal.

8.2 Headlight beam adjuster locations

9.3a Slacken and remove the left- and right-hand retaining screws . . .

9.4a Undo the retaining screws (arrowed) . . .

2 Carefully prise the left- and right-hand front speaker grilles out from the top of the facia panel to gain access to the facia top panel retaining screws.
3 Slacken and remove the retaining screw from the left- and right-hand end then unclip the top panel from the main facia and remove it from the vehicle **(see illustrations)**.
4 Slacken and remove the three retaining screws and remove the instrument panel, disconnecting the wiring connectors as they become accessible **(see illustrations)**.

Refitting

5 Securely reconnect the wiring connectors then clip the instrument panel back into position in the facia. Refit the panel retaining screws and tighten securely.
6 Manoeuvre the facia top panel into position and clip it firmly onto the main facia assembly.
7 Refit the top panel retaining screws, tightening them securely, and refit the loudspeaker grilles.
8 Reconnect the battery and check the operation of the panel warning lights to ensure that they are functioning correctly.

10 Instrument panel components - removal and refitting

At the time of writing, it was unclear if any of the instrument panel components were available separately. Refer to your Renault dealer for the latest parts information; they will

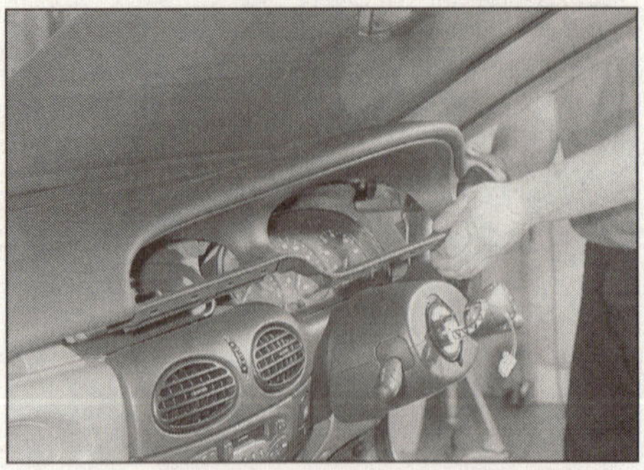

9.3b . . . then unclip the top panel and remove it from the facia

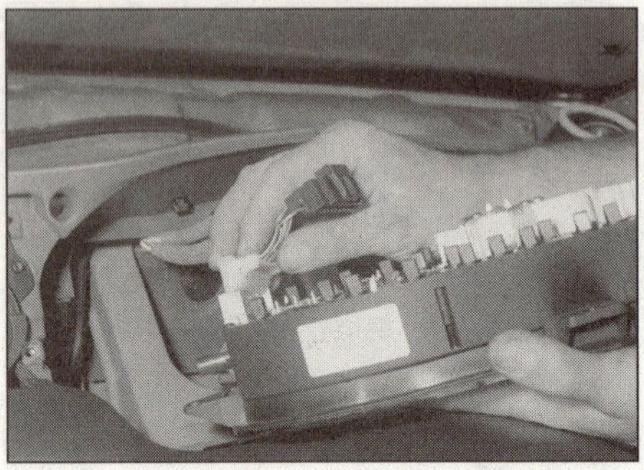

9.4b . . . and remove the instrument panel, disconnecting the wiring connectors as they become accessible

be able to advise you on the best course of action should the instrument(s) develop a fault.

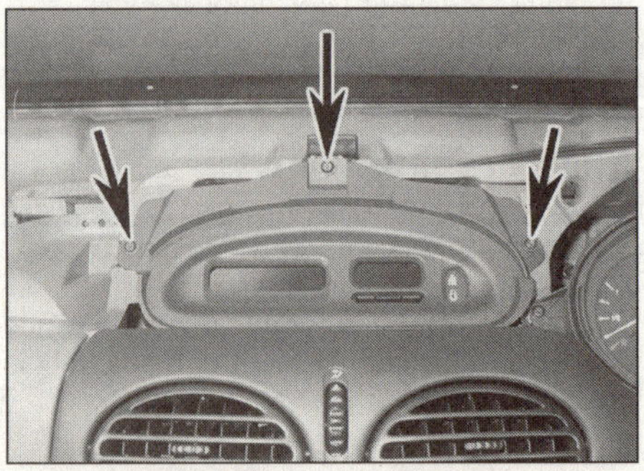

11.2 Multi-function display retaining screws (arrowed)

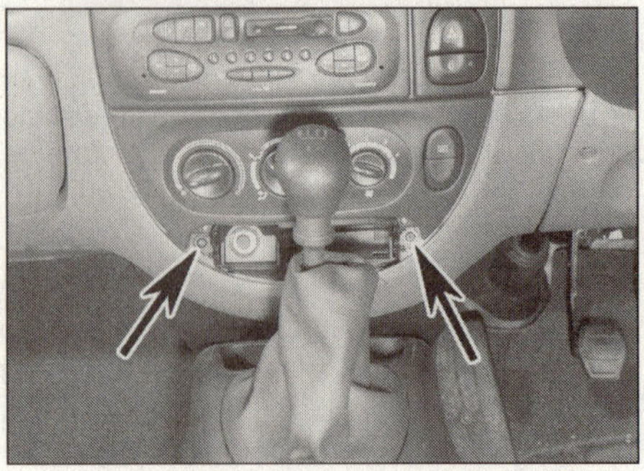

12.2b . . . then slacken and remove the ashtray retaining screws (arrowed)

11 Multi-function display - removal and refitting

Removal

1 Remove the facia top panel as described in paragraphs 1 to 3 of Section 9.
2 Slacken and remove the retaining screws and remove the multi-function display from the facia, disconnecting its wiring connectors as they become accessible **(see illustration)**.

Refitting

3 Refitting is the reverse of removal.

12 Cigarette lighter - removal and refitting

Removal

1 Undo the retaining screw and remove the battery cover then disconnect the battery negative terminal.
2 Remove the ashtray insert then slacken and remove the ashtray mounting bracket retaining screws **(see illustrations)**.
3 Remove the ashtray mounting bracket from the facia and disconnect its wiring connectors **(see illustration)**.
4 Release the retaining clips and slide out the cigarette lighter insert then push the plastic surround out of position **(see illustrations)**.

12.2a Remove the ashtray insert . . .

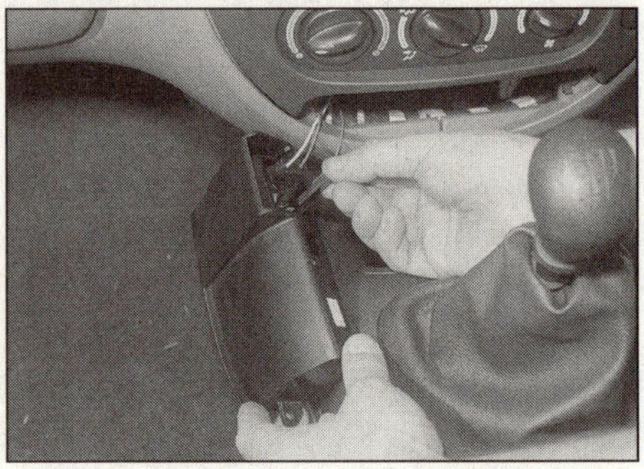

12.3 Withdraw the ashtray from the facia and disconnect its wiring connectors

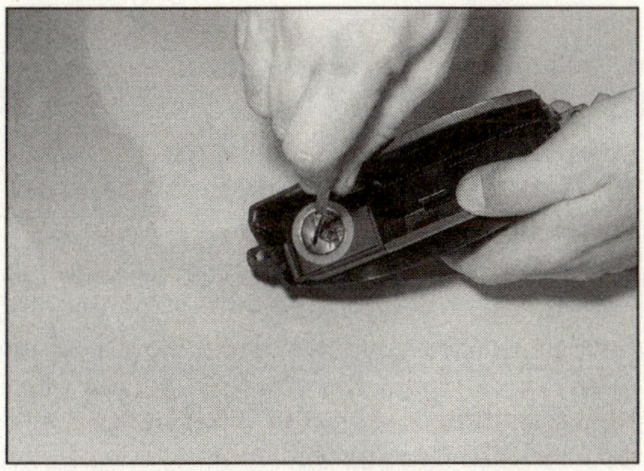

12.4a Release the retaining clips . . .

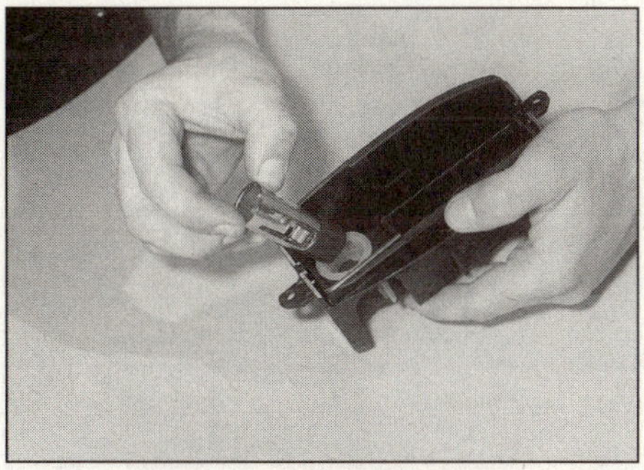

12.4b . . . then slide out the metal insert . . .

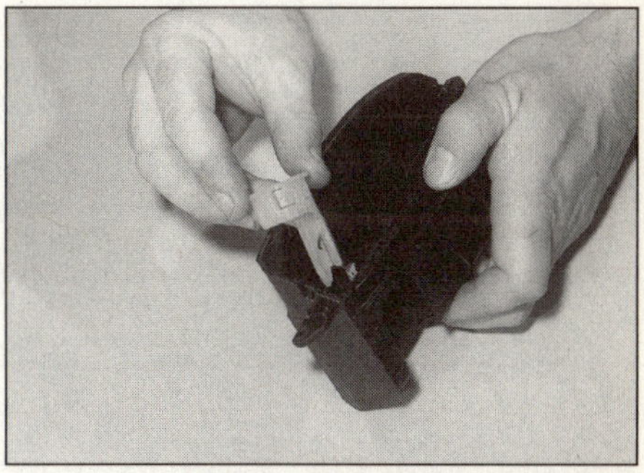

12.4c . . . and remove the plastic surround

Refitting

5 Refitting is a reversal of the removal procedure, ensuring that all the wiring connectors are securely reconnected.

13 Clock - removal and refitting

Removal

1 Remove the multi-function display as described in Section 11.

2 Carefully unclip the clock and remove it from the display unit **(see illustration)**.

Refitting

3 Refitting is the reverse of removal.

14 Horn(s) - removal and refitting

Removal

1 The horns are located behind the front bumper, one is fitted on the right-hand side and the other on the left-hand side.

2 To gain access to the assembly, firmly apply the handbrake then jack up the front of the vehicle and support it on axle stands (refer to "*Jacking and vehicle support*"). Where necessary, undo the retaining screws and free the wheelarch liner/cover from the underside of the bumper and wheelarch (as applicable).

3 Disconnect the wiring connector(s) then slacken and remove the mounting bolt(s) and remove the horn(s) from the vehicle.

Refitting

4 Refitting is the reverse of removal.

15 Wiper arm - removal and refitting

Note: *The wiper arms are a very tight fit on the spindles and it is likely that a puller will be needed to remove them safely, without damage.*

Removal

1 Operate the wiper motor then switch it off so that the wiper arm returns to the at rest position.

2 Lift the wiper arm retaining nut cover then slacken and remove the nut **(see illustration)**.

3 Lift the blade off the glass and pull the wiper arm off the motor and carefully lever the arm off the spindle using a large, flat-bladed screwdriver. If the arm is very tight, free it from the spindle using a suitable puller.

Refitting

4 Ensure that the wiper and spindle are clean and dry then refit the arm. Ensure that the arm is correctly positioned then securely tighten the retaining nut and refit the spindle nut cover.

16 Windscreen wiper motor and linkage - removal and refitting

Wiper motor

Removal

1 Remove the battery as described in Chapter 5A.

2 Working through the battery aperture, disconnect the wiring connector from the wiper motor.

3 Mark the relative positions of the motor shaft and crank then unscrew the retaining nut and washer and free the wiper linkage from the motor spindle. Unscrew the motor retaining bolts and manoeuvre the motor out of position and remove it from the vehicle **(see illustration)**.

Refitting

4 Manoeuvre the motor into position and refit its mounting bolts, tightening them securely.

5 Engage the crank with the motor spindle, aligning the marks made prior to removal, and securely tighten the spindle nut.

6 Reconnect the motor wiring connector then refit the battery.

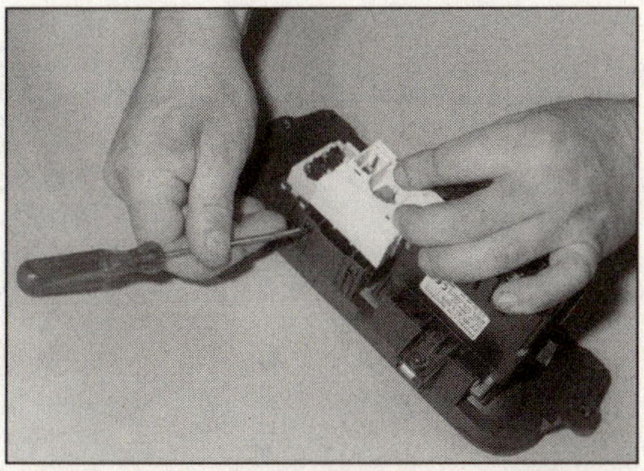

13.2 Release the retaining clips and unclip the clock unit from the multi-function display

15.2 Lift up the cover then slacken and remove the wiper arm retaining nut

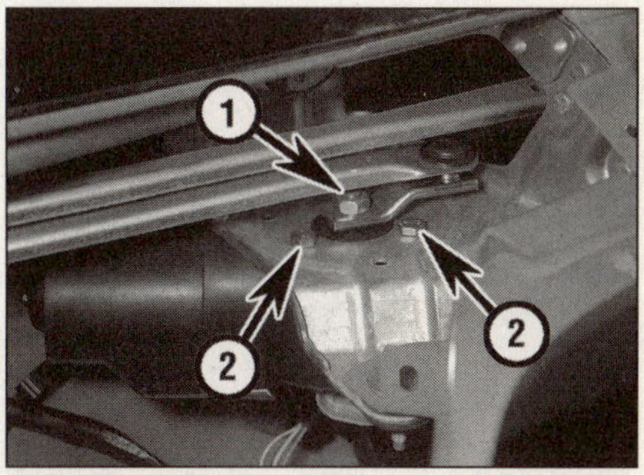

16.3 Slacken and remove the spindle nut (1) then unscrew the mounting bolts (2) - one hidden - and remove the wiper motor

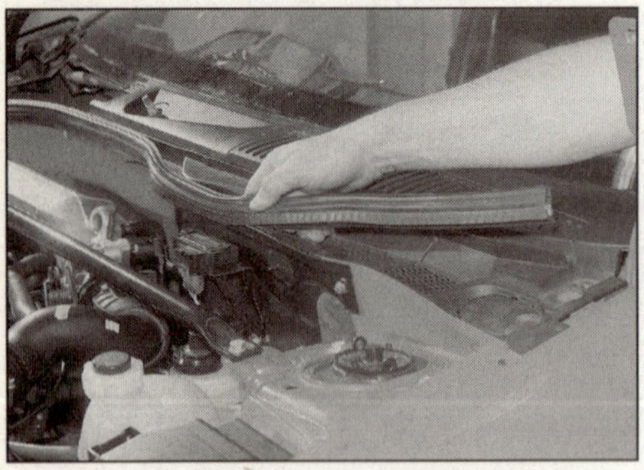

16.9 Peel off the rubber seal from the engine compartment bulkhead

16.10b . . . and the large vent panel from the left-hand side

Wiper motor and linkage - right-hand drive models

Removal

7 Undo the retaining screw and remove the battery cover then disconnect the battery negative terminal.

8 Remove the wiper arms as described in the previous Section.

9 Open up the bonnet and remove the rubber seal from the top of the engine compartment bulkhead **(see illustration)**.

10 Undo the retaining screws then unclip and remove the remaining inlet vent panel sections from beneath the windscreen **(see illustrations)**.

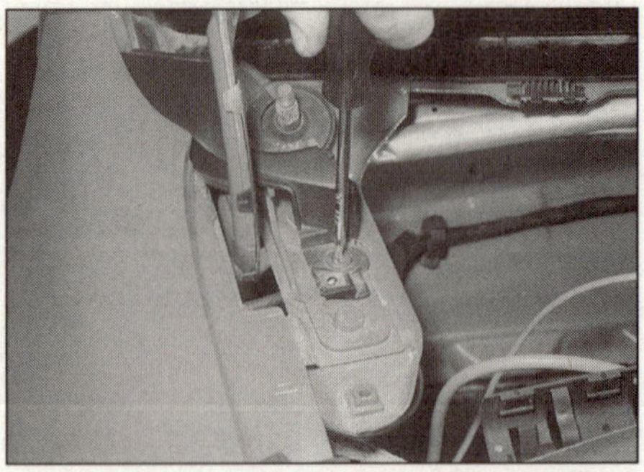

16.10a Undo the retaining screws and small inlet vent panel section from the right-hand side . . .

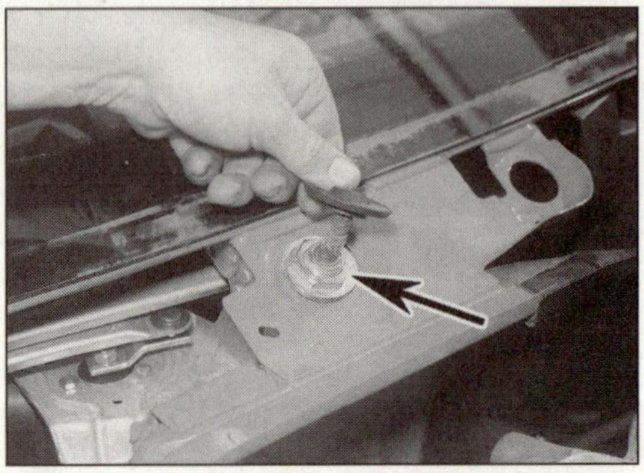

16.11 Remove the rubber covers then slacken and remove the spindle nuts (arrowed) and washers

11 Remove the covers from the wiper arm spindles then slacken and remove the spindle nuts and washers **(see illustration)**.

12 Disconnect the wiring connector from the wiper motor then unscrew the mounting bolts and manoeuvre the assembly out of position and away from the car **(see illustrations)**. If necessary the motor and linkage can then be separated as described in paragraph 3.

Refitting

13 Refitting is a reverse of removal. Prior to refitting ensure that the rubber mounting grommets fitted to the spindle holes are in good condition. Ensure that all fasteners are secure and make sure that the inlet vent panels are correctly clipped in position prior to refitting the wiper arms.

Wiper motor and linkage - left-hand drive models not fitted with air conditioning

14 Removal and refitting is as described in paragraphs 7 to 13 noting that it will also be necessary to remove the heater blower motor assembly to gain the necessary clearance required to manoeuvre the wiper motor out of position (see Chapter 3).

Wiper motor and linkage - left-hand drive models with air conditioning

15 On left-hand drive models with air conditioning, removal of the wiper motor and linkage assembly requires the air conditioning system evaporator to be removed. This means that the refrigerant circuit will have to be discharged/recharged and therefore this task will have to be entrusted to a Renault dealer or other suitably equipped workshop (see Chapter 3).

17 Tailgate/rear window wiper motor - removal and refitting

Removal

1 Disconnect the battery negative terminal then remove the wiper arm as described in Section 15.

2 Prise off the spindle cover then unscrew the nut and lift off the spacer from the tailgate wiper motor spindle **(see illustrations)**.

3 On Hatchback models, open up the tailgate, then slacken and remove the retaining screws and unclip the tailgate inner trim panel. Disconnect the wiring connector from the motor then undo the retaining bolts and remove the motor from the tailgate. Remove the rubber sealing grommet from the tailgate **(see illustrations)**.

4 On Coupe models, open up the boot lid and

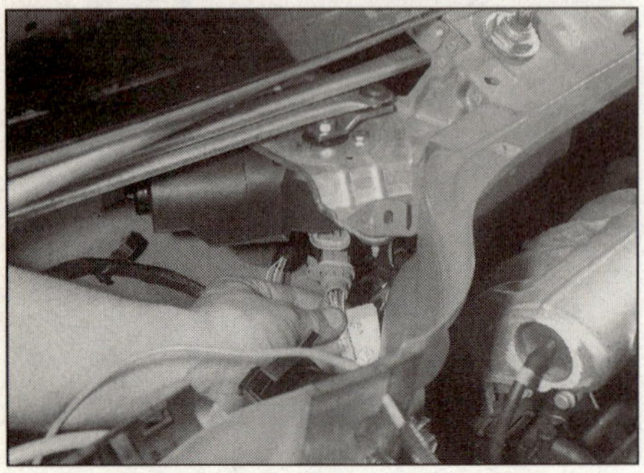

16.12a Disconnect the wiring connector . . .

16.12b . . . then undo the mounting bolts (arrowed) . . .

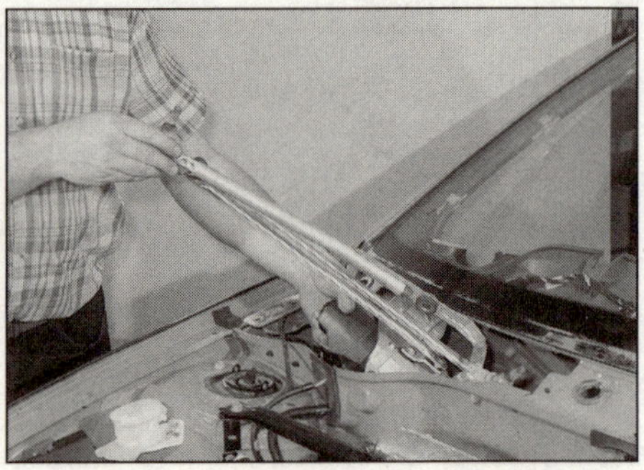

16.12c . . . and manoeuvre the wiper motor and linkage out of position

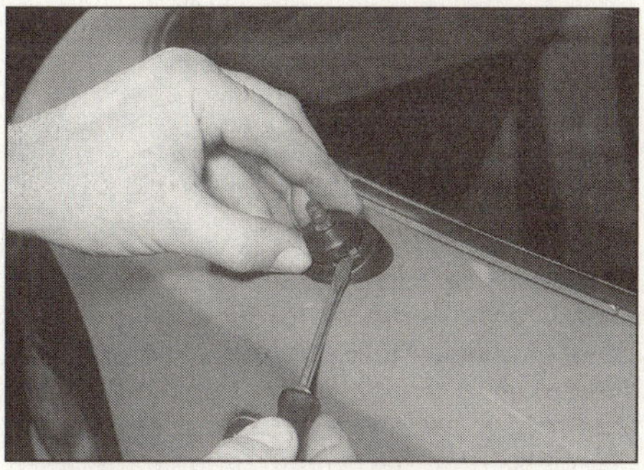

17.2a Prise off the spindle cover . . .

17.2b . . . then unscrew the nut . . .

17.2c . . . and lift off the spacer (Hatchback model shown)

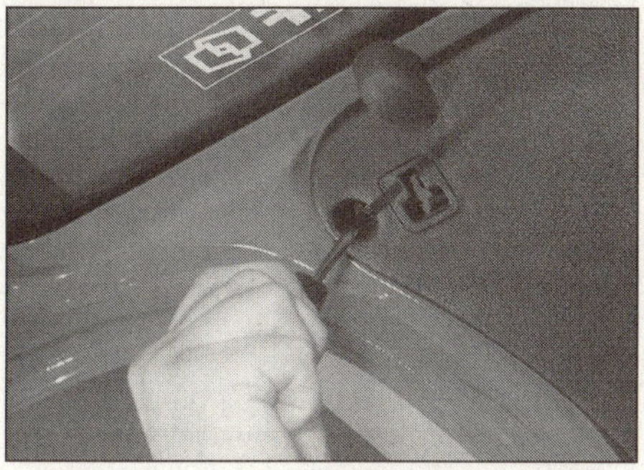

17.3a On Hatchback models, undo the retaining screws and unclip the inner trim panel from the tailgate . . .

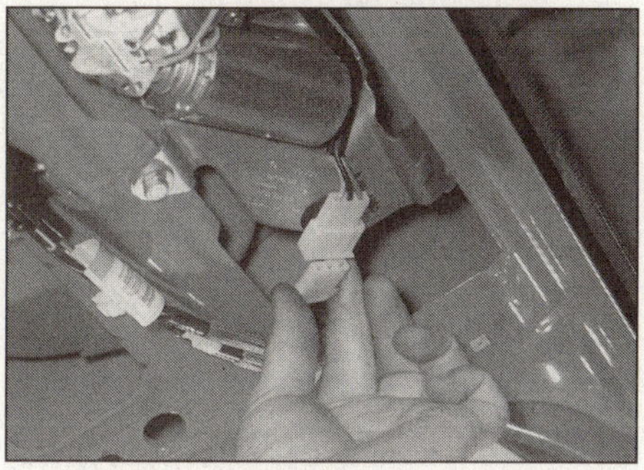

17.3b . . . disconnect the wiring connector . . .

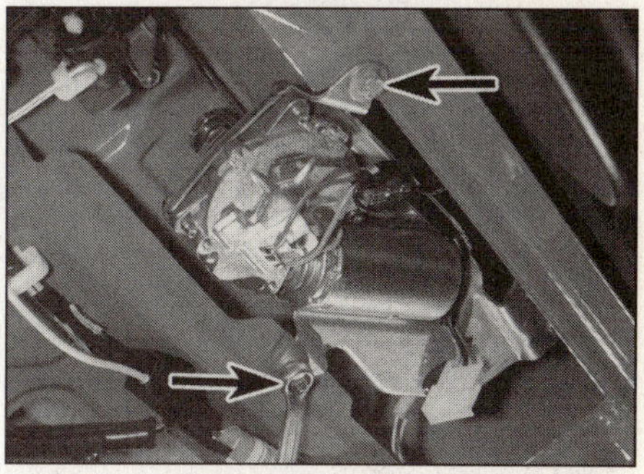

17.3c . . . then undo the retaining bolts (arrowed) and remove the wiper motor from the tailgate

disconnect the wiring connector from the motor. Slacken and remove the mounting bolts and remove the wiper motor from the vehicle, taking care not to lose the motor mounting rubbers **(see illustration)**. Remove the rubber sealing grommet from the tailgate.

Refitting

5 Refitting is the reverse of removal. Prior to refitting, inspect the sealing grommet for signs of damage or deterioration and renew if necessary **(see illustration)**.

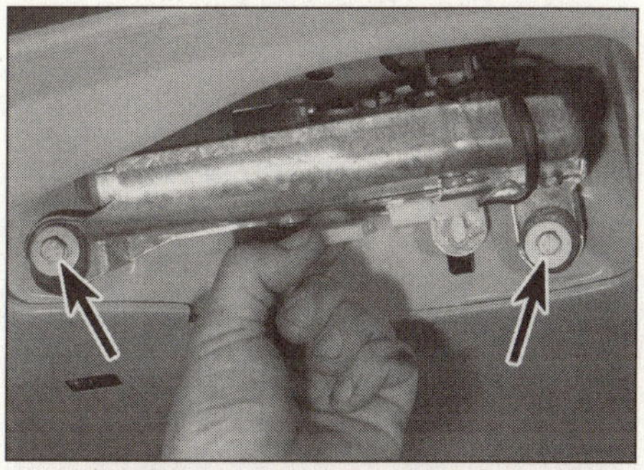

17.4 On Coupe models, disconnect the wiring then undo the mounting bolts (arrowed) and remove the wiper motor

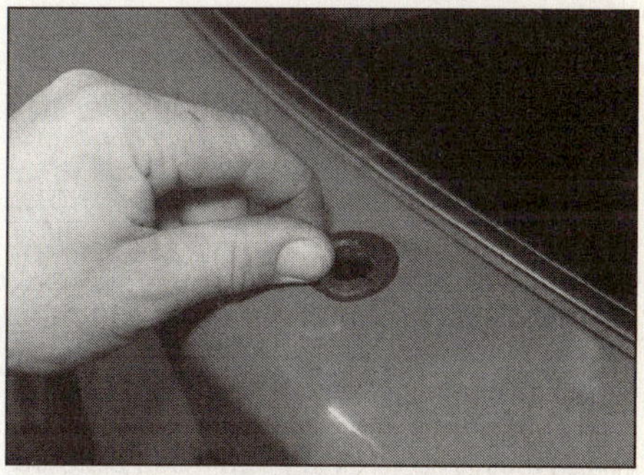

17.5 Prior to refitting ensure that the wiper motor spindle sealing grommet is in good condition

18 Windscreen/tailgate washer system components - removal and refitting

Windscreen/tailgate washer reservoir

1 Working in the engine compartment, undo the retaining nut/screw (as applicable) and remove the washer reservoir filler neck.

2 Firmly apply the handbrake then jack up the front of the vehicle and support it on axle stands (see "*Jacking and vehicle support*"). Remove the right-hand front roadwheel then remove the retaining screws and fasteners and remove the wheelarch liner sections to reveal the reservoir.

3 Release the retaining clip(s) and disconnect the hose(s) from the washer pump union(s) (as applicable). Where there is more than one hose connection, make alignment marks to ensure the hoses are correctly reconnected on refitting.

4 Trace the wiring back from the washer pump to the connector and disconnect it from the main harness; on some models it will also be necessary to release the side repeater bulbholder from the light unit to allow the bulbholder to be removed with the reservoir.

5 Slacken and remove the retaining nut from the front of the reservoir then manoeuvre the reservoir out from underneath the wing.

6 Refitting is the reverse of removal ensuring that the hose(s) are securely reconnected. Refill the reservoir and check for leakage.

Washer pump

7 Remove the reservoir as described earlier in this Section.

8 Carefully ease the pump out from the reservoir and recover its sealing grommet. Wash off any spilt fluid with cold water.

9 Refitting is the reverse of removal, using a new sealing grommet if necessary. Refill the reservoir and check the pump grommet for leaks.

Washer reservoir level switch

10 To gain access to the level switch (where fitted), firmly apply the handbrake then jack up the front of the vehicle and support it on axle stands (see "*Jacking and vehicle support*"). Remove the front roadwheel then remove the retaining screws and fasteners and remove the rear section of the wheelarch liner.

11 Disconnect the wiring connector from the level switch and carefully ease the switch out

19.1 Release the retaining clips by inserting welding rods into the holes provided . . .

from the reservoir and catch the reservoir contents in a suitable container. Recover the sealing grommet and wash off any spilt fluid with cold water.

12 Refitting is the reverse of removal, using a new sealing grommet if the original one shows signs of damage or deterioration.

Windscreen washer jets

13 Open up the bonnet then carefully prise out the retaining clips and peel back the bonnet insulation panel to gain access to the washer jets. Disconnect the washer hoses from the relevant jet then depress the retaining clips and carefully ease the jet away from the bonnet.

14 On refitting, securely connect the jet to the hose and clip it into position in the bonnet. Check the operation of the jet. If necessary adjust the nozzles using a pin, aiming one nozzle to a point slightly above the centre of the swept area and the other to slightly below the centre point to ensure complete coverage.

Tailgate washer jet

15 Open up the tailgate and peel the sealing strip away from the top of the tailgate aperture so that it is clear of the headlining.

16 Depress the retaining tangs and slide out the retaining clips from the rear of the headlining.

17 Prise out the tailgate hinge access covers from the headlining then carefully peel back the headlining until access can be gained to the washer jet.

18 Unscrew the retaining nut then lift the washer jet and seal away from the roof, disconnecting it from the hose.

19 Refitting is the reverse of removal. If necessary adjust the nozzle using a pin.

19 Radio/cassette player - removal and refitting

Note: *The following removal and refitting procedure is for the range of equipment fitted by Renault.*

Removal

1 In order to release the radio retaining clips,

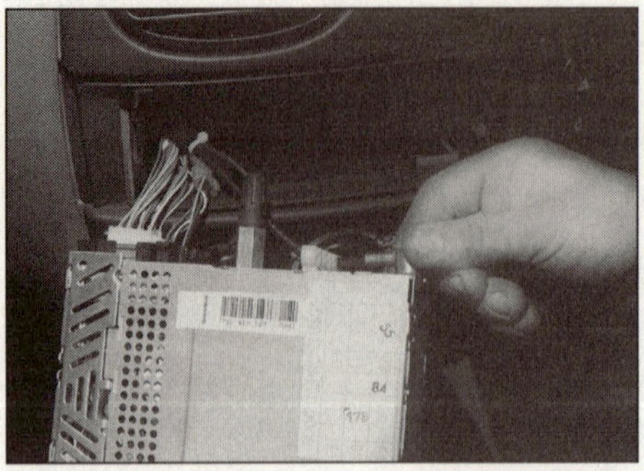

19.2 . . . then slide out the radio/cassette player and disconnect the wiring connectors and the aerial lead

two lengths of welding rod will be required. Insert the rods into the holes in each side of the unit until the retaining clips are felt to release **(see illustration)**.

2 Carefully slide the audio unit out of position whilst ensuring that the wiring does not become trapped or caught. Disconnect the wiring connectors and the aerial lead and remove the unit from the vehicle **(see illustration)**.

Refitting

3 Refitting is the reverse of removal.

20 Loudspeakers - removal and refitting

Facia loudspeaker

1 Carefully prise the speaker out of position, taking great care not to damage the facia panel or speaker grille **(see illustration)**.

2 Slacken and remove the retaining screws then manoeuvre the speaker out of position, disconnecting the wiring connector as it becomes accessible **(see illustrations)**.

3 Refitting is the reverse of removal.

Front door speaker

4 On models with electric front windows, carefully prise the front of the speaker grille out of position, and remove the grille from the door **(see illustration)**.

5 On models with manual front windows, rotate the speaker grille anti-clockwise and remove it from the door.

6 Slacken and remove the retaining screws then remove the speaker from the door, disconnecting the wiring connector as it becomes accessible **(see illustrations)**.

7 Refitting is the reverse of removal.

Rear speaker

Hatchback models

8 From within the luggage compartment, disconnect the wiring connector from the base of the speaker then slacken and remove

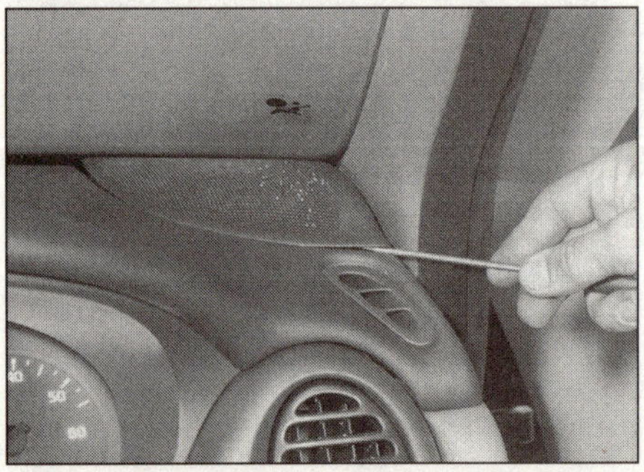

20.1 Carefully prise the speaker grille out from the top of the facia . . .

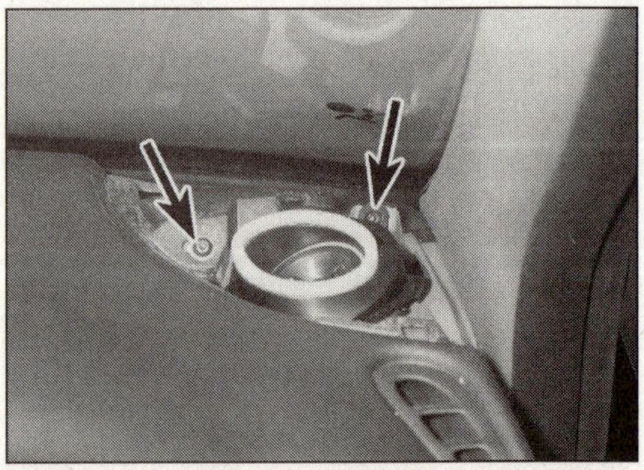

20.2a . . . then undo the retaining screws (arrowed) . . .

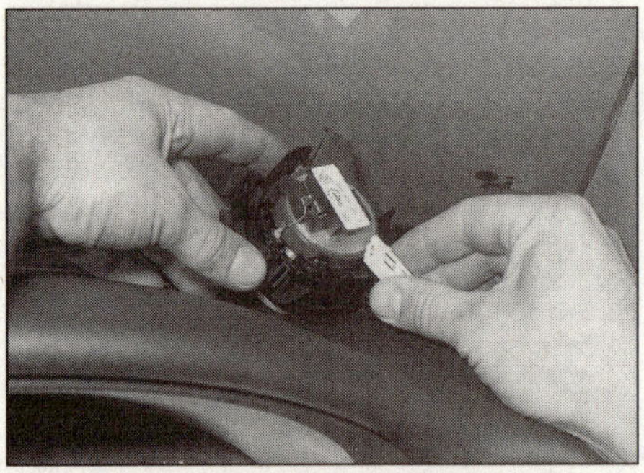

20.2b . . . and lift out the speaker, disconnecting its wiring connector

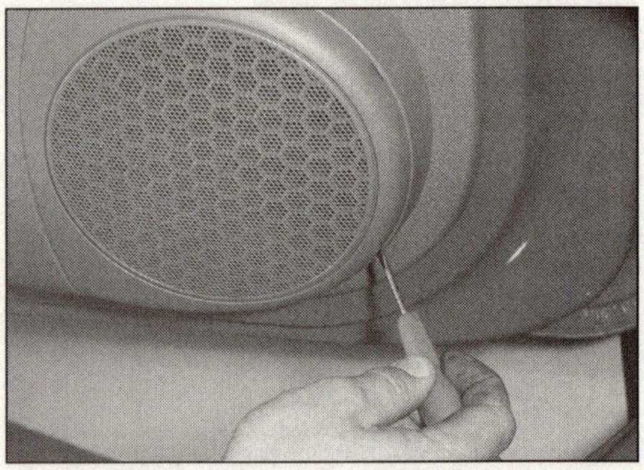

20.4 On models with electric windows, unclip the speaker grille from the door trim panel

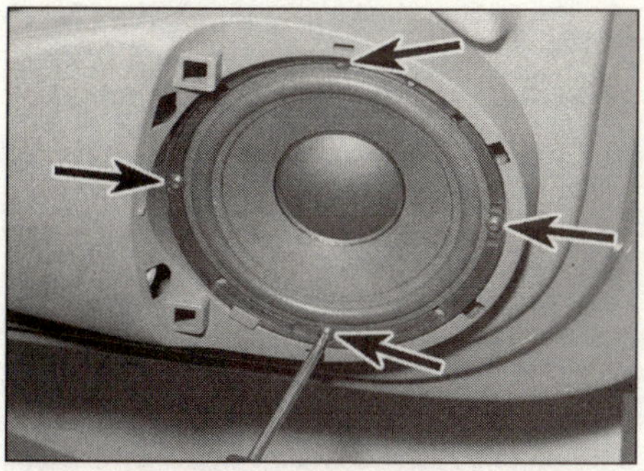

20.6a Slacken and remove the retaining screws (arrowed) . . .

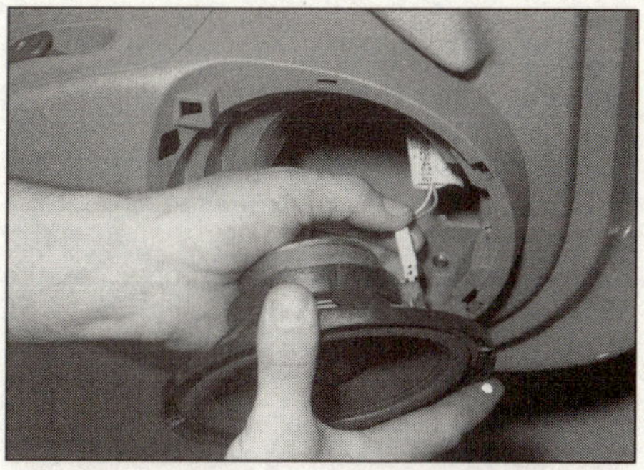

20.6b . . . then remove the speaker and disconnect its wiring connector

the retaining screws and remove the speaker from the vehicle.
9 Refitting is the reverse of removal.

Coupe and Saloon models

10 Carefully twist and unclip the speaker cover from the rear parcel shelf.
11 Undo the retaining screws and remove the speaker, disconnecting its wiring connectors as they become accessible.
12 Refitting is the reverse of removal.

21 Radio aerial - removal and refitting

Removal

1 The aerial mast can is a screw-fit in its base and is easily removed.
2 To remove the complete aerial it is necessary to remove the overhead console. To do this, slide the front cover off the console in a forwards direction and disconnect it from the wiring connector. Slacken and remove the console retaining screws then slide the console assembly towards the front of the vehicle to release it from the roof. Disconnect the wiring connectors and remove the console from the vehicle. Remove the insulating cap (where fitted) then slacken and remove the aerial lead nut **(see illustration)**. Detach the aerial lead and lift the aerial off from the roof, noting its rubber seal.

Refitting

3 Refitting is the reverse of removal.

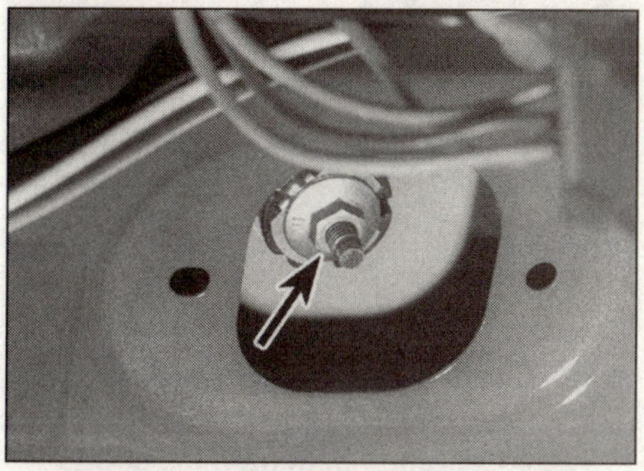

21.2 Aerial retaining nut (arrowed) is located behind the overhead console

22 Anti-theft alarm and engine immobiliser - general information

Note: *This information is applicable only to the anti-theft alarm/engine immobiliser systems fitted by Renault as standard equipment.*

General

1 Most models in the range are fitted with an anti-theft alarm system as standard equipment. The alarm is automatically armed and disarmed using the remote central locking transmitter (where applicable). When the system is activated, the alarm indicator light, located on the roof console, will flash continuously. In addition to the alarm function, the system also incorporates an engine immobiliser.
2 Most models are also fitted with an engine immobiliser device. On models with an anti-theft alarm the immobiliser is also armed and disarmed at the same time as the alarm. On models not equipped with remote central locking, the immobiliser is activated by the coded ignition key. When the immobiliser is armed, the indicator light on the instrument panel will flash continuously.

Anti-theft alarm system

Note: *If the doors are operated using the key, the alarm will not be armed or disarmed (as applicable).*
3 The alarm system has switches on the bonnet, tailgate/boot and each of the doors. It also has ultrasonic sensing, which detects movement inside the vehicle, via sensors mounted on the roof console. If, for some reason the remote central locking transmitter fails whilst the alarm is armed, the alarm can be disarmed using the key. To do this, open the door with the key, then enter the vehicle, noting that the alarm will sound as the door is opened. Open up the glovebox and switch the alarm system isolation switch to the "OFF" position.
4 When the alarm is set using the remote central locking transmitter, the hazard warning light should flash and the alarm siren should "bleep". If there is no sound and the hazard warning lights do not flash as the doors are locked, check that the isolation switch is not turned off. If the alarm "bleeps" but the hazard warning lights do not flash, this indicates that a door or the tailgate/boot/ bonnet are not properly shut; as soon as the relevant item is securely closed the hazard warning lights will flash to indicate the alarm system is now active.
5 To de-activate the alarm system, unlock the doors with the remote central locking transmitter then open up the glovebox and, using the ignition key, turn the alarm system isolation switch to the "OFF" position. To re-activate the system, turn the isolation switch back "ON" then lock then unlock the vehicle using the remote transmitter; the alarm will be functional again the next time the vehicle is locked with the remote control.
6 Should the alarm system become faulty, the vehicle should be taken to a Renault dealer for examination.

Engine immobiliser

Models with remote central locking

Caution: If the central locking remote transmitter becomes faulty whilst the doors are locked, it will be necessary to use the spare central locking remote transmitter (supplied with the vehicle) to unlock the doors and disarm the immobiliser. If the spare transmitter is not available, then it will be necessary to seek the advice of your Renault dealer; it will not be possible to start the engine without a functioning transmitter.
7 On models with remote central locking, the engine immobiliser is turned on and off using the remote central locking transmitter.
8 When the doors are locked using the transmitter, the immobiliser is switched on and the warning light in the instrument panel flashes continuously. When the doors are unlocked using the transmitter, the warning light will go out. When the ignition switch is turned on, the light should illuminate for a few seconds and then go out; this indicates that the immobiliser has received the correct signal and the engine can be started.
9 If the immobiliser warning light fails to go out when the ignition key is turned or illuminates/flashes whilst the engine is running then the immobiliser system is faulty and the vehicle should be taken to a Renault dealer for testing.

Models without remote central locking

Caution: If ignition key is lost/faulty, it will be necessary to use the spare key (supplied with the vehicle) to unlock the doors and switch on the ignition. If the spare key is not available, then it will be necessary to seek the advice of your Renault dealer; it will not be possible to start the engine without a functioning transmitter.
10 On models not equipped with remote central locking, the engine immobiliser is activated by the coded ignition key.

11 When the ignition switch is turned off and the key is removed, the immobiliser is activated and the warning light in the instrument panel flashes continuously. When the key is inserted and the ignition switch is turned on, the light should illuminate for a few seconds and then go out; this indicates that the immobiliser has received the correct signal from the ignition key and the engine can be started.

12 If the immobiliser warning light illuminates or flashes whilst the engine is running then the immobiliser system is faulty and the vehicle should be taken to a Renault dealer for testing.

Disconnecting the battery

13 Prior to disconnecting the battery, the alarm system must be de-activated by turning the isolation switch in the glovebox off (see paragraph 5). Failure to de-activate the alarm will lead to the siren sounding when the battery is disconnected. Once the battery is reconnected, reactivate the alarm by switching the isolation switch back on again.

23 Electric front seat components - removal and refitting

Renewal of the front seat heater pads and/or electric motors should be entrusted to a Renault dealer. Renewal involves dismantling of the complex seat assembly. Heater pad renewal is especially difficult to achieve successfully. In practice, it will be very difficult for the home mechanic to carry out the job without ruining the upholstery. The only item which is easily removed/refitted is the operating switches (see Section 4).

24 Airbag system - general information and precautions

Both a driver and passenger's airbag were fitted as standard to some models in the range; on other models they were available as an optional extra. Models fitted with a driver's side airbag have the word AIRBAG stamped on the airbag unit, which is fitted to the centre of the steering wheel. Models also equipped with a passenger side airbag also have the word AIRBAG stamped on the passenger airbag unit which is fitted to the passenger side of the facia. The airbag system comprises of the airbag unit(s) (complete with gas generators), the control unit (with an integral impact sensor) and a warning light in the instrument panel.

The airbag system is triggered in the event of a heavy frontal impact above a predetermined force; depending on the point of impact. The airbag is inflated within milliseconds and forms a safety cushion between the driver and steering wheel and (where fitted) the passenger and facia. This prevents contact between the upper body and wheel/facia and therefore greatly reduces the risk of injury. The airbag then deflates almost immediately. The control unit also operates the front seat belt tensioner mechanisms at the same time as the airbag(s) (see Chapter 11).

Every time the ignition is switched on, the airbag control unit performs a self-test. The self-test takes approximately 3 seconds and during this time the airbag warning light in the instrument panel is illuminated. After the self-test has been completed the warning light should go out. If the warning light fails to come on, remains illuminated after the initial period or comes on at any time when the vehicle is being driven, there is a fault in the airbag system. The vehicle should be taken to a Renault dealer for examination at the earliest possible opportunity.

Warning: Before carrying out any operations on the airbag system, to prevent the risk of injury if the system is triggered inadvertently when working on the vehicle, disconnect the battery and disable the system. To do this, remove the seat belt tensioner/airbag system fuse (see "Wiring diagrams") from the fusebox and wait for at least five minutes. This will allow the reserve power capacitors in the control unit to discharge. When operations are complete, refit the fuse. Make sure no one is inside the vehicle when the battery is reconnected then, with the driver's door open, switch the ignition on from outside the vehicle.

Warning: Note that the airbag(s) must not be subjected to excess temperatures. When the airbag is removed, ensure that it is stored the correct way up to prevent possible inflation.

Warning: Do not allow any solvents or cleaning agents to contact the airbag assemblies. They must be cleaned using only a damp cloth.

Warning: The airbags and control unit are both sensitive to impact. If either is dropped or damaged they should be renewed.

25.3 Lift the driver's side airbag away from the steering wheel and disconnect its wiring connector

Warning: Disconnect the airbag control unit wiring plug prior to using arc-welding equipment on the vehicle.

25 Airbag system components - removal and refitting

Note: *Refer to the warnings in Section 24 before carrying out the following operations.*

1 Disconnect the battery negative terminal, remove the seat belt tensioner/airbag system fuse (see "*Wiring diagrams*") from the fusebox and wait for at least five minutes. This will allow the reserve power capacitors in the control unit to discharge and disable the airbag system (see Section 24).

Driver's side airbag

Removal

2 Slacken and remove the two airbag retaining screws from the rear of the steering wheel, rotating the wheel as necessary to gain access to the screws.

3 Return the steering wheel to the straight-ahead position then carefully lift the airbag assembly away from the steering wheel and disconnect the wiring connector from the rear of the unit **(see illustration)**. Note that the airbag must not be knocked or dropped and should be stored the correct way up with its padded surface uppermost.

Refitting

4 Ensure that the wiring connector is securely reconnected and seat the airbag unit centrally in the steering wheel, making sure the wire does not become trapped. Fit the retaining screws and tighten them to the specified torque setting **(see illustration)**.

5 Refit the fuse and, ensuring no one is inside the vehicle, reconnect the battery. With the driver's door open, turn on the ignition switch and check the operation of the airbag warning light.

Passenger side airbag

Removal

6 Remove the facia, referring to Chapter 11.

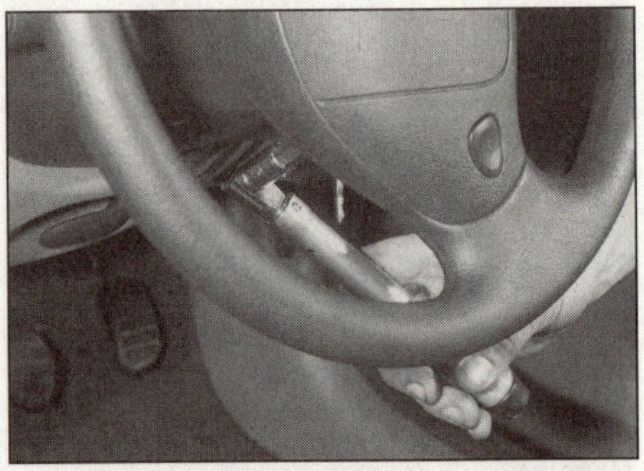

25.4 On refitting, tighten the airbag retaining screws to the specified torque

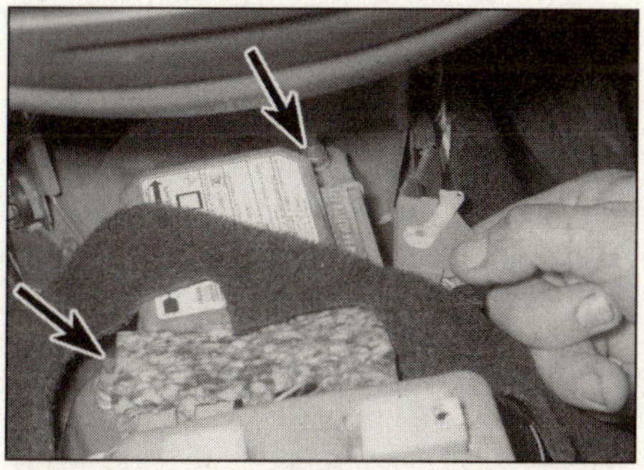

25.12 Disconnect the wiring connector then undo the retaining nuts (arrowed) and remove the airbag control unit

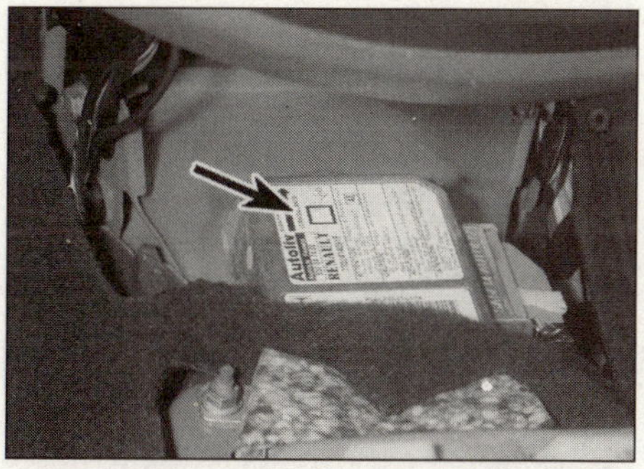

25.13 Ensure that the control unit is fitted with its arrow (arrowed) pointing towards the front of the vehicle

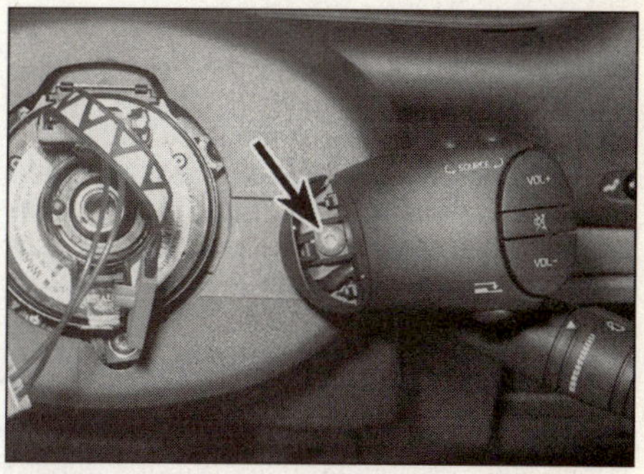

25.17 Undo the retaining screw (arrowed) and free the radio control switch from the steering column

7 Disconnect the wiring connector(s) then slacken and remove the retaining nuts/bolts. Release the airbag unit from its mountings and remove it from the facia.

Refitting

8 Manoeuvre the airbag into position, making sure it is correctly engaged with its mounting tangs, then refit the mounting nuts and tighten them securely. Reconnect the wiring connector(s), ensuring that the wiring is correctly routed and retained by all the necessary clips and ties.

9 Refit the facia as described in Chapter 11.

10 Refit the fuse and, ensuring no one is inside the vehicle, reconnect the battery. With the driver's door open, turn on the ignition switch and check the operation of the airbag warning light.

Airbag control unit

Removal

11 Remove the centre console as described in Chapter 11 to gain access to the control unit which is mounted in front of the gearchange/selector lever.

12 Release the retaining clip and disconnect the wiring connector then unscrew the retaining nuts and remove the control unit from the vehicle **(see illustration)**.

Refitting

13 Refit the control unit, making sure the arrow on the top of the unit is pointing towards the front of the vehicle **(see illustration)**. Refit the mounting nuts and tighten them securely.

14 Reconnect the wiring connector and refit the centre console as described in the relevant Section of Chapter 11.

15 Refit the fuse and, ensuring no one is inside the vehicle, reconnect the battery. With the driver's door open, turn on the ignition switch and check the operation of the airbag warning light.

Airbag wiring contact unit

Removal

16 Remove the driver's side airbag unit then remove the steering wheel as described in Chapter 10. Ensure that the front wheels are pointing in the straight-ahead position.

17 Unclip the cover from the inner end of the radio control switch assembly and slacken the switch retaining screw by a few turns **(see illustration)**.

18 Slacken and remove the retaining screws and remove the lower half of the steering column shroud. Undo the two retaining screws and remove the upper half of the steering column shroud **(see illustrations)**.

19 Disconnect the wiring connector then slacken and remove the retaining screws and remove the contact unit from the top of the steering column **(see illustrations)**.

Refitting

20 Prior to refitting it is necessary to ensure that the contact unit is correctly centralised, with the wiring connector at the top of the unit, and that the front wheels are pointing in the straight-ahead position. If there is any doubt about the contact unit position, rotate the contact unit insert in a clockwise direction until resistance is felt. From this point, rotate the insert back until its wiring connector is positioned at the top of the unit, then rotate it through a further two rotations. Do not rotate the unit from now on **(see illustration)**.

21 Slide the contact unit into position and engage it with the switch housing. Ensure that

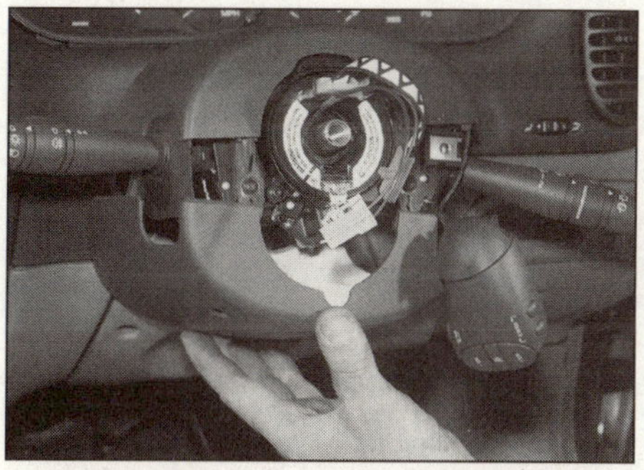

25.18a Undo the retaining screws and remove the lower shroud . . .

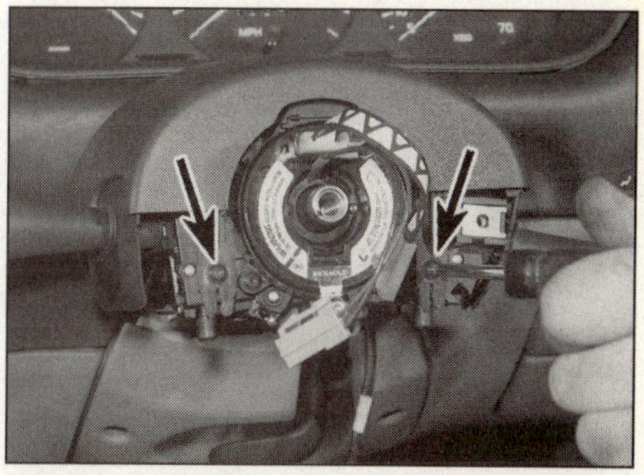

25.18b . . . then undo the retaining screws (arrowed) and lift off the upper shroud

25.19a Disconnect the wiring connector . . .

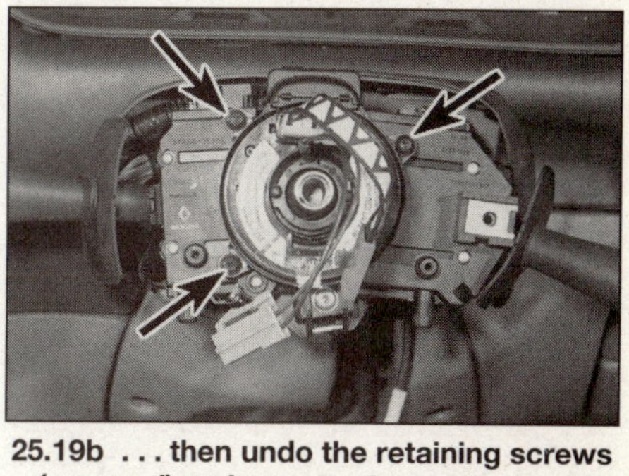

25.19b . . . then undo the retaining screws (arrowed) and remove the contact unit from the steering column

the contact ring is correctly seated and securely tighten its retaining screws.

22 Reconnect the contact unit wiring connector and refit the steering column upper shroud, tighten its retaining screws securely. Refit the lower shroud, tightening its screws securely, then retighten the radio switch stalk screw.

23 Refit the steering wheel as described in Chapter 10 then refit the airbag unit as described in paragraphs 4 and 5.

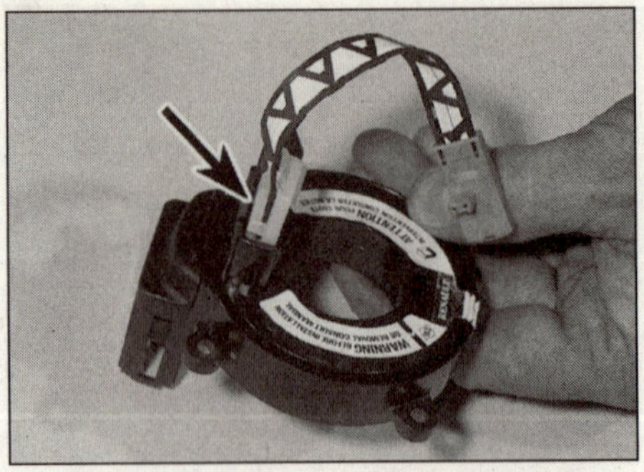

25.20 Centralise the contact unit as described in the text so that the wiring connector (arrowed) is at the top

Key to symbols

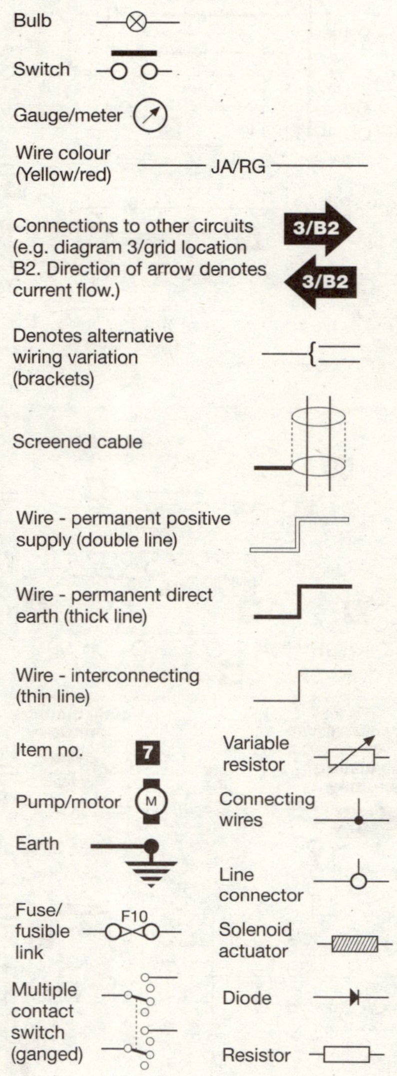

Typical passenger fusebox

Fuse	Rating	Circuit protected
F1	25A	Heated rear window
F2	7.5A	Heated mirrors
F3	7.5A	LH sidelights, glovebox
F4	7.5A	RH sidelights, instrument panel
F5	10A	RH main beam
F6	10A	LH main beam
F7	15A	Horn
F8	5A	Rear foglights
F9	15A	Interior and luggage compartment lighting
F10	15A	Clock, instrument panel, radio and memory seats
F12	10A	Headlight washer
F13	20A	Central locking
F14	15A	Front foglights
F15	25A	Memory seats
F16	20A	Cooling fan
F17	15A	Heated seats
F18	25A	Driver's electric window
F19	25A	Passenger's electric window
F20	25A	LH rear electric window
F21	25A	RH rear electric window
F22	20A	Sunroof
F23	10A	Spare
F24	20A	Electric seats
F25	10A	LH dipped beam
F26	10A	RH dipped beam
F27	20A	Heater
F28	15A	Cigar lighter, radio, clock and switch illumination
F30	15A	Windscreen wash/wipe
F31	15A	Instrument panel, airbag, pre-tensioners, decoder unit, alarm and memory seats
F33	5A	Anti-lock brake system
F34	5A	Spare
F35	7.5A	Suspension height control
F36	15A	Rear screen wash/wipe, reversing light, heated windscreen
F37	15A	Direction indicators and hazard warning lights
F38	20A	Electric seats
F39	5A	Electronic unit, remote control, alarm immobiliser
F41	2A	Injection memory
F43	30A	Accessories cut-off
F44	7.5A	Rear accessories
F45	20A	Twin headlight shunt

Passenger compartment fusebox (low specification models)

Passenger compartment fusebox (high specification models)

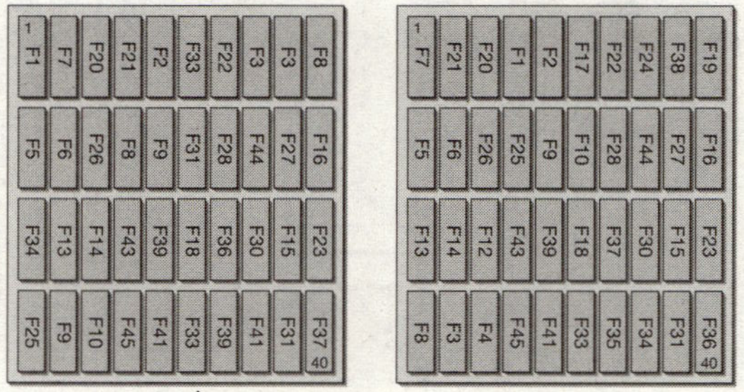

Engine compartment fusebox (typical)

Engine compartment fusebox (typical)

A Reversing light fuse
B Starter relay
C Not used
D Not used
E Fan high speed relay
F Air conditioning fan unit fuse
G Automatic transmission fuse
H Not used
I Inlet manifold heater
J Injection locking relay
K Fan low speed relay
L Diesel fuel filter heater/injection pump relay
M Engine protection fuse
N Engine functions fuse
O Air conditioning diode

H29552

Diagram 1 : Information for wiring diagrams

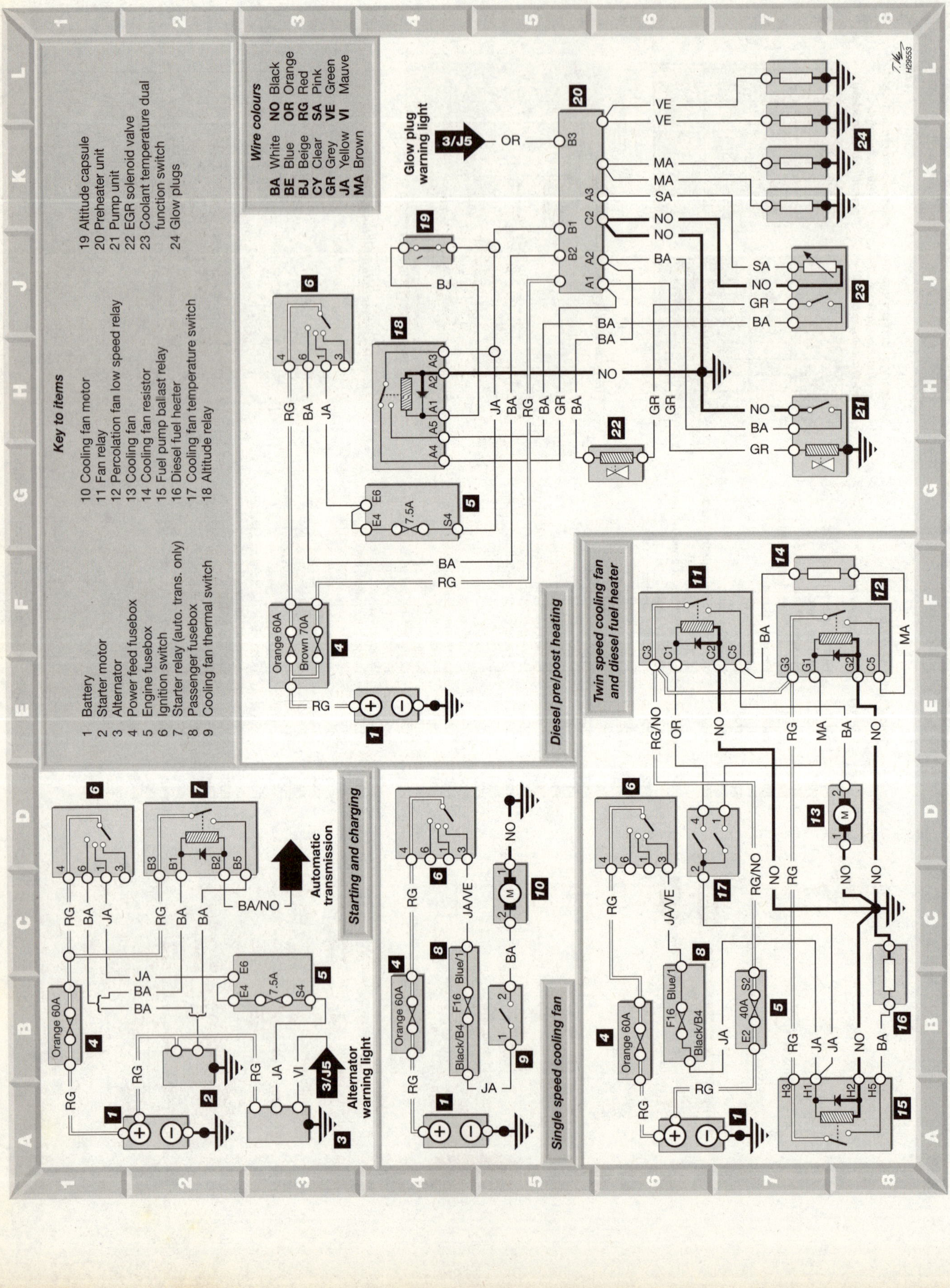

Diagram 2 : Starting, charging, engine cooling fan, diesel fuel heater, pre/post heating

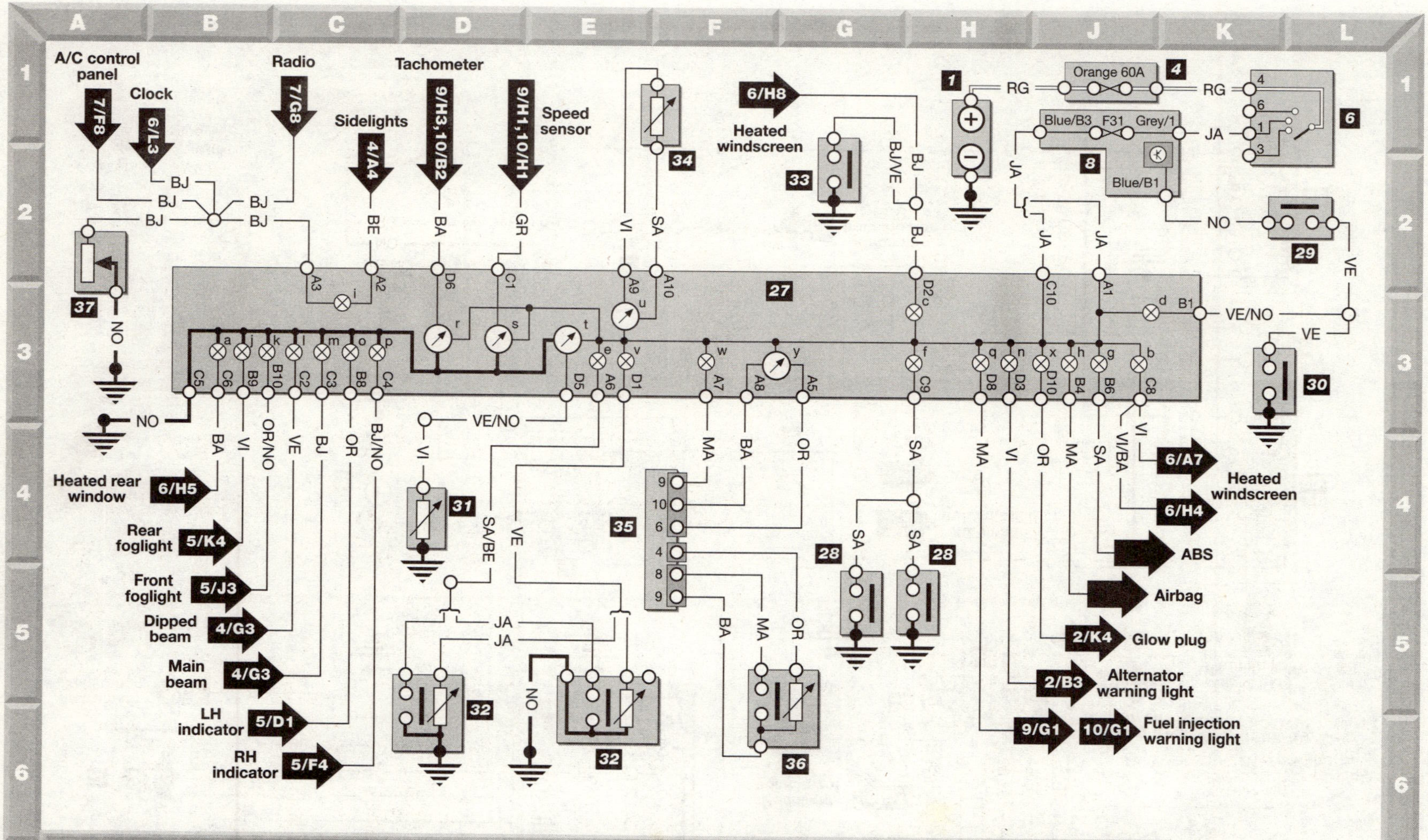

Key to items

1 Battery
4 Power feed fusebox
6 Ignition switch
8 Passenger fuse box
27 Instrument cluster
a = heated rear window warning light
b = heated windscreen warning light
c = oil pressure warning light
d = handbrake/brake system warning light
e = coolant temperature warning light
f = pad wear warning light
g = ABS warning light
h = airbag warning light
i = instrument illumination
j = rear foglight warning light
k = front foglight warning light
l = dipped beam warning light
m = high beam warning light
n = no charge warning light
o = LH indicator warning light
p = RH indicator warning light
q = electronic incident warning light
r = tachometer
s = speedometer
t = oil pressure gauge
u = oil level gauge
v = temperature gauge
w = low fuel warning light
x = preheat warning light
y = fuel gauge
28 Brake pad wear sensor
29 Low brake fluid switch
30 Handbrake switch
31 Oil pressure sensor
32 Coolant temp. sender unit
33 Oil pressure switch
34 Oil level sensor
35 Fuel gauge control unit
36 Fuel gauge sender unit
37 Interior lighting rheostat

Wire colours

BA	White	**NO**	Black
BE	Blue	**OR**	Orange
BJ	Beige	**RG**	Red
CY	Clear	**SA**	Pink
GR	Grey	**VE**	Green
JA	Yellow	**VI**	Mauve
MA	Brown		

Diagram 3 : Warning lights and gauges

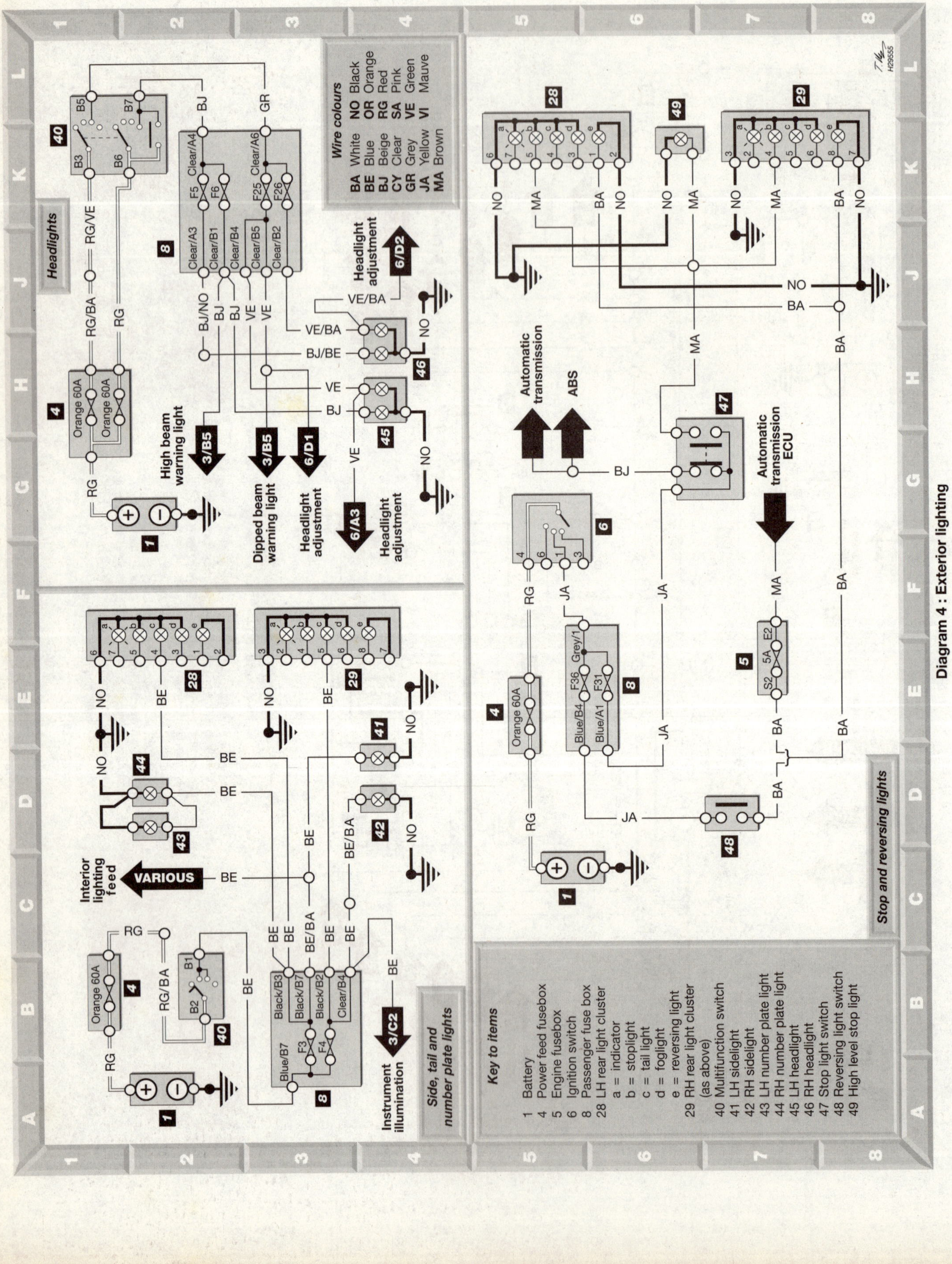
Headlights
Side, tail and number plate lights
Stop and reversing lights
Wire colours
BA White
BE Blue
BJ Beige
CY Clear
GR Grey
JA Yellow
MA Brown
NO Black
OR Orange
RG Red
SA Pink
VE Green
VI Mauve
Key to items
1 Battery
4 Power feed fusebox
5 Engine fusebox
6 Ignition switch
8 Passenger fuse box
28 LH rear light cluster
a = indicator
b = stoplight
c = tail light
d = foglight
e = reversing light
29 RH rear light cluster (as above)
40 Multifunction switch
41 LH sidelight
42 RH sidelight
43 LH number plate light
44 RH number plate light
45 LH headlight
46 RH headlight
47 Stop light switch
48 Reversing light switch
49 High level stop light
High beam warning light
Dipped beam warning light
Headlight adjustment
Interior lighting feed
VARIOUS
Instrument illumination
Automatic transmission
ABS
Automatic transmission ECU
H29555
Diagram 4 : Exterior lighting

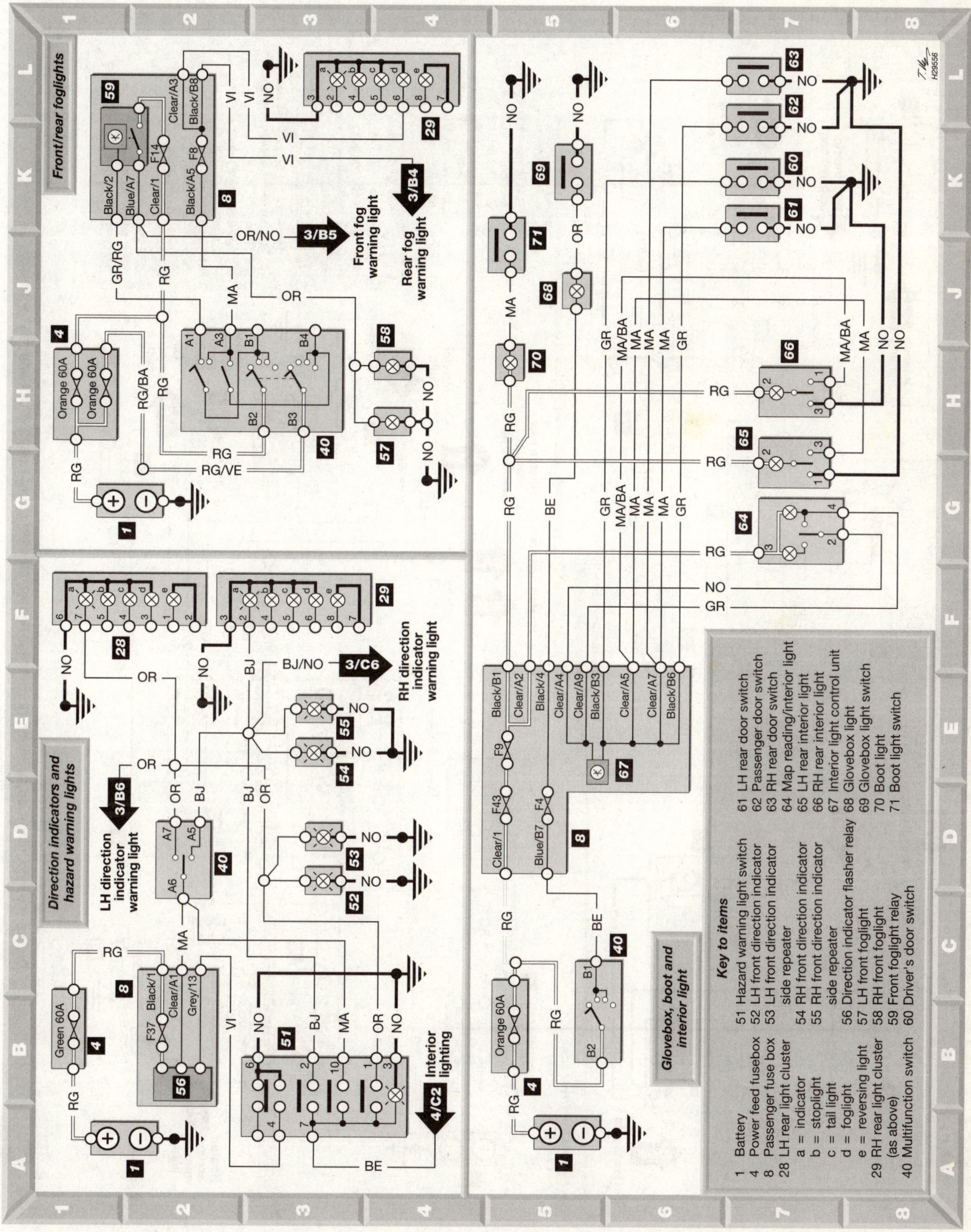

Diagram 5 : Exterior lighting continued, and interior lighting

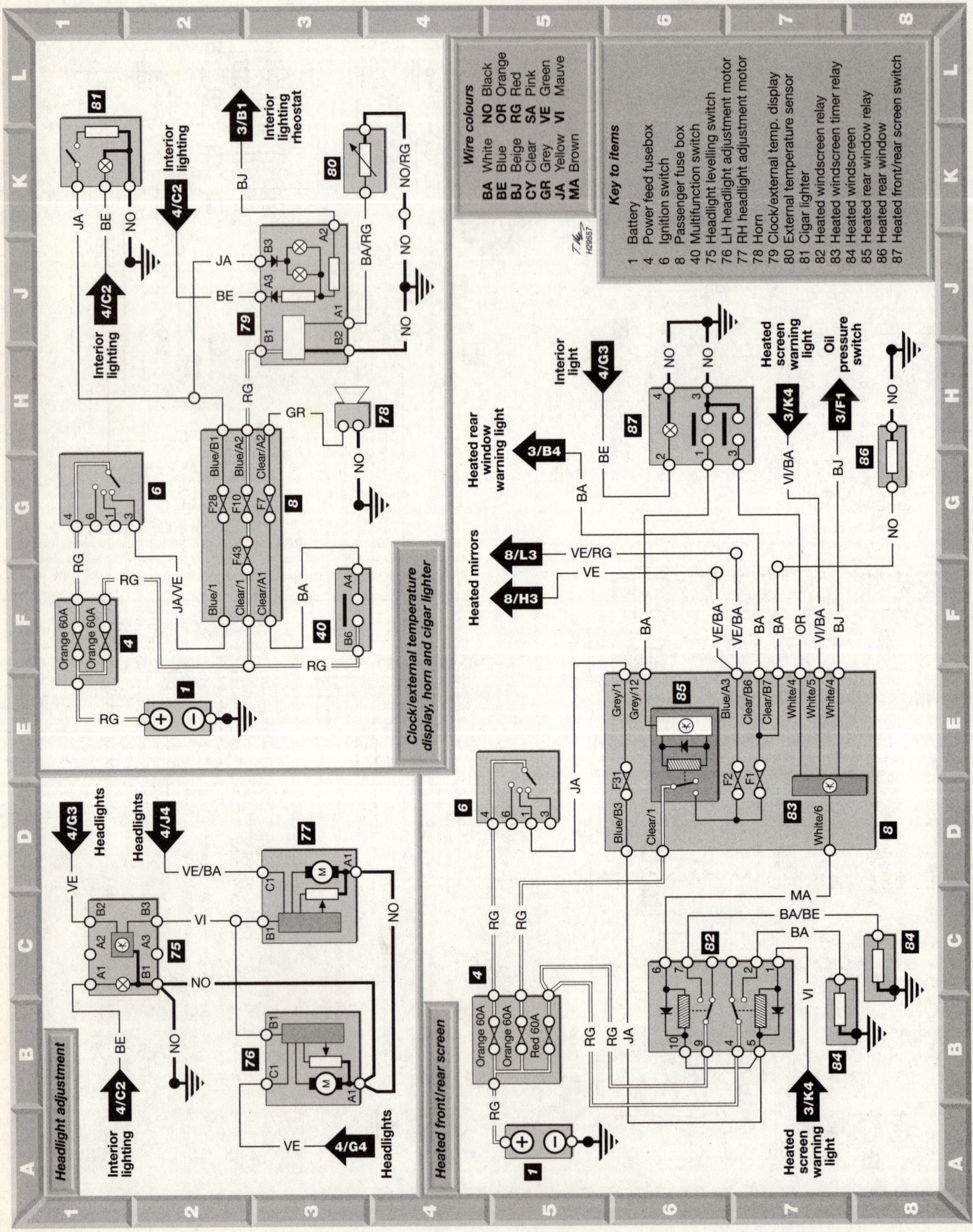

Diagram 6 : Headlight adjustment, horn, cigar lighter, clock/display, heated front and rear screens

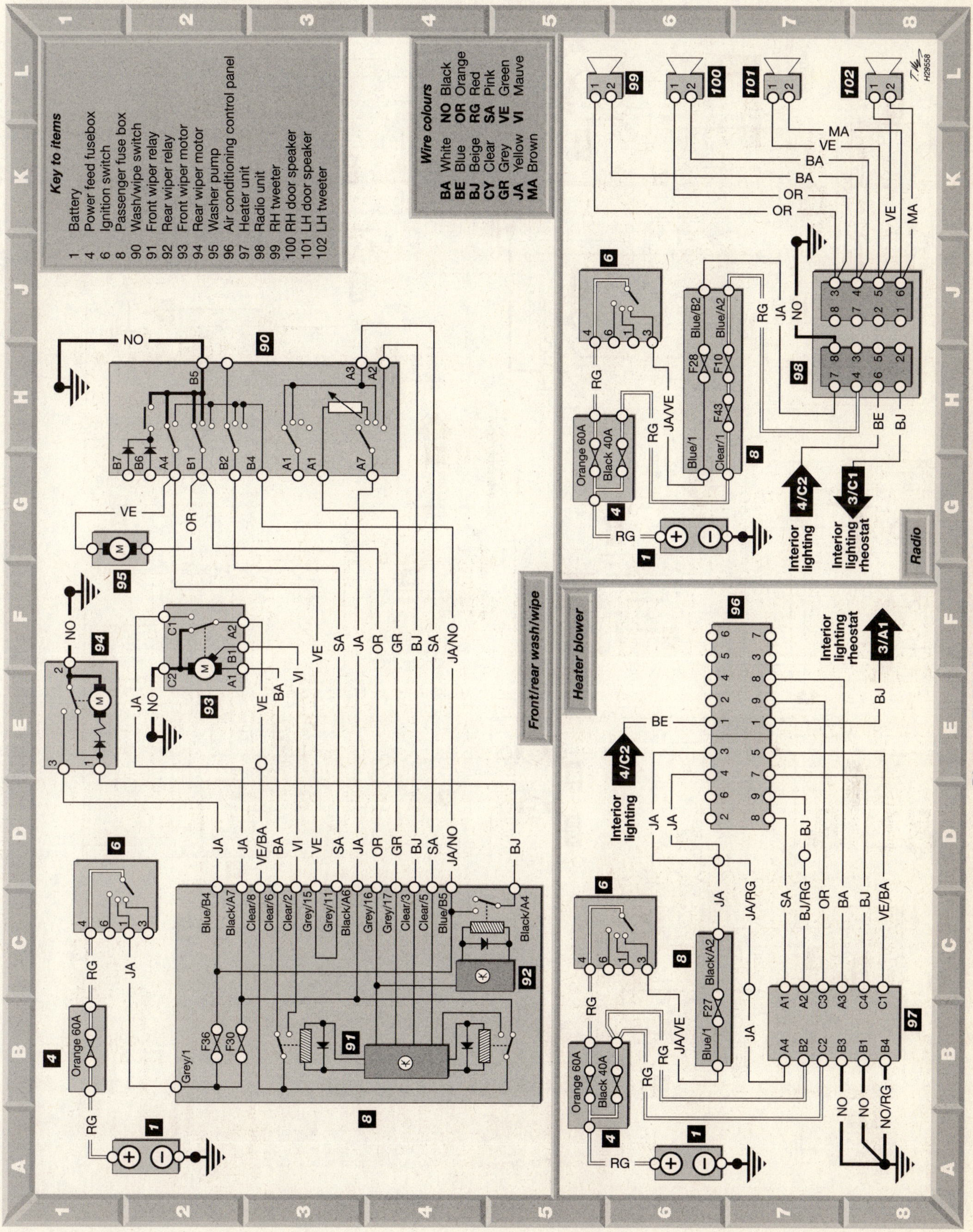

Diagram 7 : Wash/wipe, heater blower and radio

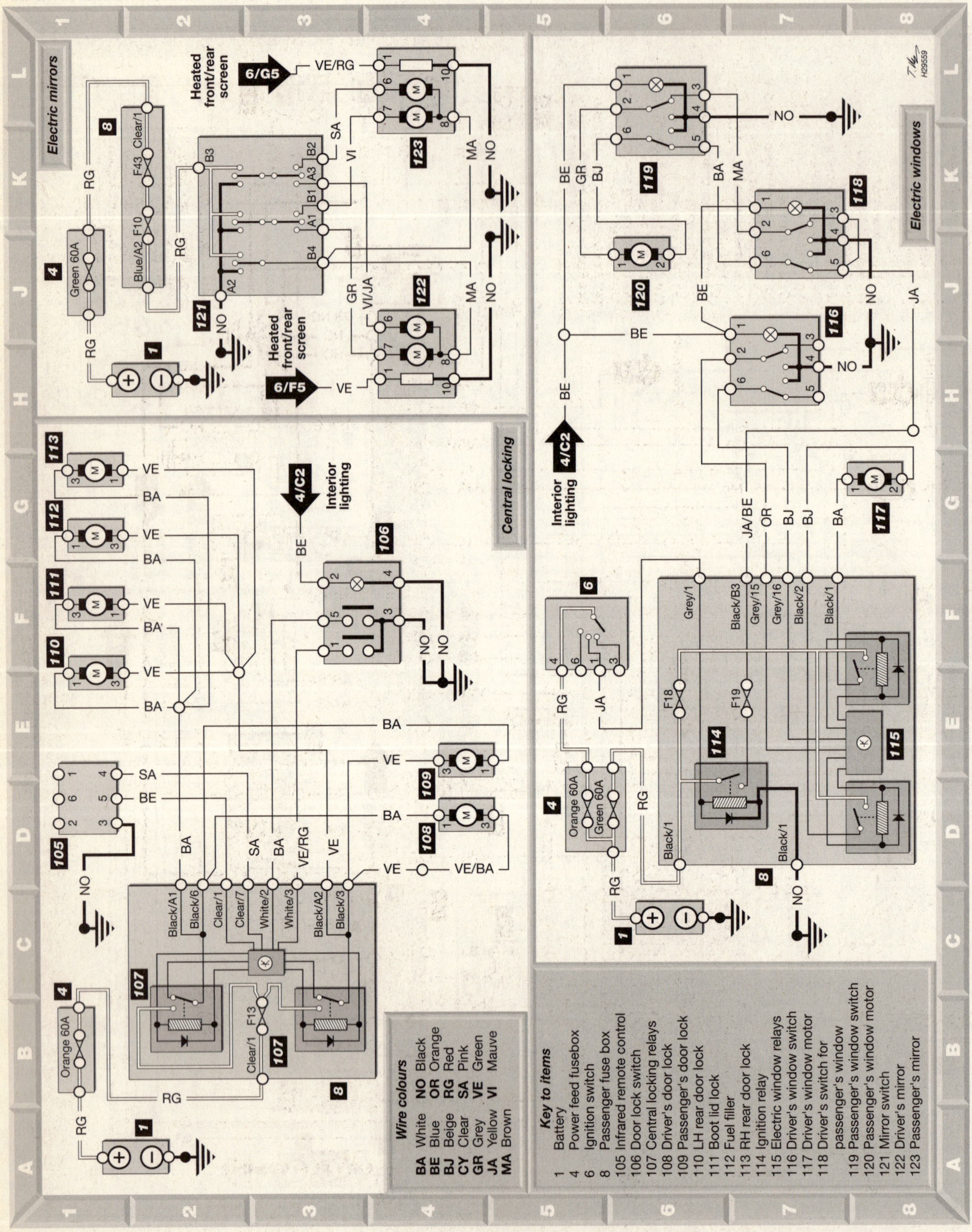

Diagram 8 : Central locking, electric windows and mirrors

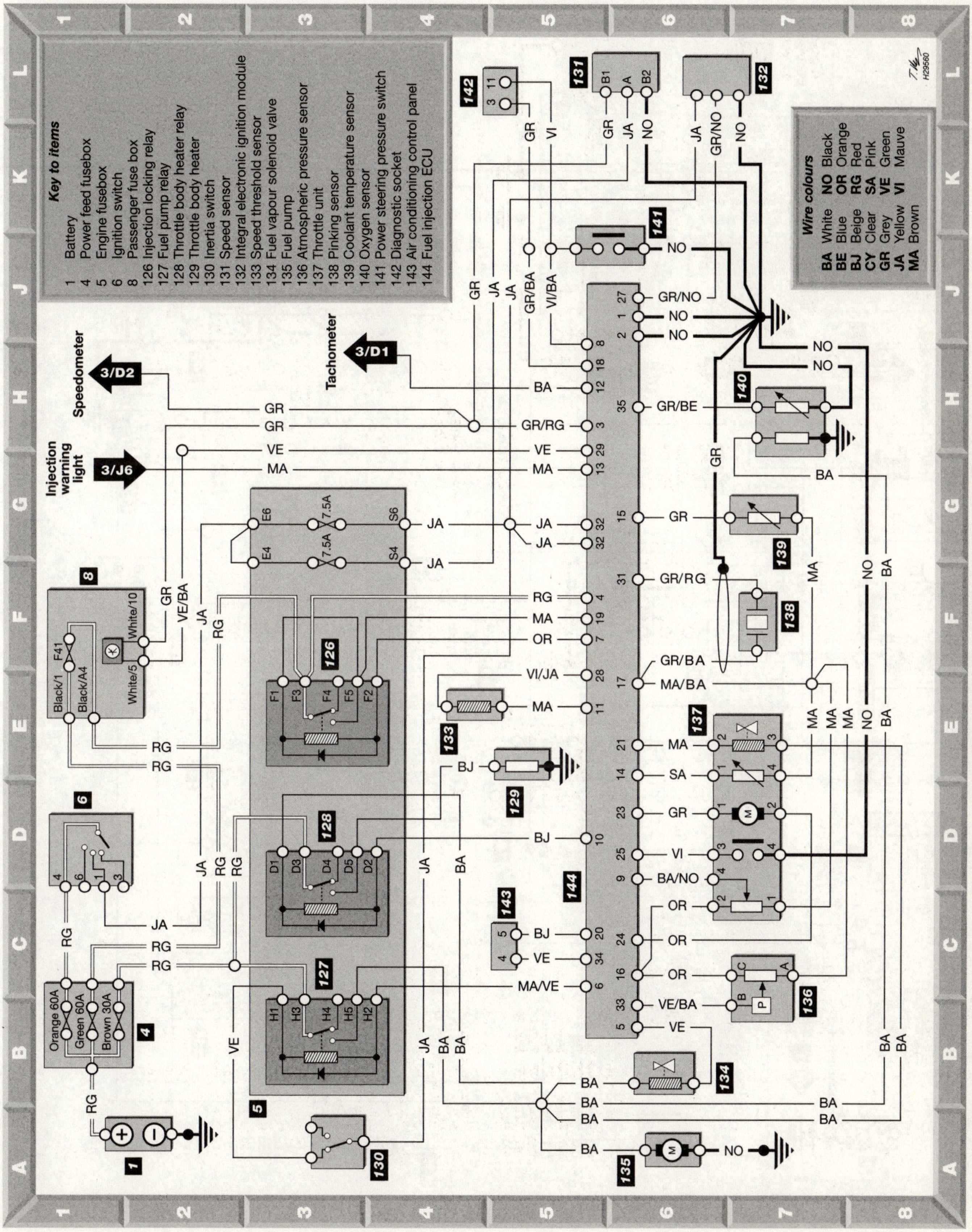

Diagram 9 : Typical single-point fuel injection system

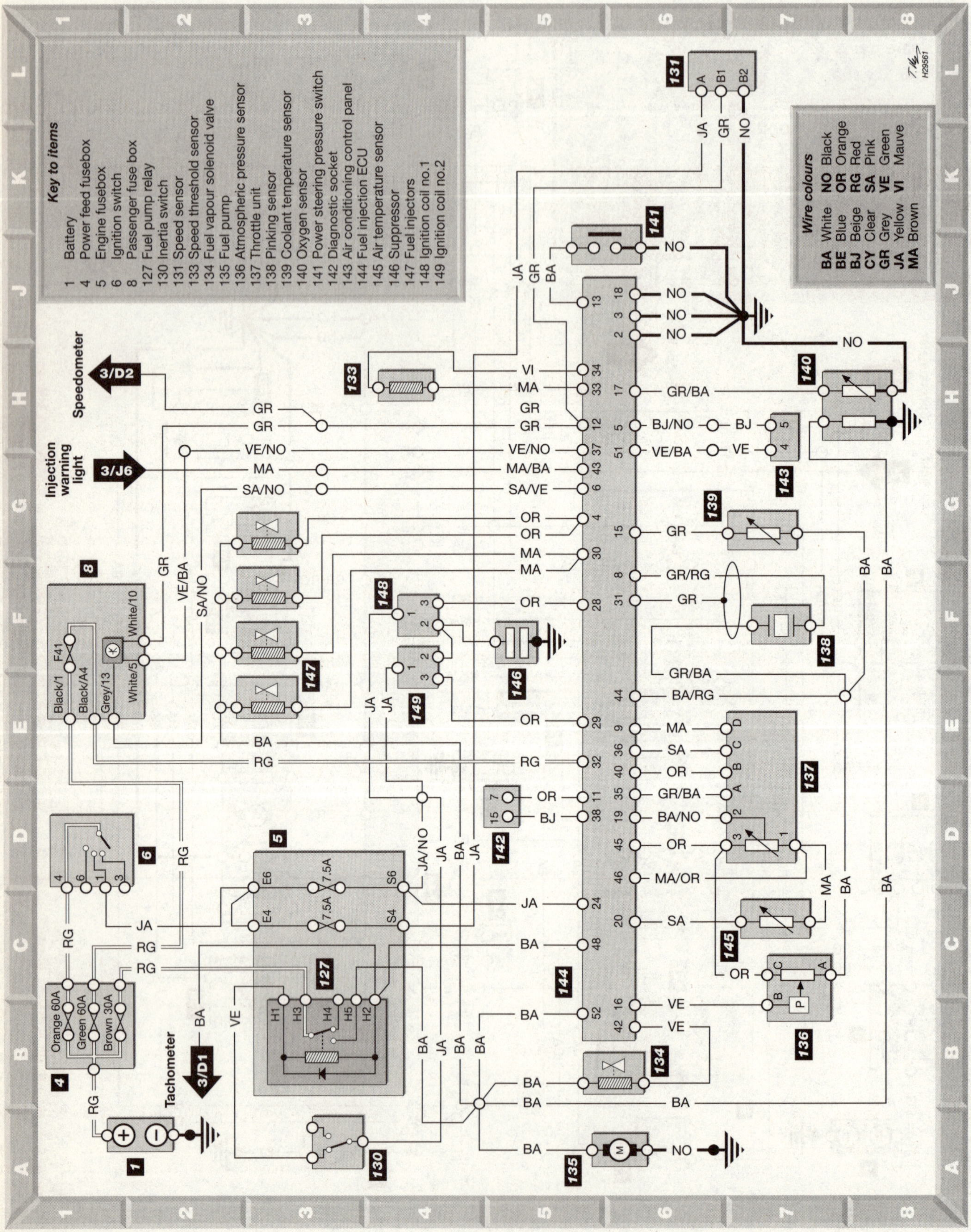

Diagram 10 : Typical multi-point fuel injection system

Dimensions and weights

Note: *All figures are approximate, and may vary according to model. Refer to manufacturer's data for exact figures.*

Dimensions

Overall length .... 4129 mm
Overall width (excluding wing mirrors) .... 1699 mm
Overall height (unladen) .... 1420 mm
Wheelbase .... 2580 mm

Weights

Kerb weight:
- 1.4 litre petrol engine models .... 1015 to 1025 kg
- 1.6 litre petrol engine models:
  - Manual transmission models .... 1055 to 1065 kg
  - Automatic transmission models .... 1085 to 1095 kg
- 1.9 litre diesel engine models (except Van):
  - Non-turbo models .... 1110 to 1115 kg
  - Turbo models .... 1130 to 1140 kg

Maximum roof rack load .... 80 kg

Maximum towing weight:
- Unbraked trailer:
  - 1.4 litre petrol engine models .... 545 kg
  - 1.6 litre petrol engine models:
    - Manual transmission models .... 565 kg
    - Automatic transmission models .... 580 kg
  - 1.9 litre diesel engine models:
    - Non-turbo models .... 590 kg
    - Turbo models .... 600 kg
- Braked trailer:
  - 1.4 litre petrol engine models .... 950 kg
  - 1.6 litre petrol engine models:
    - Manual transmission models .... 1000 kg
    - Automatic transmission models .... 1000 kg
  - 1.9 litre diesel engine models:
    - Non-turbo models .... 925 kg
    - Turbo models .... 1000 kg

Conversion factors

Length (distance)

Inches (in)	x 25.4 =	Millimetres (mm)	x 0.0394 =	Inches (in)
Feet (ft)	x 0.305 =	Metres (m)	x 3.281 =	Feet (ft)
Miles	x 1.609 =	Kilometres (km)	x 0.621 =	Miles

Volume (capacity)

Cubic inches (cu in; in^3)	x 16.387 =	Cubic centimetres (cc; cm^3)	x 0.061 =	Cubic inches (cu in; in^3)
Imperial pints (Imp pt)	x 0.568 =	Litres (l)	x 1.76 =	Imperial pints (Imp pt)
Imperial quarts (Imp qt)	x 1.137 =	Litres (l)	x 0.88 =	Imperial quarts (Imp qt)
Imperial quarts (Imp qt)	x 1.201 =	US quarts (US qt)	x 0.833 =	Imperial quarts (Imp qt)
US quarts (US qt)	x 0.946 =	Litres (l)	x 1.057 =	US quarts (US qt)
Imperial gallons (Imp gal)	x 4.546 =	Litres (l)	x 0.22 =	Imperial gallons (Imp gal)
Imperial gallons (Imp gal)	x 1.201 =	US gallons (US gal)	x 0.833 =	Imperial gallons (Imp gal)
US gallons (US gal)	x 3.785 =	Litres (l)	x 0.264 =	US gallons (US gal)

Mass (weight)

Ounces (oz)	x 28.35 =	Grams (g)	x 0.035 =	Ounces (oz)
Pounds (lb)	x 0.454 =	Kilograms (kg)	x 2.205 =	Pounds (lb)

Force

Ounces-force (ozf; oz)	x 0.278 =	Newtons (N)	x 3.6 =	Ounces-force (ozf; oz)
Pounds-force (lbf; lb)	x 4.448 =	Newtons (N)	x 0.225 =	Pounds-force (lbf; lb)
Newtons (N)	x 0.1 =	Kilograms-force (kgf; kg)	x 9.81 =	Newtons (N)

Pressure

Pounds-force per square inch (psi; lbf/in^2; lb/in^2)	x 0.070 =	Kilograms-force per square centimetre (kgf/cm^2; kg/cm^2)	x 14.223 =	Pounds-force per square inch (psi; lbf/in^2; lb/in^2)
Pounds-force per square inch (psi; lbf/in^2; lb/in^2)	x 0.068 =	Atmospheres (atm)	x 14.696 =	Pounds-force per square inch (psi; lbf/in^2; lb/in^2)
Pounds-force per square inch (psi; lbf/in^2; lb/in^2)	x 0.069 =	Bars	x 14.5 =	Pounds-force per square inch (psi; lbf/in^2; lb/in^2)
Pounds-force per square inch (psi; lbf/in^2; lb/in^2)	x 6.895 =	Kilopascals (kPa)	x 0.145 =	Pounds-force per square inch (psi; lbf/in^2; lb/in^2)
Kilopascals (kPa)	x 0.01 =	Kilograms-force per square centimetre (kgf/cm^2; kg/cm^2)	x 98.1 =	Kilopascals (kPa)
Millibar (mbar)	x 100 =	Pascals (Pa)	x 0.01 =	Millibar (mbar)
Millibar (mbar)	x 0.0145 =	Pounds-force per square inch (psi; lbf/in^2; lb/in^2)	x 68.947 =	Millibar (mbar)
Millibar (mbar)	x 0.75 =	Millimetres of mercury (mmHg)	x 1.333 =	Millibar (mbar)
Millibar (mbar)	x 0.401 =	Inches of water (inH$_2$O)	x 2.491 =	Millibar (mbar)
Millimetres of mercury (mmHg)	x 0.535 =	Inches of water (inH$_2$O)	x 1.868 =	Millimetres of mercury (mmHg)
Inches of water (inH$_2$O)	x 0.036 =	Pounds-force per square inch (psi; lbf/in^2; lb/in^2)	x 27.68 =	Inches of water (inH$_2$O)

Torque (moment of force)

Pounds-force inches (lbf in; lb in)	x 1.152 =	Kilograms-force centimetre (kgf cm; kg cm)	x 0.868 =	Pounds-force inches (lbf in; lb in)
Pounds-force inches (lbf in; lb in)	x 0.113 =	Newton metres (Nm)	x 8.85 =	Pounds-force inches (lbf in; lb in)
Pounds-force inches (lbf in; lb in)	x 0.083 =	Pounds-force feet (lbf ft; lb ft)	x 12 =	Pounds-force inches (lbf in; lb in)
Pounds-force feet (lbf ft; lb ft)	x 0.138 =	Kilograms-force metres (kgf m; kg m)	x 7.233 =	Pounds-force feet (lbf ft; lb ft)
Pounds-force feet (lbf ft; lb ft)	x 1.356 =	Newton metres (Nm)	x 0.738 =	Pounds-force feet (lbf ft; lb ft)
Newton metres (Nm)	x 0.102 =	Kilograms-force metres (kgf m; kg m)	x 9.804 =	Newton metres (Nm)

Power

Horsepower (hp)	x 745.7 =	Watts (W)	x 0.0013 =	Horsepower (hp)

Velocity (speed)

Miles per hour (miles/hr; mph)	x 1.609 =	Kilometres per hour (km/hr; kph)	x 0.621 =	Miles per hour (miles/hr; mph)

Fuel consumption*

Miles per gallon (mpg)	x 0.354 =	Kilometres per litre (km/l)	x 2.825 =	Miles per gallon (mpg)

Temperature

Degrees Fahrenheit = (°C x 1.8) + 32

Degrees Celsius (Degrees Centigrade; °C) = (°F - 32) x 0.56

** It is common practice to convert from miles per gallon (mpg) to litres/100 kilometres (l/100km), where mpg x l/100 km = 282*

Buying spare parts

Spare parts are available from many sources, including maker's appointed garages, accessory shops, and motor factors. To be sure of obtaining the correct parts, it will sometimes be necessary to quote the vehicle identification number. If possible, it can also be useful to take the old parts along for positive identification. Items such as starter motors and alternators may be available under a service exchange scheme - any parts returned should be clean.

Our advice regarding spare parts is as follows.

Officially appointed garages

This is the best source of parts which are peculiar to your car, and which are not otherwise generally available (eg, badges, interior trim, certain body panels, etc). It is also the only place at which you should buy parts if the vehicle is still under warranty.

Accessory shops

These are very good places to buy materials and components needed for the maintenance of your car (oil, air and fuel filters, light bulbs, drivebelts, greases, brake pads, tough-up paint, etc). Components of this nature sold by a reputable shop are of the same standard as those used by the car manufacturer.

Besides components, these shops also sell tools and general accessories, usually have convenient opening hours, charge lower prices, and can often be found close to home. Some accessory shops have parts counters where components needed for almost any repair job can be purchased or ordered.

Motor factors

Good factors will stock all the more important components which wear out comparatively quickly, and can sometimes supply individual components needed for the overhaul of a larger assembly (eg, brake seals and hydraulic parts, bearing shells, pistons, valves). They may also handle work such as cylinder block reboring, crankshaft regrinding, etc.

Tyre and exhaust specialists

These outlets may be independent, or members of a local or national chain. They frequently offer competitive prices when compared with a main dealer or local garage, but it will pay to obtain several quotes before making a decision. When researching prices, also ask what "extras" may be added - for instance fitting a new valve and balancing the wheel are both commonly charged on top of the price of a new tyre.

Other sources

Beware of parts or materials obtained from market stalls, car boot sales or similar outlets. Such items are not invariably sub-standard, but there is little chance of compensation if they do prove unsatisfactory. In the case of safety-critical components such as brake pads, there is the risk not only of financial loss, but also of an accident causing injury or death.

Second-hand components or assemblies obtained from a car breaker or "scrapyard" can be a good buy in some circumstances, but this sort of purchase is best made by the experienced DIY mechanic.

Vehicle identification

Modifications are a continuing and unpublicised process in vehicle manufacture, quite apart from major model changes. Spare parts manuals and lists are compiled upon a numerical basis, the individual vehicle identification numbers being essential to correct identification of the component concerned.

When ordering spare parts, always give as much information as possible. Quote the car model, year of manufacture, body and engine numbers as appropriate.

The *Vehicle Identification Number (VIN)* adhesive label is located either on the right-hand front suspension turret in the engine compartment, or in the spare wheel well - an additional label is also located on the right-hand centre pillar **(see illustrations)**.

The *engine number plate* is riveted to the front left-hand corner of the cylinder block **(see illustrations)**. **Note:** *The first part of the engine number is the engine code - eg, "K7M".*

The *chassis number* is stamped into the top of the right-hand front suspension turret **(see illustration)**.

Vehicle identification plate numbers

1 *Vehicle type and chassis number*
2 *Max. all-up weight*
3 *Max. total train weight*
4 *Max. front axle load*
5 *Max. rear axle load*
6 *Trim code*
7 *Special points*
8 *Type of vehicle*
9 *Paint reference*
10 *Equipment level*
11-14 *Fabrication no.*
12 *Options*
13 *Version*

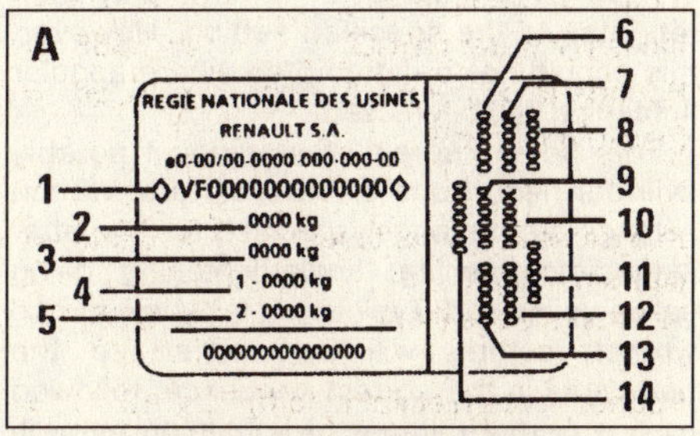

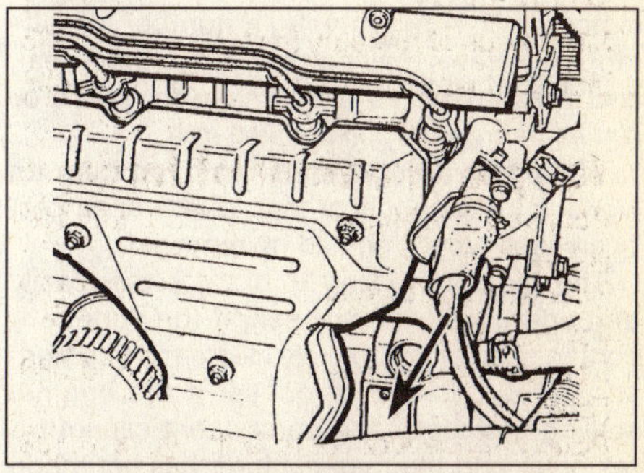

Engine number plate location on E7J and K7M petrol engines

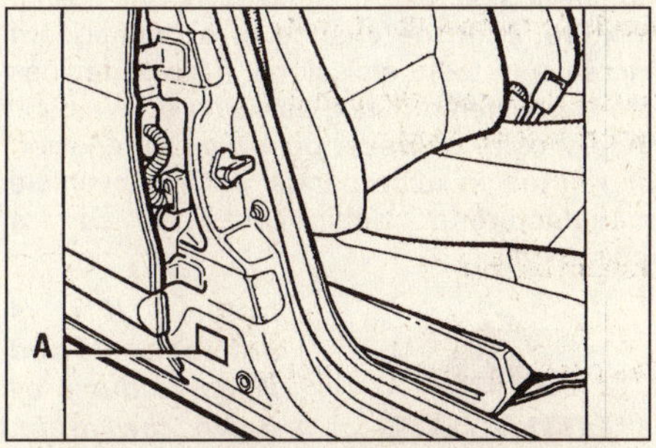

Additional identification label (A) located on the right-hand centre pillar

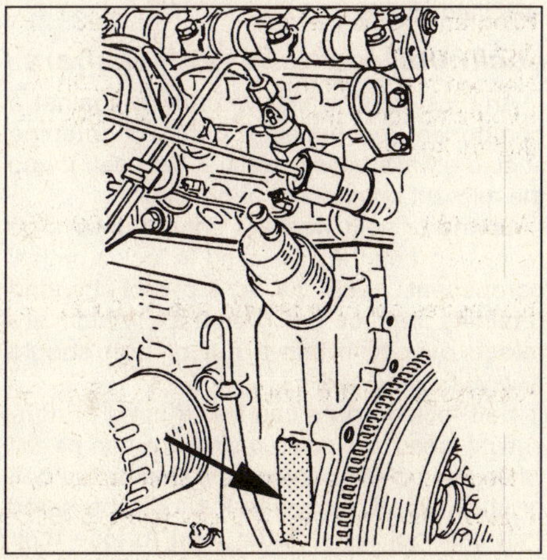

Engine number plate location on the F8Q diesel engine

The chassis number is stamped onto the top of the right-hand front suspension turret

Whenever servicing, repair or overhaul work is carried out on the car or its components, it is necessary to observe the following procedures and instructions. This will assist in carrying out the operation efficiently and to a professional standard of workmanship.

Joint mating faces and gaskets

When separating components at their mating faces, never insert screwdrivers or similar implements into the joint between the faces in order to prise them apart. This can cause severe damage which results in oil leaks, coolant leaks, etc upon reassembly. Separation is usually achieved by tapping along the joint with a soft-faced hammer in order to break the seal. However, note that this method may not be suitable where dowels are used for component location.

Where a gasket is used between the mating faces of two components, ensure that it is renewed on reassembly, and fit it dry unless otherwise stated in the repair procedure. Make sure that the mating faces are clean and dry, with all traces of old gasket removed. When cleaning a joint face, use a tool which is not likely to score or damage the face, and remove any burrs or nicks with an oilstone or fine file.

Make sure that tapped holes are cleaned with a pipe cleaner, and keep them free of jointing compound, if this is being used, unless specifically instructed otherwise.

Ensure that all orifices, channels or pipes are clear, and blow through them, preferably using compressed air.

Oil seals

Oil seals can be removed by levering them out with a wide flat-bladed screwdriver or similar tool. Alternatively, a number of self-tapping screws may be screwed into the seal, and these used as a purchase for pliers or similar in order to pull the seal free.

Whenever an oil seal is removed from its working location, either individually or as part of an assembly, it should be renewed.

The very fine sealing lip of the seal is easily damaged, and will not seal if the surface it contacts is not completely clean and free from scratches, nicks or grooves. If the original sealing surface of the component cannot be restored, and the manufacturer has not made provision for slight relocation of the seal relative to the sealing surface, the component should be renewed.

Protect the lips of the seal from any surface which may damage them in the course of fitting. Use tape or a conical sleeve where possible. Lubricate the seal lips with oil before fitting and, on dual-lipped seals, fill the space between the lips with grease.

Unless otherwise stated, oil seals must be fitted with their sealing lips toward the lubricant to be sealed.

Use a tubular drift or block of wood of the appropriate size to install the seal and, if the seal housing is shouldered, drive the seal down to the shoulder. If the seal housing is unshouldered, the seal should be fitted with its face flush with the housing top face (unless otherwise instructed).

Screw threads and fastenings

Seized nuts, bolts and screws are quite a common occurrence where corrosion has set in, and the use of penetrating oil or releasing fluid will often overcome this problem if the offending item is soaked for a while before attempting to release it. The use of an impact driver may also provide a means of releasing such stubborn fastening devices, when used in conjunction with the appropriate screwdriver bit or socket. If none of these methods works, it may be necessary to resort to the careful application of heat, or the use of a hacksaw or nut splitter device.

Studs are usually removed by locking two nuts together on the threaded part, and then using a spanner on the lower nut to unscrew the stud. Studs or bolts which have broken off below the surface of the component in which they are mounted can sometimes be removed using a stud extractor. Always ensure that a blind tapped hole is completely free from oil, grease, water or other fluid before installing the bolt or stud. Failure to do this could cause the housing to crack due to the hydraulic action of the bolt or stud as it is screwed in.

When tightening a castellated nut to accept a split pin, tighten the nut to the specified torque, where applicable, and then tighten further to the next split pin hole. Never slacken the nut to align the split pin hole, unless stated in the repair procedure.

When checking or retightening a nut or bolt to a specified torque setting, slacken the nut or bolt by a quarter of a turn, and then retighten to the specified setting. However, this should not be attempted where angular tightening has been used.

For some screw fastenings, notably cylinder head bolts or nuts, torque wrench settings are no longer specified for the latter stages of tightening, "angle-tightening" being called up instead. Typically, a fairly low torque wrench setting will be applied to the bolts/nuts in the correct sequence, followed by one or more stages of tightening through specified angles.

Locknuts, locktabs and washers

Any fastening which will rotate against a component or housing during tightening should always have a washer between it and the relevant component or housing.

Spring or split washers should always be renewed when they are used to lock a critical component such as a big-end bearing retaining bolt or nut. Locktabs which are folded over to retain a nut or bolt should always be renewed.

Self-locking nuts can be re-used in non-critical areas, providing resistance can be felt when the locking portion passes over the bolt or stud thread. However, it should be noted that self-locking stiffnuts tend to lose their effectiveness after long periods of use, and should be renewed as a matter of course.

Split pins must always be replaced with new ones of the correct size for the hole.

When thread-locking compound is found on the threads of a fastener which is to be re-used, it should be cleaned off with a wire brush and solvent, and fresh compound applied on reassembly.

Special tools

Some repair procedures in this manual entail the use of special tools such as a press, two or three-legged pullers, spring compressors, etc. Wherever possible, suitable readily-available alternatives to the manufacturer's special tools are described, and are shown in use. In some instances, where no alternative is possible, it has been necessary to resort to the use of a manufacturer's tool, and this has been done for reasons of safety as well as the efficient completion of the repair operation. Unless you are highly-skilled and have a thorough understanding of the procedures described, never attempt to bypass the use of any special tool when the procedure described specifies its use. Not only is there a very great risk of personal injury, but expensive damage could be caused to the components involved.

Environmental considerations

When disposing of used engine oil, brake fluid, antifreeze, etc, give due consideration to any detrimental environmental effects. Do not, for instance, pour any of the above liquids down drains into the general sewage system, or onto the ground to soak away. Many local council refuse tips provide a facility for waste oil disposal, as do some garages. If none of these facilities are available, consult your local Environmental Health Department, or the National Rivers Authority, for further advice.

With the universal tightening-up of legislation regarding the emission of environmentally-harmful substances from motor vehicles, most current vehicles have tamperproof devices fitted to the main adjustment points of the fuel system. These devices are primarily designed to prevent unqualified persons from adjusting the fuel/air mixture, with the chance of a consequent increase in toxic emissions. If such devices are encountered during servicing or overhaul, they should, wherever possible, be renewed or refitted in accordance with the vehicle manufacturer's requirements or current legislation.

OIL CARE
FOLLOW THE CODE
OIL BANK LINE
0800 66 33 66

***Note:** It is antisocial and illegal to dump oil down the drain. To find the location of your local oil recycling bank, call this number free.*

The jack supplied with the vehicle tool kit should only be used for changing the roadwheels - see *"Wheel changing"* at the front of this manual. When carrying out any other kind of work, raise the vehicle using a hydraulic (or "trolley") jack, and always supplement the jack with axle stands positioned under the vehicle jacking points.

When using a hydraulic jack or axle stands, always position the jack head or axle stand head under, or adjacent to one of the relevant wheel changing jacking points under the sills. Use a block of wood between the jack or axle stand and the sill - the block of wood should have a groove cut into it, in which the weld flange of the sill will locate **(see illustrations)**.

The front of the vehicle can be raised using a substantial wooden block located across both body side-members, just behind the front crossmember. Place the jack under the wooden block.

Do not attempt to jack the vehicle under the front crossmember, the sump, or any of the suspension components.

The jack supplied with the vehicle locates in the jacking points on the underside of the sills - see *"Wheel changing"* at the front of this manual. Ensure that the jack head is correctly engaged before attempting to raise the vehicle.

Never work under, around, or near a raised vehicle, unless it is adequately supported in at least two places.

Use a wooden block between the jack or axle stands and sill

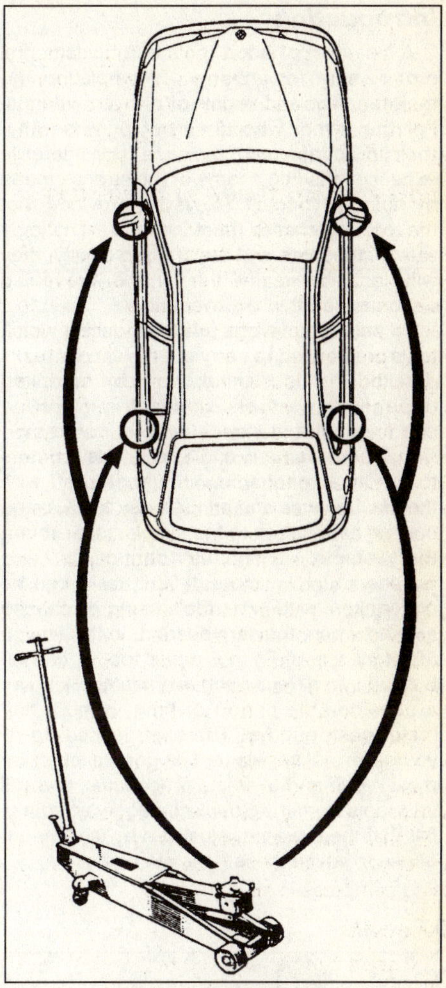

Vehicle jacking points

Radio/cassette unit anti-theft system - precaution

The radio/cassette unit fitted as standard equipment by Renault may be equipped with a built-in security code, to deter thieves. If the power source to the unit is cut, the anti-theft system will activate. Even if the power source is immediately reconnected, the radio/cassette unit will not function until the correct security code has been entered. Therefore if you do not know the correct security code for the unit, **do not** disconnect the battery negative lead, or remove the radio/cassette unit from the car.

A several different types of radio/cassette player may be fitted, with different methods of entering the security code.

If the incorrect code is entered a number of times, the unit will lock.

If the security code is lost or forgotten, seek the advice of your Renault dealer. On presentation of proof of ownership, a Renault dealer will be able to provide you with a new security code.

Tools and working facilities

Introduction

A selection of good tools is a fundamental requirement for anyone contemplating the maintenance and repair of a motor vehicle. For the owner who does not possess any, their purchase will prove a considerable expense, offsetting some of the savings made by doing-it-yourself. However, provided that the tools purchased meet the relevant national safety standards and are of good quality, they will last for many years and prove an extremely worthwhile investment.

To help the average owner to decide which tools are needed to carry out the various tasks detailed in this manual, we have compiled three lists of tools under the following headings: *Maintenance and minor repair*, *Repair and overhaul*, and *Special*. Newcomers to practical mechanics should start off with the *Maintenance and minor repair* tool kit, and confine themselves to the simpler jobs around the vehicle. Then, as confidence and experience grow, more difficult tasks can be undertaken, with extra tools being purchased as, and when, they are needed. In this way, a *Maintenance and minor repair* tool kit can be built up into a *Repair and overhaul* tool kit over a considerable period of time, without any major cash outlays. The experienced do-it-yourselfer will have a tool kit good enough for most repair and overhaul procedures, and will add tools from the *Special* category when it is felt that the expense is justified by the amount of use to which these tools will be put.

Maintenance and minor repair tool kit

The tools given in this list should be considered as a minimum requirement if routine maintenance, servicing and minor repair operations are to be undertaken. We recommend the purchase of combination spanners (ring one end, open-ended the other); although more expensive than open-ended ones, they do give the advantages of both types of spanner.

- ☐ *Combination spanners:*
  Metric - 8 to 19 mm inclusive
- ☐ *Adjustable spanner - 35 mm jaw (approx.)*
- ☐ *Spark plug spanner (with rubber insert) - petrol models*
- ☐ *Spark plug gap adjustment tool - petrol models*
- ☐ *Set of feeler blades*
- ☐ *Brake bleed nipple spanner*
- ☐ *Screwdrivers:*
  Flat blade - 100 mm long x 6 mm dia
  Cross blade - 100 mm long x 6 mm dia
- ☐ *Combination pliers*
- ☐ *Hacksaw (junior)*
- ☐ *Tyre pump*
- ☐ *Tyre pressure gauge*
- ☐ *Oil can*
- ☐ *Oil filter removal tool*
- ☐ *Fine emery cloth*
- ☐ *Wire brush (small)*
- ☐ *Funnel (medium size)*

Repair and overhaul tool kit

These tools are virtually essential for anyone undertaking any major repairs to a motor vehicle, and are additional to those given in the *Maintenance and minor repair* list. Included in this list is a comprehensive set of sockets. Although these are expensive, they will be found invaluable as they are so versatile - particularly if various drives are included in the set. We recommend the half-inch square-drive type, as this can be used with most proprietary torque wrenches.

The tools in this list will sometimes need to be supplemented by tools from the *Special* list:

- ☐ *Sockets (or box spanners) to cover range in previous list (including Torx sockets)*
- ☐ *Reversible ratchet drive (for use with sockets)*
- ☐ *Extension piece, 250 mm (for use with sockets)*
- ☐ *Universal joint (for use with sockets)*
- ☐ *Torque wrench (for use with sockets)*
- ☐ *Self-locking grips*
- ☐ *Ball pein hammer*
- ☐ *Soft-faced mallet (plastic/aluminium or rubber)*
- ☐ *Screwdrivers:*
  Flat blade - long & sturdy, short (chubby), and narrow (electrician's) types
  Cross blade – Long & sturdy, and short (chubby) types
- ☐ *Pliers:*
  Long-nosed
  Side cutters (electrician's)
  Circlip (internal and external)
- ☐ *Cold chisel - 25 mm*
- ☐ *Scriber*
- ☐ *Scraper*
- ☐ *Centre-punch*
- ☐ *Pin punch*
- ☐ *Hacksaw*
- ☐ *Brake hose clamp*
- ☐ *Brake/clutch bleeding kit*
- ☐ *Selection of twist drills*
- ☐ *Steel rule/straight-edge*
- ☐ *Allen keys (inc. splined/Torx type)*
- ☐ *Selection of files*
- ☐ *Wire brush*
- ☐ *Axle stands*
- ☐ *Jack (strong trolley or hydraulic type)*
- ☐ *Light with extension lead*

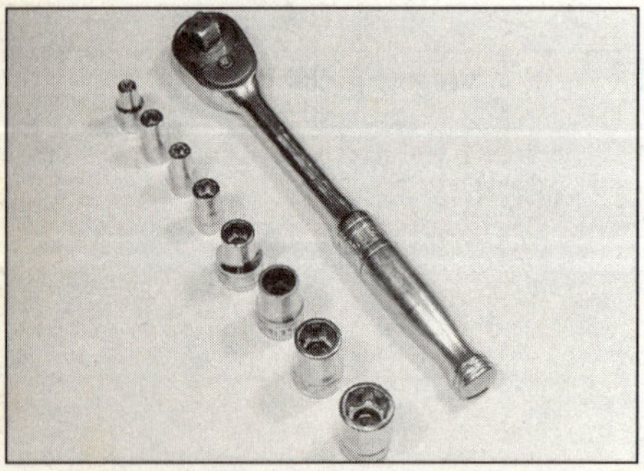

Sockets and reversible ratchet drive

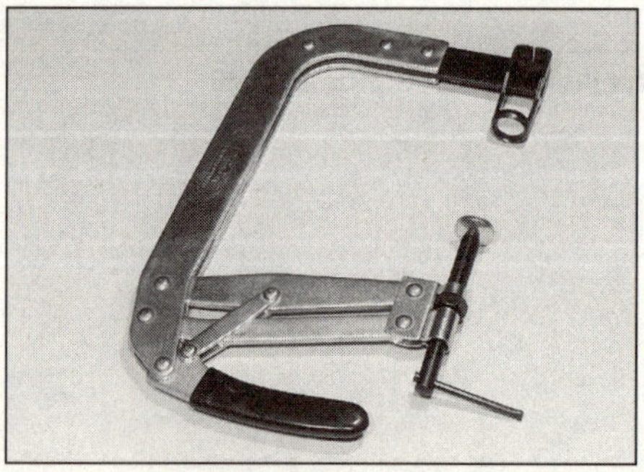

Valve spring compressor

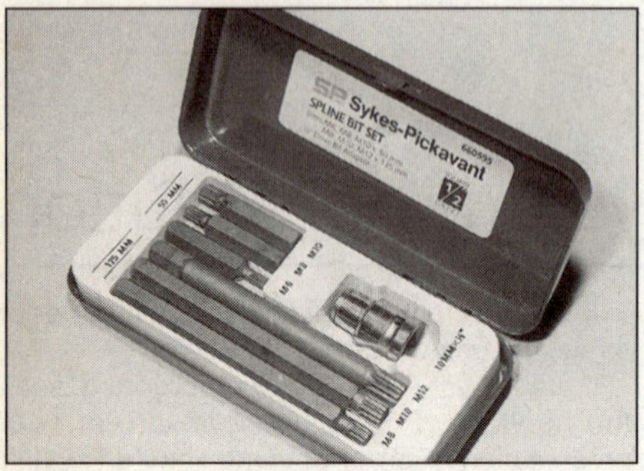

Spline bit set

Piston ring compressor

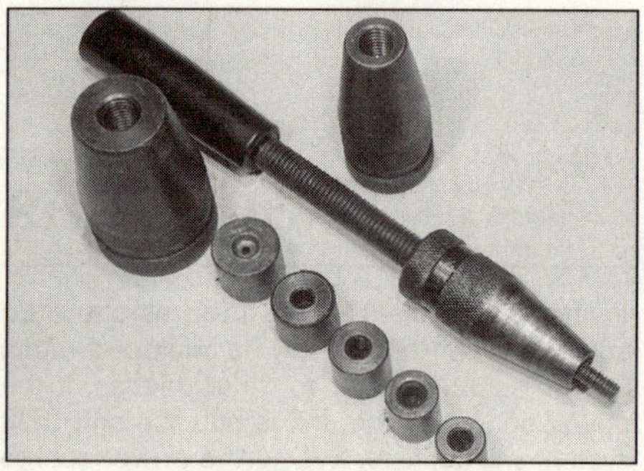

Clutch plate alignment set

Special tools

The tools in this list are those which are not used regularly, are expensive to buy, or which need to be used in accordance with their manufacturers' instructions. Unless relatively difficult mechanical jobs are undertaken frequently, it will not be economic to buy many of these tools. Where this is the case, you could consider clubbing together with friends (or joining a motorists' club) to make a joint purchase, or borrowing the tools against a deposit from a local garage or tool hire specialist. It is worth noting that many of the larger DIY superstores now carry a large range of special tools for hire at modest rates.

The following list contains only those tools and instruments freely available to the public, and not those special tools produced by the vehicle manufacturer specifically for its dealer network. You will find occasional references to these manufacturers' special tools in the text of this manual. Generally, an alternative method of doing the job without the vehicle manufacturers' special tool is given. However, sometimes there is no alternative to using them. Where this is the case and the relevant tool cannot be bought or borrowed, you will have to entrust the work to a dealer.

- ☐ *Valve spring compressor*
- ☐ *Valve grinding tool*
- ☐ *Piston ring compressor*
- ☐ *Piston ring removal/installation tool*
- ☐ *Cylinder bore hone*
- ☐ *Balljoint separator*
- ☐ *Coil spring compressors (where applicable)*
- ☐ *Two/three-legged hub and bearing puller*
- ☐ *Impact screwdriver*
- ☐ *Micrometer and/or vernier calipers*
- ☐ *Dial gauge*
- ☐ *Stroboscopic timing light*
- ☐ *Dwell angle meter/tachometer*
- ☐ *Universal electrical multi-meter*
- ☐ *Cylinder compression gauge*
- ☐ *Hand-operated vacuum pump and gauge*
- ☐ *Clutch plate alignment set*
- ☐ *Brake shoe steady spring cup removal tool*
- ☐ *Bush and bearing removal/installation set*
- ☐ *Stud extractors*
- ☐ *Tap and die set*
- ☐ *Lifting tackle*
- ☐ *Trolley jack*

Buying tools

Reputable motor accessory shops and superstores often offer excellent quality tools at discount prices, so it pays to shop around.

Remember, you don't have to buy the most expensive items on the shelf, but it is always advisable to steer clear of the very cheap tools. Beware of 'bargains' offered on market stalls or at car boot sales. There are plenty of good tools around at reasonable prices, but always aim to purchase items which meet the relevant national safety standards. If in doubt, ask the proprietor or manager of the shop for advice before making a purchase.

Care and maintenance of tools

Having purchased a reasonable tool kit, it is necessary to keep the tools in a clean and serviceable condition. After use, always wipe off any dirt, grease and metal particles using a clean, dry cloth, before putting the tools away. Never leave them lying around after they have been used. A simple tool rack on the garage or workshop wall for items such as screwdrivers and pliers is a good idea. Store all normal spanners and sockets in a metal box. Any measuring instruments, gauges, meters, etc, must be carefully stored where they cannot be damaged or become rusty.

Take a little care when tools are used. Hammer heads inevitably become marked, and screwdrivers lose the keen edge on their blades from time to time. A little timely attention with emery cloth or a file will soon restore items like this to a good finish.

Working facilities

Not to be forgotten when discussing tools is the workshop itself. If anything more than routine maintenance is to be carried out, a suitable working area becomes essential.

It is appreciated that many an owner-mechanic is forced by circumstances to remove an engine or similar item without the benefit of a garage or workshop. Having done this, any repairs should always be done under the cover of a roof.

Wherever possible, any dismantling should be done on a clean, flat workbench or table at a suitable working height.

Any workbench needs a vice; one with a jaw opening of 100 mm is suitable for most jobs. As mentioned previously, some clean dry storage space is also required for tools, as well as for any lubricants, cleaning fluids, touch-up paints etc, which become necessary.

Another item which may be required, and which has a much more general usage, is an electric drill with a chuck capacity of at least 8 mm. This, together with a good range of twist drills, is virtually essential for fitting accessories.

Last, but not least, always keep a supply of old newspapers and clean, lint-free rags available, and try to keep any working area as clean as possible.

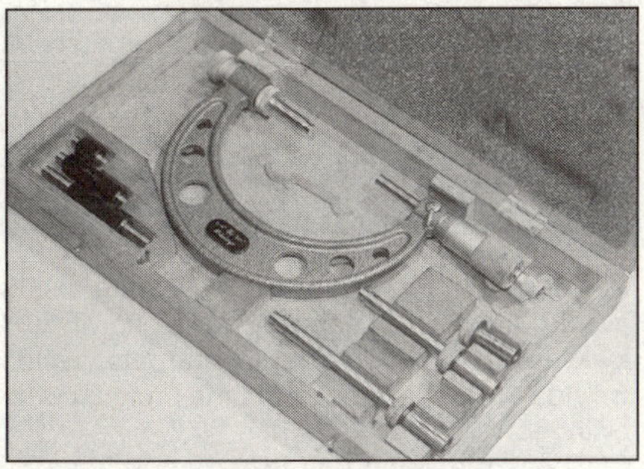

Micrometer set

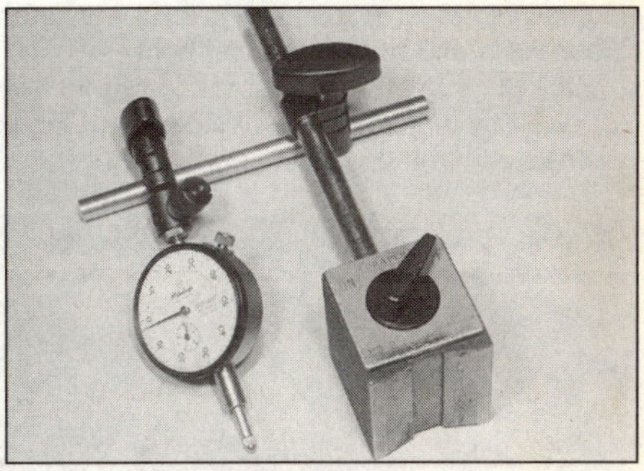

Dial test indicator ("dial gauge")

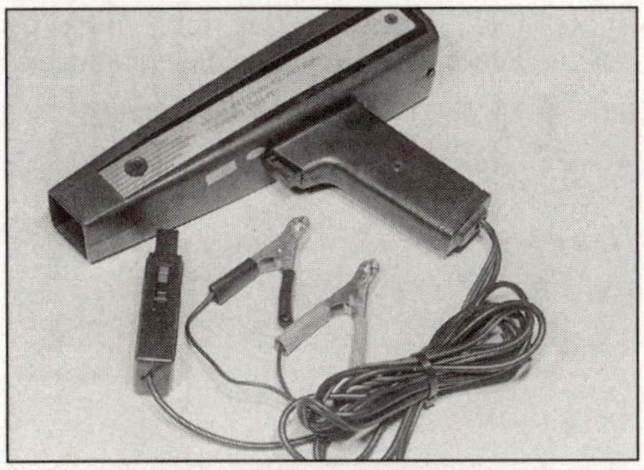

Stroboscopic timing light

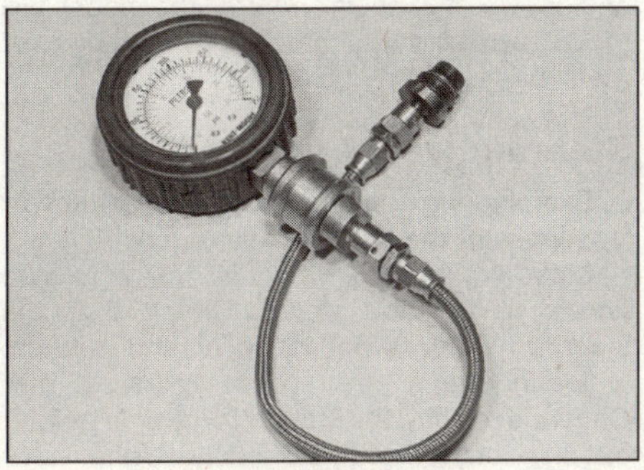

Compression tester

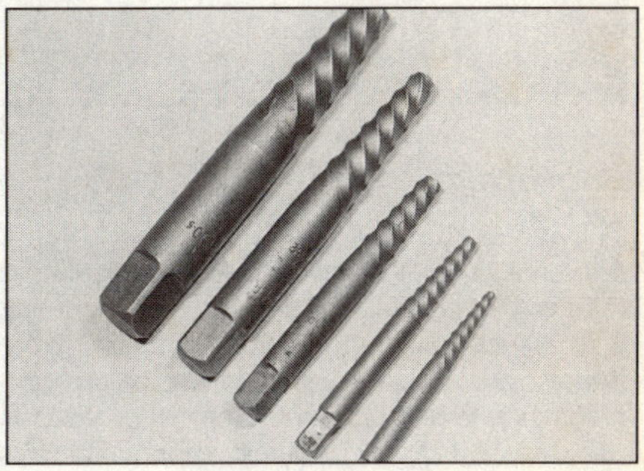

Stud extractor set

This is a guide to getting your vehicle through the MOT test. Obviously it will not be possible to examine the vehicle to the same standard as the professional MOT tester. However, working through the following checks will enable you to identify any problem areas before submitting the vehicle for the test.

Where a testable component is in borderline condition, the tester has discretion in deciding whether to pass or fail it. The basis of such discretion is whether the tester would be happy for a close relative or friend to use the vehicle with the component in that condition. If the vehicle presented is clean and evidently well cared for, the tester may be more inclined to pass a borderline component than if the vehicle is scruffy and apparently neglected.

It has only been possible to summarise the test requirements here, based on the regulations in force at the time of printing. Test standards are becoming increasingly stringent, although there are some exemptions for older vehicles. For full details obtain a copy of the Haynes publication Pass the MOT! (available from stockists of Haynes manuals).

An assistant will be needed to help carry out some of these checks.

The checks have been sub-divided into four categories, as follows:

1 Checks carried out **FROM THE DRIVER'S SEAT**

2 Checks carried out **WITH THE VEHICLE ON THE GROUND**

3 Checks carried out **WITH THE VEHICLE RAISED AND THE WHEELS FREE TO TURN**

4 Checks carried out on **YOUR VEHICLE'S EXHAUST EMISSION SYSTEM**

1 Checks carried out FROM THE DRIVER'S SEAT

Handbrake

☐ Test the operation of the handbrake. Excessive travel (too many clicks) indicates incorrect brake or cable adjustment.

☐ Check that the handbrake cannot be released by tapping the lever sideways. Check the security of the lever mountings.

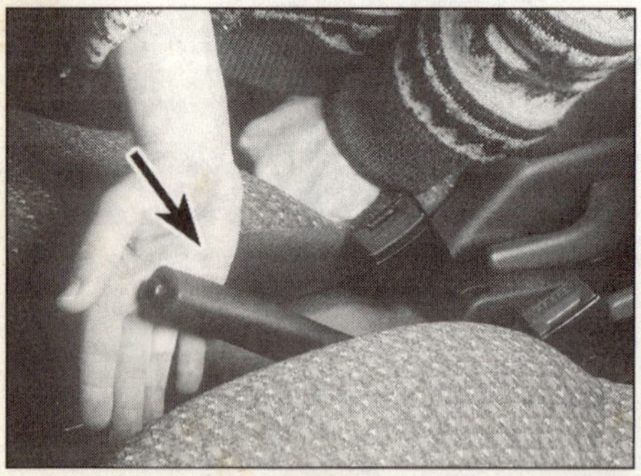

Footbrake

☐ Depress the brake pedal and check that it does not creep down to the floor, indicating a master cylinder fault. Release the pedal, wait a few seconds, then depress it again. If the pedal travels nearly to the floor before firm resistance is felt, brake adjustment or repair is necessary. If the pedal feels spongy, there is air in the hydraulic system which must be removed by bleeding.

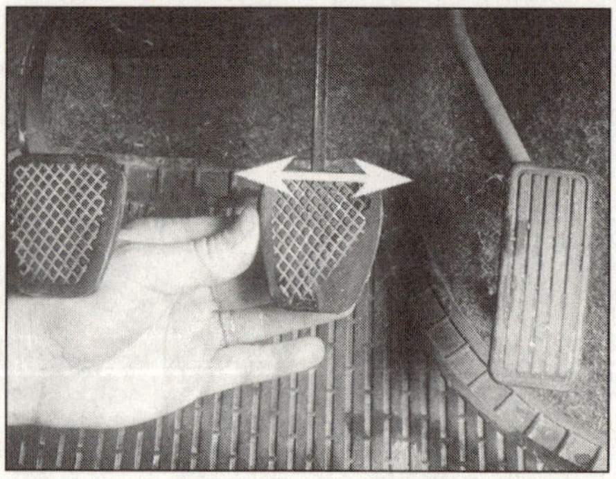

☐ Check that the brake pedal is secure and in good condition. Check also for signs of fluid leaks on the pedal, floor or carpets, which would indicate failed seals in the brake master cylinder.

☐ Check the servo unit (when applicable) by operating the brake pedal several times, then keeping the pedal depressed and starting the engine. As the engine starts, the pedal will move down slightly. If not, the vacuum hose or the servo itself may be faulty.

Steering wheel and column

☐ Examine the steering wheel for fractures or looseness of the hub, spokes or rim.

☐ Move the steering wheel from side to side and then up and down. Check that the steering wheel is not loose on the column, indicating wear or a loose retaining nut. Continue moving the steering wheel as before, but also turn it slightly from left to right.

☐ Check that the steering wheel is not loose on the column, and that there is no abnormal

movement of the steering wheel, indicating wear in the column support bearings or couplings.

Windscreen and mirrors

☐ The windscreen must be free of cracks or other significant damage within the driver's field of view. (Small stone chips are acceptable.) Rear view mirrors must be secure, intact, and capable of being adjusted.

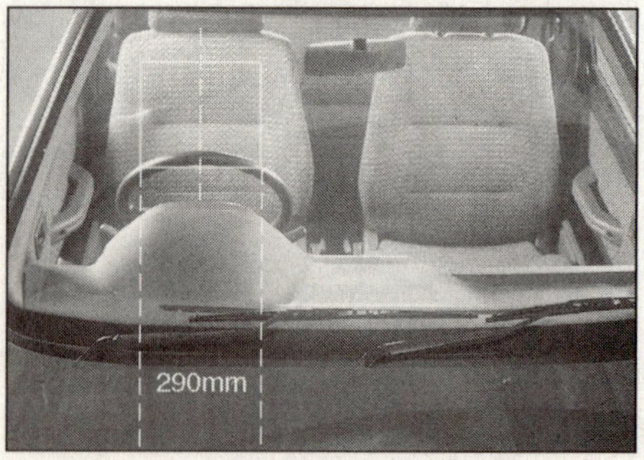

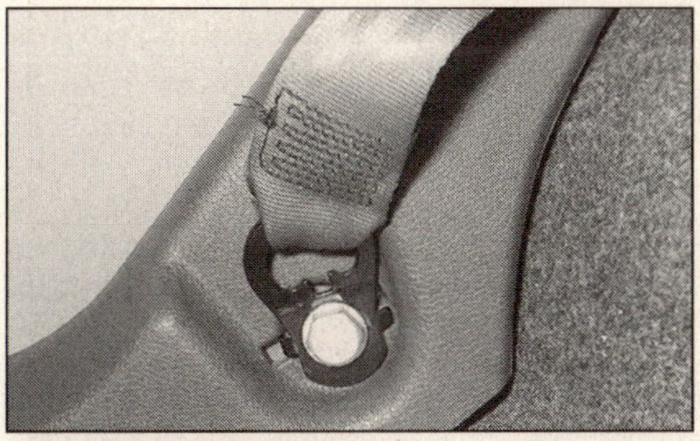

Seat belts and seats

Note: *The following checks are applicable to all seat belts, front and rear.*

□ Examine the webbing of all the belts (including rear belts if fitted) for cuts, serious fraying or deterioration. Fasten and unfasten each belt to check the buckles. If applicable, check the retracting mechanism. Check the security of all seat belt mountings accessible from inside the vehicle.

□ The front seats themselves must be securely attached and the backrests must lock in the upright position.

Doors

□ Both front doors must be able to be opened and closed from outside and inside, and must latch securely when closed.

2 Checks carried out **WITH THE VEHICLE ON THE GROUND**

Vehicle identification

□ Number plates must be in good condition, secure and legible, with letters and numbers correctly spaced – spacing at (A) should be twice that at (B).

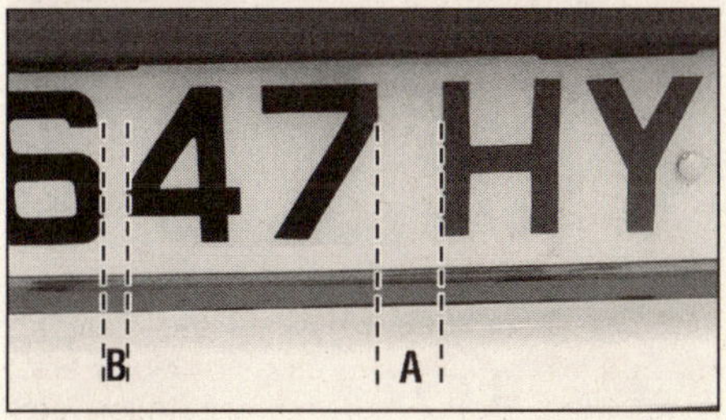

□ The VIN plate and/or homologation plate must be legible.

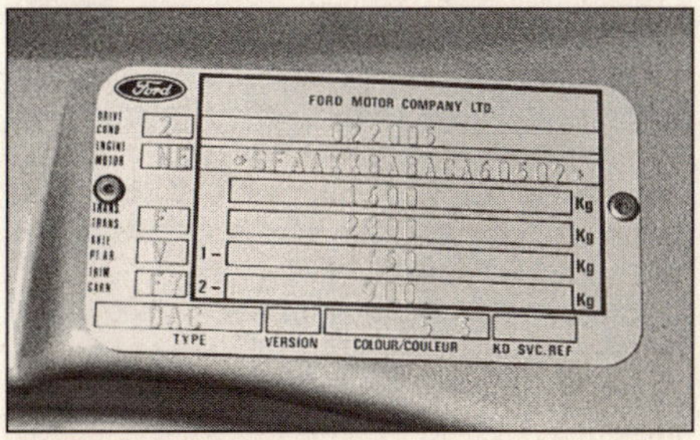

Electrical equipment

□ Switch on the ignition and check the operation of the horn.

□ Check the windscreen washers and wipers, examining the wiper blades; renew damaged or perished blades. Also check the operation of the stop-lights.

□ Check the operation of the sidelights and number plate lights. The lenses and reflectors must be secure, clean and undamaged.

□ Check the operation and alignment of the headlights. The headlight reflectors must not be tarnished and the lenses must be undamaged.

□ Switch on the ignition and check the operation of the direction indicators (including the instrument panel tell-tale) and the hazard warning lights. Operation of the sidelights and stop-lights must not affect the indicators - if it does, the cause is usually a bad earth at the rear light cluster.

□ Check the operation of the rear foglight(s), including the warning light on the instrument panel or in the switch.

Footbrake

□ Examine the master cylinder, brake pipes and servo unit for leaks, loose mountings, corrosion or other damage.

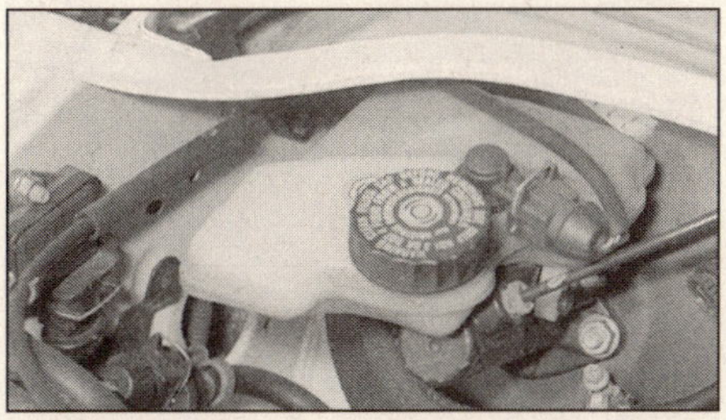

□ The fluid reservoir must be secure and the fluid level must be between the upper (**A**) and lower (**B**) markings.

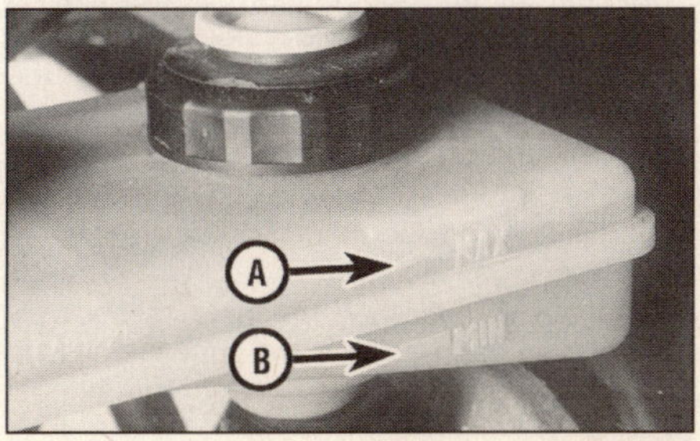

□ Inspect both front brake flexible hoses for cracks or deterioration of the rubber. Turn the steering from lock to lock, and ensure that the hoses do not contact the wheel, tyre, or any part of the steering or suspension mechanism. With the brake pedal firmly depressed, check the hoses for bulges or leaks under pressure.

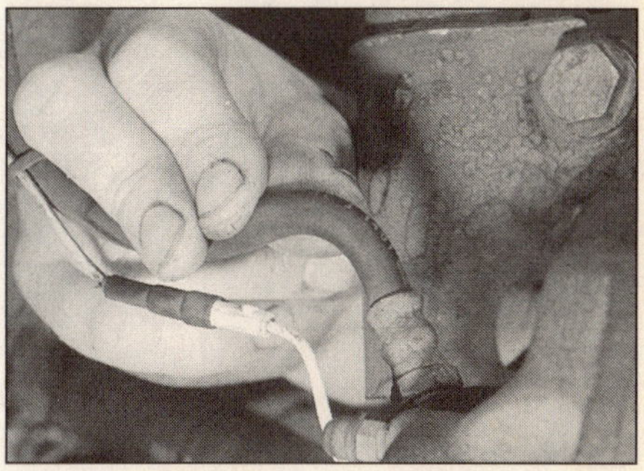

Steering and suspension

□ Have your assistant turn the steering wheel from side to side slightly, up to the point where the steering gear just begins to transmit this movement to the roadwheels. Check for excessive free play between the steering wheel and the steering gear, indicating wear or insecurity of the steering column joints, the column-to-steering gear coupling, or the steering gear itself.

□ Have your assistant turn the steering wheel more vigorously in each direction, so that the roadwheels just begin to turn. As this is done, examine all the steering joints, linkages, fittings and attachments. Renew any component that shows signs of wear or damage. On vehicles with power steering, check the security and condition of the steering pump, drivebelt and hoses.

□ Check that the vehicle is standing level, and at approximately the correct ride height.

Shock absorbers

□ Depress each corner of the vehicle in turn, then release it. The vehicle should rise and then settle in its normal position. If the vehicle continues to rise and fall, the shock absorber is defective. A shock absorber which has seized will also cause the vehicle to fail.

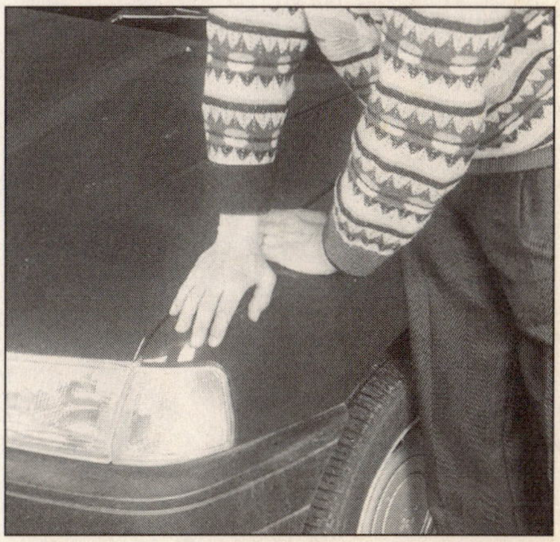

Exhaust system

☐ Start the engine. With your assistant holding a rag over the tailpipe, check the entire system for leaks. Repair or renew leaking sections.

3 Checks carried out WITH THE VEHICLE RAISED AND THE WHEELS FREE TO TURN

Jack up the front and rear of the vehicle, and securely support it on axle stands. Position the stands clear of the suspension assemblies. Ensure that the wheels are clear of the ground and that the steering can be turned from lock to lock.

Steering mechanism

☐ Have your assistant turn the steering from lock to lock. Check that the steering turns smoothly, and that no part of the steering mechanism, including a wheel or tyre, fouls any brake hose or pipe or any part of the body structure.

☐ Examine the steering rack rubber gaiters for damage or insecurity of the retaining clips. If power steering is fitted, check for signs of damage or leakage of the fluid hoses, pipes or connections. Also check for excessive stiffness or binding of the steering, a missing split pin or locking device, or severe corrosion of the body structure within 30 cm of any steering component attachment point.

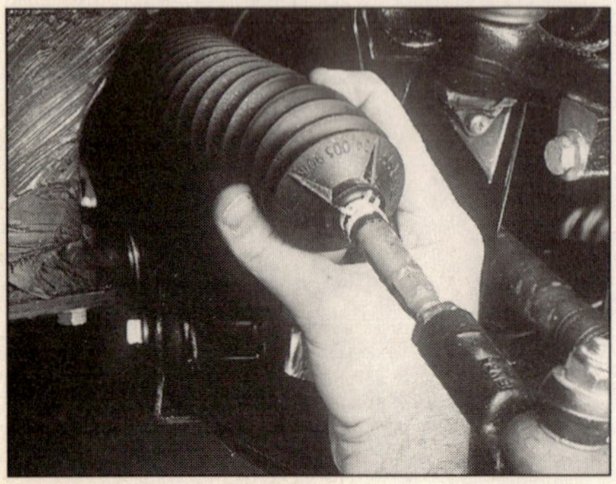

Front and rear suspension and wheel bearings

☐ Starting at the front right-hand side, grasp the roadwheel at the 3 o'clock and 9 o'clock positions and shake it vigorously. Check for free play or insecurity at the wheel bearings, suspension balljoints, or suspension mountings, pivots and attachments.

☐ Now grasp the wheel at the 12 o'clock and 6 o'clock positions and repeat the previous inspection. Spin the wheel, and check for roughness or tightness of the front wheel bearing.

☐ If excess free play is suspected at a component pivot point, this can be confirmed by using a large screwdriver or similar tool and levering between the mounting and the component attachment. This will confirm whether the wear is in the pivot bush, its retaining bolt, or in the mounting itself (the bolt holes can often become elongated).

☐ Carry out all the above checks at the other front wheel, and then at both rear wheels.

Springs and shock absorbers

☐ Examine the suspension struts (when applicable) for serious fluid leakage, corrosion, or damage to the casing. Also check the security of the mounting points.

☐ If coil springs are fitted, check that the spring ends locate in their seats, and that the spring is not corroded, cracked or broken.

☐ If leaf springs are fitted, check that all leaves are intact, that the axle is securely attached to each spring, and that there is no deterioration of the spring eye mountings, bushes, and shackles.

☐ The same general checks apply to vehicles fitted with other suspension types, such as torsion bars, hydraulic displacer units, etc. Ensure that all mountings and attachments are secure, that there are no signs of excessive wear, corrosion or damage, and (on hydraulic types) that there are no fluid leaks or damaged pipes.

☐ Inspect the shock absorbers for signs of serious fluid leakage. Check for wear of the mounting bushes or attachments, or damage to the body of the unit.

Driveshafts (fwd vehicles only)

☐ Rotate each front wheel in turn and inspect the constant velocity joint gaiters for splits or damage. Also check that each driveshaft is straight and undamaged.

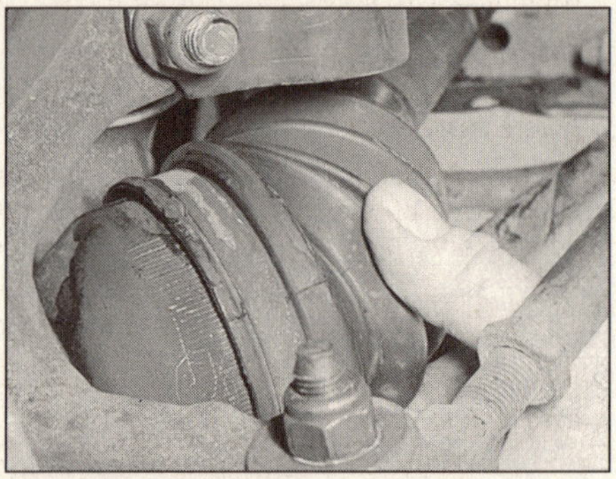

Braking system

☐ If possible without dismantling, check brake pad wear and disc condition. Ensure that the friction lining material has not worn excessively, (A) and that the discs are not fractured, pitted, scored or badly worn (B).

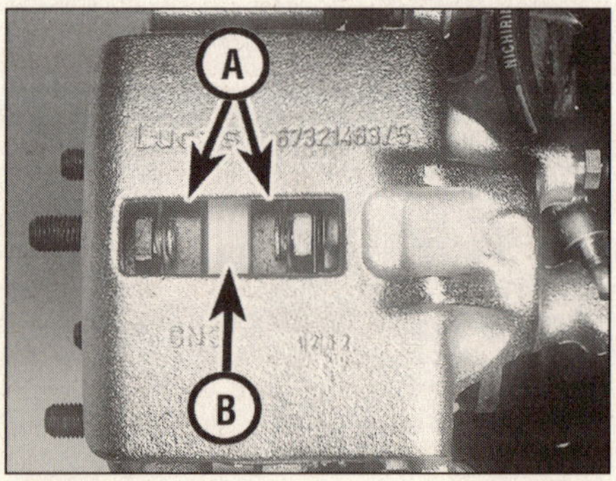

☐ Examine all the rigid brake pipes underneath the vehicle, and the flexible hose(s) at the rear. Look for corrosion, chafing or insecurity of the pipes, and for signs of bulging under pressure, chafing, splits or deterioration of the flexible hoses.

☐ Look for signs of fluid leaks at the brake calipers or on the brake backplates. Repair or renew leaking components.

☐ Slowly spin each wheel, while your assistant depresses and releases the footbrake. Ensure that each brake is operating and does not bind when the pedal is released.

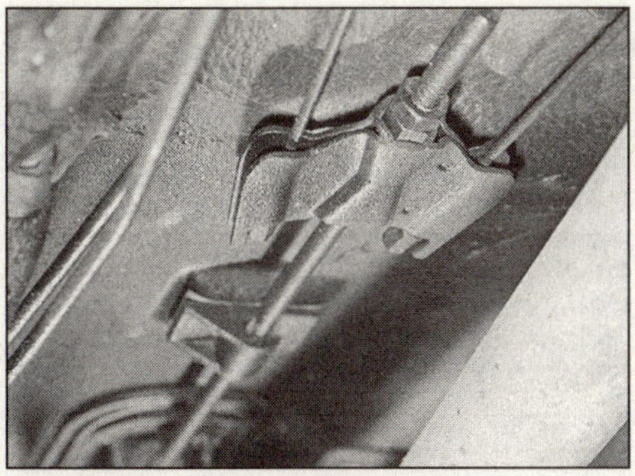

☐ Examine the handbrake mechanism, checking for frayed or broken cables, excessive corrosion, or wear or insecurity of the linkage. Check that the mechanism works on each relevant wheel, and releases fully, without binding.

☐ It is not possible to test brake efficiency without special equipment, but a road test can be carried out later to check that the vehicle pulls up in a straight line.

Fuel and exhaust systems

☐ Inspect the fuel tank (including the filler cap), fuel pipes, hoses and unions. All components must be secure and free from leaks.

☐ Examine the exhaust system over its entire length, checking for any damaged, broken or missing mountings, security of the retaining clamps and rust or corrosion.

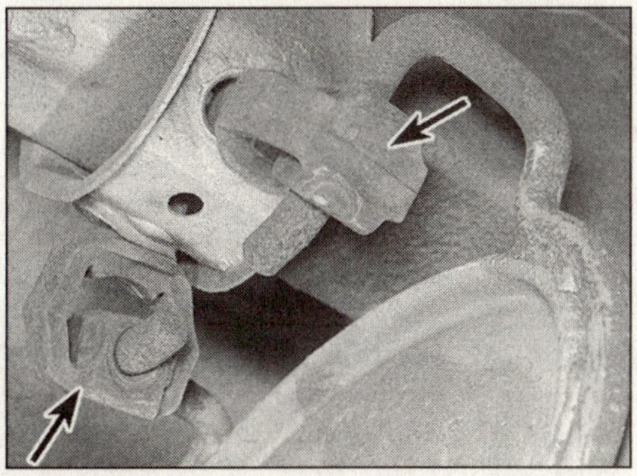

Wheels and tyres

☐ Examine the sidewalls and tread area of each tyre in turn. Check for cuts, tears, lumps, bulges, separation of the tread, and exposure of the ply or cord due to wear or damage. Check that the tyre bead is correctly seated on the wheel rim, that the valve is sound and

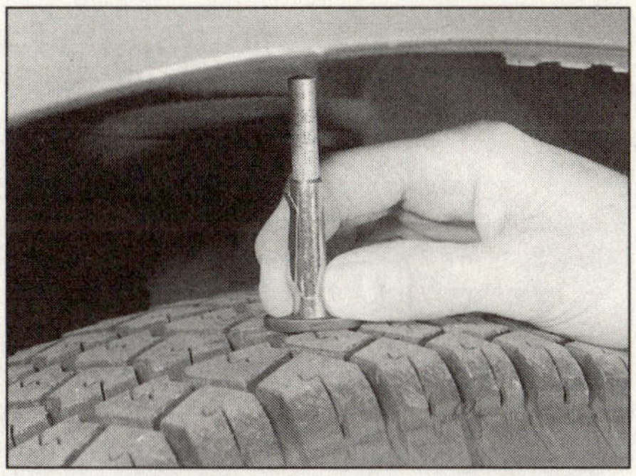

properly seated, and that the wheel is not distorted or damaged.

☐ Check that the tyres are of the correct size for the vehicle, that they are of the same size and type on each axle, and that the pressures are correct.

☐ Check the tyre tread depth. The legal minimum at the time of writing is 1.6 mm over at least three-quarters of the tread width. Abnormal tread wear may indicate incorrect front wheel alignment.

Body corrosion

☐ Check the condition of the entire vehicle structure for signs of corrosion in load-bearing areas. (These include chassis box sections, side sills, cross-members, pillars, and all suspension, steering, braking system and seat belt mountings and anchorages.) Any corrosion which has seriously reduced the thickness of a load-bearing area is likely to cause the vehicle to fail. In this case professional repairs are likely to be needed.

☐ Damage or corrosion which causes sharp or otherwise dangerous edges to be exposed will also cause the vehicle to fail.

4 Checks carried out on **YOUR VEHICLE'S EXHAUST EMISSION SYSTEM**

Petrol models

☐ Have the engine at normal operating temperature, and make sure that it is in good tune (ignition system in good order, air filter element clean, etc).

☐ Before any measurements are carried out, raise the engine speed to around 2500 rpm, and hold it at this speed for 20 seconds. Allow the engine speed to return to idle, and watch for smoke emissions from the exhaust tailpipe. If the idle speed is obviously much too high, or if dense blue or clearly-visible black smoke comes from the tailpipe for more than 5 seconds, the vehicle will fail. As a rule of thumb, blue smoke signifies oil being burnt (engine wear) while black smoke signifies unburnt fuel (dirty air cleaner element, or other carburettor or fuel system fault).

☐ An exhaust gas analyser capable of measuring carbon monoxide (CO) and hydrocarbons (HC) is now needed. If such an instrument cannot be hired or borrowed, a local garage may agree to perform the check for a small fee.

CO emissions (mixture)

☐ At the time of writing, the maximum CO level at idle is 3.5% for vehicles first used after August 1986 and 4.5% for older vehicles. From January 1996 a much tighter limit (around 0.5%) applies to catalyst-equipped vehicles first used from August 1992. If the CO level cannot be reduced far enough to pass the test (and the fuel and ignition systems are otherwise in good condition) then the carburettor is badly worn, or there is some problem in the fuel injection system or catalytic converter (as applicable).

HC emissions

☐ With the CO emissions within limits, HC emissions must be no more than 1200 ppm (parts per million). If the vehicle fails this test at idle, it can be re-tested at around 2000 rpm; if the HC level is then 1200 ppm or less, this counts as a pass.

☐ Excessive HC emissions can be caused by oil being burnt, but they are more likely to be due to unburnt fuel.

Diesel models

☐ The only emission test applicable to Diesel engines is the measuring of exhaust smoke density. The test involves accelerating the engine several times to its maximum unloaded speed.

Note: *It is of the utmost importance that the engine timing belt is in good condition before the test is carried out.*

☐ Excessive smoke can be caused by a dirty air cleaner element. Otherwise, professional advice may be needed to find the cause.

Introduction

The vehicle owner who does his or her own maintenance according to the recommended service schedules should not have to use this section of the manual very often. Modern component reliability is such that, provided those items subject to wear or deterioration are inspected or renewed at the specified intervals, sudden failure is comparatively rare. Faults do not usually just happen as a result of sudden failure, but develop over a period of time. Major mechanical failures in particular are usually preceded by characteristic symptoms over hundreds or even thousands of miles. Those components which do occasionally fail without warning are often small and easily carried in the vehicle.

With any fault-finding, the first step is to decide where to begin investigations. This may be obvious, but some detective work may be necessary. The owner who makes half a dozen haphazard adjustments or replacements may be successful in curing a fault (or its symptoms), but will be none the wiser if the fault recurs, and ultimately may have spent more time and money than was necessary. A calm and logical approach will be found to be more satisfactory in the long run. Always take into account any warning signs that may have been noticed in the period preceding the fault - power loss, high or low gauge readings, unusual smells, etc - and remember - failure of components such as fuses or spark plugs may only be pointers to some underlying fault.

The pages which follow provide an easy-reference guide to the more common problems which may occur during the operation of the vehicle. These problems and their possible causes are grouped under headings denoting various components or systems, such as Engine, Cooling system, etc. The Chapter and/or Section which deals with the problem is also shown in brackets. Whatever the fault, certain basic principles apply. These are as follows:

Verify the fault. This is simply a matter of being sure that you know what the symptoms are before starting work. This is particularly important if you are investigating a fault for someone else, who may not have described it very accurately.

Don't overlook the obvious. For example, if the vehicle won't start, is there fuel in the tank? (Don't take anyone else's word on this particular point, and don't trust the fuel gauge either!) If an electrical fault is indicated, look for loose or broken wires before digging out the test gear.

Cure the disease, not the symptom. Substituting a flat battery with a fully-charged one will get you off the hard shoulder, but if the underlying cause is not attended to, the new battery will go the same way. Similarly, changing oil-fouled spark plugs for a new set will get you moving again, but remember that the reason for the fouling (if it wasn't simply an incorrect grade of plug) will have to be established and corrected.

Don't take anything for granted. Particularly, don't forget that a "new" component may itself be defective (especially if it's been rattling around in the boot for months), and don't leave components out of a fault diagnosis sequence just because they are new or recently-fitted. When you do finally diagnose a difficult fault, you'll probably realise that all the evidence was there from the start.

1 Engine

Engine fails to rotate when attempting to start

- ☐ Battery terminal connections loose or corroded (*"Weekly checks"*).
- ☐ Battery discharged or faulty (Chapter 5A).
- ☐ Broken, loose or disconnected wiring in the starting circuit (Chapter 5A).
- ☐ Defective starter solenoid or switch (Chapter 5A).
- ☐ Defective starter motor (Chapter 5A).
- ☐ Starter pinion or flywheel ring gear teeth loose or broken (Chapters 2A, 2B, 2C and 5A).
- ☐ Engine earth strap broken or disconnected (Chapter 5A).

Engine rotates, but will not start

- ☐ Fuel tank empty.
- ☐ Battery discharged (engine rotates slowly) (Chapter 5A).
- ☐ Battery terminal connections loose or corroded (*"Weekly checks"*).
- ☐ Ignition components damp or damaged - petrol models (Chapters 1A and 5B).
- ☐ Broken, loose or disconnected wiring in the ignition circuit - petrol models (Chapters 1A and 5B).
- ☐ Worn, faulty or incorrectly-gapped spark plugs - petrol models (Chapter 1A).
- ☐ Preheating system faulty - diesel models (Chapter 5C).
- ☐ Fuel injection system fault - petrol models (Chapter 4A).
- ☐ Stop solenoid faulty - diesel models (Chapter 4B).
- ☐ Air in fuel system - diesel models (Chapter 4B).
- ☐ Major mechanical failure (eg camshaft drive) (Chapter 2A, 2B or 2C).

Engine difficult to start when cold

- ☐ Battery discharged (Chapter 5A).
- ☐ Battery terminal connections loose or corroded (*"Weekly checks"*).
- ☐ Worn, faulty or incorrectly-gapped spark plugs - petrol models (Chapter 1A).
- ☐ Preheating system faulty - diesel models (Chapter 5C).
- ☐ Fuel injection system fault - petrol models (Chapter 4A).
- ☐ Other ignition system fault - petrol models (Chapters 1A and 5B).
- ☐ Fast idle valve incorrectly adjusted - diesel models (Chapter 4B).
- ☐ Low cylinder compressions (Chapter 2A or 2B).

Engine difficult to start when hot

- ☐ Air filter element dirty or clogged (Chapter 1A or 1B).
- ☐ Fuel injection system fault - petrol models (Chapter 4A).
- ☐ Low cylinder compressions (Chapter 2A or 2B).

Starter motor noisy or excessively rough in engagement

- ☐ Starter pinion or flywheel ring gear teeth loose or broken (Chapters 2A, 2B, 2C and 5A).
- ☐ Starter motor mounting bolts loose or missing (Chapter 5A).
- ☐ Starter motor internal components worn or damaged (Chapter 5A).

Engine starts, but stops immediately

- ☐ Loose or faulty electrical connections in the ignition circuit - petrol models (Chapters 1A and 5B).
- ☐ Vacuum leak at the throttle body or inlet manifold - petrol models (Chapter 4A).
- ☐ Blocked injector/fuel injection system fault - petrol models (Chapter 4A).

Engine idles erratically

- ☐ Air filter element clogged (Chapter 1A or 1B).
- ☐ Vacuum leak at the throttle body, inlet manifold or associated hoses - petrol models (Chapter 4A).
- ☐ Worn, faulty or incorrectly-gapped spark plugs - petrol models (Chapter 1A).
- ☐ Uneven or low cylinder compressions (Chapter 2A or 2B).
- ☐ Camshaft lobes worn (Chapter 2A, 2B or 2C).
- ☐ Timing belt incorrectly tensioned (Chapter 2A or 2B).
- ☐ Blocked injector/fuel injection system fault - petrol models (Chapter 4A).
- ☐ Faulty injector(s) - diesel models (Chapter 4B).

Engine misfires at idle speed

- ☐ Worn, faulty or incorrectly-gapped spark plugs - petrol models (Chapter 1A).
- ☐ Faulty spark plug HT leads - petrol models (Chapter 1A).
- ☐ Vacuum leak at the throttle body, inlet manifold or associated hoses - petrol models (Chapter 4A).
- ☐ Blocked injector/fuel injection system fault - petrol models (Chapter 4A).
- ☐ Faulty injector(s) - diesel models (Chapter 4B).
- ☐ Distributor cap cracked or tracking internally - petrol models (where applicable) (Chapter 5B).
- ☐ Uneven or low cylinder compressions (Chapter 2A or 2B).
- ☐ Disconnected, leaking, or perished crankcase ventilation hoses (Chapter 4C).

1 Engine (continued)

Engine misfires throughout the driving speed range

- ☐ Fuel filter choked (Chapter 1A or 1B).
- ☐ Fuel pump faulty, or delivery pressure low - petrol models (Chapter 4A).
- ☐ Fuel tank vent blocked, or fuel pipes restricted (Chapter 4A or 4B).
- ☐ Vacuum leak at the throttle body, inlet manifold or associated hoses - petrol models (Chapter 4A).
- ☐ Worn, faulty or incorrectly-gapped spark plugs - petrol models (Chapter 1A).
- ☐ Faulty spark plug HT leads - petrol models (Chapter 1A).
- ☐ Faulty injector(s) - diesel models (Chapter 4B).
- ☐ Distributor cap cracked or tracking internally - petrol models (where applicable) (Chapter 5B).
- ☐ Faulty ignition coil - petrol models (Chapter 5B).
- ☐ Uneven or low cylinder compressions (Chapter 2A or 2B).
- ☐ Blocked injector/fuel injection system fault - petrol models (Chapter 4A).

Engine hesitates on acceleration

- ☐ Worn, faulty or incorrectly-gapped spark plugs - petrol models (Chapter 1A).
- ☐ Vacuum leak at the throttle body, inlet manifold or associated hoses - petrol models (Chapter 4A).
- ☐ Blocked injector/fuel injection system fault - petrol models (Chapter 4A).
- ☐ Faulty injector(s) - diesel models (Chapter 4B).

Engine stalls

- ☐ Vacuum leak at the throttle body, inlet manifold or associated hoses - petrol models (Chapter 4A).
- ☐ Fuel filter choked (Chapter 1A or 1B).
- ☐ Fuel pump faulty, or delivery pressure low - petrol models (Chapter 4A).
- ☐ Fuel tank vent blocked, or fuel pipes restricted (Chapter 4A or 4B).
- ☐ Blocked injector/fuel injection system fault - petrol models (Chapter 4A).
- ☐ Faulty injector(s) - diesel models (Chapter 4B).

Engine lacks power

- ☐ Timing belt incorrectly fitted or tensioned (Chapter 2A or 2B).
- ☐ Fuel filter choked (Chapter 1A or 1B).
- ☐ Fuel pump faulty, or delivery pressure low - petrol models (Chapter 4A).
- ☐ Uneven or low cylinder compressions (Chapter 2A or 2B).
- ☐ Worn, faulty or incorrectly-gapped spark plugs - petrol models (Chapter 1A).
- ☐ Vacuum leak at the throttle body, inlet manifold or associated hoses - petrol models (Chapter 4A).
- ☐ Blocked injector/fuel injection system fault - petrol models (Chapter 4A).
- ☐ Faulty injector(s) - diesel models (Chapter 4B).
- ☐ Injection pump timing incorrect - diesel models (Chapter 4B).
- ☐ Brakes binding (Chapters 1A or 1B and 9).
- ☐ Clutch slipping (Chapter 6).

Engine backfires

- ☐ Timing belt incorrectly fitted or tensioned (Chapter 2A or 2B).
- ☐ Vacuum leak at the throttle body, inlet manifold or associated hoses - petrol models (Chapter 4A).
- ☐ Blocked injector/fuel injection system fault - petrol models (Chapter 4A).

Oil pressure warning light illuminated with engine running

- ☐ Low oil level, or incorrect oil grade (*"Weekly checks"*).
- ☐ Faulty oil pressure sensor (Chapter 5A).
- ☐ Worn engine bearings and/or oil pump (Chapter 2A, 2B or 2C).
- ☐ High engine operating temperature (Chapter 3).
- ☐ Oil pressure relief valve defective (Chapter 2A or 2B).
- ☐ Oil pick-up strainer clogged (Chapter 2A or 2B).

Engine runs-on after switching off

- ☐ Excessive carbon build-up in engine (Chapter 2A, 2B or 2C).
- ☐ High engine operating temperature (Chapter 3).
- ☐ Fuel injection system fault - petrol models (Chapter 4A).
- ☐ Faulty stop solenoid - diesel models (Chapter 4B).

Engine noises

Pre-ignition (pinking) or knocking during acceleration or under load

- ☐ Ignition timing incorrect/ignition system fault - petrol models (Chapters 1A and 5B).
- ☐ Incorrect grade of spark plug - petrol models (Chapter 1A).
- ☐ Incorrect grade of fuel (Chapter 1A or 1B).
- ☐ Vacuum leak at the throttle body, inlet manifold or associated hoses - petrol models (Chapter 4A).
- ☐ Excessive carbon build-up in engine (Chapter 2A, 2B or 2C).
- ☐ Blocked injector/fuel injection system fault - petrol models (Chapter 4A).

Whistling or wheezing noises

- ☐ Leaking inlet manifold or throttle body gasket - petrol models (Chapter 4A).
- ☐ Leaking exhaust manifold gasket or pipe-to-manifold joint (Chapter 4A or 4B).
- ☐ Leaking vacuum hose (Chapters 4A, 4B or 4C, 5B and 9).
- ☐ Blowing cylinder head gasket (Chapter 2A or 2B).

Tapping or rattling noises

- ☐ Worn valve gear or camshaft (Chapter 2A, 2B or 2C).
- ☐ Ancillary component fault (water pump, alternator, etc) (Chapters 3, 5A, etc).

Knocking or thumping noises

- ☐ Worn big-end bearings (regular heavy knocking, perhaps less under load) (Chapter 2C).
- ☐ Worn main bearings (rumbling and knocking, perhaps worsening under load) (Chapter 2C).
- ☐ Piston slap (most noticeable when cold) (Chapter 2C).
- ☐ Ancillary component fault (water pump, alternator, etc) (Chapters 3, 5A, etc).

2 Cooling system

Overheating

- ☐ Insufficient coolant in system (*"Weekly checks"*).
- ☐ Thermostat faulty (Chapter 3).
- ☐ Radiator core blocked, or grille restricted (Chapter 3).
- ☐ Electric cooling fan or thermostatic switch faulty (Chapter 3).
- ☐ Inaccurate temperature gauge sender unit (Chapter 3).
- ☐ Airlock in cooling system (Chapter 3).
- ☐ Expansion tank pressure cap faulty (Chapter 3).

Overcooling

- ☐ Thermostat faulty (Chapter 3).
- ☐ Inaccurate temperature gauge sender unit (Chapter 3).

External coolant leakage

- ☐ Deteriorated or damaged hoses or hose clips (Chapter 1A or 1B).
- ☐ Radiator core or heater matrix leaking (Chapter 3).
- ☐ Pressure cap faulty (Chapter 3).
- ☐ Coolant pump internal seal leaking (Chapter 3).
- ☐ Coolant pump-to-block seal leaking (Chapter 3).
- ☐ Boiling due to overheating (Chapter 3).
- ☐ Core plug leaking (Chapter 2A, 2B or 2C).

Internal coolant leakage

- ☐ Leaking cylinder head gasket (Chapter 2A or 2B).
- ☐ Cracked cylinder head or cylinder block (Chapter 2A, 2B or 2C).

Corrosion

- ☐ Infrequent draining and flushing (Chapter 1A or 1B).
- ☐ Incorrect coolant mixture or inappropriate coolant (*"Weekly checks"*).

3 Fuel and exhaust systems

Excessive fuel consumption

- ☐ Air filter element dirty or clogged (Chapter 1A or 1B).
- ☐ Fuel injection system fault - petrol models (Chapter 4A).
- ☐ Faulty injector(s) - diesel models (Chapter 4B).
- ☐ Ignition timing incorrect/ignition system fault - petrol models (Chapters 1A and 5B).
- ☐ Tyres under-inflated (*"Weekly checks"*).

Fuel leakage and/or fuel odour

- ☐ Damaged or corroded fuel tank, pipes or connections (Chapter 4A or 4B).

Excessive noise or fumes from exhaust system

- ☐ Leaking exhaust system or manifold joints (Chapters 1A or 1B and 4A or 4B).
- ☐ Leaking, corroded or damaged silencers or pipe (Chapters 1A or 1B and 4A or 4B).
- ☐ Broken mountings causing body or suspension contact (Chapter 1A or 1B).

4 Clutch

Pedal travels to floor - no pressure or very little resistance

- ☐ Badly stretched or broken cable (Chapter 6).
- ☐ Incorrect clutch adjustment (Chapter 6).
- ☐ Broken clutch release bearing or arm (Chapter 6).
- ☐ Broken diaphragm spring in clutch pressure plate (Chapter 6).

Clutch fails to disengage (unable to select gears)

- ☐ Incorrect clutch adjustment (Chapter 6).
- ☐ Clutch friction plate sticking on gearbox input shaft splines (Chapter 6).
- ☐ Clutch friction plate sticking to flywheel or pressure plate (Chapter 6).
- ☐ Faulty pressure plate assembly (Chapter 6).
- ☐ Clutch release mechanism worn or badly assembled (Chapter 6).

Clutch slips (engine speed increases, with no increase in vehicle speed)

- ☐ Clutch friction plate linings excessively worn (Chapter 6).
- ☐ Clutch friction plate linings contaminated with oil or grease (Chapter 6).
- ☐ Faulty pressure plate or weak diaphragm spring (Chapter 6).

Judder as clutch is engaged

- ☐ Clutch friction plate linings contaminated with oil or grease (Chapter 6).
- ☐ Clutch friction plate linings excessively worn (Chapter 6).
- ☐ Faulty or distorted pressure plate or diaphragm spring (Chapter 6).
- ☐ Worn or loose engine or gearbox mountings (Chapter 2A).
- ☐ Clutch friction plate hub or gearbox input shaft splines worn (Chapter 6).

Noise when depressing or releasing clutch pedal

- ☐ Worn clutch release bearing (Chapter 6).
- ☐ Worn or dry clutch pedal pivot (Chapter 6).
- ☐ Faulty pressure plate assembly (Chapter 6).
- ☐ Pressure plate diaphragm spring broken (Chapter 6).
- ☐ Broken clutch friction plate cushioning springs (Chapter 6).

5 Manual transmission

Noisy in neutral with engine running

- ☐ Input shaft bearings worn (noise apparent with clutch pedal released, but not when depressed) (Chapter 7A).*
- ☐ Clutch release bearing worn (noise apparent with clutch pedal depressed, possibly less when released) (Chapter 6).

Noisy in one particular gear

- ☐ Worn, damaged or chipped gear teeth (Chapter 7A).*

Difficulty engaging gears

- ☐ Clutch fault (Chapter 6).
- ☐ Worn or damaged gear linkage (Chapter 7A).
- ☐ Worn synchroniser units (Chapter 7A).*

Jumps out of gear

- ☐ Worn or damaged gear linkage (Chapter 7A).
- ☐ Worn synchroniser units (Chapter 7A).*
- ☐ Worn selector forks (Chapter 7).*

Vibration

- ☐ Lack of oil (Chapter 1A or 1B).
- ☐ Worn bearings (Chapter 7A).*

Lubricant leaks

- ☐ Leaking oil seal (Chapter 7A).
- ☐ Leaking housing joint (Chapter 7A).*

**Although the corrective action necessary to remedy the symptoms described is beyond the scope of the home mechanic, the above information should be helpful in isolating the cause of the condition, so that the owner can communicate clearly with a professional mechanic.*

6 Automatic transmission

Note: *Due to the complexity of the automatic transmission, it is difficult for the home mechanic to properly diagnose and service this unit. For problems other than the following, the vehicle should be taken to a dealer service department or automatic transmission specialist. Do not be too hasty in removing the transmission if a fault is suspected, as most of the testing is carried out with the unit still fitted.*

Fluid leakage

- ☐ Automatic transmission fluid is usually dark in colour. Fluid leaks should not be confused with engine oil, which can easily be blown onto the transmission by airflow.
- ☐ To determine the source of a leak, first remove all built-up dirt and grime from the transmission housing and surrounding areas using a degreasing agent, or by steam-cleaning. Drive the vehicle at low speed, so airflow will not blow the leak far from its source. Raise and support the vehicle, and determine where the leak is coming from. The following are common areas of leakage:
  - *a) Oil pan (Chapter 1A and 7B).*
  - *b) Dipstick tube (Chapter 1A and 7B).*
  - *c) Transmission-to-fluid cooler pipes/unions (Chapter 7B).*

Transmission fluid brown, or has burned smell

- ☐ Transmission fluid level low, or fluid in need of renewal (Chapter 1A and 7B).

General gear selection problems

- ☐ Chapter 7B deals with checking and adjusting the selector cable on automatic transmissions. The following are common problems which may be caused by a poorly-adjusted cable:
  - *a) Engine starting in gears other than Park or Neutral.*
  - *b) Indicator panel indicating a gear other than the one actually being used.*
  - *c) Vehicle moves when in Park or Neutral.*
  - *d) Poor gear shift quality or erratic gear changes.*
- ☐ Refer to Chapter 7B for the selector cable adjustment procedure.

Transmission will not downshift (kickdown) with accelerator pedal fully depressed

- ☐ Low transmission fluid level (Chapter 1A and 7B).
- ☐ Incorrect selector cable adjustment (Chapter 7B).

Engine will not start in any gear, or starts in gears other than Park or Neutral

- ☐ Incorrect starter/inhibitor switch adjustment (Chapter 7B).
- ☐ Incorrect selector cable adjustment (Chapter 7B).

Transmission slips, shifts roughly, is noisy, or has no drive in forward or reverse gears

- ☐ There are many probable causes for the above problems, but the home mechanic should be concerned with only one possibility - fluid level. Before taking the vehicle to a dealer or transmission specialist, check the fluid level and condition of the fluid as described in Chapter 1A, or 7B, as applicable. Correct the fluid level as necessary, or change the fluid and filter if needed. If the problem persists, professional help will be necessary.

7 Driveshafts

Clicking or knocking noise on turns (at slow speed on full-lock)

- ☐ Lack of constant velocity joint lubricant, possibly due to damaged gaiter (Chapter 8).
- ☐ Worn outer constant velocity joint (Chapter 8).

Vibration when accelerating or decelerating

- ☐ Worn inner constant velocity joint (Chapter 8).
- ☐ Bent or distorted driveshaft (Chapter 8).
- ☐ Worn right-hand driveshaft intermediate bearing - where applicable (Chapter 8).

8 Braking system

Note: *Before assuming that a brake problem exists, make sure that the tyres are in good condition and correctly inflated, that the front wheel alignment is correct, and that the vehicle is not loaded with weight in an unequal manner. Apart from checking the condition of all pipe and hose connections, any faults occurring on the anti-lock braking system should be referred to a Renault dealer for diagnosis.*

Vehicle pulls to one side under braking

- ☐ Worn, defective, damaged or contaminated front or rear brake pads/shoes on one side (Chapters 1A or 1B and 9).
- ☐ Seized or partially-seized front or rear brake caliper/wheel cylinder piston (Chapter 9).
- ☐ A mixture of brake pad/shoe lining materials fitted between sides (Chapter 9).
- ☐ Brake caliper or rear brake backplate mounting bolts loose (Chapter 9).
- ☐ Worn or damaged steering or suspension components (Chapters 1A or 1B and 10).

Noise (grinding or high-pitched squeal) when brakes applied

- ☐ Brake pad or shoe friction lining material worn down to metal backing (Chapters 1A or 1B and 9).
- ☐ Excessive corrosion of brake disc or drum - may be apparent after the vehicle has been standing for some time (Chapters 1A or 1B and 9).

Excessive brake pedal travel

- ☐ Faulty rear drum brake self-adjust mechanism (Chapter 9).
- ☐ Faulty master cylinder (Chapter 9).
- ☐ Air in hydraulic system (Chapter 9).
- ☐ Faulty vacuum servo unit (Chapter 9).
- ☐ Faulty vacuum pump - diesel models (Chapter 9).

Brake pedal feels spongy when depressed

- ☐ Air in hydraulic system (Chapter 9).
- ☐ Deteriorated flexible rubber brake hoses (Chapters 1A or 1B and 9).
- ☐ Master cylinder mountings loose (Chapter 9).
- ☐ Faulty master cylinder (Chapter 9).

Excessive brake pedal effort required to stop vehicle

- ☐ Faulty vacuum servo unit (Chapter 9).
- ☐ Disconnected, damaged or insecure brake servo vacuum hose (Chapters 1A or 1B and 9).
- ☐ Faulty vacuum pump - diesel models (Chapter 9).
- ☐ Primary or secondary hydraulic circuit failure (Chapter 9).
- ☐ Seized brake caliper or wheel cylinder piston(s) (Chapter 9).
- ☐ Brake pads or brake shoes incorrectly fitted (Chapter 9).
- ☐ Incorrect grade of brake pads or brake shoes fitted (Chapter 9).
- ☐ Brake pads or brake shoe linings contaminated (Chapter 9).

Judder felt through brake pedal or steering wheel when braking

- ☐ Excessive run-out/distortion of brake disc(s) or drum(s) (Chapter 9).
- ☐ Brake pad or brake shoe linings worn (Chapters 1A or 1B and 9).
- ☐ Brake caliper or rear brake backplate mounting bolts loose (Chapter 9).
- ☐ Wear in suspension or steering components or mountings (Chapters 1A or 1B and 10).

Pedal pulsates when braking hard

- ☐ Normal feature of ABS - no fault

Brakes binding

- ☐ Seized brake caliper piston(s) or wheel cylinder piston(s) (Chapter 9).
- ☐ Incorrectly-adjusted handbrake mechanism or linkage (Chapter 9).
- ☐ Faulty master cylinder (Chapter 9).

Rear wheels locking under normal braking

- ☐ Seized brake caliper piston(s) or wheel cylinder piston(s) (Chapter 9).
- ☐ Faulty brake pressure regulator (Chapter 9).

9 Steering and suspension

Note: *Before diagnosing suspension or steering faults, be sure that the trouble is not due to incorrect tyre pressures, mixtures of tyre types, or binding brakes.*

Vehicle pulls to one side

- ☐ Defective tyre (*"Weekly checks"*).
- ☐ Excessive wear in suspension or steering components (Chapters 1A or 1B and 10).
- ☐ Incorrect front wheel alignment (Chapter 10).
- ☐ Accident damage to steering or suspension components (Chapters 1A or 1B and 10).

Wheel wobble and vibration

- ☐ Front roadwheels out of balance (vibration felt mainly through the steering wheel) (Chapter 10).
- ☐ Rear roadwheels out of balance (vibration felt throughout the vehicle) (Chapter 10).
- ☐ Roadwheels damaged or distorted (Chapter 10).
- ☐ Faulty or damaged tyre (*"Weekly checks"*).
- ☐ Worn steering or suspension joints, bushes or components (Chapters 1A or 1B and 10).
- ☐ Wheel bolts loose (Chapter 10).

Excessive pitching and/or rolling around corners, or during braking

- ☐ Defective shock absorbers (Chapters 1A or 1B and 10).
- ☐ Broken or weak coil spring and/or suspension component (Chapters 1A or 1B and 10).
- ☐ Worn or damaged anti-roll bar or mountings (Chapter 10).

Wandering or general instability

- ☐ Incorrect front wheel alignment (Chapter 10).
- ☐ Worn steering or suspension joints, bushes or components (Chapters 1A or 1B and 10).
- ☐ Roadwheels out of balance (Chapter 10).
- ☐ Faulty or damaged tyre (*"Weekly checks"*).
- ☐ Wheel bolts loose (Chapter 10).
- ☐ Defective shock absorbers (Chapters 1A or 1B and 10).

Excessively stiff steering

- ☐ Lack of steering gear lubricant (Chapter 10).
- ☐ Seized track rod end balljoint or suspension balljoint (Chapters 1A or 1B and 10).
- ☐ Broken or incorrectly adjusted auxiliary drivebelt (Chapter 1).
- ☐ Incorrect front wheel alignment (Chapter 10).
- ☐ Steering rack or column bent or damaged (Chapter 10).

9 Steering and suspension

Excessive play in steering

- ☐ Worn steering column universal joint(s) (Chapter 10).
- ☐ Worn steering track rod end balljoints (Chapters 1A or 1B and 10).
- ☐ Worn rack-and-pinion steering gear (Chapter 10).
- ☐ Worn steering or suspension joints, bushes or components (Chapters 1A or 1B and 10).

Lack of power assistance

- ☐ Broken or incorrectly-adjusted auxiliary drivebelt (Chapter 1).
- ☐ Incorrect power steering fluid level (*"Weekly checks"*).
- ☐ Restriction in power steering fluid hoses (Chapter 10).
- ☐ Faulty power steering pump (Chapter 10).
- ☐ Faulty rack-and-pinion steering gear (Chapter 10).

Tyre wear excessive

Tyres worn on inside or outside edges

- ☐ Tyres under-inflated (wear on both edges) (*"Weekly checks"*).
- ☐ Incorrect camber or castor angles (wear on one edge only) (Chapter 10).
- ☐ Worn steering or suspension joints, bushes or components (Chapters 1A or 1B and 10).
- ☐ Excessively-hard cornering.
- ☐ Accident damage.

Tyre treads exhibit feathered edges

- ☐ Incorrect toe setting (Chapter 10).

Tyres worn in centre of tread

- ☐ Tyres over-inflated (*"Weekly checks"*).

Tyres worn on inside and outside edges

- ☐ Tyres under-inflated (*"Weekly checks"*).
- ☐ Worn shock absorbers (Chapters 1A or 1B and 10).

Tyres worn unevenly

- ☐ Tyres out of balance (*"Weekly checks"*).
- ☐ Excessive wheel or tyre run-out (Chapter 1A or 1B).
- ☐ Worn shock absorbers (Chapters 1A or 1B and 10).
- ☐ Faulty tyre (*"Weekly checks"*).

10 Electrical system

Note: *For problems associated with the starting system, refer to the faults listed under "Engine" earlier in this Section.*

Battery will only hold a charge for a few days

- ☐ Battery defective internally (Chapter 5A).
- ☐ Battery electrolyte level low - where applicable (*"Weekly checks"*).
- ☐ Battery terminal connections loose or corroded (*"Weekly checks"*).
- ☐ Auxiliary drivebelt worn - or incorrectly adjusted, where applicable (Chapter 1A or 1B).
- ☐ Alternator not charging at correct output (Chapter 5A).
- ☐ Alternator or voltage regulator faulty (Chapter 5A).
- ☐ Short-circuit causing continual battery drain (Chapters 5A and 12).

Ignition/no-charge warning light remains illuminated with engine running

- ☐ Auxiliary drivebelt broken, worn, or incorrectly adjusted (Chapter 1A or 1B).
- ☐ Alternator brushes worn, sticking, or dirty (Chapter 5A).
- ☐ Alternator brush springs weak or broken (Chapter 5A).
- ☐ Internal fault in alternator or voltage regulator (Chapter 5A).
- ☐ Broken, disconnected, or loose wiring in charging circuit (Chapter 5A).

Ignition/no-charge warning light fails to come on

- ☐ Warning light bulb blown (Chapter 12).
- ☐ Broken, disconnected, or loose wiring in warning light circuit (Chapter 12).
- ☐ Alternator faulty (Chapter 5A).

Lights inoperative

- ☐ Bulb blown (Chapter 12).
- ☐ Corrosion of bulb or bulbholder contacts (Chapter 12).
- ☐ Blown fuse (Chapter 12).
- ☐ Faulty relay (Chapter 12).
- ☐ Broken, loose, or disconnected wiring (Chapter 12).
- ☐ Faulty switch (Chapter 12).

Instrument readings inaccurate or erratic

Instrument readings increase with engine speed

- ☐ Faulty voltage regulator (Chapter 12).

Fuel or temperature gauges give no reading

- ☐ Faulty gauge sender unit (Chapters 3 and 4).
- ☐ Wiring open-circuit (Chapter 12).
- ☐ Faulty gauge (Chapter 12).

Fuel or temperature gauges give continuous maximum reading

- ☐ Faulty gauge sender unit (Chapters 3 and 4).
- ☐ Wiring short-circuit (Chapter 12).
- ☐ Faulty gauge (Chapter 12).

Horn inoperative, or unsatisfactory in operation

Horn operates all the time

- ☐ Horn contacts permanently bridged or horn push stuck down (Chapter 12).

Horn fails to operate

- ☐ Blown fuse (Chapter 12).
- ☐ Cable or cable connections loose, broken or disconnected (Chapter 12).
- ☐ Faulty horn (Chapter 12).

Horn emits intermittent or unsatisfactory sound

- ☐ Cable connections loose (Chapter 12).
- ☐ Horn mountings loose (Chapter 12).
- ☐ Faulty horn (Chapter 12).

10 Electrical system (continued)

Windscreen/tailgate wipers inoperative, or unsatisfactory in operation

Wipers fail to operate, or operate very slowly

- ☐ Wiper blades stuck to screen, or linkage seized or binding (*"Weekly checks"* and Chapter 12).
- ☐ Blown fuse (Chapter 12).
- ☐ Cable or cable connections loose, broken or disconnected (Chapter 12).
- ☐ Faulty relay (Chapter 12).
- ☐ Faulty wiper motor (Chapter 12).

Wiper blades sweep over too large or too small an area of the glass

- ☐ Wiper arms incorrectly positioned on spindles (Chapter 12).
- ☐ Excessive wear of wiper linkage (Chapter 12).
- ☐ Wiper motor or linkage mountings loose or insecure (Chapter 12).

Wiper blades fail to clean the glass effectively

- ☐ Wiper blade rubbers worn or perished (*"Weekly checks"*).
- ☐ Wiper arm tension springs broken, or arm pivots seized (Chapter 12).
- ☐ Insufficient windscreen washer additive to adequately remove road film (*"Weekly checks"*).

Windscreen/tailgate washers inoperative, or unsatisfactory in operation

One or more washer jets inoperative

- ☐ Blocked washer jet (Chapter 12).
- ☐ Disconnected, kinked or restricted fluid hose (Chapter 12).
- ☐ Insufficient fluid in washer reservoir (*"Weekly checks"*).

Washer pump fails to operate

- ☐ Broken or disconnected wiring or connections (Chapter 12).
- ☐ Blown fuse (Chapter 12).
- ☐ Faulty washer switch (Chapter 12).
- ☐ Faulty washer pump (Chapter 12).

Washer pump runs for some time before fluid is emitted from jets

- ☐ Faulty one-way valve in fluid supply hose (Chapter 12).

Electric windows inoperative, or unsatisfactory in operation

Window glass will only move in one direction

- ☐ Faulty switch (Chapter 12).

Window glass slow to move

- ☐ Regulator seized or damaged, or in need of lubrication (Chapter 11).
- ☐ Door internal components or trim fouling regulator (Chapter 11).
- ☐ Faulty motor (Chapter 11).

Window glass fails to move

- ☐ Blown fuse (Chapter 12).
- ☐ Faulty relay (Chapter 12).
- ☐ Broken or disconnected wiring or connections (Chapter 12).
- ☐ Faulty motor (Chapter 12).

Central locking system inoperative, or unsatisfactory in operation

Complete system failure

- ☐ Blown fuse (Chapter 12).
- ☐ Faulty relay (Chapter 12).
- ☐ Broken or disconnected wiring or connections (Chapter 12).

Latch locks but will not unlock, or unlocks but will not lock

- ☐ Faulty switch (Chapter 12).
- ☐ Broken or disconnected latch operating rods or levers (Chapter 11).
- ☐ Faulty relay (Chapter 12).

One motor fails to operate

- ☐ Broken or disconnected wiring or connections (Chapter 12).
- ☐ Faulty motor (Chapter 11).
- ☐ Broken, binding or disconnected lock operating rods or levers (Chapter 11).
- ☐ Fault in door lock (Chapter 11).

A

ABS (Anti-lock brake system) A system, usually electronically controlled, that senses incipient wheel lockup during braking and relieves hydraulic pressure at wheels that are about to skid.
Air bag An inflatable bag hidden in the steering wheel (driver's side) or the dash or glovebox (passenger side). In a head-on collision, the bags inflate, preventing the driver and front passenger from being thrown forward into the steering wheel or windscreen.
Air cleaner A metal or plastic housing, containing a filter element, which removes dust and dirt from the air being drawn into the engine.
Air filter element The actual filter in an air cleaner system, usually manufactured from pleated paper and requiring renewal at regular intervals.

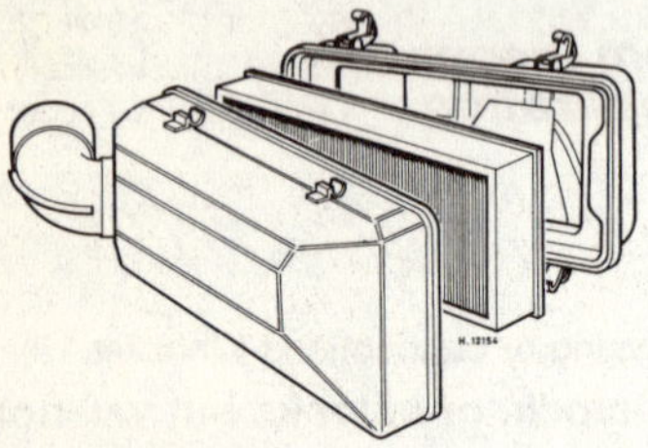

Air filter

Allen key A hexagonal wrench which fits into a recessed hexagonal hole.
Alligator clip A long-nosed spring-loaded metal clip with meshing teeth. Used to make temporary electrical connections.
Alternator A component in the electrical system which converts mechanical energy from a drivebelt into electrical energy to charge the battery and to operate the starting system, ignition system and electrical accessories.
Ampere (amp) A unit of measurement for the flow of electric current. One amp is the amount of current produced by one volt acting through a resistance of one ohm.
Anaerobic sealer A substance used to prevent bolts and screws from loosening. Anaerobic means that it does not require oxygen for activation. The Loctite brand is widely used.
Antifreeze A substance (usually ethylene glycol) mixed with water, and added to a vehicle's cooling system, to prevent freezing of the coolant in winter. Antifreeze also contains chemicals to inhibit corrosion and the formation of rust and other deposits that would tend to clog the radiator and coolant passages and reduce cooling efficiency.
Anti-seize compound A coating that reduces the risk of seizing on fasteners that are subjected to high temperatures, such as exhaust manifold bolts and nuts.
Asbestos A natural fibrous mineral with great heat resistance, commonly used in the composition of brake friction materials. Asbestos is a health hazard and the dust created by brake systems should never be inhaled or ingested.
Axle A shaft on which a wheel revolves, or which revolves with a wheel. Also, a solid beam that connects the two wheels at one end of the vehicle. An axle which also transmits power to the wheels is known as a live axle.
Axleshaft A single rotating shaft, on either side of the differential, which delivers power from the final drive assembly to the drive wheels. Also called a driveshaft or a halfshaft.

B

Ball bearing An anti-friction bearing consisting of a hardened inner and outer race with hardened steel balls between two races.
Bearing The curved surface on a shaft or in a bore, or the part assembled into either, that permits relative motion between them with minimum wear and friction.

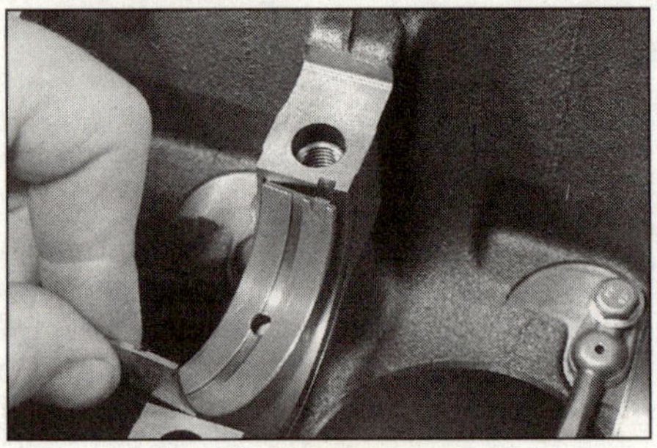

Bearing

Big-end bearing The bearing in the end of the connecting rod that's attached to the crankshaft.
Bleed nipple A valve on a brake wheel cylinder, caliper or other hydraulic component that is opened to purge the hydraulic system of air. Also called a bleed screw.
Brake bleeding Procedure for removing air from lines of a hydraulic brake system.

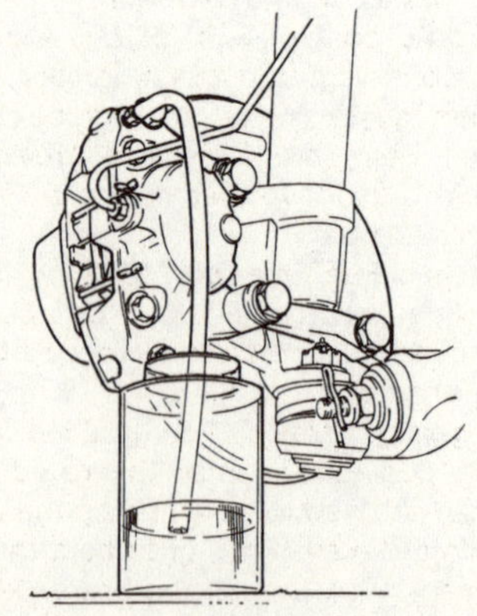

Brake bleeding

Brake disc The component of a disc brake that rotates with the wheels.
Brake drum The component of a drum brake that rotates with the wheels.
Brake linings The friction material which contacts the brake disc or drum to retard the vehicle's speed. The linings are bonded or riveted to the brake pads or shoes.
Brake pads The replaceable friction pads that pinch the brake disc when the brakes are applied. Brake pads consist of a friction material bonded or riveted to a rigid backing plate.
Brake shoe The crescent-shaped carrier to which the brake linings are mounted and which forces the lining against the rotating drum during braking.
Braking systems For more information on braking systems, consult the *Haynes Automotive Brake Manual*.
Breaker bar A long socket wrench handle providing greater leverage.
Bulkhead The insulated partition between the engine and the passenger compartment.

C

Caliper The non-rotating part of a disc-brake assembly that straddles the disc and carries the brake pads. The caliper also contains the hydraulic components that cause the pads to pinch the disc when the brakes are applied. A caliper is also a measuring tool that can be set to measure inside or outside dimensions of an object.
Camshaft A rotating shaft on which a series of cam lobes operate the valve mechanisms. The camshaft may be driven by gears, by sprockets and chain or by sprockets and a belt.
Canister A container in an evaporative emission control system; contains activated charcoal granules to trap vapours from the fuel system.

Canister

Carburettor A device which mixes fuel with air in the proper proportions to provide a desired power output from a spark ignition internal combustion engine.
Castellated Resembling the parapets along the top of a castle wall. For example, a castellated balljoint stud nut.
Castor In wheel alignment, the backward or forward tilt of the steering axis. Castor is positive when the steering axis is inclined rearward at the top.

Catalytic converter A silencer-like device in the exhaust system which converts certain pollutants in the exhaust gases into less harmful substances.

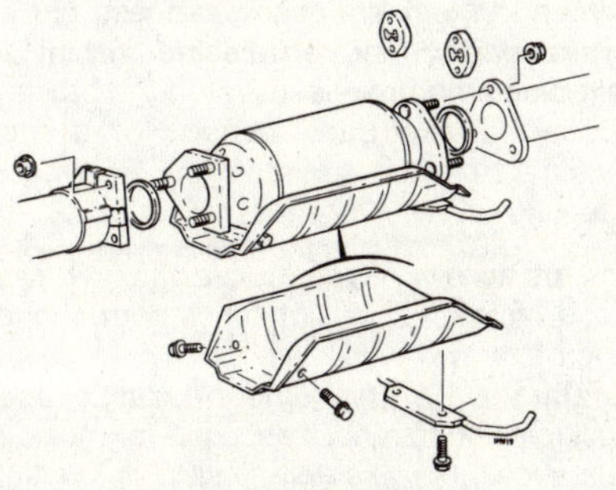

Catalytic converter

Circlip A ring-shaped clip used to prevent endwise movement of cylindrical parts and shafts. An internal circlip is installed in a groove in a housing; an external circlip fits into a groove on the outside of a cylindrical piece such as a shaft.

Clearance The amount of space between two parts. For example, between a piston and a cylinder, between a bearing and a journal, etc.

Coil spring A spiral of elastic steel found in various sizes throughout a vehicle, for example as a springing medium in the suspension and in the valve train.

Compression Reduction in volume, and increase in pressure and temperature, of a gas, caused by squeezing it into a smaller space.

Compression ratio The relationship between cylinder volume when the piston is at top dead centre and cylinder volume when the piston is at bottom dead centre.

Constant velocity (CV) joint A type of universal joint that cancels out vibrations caused by driving power being transmitted through an angle.

Core plug A disc or cup-shaped metal device inserted in a hole in a casting through which core was removed when the casting was formed. Also known as a freeze plug or expansion plug.

Crankcase The lower part of the engine block in which the crankshaft rotates.

Crankshaft The main rotating member, or shaft, running the length of the crankcase, with offset "throws" to which the connecting rods are attached.

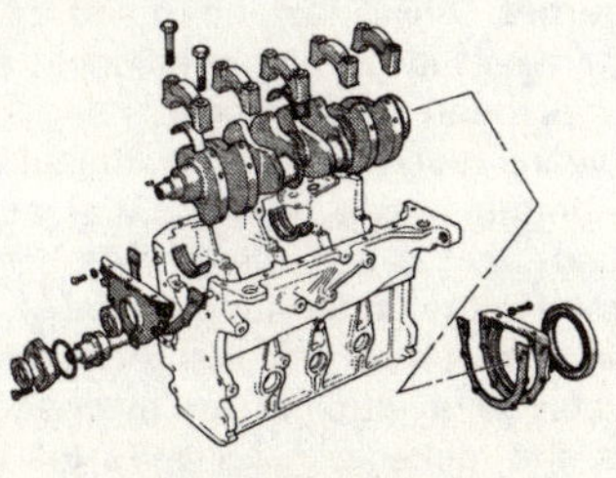

Crankshaft assembly

Crocodile clip See Alligator clip

D

Diagnostic code Code numbers obtained by accessing the diagnostic mode of an engine management computer. This code can be used to determine the area in the system where a malfunction may be located.

Disc brake A brake design incorporating a rotating disc onto which brake pads are squeezed. The resulting friction converts the energy of a moving vehicle into heat.

Double-overhead cam (DOHC) An engine that uses two overhead camshafts, usually one for the intake valves and one for the exhaust valves.

Drivebelt(s) The belt(s) used to drive accessories such as the alternator, water pump, power steering pump, air conditioning compressor, etc. off the crankshaft pulley.

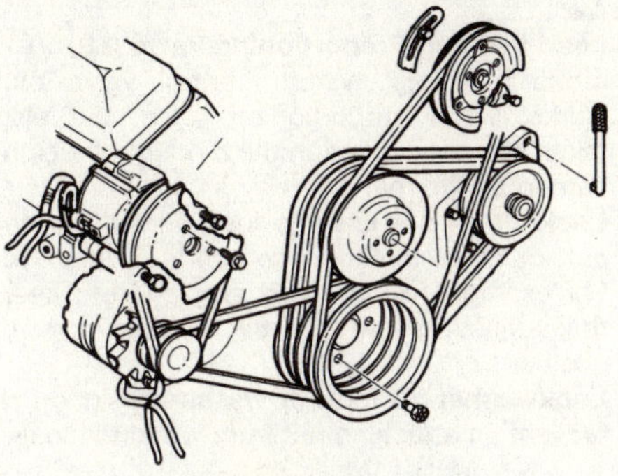

Accessory drivebelts

Driveshaft Any shaft used to transmit motion. Commonly used when referring to the axleshafts on a front wheel drive vehicle.

Drum brake A type of brake using a drum-shaped metal cylinder attached to the inner surface of the wheel. When the brake pedal is pressed, curved brake shoes with friction linings press against the inside of the drum to slow or stop the vehicle.

E

EGR valve A valve used to introduce exhaust gases into the intake air stream.

Electronic control unit (ECU) A computer which controls (for instance) ignition and fuel injection systems, or an anti-lock braking system. For more information refer to the *Haynes Automotive Electrical and Electronic Systems Manual.*

Electronic Fuel Injection (EFI) A computer controlled fuel system that distributes fuel through an injector located in each intake port of the engine.

Emergency brake A braking system, independent of the main hydraulic system, that can be used to slow or stop the vehicle if the primary brakes fail, or to hold the vehicle stationary even though the brake pedal isn't depressed. It usually consists of a hand lever that actuates either front or rear brakes mechanically through a series of cables and linkages. Also known as a handbrake or parking brake.

Endfloat The amount of lengthwise movement between two parts. As applied to a crankshaft, the distance that the crankshaft can move forward and back in the cylinder block.

Engine management system (EMS) A computer controlled system which manages the fuel injection and the ignition systems in an integrated fashion.

Exhaust manifold A part with several passages through which exhaust gases leave the engine combustion chambers and enter the exhaust pipe.

F

Fan clutch A viscous (fluid) drive coupling device which permits variable engine fan speeds in relation to engine speeds.

Feeler blade A thin strip or blade of hardened steel, ground to an exact thickness, used to check or measure clearances between parts.

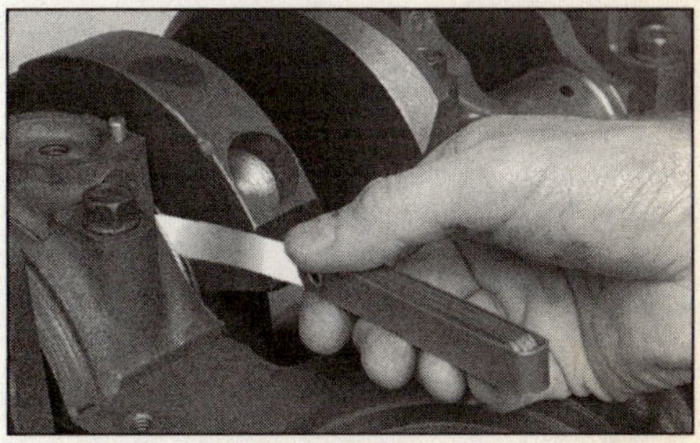

Feeler blade

Firing order The order in which the engine cylinders fire, or deliver their power strokes, beginning with the number one cylinder.

Flywheel A heavy spinning wheel in which energy is absorbed and stored by means of momentum. On cars, the flywheel is attached to the crankshaft to smooth out firing impulses.

Free play The amount of travel before any action takes place. The "looseness" in a linkage, or an assembly of parts, between the initial application of force and actual movement. For example, the distance the brake pedal moves before the pistons in the master cylinder are actuated.

Fuse An electrical device which protects a circuit against accidental overload. The typical fuse contains a soft piece of metal which is calibrated to melt at a predetermined current flow (expressed as amps) and break the circuit.

Fusible link A circuit protection device consisting of a conductor surrounded by heat-resistant insulation. The conductor is smaller than the wire it protects, so it acts as the weakest link in the circuit. Unlike a blown fuse, a failed fusible link must frequently be cut from the wire for replacement.

G

Gap The distance the spark must travel in jumping from the centre electrode to the side electrode in a spark plug. Also refers to the spacing between the points in a contact breaker assembly in a conventional points-type ignition, or to the distance between the reluctor or rotor and the pickup coil in an electronic ignition.

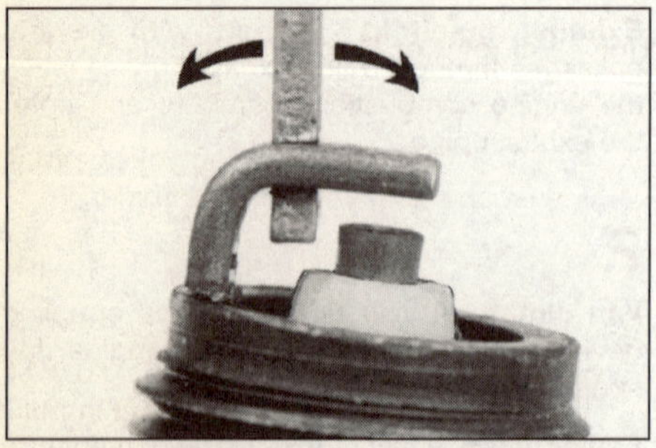

Adjusting spark plug gap

Gasket Any thin, soft material - usually cork, cardboard, asbestos or soft metal - installed between two metal surfaces to ensure a good seal. For instance, the cylinder head gasket seals the joint between the block and the cylinder head.

Gasket

Gauge An instrument panel display used to monitor engine conditions. A gauge with a movable pointer on a dial or a fixed scale is an analogue gauge. A gauge with a numerical readout is called a digital gauge.

H

Halfshaft A rotating shaft that transmits power from the final drive unit to a drive wheel, usually when referring to a live rear axle.

Harmonic balancer A device designed to reduce torsion or twisting vibration in the crankshaft. May be incorporated in the crankshaft pulley. Also known as a vibration damper.

Hone An abrasive tool for correcting small irregularities or differences in diameter in an engine cylinder, brake cylinder, etc.

Hydraulic tappet A tappet that utilises hydraulic pressure from the engine's lubrication system to maintain zero clearance (constant contact with both camshaft and valve stem). Automatically adjusts to variation in valve stem length. Hydraulic tappets also reduce valve noise.

I

Ignition timing The moment at which the spark plug fires, usually expressed in the number of crankshaft degrees before the piston reaches the top of its stroke.

Inlet manifold A tube or housing with passages through which flows the air-fuel mixture (carburettor vehicles and vehicles with throttle body injection) or air only (port fuel-injected vehicles) to the port openings in the cylinder head.

J

Jump start Starting the engine of a vehicle with a discharged or weak battery by attaching jump leads from the weak battery to a charged or helper battery.

L

Load Sensing Proportioning Valve (LSPV) A brake hydraulic system control valve that works like a proportioning valve, but also takes into consideration the amount of weight carried by the rear axle.

Locknut A nut used to lock an adjustment nut, or other threaded component, in place. For example, a locknut is employed to keep the adjusting nut on the rocker arm in position.

Lockwasher A form of washer designed to prevent an attaching nut from working loose.

M

MacPherson strut A type of front suspension system devised by Earle MacPherson at Ford of England. In its original form, a simple lateral link with the anti-roll bar creates the lower control arm. A long strut - an integral coil spring and shock absorber - is mounted between the body and the steering knuckle. Many modern so-called MacPherson strut systems use a conventional lower A-arm and don't rely on the anti-roll bar for location.

Multimeter An electrical test instrument with the capability to measure voltage, current and resistance.

N

NOx Oxides of Nitrogen. A common toxic pollutant emitted by petrol and diesel engines at higher temperatures.

O

Ohm The unit of electrical resistance. One volt applied to a resistance of one ohm will produce a current of one amp.

Ohmmeter An instrument for measuring electrical resistance.

O-ring A type of sealing ring made of a special rubber-like material; in use, the O-ring is compressed into a groove to provide the sealing action.

Overhead cam (ohc) engine An engine with the camshaft(s) located on top of the cylinder head(s).

Overhead valve (ohv) engine An engine with the valves located in the cylinder head, but with the camshaft located in the engine block.

Oxygen sensor A device installed in the engine exhaust manifold, which senses the oxygen content in the exhaust and converts this information into an electric current. Also called a Lambda sensor.

P

Phillips screw A type of screw head having a cross instead of a slot for a corresponding type of screwdriver.

Plastigage A thin strip of plastic thread, available in different sizes, used for measuring clearances. For example, a strip of Plastigage is laid across a bearing journal. The parts are assembled and dismantled; the width of the crushed strip indicates the clearance between journal and bearing.

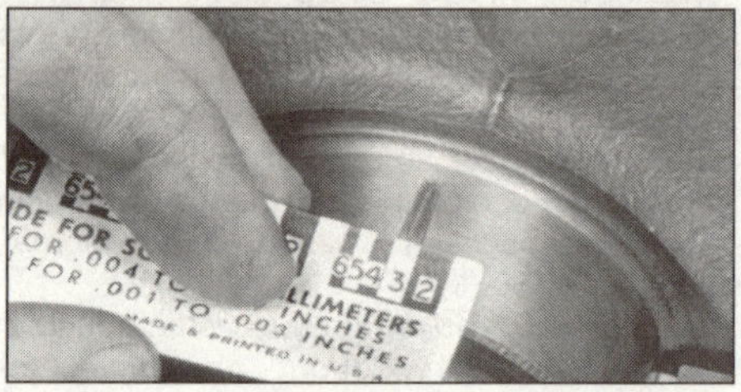

Plastigage

Propeller shaft The long hollow tube with universal joints at both ends that carries power from the transmission to the differential on front-engined rear wheel drive vehicles.

Proportioning valve A hydraulic control valve which limits the amount of pressure to the rear brakes during panic stops to prevent wheel lock-up.

R

Rack-and-pinion steering A steering system with a pinion gear on the end of the steering shaft that mates with a rack (think of a geared wheel opened up and laid flat). When the steering wheel is turned, the pinion turns, moving the rack to the left or right. This movement is transmitted through the track rods to the steering arms at the wheels.

Radiator A liquid-to-air heat transfer device designed to reduce the temperature of the coolant in an internal combustion engine cooling system.

Refrigerant Any substance used as a heat transfer agent in an air-conditioning system. R-12 has been the principle refrigerant for many years; recently, however, manufacturers have begun using R-134a, a non-CFC substance that is considered less harmful to the ozone in the upper atmosphere.

Rocker arm A lever arm that rocks on a shaft or pivots on a stud. In an overhead valve engine, the rocker arm converts the upward movement of the pushrod into a downward movement to open a valve.

Rotor In a distributor, the rotating device inside the cap that connects the centre electrode and the outer terminals as it turns, distributing the high voltage from the coil secondary winding to the proper spark plug. Also, that part of an alternator which rotates inside the stator. Also, the rotating assembly of a turbocharger, including the compressor wheel, shaft and turbine wheel.

Runout The amount of wobble (in-and-out movement) of a gear or wheel as it's rotated. The amount a shaft rotates "out-of-true." The out-of-round condition of a rotating part.

S

Sealant A liquid or paste used to prevent leakage at a joint. Sometimes used in conjunction with a gasket.

Sealed beam lamp An older headlight design which integrates the reflector, lens and filaments into a hermetically-sealed one-piece unit. When a filament burns out or the lens cracks, the entire unit is simply replaced.

Serpentine drivebelt A single, long, wide accessory drivebelt that's used on some newer vehicles to drive all the accessories, instead of a series of smaller, shorter belts. Serpentine drivebelts are usually tensioned by an automatic tensioner.

Serpentine drivebelt

Shim Thin spacer, commonly used to adjust the clearance or relative positions between two parts. For example, shims inserted into or under bucket tappets control valve clearances. Clearance is adjusted by changing the thickness of the shim.

Slide hammer A special puller that screws into or hooks onto a component such as a shaft or bearing; a heavy sliding handle on the shaft bottoms against the end of the shaft to knock the component free.

Sprocket A tooth or projection on the periphery of a wheel, shaped to engage with a chain or drivebelt. Commonly used to refer to the sprocket wheel itself.

Starter inhibitor switch On vehicles with an automatic transmission, a switch that prevents starting if the vehicle is not in Neutral or Park.

Strut See MacPherson strut.

T

Tappet A cylindrical component which transmits motion from the cam to the valve stem, either directly or via a pushrod and rocker arm. Also called a cam follower.

Thermostat A heat-controlled valve that regulates the flow of coolant between the cylinder block and the radiator, so maintaining optimum engine operating temperature. A thermostat is also used in some air cleaners in which the temperature is regulated.

Thrust bearing The bearing in the clutch assembly that is moved in to the release levers by clutch pedal action to disengage the clutch. Also referred to as a release bearing.

Timing belt A toothed belt which drives the camshaft. Serious engine damage may result if it breaks in service.

Timing chain A chain which drives the camshaft.

Toe-in The amount the front wheels are closer together at the front than at the rear. On rear wheel drive vehicles, a slight amount of toe-in is usually specified to keep the front wheels running parallel on the road by offsetting other forces that tend to spread the wheels apart.

Toe-out The amount the front wheels are closer together at the rear than at the front. On front wheel drive vehicles, a slight amount of toe-out is usually specified.

Tools For full information on choosing and using tools, refer to the *Haynes Automotive Tools Manual*.

Tracer A stripe of a second colour applied to a wire insulator to distinguish that wire from another one with the same colour insulator.

Tune-up A process of accurate and careful adjustments and parts replacement to obtain the best possible engine performance.

Turbocharger A centrifugal device, driven by exhaust gases, that pressurises the intake air. Normally used to increase the power output from a given engine displacement, but can also be used primarily to reduce exhaust emissions (as on VW's "Umwelt" Diesel engine).

U

Universal joint or U-joint A double-pivoted connection for transmitting power from a driving to a driven shaft through an angle. A U-joint consists of two Y-shaped yokes and a cross-shaped member called the spider.

V

Valve A device through which the flow of liquid, gas, vacuum, or loose material in bulk may be started, stopped, or regulated by a movable part that opens, shuts, or partially obstructs one or more ports or passageways. A valve is also the movable part of such a device.

Valve clearance The clearance between the valve tip (the end of the valve stem) and the rocker arm or tappet. The valve clearance is measured when the valve is closed.

Vernier caliper A precision measuring instrument that measures inside and outside dimensions. Not quite as accurate as a micrometer, but more convenient.

Viscosity The thickness of a liquid or its resistance to flow.

Volt A unit for expressing electrical "pressure" in a circuit. One volt that will produce a current of one ampere through a resistance of one ohm.

W

Welding Various processes used to join metal items by heating the areas to be joined to a molten state and fusing them together. For more information refer to the *Haynes Automotive Welding Manual*.

Wiring diagram A drawing portraying the components and wires in a vehicle's electrical system, using standardised symbols. For more information refer to the *Haynes Automotive Electrical and Electronic Systems Manual*.

Note: *References throughout this index are in the form - "Chapter number" • "page number"*

10,000 - inspected brakes, tyres, battery, changed oil + filter,

20,000 - (29th Sept 99) changed oil + filter, changed spark plugs, changed air filter, changed coolant, checked brakes (front driver side disc scored on the inside-engine side). rotated tyres, checked drive belts, checked battery levels,

Haynes Manuals – The Complete List

Title	Book No.
ALFA ROMEO	
Alfa Romeo Alfasud/Sprint (74 - 88)	0292
Alfa Romeo Alfetta (73 - 87)	0531
AUDI	
Audi 80 (72 - Feb 79)	0207
Audi 80, 90 (79 - Oct 86) & Coupe (81 - Nov 88)	0605
Audi 80, 90 (Oct 86 - 90) & Coupe (Nov 88 - 90)	1491
Audi 100 (Oct 76 - Oct 82)	0428
Audi 100 (Oct 82 - 90) & 200 (Feb 84 - Oct 89)	0907
AUSTIN	
Austin Ambassador (82 - 84)	0871
Austin/MG Maestro 1.3 & 1.6 (83 - 95)	0922
Austin Maxi (69 - 81)	0052
Austin/MG Metro (80 - May 90)	0718
Austin Montego 1.3 & 1.6 (84 - 94)	1066
Austin/MG Montego 2.0 (84 - 95)	1067
Mini (59 - 69)	0527
Mini (69 - Oct 96)	0646
Austin/Rover 2.0 litre Diesel Engine (86 - 93)	1857
BEDFORD	
Bedford CF (69 - 87)	0163
Bedford Rascal (86 - 93)	3015
BL	
BL Princess & BLMC 18-22 (75 - 82)	0286
BMW	
BMW 316, 320 & 320i (4-cyl) (75 - Feb 83)	0276
BMW 320, 320i, 323i & 325i (6-cyl) (Oct 77 - Sept 87)	0815
BMW 3-Series (Apr 91 - 96)	3210
BMW 3-Series (sohc) (83 - 91)	1948
BMW 520i & 525e (Oct 81 - June 88)	1560
BMW 525, 528 & 528i (73 - Sept 81)	0632
BMW 5-Series (sohc) (81 - 93)	1948
BMW 1500, 1502, 1600, 1602, 2000 & 2002 (59 - 77)	0240
CITROEN	
Citroen 2CV, Ami & Dyane (67 - 90)	0196
Citroen AX Petrol & Diesel (87 - 94)	3014
Citroen BX (83 - 94)	0908
Citroen CX (75 - 88)	0528
Citroen Visa (79 - 88)	0620
Citroen Xantia Petrol & Diesel (93 - Oct 95)	3082
Citroen ZX Diesel (91 - 93)	1922
Citroen ZX Petrol (91 - 94)	1881
Citroen 1.7 & 1.9 litre Diesel Engine (84 - 96)	1379
COLT	
Colt 1200, 1250 & 1400 (79 - May 84)	0600
DAIMLER	
Daimler Sovereign (68 - Oct 86)	0242
Daimler Double Six (72 - 88)	0478
DATSUN *(see also Nissan)*	
Datsun 120Y (73 - Aug 78)	0228
Datsun 1300, 1400 & 1600 (69 - Aug 72)	0123
Datsun Cherry (71 - 76)	0195
Datsun Pick-up (75 - 78)	0277
Datsun Sunny (Aug 78 - May 82)	0525
Datsun Violet (78 - 82)	0430

Title	Book No.
FIAT	
Fiat 126 (73 - 87)	0305
Fiat 127 (71 - 83)	0193
Fiat 500 (57 - 73)	0090
Fiat 850 (64 - 81)	0038
Fiat Panda (81 - 95)	0793
Fiat Punto (94 - 96)	3251
Fiat Regata (84 - 88)	1167
Fiat Strada (79 - 88)	0479
Fiat Tipo (88 - 91)	1625
Fiat Uno (83 - 95)	0923
Fiat X1/9 (74 - 89)	0273
FORD	
Ford Capri II (& III) 1.6 & 2.0 (74 - 87)	0283
Ford Capri II (& III) 2.8 & 3.0 (74 - 87)	1309
Ford Cortina Mk IV (& V) 1.6 & 2.0 (76 - 83)	0343
Ford Cortina Mk IV (& V) 2.3 V6 (77 - 83)	0426
Ford Escort (75 - Aug 80)	0280
Ford Escort (Sept 80 - Sept 90)	0686
Ford Escort (Sept 90 - 97)	1737
Ford Escort Mk II Mexico, RS 1600 & RS 2000 (75 - 80)	0735
Ford Fiesta (inc. XR2) (76 - Aug 83)	0334
Ford Fiesta (inc. XR2) (Aug 83 - Feb 89)	1030
Ford Fiesta (Feb 89 - Oct 95)	1595
Ford Granada (Sept 77 - Feb 85)	0481
Ford Granada (Mar 85 - 94)	1245
Ford Mondeo 4-cyl (93 - 96)	1923
Ford Orion (83 - Sept 90)	1009
Ford Orion (Sept 90 - 93)	1737
Ford Sierra 1.3, 1.6, 1.8 & 2.0 (82 - 93)	0903
Ford Sierra 2.3, 2.8 & 2.9 (82 - 91)	0904
Ford Scorpio (Mar 85 - 94)	1245
Ford Transit Petrol (Mk 1) (65 - Feb 78)	0377
Ford Transit Petrol (Mk 2) (78 - Jan 86)	0719
Ford Transit Petrol (Mk 3) (Feb 86 - 89)	1468
Ford Transit Diesel (Feb 86 - 95)	3019
Ford 1.6 & 1.8 litre Diesel Engine (84 - 96)	1172
Ford 2.1, 2.3 & 2.5 litre Diesel Engine (77 - 90)	1606
FREIGHT ROVER	
Freight Rover Sherpa (74 - 87)	0463
HILLMAN	
Hillman Avenger (70 - 82)	0037
Hillman Minx & Husky (56 - 66)	0009
HONDA	
Honda Accord (76 - Feb 84)	0351
Honda Accord (Feb 84 - Oct 85)	1177
Honda Civic 1300 (80 - 81)	0633
Honda Civic (Feb 84 - Oct 87)	1226
Honda Civic (Nov 91 - 96)	3199
HYUNDAI	
Hyundai Pony (85 - 94)	3398
JAGUAR	
Jaguar E Type (61 - 72)	0140
Jaguar MkI & II, 240 & 340 (55 - 69)	0098
Jaguar XJ6, XJ & Sovereign (68 - Oct 86)	0242
Jaguar XJ6 & Sovereign (Oct 86 - Sept 94)	3261
Jaguar XJ12, XJS & Sovereign (72 - 88)	0478

Title	Book No.
JEEP	
Jeep Cherokee Petrol (93 - 96)	1943
LADA	
Lada 1200, 1300, 1500 & 1600 (74 - 91)	0413
Lada Samara (87 - 91)	1610
LAND ROVER	
Land Rover 90, 110 & Defender Diesel (83 - 95)	3017
Land Rover Discovery Diesel (89 - 95)	3016
Land Rover Series IIA & III Diesel (58 - 85)	0529
Land Rover Series II, IIA & III Petrol (58 - 85)	0314
MAZDA	
Mazda 323 fwd (Mar 81 - Oct 89)	1608
Mazda 626 fwd (May 83 - Sept 87)	0929
Mazda B-1600, B-1800 & B-2000 Pick-up (72 - 88)	0267
MERCEDES-BENZ	
Mercedes-Benz 190 & 190E (83 - 87)	0928
Mercedes-Benz 200, 240, 300 Diesel (Oct 76 - 85)	1114
Mercedes-Benz 250 & 280 (68 - 72)	0346
Mercedes-Benz 250 & 280 (123 Series) (Oct 76 - 84)	0677
Mercedes-Benz 124 Series (85 - Aug 93)	3253
MG	
MGB (62 - 80)	0111
MG Maestro 1.3 & 1.6 (83 - 95)	0922
MG Metro (80 - May 90)	0718
MG Midget & AH Sprite (58 - 80)	0265
MG Montego 2.0 (84 - 95)	1067
MITSUBISHI	
Mitsubishi 1200, 1250 & 1400 (79 - May 84)	0600
Mitsubishi Shogun & L200 Pick-Ups (83 - 94)	1944
MORRIS	
Morris Ital 1.3 (80 - 84)	0705
Morris Marina 1700 (78 - 80)	0526
Morris Marina 1.8 (71 - 78)	0074
Morris Minor 1000 (56 - 71)	0024
NISSAN *(See also Datsun)*	
Nissan Bluebird 160B & 180B rwd (May 80 - May 84)	0957
Nissan Bluebird fwd (May 84 - Mar 86)	1223
Nissan Bluebird (T12 & T72) (Mar 86 - 90)	1473
Nissan Cherry (N12) (Sept 82 - 86)	1031
Nissan Micra (K10) (83 - Jan 93)	0931
Nissan Micra (93 - 96)	3254
Nissan Primera (90 - Oct 96)	1851
Nissan Stanza (82 - 86)	0824
Nissan Sunny (B11) (May 82 - Oct 86)	0895
Nissan Sunny (Oct 86 - Mar 91)	1378
Nissan Sunny (Apr 91 - 95)	3219
OPEL	
Opel Ascona & Manta (B Series) (Sept 75 - 88)	0316
Opel Ascona (81 - 88)	3215
Opel Astra (Oct 91 - 96)	3156
Opel Corsa (83 - Mar 93)	3160
Opel Corsa (Mar 93 - 94)	3159
Opel Kadett (Nov 79 - Oct 84)	0634
Opel Kadett (Oct 84 - Oct 91)	3196
Opel Omega & Senator (86 - 94)	3157

Title	Book No.
Opel Rekord (Feb 78 - Oct 86)	0543
Opel Vectra (88 - Oct 95)	3158
PEUGEOT	
Peugeot 106 Petrol & Diesel (91 - June 96)	1882
Peugeot 205 (83 - 95)	0932
Peugeot 305 (78 - 89)	0538
Peugeot 306 Petrol & Diesel (93 - 95)	3073
Peugeot 309 (86 - 93)	1266
Peugeot 405 Petrol (88 - 96)	1559
Peugeot 405 Diesel (88 - 96)	3198
Peugeot 406 Petrol & Diesel (96 - 97)	3394
Peugeot 505 (79 - 89)	0762
Peugeot 1.7 & 1.9 litre Diesel Engines (82 - 96)	0950
Peugeot 2.0, 2.1, 2.3 & 2.5 litre Diesel Engines (74 - 90)	1607
PORSCHE	
Porsche 911 (65 - 85)	0264
Porsche 924 & 924 Turbo (76 - 85)	0397
PROTON	
Proton (89 - 97)	3255
RANGE ROVER	
Range Rover V8 (70 - Oct 92)	0606
RELIANT	
Reliant Robin & Kitten (73 - 83)	0436
RENAULT	
Renault 5 (72 - Feb 85)	0141
Renault 5 (Feb 85 - 96)	1219
Renault 9 & 11 (82 - 89)	0822
Renault 12 (70 - 80)	0097
Renault 15 & 17 (72 - 79)	0763
Renault 18 (79 - 86)	0598
Renault 19 Petrol (89 - 94)	1646
Renault 19 Diesel (89 - 95)	1946
Renault 21 (86 - 94)	1397
Renault 25 (84 - 86)	1228
Renault Clio Petrol (91 - 93)	1853
Renault Clio Diesel (91 - June 96)	3031
Renault Espace (85 - 96)	3197
Renault Fuego (80 - 86)	0764
Renault Laguna (94 - 96)	3252
ROVER	
Rover 111 & 114 (95 - 96)	1711
Rover 213 & 216 (84 - 89)	1116
Rover 214 & 414 (Oct 89 - 92)	1689
Rover 216 & 416 (Oct 89 - 92)	1830
Rover 618, 620 & 623 (93 - 97)	3257
Rover 820, 825 & 827 (86 - 95)	1380
Rover 2000, 2300 & 2600 (77 - 87)	0468
Rover 3500 (76 - 87)	0365
Rover Metro (May 90 - 94)	1711
SAAB	
Saab 90, 99 & 900 (79 - Oct 93)	0765
Saab 9000 (4-cyl) (85 - 95)	1686
SEAT	
Seat Ibiza & Malaga (85 - 92)	1609
SIMCA	
Simca 1100 & 1204 (67 - 79)	0088
Simca 1301 & 1501 (63 - 76)	0199

Title	Book No.
SKODA	
Skoda Estelle 105, 120, 130 & 136 (77 - 89)	0604
Skoda Favorit (89 - 92)	1801
SUBARU	
Subaru 1600 (77 - Oct 79)	0237
Subaru 1600 & 1800 (Nov 79 - 90)	0995
SUZUKI	
Suzuki SJ Series, Samurai & Vitara (82 - 97)	1942
Suzuki Supercarry (86 - Oct 94)	3015
TALBOT	
Talbot Alpine, Solara, Minx & Rapier (75 - 86)	0337
Talbot Horizon (78 - 86)	0473
Talbot Samba (82 - 86)	0823
TOYOTA	
Toyota Carina E (May 92 - 97)	3256
Toyota Celica (Feb 82 - Sept 85)	1135
Toyota Corolla (fwd) (Sept 83 - Sept 87)	1024
Toyota Corolla (rwd) (80 - 85)	0683
Toyota Corolla (Sept 87 - 92)	1683
Toyota Corolla (Aug 92 - 97)	3259
Toyota Hi-Ace & Hi-Lux (69 - Oct 83)	0304
Toyota Starlet (78 - Jan 85)	0462
TRIUMPH	
Triumph Acclaim (81 - 84)	0792
Triumph Herald (59 - 71)	0010
Triumph Spitfire (62 - 81)	0113
Triumph Stag (70 - 78)	0441
Triumph TR7 (75 - 82)	0322
VAUXHALL	
Vauxhall Astra (80 - Oct 84)	0635
Vauxhall Astra & Belmont (Oct 84 - Oct 91)	1136
Vauxhall Astra (Oct 91 - 96)	1832
Vauxhall Carlton (Oct 78 - Oct 86)	0480
Vauxhall Carlton (Nov 86 - 94)	1469
Vauxhall Cavalier 1300 (77 - July 81)	0461
Vauxhall Cavalier 1600, 1900 & 2000 (75 - July 81)	0315
Vauxhall Cavalier (81 - Oct 88)	0812
Vauxhall Cavalier (Oct 88 - Oct 95)	1570
Vauxhall Chevette (75 - 84)	0285
Vauxhall Corsa (Mar 93 - 94)	1985
Vauxhall Nova (83 - 93)	0909
Vauxhall Rascal (86 - 93)	3015
Vauxhall Senator (Sept 87 - 94)	1469
Vauxhall Viva HB Series (ohv) (66 - 70)	0026
Vauxhall Viva & Firenza (ohc) (68 - 73)	0093
Vauxhall/Opel 1.5, 1.6 & 1.7 litre Diesel Engines (82 - 96)	1222
VOLKSWAGEN	
VW Beetle 1200 (54 - 77)	0036
VW Beetle 1300 & 1500 (65 - 75)	0039
VW Beetle 1302 & 1302S (70 - 72)	0110
VW Beetle 1303, 1303S & GT (72 - 75)	0159
VW Golf Mk 1 1.1 & 1.3 (74 - Feb 84)	0716
VW Golf Mk 1 1.5, 1.6 & 1.8 (74 - 85)	0726
VW Golf Mk 1 Diesel (78 - Feb 84)	0451
VW Golf Mk 2 (Mar 84 - Feb 92)	1081
VW Golf Mk 3 Petrol & Diesel (Feb 92 - 96)	3097

Title	Book No.
VW Jetta Mk 1 1.1 & 1.3 (80 - June 84)	0716
VW Jetta Mk 1 1.5, 1.6 & 1.8 (80 - June 84)	0726
VW Jetta Mk 1 Diesel (81 - June 84)	0451
VW Jetta Mk 2 (July 84 - 92)	1081
VW LT vans & light trucks (76 - 87)	0637
VW Passat (Sept 81 - May 88)	0814
VW Passat (May 88 - 91)	1647
VW Polo & Derby (76 - Jan 82)	0335
VW Polo (82 - Oct 90)	0813
VW Polo (Nov 90 - Aug 94)	3245
VW Santana (Sept 82 - 85)	0814
VW Scirocco Mk 1 1.5, 1.6 & 1.8 (74 - 82)	0726
VW Scirocco (82 - 90)	1224
VW Transporter 1600 (68 - 79)	0082
VW Transporter 1700, 1800 & 2000 (72 - 79)	0226
VW Transporter with air-cooled engine (79 - 82)	0638
VW Type 3 (63 - 73)	0084
VW Vento Petrol & Diesel (Feb 92 - 96)	3097
VOLVO	
Volvo 66 & 343, Daf 55 & 66 (68 - 79)	0293
Volvo 142, 144 & 145 (66 - 74)	0129
Volvo 240 Series (74 - 93)	0270
Volvo 262, 264 & 260/265 (75 - 85)	0400
Volvo 340, 343, 345 & 360 (76 - 91)	0715
Volvo 440, 460 & 480 (87 - 92)	1691
Volvo 740 & 760 (82 - 91)	1258
Volvo 850 (92 - 96)	3260
Volvo 940 (90 - 96)	3249
YUGO/ZASTAVA	
Yugo/Zastava (81 - 90)	1453

TECH BOOKS	
Automotive Brake Manual	3050
Carburettor Manual	3288
Diesel Engine Service Guide	3286
Automotive Electrical & Electronic Systems	3049
Automotive Engine Management and Fuel Injection Manual	3344
In-Car Entertainment Manual	3363
Automotive Tools Manual	3052
Automotive Welding Manual	3053

CAR BOOKS	
Automotive Fuel Injection Systems	9755
Car Bodywork Repair Manual	9864
Caravan Manual (2nd Edition)	9894
Haynes Technical Data Book (88 - 97)	1997
How to Keep Your Car Alive	9868
Japanese Vehicle Carburettors	1786
Small Engine Repair Manual	1755
SU Carburettors	0299
Weber Carburettors (to 79)	0393